# Methods in Enzymology

Volume 157
BIOMEMBRANES
Part Q
ATP-Driven Pumps and Related Transport:
Calcium, Proton, and Potassium Pumps

# METHODS IN ENZYMOLOGY

EDITORS-IN-CHIEF

John N. Abelson     Melvin I. Simon

*Methods in Enzymology*

*Volume 157*

# *Biomembranes*

*Part Q*

*ATP-Driven Pumps and Related Transport:*
*Calcium, Proton, and Potassium Pumps*

EDITED BY

*Sidney Fleischer*
*Becca Fleischer*

DEPARTMENT OF MOLECULAR BIOLOGY
VANDERBILT UNIVERSITY
NASHVILLE, TENNESSEE

*Editorial Advisory Board*

ACADEMIC PRESS, INC.
**Harcourt Brace Jovanovich, Publishers**
San Diego   New York   Berkeley   Boston
London   Sydney   Tokyo   Toronto

ACADEMIC PRESS, INC.
San Diego, California 92101

*United Kingdom Edition published by*
ACADEMIC PRESS, INC. (LONDON) LTD.
24-28 Oval Road, London NW1 7DX

LIBRARY OF CONGRESS CATALOG CARD NUMBER: 54-9110

ISBN   0-12-182058-0   (alk. paper)

PRINTED IN THE UNITED STATES OF AMERICA
88  89  90  91      9  8  7  6  5  4  3  2  1

# Table of Contents

## Section I. $Ca^{2+}$ Fluxes and Regulation

### A. Complex Systems and Subcellular Fractions Involved in $Ca^{2+}$ Regulation

## B. Characterization of $Ca^{2+}$-Pumps and Modulators from Various Sources

## C. $Ca^{2+}$ and Other Channels

## Section II. ATP-Driven Proton Pumps

## Section III. ATP-Driven K$^+$ Pumps

## Addendum

# Contributors to Volume 157

Article numbers are in parentheses following the names of contributors.
Affiliations listed are current.

QAIS AL-AWQATI (47), *Department of Medicine, Columbia University, College of Physicians and Surgeons, New York, New York 10032*

KARLHEINZ ALTENDORF (52), *Fachbereich Biologie/Chemie, Universität Osnabrück, D-4500 Osnabrück, Federal Republic of Germany*

ANTOINE AMORY (38), *Laboratory of Enzymology, Catholic University of Louvain, B-1348 Louvain-la-Neuve, Belgium*

JENS P. ANDERSEN (20), *Institute of Physiology, University of Aarhus, 8000 Aarhus C, Denmark*

YASUHIRO ANRAKU (41), *Department of Biology, Faculty of Science, University of Tokyo, Hongo, Bunkyo-ku, Tokyo 113, Japan*

BOZENA ANTONIU (34), *Department of Muscle Research, Boston Biomedical Research Institute, Boston, Massachusetts 02114, and Department of Neurology, Harvard Medical School, Boston, Massachusetts 02115*

AKIKO BABA (19), *Department of Pure and Applied Sciences, College of Arts and Sciences, The University of Tokyo, Komaba, Meguro-ku, Tokyo 153, Japan*

ALAN B. BENNETT (44), *Mann Laboratory, Department of Vegetable Crops, University of California, Davis, California 95616*

BARRY J. BOWMAN (42), *Department of Biology, University of California, Santa Cruz, California 95064*

EMMA JEAN BOWMAN (42), *Department of Biology, University of California, Santa Cruz, California 95064*

CHRISTOPHER J. BRANDL (22), *Department of Biochemistry, Harvard Medical School, Boston, Massachusetts 02115*

NEIL R. BRANDT (7), *Department of Pharmacology, University of Miami School of Medicine, Miami, Florida 33101*

J-P. BRUNSCHWIG (7), *Department of Pharmacology, University of Miami School of Medicine, Miami, Florida 33101*

ERNESTO CARAFOLI (1, 27), *Laboratory of Biochemistry, Swiss Federal Institute of Technology (ETH), CH-8092 Zurich, Switzerland*

ANTHONY H. CASWELL (7), *Department of Pharmacology, University of Miami School of Medicine, Miami, Florida 33101*

CHRISTOPHER CHADWICK (5), *Department of Molecular Biology, Vanderbilt University, Nashville, Tennesse 37235*

BRIAN K. CHAMBERLAIN (9), *Research and Development, Norwich Eaton Pharmaceuticals Incorporated, Norwich, New York 13815*

JEAN CHIN (45), *Department of Biochemistry, Dartmouth Medical School, Hanover, New Hampshire 03756*

ALICE CHU (4), *Cardiovascular Sciences Section, Department of Medicine, Baylor College of Medicine, Houston, Texas 77030*

SHULAMIT CIDON (48), *Laboratory of Molecular and Cellular Neuroscience, The Rockefeller University, New York, New York 10021*

ROBERTO CORONADO (35), *Department of Physiology and Molecular Biophysics, Baylor College of Medicine, Houston, Texas 77030*

BRIAN COSTELLO (5), *Department of Cell Biology, Vanderbilt University Medical School, Nashville, Tennessee 37235*

GARY E. DEAN (46), *Department of Microbiology and Molecular Genetics, University of Cincinnati, Cincinnati, Ohio 45267*

WILLIAM L. DEAN (28), *Department of Biochemistry, University of Louisville School of Medicine, Louisville, Kentucky 40292*

PAUL DeFOOR (30), *Eastern Regional Research Center, United States Department of Agriculture, Philadelphia, Pennsylvania 19118*

LEOPOLDO DE MEIS (13), *Instituto de Ciências Biomédicas, Departmento de Bioquímica, Universidade Federal do Rio de Janeiro, Rio de Janeiro 21910, Brazil*

MARK C. DIXON (4), *University of Tennessee, Memphis Medical School, Memphis, Tennessee*

JEAN-PIERRE DUFOUR (38, 39), *Laboratory of Brewing Sciences, Catholic University of Louvain, B-1348 Louvain-la-Neuve, Belgium*

Y. DUPONT (14), *Laboratoire de Biophysique Moléculaire et Cellulaire, Département de Recherches Fondamentales, Centre d'Etudes Nucléaires de Grenoble, BP-85X, 38041 Grenoble, France*

L. DUX (21), *Szent-Gyorgyi Albert University Medical School, Institute of Biochemistry, H-6701 Szeged, Hungary*

M. ENDO (2), *Department of Pharmacology, Faculty of Medicine, University of Tokyo, Bunkyo-ku, Tokyo 113, Japan*

WOLFGANG EPSTEIN (51), *Department of Molecular Genetics and Cell Biology, University of Chicago, Chicago, Illinois 60637*

ALEXANDRE FABIATO (31), *Department of Physiology and Biophysics, Medical College of Virginia, Richmond, Virginia 23298*

E. FASSOLD (15), *Department of Physiology, Max-Planck-Institute for Medical Research, 6900 Heidelberg, Federal Republic of Germany*

ADELAIDA FILOTEO (27), *Section of Biochemistry, Mayo Clinic/Foundation, Rochester, Minnesota 55905*

SIDNEY FLEISCHER (3, 4, 5, 6, 9, 10, 23, 24, 25, 30, 36), *Department of Molecular Biology, Vanderbilt University, Nashville, Tennessee 37235*

LARRY FLIEGEL (22), *Department of Pediatrics, University of Alberta, Edmonton, Alberta T6G 2R7, Canada*

RENATA FUCHS (46), *Department of Experimental Pathology, University of Vienna, A-1090 Vienna, Austria*

JUN-ICHI FUJII (11), *Division of Cardiology, Departments of Medicine and Pathophysiology, Osaka University School of Medicine, Fukushima-ku, Osaka 553, Japan*

PETER FÜRST (53), *National Institutes of Health, Bethesda, Maryland 20205*

CYNTHIA J. GALLOWAY (46), *Department of Pharmacy, University of California, San Francisco, San Francisco, California 94720*

ANDRÉ GOFFEAU (38, 39), *Laboratory of Enzymology, Catholic University of Louvain, B-1348 Louvain-la-Neuve, Belgium*

F. GUILLAIN (14), *Service de Biophysique, Département de Biologie, Centre d'Etudes Nucléaires de Saclay, 91190 Gif-Sur-Yvette, France*

WILHELM HASSELBACH (15), *Department of Physiology, Max-Planck-Institute for Medical Research, 6900 Heidelberg, Federal Republic of Germany*

LIN HYMEL (23), *Institute for Biophysics, University of Linz, A-4040 Linz, Austria*

M. IINO (2), *Department of Pharmacology, Faculty of Medicine, University of Tokyo, Bunkyo-ku, Tokyo 113, Japan*

NORIAKI IKEMOTO (34), *Department of Muscle Research, Boston Biomedical Research Institute, Boston, Massachusetts 02114, and Department of Neurology, Harvard Medical School, Boston, Massachusetts 02115*

WHA BIN IM (50), *Experimental Biology, Upjohn Company, Kalamazoo, Michigan 49001*

GIUSEPPE INESI (12), *Department of Biological Chemistry, University of Maryland,*

School of Medicine, Baltimore, Maryland 21201

MAKOTO INUI (10, 11, 24, 36), Department of Molecular Biology, Vanderbilt University, Nashville, Tennessee 37235

LARRY R. JONES (8, 29), Krannert Institute of Cardiology, Indiana University School of Medicine, Indianapolis, Indiana 46202

MASAAKI KADOMA (11), Division of Cardiology, First Department of Medicine, Osaka University School of Medicine, Fukushima-ku, Osaka 553, Japan

MICHIKI KASAI (33), Department of Biophysical Engineering, Faculty of Engineering Science, Osaka University, Toyonaka, Osaka 560, Japan

MASAO KAWAKITA (19), Department of Pure and Applied Sciences, College of Arts and Sciences, The University of Tokyo, Komaba, Meguro-ku, Tokyo 153, Japan

RICHARD M. KAWAMOTO (7), Norwich Eaton Pharmaceuticals Incorporated, Woods Corners, Norwich, New York 13815

DO HAN KIM (34), Department of Medicine, Division of Cardiology, University of Connecticut Health Center, Farmington, Connecticut 06032

MARK KURZMACK (12), Department of Biological Chemistry, University of Maryland, School of Medicine, Baltimore, Maryland 21201

J. J. LACAPERE (14), Département de Biologie, Centre d'Etudes Nucléaires de Saclay, 91190 Gif-Sur-Yvette, France

ROGER A. LEIGH (44), Rothamsted Experimental Station, Harpenden, Hertfordshire AL5 2JQ, England

MARC LE MAIRE (20), Centre de Génétique Moléculaire, Laboratoire propre du Centre National de la Recherche Scientifique, Associíe a l'Université de Paris VI, 91190 Gif-Sur-Yvette, France

DAVID LEWIS (12), Department of Biological Chemistry, University of Maryland, School of Medicine, Baltimore, Maryland 21201

ANSON LOWE (47), Department of Physiology and Biochemistry, University of California, San Francisco, San Francisco, California 94109

THOMAS J. LUKAS (26), Howard Hughes Medical Institute, and Department of Pharmacology, Vanderbilt University School of Medicine, Nashville, Tennessee 37232

JOËL LUNARDI (30), Laboratoire de Biochemie, CEN Grenoble 85X, 38041 Grenoble Cedex, France

DAVID H. MACLENNAN (22), Banting and Best Department of Medical Research, Charles H. Best Institute, University of Toronto, Toronto, Ontario M5G 1L6, Canada

MADOKA MAKINOSE (16), Department of Physiology, Max-Planck-Institute for Medical Research, 6900 Heidelberg, Federal Republic of Germany

A. MARTONOSI (21), State University of New York, Health Science Center, Department of Biochemistry and Molecular Biology, Syracuse, New York 13210

ANDREAS MAURER (25), Department Dia-Fa, F. Hoffman–LaRoche and Company, Ltd., Kaiseraugst, Switzerland

CAROL S. MCDONOUGH (27), Section of Biochemistry, Mayo Clinic/Foundation, Rochester, Minnesota 55905

GERHARD MEISSNER (32, 35), Departments of Biochemistry and Nutrition, and Pharmacology, University of North Carolina, Chapel Hill, North Carolina 27599

IRA MELLMAN (46), Department of Cell Biology, Yale University School of Medicine, New Haven, Connecticut 06510

R. D. MITCHELL (6), Department of Medicine, Krannert Institute of Cardiology, Indiana University, Indianapolis, Indiana 46202

JESPER V. MØLLER (20), Institute of Medical Biochemistry, University of Aarhus, 8000 Aarhus C, Denmark

YOSHINORI MORIYAMA (48), Roche Institute of Molecular Biology, Roche Research Center, Nutley, New Jersey 07110

NATHAN NELSON (48), *Roche Institute of Molecular Biology, Roche Research Center, Nutley, New Jersey 07110*

KAZUKI NUNOGAKI (33), *Department of Biophysical Engineering, Faculty of Engineering Science, Osaka University, Toyonaka, Osaka 560, Japan*

YOSHINORI OHSUMI (41), *Department of Biology, Faculty of General Education, University of Tokyo, Komaba, Meguro-ku, Tokyo 153, Japan*

TAKAYUKI OZAWA (25), *Department of Biomedical Chemistry, University of Nagoya, Tsuruma-Cho 65, Showa-ku, Japan*

PHILIP PALADE (6), *Department of Physiology and Biophysics, University of Texas Medical Branch, Galveston, Texas 77550*

JOHN T. PENNISTON (27), *Section of Biochemistry, Mayo Clinic/Foundation, Rochester, Minnesota 55905*

JAMES W. POLAREK (51), *Department of Biochemistry and Molecular Biology, University of Chicago, Chicago, Illinois 60637*

EDD C. RABON (50), *Department of Medicine, Center for Ulcer Research and Education, University of California, Los Angeles, Los Angeles, Califronia 90024*

EFRAIM RACKER (49), *Department of Biochemistry, Cell and Molecular Biology, Cornell University, Ithaca, New York 14853*

JOHN P. REEVES (37), *Roche Institute of Molecular Biology, Roche Research Center, Nutley, New Jersey 07110*

GEORGE SACHS (50), *Department of Medicine, Center for Ulcer Research and Education, University of California, Los Angeles, Los Angeles, California 90024*

AKITSUGU SAITO (4, 6, 10), *Department of Molecular Biology, Vanderbilt University, Nashville, Tennessee 37235*

KIMIKO SAITO-NAKATSUKA (19), *Department of Pure and Applied Sciences, College of Arts and Sciences, The University of Tokyo, Komaba, Meguro-ku, Tokyo 153, Japan*

GENE A. SCARBOROUGH (43), *Department of Pharmacology, University of North Carolina at Chapel Hill, Chapel Hill, North Carolina 27599*

DONALD L. SCHNEIDER (45), *Department of Biochemistry, Dartmouth Medical School, Hanover, New Hampshire 03756*

STEVEN SEILER (3, 4), *Krannert Institute of Cardiology, Department of Pharmacology/Toxicology, Indiana University School of Medicine, Indianapolis, Indiana 46202*

RAMÓN SERRANO (40), *European Molecular Biology Laboratory, 6900 Heidelberg, Federal Republic of Germany*

ANNETTE SIEBERS (52), *Fachbereich Biologie/Chemie, Universität Osnabrück, D-4500 Osnabrück, Federal Republic of Germany*

HEATHER K. B. SIMMERMAN (29), *Krannert Institute of Cardiology, Indiana University School of Medicine, Indianapolis, Indiana 46202*

JEFFREY S. SMITH (35), *Merck, Sharp and Dohme, W26-208 Research Laboratories, West Point, Pennsylvania 19486*

MARC SOLIOZ (53), *Department of Clinical Pharmacology, University of Bern, 3010 Bern, Switzerland*

ROGER M. SPANSWICK (44), *Section of Plant Biology, Cornell University, Ithaca, New York 14853*

DENNIS K. STONE (49), *Departments of Internal Medicine and Physiology, University of Texas Southwestern Medical Center at Dallas, Dallas, Texas 75235*

MICHIHIKO TADA (11), *Division of Cardiology, Departments of Medicine and Pathophysiology, Osaka University School of Medicine, Fukushima-ku, Osaka 553, Japan*

HARUHIKO TAKISAWA (16), *Department of Biology, Faculty of Science, Osaka University, Toyonaka, Osaka 560, Japan*

MASASHI TANAKA (25), *Department of Biomedical Chemistry, University of Nagoya, Tsuruma-Cho 65, Showa-ku, Japan*

K. A. Taylor (21), *Duke University Medical Center, Department of Anatomy, Durham, North Carolina 27710*

H. P. Ting-Beall (21), *Duke University Medical Center, Department of Anatomy, Durham, North Carolina 27710*

Tian Yow Tsong (18), *Department of Biological Chemistry, The Johns Hopkins University School of Medicine, Baltimore, Maryland 21205*

Etsuko Uchida (41), *Department of Neurology, Juntendo University, School of Medicine, Hongo, Bunkyo-ku, Tokyo 113, Japan*

S. Varga (21), *University Medical School, Central Research Laboratory, H-4012 Debrecen, Hungary*

Mark O. Walderhaug (51), *Department of Molecular Genetics and Cell Biology, University of Chicago, Chicago, Illinois 60637*

Sherry Wang (10), *Department of Molecular Biology, Vanderbilt University, Nashville, Tennessee 37235*

Takahide Watanabe (17), *Molecular Physiology, National Cardiovascular Center Research Institute, Fujishiro-dai, Suita, Osaka 565, Japan*

D. Martin Watterson (26), *Howard Hughes Medical Institute, and Department of Pharmacology, Vanderbilt University School of Medicine, Nashville, Tennessee 37232*

Adam D. Wegener (29), *Krannert Institute of Cardiology, Indiana University School of Medicine, Indianapolis, Indiana 46202*

Leszek Wieczorek (52), *Fachbereich Biologie/Chemie, Universität Osnabrück, D-4500 Osnabrück, Federal Republic of Germany*

Xiao-Song Xie (49), *Department of Physiology, University of Texas Southwestern Medical Center at Dallas, Dallas, Texas 75235*

Motonori Yamaguchi (17), *Nihon Schering K. K., Nishimiyahara, Yodogawa-ku, Osaka 532, Japan*

Tetsuro Yamashita (19), *Department of Pure and Applied Sciences, College of Arts and Sciences, The University of Tokyo, Komaba, Meguro-ku, Tokyo 153, Japan*

Kimiko Yasuoka-Yabe (19), *National Institute of Animal Health, Tsukuba-Science City, Ibaraki 305, Japan*

# Preface

The transport volumes of Biomembranes were initiated with Volumes 125 and 126 (Transport in Bacteria, Mitochondria, and Chloroplasts) of *Methods in Enzymology*. Biological transport represents a continuation of methodology regarding the study of membrane function, Volumes 96–98 having dealt with membrane biogenesis, assembly, targeting, and recycling.

This is a particularly good time to cover the topic of biological membrane transport because a strong conceptual basis for its understanding now exists. Membrane transport has been divided into five topics. Topic 2 is covered in Volumes 156 and 157. The remaining three topics will be covered in subsequent volumes of the Biomembranes series.

1. Transport in Bacteria, Mitochondria, and Chloroplasts
2. ATP-Driven Pumps and Related Transport
3. General Methodology of Cellular and Subcellular Transport
4. Cellular and Subcellular Transport: Eukaryotic (Nonepithelial) Cells
5. Cellular and Subcellular Transport: Epithelial Cells

We are fortunate to have the good counsel of our Advisory Board. Their input ensures the quality of these volumes. The same Advisory Board has served for the complete transport series. Valuable input on the outlines of the five volumes was also provided by Qais Al-Awqati, Ernesto Carafoli, Halvor Christensen, Isadore Edelman, Joseph Hoffman, Phil Knauf, and Hermann Passow.

The names of our board members and advisors were inadvertantly omitted in Volumes 125 and 126. When we noted the omission, it was too late to rectify the problem. For Volumes 125 and 126, we are also pleased to acknowledge the advice of Angelo Azzi, Youssef Hatefi, Dieter Oesterhelt, and Peter Pedersen.

Additional valuable input to Volumes 156 and 157 was obtained from Jens Skou, Peter Jørgensen, Steve Karlish, Gerhard Meissner, and Giuseppi Inesi. The enthusiasm and cooperation of the participants have enriched and made these volumes possible. The friendly cooperation of the staff of Academic Press is gratefully acknowledged.

These volumes are dedicated to Professor Sidney Colowick, a dear friend and colleague, who died in 1985. We shall miss his wise counsel, encouragement, and friendship.

Sidney Fleischer
Becca Fleischer

# METHODS IN ENZYMOLOGY

EDITED BY

## Sidney P. Colowick and Nathan O. Kaplan

VANDERBILT UNIVERSITY
SCHOOL OF MEDICINE
NASHVILLE, TENNESSEE

DEPARTMENT OF CHEMISTRY
UNIVERSITY OF CALIFORNIA
AT SAN DIEGO
LA JOLLA, CALIFORNIA

# METHODS IN ENZYMOLOGY

### EDITORS-IN-CHIEF

## Sidney P. Colowick and Nathan O. Kaplan

# Section I

# Ca$^{2+}$ Fluxes and Regulation

*Article 1*
A. Complex Systems and Subcellular Fractions Involved
in Ca$^{2+}$ Regulation
*Articles 2 through 11*

B. Characterization of Ca$^{2+}$-Pumps and Modulators
from Various Sources
*Articles 12 through 30*

C. Ca$^{2+}$ and Other Channels
*Articles 31 through 37*

## [1] Membrane Transport of Calcium: An Overview

By ERNESTO CARAFOLI

$Ca^{2+}$ signaling plays an important role in numerous areas of cell biochemistry and physiology, ranging from exocrine and hormonal secretion, to muscle and nonmuscle motility, to the activity and regulation of several important metabolic pathways. The central element in the signaling function of $Ca^{2+}$ is its reversible complexation by specific proteins, which bind $Ca^{2+}$ in micromolar or submicromolar concentrations with optimal affinity in the presence of much larger concentrations of other divalent and monovalent cations such as $Mg^{2+}$, $Na^+$, and $K^+$. Many of these proteins are dissolved in the cytosol or organized in nonmembrane structures such as the myofibrils. The structure of these proteins repeats a standardized model first described for the muscle $Ca^{2+}$-binding protein parvalbumin,[1] and now referred to as the E/F-hand structure. $Ca^{2+}$ buffering and thus the regulation of the $Ca^{2+}$ messenger function by the cytosolic (or nonmembranous) $Ca^{2+}$-binding proteins, albeit important, are quantitatively limited by the amount of protein present in the cell. Cell $Ca^{2+}$, however, is also specifically and reversibly complexed by a second class of binding proteins, intrinsic to membranes, and apparently different from the E/F-hand structural pattern.[2] The quantitative limitation in the total $Ca^{2+}$-handling capacity linked to the amount of binding protein present may be overcome by moving $Ca^{2+}$ in and out of the cell and across intracellular membrane boundaries. Thus, intrinsic membrane proteins can be (they indeed are) present in minute amounts and still perform the $Ca^{2+}$-buffering function efficiently, since they bind $Ca^{2+}$ on one side of a membrane, transport it across the membrane, release it at the other side, and come back uncomplexed for the next binding and transport cycle. Thus, membrane transport performs a role of paramount importance in the regulation of the signaling function of $Ca^{2+}$. This concept is underscored by the fact that numerous $Ca^{2+}$-transporting systems have been identified in all (eukaryotic) cells. They are located in the plasma membrane and in the intracellular organelles, and are characterized by different functional properties (e.g., affinity for $Ca^{2+}$, total transport capacity, sensitivity to modulating agents). They provide cells with the means to regulate the ionic concentration of $Ca^{2+}$ in various compartments according to the

[1] P. C. Moews and R. H. Kretsinger, J. Mol. Biol. 91, 201 (1976).
[2] D. H. MacLennan, C. J. Brandl, B. Korczak, and N. M. Green, Nature (London) 316, 696 (1985).

different demands of the physiological cycle. The transport systems will respond to situations which require rapid and fine regulation of $Ca^{2+}$, as well as to situations where longer term and larger scale buffering of $Ca^{2+}$ is demanded.

At least seven $Ca^{2+}$-transporting systems can be recognized in normal eukaryotic cells: three are present in the plasma membrane (a $Ca^{2+}$-ATPase, a $Na^+/Ca^{2+}$ exchanger, and a $Ca^{2+}$ channel), two in the inner membrane of mitochondria (an electrophoretic uniporter and a $Na^+/Ca^{2+}$ exchanger), and two in the endo(sarco)plasmic reticulum (a $Ca^{2+}$-ATPase and a $Ca^{2+}$ channel). Additional transporting systems have been suggested in other membranes as well (e.g., the lysosomes and the Golgi body), but have not yet been characterized to an extent which would warrant their inclusion among the certified $Ca^{2+}$ transporting systems. All of these systems can be simplified to four basic transport modes: ATPases, exchangers, channels, and electrophoretic uniporters (Table I). The ATPase mode is the only one which confers on the system the ability to interact with $Ca^{2+}$ with high affinity, and is therefore used [by plasma membranes and endo(sarco)plasmic reticulum] whenever the situation demands the fine regulation of $Ca^{2+}$ in submicromolar concentrations. The other three modes operate with lower $Ca^{2+}$ affinity ($K_m$, $>1 \mu M$), and normally have larger transporting capacity. They are used as long-term $Ca^{2+}$ buffers. In $Ca^{2+}$ exchange reactions $Na^+$ is a preferred partner because of the chemical similarities between $Na^+$ and $Ca^{2+}$, which permit these ions to fit into the same basic binding sites (cavities) on protein ligands.[3] In the chapters that follow, these transport systems will be discussed in detail. In this overview, a succinct description of the properties of transport systems will be offered, with the aim of providing an integrated view of their role in the overall process of intracellular $Ca^{2+}$ homeostasis.

## Plasma Membrane $Ca^{2+}$-Transporting Systems

The $Ca^{2+}$ channel mediates the penetration of $Ca^{2+}$ into cells. Although the channel is probably present in all cell types,[4] it is normally studied in excitable plasma membranes, where $Ca^{2+}$-dependent electrical currents can be measured.[5,6] Considerable advances in the characterization of the $Ca^{2+}$ channels have been made possible by the use of the so-

[3] R. J. P. Williams, *Symp. Soc. Exp. Biol.* **30**, 1 (1976).
[4] L. Varecka and E. Carafoli, *J. Biol. Chem.* **257**, 7414 (1982).
[5] P. G. Kostyuk, *Biochim. Biophys. Acta* **650**, 128 (1981).
[6] H. Reuter, *Annu. Rev. Physiol.* **46**, 473 (1984).

TABLE I
Ca$^{2+}$-TRANSPORTING SYSTEMS OF CELL MEMBRANES

| Transporting mode | Membrane | Ca$^{2+}$ affinity |
| --- | --- | --- |
| ATPases | Plasma membranes | |
| | Endo(sarco)plasmic reticulum | High |
| Exchangers (Na$^+$/Ca$^{2+}$) | Plasma membrane | |
| | Inner mitochondrial membrane | Low |
| Channels | Plasma membranes | |
| | Endo(sarco)plasmic reticulum? | Low[a] |
| Electrophoretic uniporters | Inner mitochondrial membrane | Low |

[a] The calcium release channel of sarcoplasmic reticulum has recently been isolated, this volume [36]. It is activated by submicromolar concentrations of calcium ion.

called Ca$^{2+}$ antagonists (or Ca$^{2+}$-entry blockers),[7] and by the introduction of the patch-clamp technique, which permits the study of individual Ca$^{2+}$ channels.[8] The Ca$^{2+}$ channels are not completely selective; in fact, they transport Ba$^{2+}$ and Sr$^{2+}$ in preference to Ca$^{2+}$, and even Na$^+$ is readily transported if Ca$^{2+}$ is excluded from the experimental system. The conductance of the channel is of the order of 15–25 pS, corresponding to the transfer of about 3000 Ca$^{2+}$ ions per millisecond, which is the average duration of the opening event. It is important to emphasize here that only one Ca$^{2+}$ atom at a time is allowed to pass through the open channel. In excitable cells, (e.g., heart) the opening of the channels is triggered by the depolarization of the plasma membrane, which is in turn produced by the stimulation-induced influx of Na$^+$ through the Na$^+$ channels. The relationship linking transmembrane potential to the probability of Ca$^{2+}$ channel-opening saturates at considerably less than one. However, the cAMP-mediated phosphorylation of a component of the Ca$^{2+}$ channel greatly increases the probability of channel opening.[8] Work aimed at the molecular resolution of the Ca$^{2+}$ channel is currently in progress in several laboratories and it now appears that the channel consists of multiple protein subunits. Four subunits having $M_r$ values of 175,000, 170,000, 52,000, and 30,000 are now found.

[7] A. Fleckenstein, H. Tritthard, B. Fleckenstein, A. Herbst, and G. Gründ, *Pfluegers Arch. Gesamte Physiol. Menschen Tiere* **307**, R25 (1969).
[8] H. Reuter, C. F. Stevens, R. W. Tsien, and G. Yellen, *Nature (London)* **297,** 501 (1982).

The $Ca^{2+}$-pumping ATPase, first described in 1966,[9] ejects $Ca^{2+}$ from all (eukaryotic) cells studied. It is one of the ion-motive ATPases which forms an acyl phosphate during the reaction cycle and is sensitive to vanadate. It interacts with $Ca^{2+}$ with high affinity ($K_m$, $<0.5 \mu M$), and is thus able to pump $Ca^{2+}$ out of cells even when its level in the cytosol becomes very low, e.g., during the diastole of heart cells. However, the amount of ATPase present in most plasma membranes is minute, resulting in a low total transport capacity. One important property of the ATPase is the stimulation by calmodulin, which shifts the ATPase to its appropriate high $Ca^{2+}$ affinity state. However, in the absence of calmodulin, the ATPase can be shifted to the high $Ca^{2+}$ affinity state by acidic phospholipids or polyunsaturated fatty acids (particularly active are the phosphorylated metabolites of phosphatidylinositol) or by a controlled proteolytic treatment. Calmodulin stimulation is due to the direct interaction of the activator with the ATPase, and not to the stimulation of a calmodulin-dependent kinase, which is the case for the analogous ATPase of sarcoplasmic reticulum. The direct interaction with the ATPase has been exploited to purify the enzyme from membrane environments: first, in erythrocytes and, then, in a number of other cells, using calmodulin columns.[10] The purified enzyme is a single polypeptide of $M_r$ 138K, which repeats the properties of the enzyme *in situ*. It can be reconstituted into liposomes,[11] where it pumps $Ca^{2+}$ with a stoichiometry to ATP of 1 : 1 in exchange for $H^+$. Recent work[12] on trypsin fragments of the purified ATPase has shown that polypeptides of $M_r$ 90K, 85K, and 81K, reconstituted into liposomes, are still capable of $Ca^{2+}$ transport. Another interesting observation on the $Ca^{2+}$-ATPase of plasma membrane is its stimulation by a cAMP-dependent phosphorylation process. The ATPase molecule itself is the target of the phosphorylation process.[13]

The $Na^+/Ca^{2+}$ exchanger is a lower $Ca^{2+}$ affinity system, which is particularly active in excitable plasma membranes.[14,15] It functions in both directions, i.e., it mediates both the influx and the efflux of $Ca^{2+}$. The direction of the $Ca^{2+}$ movements is affected not only by the magnitude of the trans plasma membrane $Na^+$ and $Ca^{2+}$ gradients, but also by the magnitude and sign of the trans plasma membrane electrical potential. This is due to the fact that the exchanger operates electrogenically, ex-

[9] H. Schatzmann, *Experientia* **22**, 364 (1966).
[10] V. Niggli, J. T. Penniston, and E. Carafoli, *J. Biol. Chem.* **254**, 9955 (1979).
[11] V. Niggli, E. Sigel, and E. Carafoli, *J. Biol. Chem.* **257**, 2350 (1982).
[12] G. Benaim, M. Zurini, and E. Carafoli, *J. Biol. Chem.* **259**, 8471 (1984).
[13] L. Neyses, L. Reinlib, and E. Carafoli, *J. Biol. Chem.* **260**, 10283 (1985).
[14] H. Reuter and N. Seitz, *J. Physiol. (London)* **195**, 451 (1968).
[15] M. P. Blaustein and A. L. Hodgkin, *J. Physiol. (London)* **200**, 497 (1969).

changing three $Na^+$ for one $Ca^{2+}$. The electrogenicity of the exchanger is indeed a necessity, if the system is to maintain the steep gradient of $Ca^{2+}$ ($\sim 10^4$) normally present across plasma membranes. Important advances in the definition of some of the properties of the exchanger, at least in heart cells, have been made possible by the introduction of plasma membrane vesicles as an experimental tool.[16] Thus, it has been possible to establish that the system interacts with $Ca^{2+}$ with low affinity ($K_m$, 1–5 $\mu M$), but transports $Ca^{2+}$ with a much larger total capacity than the ATPase. The affinity of the exchanger for $Ca^{2+}$ is increased somewhat by a calmodulin-directed phosphorylation system,[17] but not to a point where the exchanger could be considered as a high-affinity system. At the molecular level, very little is known about the exchanger. Recent solubilization and reconstitution studies have tentatively identified it as a protein of $M_r$ 82K, 70K, or 33K.[18–20]

## $Ca^{2+}$-Transporting Systems in Endo(sarco)plasmic Reticulum

Although interest in and knowledge of the endoplasmic reticulum as a $Ca^{2+}$-transporting system have increased considerably in recent years, most of the studies on $Ca^{2+}$ uptake have so far been carried out on the sarcoplasmic reticulum. The enzyme responsible is an ATPase of the same type as that of the plasma membrane, which interacts with $Ca^{2+}$ with high affinity ($K_m$, 0.5 $\mu M$) and transports it with a stoichiometry to ATP of 2:1. Since the ATPase is very abundant in the membrane of the sarcoplasmic reticulum (up to 90% of the total protein in some skeletal muscles), the organelle can rapidly remove very large amounts of $Ca^{2+}$ from the cytosol (20–40 nmol per milligram of membrane protein per second at 25° in some skeletal muscles, but only 5–10 times less in heart). The $Ca^{2+}$-ATPase has been isolated as a single polypeptide of $M_r$ 115K[21] that can be reconstituted in liposomes with reasonable pumping efficiency.[22] In heart sarcoplasmic reticulum, an acidic proteolipid termed phospholamban,[23] which is phosphorylated by both a cAMP-dependent and a calmodulin-dependent kinase, interacts with the ATPase and stimu-

[16] J. P. Reeves and J. L. Sutko, *Proc. Natl. Acad. Sci. U.S.A.* **76**, 590 (1979).

[17] P. Caroni and E. Carafoli, *Eur. J. Biochem.* **132**, 451 (1983).

[18] C. C. Hale, R. S. Slaughter, D. C. Ahrens, and J. P. Reeves, *Proc. Natl. Acad. Sci. U.S.A.* **81**, 6569 (1984).

[19] A. Barzilai, R. Spanier, and H. Rahamimoff, *Proc. Natl. Acad. Sci. U.S.A.* **81**, 6521 (1984).

[20] L. Soldati, S. Longoni, and E. Carafoli, *J. Biol. Chem.* **260**, 13321 (1985).

[21] C. J. Brandl, Larry Fliegel, and D. H. MacLennan, this volume [22].

[22] G. Inesi, R. Nakamoto, L. Hymel, and S. Fleischer, *J. Biol. Chem.* **258**, 14804 (1983).

[23] M. Tada, M. A. Kirchberger, and A. M. Katz, *J. Biol. Chem.* **249**, 6178 (1974).

lates both its hydrolytic and its Ca$^{2+}$-transporting activity. A very important advance in our understanding of the molecular properties of the ATPase has been the clarification of its primary structure.[2] As mentioned above, the amino acid sequence of the ATPase bears no resemblance to that of the soluble Ca$^{2+}$-binding proteins of the E/F-hand type.

The path for Ca$^{2+}$ release from sarcoplasmic reticulum is still not understood, but a specific channel is generally postulated. The opening of the channel must be somehow connected to the depolarization of the plasma membrane which, at least in the heart, would spread from the T system to the terminal cisternae of sarcoplasmic reticulum. Interesting studies on skinned heart fibers[24] have characterized in detail the phenomenon of Ca$^{2+}$-induced Ca$^{2+}$ release, i.e., the opening of the putative Ca$^{2+}$ release channels by the addition of small amounts of Ca$^{2+}$ to the spaces surrounding sarcoplasmic reticulum.

Interest in endoplasmic reticulum as a Ca$^{2+}$ reservoir has been heightened by the finding that inositol trisphosphate, one of the two important catabolites of phosphatidylinositol diphosphate, specifically releases Ca$^{2+}$ from endoplasmic reticulum in response to a number of plasma membrane stimuli that activate a phospholipase. The release was originally demonstrated in permeabilized exocrine pancreas cells[25] but, more recently, has been shown in isolated endoplasmic reticulum vesicles. In the latter system, however, particular conditions must be applied.[26]

## Ca$^{2+}$-Transporting Systems of Mitochondria

Mitochondria possess separate pathways for Ca$^{2+}$ influx and efflux.[27] The uptake system is powered by the electrical component of the proton-motive force maintained by respiration across the inner membrane, and transfers Ca$^{2+}$ into the matrix space without charge compensation. Although a proteinaceous carrier defined as an electrophoretic uniporter is generally postulated, its molecular definition has not been determined despite intensive efforts by several laboratories. It has been established, however, that the route interacts with Ca$^{2+}$ with low affinity ($K_m$, 1–10 $\mu M$) and that it can lead to the accumulation of very large amounts of Ca$^{2+}$ in the matrix space, provided that mitochondria are permitted to take up inorganic phosphate on one of their phosphate carriers. The accumulated phosphate precipitates Ca$^{2+}$ into the matrix, thus making room for the influx of more Ca$^{2+}$ and permitting mitochondria to function as

[24] A. Fabiato and F. Fabiato, *J. Physiol.* (*London*) **249**, 457 (1975).
[25] H. Streb, R. F. Irvine, M. J. Berridge, and I. Schulz, *Nature* (*London*) **306**, 67 (1983).
[26] A. P. Dawson and R. F. Irvine, *Biochem. Biophys. Res. Commun.* **120**, 858 (1984).
[27] E. Carafoli, *FEBS Lett.* **104**, 1 (1979).

large-capacity buffers of cell $Ca^{2+}$. Since the uptake route is energy dependent, it is naturally inhibited by agents that interfere with the production of the trans inner membrane electrical potential, like uncouplers or respiration inhibitors. The most useful inhibitor, however, is a polycation termed ruthenium red,[28] which blocks the uptake route at submicromolar concentrations without interfering with other mitochondrial reactions. Ruthenium red has been instrumental in helping to identify the route for $Ca^{2+}$ release from mitochondria, which is now known to be a $Na^+/Ca^{2+}$ exchanger.[29] An independent $Ca^{2+}$ release route was demanded by the fact that the electrophoretic uptake route, driven by a continuously present transmembrane potential of the order of $-200$ mV, is essentially irreversible, and never mediates $Ca^{2+}$ efflux from the matrix. The independent release route was discovered by blocking the electrophoretic uniporter with ruthenium red to prevent the reuptake of the lost $Ca^{2+}$, and by adding $Na^+$ to the mitochondria after the accumulation of a $Ca^{2+}$ pulse to successfully initiate a $Na^+/Ca^{2+}$ exchange reaction that allowed $Ca^{2+}$ to flow out of mitochondria.[29] Later studies[30-32] have established the properties of the mitochondrial $Na^+/Ca^{2+}$ exchange system, among them its expected electroneutrality, its predominance in mitochondria from excitable tissues (e.g., heart), the specificity of $Na^+$ as a partner in the exchange process, and the fact that the route is half-maximally activated by $Na^+$ concentrations which are in the range of those presumed to exist in most cytoplasms. The route has low $Ca^{2+}$ affinity ($K_m$, $\sim 10 \mu M$)[33] and has rather low maximal transport capacity (about 0.2 nmol per milligram of mitochondrial protein per second at 25° in heart, which is a tissue where the exchanger is very active). An interesting and potentially useful development has been the discovery that one of the $Ca^{2+}$ antagonists, a benzothiazepine termed diltiazem, effectively inhibits the exchanger.[34]

Integrated Role of the $Ca^{2+}$-Transporting Systems in the Cellular
   Homeostasis of $Ca^{2+}$ and in the Regulation of Its Signaling Function

The kinetic parameters of the various systems described above are presented in Table II. Clearly, the mitochondrion is a low-$Ca^{2+}$-affinity

[28] C. L. Moore, *Biochem. Biophys. Res. Commun.* **42,** 298 (1971).
[29] E. Carafoli, R. Tiozzo, G. Lugli, F. Crovetti, and C. Kratzing, *J. Mol. Cell. Cardiol.* **6,** 361 (1974).
[30] M. Crompton, M. Capano, and E. Carafoli, *Eur. J. Biochem.* **69,** 453 (1976).
[31] M. Crompton, R. Moser, H. Lüdi, and E. Carafoli, *Eur. J. Biochem.* **8,** 25 (1978).
[32] H. Affolter and E. Carafoli, *Biochem. Biophys. Res. Commun.* **95,** 193 (1980).
[33] M. Crompton, M. Künzi, and E. Carafoli, *Eur. J. Biochem.* **79,** 549 (1977).
[34] P. L. Vaghy, J. D. Johnson, M. A. Matlib, T. Wong, and A. Schwartz, *J. Biol. Chem.* **257,** 6000 (1982).

TABLE II

KINETIC PROPERTIES OF Ca²⁺ TRANSPORTING SYSTEMS IN
CELL MEMBRANES[a]

| Transporting system | $K_m$ (Ca²⁺) ($\mu M$) | $V_{max}$ of transport (nmol Ca²⁺ per mg of membrane protein per sec, at 25°) |
|---|---|---|
| Ca²⁺-ATPase of plasma membranes | 0.5 | ~0.5 |
| Na⁺/Ca²⁺ exchanger of plasma membranes | 1–5 | 15–30 |
| Ca²⁺-ATPase of sarco-plasmic reticulum | 0.5 | 20–30 |
| Release route of sarco-plasmic reticulum[a] | <1 | — |
| Uptake route in mito-chondria | 1–10 | ~10 |
| Na⁺/Ca²⁺ exchanger of mitochondria | ~10 | ~0.2 |

[a] The calcium release channel of sarcoplasmic reticulum has recently been isolated (cf. this volume [36]). The data presented are from the membrane type where the process is most active (e.g., heart for the case of the Na⁺/Ca²⁺ exchanger of mitochondria). They refer to the transporting protein in the membrane environment. However, since the membranes are normally studied in the isolated state, the data in the table do not necessarily correspond to those which would prevail in the cellular environment *in vivo*. This is particularly striking for the case of the uptake route of mitochondria, which is slowed down considerably by physiological concentrations of Mg²⁺.

structure, whereas the endo(sarco)plasmic reticulum is the cellular locus for high-affinity Ca²⁺ interaction. It is logical to conclude from this that the mitochondrion is mainly a long-term buffer for cellular Ca²⁺, whereas the reticulum is its rapid and fine tuner. The plasma membrane, at least in excitable cells, contains both a high- and a low-affinity Ca²⁺ ejection system, and it seems obvious to postulate that the two operate in parallel, the former (the ATPase) acting as a continuously active export system, the latter (the exchanger) being reasonably active and predominant only under conditions that lead to the increase of the cytosolic free Ca²⁺ concentration (e.g., at peak activation in heart).

In considering the role of mitochondria in detail, it now becomes clear that the energy-driven uptake process, coupled to the energy-independent $Na^+/Ca^{2+}$ exchanger, corresponds to an energy-dissipating operation that has indeed been termed the "mitochondrial $Ca^{2+}$ cycle."[27] It may be assumed that the release leg of the cycle (the exchanger) operates at a relatively constant slow rate, whereas the electrophoretic uptake leg oscillates below and above the rate of the release exchanger, responding essentially to the fluctuations in the cytosolic $Ca^{2+}$. It is important to bear in mind, however, that even under the most favorable conditions that may be expected during the physiological life of cells, the uptake route is heavily damped by cytosolic $Mg^{2+}$ (see Table II) and by the fact that cytosolic $Ca^{2+}$ will never reach values that would fully activate it. Thus, under normal conditions the futile $Ca^{2+}$ cycle in mitochondria operates at a low speed, and contributes less than the reticulum and the plasma membrane to the buffering of the cytosolic $Ca^{2+}$. Under normal cell conditions, the rationale for the existence of the sophisticated machinery for the handling of $Ca^{2+}$ by mitochondria must be sought elsewhere, i.e., in the existence of at least 3 $Ca^{2+}$-dependent dehydrogenases in the matrix.[35] It has become clear that the correct functioning of the tricarboxylic acid cycle requires that matrix-free $Ca^{2+}$ be properly balanced at the micromolar level. In normal cells, mitochondria must be regarded more as organelles that control their own internal $Ca^{2+}$ than as organelles that play a major role in cytosolic $Ca^{2+}$ homeostasis. It must not be forgotten, however, that mitochondria can store very large amounts of $Ca^{2+}$ in the matrix without essential variations in its ionic concentration, due to the simultaneous accumulation of phosphate and to the precipitation of insoluble calcium phosphate deposits (see above). This may become very important if pathological conditions allow the cytosolic concentration of $Ca^{2+}$ to increase to levels where a significant activation of the electrophoretic uptake route may be expected. Under these conditions the mitochondrial uptake system becomes important for the buffering of cytosolic $Ca^{2+}$, removing it to the matrix until it can be released slowly, as expected, on the $Na^+/Ca^{2+}$ exchanger.

[35] J. G. McCormack and R. M. Denton, *Biochem. J.* **10,** 151 (1978).

## [2] Measurement of Ca²⁺ Release in Skinned Fibers from Skeletal Muscle

*By* M. ENDO and M. IINO

### Introduction

Ca²⁺ release from the sarcoplasmic reticulum (SR) is one of the most important steps in excitation–contraction coupling of skeletal muscle. However, while various kinds of stimuli experimentally applied to the SR can be shown to cause Ca²⁺ release, the mechanism of *physiological* Ca²⁺ release from the SR is not yet known.[1] We have been trying to find out what kinds of stimuli directly applied to the SR of skinned muscle fibers cause Ca²⁺ release in the hope that, from the effective stimuli, we might be able to determine the mechanism utilized in physiological Ca²⁺ release. In this chapter, we will describe in detail the methods for measurement of Ca²⁺ release in skinned fibers that are used in our laboratory.

### General Considerations

In the experiments on Ca²⁺ release from the SR, one must take the following facts into account. (1) There are several Ca²⁺ release mechanisms in the SR, which are different at least in their activation mechanisms but probably also in their Ca²⁺ channels.[1] Therefore, determinations regarding one of the release mechanisms can not be applied automatically to Ca²⁺ release in general. (2) SR membrane also has a strong Ca²⁺ uptake activity through the calcium pump ATPase.[2,3] Therefore, unless the Ca²⁺-pump is inhibited by the experimental condition, Ca²⁺ released by any means tends to activate the pump and partly be taken up again. (3) Ca²⁺ also activates one of the Ca²⁺ release mechanisms present in the SR, the so-called Ca²⁺-induced Ca²⁺ release mechanism.[4,5] The secondary Ca²⁺ release caused by Ca²⁺ released by a primary process must be strictly distinguished from the primary release, if the properties of the latter are to be examined. For this reason, simple monitoring of

[1] M. Endo, *in* "Regulation of Calcium Transport across Muscle Membranes" (A. E. Shamoo, ed.), Current Topics in Membranes and Transport, Vol. 25, p. 181. Academic Press, New York, 1985.
[2] S. Ebashi and F. Lipmann, *J. Cell Biol.* **14,** 389 (1962).
[3] W. Hasselbach and M. Makinose, *Biochem. Z.* **333,** 518 (1961).
[4] L. E. Ford and R. J. Podolsky, *Science* **167,** 58 (1970).
[5] M. Endo, M. Tanaka, and Y. Ogawa, *Nature (London)* **228,** 34 (1970).

Ca$^{2+}$ concentrations outside the SR is not suitable for the measurement of the primary Ca$^{2+}$ release process. Instead Ca$^{2+}$ concentrations outside the SR should be fixed as far as possible by the use of a Ca$^{2+}$ buffer. Under this condition, the Ca$^{2+}$ release can be estimated by the time course of decrease in the amount of Ca$^{2+}$ in the SR.

Ca$^{2+}$ release from the SR can be studied in skinned fibers as described here or in vesicles of fragmented sarcoplasmic reticulum (FSR) as described in [32–34], this volume. Skinned fibers have a physiological advantage over the FSR and, probably for this reason, Ca$^{2+}$ release is easier to evoke in skinned fibers than in FSR. In fact, many important modes of Ca$^{2+}$ release have been found in skinned fibers first and then confirmed with FSR.[4–6] It is also easier to exchange the environmental solutions quickly in skinned fibers, whereas in FSR an exchange of solutions can only be made by filtration or centrifugation or, less precisely, by dilution. Another advantage (or possible disadvantage) of skinned fibers is the fact that the lumen of the whole SR is most probably continuous in skinned fibers, so that even if a single Ca$^{2+}$ channel throughout the whole SR is open, the Ca$^{2+}$ release continues until the electrochemical potential gradient of Ca$^{2+}$ disappears between the inside and the outside of the SR lumen. In contrast, some FSR vesicles may not have a certain kind of Ca$^{2+}$ channel and thus may not respond to stimuli that activate that kind of channel. This heterogeneity of the vesicles must be kept in mind in interpreting results of FSR experiments. On the other hand, it is easier with FSR than with skinned fibers to follow Ca$^{2+}$ movements precisely and to conduct a large number of experiments by changing experimental conditions. Another disadvantage of skinned fibers is that while an exchange of solutions in the extrafiber space could be immediate, the change in composition in the real environment of the SR is rather slow due to a relatively long diffusion distance.

### Preparation of Skinned Fibers

Preparing skinned fibers from muscles requires two steps: (1) isolation of a single fiber or a sufficiently thin bundle of fibers and (2) skinning or destruction of the semipermeability of the surface membrane. Skinning can be carried out before the isolation of single fibers or bundles if chemical skinning is used. Thicker bundles, of diameter larger than about 150 $\mu$m, may be used for qualitative studies but they are not recommended for quantitative experiments because the diffusion time is too long.

The best way to prepare a skinned fiber is first to isolate an *intact*

---

[6] M. Endo and Y. Nakajima, *Nature (London)*, *New Biol.* **246**, 216 (1973).

single fiber from tendon to tendon in a physiological extracellular solution and then to skin it. This method assures that only damage will be due to skinning. Whenever possible, therefore, this procedure is recommended, but since isolating intact single fibers requires considerable skill and is time-consuming, an alternative is to isolate single fibers or a segment of fibers in a relaxing solution either before or after skinning. In this case, fibers should be examined under a microscope of a sufficient power to select parts with uniform striations for experimentation.

The isolation of single fibers is done under a stereomicroscope of 40× – 80× magnification, with the aid of forceps and small scissors, knives, or needles. Tips of these instruments should be sharpened on an oil stone.

The semitendinosus, iliofibularis, and tibialis anterior of amphibia are the muscles used most frequently for intact single fiber isolation. Slow tonic fibers can be obtained from amphibian iliofibularis muscles.[7] Mammalian fast twitch and slow twitch fibers can be obtained from extensor digitorum longus and soleus muscles, respectively. The strontium sensitivity of the contractile system is about one order of magnitude higher in both slow twitch[8,9] and slow tonic fibers[7] than in fast fibers and can be used to confirm fiber type. However, any kind of animal muscles can be used. Skinning can be carried out either mechanically or chemically.

*Mechanical Skinning*

Mechanical skinning can be performed in either an oil[10] or a relaxing solution.[6] The entire surface membrane can be removed[10] or the fibers can be split into two (or more) longitudinal pieces to get partially skinned fibers.[6] The main difference between completely skinned fibers and partially skinned fibers is that, whereas in completely skinned fibers the disrupted T system membrane is likely to be sealed off and a potential difference across the T membrane may be reestablished with the aid of active sodium transport, in partially skinned fibers (if appropriately prepared) the mouths of the T tubules on the remaining sarcolemma may still be open to the outside and no ionic gradient across the T membrane can be established.[6] In the former, therefore, the real physiological stimulus to cause Ca$^{2+}$ release, the depolarization of the T membrane, can still be effective, but in the latter, since the T membrane is completely depolarized from the beginning, it is extremely difficult, if not impossible, to

[7] K. Horiuti, *J. Physiol.* (*London*) **373**, 1 (1986).
[8] S. K. B. Donaldson and W. G. L. Kerrick, *J. Gen. Physiol.* **66**, 427 (1975).
[9] A. Takagi and M. Endo, *Exp. Neurol.* **55**, 95 (1977).
[10] R. Natori, *Jikeikai Med. J.* **1**, 119 (1954).

cause Ca$^{2+}$ release by manipulating the T membrane potential. To obtain well-resealed T tubules, skinning in an oil with a minimal quantity of extracellular fluid around the fiber is recommended, since Ca$^{2+}$ seems to be required for the fusion of the disrupted membranes.

*Chemical Skinning*

Chemical skinning for Ca$^{2+}$ release experiments should be performed with an agent that destroys the semipermeability of the surface membrane but not that of the SR. Two techniques have been reported for skeletal muscles: (1) ethylenediaminetetraacetic acid (EDTA) or ethylene glycol bis($\beta$-aminoethyl ether)-$N,N'$-tetraacetic acid (EGTA) treatment of human fibers and (2) treatment with saponin.

EDTA-treated chemically skinned fibers were first described by Winegrad.[11] He showed that, when bundles of cardiac fibers are treated with a solution containing 3 m$M$ EDTA for longer than 15 min at room temperature, the surface membrane became permeable to Ca$^{2+}$, EGTA, ATP, and other small molecules but not to large molecules such as proteins. This method does not work for skeletal muscle fibers of amphibia, but human skeletal muscle fibers are reported to be skinned by a similar EGTA treatment.[12] However, this might not be purely chemical skinning but EGTA-assisted disruption of the surface membrane, unlike EDTA-skinned cardiac fibers.[12]

Treatment with 5–50 $\mu$g/ml saponin for 30 min specifically perforates the surface membrane in amphibian skeletal fast muscle fibers.[13] A concentration of saponin higher than 150 $\mu$g/ml destroys the function of the SR as well.[13] Essentially the same applies to other kinds of fibers, but the concentration of saponin affecting the SR function differs in different kinds of fibers. For slow tonic fibers of amphibia and for mammalian skeletal muscle fibers, a lower concentration such as 20 $\mu$g/ml is recommended. In precise quantitative studies the possible effect of saponin on the SR should be checked under experimental conditions. The specificity of saponin comes from the fact that it acts on cholesterol molecules[14] and that the cholesterol content of the surface membrane is much higher than that of the SR membrane.[15,16] Therefore, besides saponin, other agents

[11] S. Winegrad, *J. Gen. Physiol.* **58,** 71 (1971).
[12] D. S. Wood, J. Zollman, J. P. Reuben, and P. W. Brandt, *Science* **187,** 1075 (1975).
[13] M. Endo and M. Iino, *J. Muscle Res. Cell Motil.* **1,** 89 (1980).
[14] I. Ohtsuki, R. M. Manzi, G. E. Palade, and J. D. Jamieson, *Biol. Cell.* **31,** 119 (1978).
[15] A. Martonosi, *Biochim. Biophys. Acta* **150,** 694 (1968).
[16] K. Waku, Y. Uda, and Y. Nakazawa, *J. Biochem. (Tokyo)* **69,** 483 (1971).

which affect cholesterol, such as digitonin, may have a similar effect. Such agents could also be used for chemical skinning for $Ca^{2+}$ release measurement, if their effects on the SR are negligible.

In chemical skinning, the T tubule membrane is also permeabilized along with the surface membrane, so that no potential gradient can be established across the T membrane, as in the case of partially skinned fibers.

### $Ca^{2+}$ Measurement

As described under General Considerations, measurement of $Ca^{2+}$ release from the SR should be made by determining the time course of decrease in the amount of $Ca^{2+}$ in the SR at a fixed $Ca^{2+}$ concentration, to avoid the secondary $Ca^{2+}$-induced $Ca^{2+}$ release. Measurement of the amount of $Ca^{2+}$ in the SR can be made directly by using $^{45}Ca$ or by discharging all of $Ca^{2+}$ and measuring the amount discharged.

A more-or-less continuous monitoring of $Ca^{2+}$ content of the SR in skinned fibers with $^{45}Ca$ is probably possible, since such experiments in intact single fibers have been reported.[17,18] However, it requires $^{45}Ca$ of a very high specific activity and very careful experimental apparatus and design because counts of $\beta$-emission of $^{45}Ca$ that is of low energy are sharply diminished by small increases in distance between fiber and counter. Alternatively, if $^{45}Ca$ is extracted from the fiber for counting,[19,20] the measurement is reliable. However, a single time course cannot be determined with a single fiber, but requires a large number of fibers, which makes the experiment cumbersome.

On the other hand, under appropriate conditions a high concentration of caffeine causes an almost complete release of $Ca^{2+}$ from the SR without any damage to the SR, so that many time courses can be determined in one skinned fiber. Since this kind of experiment can be more conveniently conducted than the $^{45}Ca$ experiments mentioned above, this is the method of choice in general $Ca^{2+}$ release measurements.

The amount of $Ca^{2+}$ discharged from the SR can be determined by using (1) the size of resulting contracture of the skinned fiber (bioassay), (2) $Ca^{2+}$ indicator dyes, (3) a luminescent protein, aequorin,[21] and (4) probably $^{45}Ca$. Only the first two methods will be described.

[17] B. A. Curtis, J. Gen. Physiol. **50**, 255 (1966).
[18] B. A. Curtis and R. S. Eisenberg, J. Gen. Physiol. **85**, 383 (1982).
[19] L. E. Ford and R. J. Podolsky, J. Physiol. (London) **223**, 1 (1972).
[20] E. W. Stephenson, J. Gen. Physiol. **71**, 411 (1978).
[21] J. R. Blinks, W. G. Wier, P. Hess, and F. G. Prendergast, Prog. Biophys. Mol. Biol. **40**, 1 (1982).

*Ca²⁺ Measurement by Utilizing Contraction of the Skinned Fiber*

Under appropriate conditions, application of a high concentration of caffeine causes contraction of skinned fibers in a relaxing solution due to the release of $Ca^{2+}$ from the SR.[22] Caffeine causes this by stimulating the $Ca^{2+}$-induced $Ca^{2+}$ release.[23] The contractions are transient because the $Ca^{2+}$ released in the fiber space quickly diffuses into the relatively large volume of medium bathing the fiber. Free EGTA in the relaxing solution in turn diffuses into the fiber space. In order to obtain a good estimate of $Ca^{2+}$ release from the size of contraction, the time course of $Ca^{2+}$ released by caffeine must be rapid compared with that of diffusion. After a single caffeine treatment, the SR becomes practically empty of $Ca^{2+}$. Reapplication of caffeine (after the drug is washed out once) no longer causes any response until the SR is allowed to reaccumulate $Ca^{2+}$. The completeness of the $Ca^{2+}$ release by caffeine can be checked either directly by determining the amount of $^{45}Ca$ present in the SR before and after the caffeine treatment or indirectly and less precisely by showing that incubation with a low concentration of $Ca^{2+}$, e.g., 0.03 $\mu M$, can cause a recovery of detectable caffeine contracture. This indicates that caffeine reduces the level of $Ca^{2+}$ in the SR below the equilibrium level with the low $Ca^{2+}$ concentration.[22]

To obtain rapid and complete $Ca^{2+}$ release, the effect of caffeine should be strong. To achieve this, (1) a high concentration of caffeine must be used (25–50 m$M$). (2) EGTA concentration in a relaxing solution during caffeine application should be appropriate. If the EGTA concentration is too high, the $Ca^{2+}$ concentration will not be raised sufficiently as a result of the $Ca^{2+}$ release, which makes (a) $Ca^{2+}$ release rather slow,[24] and (b) contractile response too small. If the concentration of EGTA is too low, distortion of the linearity of the contractile response with the amount of $Ca^{2+}$ released becomes greater due to saturation of contraction and fibers run down rapidly because of long intensive contractions. (3) Free $Mg^{2+}$ ion concentration must also be adjusted. $Ca^{2+}$-induced $Ca^{2+}$ release and hence caffeine-induced $Ca^{2+}$ release are inhibited by raising $Mg^{2+}$ concentration in the medium.[25] Therefore, lower $Mg^{2+}$ concentrations are preferable for rapid and complete $Ca^{2+}$ release by caffeine. On the other hand, since $Ca^{2+}$ is assayed by contraction, a $Mg^{2+}$ concentra-

[22] M. Endo, *Physiol. Rev.* **57**, 71 (1977).
[23] M. Endo, *Proc. Jpn. Acad.* **51**, 479 (1975).
[24] Because caffeine-induced $Ca^{2+}$ release is stronger at higher $Ca^{2+}$ concentrations, a very rapid $Ca^{2+}$ release is obtained when the $Ca^{2+}$ ion released in turn accelerates a further release of $Ca^{2+}$.
[25] M. Endo, *Proc. Jpn. Acad.* **51**, 467 (1975).

TABLE I

MAIN COMPOSITIONS OF SOLUTIONS USED FOR MEASUREMENT OF $Ca^{2+}$-INDUCED $Ca^{2+}$ RELEASE
BY UTILIZING CONTRACTION OF SKINNED FIBERS[a]

| Solutions | $[Mg^{2+}]$[b] (m$M$) | $[MgATP^{2-}]$[b] (m$M$) | [EGTA] total (m$M$) | $[Ca^{2+}]$[b] ($M$) | [Caffeine] (m$M$) | [Procaine⁺] (m$M$) |
|---|---|---|---|---|---|---|
| For Step 1 | | | | | | |
| G2 relaxing solution | 1 | 4 | 2 | 0 | 0 | 0 |
| G10 relaxing solution | 1 | 4 | 10 | 0 | 0 | 0 |
| Loading solution | 1 | 4 | 10 | $1-5 \times 10^{-7}$ | 0 | 0 |
| For Step 2 | | | | | | |
| Rigor solution | 1 | 0 | 2 | 0 | 0 | 0 |
| Prereleasing solution[c] | 0 | 0 | 2 | 0 | 0 | 0 |
| Releasing solution[c] | 0 | 0 | 10 | Variable | 0 | 0 |
| Stopping solution | 10 | 0 | 10 | 0 | 0 | 10 |
| For Step 3 | | | | | | |
| Preassay solution A[d] | 1 | 4 | 0.5–2 | 0 | 0 | 0 |
| Preassay solution B[e] | 1 | 4 | 0.1 | 0 | 0 | 0–5 |
| Assay solution A[d] | 1 | 4 | 0.5–2 | 0 | 25–50 | 0 |
| Assay solution B[e] | 0.02 | 1[f] | 0.1 | 0 | 25–50 | 0 |

[a] All solutions contain total 20 m$M$ piperazine-$N,N'$-bis(2-ethanesulfonic acid) (PIPES), pH 7.0, at 0–5° for amphibian fibers and 20–25° for mammalian fibers. Ionic strength of all solutions is adjusted to 0.17 $M$ and 0.2 $M$ for amphibian and mammalian fibers, respectively, by adding an appropriate amount of potassium methanesulfonate.

[b] To calculate free $Ca^{2+}$ and $Mg^{2+}$ concentrations etc., refer to some original papers, e.g., M. Iino, *J. Physiol.* (*London*) **320**, 513 (1981) or K. Horiuti, *J. Physiol.* (*London*) **373**, 1 (1986).

[c] Depending on the purpose of the experiment, necessary modifications must of course be made. For example, if the effect of $Mg^{2+}$ ion is to be examined, an appropriate concentration of $Mg^{2+}$ ions should be added.

[d] A is for amphibian fast fibers.

[e] B is for amphibian slow fibers or mammalian fibers.

[f] Total ATP is nearly 5 m$M$.

tion is required to obtain a sufficient concentration of MgATP. Values for these three concentration factors should be chosen according to fiber types. The $Ca^{2+}$-accumulating capacity of the SR per unit fiber volume is dependent on fiber type and the time of diffusion is strongly dependent on the diameter of the skinned fiber preparation. Typical figures are given in Table I.

The reproducibility of the assay system can be demonstrated by the fact that repeated caffeine tests after the same procedure always give the same magnitude of contracture if slow run-down of the fiber is taken into account.[26]

[26] At the beginning of experiments in each fiber, contractile responses may not decrease but may *increase* slightly for the first two or three trials.

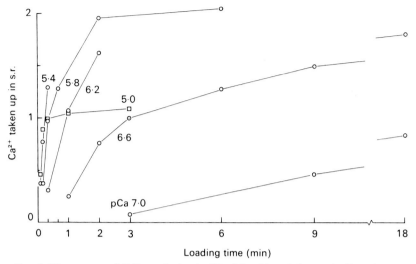

FIG. 1. Time course of Ca²⁺ uptake by the SR of the skinned slow tonic fiber of *Xenopus* at various Ca²⁺ concentrations. pCa values are identified on each curve. The Ca²⁺ ion in the SR is discharged by caffeine, and the area under the resulting contracture tension curve (tension-time integral) is plotted after normalization to that at pCa 6.6 and 3 min. The maximum level of loading at pCa 5.0 is lower than at higher pCa values, probably because of the operation of the Ca²⁺-induced Ca²⁺ release mechanism. Results for three slow fibers are combined. [Reproduced from K. Horiuti, *J. Physiol. (London)* **373**, 1 (1986).]

Areas under the contracture curve (tension-time integrals) but not peak tensions are recommended as the index of the magnitude of contracture and hence of the amount of Ca²⁺ released. With an increase in the amount of Ca²⁺ released peak tension does reach saturation at some point, but the tension-time integral does not because the duration of the contracture can still increase. For this reason the distortion of linearity is less if the tension-time integral is used. The approximate linearity of the assay system using tension-time integrals of caffeine contracture is supported by the fact that the time course of Ca²⁺ uptake is approximately exponential (Fig. 1).[27] As shown later, the time course of Ca²⁺ release is also approximately exponential.

Both ends of a skinned fiber segment, 5 mm long, are tied with a single silk thread to hooks, one of which is connected to a tension transducer. The output from the strain-gauge transducer (UL-2, UL-20 of NMB, Ja-

---

[27] A closer examination of Fig. 1 reveals that the initial parts of Ca²⁺ uptake curves have a concave form. This is due to the fact that when the amount of Ca²⁺ released is small enough, it only binds to EGTA in the solution and cannot evoke any contraction at all. With lower EGTA concentrations, this distortion of linearity is smaller.

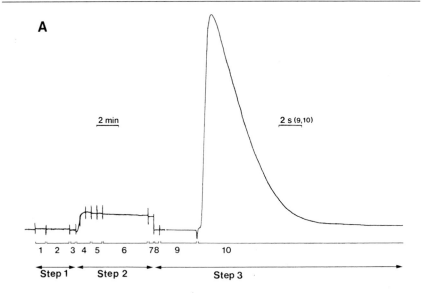

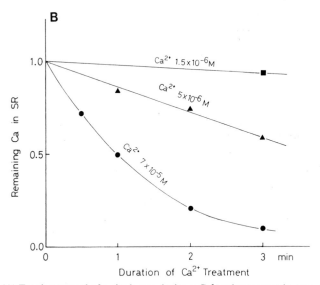

FIG. 2. (A) Tension record of a single run during a Ca²⁺ release experiment on a skinned guinea pig skeletal muscle fiber. The procedure consists of three steps: (1) Ca²⁺ loading to a fixed level, (2) Ca²⁺ release, and (3) assay. In each step several solution exchanges are made at the time indicated by the artifacts on the tension record. The composition of the solutions is given in Table I. Solution 1, G2 relaxing solution; 2, loading solution; 3, G10 relaxing solution; 4, rigor solution; 5, prereleasing solution; 6, releasing solution; 7, stopping solution; 8, 9, preassay solution; 10, assay solution. Note that the time scale is different for

pan or AE801, AME, Norway) is amplified by a strain meter (DSA-601-B, NMB, Japan) and recorded on a pen recorder (Recticorder, Nihon-Kohden, Japan). The length of the skinned fiber is set, using a microscope or laser diffraction, at a longer-than-optimum length (around 2.8 $\mu$m per sarcomere) because rundown of the fiber is slower at longer lengths. By exchanging solutions in a logical order, Ca$^{2+}$ release can be measured. To exchange solutions, a volume of desired solution, which is several times that of the experimental trough, is rapidly injected from a syringe through a thin short tubing and the overflow is aspirated. The following method may also be used. Solutions are placed in small wells (about 0.5 ml in volume), which are drilled in aluminum, brass, or plastic plate and resin-coated, so that the surface of the solution is about 2–3 mm above plate surface. The skinned fiber is set horizontally in the convex solution. By moving the plate horizontally, the solution exchange can easily be made. The temperature of the solution is kept constant by circulating water just underneath the trough or the plate. In the latter case a small magnetic bar is put in the solution and a magnetic stirrer is placed under the stage holding the plate. Temperatures of 0°–5° for amphibian muscles and 20°–25° for mammalian and avian muscles are recommended. Higher temperatures can be used, but they tend to cause a very rapid rundown of skinned fibers.

Protocols for the measurement of Ca$^{2+}$ release in skinned fibers essentially consist of three steps. Starting from the empty SR, (1) load the SR with Ca$^{2+}$ to a fixed level by immersing the fiber in a medium containing Ca$^{2+}$ and MgATP for a period of time, and allow the SR to accumulate Ca$^{2+}$ by Ca$^{2+}$-pump ATPase. (2) Stimulate Ca$^{2+}$ release for various periods of time by suppressing the Ca$^{2+}$-pump, and by preventing secondary Ca$^{2+}$-induced Ca$^{2+}$ release. (3) Totally discharge Ca$^{2+}$ remaining in the SR and measure the amount. An example of the detailed protocol for Ca$^{2+}$-induced Ca$^{2+}$ release is given in Fig. 2A. Here the run starts with the empty SR since Step 3 immediately preceding the run should have discharged all the Ca$^{2+}$ in the SR. The assay solution used in Step 3 is replaced by a relaxing solution with 2 m$M$ EGTA. If the EGTA concen-

solutions 9 and 10. Peak tension is 20 mg. Room temperature. (B) Time course of decrease in the amount of Ca$^{2+}$ in the SR of a *Xenopus* fast fiber during Ca$^{2+}$ release stimulated by various Ca$^{2+}$ concentrations. Experiments were performed as in (A) and described in the text. The tension-time integral during the application of solution 10 was plotted after normalization to that of the control run, which was done in exactly the same way except that the application of the releasing solution was omitted. [Reproduced from M. Endo, *in* "Regulation of Calcium Transport across Muscle Membranes" (A. E. Shamoo, ed.), Current Topics in Membranes and Transport, Vol. 25, p. 181. Academic Press, New York, 1985.]

tration in this relaxing solution is too high, the start of $Ca^{2+}$ loading is delayed by the diffusion time. If it is too low, $Ca^{2+}$ loading may already start in the relaxing solution because of contaminating $Ca^{2+}$.

$Ca^{2+}$ loading (Step 1) starts by applying the loading solution (typical composition is given in Table I). A $Ca^{2+}$ concentration that does not directly activate the contractile system is usually preferred. The $Ca^{2+}$ concentration should be buffered with a high concentration of a $Ca^{2+}$-buffer. After an appropriate period of time, the loading solution is replaced by a relaxing solution having a high concentration of EGTA to stop the loading immediately. The loading period is chosen so as to obtain a 30–70% maximum level of loading. With excessive loading $Ca^{2+}$ tends to be released very easily or even spontaneously.[28] On the other hand, the time course of $Ca^{2+}$ decrease in the SR due to a stimulus may be difficult to determine accurately with too small a level of initial loading.

Before the $Ca^{2+}$-releasing stimulus is applied in Step 2, the solution must be changed appropriately. First, ATP is removed to stop the $Ca^{2+}$ pump. The removal of ATP takes a relatively long time because of its slow diffusion due to the presence of concentrated ATP-binding sites in the fiber. In the example given in Fig. 2A, several exchanges of ATP-free rigor solution (Table I) are made. Removal of MgATP is indicated by rigor tension development. If a sufficient amount of $Mg^{2+}$ ion is present as in the rigor solution of this example, this can approximately be equated with the removal of total ATP. Prereleasing solution (Table I), which is the same as releasing solution except that it does not contain $Ca^{2+}$-releasing stimulus, was then applied. Sufficient time must be allowed so that each constituent in the prereleasing solution can reach equilibrium. Then, the $Ca^{2+}$-releasing stimulus, in this example, the $Ca^{2+}$ ion, (releasing solution, Table I), is applied for a predetermined period of time. Care should be taken to initiate and to terminate the releasing stimulus as quickly as possible, especially if the period of stimulation is short. In the case of $Ca^{2+}$-induced $Ca^{2+}$ release, a $Ca^{2+}$ solution with a high concentration of $Ca^{2+}$ buffer is used to initiate $Ca^{2+}$ stimulus quickly. A special stopping solution, which is not only devoid of $Ca^{2+}$, but also contains inhibitors of $Ca^{2+}$-induced $Ca^{2+}$ release, $Mg^{2+}$ ion, and procaine[19,29] (Table I), is used to shut off the $Ca^{2+}$-induced $Ca^{2+}$ release as quickly as possible.

In Step 3, before the assay solution is applied, EGTA concentration must be reduced to the level of that in the preassay solution by applying preassay solution (Table I). However, if the skinned fiber is kept in a

---

[28] This is probably ascribable to secondary $Ca^{2+}$ release due to activation of $Ca^{2+}$-induced $Ca^{2+}$ release by $Ca^{2+}$ leakage from the SR. The rate of $Ca^{2+}$ leakage should be nearly proportional to the level of $Ca^{2+}$ loaded.

[29] S. Thorens and M. Endo, *Proc. Jpn. Acad.* **51**, 473 (1975).

relaxing solution containing a low concentration of EGTA with minimal buffering capacity, a small amount of Ca$^{2+}$ leaking out of the SR may not be effectively buffered and may stimulate the Ca$^{2+}$-induced Ca$^{2+}$ release before the assay is made. To avoid this, the Mg$^{2+}$ concentration in the preassay solution should be increased and, if necessary, procaine should be added. After equilibration with the preassay solution, the assay solution is applied and the resulting contraction is recorded. In this assay, the decrease in Mg$^{2+}$ and procaine concentrations to the level of the assay solution is obviously not immediate. However, unlike Ca$^{2+}$ release by the releasing solution (the main purpose of the experiment), what is important in the assay is to apply the assay solution under exactly the same conditions and to obtain a rapid enough Ca$^{2+}$ release. An accurate time course of Ca$^{2+}$ release by the assay solution is not necessary.

Experiments as shown in Fig. 2A are repeated by changing the Ca$^{2+}$ concentration and the application time of each releasing solution. To account for a slow run down of the skinned fiber, controls are run after every 4–6 experiments. The same series of solution exchanges is made without any releasing solution for the controls. The result of experiments is presented as a value relative to that of the controls in Fig. 2B. The amount of Ca$^{2+}$ in the control runs is taken as 100%. Figure 2B shows that the time courses are roughly exponential, and, therefore, the rate constant of the decline can be taken as the index of magnitude of Ca$^{2+}$ release under these conditions.

Essentially the same procedures can be followed for amphibian and mammalian (or avian) muscles. The main differences in procedures for these classes occur in experimental temperature, ionic strength of solutions, and composition of the assay solution. (For amphibian slow tonic fibers, assay solution similar to that for mammalian muscle is suitable, probably due to the smaller Ca$^{2+}$-accumulating capacity of the SR of these fibers than amphibian fast fibers.)

For other kinds of Ca$^{2+}$-releasing stimuli, essentially the same procedures can be used with appropriate modifications.

## Ca$^{2+}$ Measurement by a Fluorescent Ca$^{2+}$ Indicator

Amount of Ca$^{2+}$ discharged from the SR of skinned fiber can also be measured by the change in the optical properties of a Ca$^{2+}$ indicator dye. Fura-2, a fluorescent Ca$^{2+}$ indicator,[30] is an EGTA analog which has a dissociation constant for Ca$^{2+}$ as low as 200 n$M$. Therefore fura-2 can bind most of the Ca$^{2+}$ discharged from the SR, when it is the major Ca$^{2+}$

[30] G. Grynkiewicz, M. Poenie, and R. Y. Tsien, *J. Biol. Chem.* **260**, 3440 (1985).

buffer in the assay solution. When this dye is excited by 340 nm ultraviolet light, the fluorescence intensity increases 3-fold upon binding of Ca$^{2+}$.

For our experiments, we use a glass capillary as a cuvette and a fluorescence microscope equipped with a photomultiplier tube as a fluorometer. Conventional cuvettes and spectrofluorometers should not be used for the following reasons: (1) The maximum amount of Ca$^{2+}$ released from a skinned skeletal muscle fiber with a diameter of 100 $\mu$m is of the order of 10 pmol/mm of fiber length. If the amount of Ca$^{2+}$ is then diluted with 1 ml of the assay solution, it will result in only a 0.1 $\mu M$ increase in Ca$^{2+}$ concentration even if a 10-mm-long skinned fiber is used. On the other hand, in a capillary with an internal diameter of 400 $\mu$m, several tens of a micromolar change in Ca$^{2+}$ concentration can be produced. (2) Solution changes cannot be carried out rapidly using a conventional cuvette and large quantities of solutions containing fura-2 (which is rather expensive) are discarded. This problem can be circumvented if a capillary cuvette is used.

Using silk filaments, skinned fiber, 5 mm in length, is tied to a resin-coated tungsten wire with a diameter of 100 $\mu$m. It is then inserted into and fixed in a glass capillary (internal diameter 400 $\mu$m, length 32 mm) and mounted on the stage of a microscope. Both ends of the capillary are linked to silicone tubing. One of the tubings is connected to a peristaltic pump, so that, by placing the free end of the silicone tubing in solution and running the pump, it is possible to rapidly change and to perfuse the solution in the capillary cuvette.

A fluorescence microscope is used for the detection of fura-2 fluorescence. Excitation light (340 nm) is provided by a Xenon lamp via a narrow bandwidth interference filter. A 0.8-mm length of the cuvette is epi-illuminated through a 20× objective. Emitted light is collected by the same objective, and focused on the photomultiplier tube through 500 nm band-pass filter. Calibration of the system using 40–100 $\mu M$ of fura-2 gives a linear relationship between the total Ca$^{2+}$ concentration and the change in fluorescence intensity up to one-half of the maximum change as theoretically expected from one-to-one binding between Ca$^{2+}$ and the dye.

Since ultraviolet light is known to be harmful to the SR,[31] it is desirable to have a shutter in the excitation light path and to keep the exposure time as short as possible. In our system, the shutter is pneumatically opened for 50 msec every 5 sec. The shutter operation, as well as the peristaltic pump operation and data collection, is controlled by a microcomputer.

The procedure for the Ca$^{2+}$ assay is essentially the same as that de-

[31] T. Nagai, M. Makinose, and W. Hasselbach, *Biochim. Biophys. Acta* **43,** 223 (1960).

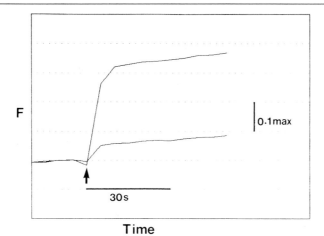

FIG. 3. Fluorescence change of fura-2 during $Ca^{2+}$ assay. Caffeine-containing assay solution is injected into the capillary cuvette at the time marked by the arrow. Upper trace is obtained after $Ca^{2+}$ loading at pCa 6.7 for 60 sec. No $Ca^{2+}$ loading is carried out for the lower trace. Fura-2 concentration is 40 $\mu M$, and, therefore, the vertical bar corresponds to 4 $\mu M$ of $Ca^{2+}$. Guinea pig EDL; fiber width, 40 $\mu m$.

scribed in the previous section, and 50 m$M$ caffeine is used to discharge $Ca^{2+}$ from the SR. There are, however, a few modifications. ATP is withdrawn prior to the assay and replaced by 25 m$M$ AMP during the assay. This will ensure that the fiber remains unmoved on increasing $Ca^{2+}$ concentration. AMP is introduced because adenine nucleotides are known to enhance $Ca^{2+}$-induced $Ca^{2+}$ release.[32,33] Both $Mg^{2+}$ and EGTA are also removed prior to the assay in order to enhance the $Ca^{2+}$ release by caffeine and to allow fura-2 to bind all the $Ca^{2+}$ discharged.

The change in fluorescence intensity takes place in two phases after the application of the caffeine solution (Fig. 3). First, it increases rapidly within 10 to 15 sec (step), then there is a slow, more-or-less linear increase (creep). A part of the step change is due to the effect of caffeine on fura-2 fluorescence and is $Ca^{2+}$ independent. The slope of the creep is dependent on the duration and $Ca^{2+}$ concentration of the $Ca^{2+}$ treatment before the assay, but is present even if ATP is withdrawn during the $Ca^{2+}$ treatment. Therefore, the creep seems to be due to very slow release of $Ca^{2+}$ passively trapped or bound by the skinned fiber. *Step amplitude* minus *$Ca^{2+}$-independent step amplitude* is both $Ca^{2+}$- and ATP-dependent, and should

[32] M. Endo and T. Kitazawa, *Proc. Jpn. Acad.* **52,** 595 (1976).
[33] Y. Kakuta, *Pfluegers Arch.* **400,** 72 (1984).

represent the amount of $Ca^{2+}$ which has been stored in the SR and is discharged by caffeine.

Since the $Ca^{2+}$ signal is obtained from the center of a 5-mm-long skinned fiber, any distortion of the fluorescence change due to diffusion of $Ca^{2+}$ along the fiber length seems minimal. The concentration of fura-2 chosen is such that the size of the step change in fluorescence intensity does not exceed one-half of the maximum response. The time course of both $Ca^{2+}$ uptake by and $Ca^{2+}$-induced $Ca^{2+}$ release from the SR of skinned fibers appears to be exponential. This provides another test for the linearity of the $Ca^{2+}$-measuring system. This $Ca^{2+}$-measuring system has also been successfully applied to cardiac and smooth muscle chemically skinned fibers in our laboratory.[34]

[34] M. Iino, *Biochem. Biophys. Res. Commun.* **142**, 47 (1987).

# [3] Isolation and Characterization of Sarcolemmal Vesicles from Rabbit Fast Skeletal Muscle

By STEVEN SEILER and SIDNEY FLEISCHER

The study of sarcolemmal ion transport using whole muscle cells is complicated by contributions of the internal membrane systems and their respective transport activities. The availability of purified sarcolemmal membrane vesicles simplifies the study of ion transport referable to sarcolemma. This report describes the preparation of skeletal muscle sarcolemma in the form of sealed, predominantly inside-out vesicles, which are suitable for transport studies.

Overview of the procedure—the sarcolemmal vesicle isolation procedure includes subjecting ground fast skeletal muscle to several limited blendings and washings in 0.6 *M* KCl. A low-speed sediment is rehomogenized in buffered sucrose and a microsomal fraction is obtained by differential centrifugation. The microsomes are then fractionated using sequential isopycnic sucrose and dextran T-10 density gradient centrifugations.[1,2]

[1] These studies were supported by a grant from the National Institutes of Health DK 14632 and the Muscular Dystrophy Association of America, and a Biomedical Research Support Grant from the National Institutes of Health administered by Vanderbilt University (SF).
[2] S. Seiler and S. Fleischer, *J. Biol. Chem.* **257**, 13862 (1982).

Preparation Procedure

*Reagents*

0.75 *M* KCl, 5 m*M* imidazole, pH 7.4 (homogenizing medium)
0.3 *M* sucrose, 5 m*M* imidazole, pH 7.4
0.6 *M* KCl, 4 m*M* imidazole, pH 7.4
5 m*M* imidazole, pH 7.4
Sucrose gradient:
    (a) 17% sucrose (w/w), 5 m*M* imidazole, pH 7.4
    (b) 23% sucrose (w/w), 5 m*M* imidazole, pH 7.4
    (c) 50% sucrose (w/w), 5 m*M* imidazole, pH 7.4
Dextran T10 gradient:
    (a) 8.7% dextran T10 (w/w) (obtained from Pharmacia) in 5 m*M*
      imidazole, pH 7.4
    (b) 18% dextran T10 in 5 m*M* imidazole, pH 7.4

Typically, three female New Zealand white rabbits (1–1.8 kg weight per rabbit) are killed by cervical dislocation. All subsequent steps are carried out at 0–4°. The predominantly white muscles [sartorius, gracilis, vastus lateralis, vastus medialis, adductor magnus, glutens (maxima, medius, minimus), biceps femoris, and gastrocnemius] of the hind legs are dissected from red muscle, large blood vessels, nerves, tendon, and fascia. The dissection is performed in a cold room on a glass tray seated on crushed ice. The muscle is then ground in a meat grinder (General Model A meat grinder). Typically, approximately 40 g of trimmed muscle is obtained from a single rabbit leg and approximately 200 g of ground muscle is obtained from three rabbits. The ground muscle is then divided into 50-g portions.

The ground muscle (50-g portion) is combined with 250 ml of 0.75 *M* KCl, 5 m*M* imidazole, pH 7.4, and briefly homogenized using a Waring blender (Waring Products Div. New Hartford, CT) at half-maximal speed for 5 sec. This initial homogenate is then centrifuged for 20 min at 9000 rpm in a JA-10 rotor (Beckman model J-21 centrifuge). The supernatant is discarded. The pellet is resuspended in another 250 ml of 0.75 *M* KCl, 5 m*M* imidazole, pH 7.4, by brisk stirring with a glass rod. The resuspended pellet is again centrifuged in the JA-10 rotor for 2 min at 9000 rpm. The supernatant is discarded, and the pellet is resuspended in the same solution and centrifuged once again. The pellet is then suspended as above in 0.3 *M* sucrose, 5 m*M* imidazole, pH 7.4, and centrifuged as before. The supernatant is discarded and the pellet is rehomogenized in 250 ml of 0.3 *M* sucrose, 5 m*M* imidazole, pH 7.4, using a Waring blender at full speed for 1 min. This rehomogenate is centrifuged in a JA-10 rotor at 9000 rpm for 20 min and the pellet is discarded. The supernatant is then filtered through several layers of cheesecloth to remove floating particles, and the filtrate is centrifuged at 30,000 rpm for 2 hr in a Beckman Type 35 rotor to obtain a microsomal fraction. The pellets are suspended in 40–60 ml of ice-cold 0.6 $M$ KCl, 4 m$M$ imidazole, pH 7.4, using a loose-fitting Dounce homogenizer, and incubated on ice for 1–3 hr.

The microsomes (approximately 125 mg of protein in 10–12 ml per centrifuge tube) are layered onto a discontinuous sucrose gradient consisting of 17% (w/w) sucrose (10 ml), 23% sucrose (10 ml), and 50% sucrose (7 ml), all buffered with 5 m$M$ imidazole, pH 7.4. The tubes are centrifuged in a Beckman SW27 rotor at 20,000 rpm overnight (12–16 hr). The sarcolemmal vesicles migrate to the 17/23% sucrose interface (Fig. 1A, band 2). This sucrose gradient fraction is diluted with an equal vol-

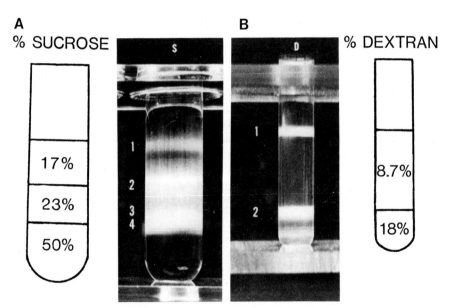

A                            B

% SUCROSE                      % DEXTRAN

17%

23%

50%

8.7%

18%

Fig. 1. Purification of sarcolemmal vesicles on sucrose and dextran density gradients. Separation of muscle microsomes is performed using a discontinuous sucrose gradient (A). The muscle microsomes (in 10 ml of 0.6 $M$ KCl) are layered on top of a step gradient containing 17% (10 ml), 23% (10 ml), and 50% (10 ml) sucrose and centrifuged overnight in an SW 27 rotor. The fraction containing sarcolemmal vesicles is located at the 17/23% sucrose interface (band 2). Band 1 also contains some sarcolemma, but contains aggregated contractile protein and is not normally used for further purification. The sarcolemmal vesicles from band 2 of the sucrose gradient can be further purified using a dextran T10 gradient (B). The sarcolemmal vesicles (2 ml in 5 m$M$ imidazole, pH 7.4) are layered onto a discontinuous gradient containing 8.7% (2 ml) and 18% (1 ml) dextran T10. The gradient is centrifuged overnight and the sarcolemmal vesicles do not enter the 8.7% dextran T10 (band 1). This fraction is recovered by dilution with 5 m$M$ imidazole, pH 7.4, and sedimentation.

ume of 5 m$M$ imidazole, pH 7.4, and the sarcolemmal vesicles are recovered by sedimentation at 35,000 rpm for 2 hr in a type 40.1 rotor. This is referred to as sucrose-purified plasma membrane vesicles (or PMs). The sarcolemmal vesicles are then suspended in 5 m$M$ imidazole, pH 7.4.

The sucrose-purified sarcolemmal vesicles (PMs) are usually further enriched with respect to sealed sarcolemmal vesicles using a discontinuous dextran T10 gradient. Dextran does not penetrate into sealed vesicles, therefore their buoyant density is lower than leaky membranes.[3] PMs (typically 2–3 mg protein) suspended in 5 m$M$ imidazole, pH 7.4, are layered onto a discontinuous dextran T10 gradient (buffered with 5 m$M$ imidazole, pH 7.4). The gradient contains steps of 8.7% (w/w) and 18% dextran T10 using either a Beckman SW56 or SW41 swinging bucket rotor. The gradients are centrifuged overnight (16–20 hr) at 35,000 rpm. The fraction at the 5 m$M$ imidazole and 8.7% dextran T10 interface (Fig. 1B, band 1) contained the highly purified sarcolemmal vesicles. The sarcolemmal vesicles are diluted with an equal volume of 5 m$M$ imidazole, pH 7.4, and recovered by centrifugation at 35,000 rpm for 2 hr in a Beckman 75Ti rotor. This fraction is referred to as the dextran-purified sarcolemmal vesicles (PMd).

The sarcolemmal membrane vesicles are suspended in 0.3 $M$ sucrose, 5 m$M$ imidazole, pH 7.4, to a protein concentration of 3–10 mg/ml and are frozen in liquid nitrogen and stored at −70° until used. The yield of sarcolemma obtained was approximately 3 mg of sucrose-gradient-purified sarcolemma and 0.5–1.0 mg of dextran-purified vesicles from 50 g of ground muscle (Table I).[2] See footnotes in Table I for relevant methodology.[4–7]

## Notes Regarding the Preparation Procedure

The use of young rabbits (1.8 kg or less) yields sarcolemmal preparations with consistently higher specific activity of sarcolemmal marker enzymes (Na$^+$,K$^+$-ATPase and acetylcholinesterase). Sarcolemmal preparations from larger rabbits appear to be contaminated by large multilamellar vesicles, as observed in thin-section electron microscopy,[8] possibly originating from peripheral nerve myelin.

The discontinuous dextran T10 gradient is readily overloaded. When

[3] T. Steck, *Methods Membr. Biol.* **2**, 245 (1975).
[4] R. Mitchell, P. Volpe, P. Palade, and S. Fleischer, *J. Biol. Chem.* **258**, 9867 (1983).
[5] G. Meissner, G. E. Conner, and S. Fleischer, *Biochim. Biophys. Acta* **298**, 246 (1973).
[6] J. Comte and D. C. Gautheron, this series, Vol. 55, p. 98.
[7] S. Fleischer, G. Brierley, H. Klouwen, and D. B. Slatterbach, *J. Biol. Chem.* **237**, 3264 (1962).
[8] S. Seiler, Ph.D. Thesis, Vanderbilt Univ., 1982.

TABLE I

CHARACTERISTICS OF FRACTIONS IN THE PURIFICATION OF SKELETAL MUSCLE PLASMA MEMBRANES[a]

| Step | Microsomes | PMs | PMd |
|---|---|---|---|
| Yield (mg protein/50 g of skeletal muscle) | 125 | 2.7 | 0.5–1.0 |
| $\mu$g P/mg protein | 24.9 | 53.3 | 73.4 ± 4.6 (3) |
| Total ATPase[b] ($\mu$mol/mg · hr) | 8.5 | 74.5 | 73.5 ± 11.1 (6) |
| Total ATPase[b] + 1 m$M$ oubain ($\mu$mol/mg · hr) | 5.7 | 36.7 | 18.6 ± 3.1 (6) |
| Ouabain-sensitive Na$^+$,K$^+$-ATPase[b,c] ($\mu$mol/mg · hr) | 2.8 | 37.8 | 56.5 ± 9.1 (6) |
| Acetylcholinesterase[b] (nmol/mg · min) | 63.5 | 154 (2) | 271 (3) |
| ATP-dependent Na$^+$ uptake[d] (capacity, nmol/mg protein) | — | 139 (2) | 218 (2) |
| Ca$^{2+}$-ATPase[e] ($\mu$mol/mg · min) | 2.01 | 0.26 | <0.03 (2) |
| Monoamine oxidase[f] (pmol/mg · min) | 27.5 | 222 | 48.2 |
| Succinate–cytochrome $c$ reductase[g] (nmol/mg · min) | 47.2 (2) | 10.0 | 3.0 |

[a] PMs and PMd refer to plasma membrane fractions obtained from sucrose gradient and dextran T10 gradient, respectively. The numbers in parentheses represent the number of preparations analyzed. From Ref. 2.

[b] Activity was measured after preincubation with SDS.

[c] Ouabain-sensitive ATPase as obtained by subtracting the values obtained in the presence of ouabain from the total ATPase (above). See also Ref. 4.

[d] Ten-minute point, measuring uptake capacity at 25° in a medium containing 5 m$M$ NaCl, 50 m$M$ KCl, 5 m$M$ MgCl$_2$, 30 m$M$ imidazole, 1 m$M$ EGTA, and 5 m$M$ Tris/ATP.

[e] Purified sarcoplasmic reticulum[5] has Ca$^{2+}$-ATPase activity of 3.0 $\mu$mol/mg · min measured under the conditions described. This value was used to estimate the upper limit of SR contamination. The microsome fraction contained 65–70% sarcoplasmic reticulum, the sucrose gradient fraction contained 10%, and the dextran T10-purified fraction contained less than 1% sarcoplasmic reticulum contamination.

[f] Data are not available on the specific activity of monoamine oxidase activity of purified skeletal muscle outer mitochondrial membrane. The pig heart outer mitochondrial membrane has an activity of 12.8 nmol/mg · min using benzylamine as substrate.[6]

[g] Purified bovine heart mitochondria have a succinate–cytochrome-$c$ reductase rate of approximately 800–900 nmol/mg · min.[7] Using this value, the microsomal fraction and PMs contained approximately 5 and 1% mitochondria, respectively. The mitochondrial contamination was reduced to 0.3% in the PMd plasma membrane fraction.

this occurs, the SR does not cleanly separate from the sarcolemmal vesicles.

A major problem in isolating sarcolemmal vesicles is that sarcolemma is not readily released from muscle by homogenization. The sticking of the membrane to fibrillar elements is reduced by the presence of high ionic strength during fiber disruption. However, high salt, which serves to disaggregate muscle contractile elements, also cause changes in the morphology of the muscle sarcolemma.[8,9]

### Characterization of Sarcolemmal Vesicles

The dextran-purified sarcolemma preparation is highly enriched in plasmalemma marker activities (ouabain-sensitive $Na^+,K^+$-ATPase, adenylate cyclase, and acetylcholinesterase) (Table I). Negligible contamination by sarcoplasmic reticulum and mitochondria is indicated by low $Ca^{2+}$-ATPase, succinate–cytochrome $c$ reductase, and monoamine oxidase enzyme activities. Thin-section and negative-staining electron microscopy confirms the absence of sarcoplasmic reticulum and mitochondrial contamination.[2]

The sarcolemmal preparation, as observed by thin-section electron microscopy, consists mainly of large, irregularly shaped vesicles varying in size from 0.2–0.5 $\mu$m in diameter (Fig. 2). Relatively few contaminating transverse tubules are observed. Transverse tubules are observed in other fractions (especially in band 3 of the sucrose gradient, see Fig. 1A), as indicated by characteristic morphology of the isolated transverse tubules and the higher content of nitrendipine receptor sites than for the sarcolemmal vesicle fraction (unpublished observations).

The sarcolemmal preparation consists of sealed vesicles, as indicated by thin-section electron microscopy and latency of enzymatic activities (Table II). The vesicle sidedness can be estimated from latency of ouabain-sensitive $Na^+,K^+$-ATPase and acetylcholinesterase,[10] as well as from latency of ouabain binding.[2] Estimates suggest that 50–66% of the vesicles are sealed and inside-out, 19–25% of the vesicles are sealed and right-side-out, and the remaining 15–25% are leaky.

The sarcolemma vesicles are useful for studying ion transport as carried out by the $Na^+,K^+$-ATPase,[2] sarcolemmal $Ca^{2+}$-ATPase,[11] $Na^+/Ca^{2+}$ exchange,[8] and $Na^+/H^+$ exchange (S. Seiler, unpublished observations).

[9] A. Saito, S. Seiler, and S. Fleischer, *J. Ultrastruct. Res.* **86,** 277 (1984).
[10] T. Steck and J. Kant, this series, Vol. 31, p. 172.
[11] M. Michalak, K. Famulski, and E. Carafoli, *J. Biol. Chem.* **259,** 15540 (1984).

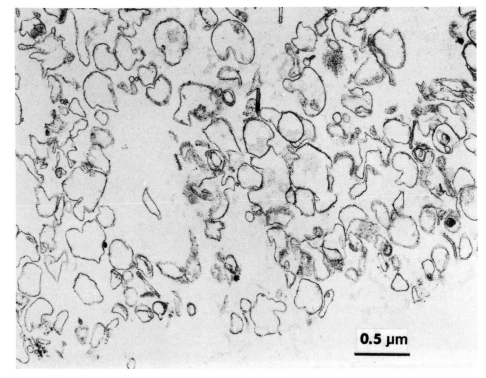

FIG. 2. Electron micrograph of sarcolemmal vesicles. Purified sarcolemmal vesicles obtained from dextran gradient were prepared for thin-section electron microscopy, as described in Ref.2. Most of the preparation contains sealed vesicles 0.2–0.5 $\mu$m in diameter. Magnification: 30,000×.

## Assay of Marker Enzymes

The most commonly used plasma membrane marker is the Na⁺,K⁺-ATPase, which can also be used to estimate the sidedness of the vesicles. Acetylcholinesterase copurifies with the Na⁺,K⁺-ATPase, and this enzyme can also be used to estimate the sidedness characteristics of the vesicles.[2]

Measurement of the total Na⁺,K⁺-ATPase activity in sealed membrane vesicles requires exposure of all the active sites of the enzyme. Therefore, preincubation with detergent is carried out to make all of the vesicles leaky. The preincubation conditions used to expose maximal Na⁺,K⁺-ATPase and acetylcholinesterase are the same. The vesicles are incubated in a small volume containing 1 mg/ml of sample protein, 40 m$M$

TABLE II

LEAKINESS AND MEMBRANE SIDEDNESS OF PURIFIED PLASMA MEMBRANE VESICLES[a]

| | PMs | | | PMd | | |
|---|---|---|---|---|---|---|
| | Enzyme activity | | Latency[c] (%) | Enzyme activity | | Latency[c] (%) |
| Assay | No SDS | + SDS[b] | | No SDS | + SDS[b] | |
| Ouabain-sensitive Na$^+$,K$^+$-ATPase ($\mu$mol/mg · hr) | 11.0(1) | 37.8(1) | 71 | 8.3 ± 2.6(6) | 56.5 ± 1(6) | 85 ± 4[d](6) |
| Acetylcholinesterase (nmol/mg · min) | 73.0(2) | 153 (2) | 53 | 96.0 ± 6.4(3) | 275 ± 13(3) | 66 ± 1[e](3) |

[a] The values in parentheses denote the number of preparations analyzed. From Ref. 2.

[b] Measured after preincubation in SDS (see Methods and Ref. 2 for details).

[c] Percent latent activity = [total activity (SDS pretreated) − activity (no SDS)]/[total activity (SDS pretreated)].

[d] The latency of ouabain-sensitive Na$^+$,K$^+$-ATPase is a measure of sealed vesicles both right-side-out and inside-out (85%). The difference from 100% gives the percentage of leaky vesicles (15%).

[e] The latency of acetylcholinesterase is a measure of inside-out vesicles (66%) and, when subtracted from the percentage of sealed vesicles (85%), gives the percentage of right-side-out vesicles (19%).

imidazole, 40 m$M$ HEPES ($N$-2-hydroxyethylpiperazine-$N'$-2-ethane-sulfonic acid), pH 7.3, 2 m$M$ Tris-EGTA, and from 0 to 0.5 mg/ml sodium dodecyl sulfate (SDS). The SDS concentration required for maximal Na$^+$,K$^+$-ATPase and acetylcholinesterase activity depends on the purity of the sarcolemmal vesicles. Less SDS is required to maximally activate enzyme activity in crude microsomes than is required for purified sarcolemmal vesicles.

The sample is preincubated with detergent at room temperature 20–30 min prior to 50-fold dilution of the sample into temperature-equilibrated assay media (37°). For the measurement of Na$^+$,K$^+$-ATPase, the assay medium consists of 120 m$M$ NaCl, 20 m$M$ KCl, 3 m$M$ MgCl$_2$, 3 m$M$ Na$_2$ATP, 0.5 m$M$ EGTA, 5 m$M$ NaN$_3$, 30 m$M$ imidazole/HCl, pH 7.5, with or without 1 m$M$ ouabain. The reaction is started by addition of detergent-treated sample and then the sample is incubated for 5–20 min before being stopped by the addition of 0.5 ml of 5% (w/v) SDS, 10 m$M$ Na$_2$EGTA. The advantages of SDS in the stop solutions are that deproteination of the sample is not required and that the total amount of phosphate liberated by the sample can be measured. After stopping the reaction, the sample is kept on ice at neutral pH until assayed for phosphate.

The samples can be kept on ice several hours to overnight without appreciable nonenzymatic release of phosphate from ATP. SDS precipitates when the sample is cold. However, the SDS redissolves when warmed to room temperature. The inorganic phosphate liberated is measured, as previously described.[2]

Measurement of acetylcholinesterase activity in sarcolemmal vesicles is as described,[10] except that pretreatment of the vesicles with SDS to obviate latency is carried out as above. Table II illustrates such treatment to expose latent enzyme activity and its use in the estimation of the vesicle sidedness.

### Measurement of Ion Transport in the Sarcolemmal Vesicles

The inside-out vesicles accumulate $Na^+$ ions energized by ATP in the presence of $K^+$ or $NH_4^+$. $Na^+$ uptake can be prevented by low concentrations of vanadate or digitoxin,[2] as would be expected for the $Na^+,K^+$-pump. Optimal transport of $Na^+$ by the $Na^+,K^+$-pump requires the counter transport of $K^+$, $Rb^+$, or $NH_4^+$. Therefore, the vesicles are preincubated in solutions containing these ions to allow equilibration with the intravesicular space. The following protocol typifies a $Na^+$ uptake experiment using these vesicles (see Fig. 3).

The sarcolemmal membrane vesicles are preincubated on ice for 3 hr to overnight in a medium containing 10 m$M$ NaCl ($2.3 \times 10^6$ cpm/ml of

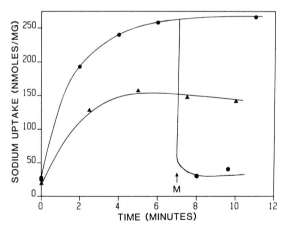

FIG. 3. ATP-dependent sodium uptake into sarcolemmal vesicles. Sodium uptake was measured by accumulation of $^{22}$Na into the vesicle. Sucrose-gradient-purified (▲) and dextran-gradient-purified (●) sarcolemmal vesicles are first preincubated in 5 m$M$ MgCl$_2$, 1 m$M$ EGTA, 50 m$M$ KCl, 5 m$M$ NaCl ($2 \times 10^6$ cpm/ml), 30 m$M$ imidazole, pH 7.2. The uptake reaction is started by addition of concentrated Tris-ATP to a final concentration of 5 m$M$. At the indicated time points, 0.5 ml of sample is withdrawn and filtered through 0.22-$\mu$m Millipore filters. The filters are then washed twice with 2 ml of 0.5 m$M$ NaCl. The arrow indicates the addition of 1 $\mu M$ monensin, a $Na^+$ ionophore, which releases the sodium accumulated.

$^{22}$Na), 100 m$M$ KCl, 5 m$M$ MgCl$_2$, 40 m$M$ imidazole, 40 m$M$ HEPES, pH 7.3, and a protein concentration of 40 $\mu$g/ml. The sample is then further preincubated for 30 min at 37°. After such preequilibration, an aliquot (usually 0.5 ml per point) is withdrawn (zero time point) to determine the amount of Na$^+$ in the vesicles prior to addition of 5 m$M$ ATP (diTris salt, pH 7.0–7.2), which initiates the uptake reaction. At appropriate times, aliquots (0.5 ml containing 20–50 $\mu$g of sample) are withdrawn and filtered on 0.22-$\mu$m Millipore filter disks (GSWP) (Bedford, MA). The filters are immediately washed three times (2 ml/wash) with wash medium. The wash medium can be 0.5 $M$ NaCl, 0.5 $M$ Na$_2$SO$_4$, or the reaction medium minus ATP and $^{22}$Na. Similar results are obtained with either of these wash media.

The filters are placed on absorbent tissue to remove excess retained filtrate and then dissolved in 2 ml of ethylene glycol monoethyl ether (Cellosolve). Liquid scintillant (10 ml of ACS, Beckman Instruments) is added and the radioactivity measured by liquid scintillation counting.

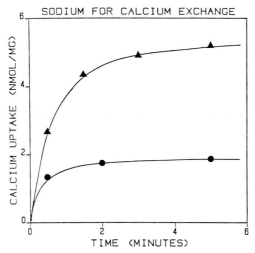

FIG. 4. Calcium uptake into skeletal muscle sarcolemmal vesicles by Na$^+$/Ca$^{2+}$ exchange. Dextran-gradient-purified sarcolemmal vesicles are preequilibrated 3 hr to overnight on ice in 160 m$M$ NaCl, 40 m$M$ MOPS/Tris, pH 7.4, sedimented, and then resuspended in a small volume to 3 mg/ml. The reaction is initiated by dilution (at least 20-fold) into the 160 m$M$ KCl, 40 m$M$ MOPS/Tris, 44 $\mu$M CaCl$_2$ (3 $\mu$C/ml $^{45}$Ca), and 1 $\mu$M valinomycin (▲). The control (●), with no Na$^+$ gradient, replaced KCl in the uptake medium with NaCl. Following dilution of the vesicles into uptake medium, the sarcolemmal vesicles (200 $\mu$l aliquots, 8–10 $\mu$g of protein) are withdrawn and rapidly filtered through 0.22-$\mu$m Millipore filters (GSWP). The filtered sample is washed three times with 2 ml wash of 200 m$M$ KCl, 1 m$M$ La(NO$_3$)$_3$, 5 m$M$ HEPES, pH 7.4, blotted dry, and radioactivity is measured by scintillation counting, as described earlier.

## $Ca^{2+}$ Transport by $Na^+/Ca^{2+}$ Exchange

The skeletal muscle sarcolemmal vesicles are also capable of $Ca^{2+}$ transport via the $Na^+/Ca^{2+}$ exchanger.[8] When the vesicles have a $Na^+$ gradient established across the vesicle, $Ca^{2+}$ will accumulate in response to the transmembrane $Na^+$ gradient. The outwardly directed $Na^+$ gradient is developed by preequilibrating the sarcolemmal vesicles in $Na^+$-containing medium and then diluting them in isotonic Na-free medium containing $^{45}Ca^{2+}$ (Fig. 4).

The vesicles are preequilibrated 3 hr to overnight on ice in 160 m$M$ NaCl, 40 m$M$ MOPS/Tris, pH 7.4, in a large volume, then sedimented. The pellet is resuspended in a small volume to a protein concentration of approximately 3 mg/ml (protein determination is later performed to obtain the precise protein concentration).

The reaction is initiated by dilution of the presoaked vesicles at least 20-fold into 160 m$M$ KCl, 40 m$M$ MOPS/Tris, 44 $\mu M$ CaCl₂ (3 $\mu$Ci/ml $^{45}$Ca), and 1 $\mu M$ valinomycin. The uptake (dilution) medium is temperature-equilibrated and maintained at 37°. The control, with no $Na^+$ gradient, contains $Na^+$ in the uptake medium in place of $K^+$.

After the membrane samples are added, 200 $\mu$l aliquots (8–10 $\mu$g of protein) are withdrawn and rapidly filtered through 0.22-$\mu$m Millipore filters (GSWP). The filtered samples are washed three times with 2 ml of 200 m$M$ KCl, 1 m$M$ La(NO₃)₃, 5 m$M$ HEPES, pH 7.4. The filters are blotted dry and prepared for scintillation counting, as described for $Na^+$ transport activities.

## [4] Isolation of Sarcoplasmic Reticulum Fractions Referable to Longitudinal Tubules and Junctional Terminal Cisternae from Rabbit Skeletal Muscle[1]

By ALICE CHU, MARK C. DIXON, AKITSUGU SAITO, STEVEN SEILER, and SIDNEY FLEISCHER

The sarcoplasmic reticulum (SR) of fast skeletal muscle fibers is a sleevelike network surrounding the myofibrils that controls the cytoplasmic calcium concentration and thereby regulates muscle contraction and relaxation. The membranous system consists of mainly two regions: (1)

[1] Supported by National Institutes of Health postdoctoral fellowship (AM 07016) to A.C., in part by grants from the National Institutes of Health (DK 14632) and the Muscular Dystrophy Association to S.F.

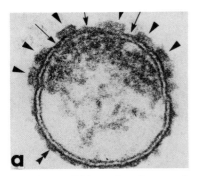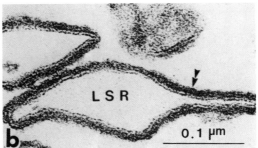

FIG. 1. Thin-section electron micrographs of junctional terminal cisternae and light SR. The samples were fixed with tannic acid/glutaraldehyde. (a) A junctional terminal cisternae vesicle; the junctional face membrane (arrow) containing "feet" structures (arrowhead) and calcium-pump membrane (double arrowhead) are indicated. (Magnification: 221,000×. (b) Light SR, referable to longitudinal tubule, consists only of the calcium-pump membrane (double arrowhead). (Magnification: 221,000× (From Ref. 2.)

tubular longitudinal elements and (2) bulbous terminal cisternae. The latter is junctionally associated with the tranverse tubule via bridging structures referred to as "feet." Recently, we have reported the preparation of a highly enriched SR fraction of junctional terminal cisternae with morphologically intact feet structures.[2] The isolation procedures for small- and large-scale preparations are described here. The small-scale preparation uses swinging bucket rotors while the large-scale preparation uses zonal rotors. SR microsomes are first isolated by homogenization of muscle and differential centrifugation. Subfractionation by isopycnic density centrifugation gives rise to a light SR fraction at lower sucrose density and a heavy SR fraction at higher sucrose density, referable to longitudinal tubules and terminal cisternae, respectively (Fig. 1).

## Preparation of Defined Sarcoplasmic Reticulum Fractions

### Reagents

Imidazole buffer, 5 m$M$, pH 7.4, adjusted with HCl at room temperature

Homogenization medium consists of 5 m$M$ imidazole-HCl (pH 7.4) and 0.3 $M$ sucrose

Buffered sucrose solutions of 5 m$M$ imidazole-HCl (pH 7.4) and various sucrose concentrations (ultrapure grade, Schwarz/Mann, Orangeburg, NY) of 24% (w/w), 27.0% (w/w), 32.0% (w/w), 34.0%

[2] A. Saito, S. Seiler, A. Chu, and S. Fleischer, *J. Cell Biol.* **99,** 875 (1984).

(w/w), 38.0% (w/w), 45.0% (w/w), and 60.0% (w/w). The % sucrose is checked by a refractometer (Abbe Type 3L, Bausch and
Lomb) and adjusted to within ±0.2%
Sucrose 60.0% (w/w) as cushion (need not be ultrapure grade, enzyme grade or table sugar is sufficient)

*Small-Scale Preparation*

One female New Zealand white rabbit, about 3 kg, is sacrificed by
cervical dislocation. All remaining procedures are carried out at 0–4°. The
leg muscles are quickly excised, rinsed in homogenizing medium,
trimmed of fat, connective tissue, and red muscle, and the predominantly
white muscles[3] are selected and passed through a meat grinder of face
plate of 2- to 4-mm-size holes. Portions of 50 g of ground muscle are
homogenized with 250 ml of homogenizing medium in a blender of 40 oz
or 1200 ml size (model 31BL46, Waring Products Division, New Hartford,
CT) for 1 min at maximum speed, which is controlled by an autotrans-
former set at 120–140 V (see Table I). The homogenates are centrifuged in
a Beckman J21 centrifuge using a JA-10 rotor or equivalent (Spinco Division, Palo Alto, CA) for 20 min at 8000 rpm (11,000 $g_{max}$), one homogenate
per rotor bottle. Each resulting pellet is rehomogenized with 250 ml of
homogenization medium in the blender as described above, and centrifuged again in a JA-10 rotor. The second supernatant (SUP) is filtered
through 4–6 layers of cheesecloth and recentrifuged in a Beckman preparative ultracentrifuge for 60 min in a Type 35 or 45Ti rotor at 30,000 rpm
(110,000 $g_{max}$). The pellets are combined and resuspended in about a total
of 60 ml homogenizing medium (see below), and about 10 ml each of this
microsomal fraction is loaded on top of a sucrose (w/w) step gradient
constructed of 4 ml of 45% (bottom step), 7 ml of 38%, 7 ml of 34%, 7 ml
of 32%, and 4 ml of 27% sucrose in 5 m$M$ imidazole-HCl (pH 7.4). The
densities of the sucrose layers are 1.203, 1.166, 1.146, 1.137, and 1.113 g/
cm$^3$, respectively. It has been our experience that about 60 mg microsomal protein can be loaded on top of each gradient tube with good
separation. The time involved up to this stage is 4–5 hr. Centrifugation is
carried out overnight in a Beckman swinging bucket SW28 rotor or equivalent for at least 12–16 hr at 20,000 rpm (70,000 $g_{max}$). (See flow diagram in
Fig. 2.)

The fractions band at the interfaces of the steps of the gradient as
shown in Figs. 2 and 3. Fractions 2 and 4 are collected by the careful use
of Pasteur pipets or plastic hyperdermic syringes with long needles and

---

[3] The white portions of leg muscles include sartorius, gracilis, vastus lateralis, vastus medialis, adductor magnus, gluteus (maximus, medius, minimus), biceps femoralis, and gas-
trocnemius.

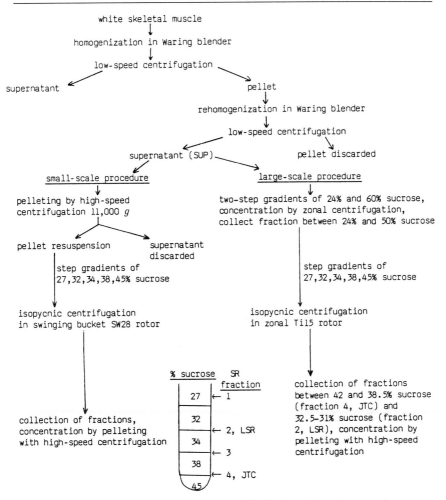

FIG. 2. Flow chart of isolation procedures of SR. Both small- and large-scale preparations are presented. LSR, light SR (fraction 2); JTC, junctional terminal cisternae (fraction 4).

diluted with 5 m$M$ imidazole-HCl (pH 7.4) to a sucrose concentration of about 10–15%, as assessed by a refractometer at room temperature. The fractions are then sedimented for 1 hr in a Beckman 35 or 45Ti rotor at 32,000 rpm (125,000 $g_{max}$). The pellets are resuspended in homogenizing medium to give a protein concentration of about 10–30 mg/ml, divided into appropriate aliquots, quick-frozen in liquid nitrogen, and stored at −70°.

TABLE I

PREPARATION AND FUNCTIONAL PROPERTIES OF SARCOPLASMIC RETICULUM VESICLE SUBFRACTIONS

| Preparation | Small scale | | Large scale | |
|---|---|---|---|---|
| Number of rabbits | 1 | | 3 | |
| Net weight of muscle (g) | 143 (125–165)[a] | | 506 (400–600) | |
| Crude microsomes (mg) | 173 (100–250) | | 690 (488–900) | |
| Microsomal yield (mg protein/g muscle) | 1.3 (0.8–1.65) | | 1.3 (0.8–1.8) | |

| Property | Small scale | | Large scale | |
| | Fraction 2[b] | Fraction 4[b] | Fraction 2 | Fraction 4 |
|---|---|---|---|---|
| SR yield (mg) | 15 (6–21) | 17 (12–26) | 29 (12–82) | 58 (23–94) |
| SR% recovery (mg protein/mg microsomes) | 9 (6–11) | 11 (9–13) | 5 (2–13) | 9 (4–17) |
| Ca²⁺,Mg²⁺-dependent rate ATPase activity[c] (μmol P$_i$/mg protein · min) | | | | |
| No A23187 | 1.9 (1.3–2.6) | 1.9 (1.6–2.0) | 1.1 (0.8–1.6) | 1.2 (0.8–1.8) |
| + A23187 | 3.7 (3.5–4.1) | 2.5 (2.3–2.8) | 3.1 (2.1–4.3) | 1.6 (0.6–2.4) |
| Ratio of (+A23187/No A23187) | 2.1 (2.0–3.2) | 1.3 (1.2–1.5) | 2.7 (1.9–3.6) | 1.3 (0.8–1.5) |
| Mg²⁺-dependent (basal) rate | | | | |
| No A23187 | 0.18 (0.10–0.28) | 0.08 (0.05–0.12) | 0.26 (0.20–0.39) | 0.15 (0.10–0.18) |
| + A23187 | 0.11 (0.10–0.15) | 0.11 (0.04–0.14) | 0.28 (0.23–0.36) | 0.16 (0.13–0.21) |
| Ca²⁺-loading rates (nmol Ca²⁺/mg protein · min) | | | | |
| no RR[d] | 1.58 (0.63–2.32) | 0.14 (0.04–0.30) | 1.85 (0.80–2.50) | 0.21 (0.02–0.44) |
| + RR | 1.54 (0.62–3.06) | 0.90 (0.32–2.05) | 1.95 (0.91–2.86) | 1.07 (0.40–1.95) |
| Ratio of (+RR/–RR) | 1.1 (0.9–1.3) | 7.2 (3.2–16.7) | 1.0 (0.9–1.2) | 8.1 (2.8–13.7) |

[a] The values reported here represent an average of at least five separate small- and large-scale preparations. Due to the variability, the range of values is given in parentheses. Note particularly the values of Ca²⁺ loading for fraction 4.

[b] Fractions 2 and 4 refer to the longitudinal tubules and junctional terminal cisternae, respectively.

[c] ATPase activity is measured in the presence and absence of Ca²⁺ as described in the text. The Ca²⁺,Mg²⁺-dependent rate is the difference between the total ATPase rate (in the presence of Ca²⁺) and the Ca²⁺-independent, Mg²⁺-dependent (basal) rate (in the absence of Ca²⁺ and in the presence of EGTA).

[d] Ruthenium red concentration used was 20 μM.

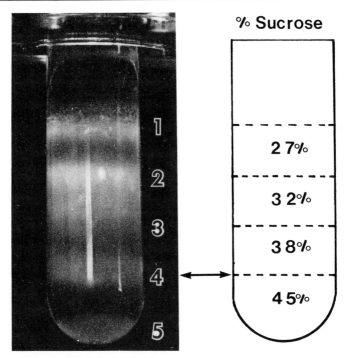

FIG. 3. Discontinuous sucrose gradient for subfractionation of SR microsomes.[2] Fractions 2 and 4 give rise to light SR and junctional terminal cisternae. The 34% sucrose step is not included in this illustration.

## Large-Scale Preparation

Fast skeletal muscle from the hind legs of three rabbits is used. The selection of predominantly white muscle and homogenization of ground muscle are carried out as described above for the small-scale preparation to the supernatant (SUP) step (see Fig. 2 and Table I). The large volume of supernatant is filtered through 4–6 layers of cheesecloth, diluted to 1300 ml and poured directly into a Beckman Ti15 zonal rotor equipped with a B-29 core plus liner. The rotor is assembled and accelerated to 2000 rpm. A two-step sucrose gradient is established at the outer edge by pumping in 100 ml of 24% (w/w) buffered sucrose and about 100 ml of 60% (w/w) buffered sucrose (to fill the rotor). A peristaltic pump (e.g., Minipuls 2, Gilson Medical Electronics, Middleton, WI) (tubing size of about 2.2 mm inner diameter) is used at a speed of about 13 ml per minute. The rotor is then accelerated to 25,000 rpm and centrifuged for 90 min. Then it is decelerated to 2000 rpm. Distilled water is pumped into the center of the rotor at 13 ml per minute to collect the microsomal fraction. The turbid material banding between 50 and 24% sucrose is collected by monitoring the concentration of sucrose with a refractometer. The microsomes collected from two zonal rotors, carried out either sequentially or in parallel, are pooled and diluted to 650 ml with 5 m$M$ imidazole-HCl (pH 7.4) to about 10–15% sucrose.

The sample is poured into a Ti15 rotor (containing the B29 core and liner), and the rotor accelerated to 2000 rpm. A discontinuous gradient, consisting of 125 ml each of 27%, 32%, 34%, 38%, 45% (w/w) sucrose, and buffered 5 m$M$ imidazole-HCl (pH 7.4), is pumped in to the periphery of the rotor. The remaining volume is filled with 60% sucrose as cushion. The time involved up to this stage is about 6–7 hr. The centrifugation is carried out at 20,000 rpm (48,000 $g_{max}$) overnight for about 12–16 hr. For unloading, the rotor is decelerated to 2000 rpm, and the fractions are collected as described below. The junctional terminal cisternae (fraction 4) is collected between 42 and 38.5% and the light SR (fraction 2) between 33 and 31% sucrose by pumping distilled water into the center of the rotor. Fraction 3, which is a mixture of light SR and junctional terminal cisternae, may also be collected, if desired, between 37.5% and 34% sucrose. The fractions are then diluted with 5 m$M$ imidazole-HCl (pH 7.4) to about 10–15% sucrose, and pelleted in a Beckman 35 or 45Ti rotor for 60 min at 32,000 rpm (125,000 $g_{max}$). The pellets are resuspended in homogenizing medium as described for the small-scale preparation, quick-frozen in liquid nitrogen, and stored at $-70°$.

## Functional Characterization of Sarcoplasmic Reticulum Vesicles

There are a number of criteria to determine the purity and function of the SR preparations, including Ca$^{2+}$,Mg$^{2+}$-dependent ATPase (CaATPase) activity of the calcium pump and energized Ca$^{2+}$ loading. The fractions have characteristic protein profiles using SDS-polyacrylamide gel electrophoresis and morphology as viewed by electron microscopy, respectively.[2] A coupled-enzyme assay utilizing an ATP regenerating system is described for measuring the ATPase activity.[4] A Ca$^{2+}$ ionophore is used during the assay to dissipate the buildup of a Ca$^{2+}$ gradient so that the CaATPase activity is maximal. EGTA is added after the initial rate is obtained to chelate the Ca$^{2+}$ so that the basal (Ca$^{2+}$-independent) ATPase rate can be subtracted. CaATPase is also frequently measured by the production of inorganic phosphate colorimetrically.[5]

---

[4] A. Schwartz, J. C. Allen, and S. Harigaya, *J. Pharmacol. Exp. Ther.* **168,** 31 (1969).
[5] P. Ottolenghi, *Biochem. J.* **151,** 61 (1975).

ATP-dependent $Ca^{2+}$ transport is readily measured by $Ca^{2+}$ loading (in the presence of a calcium-precipitating anion, oxalate or phosphate). A characteristic of the junctional terminal cisternae is the low $Ca^{2+}$-loading rates which can be stimulated by ruthenium red (Table I).[6,7] We measure $Ca^{2+}$ loading in the absence and presence of ruthenium red, which stimulates $Ca^{2+}$ loading. The $Ca^{2+}$-loading rate of light SR is already optimal and is not stimulated by ruthenium red. The assay for $Ca^{2+}$ loading using $Ca^{2+}$-sensitive metallochromic dyes for measuring $Ca^{2+}$ transport is described.[8] An alternative method is to measure radioactive $^{45}Ca^{2+}$ loading by filtration.[9] (See Meissner, this series.[10])

## Assays for Sarcoplasmic Reticulum Characterization

### CaATPase Activity

#### Reagents

Assay medium: HEPES (*N*-2-hydroxyethylpiperazine-*N'*-2-ethanesulfonic acid) buffer, 7 m*M* (pH 7.0), with 143 m*M* KCl, 7 m*M* $MgCl_2$, 143 μ*M* $CaCl_2$, 85.6 μ*M* EGTA (ethylene glycol bis(β-aminoethyl ether)-*N,N,N',N'*-tetraacetic acid), and 0.43 m*M* sucrose SR vesicles, about 1 mg protein/ml

$Na_2ATP$, 50 m*M* (adjusted to pH 7.0 with NaOH)

NADH, 10 m*M* (Sigma Chemical Co., St. Louis, MO)

Phosphoenolpyruvate 100 m*M* (Sigma), tricyclohexylammonium salt

Coupling enzyme mixture of pyruvate kinase and lactate dehydrogenase 700 and 1000 units/ml, respectively, in 2 m*M* $(NH_4)_2SO_4$ (Sigma)

A23187, 0.5 mg/ml in 95% ethanol (Calbiochem-Behring, La Jolla, CA)

EGTA, 100 m*M* (adjusted to pH 7.0 with KOH) ($K_2EGTA$)

#### Procedure

The oxidation of NADH is continuously monitored by the decreased absorbance at 340 nm using a recording spectrophotometer.[4] The assay is based on the following coupled reactions using the enzyme mixture:

[6] A. Chu, P. Volpe, B. Costello, and S. Fleischer, *Biochemistry* **25**, 8315 (1986).

[7] A. Chu, A. Saito, and S. Fleischer, *Arch. Biochem. Biophys.* **258**, 13 (1987).

[8] A. Scarpa, this series, Vol. 56, p. 301.

[9] A. Martonosi and R. Feretos, *J. Biol. Chem.* **239**, 648 (1964).

[10] G. Meissner, this series, Vol. 31, p. 238.

$$ATP \xrightarrow{\text{SR CaATPase}} ADP + P_i$$

$$ADP + \text{phosphoenolpyruvate} \xrightarrow{\text{pyruvate kinase}} ATP + \text{pyruvate}$$

$$\text{Pyruvate} + NADH \xrightarrow{\text{lactate dehydrogenase}} \text{lactate} + NAD^+$$

To a final volume of 1.0 ml, 0.7 ml of assay medium (pH 7.0), 40 $\mu$l NADH, 20 $\mu$l phosphoenolpyruvate, 12 $\mu$l of coupling enzyme mixture, and 10 $\mu$l SR are added to give a final concentration of 5 m$M$ HEPES, 100 m$M$ KCl, 5 m$M$ MgCl$_2$, 100 $\mu M$ CaCl$_2$, 60 $\mu M$ EGTA, 0.4 m$M$ NADH, 2 m$M$ phosphoenolpyruvate, 8.4 units of pyruvate kinase and 12 units of lactate dehydrogenase/ml, and 10 $\mu$g SR protein/ml. A baseline is established and the temperature equilibrated at 25° for 2–4 min. The reaction is started by adding 20 $\mu$l of Na$_2$ATP to give a final concentration of 1 m$M$. The reaction is run for several minutes to obtain a linear slope (a chart speed of about 4 min/inch is convenient), of the total ATPase activity. Then 40 $\mu$l of 100 m$M$ K$_2$EGTA (pH 7.0) is added to give a final concentration of 4 m$M$ K$_2$EGTA to obtain the basal (Ca$^{2+}$-independent) ATPase rate. In another cuvette, the same reaction is repeated in the presence of 3 $\mu$l of the Ca$^{2+}$ ionophore A23187 (final concentration of 1.5 $\mu$g/ml). If the rate is rapid, the SR protein is reduced to obtain a readily measurable slope. When less pure fractions are used, an inhibitor of the mitochondrial F$_1$ ATPase should be included in the assay (e.g., 5 m$M$ sodium azide).

The amount of NADH oxidized is equivalent to the ATP hydrolyzed. Based on the extinction coefficient for NADH, 6.22 cm$^2$ $\mu$mol$^{-1}$, the following general formula can be used to calculate the rate of ATP hydrolysis, in $\mu$mol/mg · min:

$$\text{Rate} = \frac{\Delta \text{ absorbance}}{6.22 \times \text{protein (mg)} \times \text{time (min)}}$$

*Ca$^{2+}$ Loading Assay*

*Reagents*

Phosphate buffer (pH 7.0), 125 m$M$ phosphoric acid neutralized with concentrated KOH

MgCl$_2$, 100 m$M$

Antipyrylazo III, 5 m$M$ (neutralized to about pH 8.0 with HCl) (Sigma)

SR vesicles, about 10 mg/ml

CaCl$_2$, 5 m$M$

Na$_2$ATP, 50 m$M$ (pH 7.0)

Ruthenium red, 1 m$M$ (Sigma). The concentration is adjusted for the purity as indicated on the product label

*Procedure*

A dual-wavelength spectrophotomer is used for the continuous monitoring of $Ca^{2+}$ transport with the metallochromic indicator antipyrylazo III, at the wavelength pair of 710–790 nm. We use a Hewlett Packard UV/ VIS spectrophotometer (Model 8450A, Palo Alto, CA). In a cuvette of 1 ml final volume, 40 $\mu$l antipyrylazo III, 10 $\mu$l $MgCl_2$, and 800 $\mu$l phosphate buffer are added to give a final concentration of 200 $\mu M$ antipyrylazo III, 1 m$M$ $MgCl_2$, and 100 m$M$ phosphate (pH 7.0). SR vesicles (~5 $\mu$l) and $Na_2ATP$ (20 $\mu$l) are added to give final concentrations of ~50 $\mu$g SR protein/ml and 1 m$M$. The temperature is equilibrated to 25° for about 2 min and baseline is established. After the uptake of contaminating $Ca^{2+}$ in the medium, 10 $\mu$l of $CaCl_2$ (final concentration of 50 $\mu M$) is added to start the reaction, and a loading rate is obtained from the linear slope. Ruthenium red (5–7 $\mu M$) can then be pulsed in to obtain a stimulated rate in the case of terminal cisternae (see Fig. 4).[11] The change in absorbance upon the addition of $Ca^{2+}$ is used for the calibration of the amount of external $Ca^{2+}$ added to the medium. When ruthenium red has been added, a second calibration with $Ca^{2+}$ (10 $\mu$l of 5 m$M$ $CaCl_2$) in the presence of ruthenium red is required. Alternatively, the enhanced rate with ruthenium red can be measured in a separate cuvette using the initial $Ca^{2+}$ calibration. The amount of SR can be adjusted so that the $Ca^{2+}$-loading rate is readily measurable.

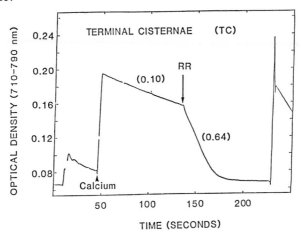

Fig. 4. $Ca^{2+}$ loading of junctional terminal cisternae is measured in the presence of the metallochromic dye, antipyrylazo III. At the arrow, 7 $\mu M$ ruthenium red is added to give an enhanced $Ca^{2+}$-loading rate. The values in parentheses represent the loading rates in $\mu$mol/ mg · min.[11]

[11] S. Fleischer, E. M. Ogunbunmi, M. C. Dixon, and E. A. M. Fleer, *Proc. Natl. Acad. Sci. U.S.A.* **82,** 7256 (1985).

A general formula can be used to calculate the rate of $Ca^{2+}$ loading, in $\mu$mol $Ca^{2+}$/mg · min:

$$\text{Slope of } Ca^{2+} \text{ loading} = \frac{\Delta \text{ absorbance}}{\text{time (min)}}$$

$$Ca^{2+} \text{ calibration} = \frac{50 \text{ nmol } Ca^{2+}}{\Delta \text{ absorbance}}$$

$$\text{Rate of } Ca^{2+} \text{ loading} = \frac{\text{slope of } Ca^{2+} \text{ loading} \times Ca^{2+} \text{ calibration}}{\text{protein (mg)}}$$

The SR serves a central role in $Ca^{2+}$ uptake, storage, and release. Therefore, fractions of junctional terminal cisternae and light SR can be studied in this context. Both SR fractions are obtained from the same gradient and have undergone similar treatment. The light SR is essentially only the calcium pump membrane, capable of energized $Ca^{2+}$ uptake. The junctional terminal cisternae contains the calcium pump membrane (80–85% of the membrane) as well as the junctional face membrane with well-defined junctional feet structures (15–20% of the membrane surface area).[2] Junctional terminal cisternae are highly permeable to $Ca^{2+}$.[6,7,11] Enhanced $Ca^{2+}$ loading is obtained with ruthenium red. This high permeability is due to the $Ca^{2+}$-release channels which are localized in this portion of the SR.[11]

# [5] Isolation of the Junctional Face Membrane of Sarcoplasmic Reticulum

By BRIAN COSTELLO, CHRISTOPHER CHADWICK, and SIDNEY FLEISCHER

The sarcoplasmic reticulum (SR)[1,2] of skeletal muscle controls the myoplasmic free calcium concentration and thereby mediates muscle contraction and relaxation. The molecular machinery involved in the release of calcium from the SR remains the key unknown link in the chain of events

[1] Supported in part by grants from National Institutes of Health DK 14632 and the Muscular Dystrophy Association to SF; NIH National Research Service Award to BC (5 F32 GM 08198); and a Biomedical Research Support Grant from the National Institutes of Health administered by Vanderbilt University.
[2] Abbreviations: CC, compartmental contents; EDTA, ethylenediaminetetraacetic acid; HEPES, N-2-hydroxyethylpiperazine-N'-2-ethanesulfonic acid; JFM, junctional face membrane; JFM-CC, junctional face membrane-compartmental contents; JTC, junctional terminal cisternae; SR, sarcoplasmic reticulum; T tubule, transverse tubule.

underlying excitation–contraction coupling in skeletal muscle. It has long been inferred that the immediate signal for calcium release is transmitted across the triad junction between the transverse tubule and the terminal cisternae of SR in response to the depolarization of the transverse tubule with release of $Ca^{2+}$ from the terminal cisternae of SR. Transverse tubules are narrow invaginations of the plasma membrane of the muscle fiber which are situated in register with the underlying sarcomeres. The transverse tubules are junctionally associated with the terminal cisternae by way of the junctional face membrane (JFM) containing the junctional feet structures. The latter are observed by electron microscopy as electron-dense structures, spanning the gap between the transverse tubule and the terminal cisternae.[3,4] Terminal cisternae consist of two types of membranes, the calcium-pump membrane and the JFM. The former membrane is involved in energized $Ca^{2+}$ uptake, the latter we believe mediates $Ca^{2+}$ release, which triggers muscle contraction.[5,6,6a] It is of obvious interest to be able to isolate and study the JFM. A method is presented for the isolation of the JFM of the junctional terminal cisternae (JTC) of sarcoplasmic reticulum.

The starting material for the JFM isolation (see Fig. 1) is JTC vesicles of rabbit fast twitch skeletal muscle.[7] This fraction derives from the terminal cisternae region of the SR and, in contrast to other heavy SR fractions,[8] contains a high percentage (~20%) of JFM that contain well-defined junctional feet structures[7] (Fig. 2a). A prominent feature of the JTC that is shared with other heavy SR preparations is the electron-opaque internal contents (Fig. 2a), which consist mainly of calcium-binding protein (CBP),[9] also referred to as calsequestrin.[10] For the JFM isolation, we use a fraction of JTC obtained from fresh rabbit leg muscle[7] or from frozen back muscle.[11] The protein composition of the JTC is shown in Fig. 3 (lane 1). The two predominant proteins are the calcium-pump protein

[3] S. Fleischer, in "Structure and Function of Sarcoplasmic Reticulum" (S. Fleischer and Y. Tonomura, eds.), pp. 119–145. Academic Press, New York, 1985.

[4] A. Martonosi, *Physiol. Rev.* **64,** 1240 (1984).

[5] S. Fleischer, E. M. Ogunbunmi, M. C. Dixon, and E. A. M. Fleer, *Proc. Natl. Acad. Sci. U.S.A.* **82,** 7256 (1985).

[6] M. Inui, A. Saito, and S. Fleischer, *J. Biol. Chem.* **262,** 1740 (1987).

[6a] L. Hymel, M. Inui, S. Fleischer, and H. G. Schindler, *Proc. Natl. Acad. Sci. U.S.A.* **85,** 441 (1988).

[7] A. Saito, S. Seiler, A. Chu, and S. Fleischer, *J. Cell Biol.* **99,** 875 (1984).

[8] G. Meissner, *Biochim. Biophys. Acta* **389,** 51 (1975).

[9] G. Meissner, G. E. Conner, and S. Fleischer, *Biochim. Biophys. Acta* **298,** 246 (1973).

[10] D. H. MacLennan and P. T. S. Wong, *Proc. Natl. Acad. Sci. U.S.A.* **68,** 1231 (1971).

[11] B. Costello, C. Chadwick, A. Saito, A. Chu, A. Maurer, and S. Fleischer, *J. Cell Biol.* **103,** 741 (1986).

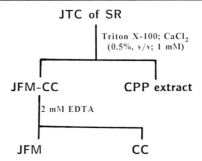

FIG. 1. Scheme for the preparation of the junctional face membrane.[11]

(CPP) and CBP. A characteristic of JTC is a band at 340 kDa which sometimes appears as a doublet.

## Procedure for the Isolation of Junctional Face Membrane

The necessary stock solutions are as follows:
  Buffered sucrose (enzyme grade, Schwarz-Mann Biotech, Cleveland, OH): 0.3 $M$ sucrose and 5 m$M$ K HEPES, pH 7.4
  CaCl₂ stock solution: 100 m$M$ CaCl₂
  Triton X-100 stock solution: 10% (v/v) Triton X-100 (Pierce Chemical Co., Rockford, IL)
  EDTA stock solution: 100 m$M$ NaEDTA, pH 7.4

The procedure is as follows. All operations are performed at 0–4° (see Fig. 1) and pellets are resuspended with a Dounce glass homogenizer (loose pestle B).

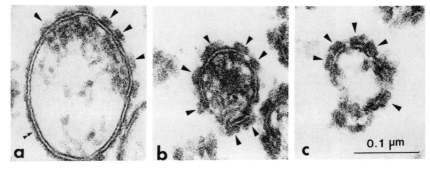

FIG. 2. Electron micrographs comparing a typical vesicle of (a) JTC, (b) JFM-CC, and (c) JFM. For each, the JFM is approximately of similar size and contains unidirectionally aligned feet structures. The feet structures are indicated by arrowheads and the calcium-pump membrane by the double arrowhead.[11]

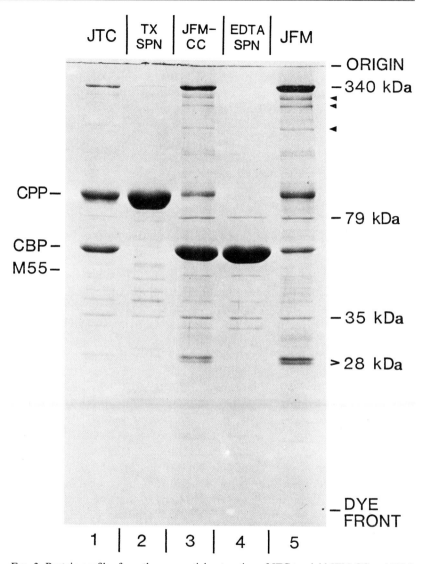

FIG. 3. Protein profiles from the sequential extraction of JTC to yield JFM-CC and JFM (see Fig. 1). TX, Triton X-100; SPN, supernatant.[11] CPP, calcium pump protein; CBP, calcium binding protein.

1. The JTC fraction in buffered sucrose is adjusted to a final concentration of 3.3 mg protein/ml and 1 m$M$ CaCl$_2$ using buffered sucrose and stock CaCl$_2$ solution and maintained for 10 min.

2. Triton X-100 stock solution is added to a final concentration of

0.5% (v/v), mixed thoroughly for 5 sec using a vortex mixer, and incubated for 20 min.

3. The mixture is centrifuged in either (1) a Beckman Type 75Ti rotor at 35,000 rpm (110,000 $g_{max}$) for 60 min in a Beckman L-8 ultracentrifuge, (2) an A95 or A110 rotor at 29 psi (160,000 $g_{max}$) for 30 min in a Beckman Airfuge, or (3) a TLA 100.2 rotor at 55,000 rpm (109,000 $g_{max}$) for 15 min using a Beckman TL-100 Tabletop Ultracentrifuge.

4. The supernatant containing the calcium pump protein is decanted.

5. The pellet (JFM-CC) is resuspended in buffered sucrose to 1.0 mg protein/ml and EDTA stock solution is added to a final concentration of 2 m$M$. The mixture is incubated for 10 min.

6. Centrifuge as in step 5.

7. The supernatant containing mainly calcium binding protein is decanted.

8. The pellet containing JFM is resuspended in buffered sucrose. The sample is quick-frozen using liquid nitrogen and stored at −80°.

Treatment of JTC in the presence of Ca²⁺ and Triton X-100 selectively extracts the calcium pump membrane (Fig. 3, lane 2) yielding an insoluble pellet consisting of JFM with tightly bound CC (Fig. 2b). This JFM-CC pellet (~50% of the starting JTC protein) contains a small amount of protein migrating with a mobility equivalent to CPP (Fig. 3, lane 3). It is likely that this band is referable to a protein other than the calcium pump protein. All of the CBP remains in the pellet after Triton extraction in the presence of 1 m$M$ CaCl₂. Treatment of the JFM-CC with EDTA extracts most of the CC from the JFM-CC, thereby yielding intact JFM which is recovered by sedimentation (Fig. 2C) (~25% of the JTC protein). The SDS-PAGE profile of the JFM pellet shows that EDTA treatment of the JFM-CC extracts most of the CBP and little else (Fig. 3, lane 5). Hence, the EDTA supernatant consists almost entirely of solubilized CBP with trace amounts of other protein components (Fig. 3, lane 4). Characteristically, the JFM is enriched in the 340-kDa protein and in the protein doublet at 28 kDa. Also present in this JFM fraction is some of the 79-kDa protein as well as a small fraction (approximately 10%) of the calcium-binding protein. This small fraction is quite resistant to EDTA extraction.[11] The JFM of SR, now available in the test tube, should facilitate studies on junctional association and the calcium release process.

# [6] Isolation of Triads from Skeletal Muscle[1]

*By* R. D. MITCHELL, PHILIP PALADE, AKITSUGU SAITO,
and SIDNEY FLEISCHER

Excitation–contraction coupling in skeletal muscle is initiated at the neuromuscular junction. The action potential then proceeds longitudinally along the length of the fiber and transversely to within the fiber by way of invaginations from the plasmalemma, referred to as transverse tubules. The latter are junctionally associated with the terminal cisternae of sarcoplasmic reticulum via "feet" structures. By an as yet unknown mechanism, $Ca^{2+}$ is then released from the terminal cisternae, elevating the myoplasmic calcium concentration, and thereby triggering muscle contraction. The structure formed by the junctional association of the transverse tubule with the terminal cisternae is referred to as the "triad," and the junction is the triad junction. The triad[2] and the triad junction serve a key role in excitation–contraction coupling. We describe methodology for the isolation of morphologically intact triad structures.[3]

The methodology is presented in two sections: (1) procedures for triad isolation, and (2) characterization of the isolated triads.

## Isolation Procedure

All operations are carried out in the cold ($\sim 4°$). The pH of the solutions is adjusted at room temperature.

The flow diagram for the preparation of triads is given in Fig. 1. Two variants are described. The Standard variant provides a triad fraction with somewhat lesser purity and integrity of the transverse tubule compared with the Pyrophosphate variant. The Pyrophosphate variant pro-

[1] These studies were supported by grants from NIH DK 14632 and the Muscular Dystrophy Association of America, and a postdoctoral fellowship from the Muscular Dystrophy Association (RM) and from Public Health Service (PP), and a Biomedical Research Support Grant from the National Institutes of Health, administered by Vanderbilt University (SF). We thank Dr. Christopher Chadwick for his comments on the manuscript and Ms. Laura Taylor for her secretarial skills.

[2] The term "triad" derives from electron microscopy of fast twitch skeletal muscle in which the transverse tubule is apposed between two terminal cisternae of sarcoplasmic reticulum, each in junctional association. The term "triad" is used to denote isolated structures consisting of one or more junctional associations of transverse tubule with terminal cisternae.

[3] R. Mitchell, P. Palade, and S. Fleischer, *J. Cell Biol.* **96**, 1008 (1983).

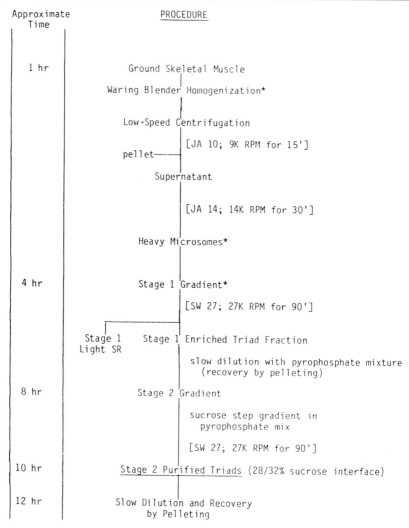

Approximate
Time

PROCEDURE

1 hr

Ground Skeletal Muscle

Waring Blender Homogenization*

Low-Speed Centrifugation

pellet—          [JA 10; 9K RPM for 15']

Supernatant

[JA 14; 14K RPM for 30']

Heavy Microsomes*

4 hr

Stage 1 Gradient*

[SW 27; 27K RPM for 90']

Stage 1          Stage 1 Enriched Triad Fraction
Light SR

slow dilution with pyrophosphate mixture
(recovery by pelleting)

8 hr

Stage 2 Gradient

sucrose step gradient in
pyrophosphate mix

[SW 27; 27K RPM for 90']

10 hr

Stage 2 Purified Triads (28/32% sucrose interface)

12 hr

Slow Dilution and Recovery
by Pelleting

FIG. 1. Flow diagram for the isolation of morphologically intact triads from skeletal muscle. Two variants are described. The Standard variant does not include pyrophosphate–$Mg^{2+}$ mixture in the solutions until the slow dilution of the Stage I gradient triads. The Pyrophosphate variant contains pyrophosphate mix in the solutions at all stages. The asterisk denotes the *absence* of the pyrophosphate mix in the solutions for the standard variant. The Stage I gradient separation is illustrated in Fig. 2. The Stage II gradient separation is shown in Fig. 3. Characterization of Stage I and Stage II fractions is summarized in Table I.

vides a triad fraction of high purity and membrane integrity, albeit with a reduced level (~50%) of adenylate cyclase activity.

The Pyrophosphate variant differs from the Standard variant in that the buffers used in the first three steps of the former are supplemented with pyrophosphate–$Mg^{2+}$ mixture, (pyrophosphate mix),[3–6] and are indicated by an asterisk in the flow diagram. That is, the pyrophosphate mix is added to each step in the purification up to the dilution of Stage I triads. At this point and beyond, each of the solutions for both procedures contain the pyrophosphate mix.

### Reagents

Sucrose (density gradient) was obtained from Schwarz/Mann (Orangebury, NJ).

### Standard Variant

Homogenization medium: 10% sucrose, 0.5 m$M$ (EDTA), pH 7.2
Gradient I: 28% sucrose (w/w) in 5 m$M$ N-2-hydroxyethylpipera-
        zine-$N'$-2-ethanesulfonic acid (HEPES), pH 7.1
        50% sucrose (w/w) in 5 m$M$ HEPES, pH 7.1
        25% sucrose (w/w) in 5 m$M$ HEPES, pH 7.1
        14% sucrose (w/w) in 5 m$M$ HEPES, pH 7.1
        10% sucrose (w/w) in 5 m$M$ HEPES, pH 7.1
Dilution buffer: 5 m$M$ HEPES, pH 7.1
*Pyrophosphate Variant.* Each of the above solutions contains pyrophosphate mix instead of 5 m$M$ HEPES.
        Pyrophosphate mix: 20 m$M$ sodium pyrophosphate, 20 m$M$
        $NaH_2PO_4$, and 1 m$M$ $MgCl_2$, pH 7.1

---

[4] Most preparations of sarcoplasmic reticulum from fast twitch skeletal muscle employ, at one step or another, a salt wash in high salt (0.6 $M$) to remove contractile constituents. Disruption of the triad structure as a result of high levels of salt has been reported by several laboratories,[3,5,6] which on dilution can lead to osmotic disruption of the transverse tubule. We employ a pyrophosphate-$Mg^{2+}$ buffer at relatively low concentrations for disaggregation to facilitate triad isolation[3] and fractions obtained from sucrose density gradients are diluted slowly to minimize osmotic damage. Higher levels of pyrophosphate (50 m$M$, 1 m$M$ $MgCl_2$) lead to disruption of the terminal cisternae similar to that observed using high salt. $MgCl_2$ is essential for the procedure to be effective, although levels higher than 1 m$M$ $MgCl_2$ were found to cause aggregation. In our experience, pyrophosphate from different batches and sources (Fisher Scientific Co. and Alfa Div., Ventron Corp.) can behave somewhat differently, so that minor adjustment in concentration with a new batch of reagent may be necessary. Optimization of morphological integrity is judged using electron microscopy.

[5] K. P. Campbell, C. Franzini-Armstrong, and A. E. Shamoo, *Biochim. Biophys. Acta* **602**, 97 (1980).

[6] J. R. Gilbert and G. Meissner, *Arch. Biochem. Biophys.* **223**, 9 (1983).

Gradient II:   45% sucrose (w/w) in pyrophosphate mix
36% sucrose (w/w) in pyrophosphate mix
34% sucrose (w/w) in pyrophosphate mix
32% sucrose (w/w) in pyrophosphate mix
28% sucrose (w/w) in pyrophosphate mix
25% sucrose (w/w) in pyrophosphate mix
15% sucrose (w/w) in pyrophosphate mix
10% sucrose (w/w) in pyrophosphate mix

Storage medium: 0.3 $M$ sucrose, 0.15 $M$ KCl, and 5 m$M$ HEPES, pH 7.1.

## Skeletal Muscle[7]

Young, New Zealand white rabbits (2–3 kg) are killed by cervical dislocation. The predominantly white muscle of the hind legs is separated from red muscle, large blood vessels, nerves, tendons, and fascia. The hind leg muscle from one rabbit is about 200–250 g. Back muscle (100–150 g) per rabbit can also be used. The dissection is performed in a cold room, and the muscle is then temperature equilibrated (approximately 5 min) before passing it through a prechilled meat grinder. This limits the amount of tissue sticking to the already chilled metal of the meat grinder. The meat grinder (General Model A meat grinder) uses a mincing plate with 2-mm holes. Approximately 300 g of fresh muscle is ground, providing sufficient ground meat for four packets of 60 g (Standard variant) or six 40-g packets (Pyrophosphate variant).

## Homogenization and Preparation of Muscle Microsomes

One packet is placed into the Waring blender container (Waring Commercial Blender, 1182 ml capacity and 590 ml water load, Allied Fischer catalog 1986, 14-509-10) together with 300 ml of homogenization medium (10% sucrose (w/w), 0.5 m$M$ EDTA, pH 7.2, Standard variant) or homogenizing medium containing pyrophosphate mix (Pyrophosphate variant) and homogenized for 60 sec at maximum speed. The homogenate is transferred to a Beckman JA-10 bottle and homogenization is repeated for the remaining packets of meat.

---

[7] In general, higher body weight is avoided since it correlates with increased fatty tissue contamination. The age and sex of the animals and the season of the year can also be factors. Females are more susceptible to fat buildup during the winter season, while muscle from males tends to become engorged with blood in early spring, making distinctions between fast and slow muscle types more difficult. In general, young male rabbits with body weight between 2 and 3 kg provided the best source of muscle tissue for the triad preparation.

The homogenates are centrifuged in a JA-10 rotor at 9000 rpm (8900 $g_{av}$) for 15 min in a Beckman J2-21 centrifuge. The supernatant is carefully poured through four layers of cheesecloth (layered in a funnel) into JA-14 bottles and is centrifuged in a JA-14 rotor at 14,000 rpm (18,900 $g_{av}$) for 30 min in a J2-21 centrifuge.[8] The supernatant is poured off and discarded. The pellets (microsomes) from 240 g muscle are resuspended in 20 ml total volume of 10% sucrose (w/w) containing 5 m$M$ HEPES, pH 7.1 (Standard variant), or 30 ml of 10% sucrose (total volume) containing pyrophosphate mix (Pyrophosphate variant) using a Dounce homogenizer with a loose-fitting pestle.

### Stage I Gradient Centrifugation

Five milliliters resuspended microsomes are carefully layered on top of the 14% sucrose layer of the Stage I gradient tube (see Fig. 2). The gradients are centrifuged in a swinging bucket rotor (Beckman SW27 or SW28 rotor) at 27,000 rpm (96,260 $g_{av}$) for 90 min, using reduced acceleration with no braking.[9]

The tubes are removed from the rotor. Two bands are observed (Fig. 2). In the Standard variant, the upper light SR band is spatially separated from the lower triad/mitochondria band. In the Pyrophosphate variant, the upper band sits above the triad/mitochondria band. The two bands are

---

[8] Pyrophosphate, in the initial homogenization, dissociates the contractile protein, causing the supernatant from the low-speed spin to become quite viscous. More triad material is released and hence the yield is increased (see Table I). Gravity filtration of the low-speed supernatant through cheesecloth is slow, but is necessary to remove fatty aggregates and particulates. The yield of microsomes is approximately 1.6-fold higher for the Pyrophosphate than for the Standard variant.

[9] The Stage I gradient procedure is designed to separate lower density plasmalemmal vesicles and longitudinal light SR from triads. A velocity separation is used with a gradient employing a long distance of 25% sucrose followed by a steeper linear gradient from 28 to 50% sucrose (w/w). The 90-min centrifugation permits only the densest material to approach isopycnic density, resulting in good separation of triad from the light SR and plasmalemma-derived membranes. In the Standard variant, two bands were resolved (see Fig. 2). The upper band, enriched in light SR, was white and turbid and accounted for 60% of the protein. The lower band was light-brown in color, containing visible small white aggregates, and was enriched in triads. Electron microscopy revealed that the fraction was composed primarily of intact triads (or dyads), mitochondria, free junctional heavy SR, some free transverse tubule, and contractile protein. For the Pyrophosphate variant, the separation of the two fractions for the Stage I gradient is not as complete as for the Standard variant. The recovery of light SR from the Pyrophosphate variant is low. When insufficient separation between the two occurs, some upper white material is removed and the remainder of the SR band is used. The contaminating light SR can then be separated in the Stage II separation. However, the preparation of highly enriched triads is more readily achieved when an attempt is made to remove light SR at this point.

% Sucrose

Stage I

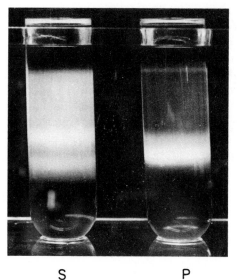

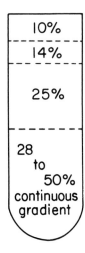

'light' SR

Triad/Mito.

S          P

FIG. 2. Stage I gradient separation of skeletal muscle microsomes. Purification of skeletal muscle microsomes was carried out using a combination step and continuous gradient in an SW27 or SW28 rotor, centrifuged at 27,000 rpm for 90 min. The right-hand diagram defines the gradient. Linear gradients from 28 to 50% are prepared on the previous day. Then, prior to use, a 12 ml step of 25% sucrose is applied atop of the linear 28 to 50% gradient and a second step (4 ml) of 14% sucrose. The sample (5 ml) in 10% sucrose is added just prior to initiating the run. The photograph shows the separation for both variants. In each case, the lowest band in the tube is the Stage I enriched triad/mitochondria fraction. The Pyrophosphate variant (P) is less effective in resolving light SR and triads into two bands than the Standard variant (S). Light SR is located above the triads and appears whiter. The triad/mitochondria fraction is immediately below. It is whitish in appearance at the upper and orange-brown at the lower portion of the band.

collected with the aid of a Pasteur pipet. It is convenient first to aspirate the fluid above the band. The samples are then slowly diluted to minimize osmotic stress. The aim is to dilute each sample to about 10% (w/w) sucrose (check with refractometer) in about 45 min. This is achieved by adding small milliliter aliquots of dilution buffer periodically (Standard variant), or pyrophosphate mix to each fraction with gentle stirring. From here on, the operation for both variants is the same. The diluted samples are placed in centrifuge tubes and centrifuged at 30,000 rpm (70,000 $g_{av}$) for 45 min in a Beckman Type 35 rotor.

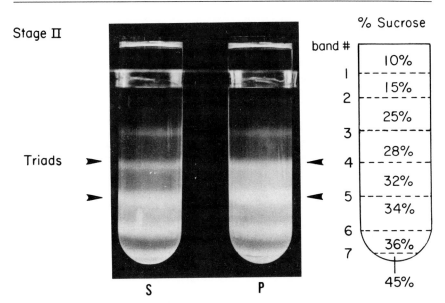

FIG. 3. Stage II gradient purification of the triads for Standard (S) and Pyrophosphate (P) variants. The sucrose step gradient that is employed is displayed to the right. The Stage I triad/mitochondria fraction, in hypertonic sucrose from the gradient, is diluted slowly (30–45 min) with pyrophosphate mix (for both variants) to a sucrose concentration of ~10%, using gentle stirring. The diluted material is sedimented (type 35 rotor, 60 min at 30,000 rpm) and resuspended in pyrophosphate mixture containing 10% sucrose and applied to the discontinuous sucrose gradient. Centrifugation is carried out at 27,000 rpm for 90 min using a Beckman SW27 rotor. The distribution of material in the gradient after centrifugation is shown. Purified triads are located in fractions 4 and 5. Fraction 4 has lower mitochondrial contamination (Table I).

### Stage II Gradient Centrifugation

After centrifugation, the supernatants are carefully aspirated and the pellets are resuspended in 10% sucrose (w/w) containing pyrophosphate mix (20 ml volume for triad/mitochondria fraction, 10 ml for the light SR fraction.[10] To each gradient tube (SW27 or SW28), 5 ml of resuspended sample is added to a Stage II gradient tube (see Fig. 3) and the tubes are

[10] A light SR fraction, referable to longitudinal tubules of SR, can be obtained as a byproduct of the Standard variant. The upper light SR band of the first stage enrichment is applied to the Stage II gradient; bands 3 and 4 are enriched in light SR. Band 3 is better light SR in that contaminating plasmalemmal activity is lower and the calcium loading is higher (Table I).

centrifuged in an SW27 (or SW28) rotor at 27,000 rpm for 90 min (reduced acceleration and no brake).

After centrifugation, the tubes are placed in a Plexiglas rack in a cold room and the upper two layers are aspirated down to the top of the 15/25% interface (band 2). The individual bands (2 to 7) are collected with the aid of a Pasteur pipet and combined to yield fractions 2 to 7, respectively (see Fig. 3).[11]

Each fraction is diluted slowly as before to ~10% sucrose in about 45 min. The diluted samples are centrifuged at 30,000 rpm for 45 min in a Type 35 rotor. The supernatants are aspirated and discarded. The pellets are resuspended in several milliliters of storage medium (0.3 $M$ sucrose containing 150 m$M$ KCl, 5 m$M$ HEPES, pH 7.1) to about 10–20 mg protein/ml. Fractions 4 and 5 are enriched in triads (see Table I).[3] Aliquots are rapidly frozen in liquid nitrogen and stored at low temperature (−80°).

### Characterization of the Triads

*Morphology*

A method of sample preparation for electron microscopy was developed to estimate the purity and morphological integrity of our subcellular fractions.[13] Dextran is added as a nonosmophilic spacer to the prefixed sample and filtration is used in the preparation for thin sectioning. The procedure requires only a small amount of sample (~100 $\mu$g protein) and ensures more representative sampling by minimizing vertical stratification of the sample. In order to minimize osmotic shock, the fractions from the gradient are fixed in suspension directly in the sucrose and salt concentration in which they are obtained from the gradient. The sample (100 to 300 $\mu$g of protein) is fixed by slow addition (~15 sec) of 0.1 volume of 25% glutaraldehyde in 200 m$M$ cacodylate, pH 7.2, with vortexing, and is kept overnight in the cold room. Shorter fixation times were sometimes inadequate in that osmotic sensitivity was occasionally observed upon dilution

---

[11] Both the Standard and Pyrophosphate variants have similar separation patterns. For both, fractions 4 and 5 are highly enriched in triads as determined by particle counting of electron micrographs (see Table II). Fraction 4, a creamy white band at the 28–32% interface, displayed peak activity of adenylate cyclase, Na⁺,K⁺-ATPase, and several other plasmalemmal markers (see Table I), while fraction 5 (32/34%), which also consisted mainly of triads, contained fourfold higher (about 3 to 7%) mitochondrial contamination as determined by particle counting.

[12] G. Meissner, G. Conner, and S. Fleischer, *Biochim. Biophys. Acta* **298**, 246 (1973).

[13] P. Palade, A. Saito, R. D. Mitchell, and S. Fleischer, *J. Histochem. Cytochem.* **31**, 971 (1983).

after fixation.[3] The prefixed sample (0.05–0.2 ml) is then admixed together with 0.05–0.3 ml of 0.1% solution of high-molecular-weight dextran (5 to 40 million, Polysciences, Inc., Warrington, PA), in 50 m$M$ sodium phosphate buffer, pH 7.0, filtered onto a Millipore filter with shiny side up (0.1 $\mu$m nominal pore size, Millipore VCWP 01300) at negative pressure of 600–700 torr. The filtration requires 5 to 40 min. The sample is then processed for thin sectioning and embedded in polymer. In practice, the combined volume of sample and dextran solution was kept at 0.4 ml or less. Filtration serves to uniformly distribute structures of different sizes and densities. The dextran increases the space between individual structures and prevents sample compression during filtration.[13]

The primary criterion for the isolation of triads is morphology, since diagnostic enzymes for plasmalemma/transverse tubule or sarcoplasmic reticulum do not reveal whether junctional association remains intact. Quantitative particle counting has been used to measure the terminal cisternae and transverse tubule which are linked in junctional association for both Stage I and Stage II triads (Table II). Stage I pyrophosphate variant is a good quality triad preparation; the stage II gradient serves mainly to reduce the mitochondrial contamination. Fractions F4 and F5 are both good quality, purified triads, although F4 has a lower level of mitochondrial contamination.[3]

The *in situ* morphology of the triad structure[14] is preserved in the isolated triads (see Fig. 4).[15] Most structures consist of a central transverse tubule, roughly 1500 to 6000 Å in length, with two apposing terminal cisternae. A number of variations of this theme are encountered, including both trans and cis attachment of doublet terminal cisternae, dyads (with a single terminal cisterna attached), and multiple terminal cisternal attachment (not shown). Occasionally the triad also retains the longitudinal cisternal portion of the SR. The junctional association of transverse tubule with terminal cisternae is by way of the feet structures which are arranged in the junctional face membrane in checkerboard array.[16]

*Osmotic Sensitivity*

Isolated triads are adversely affected both morphologically[3] and enzymatically with regard to latent enzymatic activities (see below)[17] by the presence and then rapid dilution of osmotic perturbants such as sucrose

[14] C. Franzini-Armstrong, *Fed. Proc.* **39**, 2403 (1980).
[15] R. Mitchell, A. Saito, P. Palade, and S. Fleischer, *J. Cell Biol.* **96**, 1017 (1983).
[16] A. Saito, S. Seiler, A. Chu, and S. Fleischer, *J. Cell Biol.* **99**, 875 (1984), have developed a procedure to isolate from skeletal muscle a fraction enriched in terminal cisternae from sarcoplasmic reticulum, which retain morphologically intact junctional feet structures.
[17] R. D. Mitchell, P. Volpe, P. Palade, and S. Fleischer, *J. Biol. Chem.* **258**, 9867 (1983).

TABLE I

PURIFICATION OF TRIADS AND LIGHT SARCOPLASMIC RETICULUM[a]

| | | Protein | | Enzymatic activity | | | | | |
| | | | | Adenylate cyclase | | Ca²⁺-phosphate loading | | Succinate-cytochrome c reductase | |
| Fraction | Percent sucrose at interface | mg | Percentage of total | pmol/mg·min | Percentage of total | μmol/mg·min | Percentage of total | nmol/mg·min | Percentage of total |
|---|---|---|---|---|---|---|---|---|---|
| A. Summary of Stage II gradient purification of enriched triads prepared by the Standard variant | | | | | | | | | |
| Stage I | — | 16.00 | 100.0 | 28.3 | 100.0 | 1.61 | 100.0 | 118.9 | 100.0 |
| Stage II | | | | | | | | | |
| F1 | 10/15 | 0.43 | 2.7 | ND | — | ND | — | 34.3 | 0.0 |
| F2 | 15/25 | 0.51 | 3.2 | 7.0 | 1.0 | ND | — | 40.0 | 1.3 |
| F3 | 25/28 | 1.26 | 7.8 | 24.2 | 8.5 | 2.91 | 14.2 | 25.4 | 2.1 |
| F4 (triads) | 28/32 | 2.35 | 14.7 | 56.1 | 36.9 | 1.81 | 16.5 | 15.5 | 2.4 |
| F5 (triads) | 32/34 | 2.99 | 18.7 | 28.6 | 23.9 | 1.15 | 13.3 | 52.5 | 10.4 |
| F6 | 34/36 | 3.49 | 21.8 | 8.6 | 8.4 | 0.92 | 12.5 | 163.8 | 38.1 |
| F7 | 36/45 | 1.58 | 9.8 | 4.8 | 2.1 | 1.12 | 6.8 | 349.9 | 36.8 |
| Recovery | | | 78.7 | | 80.8 | | | | 91.1 |

B. Summary of Stage II gradient purification of enriched triads prepared by the Pyrophosphate variant

| Stage I | — | 68.40 | 100 | 15.7 | 100.0 | 1.0 | 100.0 | 137.5 | 100.0 |
|---|---|---|---|---|---|---|---|---|---|
| Stage II | | | | | | | | | |
| F1 | 10/15 | 2.09 | 3.0 | 33.0 | 6.5 | ND | — | 39.0 | 0.8 |
| F2 | 15/25 | 2.85 | 4.2 | 10.1 | 2.7 | ND | — | 33.0 | 1.0 |
| F3 | 25/28 | 6.52 | 9.5 | 16.8 | 10.2 | ND | 15.1 | 27.5 | 1.9 |
| F4 (triads) | 28/32 | 10.72 | 15.7 | 27.1 | 27.2 | 1.41 | 9.0 | 35.5 | 4.0 |
| F5 (triads) | 32/34 | 16.00 | 23.4 | 23.5 | 35.0 | 0.56 | 7.9 | 82.5 | 14.0 |
| F6 | 34/36 | 12.56 | 18.4 | ND | — | 0.63 | 9.2 | 302.5 | 40.4 |
| F7 | 36/45 | 8.85 | 12.9 | ND | | 0.70 | | 468.5 | 44.0 |
| Recovery | | | 87.1 | | >81.6 | | | | 106.1 |

C. Summary of Stage II gradient purification of light SR prepared from the Standard variant

| Stage I | — | 28.90 | 100.0 | 4.8 | 100.0 | 4.9 | 100.0 | 6.4 | 100.0 |
|---|---|---|---|---|---|---|---|---|---|
| Stage II | | | | | | | | | |
| L F3 | 25/28 | 9.54 | 33.0 | 2.2 | 15.1 | 5.2 | 34.9 | 8.6 | 44.0 |
| L F4 | 28/32 | 9.07 | 31.3 | 12.1 | 79.1 | 3.7 | 23.6 | 7.6 | 37.2 |

[a] Stage I triad material from Standard and Pyrophosphate variants was further purified using Stage II gradient purification (Fig. 3).[3] The amount of Stage I triad material recovered from 100 g of ground muscle is given as Stage I (protein). A total of six SW27 tubes (applying approximately 20 mg and 60 mg protein per tube for Standard and Pyrophosphate variants, respectively) is convenient for a single preparation involving one or two rabbits and one SW27 rotor for both the Standard and Pyrophosphate variants. Assays were carried out on samples that were quick-frozen and singly thawed, except for the phosphate-facilitated loading in C, which was performed on fresh material after the second stage of purification of light SR by the Standard variant. Purified triads, Stage II Standard (F4), assayed fresh without freezing and thawing, have Ca$^{2+}$-loading rates of 2.5–3 $\mu$mol/min · mg protein. The reduction in activity revealed in Table IA and B is the result of assaying material that was frozen in the absence of 0.1 $M$ KCl; the latter is required for stabilizing activity.[12] ND, not determined.

TABLE II

MORPHOLOGICAL QUANTITATION OF STAGE I AND STAGE II GRADIENT FRACTIONS OBTAINED BY STANDARD AND PYROPHOSPHATE VARIANTS OF TRIAD ISOLATION PROCEDURE[a]

| | Percentage of total vesicle population | | | | | | | | |
|---|---|---|---|---|---|---|---|---|---|
| | Junctional structures | | | | Nonjunctional structures | | Other structures | | |
| | Probable | | Possible | | | | Small empty vesicles (LSR) | Mitochondria | Large empty vesicles (PM) |
| | TC | T Tubule | TC | T Tubule | TC | T Tubule | | | |
| **A. Standard variant** | | | | | | | | | |
| Stage I | 18.9 ± 5.8 | 15.5 ± 4.1 | 13.9 ± 2.5 | 13.8 ± 3.5 | 4.7 ± 1.6 | 2.83 ± 0.77 | 16.7 ± 1.07 | 8.04 ± 2.66 | 1.06 ± 0.89 |
| Stage II | | | | | | | | | |
| Fraction 4 | 31.1 ± 3.3 | 21.4 ± 3.0 | 15.4 ± 8.0 | 7.7 ± 3.3 | 11.8 ± 3.8 | 2.28 ± 0.61 | 6.56 ± 2.61 | 1.67 ± 0.80 | 0.43 ± 0.28 |
| Fraction 5 | 28.3 ± 3.9 | 19.0 ± 2.8 | 14.4 ± 2.2 | 12.0 ± 1.3 | 11.6 ± 2.7 | 2.72 ± 1.36 | 2.94 ± 1.60 | 6.78 ± 2.05 | 0.25 ± 0.20 |
| **B. Pyrophosphate variant** | | | | | | | | | |
| Stage I | 31.8 ± 5.0 | 22.1 ± 1.6 | 8.8 ± 0.3 | 7.9 ± 0.4 | 9.4 ± 2.0 | 1.30 ± 1.05 | 7.81 ± 1.55 | 5.70 ± 1.92 | 0.47 ± 0.22 |
| Stage II | | | | | | | | | |
| Fraction 4 | 33.3 ± 3.1 | 21.8 ± 3.4 | 13.7 ± 3.3 | 10.7 ± 2.5 | 12.7 ± 3.6 | 2.57 ± 0.84 | 3.35 ± 0.69 | 0.83 ± 0.32 | 0.24 ± 0.18 |
| Fraction 5 | 34.9 ± 3.2 | 25.1 ± 2.8 | 10.1 ± 1.8 | 7.9 ± 1.2 | 12.1 ± 4.5 | 2.19 ± 1.03 | 3.43 ± 1.94 | 3.02 ± 0.65 | 0.18 ± 0.35 |

[a] Morphological quantitation of Stage I and Stage II triad fractions by particle counting.[3] Four consecutive preparations of Standard and Pyrophosphate triadic material were filtered according to Palade et al.[13] to obtain representative samples. Terminal cisternae and transverse tubule can readily be resolved morphologically. The term "junctional structures" refers to transverse tubule and terminal cisternae, oriented with respect to one another so as to suggest junctional association. Triad material is classified as junctional structures and is distinguished from nonjunctional structures [i.e., free transverse (T) tubule and terminal cisternae (TC)]. Apposed transverse tubule and TC that were clearly visible with all portions of the structure in the plane of section were tallied as "probable" junctional structures, even when bridging structures were not clearly evident. "Possible" junctional structures refers to less certain identification in which transverse tubule or terminal cisternae are identifiable and positioned near material that is out of the plane of section. Values, which appear in each category, are given as the percentage of all the vesicles counted for each fraction. Such values do not reflect the greater mass of some structures compared with others.

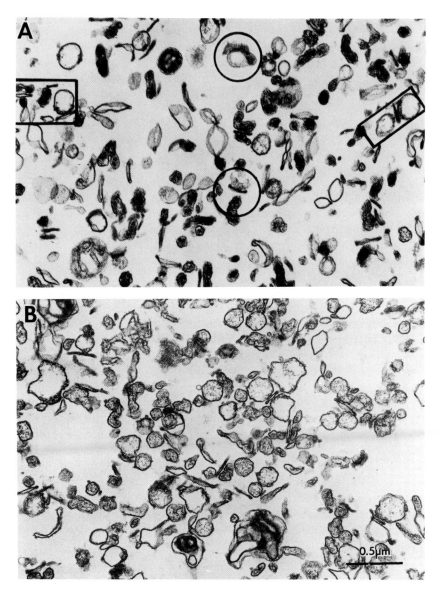

FIG. 4. Electron microscopy of a representative section of Stage II gradient fraction 4 triads. (A) Prepared according to the Standard variant. The samples were fixed with glutaraldehyde in suspension and then prepared for thin-section electron microscopy by addition of dextran and filtration[13] to ensure representative sampling. A representative field is shown (magnification: 29,000×). Material enclosed by the square is considered a "probable" junctional association (i.e., terminal cisternae and transverse tubule are juxtaposed appropriately and are in the field of focus to suggest junctional association). Material enclosed by a circle is considered a "possible" junctional association since the structure is not entirely within the same plane of focus. Quantitation of both "probable" and "possible" junctional structures is given in Table II. (B) Stage II, fraction 5, triads prepared according to the Pyrophosphate variant. (Magnification: 29,000×).

or salt. In the isolation procedure, precaution is taken to *slowly* dilute the fractions from both Stage I and Stage II gradients. Our use of relatively low levels of pyrophosphate (20 m$M$) to extract contractile material during isolation is less damaging to triad structure than is high salt. Prolonged pyrophosphate treatment has been found to reduce basal adenylate cyclase activity in the transverse tubule (see below).[3,17] SR enzymatic activity is not as adversely affected by salt treatment.

### Enzymatic Characteristics of Isolated Triad Structures

Isolated triads contain enzymatic activities representative of their component organelles, plasmalemma/transverse tubule, and sarcoplasmic reticulum. Some characteristics referable to each are provided in Table III. It is worth noting that calcium loading in the terminal cisternae of the triads can be enhanced with ruthenium red,[17] which closes the calcium release channels.[18]

Triad-associated transverse tubule contains enzymatic markers characteristic of the plasma membrane. However, transverse tubule is a specialized region of the plasma membrane and may contain a different level of some activities as compared with the surface membrane. To date, no plasmalemmal activity has been found which is a diagnostic of junctional association.

Although the specific activity appears greater for transverse tubule enzymatic markers in the Pyrophosphate preparation, the particle ratio of terminal cisternae to transverse tubule structures is similar for both (~1.5). This suggests that Pyrophosphate-variant fractions contain a higher ratio of transverse tubule to terminal cisternae membrane than those prepared by the Standard method. The specific activity ratio of a number of plasmalemmal activities for the pyrophosphate/standard variants is similar (1.5 to 1.9) for ouabain binding, cholesterol, Na$^+$,K$^+$-ATPase, and dihydroalprenolol binding (DHA, a $\beta$-agonist). Adenylate cyclase is the exception. It is 50% lower in the pyrophosphate variant[17] (Table III).

### Analysis of Membrane Sidedness and Integrity of the Transverse Tubule

Transverse tubules can exist in three configurations with regard to membrane sidedness: (1) sealed right-side-out, oriented with their extracellular face exposed; (2) sealed inside-out, oriented with their cytoplas-

[18] S. Fleischer, E. M. Ogunbunmi, M. C. Dixon, and E. A. M. Fleer, *Proc. Natl. Acad. Sci. U.S.A.* **82,** 7256 (1985).

TABLE III
BIOCHEMICAL CHARACTERIZATION OF TRIADS[a]

| Activity | Standard (F4) | Pyrophosphate (F4) | Ratio (Pyrophosphate/ Standard) |
|---|---|---|---|
| Plasmalemmal | | | |
| DHA binding $B_{max}$ (pmol/mg) | 0.258 | 0.394 | 1.53 |
| Basal adenylate cyclase (pmol/ min · mg protein) | 57.5 | 27.1 | 0.47 |
| % latent activity | <5% RSO | 12.0% RSO | — |
| Maximum ouabain binding (pmol/mg protein) | 5.07 | 9.62 | 1.90 |
| % latent (% leaky) | ~60% ISO (~38%) | ~87% ISO (~11%) | |
| $Na^+,K^+$-ATPase (max. ouabain sensitive) ($\mu$mol $P_i$/hr · mg) | 2.86 | 4.61 | 1.61 |
| % latent (% leaky) | ~70% ISO (~30%) | ~85% ISO (~15%) | |
| Cholesterol ($\mu$mol/mg protein) | 0.051 | 0.083 | 1.62 |
| Sarcoplasmic Reticulum | | | |
| $Ca^{2+}$-ATPase ($\mu$mol/min · mg protein) ($-$A23187/$+$1.5 $\mu M$ A23187) | 1.50/3.75 | 1.26/2.65 | 0.843/0.71 |
| Phosphate-facilitated $Ca^{2+}$ loading ($\mu$mol/min · mg protein) | 1.38 | 1.15 | 0.83 |
| +20 $\mu M$ ruthenium red | 2.10 | 2.31 | 1.10 |

[a] Triads (Fraction 4) were isolated by both the Standard and Pyrophosphate variants.[3] $Ca^{2+}$-loading rates were measured in the presence of 5 m$M$ potassium oxalate. Ouabain-sensitive $Na^+,K^+$-ATPase activity is the difference between the activity measured in the absence (total) and in the presence (basal) of 1 m$M$ ouabain, after preincubation with optimized detergent concentrations (0.14–0.18 mg of SDS/mg of protein or 0.4–0.6 mg of alamethicin/mg of protein). Maximal ouabain binding was determined after preincubation with optimized concentrations of alamethicin (0.4–0.6 mg/mg of protein). Similar values were obtained when SDS was used as detergent. Basal adenylate cyclase activity was measured in the presence of 5 m$M$ $MgCl_2$. The maximal number of DHA-binding sites ($B_{max}$) and $K_D$ were determined by Scatchard plot analysis. The values are reported as the mean $\pm$S.D., with the number of determinations on different triad preparations shown in parentheses. The data were taken from Mitchell et al.,[17] which also contains further details on methodology. Inside-out and right-side-out are abbreviated ISO and RSO, respectively.

mic face exposed; or (3) leaky, with both faces exposed. Sealed inside-out is the orientation referable to the triad in situ. Estimates of both membrane integrity and sidedness are generally made by determining the degree of increase in the binding of a ligand or enzymatic activity before and after the addition of detergents[19,20] or alamethicin,[19] which make latent binding sites accessible. Latency is measured by estimation of ouabain binding, adenylate cyclase, and Na$^+$,K$^+$-ATPase activities under specified conditions. With regard to the Na$^+$,K$^+$-ATPase, the ouabain-binding site is not accessible in inside-out tubules, whereas the ATP (or phosphate)-binding site is not accessible in right-side-out transverse (T) tubules.[21] With respect to adenylate cyclase, the ATP-binding site is not accessible in right-side-out transverse tubules.[19]

### Pretreatment of Triads with Alamethicin or Sodium Dodecyl Sulfate

In order to expose latent plasma membrane enzymatic activities in triads containing sealed transverse tubules, samples are preincubated with sodium dodecyl sulfate (SDS) or alamethicin. When pretreated with alamethicin, the triads (1–3 mg/ml) are preincubated for 20 min at room temperature[19] with varying concentrations of alamethicin added from a stock solution (30 mg/ml in absolute ethanol). The final concentration of alamethicin is expressed on a weight basis relative to protein. The amount of ethanol added ranged from 0.25 to 3.3% (v/v). When pretreated with SDS, the triads (1 mg/ml) are preincubated for 30 min at room temperature with varying concentrations of SDS in a medium containing 40 m$M$ imidazole-HEPES and 2 m$M$ Tris-EDTA, pH 7.5.[20]

### Ouabain Binding

Ouabain binds preferentially to the phosphoenzyme form of the catalytic subunit of the Na$^+$,K$^+$-ATPase, which can be phosphorylated by either ATP or inorganic phosphate.[22] Previous studies examining ouabain binding to transverse tubule were performed in the presence of ATP.[23] Since ATP and ouabain bind to opposite faces of the Na$^+$,K$^+$-ATPase,[20] membrane integrity can be assessed, but not sidedness.[23]

A new method for ouabain binding has been devised to determine membrane sidedness in which phosphate is used instead of ATP to form phosphoenzyme. In this procedure, triads are preincubated in the presence of 3 m$M$ MgCl$_2$, 3 m$M$ Tris-phosphate (see below for details), in a cold room for 18–20 hr. This incubation time was found to be sufficient to permit both Mg$^{2+}$ and phosphate to permeate across the membranes, enabling the Na$^+$,K$^+$-ATPase molecules to be phosphorylated, irrespec-

[19] L. R. Jones, S. W. Maddock, and H. R. Besch, Jr., J. Biol. Chem. 255, 9971 (1980).
[20] S. Seiler and S. Fleischer, J. Biol. Chem. 257, 13862 (1982).
[21] F. Shuurmans Stekhoren and S. L. Bonting, Physiol. Rev. 61, 1 (1981).
[22] A. K. Sen, T. Tobin, and R. L. Post, J. Biol. Chem. 244, 6596 (1969).
[23] Y. H. Lau, A. H. Caswell, M. Garcia, and L. Letellier, J. Gen. Physiol. 74, 335 (1979).

tive of vesicle orientation or membrane integrity. Under these conditions, ouabain binding is limited only to leaky and sealed right-side-out transverse tubules. The total number of ouabain-binding sites can be measured by exposing latent sites and by difference, the percentage of inside-out sealed transverse tubules is obtained.

### [³H]Ouabain Binding

The quantitation of the ouabain-binding sites is carried out using either of two procedures which are described below.

*1. [³H]Ouabain Binding Promoted by Phosphate and Magnesium.* Triads are equilibrated for 18–20 hr in the cold room on ice, in a medium containing 3 m$M$ Tris-PO$_4$, 3 m$M$ MgSO$_4$, 10 m$M$ Tris-Cl, pH 7.2, and 10% (w/w) sucrose. Triads are then pretreated, where applicable, with either alamethicin or SDS, as described above. The assay, in a final volume of 0.25 ml, is carried out at 37° in a medium containing 3 m$M$ Tris-PO$_4$, 3 m$M$ MgSO$_4$, 10 m$M$ Tris-Cl, pH 7.2, 1 m$M$ EGTA, 1 $\mu M$ [³H]ouabain, and 40–50 $\mu$g of protein, in the presence or absence of 0.1 m$M$ unlabeled ouabain. After a 30-min incubation, 0.22 ml is filtered through a 0.22-$\mu$m Millipore filter, which is immediately washed with 15 ml of ice-cold buffer (without ouabain). The filters are dried and then dissolved in 0.5 ml of ethylene glycol monoethyl ether (Cellosolve) for 30–45 min before adding 10 ml of aqueous counting scintillant (ACS, The Radiochemical Centre, Amersham, U.K.) and determining radioactivity content. Specific ouabain binding is defined as the difference between total and nonspecific binding, i.e., binding measured in the absence and in the presence of 0.1 m$M$ unlabeled ouabain, respectively.

*2. [³H]Ouabain Binding Promoted by ATP and Sodium Ion.* The assay, in a final volume of 0.25 ml, is performed at 37° for 30 min in a medium which contained 40 m$M$ Tris-Cl, pH 7.4, 120 m$M$ NaCl, 1 m$M$ EGTA, 10 m$M$ MgSO$_4$, 1 $\mu M$ [³H]ouabain, and 40–50 $\mu$g of protein. Leaky T-tubule membranes can be measured by ouabain binding in the presence of 10 m$M$ Na$_2$ATP, with nonspecific binding being measured with 0.1 m$M$ unlabeled ouabain, in the absence of Na$_2$ATP.[17] Total binding of ouabain can also be measured after preincubation with an optimized concentration of alamethecin. Filtration, washing of the filter, and counting are carried out as described in procedure 1.

Titration of either Pyrophosphate or Standard Triads with alamethicin revealed that most of the ouabain-binding sites are latent; thus, the transverse tubules are predominantly sealed and inside-out in orientation (Table III). The percentage of leaky vesicles is estimated by measuring the ouabain binding promoted by ATP and Na$^+$ (rather than by Mg$^{2+}$ and P$_i$) in the absence of alamethicin, divided by the total ouabain binding

(+ alamethicin). The percentage of sealed right-side-out transverse tubules is, thus, obtained by difference. Examples of such determinations are given for Standard and Pyrophosphate triads in Table III. By these criteria, the transverse tubule of the Pyrophosphate variant is 90% sealed and 87% inside-out, and is superior to the Standard variant in this respect.

*Adenylate Cyclase Activity*

Quantitation of right-side-out transverse tubules is made in a complementary manner by examining the increase in adenylate cyclase activity using alamethicin.[19] Exposure of either Standard or Pyrophosphate triads to increasing concentrations of alamethicin up to 0.2 mg/mg of protein resulted in only a small increase in adenylate cyclase activity, indicating that 12% or fewer of the transverse tubules were sealed and right-side-out (Table III). Higher concentrations of alamethicin inhibited adenylate cyclase activity. This procedure does not discriminate between sealed inside-out and leaky transverse tubules.[17]

Taken together, the evidence shows that isolated triad fractions contain transverse tubules which are largely sealed and inside-out. In this regard, the membrane integrity of the transverse tubule in the Pyrophosphate variant is superior to that of the Standard variant.

[7] Isolation of Transverse Tubule Membranes from Skeletal Muscle: Ion Transport Activity, Reformation of Triad Junctions, and Isolation of Junctional Spanning Protein of Triads

*By* Anthony H. Caswell, Neil R. Brandt, J-P. Brunschwig, and Richard M. Kawamoto

Transverse (T) tubules are invaginations of the plasma membrane which serve to carry the message of excitation into the interior of the fiber. They make physical contact with the sarcoplasmic reticulum through the triad junction and pass the message by an unknown mechanism to cause release of Ca$^{2+}$ from the sarcoplasmic reticulum.

Two basic preparations of T tubules have been described. The early stages of both preparations have involved similar protocols in each case for the production of skeletal muscle microsomes. On the one hand,

Scales and Sabbadini[1] and Rosemblatt *et al.*[2] have both observed that when skeletal muscle microsomes have been placed on a sucrose density gradient, the external membrane fraction has an isopycnic point lighter than that of longitudinal reticulum. On the other hand, Caswell *et al.*[3] and Lau *et al.*[4] found that the transverse tubular component migrated with the terminal cisternae and was indeed attached to the terminal cisternae in the form of a triad. It is pertinent to ask why in one preparation the triad has been broken—presumably during the homogenization procedure—while in the other, the triad has remained intact through both homogenization and subsequent fractionation. The differences in the isolation protocol may account for the marked difference in the integrity of the triadic junction between the two preparations.

Caswell *et al.*[3] inject the exposed back muscle of the killed rabbit with a Krebs–Ringer solution and retain the muscle on the animal for a period of 30 min prior to excision. The original purpose of this was to permit the injection of [$^3$H]ouabain as a marker for T tubules. A consequence of this treatment is that when the muscle is excised from the animal, it is flaccid and does not undergo strong contraction. Muscles immediately excised from the freshly killed animal usually go into contracture. This prior treatment of the muscle is not essential for the production of triads, although more consistent triadic preparations are obtained by the injection protocol. The second difference between the preparations is that Fernandez *et al.*[5] and Rosemblatt *et al.*[2] treated the microsomes with hypertonic KCl. We have found that hypertonic KCl does cause breakage of the triadic junction, albeit very limited, to liberate free T tubules. Therefore the time of applications of KCl rather than the treatment itself may be the important factor in maintenance of triad integrity.

The subsequent purification protocol has been markedly influenced by this difference in the initial disposition of the T tubules. In both preparations, several steps have been required in order to achieve a preparation of high purity. Scales and Sabbadini[1] and Rosemblatt *et al.*[2] have carried out iterative calcium loading followed by density gradient centrifugation to separate the calcium-loaded sarcoplasmic reticulum from the T tubules. On the other hand, Lau *et al.*[4] separated the T tubule from the terminal cisternae by passing the preparation through a French press and carrying out subsequent density gradient centrifugation. It should also be emphasized that the different protocols may give rise to preparations

[1] D. J. Scales and R. A. Sabbadini, *J. Cell Biol.* **83,** 33 (1979).
[2] M. Rosemblatt, C. Hidalgo, C. Vergara, and N. Ikemoto, *J. Biol. Chem.* **256,** 8140 (1981).
[3] A. H. Caswell, Y. H. Lau, and J.-P. Brunschwig, *Arch Biochem. Biophys.* **176,** 417 (1976).
[4] Y. H. Lau, A. H. Caswell, and J.-P. Brunschwig, *J. Biol. Chem.* **252,** 5565 (1977).
[5] J. L. Fernandez, M. Rosemblatt, and C. Hidalgo, *Biochim. Biophys. Acta* **599,** 522 (1980).

having some differences both morphologically and chemically. It is possible that the preparation of T tubules from triads is enriched in junctional T tubules. In some experiments, notably those in which the properties and qualities of the triadic junction are being investigated, it is preferable to prepare triadic vesicles first in order to evaluate features of the experimental system that are associated specifically with the junction, and distinguish them from those that are general external membane functions.

### Preparation of T Tubules from Triads

In the procedure of Caswell et al.[3] and Lau et al.,[4] a 4-lb New Zealand white rabbit is stunned with a blow on the neck and immediately bled through the carotid artery. (Alternatively, the animal may be killed with a captive bolt device.) Both back muscles are immediately exposed and injected with 10 ml each of a Krebs–Ringer solution (120 m$M$ NaCl, 4.8 m$M$ KCl, 1.5 m$M$ CaCl$_2$, 11 m$M$ glucose, and 1 m$M$ sodium phosphate, pH 7.4). The injection is performed by sliding the needle along the muscle during injection; the needle is placed at various positions along the muscle to ensure an even distribution of the fluid. At this stage it is possible to trap certain ligands in the T tubules which bind to specific receptors and which are not released after the lumen of the T tubules is sealed through the homogenization process. After 30 min at 22°, each muscle is excised from the animal, placed in 250 ml of ice-cold 250 m$M$ sucrose, 0.25 m$M$ Tris-EDTA, pH.7.0, in a Waring blender. No attempt is made to cut the muscle further before homogenization. Homogenization is carried out at full speed three times for 30 sec with 30 sec intervals. The homogenate is centrifuged at 10,000 $g$ for 20 min and the supernatant filtered through a cheesecloth to remove any debris or floating fat tissue. The supernatant should be a light-salmon color. If substantial blood has seeped into the muscle during killing, then the preparation is usually not functional. The supernatant is now centrifuged at 120,000 $g$ for 60 min. The pellet is rehomogenized with a glass Teflon homogenizer in a solution of 250 m$M$ sucrose, 2 m$M$ histidine, diluted to 60 ml per back muscle and recentrifuged under the same conditions.

The microsomal pellet is resuspended in 10 ml per back muscle of 250 m$M$ sucrose, 2 m$M$ histidine, and layered on continuous sucrose gradents in a swinging bucket rotor for isopycnic centrifugation. Eight density gradients are prepared employing a constant-volume mixing chamber containing 250 ml of 12.5% sucrose, and 65% sucrose is infused into this chamber in order to form a gradient with a volume of 35 ml per tube. We have preferred, in this and subsequent stages, to employ continuous sucrose gradients rather than step gradients because they are simple to prepare and because they have proved far superior in determination of the subcellular location of biological functions. Moreover, we have found small variations in the isopycnic density of organelles which could lead to erroneous localizations if batch techniques are employed. Visual inspection is employed in continuous gradients for determining the most appropriate place to cut the fractions for purification of the organelles. A microsomal suspension from each back muscle is loaded onto three centrifuge tubes and centrifuged for 5 hr at 120,000 $g$. This normally separates the microsomes into two distinct bands with isopycnic points at approximately 40 and 30% sucrose. In a good preparation of terminal cisternae/triads, the heavier band is usually condensed and separated from the lighter longitudinal reticulum band. We have found that the best indicator of the quality of the preparation is the degree to which the heavy band is condensed rather than its exact isopycnic position.

The heavy band is delineated from the light band visually and removed from the gradient. This band contains a mixture of free terminal cisternae and intact triadic junctions. The heavy band (terminal cisternae/triads) is diluted with water and centrifuged at 120,000 $g$ for 1 hr. It is then resuspended in 5 ml of 250 m$M$ sucrose; 2 m$M$ histidine.

The terminal cisternae/triad suspension is passed through a French press at 6000 psi. Care is taken to exclude air from the pressure cell and the extrusion is controlled to approximately 2 drops/sec by the valve at maintained pressure. Excess pressure may destroy membrane activities, while low pressures and rapid extrusion from the French press do not afford complete breakage. Centrifugation on one or two continuous gradients as described earlier is performed at 120,000 $g$ overnight. The T tubular band is distinguished as a sharp band above the sarcoplasmic reticulum. The terminal cisternae separates into two bands, more or less well distinguished, which we have designated as light and heavy terminal cisternae. The light terminal cisternae are almost devoid of internal electron-dense matter and calsequestrin. In contrast, the heavy terminal cisternae are filled with dense matter and highly enriched in calsequestrin. For many purposes, this preparation of T tubules is adequate either without further treatment or after concentration by centrifugation.

We have devised a simple protocol for further purification of T tubules where this is necessary.[6] This protocol exploits the fact that T tubules will reform a triad junction with terminal cisternae if incubated in the presence of hypertonic potassium cacodylate, pH 7.0. The original terminal cisternae/triad preparation is passed through a French press at 8000 psi and immediately made to 0.3 $M$ potassium cacodylate. The suspension is centrifuged on a continuous sucrose density gradient as described before

[6] N. R. Brandt, A. H. Caswell, and J.-P. Brunschwig, *J. Biol. Chem.* **255,** 6290 (1980).

or, more commonly, on a Sorvall TV 850 vertical gradient at 150,000 g for 1 hr. The rejoined triads are observed as the highly condensed and partially aggregated band toward the bottom of the centrifuge tube. A light diffuse band of light terminal cisternae, which we designate as "nonrejoining terminal cisternae," is seen immediately above it. The rejoined triads are removed and concentrated by centrifugation. They are then passed through a French press a second time but at 4000 psi and centrifuged on a continuous sucrose gradient as described earlier. The principle of this protocol is to remove nonrejoining light terminal cisternae from the preparation. Since the major contaminant of the transverse tubules is light terminal cisternae, it is possible to achieve a considerably purer preparation. In assays of T tubular integrity and morphological appearance, there is no indication that this procedure is at all deleterious to the T tubules.

### Preparation of T Tubules Isolated as Free Vesicles

In the method of Fernandez et al.,[5] microsomes are prepared by immediately excising back and hind leg muscles from the killed rabbit. The muscles are homogenized with 4 vol of 0.3 M sucrose, 20 mM Trismaleate, pH 7.0 in a Waring blender for 100 sec. The homogenate is centrifuged at 3000 g for 20 min and the supernatant centrifuged at 10,000 g for 20 min. The final supernatant is passed through cheesecloth and made to 0.5 M with KCl. This is then centrifuged at 150,000 g for 30 min to obtain a microsome pellet.

In the procedure of Rosemblatt et al.[2] for preparation of T tubules, the microsomes are centrifuged for 16 hr at 85,000 g in a discontinuous gradient of 5 ml 50%, 15 ml 40%, 10 ml 35%, and 5 ml 25% sucrose in Trismaleate (20 mM). The fraction from the top of the gradient is withdrawn and pelletted. This is resuspended at 0.1 mg/ml in Ca$^{2+}$-loading solution containing 50 mM potassium phosphate pH 7.4, 5 mM MgCl$_2$, 150 mM KCl, 0.3 mM CaCl$_2$, 2 mM ATP for 20 min at 22°. The vesicles are pelletted at 150,000 g and resuspended in 2 ml of loading solution. The suspension is placed on a gradient containing 4 ml 65%, 50%, and 35% sucrose in loading solution and centrifuge at 150,000 g for 25 min. The fraction at the top of the gradient is withdrawn as T tubules.

Scales and Sabbadini[1] described a similar preparation from chicken muscle. Microsomes were prepared from a muscle homogenate after treating the vesicles with 0.6 M KCl for 1 hr.[7] Contaminating SR was separated from T tubules by Ca$^{2+}$ loading in a medium containing 5 mM potassium oxalate.

[7] D. Scales, R. Sabbadini, and G. Inesi, *Biochim. Biophys. Acta* **465,** 535 (1977).

Estimation of Purity

Appropriate protocols to estimate the purity of the T tubule preparation are still incomplete and some of the approaches which have been employed are cumbersome. The estimation of the contribution of such organelles as mitochondria, Golgi apparatus, and nuclei can readily be made by conventional techniques but since none of these are likely to represent significant contaminations they are not especially helpful in evaluating the quality of the preparation. Of major concern is the contribution which the sarcoplasmic reticulum and the plasma membrane may make toward the vesicle content.

Estimation of sarcoplasmic reticulum contamination is best carried out by fluorescence assay of calcium-stimulated phosphatase using artificial high-energy phosphate sources such as 3-O-methylfluorescein phosphate.[6] This assay is simpler and more definitive than the assay of $Ca^{2+}$-stimulated ATPase since the latter activity has been discerned in some preparations of T tubules as well as plasma membrane.[4,6,8] The surface membrane $Ca^{2+}$-stimulated ATPase will not accept artificial phosphate donors.

No clean assay for plasma membrane contamination has yet been devised. Scales and Sabbadini[1] have reported a different particle distribution of freeze-fracture replicas between plasma membrane and T tubules but the employment of this distinction as a routine assay would be cumbersome, and quantitation is problematic. Barhanin et al.[9] and Jaimovich et al.[10] report the existence of toxins in scorpion venom which block the $Na^+$ channel of muscle plasma membrane but are ineffective on the $Na^+$ channel of T tubules. These authors have prepared labeled toxins and it is possible that these may be employed to estimate plasma membrane contamination.


Composition of T Tubules

It is generally agreed that the T tubule more closely resembles plasma membrane in lipid composition than that of internal organelles.[2,11,12] Cholesterol and sphingomyelin contents are both high. The lipid to protein

[8] P. V. Sulakhe, G. I. Drummond, and D. G. Ng, *J. Biol. Chem.* **248**, 4158 (1973).

[9] J. Barhanin, M. Ildefonse, O. Rougier, S. V. Sampaio, J. R. Giglio, and M. Lazdunski, *Pfluegers Arch.* **400**, 22 (1984).

[10] E. Jaimovich, R. Chicheportiche, A. Lombat, M. Lazdunski, M. Ildefonse, and O. Rougier, *Pfluegers Arch.* **397**, 1 (1983).

[11] Y. H. Lau, A. H. Caswell, J.-P. Brunschwig, R. J. Boerwald, and M. Garcia, *J. Biol. Chem.* **254**, 540 (1979).

[12] G. E. Sumnicht and R. A. Sabbadini, *Arch. Biochem. Biophys.* **215**, 628 (1982).

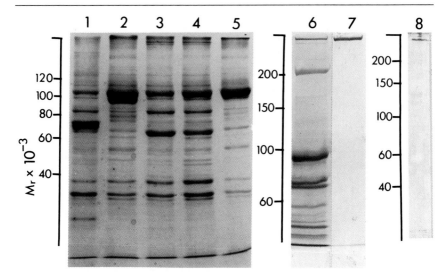

FIG. 1. SDS-PAGE of microsomal subfractions from rabbit sacrospinalis muscle comparing T tubule with sarcoplasmic reticulum and purified junctional spanning protein. Lanes: (1) T tubules purified by reforming triad junctions; (2) light microsomes (longitudinal reticulum); (3) heavy fraction from broken and reformed triads (terminal cisternae); (4) heavy microsomes (terminal cisternae/triads); (5) light fraction from reformed triads (nonrejoining light terminal cisternae). Fractions (20 $\mu$g) were run on a 7.5% Laemmli gel using Sigma Co. protein standard for gel calibration. Lane 6 is a preparation of terminal cisternae/triads used to purify the spanning protein; (7) the purified protein; (8) an ELISA development of a polyclonal antibody Western blot of terminal cisternae/triads showing the specific labeling of the junctional spanning protein. 5% Laemmli gels were run.

ratio in T tubules is approximately 1.6 $\mu$mol phospholipid/mg protein,[11] which is similar to that reported for plasma membrane.[13]

Figure 1[1-5] shows the SDS gel electrophoretic patterns of T tubules isolated from triads as well as of subfractions of sarcoplasmic reticulum. The purified T tubule is characterized by the presence of a major band of $M_r$ 72,000 which is absent from sarcoplasmic reticulum. Other T tubule-specific proteins include one of $M_r$ 200,000 and one of $M_r$ 26,000. Some protein of $M_r$ 100,000 is present in all preparations of T tubules. Hidalgo *et al.*[14] have argued that this protein is a divalent cation ATPase which is absent from sarcoplasmic reticulum. This molecular weight is very close to that of the Ca²⁺-ATPase of sarcoplasmic reticulum but the enzyme is activated either by Mg²⁺ or Ca²⁺. Beeler *et al.*[15] and Sulakhe and St.

[13] S. Seiler and S. Fleischer, *J. Biol. Chem.* **257**, 13862 (1982).
[14] C. Hildalgo, M. E. Gonzalez, and R. Lagos, *J. Biol. Chem.* **258**, 13937 (1983).
[15] T. J. Beeler, K. S. Gable, and J. M. Keffer, *Biochim. Biophys. Acta* **734**, 221 (1983).

Louis[16] have suggested that this may be an ectoenzyme. In our preparation this protein is less dominant than in that of Hidalgo *et al.*[14] Proteins of $M_r$ 34,000, 38,000, and 80,000 appear to be present in both T tubules and terminal cisternae. We have recently identified the $M_r$ 34,000 and 38,000 proteins as glyceraldehyde-3-phosphate dehydrogenase and aldolase, respectively, which are attached to the membrane as extrinsic proteins.[17] The $M_r$ 80,000 protein may be a junctional membrane protein which provides recognition sites for attachment of the spanning protein.[18] A high-molecular-weight doublet is observed predominantly in the terminal cisternae, with a low content in the T tubules. We have presented evidence that these polypeptides are subunits of the junctional spanning protein of the triad.[19]

Table I[1,2,4,6,8–15,20–32] gives estimates of the content of various enzymes, channels, pumps, and other physical properties of T tubules isolated by the two protocols as well as of plasma membrane preparations. Some caution is needed in evaluating the plasma membrane data since, in many cases, the authors have not attempted to distinguish the bulk surface membrane from T tubules and therefore some preparations may contain T tubules. The two preparations of T tubules show general agreement in the content of most proteins and ion pumps; however, some discrepancies are apparent. Most notably, the divalent cation ATPase activity in the preparation of T tubules from free vesicles is considerably higher than the value obtained in T tubules isolated from triads. The estimates of $Ca^{2+}$-stimu-

[16] P. V. Sulakhe and P. J. St. Louis, *Prog. Biophys. Mol. Biol.* **35,** 135 (1980).
[17] A. M. Corbett, A. H. Caswell, R. M. Kawamoto, F. Lugo-Gutierrez, and J.-P. Brunschwig, *Biophys. J.* **47,** 450a (1985).
[18] A. H. Caswell and J.-P. Brunschwig, *J. Cell Biol.* **99,** 929 (1984).
[19] J. J. S. Cadwell and A. H. Caswell, *J. Cell Biol.* **93,** 543 (1982).
[20] Y. H. Lau, A. H. Caswell, M. Garcia, and L. F. Lettelier, *J. Gen. Physiol.* **74,** 335 (1979).
[21] A. H. Caswell, S. P. Baker, H. Boyd, L. T. Potter, and M. Garcia, *J. Biol. Chem.* **253,** 3049 (1978).
[22] N. R. Brandt, R. M. Kawamoto, and A. H. Caswell, *J. Recept. Res.* **5,** 155 (1985).
[23] M. Fosset, E. Jaimovich, E. Delpont, and M. Lazdunski, *Eur. J. Pharmacol.* **86,** 141 (1983).
[24] B. I. Curtis and W. A. Catterall, *Biochemistry* **23,** 2113 (1984).
[25] J.-P. Galizzi, M. Fosset, and M. Lazdunski, *J. Biol. Chem.* **258,** 6086 (1984).
[26] H. Glossman, D. R. Ferry, and C. B. Boschek, *Naunyn-Schmiedeberg's Arch. Pharmacol.* **323,** 1 (1983).
[27] D. B. McNamara, P. V. Sulakha, and N. S. Dhalla, *Biochem. J.* **125,** 525 (1971).
[28] R. A. Sabbadini and V. R. Okamoto, *Arch. Biochem. Biophys.* **223,** 107 (1983).
[29] S. J. Sulakhe and P. V. Sulakhe, *Gen. Pharmacol.* **10,** 103 (1979).
[30] J. R. Gilbert and G. Meissner, *J. Membr. Biol.* **69,** 77 (1982).
[31] P. V. Sulakhe, G. I. Drummond, and D. C. Ng, *J. Biol. Chem.* **248,** 4150 (1973).
[32] N. R. Brandt, R. M. Kawamoto, and A. H. Caswell, *Biochem. Biophys. Res. Commun.* **127,** 205 (1985).

TABLE I

BIOCHEMICAL AND CHEMICAL COMPOSITION OF SKELETAL MUSCLE EXTERNAL
MEMBRANE PREPARATIONS

| Marker | T tubule/triad | T tubule/free | Sarcolemma |
|---|---|---|---|
| Ouabain receptor (pmol/mg) | 37 (20) | | |
| *Tityus serrulatus* toxin (Na$^+$ channel) (pmol/mg) | | 0.01 (9) | 0.45 (9) |
| en-Tetrodotoxin$_{II}$ (Na$^+$ channel) (pmol/mg) | | 0.7 (10) | |
| $\beta$-Adrenergic receptor (pmol/mg) | 0.61 (21) | | |
| Nitrendipine (Ca$^{2+}$ channel) (pmol/mg) | 25 (22) | 6–50 (23,24,26) | 2 (23) |
| Verapamil (Ca$^{2+}$ channel) (pmol/mg) | | 50 (25) | |
| Na$^+$,K$^+$-ATPase ($\mu$mol/min · mg) | 0.10 (4) | 0–0.2 (2,25) 0.43$^a$ (1) | 0.5–1.0 (13,27) 1.2$^a$ (28) |
| Ca$^{2+}$-Stimulated ATPase ($\mu$mol/min · mg) | 0.8 (4,6) | 0–0.6 (2,14) 0.15$^a$ (28) | 0.2–1.0 (8,29) |
| Divalent cation ATPase ($\mu$mol/min · mg) | 0.4 (4,6) | 1.4–5.5 (2,14) 0.4–4$^a$ (1,28) | 4–25 (15,30) 0.4$^a$ (28) |
| Adenylate cyclase (nmol/min · mg) | 2.3 (21) | | |
| Na$^+$ uptake (nmol/min · mg) | >45 (20) | | 270 (13) |
| Ca$^{2+}$ uptake (nmol/min · mg) | 30 (6) | 15 (14) | 400 (31) |
| Na$^+$Ca$^{2+}$ exchange (nmol/min · mg) | ND$^b$ (6) | ND (14) | 2.4 (30) |
| Cholesterol ($\mu$mol/mg) | 0.64 (11) | 0.90 (2) 0.94$^a$ (12) | 0.92 (13) 0.6$^a$ (12) |
| Phospholipid ($\mu$mol/mg) | 1.6 (11) | 1.6 (2) 1.1$^a$ (12) | 2.4 (13) 1.6$^a$ (12) |
| Orientation (%) | | | |
|   P face exposed | 90 | | 66 |
|   E face exposed | ND (32) | | 19 (13) |
|   Leaky (P + E exposed) | 10 | | 15 |

$^a$ From chicken pectoral muscle.
$^b$ ND, not detectable within experimental error.

lated ATPase are quite variable with quoted values from 0 to 1 $\mu$mol/ min · mg.

## Orientation of T Tubules

T Tubule orientation may be estimated employing [$^3$H]ouabain and [$^3$H]digoxin as impermeant and permeant ligands, respectively.[32] Both ligands bind to the extracellular (E) side of the Na pump only when it is activated by ATP (or inorganic phosphate) on the cytoplasmic (P) face. T tubule vesicles (200 $\mu$g in 1 ml) are incubated with 120 m$M$ NaCl, 10 m$M$ MgCl$_2$, 1 m$M$ Tris-EGTA, 40 m$M$ Tris-Cl, 10 m$M$ Tris-ATP, pH 7.4, and 50 n$M$ labeled ligand. Nonspecific binding is estimated by addition of 5 $\mu M$ digoxin (unlabeled). After incubation at 37° for 60 min, the suspension is filtered through Millipore GSWP filters and washed four times with binding medium to which 100 m$M$ KCl and 1 $\mu M$ ouabain (unlabeled) were added and labeled ligands were omitted. Vesicles were also rendered leaky to ATP and ouabain by addition of 0.04 mg sodium dodecyl sulfate to the binding medium. We were unable to detect any vesicles with E face only exposed. The quantity of leaky vesicles (P + E face exposed) in several assays was approximately 10%. The protocol described above is similar in principle to that of Seiler and Fleischer.[13] The absence of E-face-exposed T tubules is consonant with the preparation from triads in which only one orientation is possible.

## Morphology of T Tubules

Figure 2 shows the appearance in low magnification of T tubules and the terminal cisternae/triads from which they are prepared. The thin section of terminal cisternae/triads (Fig. 2A) shows the presence of many triadic vesicles with distinct elongate T tubules apposed by spherical vesicles which contain electron-dense matter. Several other vesicles may be of triadic origin in which the sectioning angle has not delineated the three apposed vesicles. Several triads show in higher magnification the individual junctional feet.[33] Isolated T tubules (Fig. 2B) are mainly elongate. The morphology seen in the thin section and freeze fracture (Fig. 2C) is best described as discoid or discus shaped. The pinched ends are seen in thin section to contain electron-dense material in the T tubule lumen. Intercalated particles are present in low abundance on both convex and concave faces of freeze-fracture replicas with some arrayed to

[33] J.-P. Brunschwig, N. R. Brandt, A. H. Caswell, and D. S. Lukeman, *J. Cell Biol.* **93**, 533 (1982).

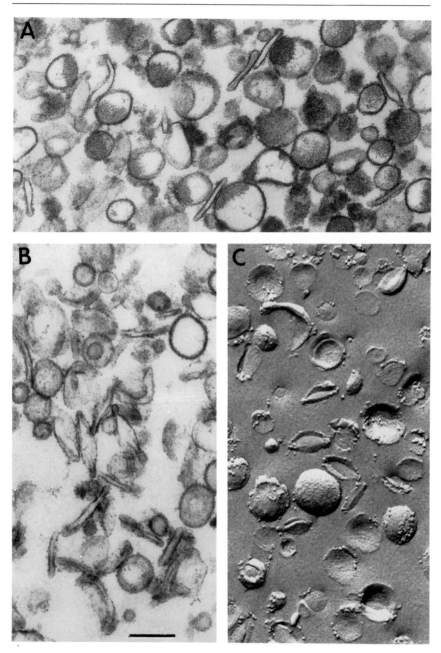

FIG. 2. Electron micrographs of terminal cisternae/triads and T tubules purified by reforming triad junctions. (A) Thin section of terminal cisternae/triad preparation showing

form lines or fused aggregates. A similar morphology has been described by Scales and Sabbadini,[1] except that these authors do not describe the retention of the discoid shape of the vesicles.

### Ion Transport in Isolated T Tubules

T Tubule preparations elicit active $Na^+$ and $Ca^{2+}$ pumping. $Na^+$ accumulation may be demonstrated by incubating the preparation in a medium containing 5 m$M$ NaCl, 50 m$M$ KCl, 5 m$M$ $MgCl_2$, 0.1 m$M$ EGTA, 30 m$M$ imidazole, and 0.2 $\mu$Ci/ml $^{22}Na^+$.[20] After 40 min of preincubation at 37°, 5 m$M$ Tris-ATP is added and the Na content of the vesicles estimated by filtration through a GSWP Millipore filter. This is washed rapidly twice or three times with 1 ml of the preincubation medium without $^{22}Na^+$. The $Na^+$ content increases from 3 nmol/mg protein to 120 nmol/mg in a period of a few seconds and a subsequent slow increase occurs up to 200 nmol/mg during 90 min of incubation. Simultaneous extrusion of $K^+$ can be demonstrated using $^{86}Rb$ as a tracer, assuming that $Rb^+$ and $K^+$ are handled by the pump in the same way. The extrusion of $Rb^+$ was estimated with much less precision. The loss of that cation from the vesicle in the presence of ATP is much less easily assayed than the enhancement of $Na^+$. The estimated turnover number of $Na^+$ by the Na pump in isolated T tubules is approximately 20/sec. This value is very close to that obtained in intact muscle. The $Na^+$ accumulation proved sensitive to digitoxin and monensin but not ouabain. The lack of effect of ouabain is associated with the inability of that ligand to reach its luminal site in the isolated T tubular preparation.

$Ca^{2+}$ accumulation by isolated T tubules was initially demonstrated by Brandt et al.[6] and subsequently by Hidalgo et al.[14] In our demonstration of this pumping activity, we were sensitive to the problem of distinguishing a low activity in the T tubules from a $Ca^{2+}$ uptake due to the presence of contaminating SR vesicles. The resolution of this problem has been three-fold: (1) We obtained a preparation of T tubules with very low SR contamination (approximately 5%); (2) the T tubular uptake did not occur when phosphate sources such as 30-methylfluorescein phosphate was employed in place of ATP; (3) $Ca^{2+}$ accumulation was not stimulated by oxalate.[6] $Ca^{2+}$ uptake may be demonstrated by employing Millipore filtra-

---

many elongate T tubules apposing spherical terminal cisternae which contains electron-dense matter. (B) Thin section of purified T tubules showing many elongate vesicles which are discus shaped. (C) Freeze-fracture profile showing discus-shaped membrane morphology. Both P and E faces show a low density of intercalated particles often associated in lines or arrays. Bar line is 200 nm.

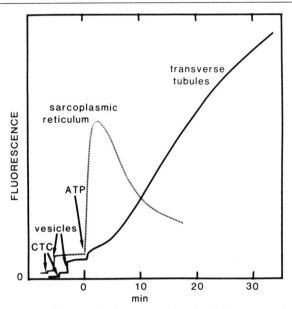

FIG. 3. Fluorescent chelate probe observation of active $Ca^{2+}$ accumulation by T tubules and sarcoplasmic reticulum. At arrows, 10 $\mu M$ chlortetracycline (CTC), vesicles, or 3 m$M$ Tris-ATP is added. The fluorimeter contains a vibrating stirrer. Excitation wavelength is 390 nm and emission 530 nm. Slit width is 10 nm. Volume of mixture is 3 ml and temperature 22°.

tion essentially as described for $Na^+$ accumulation. Alternatively, as shown in Fig. 3, the accumulation of $Ca^{2+}$ in isolated T tubules and isolated sarcoplasmic reticulum can be followed by employing the fluorescent chelate probe, chlortetracycline. Vesicles (0.15 mg vesicle protein/ml) are preincubated in 200 m$M$ sucrose, 10 m$M$ MOPS, 3 m$M$ MnCl₂, 20 $M$ CaCl₂, pH 6.8. MnCl₂ was used in the place of MgCl₂ as the divalent cation for ATP chelation. MgCl₂ chelates with chlortetracycline to form a fluorescent adduct while the adduct with MnCl₂ does not fluoresce. The choice of MgCl₂ or MnCl₂ is at the discretion of the investigator depending on the intent of the experiment. Figure 3 shows that SR gives rise to an initial rapid accumulation of $Ca^{2+}$ followed by a subsequent release. This is characteristic of many SR preparations. The chlortetracycline probably does not respond to $Ca^{2+}$ accumulation by the SR as fast as it is occurring. However, the indicator can respond very rapidly to conditions which elicit $Ca^{2+}$ release.[34] The T tubule $Ca^{2+}$ uptake differs from that of SR in that the accumulation of $Ca^{2+}$ is very slow and there is an initial lag phase

---

[34] K. Nagasaki and M. Kasai, *J. Biochem.* (*Tokyo*) **94**, 1101 (1973).

before maximal accumulation rate is achieved. The extent of accumulation is very high, being approximately equivalent to that obtained in SR. This slow accumulation but high $Ca^{2+}$ gradient is indicative of a well-sealed vesicular preparation. Quantitation by $^{45}Ca^{2+}$ and Millipore filtration has indicated that T tubules accumulate almost 100 nmol/mg protein under optimal conditions. Chlortetracycline gives a highly sensitive measure of intravesicular $Ca^{2+}$ which can readily be quantitated[35] and interpreted.

### Reformation of the Triad Junction

In certain media T tubules and terminal cisternae spontaneously reassociate to form attached vesicles which resemble morphologically native triad junctions. This ability to reassociate is labile after the triad has been broken and therefore we usually initiate triad junction reformation immediately after the junction has been broken by the French press. The most usual assay of junction integrity employs the observation of the isopycnic density of a T tubule specific marker ([$^3$H]ouabain entrapment or [$^3$H]nitrendipine binding). The isopycnic density of T tubules is 22–26% sucrose while that of triads is 38–42% sucrose. Two different agents have proved efficacious in initiating junction reformation. Hypertonic solutions of potassium cacodylate are extremely potent in promoting junction formation.[36] A more physiological initiator which we have recently described is the glycolytic enzyme, glyceraldehyde-3-phosphate dehydrogenase (GAPD).[37]

Terminal cisternae/triads are prepared from a muscle which has been injected with a Krebs–Ringer solution containing 12 $\mu$Ci [$^3$H]ouabain per back muscle. The suspension in 10 ml of 250 m$M$ sucrose, 5 m$M$ histidine, pH 7.0, is passed through a French press at 6000 psi and divided. One portion is made to 0.3 $M$ potassium cacodylate while the control portion is untreated (or treated with 0.3 $M$ potassium gluconate). After 10 min incubation the suspensions are layered on continuous sucrose gradients as described previously and centrifuged at 100,000 $g$ overnight. The appearance of the vesicles in treated and control tubes is quite different. The sharp T tubule band has disappeared in the sample gradient and a sharp band is discerned at ~40% sucrose. A diffuse band of nonrejoining light terminal cisternae overlays this. Potassium cacodylate, 0.1 to 0.6 $M$, is

[35] M. S. Millman, A. H. Caswell, and D. H. Haynes, *Membr. Biochem.* **3**, 291 (1980).
[36] A. H. Caswell, Y. H. Lau, M. Garcia, and J.-P. Brunschwig, *J. Biol. Chem.* **254**, 202 (1979).
[37] A. H. Caswell and A. M. Corbett, *J. Biol. Chem.* **260**, 6892 (1985).

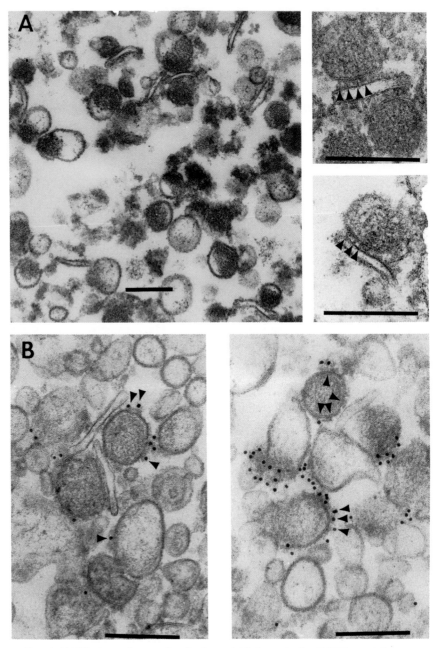

FIG. 4. (A) Electron micrographs of reformed triads employing GAPD as rejoining pro-motor. Elongate T tubules are readily distinguished attached to terminal cisternae. Little

effective at rejoining although the high concentrations of salt may give a coagulated band. Nevertheless, the T tubule is still biochemically active. It is possible to separate T tubules and terminal cisternae by centrifugation before effecting the rejoining process but rejoining is labile after French press treatment. Therefore the centrifugation must be carried out rapidly, e.g., by vertical gradient centrifugation. Samples (1.8 ml) from the continuous density gradient are withdrawn and [³H]ouabain counted in a scintillation counter in order to assay rejoining. The transfer of labeled vesicles from 22–26% sucrose to 38–42% sucrose indicates rejoining. Partly rejoined preparations have two bands at the isopycnic points of free T tubules and of triads. Rejoining between T tubules and longitudinal reticulum is very low.

The rejoining medium when GAPD is employed as promotor contains 250 m$M$ sucrose, 5 m$M$ histidine, 30 m$M$ potassium gluconate, pH 7.0. Typically, 2.8 mg GAPD is added to a suspension which contains 15 mg broken terminal cisternae/triads. The enzyme is always dialyzed overnight in a large volume of 5 m$M$ histidine, 1 m$M$ dithiothreitol before the reaction to remove salts. Rejoining occurs in the absence of potassium gluconate, but this salt prevents the tendency of this enzyme to accrete nonspecifically. Higher concentrations of salt cause progressive loss of rejoining activity.

Figure 4A shows the appearance of the reformed triads after they have been extracted from the 38–42% fraction of the sucrose density gradient. Several T tubule vesicles are distinguished, most of which are clearly attached to terminal cisternae. The morphology is hardly distinct from that of the native triads in Fig. 2A. Some triads display a characteristic pattern of individual feet or pillars. Feet appear in thin section either as solid rods or rods with a translucent core. Rejoining betwen T tubules and longitudinal reticulum or mitochondria has not been observed.

## Isolation of Spanning Protein of Triads

This protein is a constituent of the junctional feet identified previously by Caldwell and Caswell.[19] It plays the role of joining T tubules to terminal cisternae and may play an important physiological role in activation of muscle contraction. The protein may be isolated in a simple two-stage

---

loss of morphology compared with Fig. 2A is observed. Higher magnification view indicates discrete junctional processes (arrow heads), some of which have a translucent core. Bar line is 200 nm. (B) Gold-tagged indirect antibody labeling of intact terminal cisternae/triads showing protruding feet structures (arrowheads). The gold is found associated with these structures.

protocol using a monoclonal antibody column.[38] The protein is susceptible to hydrolysis by Ca$^{2+}$-stimulated protease and precautions must be taken to prevent breakdown at all stages of the preparation. We include 5 m$M$ Tris-EGTA, 100 $\mu M$ phenylmethyl sulfonyl fluoride, 1 $\mu M$ pepstatin, and 1 $\mu M$ leupeptin in our homogenizing medium and all subsequent stages in preparation of terminal cisternae/triads. The vesicle preparation (5 ml in 250 m$M$ sucrose, 2 m$M$ histidine) is dissolved in 2 mg Zwittergent 3-14/ mg protein, 1 $M$ NaCl. The antibody column is prepared by reacting 40 mg purified monoclonal antibody with 2.5 g CNBr-activated Sepharose. The dissolved vesicles are incubated with the column (1.5 cm diameter, 8 cm length) by continuous cyclic pumping for 1.5 hr at 22°. The column is washed with 4 $M$ NaCl, 0.1% Zwittergent 3-14, 1 m$M$ Tris-EGTA, 2 m$M$ histidine, pH 7.0, until no detectable absorbance at 280 nm is observed. Specific elution is achieved by replacing 4 $M$ NaCl with 2 $M$ NaSCN. The eluted band 10 ml is immediately passed onto a Fractogel TSK HW65(F) column (2.5 cm diameter, 80 cm in length) and eluted with a low-salt medium, e.g., 30 m$M$ NaCl, 4 m$M$ Tris-Cl, pH 7.0. Through this simple procedure, in excess of 500 $\mu$g of purified protein may be obtained from two rabbit back muscles as shown in Fig. 1 (lanes 6 and 7) in which the gel electrophoretic pattern of the pure protein is shown in comparison with the terminal cisternae/triads used as the starting point for the preparation.

This preparation of purified protein may be employed in its turn to produce polyclonal antibodies to study the morphology and distribution of the spanning protein in intact tissue. Fig. 1 (lane 8) shows the specificity of reaction of affinity-purified antibodies for the 300K-Da protein in a preparation of terminal cisternae/triads. Immuno-gold labeling is used to show the distribution of the protein in a preparation of terminal cisternae (Fig. 4B). The vesicles are incubated for 1 hr at 22° with rabbit antibody. This is washed by centrifugation and followed by incubation with gold-tagged goat antibody. The preparation is centrifuged in an Airfuge rotor on a cushion of 65% sucrose for 1 hr to centrifuge the dense gold into the sucrose while the vesicles are collected at the interface. The appearance of tannic acid and mordanted thin sections is shown in Fig. 4B. The gold particles are clustered around the periphery of vesicles which contain discernible foot structures (arrow heads). Most gold particles are seen to label structures protruding some distance from the membrane surface.

[38] R. M. Kawamoto, J.-P. Brunschwig, K. C. Kim, and A. H. Caswell, *J. Cell Biol.* **103**, 1405 (1986).

## [8] Rapid Preparation of Canine Cardiac Sarcolemmal Vesicles by Sucrose Flotation

*By* LARRY R. JONES

Sarcolemmal vesicles isolated from canine heart have been utilized increasingly to study the biochemical and biophysical properties of the cardiac plasma membrane. Several key activities have recently been identified in canine cardiac sarcolemmal vesicles that appear to be critically involved in regulating contractility in intact cardiac muscle. These include muscarinic–cholinergic and $\beta$-adrenergic receptors, adenylate cyclase activity, and cAMP-dependent and $Ca^{2+}$-dependent protein kinase activities[1-5]; $Na^+,K^+$-ATPase activity,[1-4] $Na^+/Ca^{2+}$ exchange activity,[6] $Na^+/H^+$ exchange activity,[7] and active $Ca^{2+}$ transport[8]; and membrane potential-regulated $Ca^{2+}$ flux.[9] Thus the utility of sarcolemmal vesicles for investigating biochemical mechanisms regulating ionic flux and contractility in the heart has been firmly established and will probably continue for some time.

Some time ago we described an isolation method for preparation of sarcolemmal vesicles from canine heart, which gave substantially purified plasma membrane vesicles in reasonable yield.[1] However, this original procedure was somewhat lengthy, in that it employed a $Ca^{2+}$ oxalate loading step in one of the final stages of purification in an attempt to reduce potential contamination of the isolated sarcolemmal vesicles with sarcoplasmic reticulum membranes. Subsequently, we modified the procedure by showing that the $Ca^{2+}$ oxalate loading step was unnecessary. It could be omitted and the yield and purity of sarcolemmal vesicles was not affected.[2-4] Sarcoplasmic reticulum contamination in such preparations was estimated at less than 5%.[2-4] Recently, we have further modified the isolation procedure by making it faster and easier, and have also suc-

[1] L. R. Jones, H. R. Besch, Jr., J. W. Fleming, M. M. McConnaughey, and A. M. Watanabe, *J. Biol. Chem.* **254**, 530 (1979).

[2] L. R. Jones, S. W. Maddock, and H. R. Besch, Jr., *J. Biol. Chem.* **255**, 9971 (1980).

[3] A. S. Manalan and L. R. Jones, *J. Biol. Chem.* **257**, 10052 (1982).

[4] L. R. Jones and H. R. Besch, Jr., *Methods Pharmacol.* **5**, 1 (1984).

[5] C. F. Presti, B. T. Scott, and L. R. Jones, *J. Biol. Chem.* **260**, 13879 (1985).

[6] J. P. Reeves and J. L. Sutko, *Science* **208**, 1461 (1980).

[7] S. M. Seiler, E. J. Cragoe, Jr., and L. R. Jones, *J. Biol. Chem.* **260**, 4869 (1985).

[8] P. Caroni and E. Carafoli, *Nature* (*London*) **283**, 765 (1980).

[9] W. P. Schilling and G. E. Lindenmayer, *J. Membr. Biol.* **79**, 163 (1984).

ceeded in doubling the yield of sarcolemmal vesicles at the same time. Approximately 25 mg of sarcolemmal vesicle protein can now be prepared from two dog hearts in ≤6 hr by this newer method, which makes use of a sucrose flotation gradient. This newer isolation method is described in detail below.

### Homogenization of Tissue and Preparation of Procedure II Canine Cardiac Microsomes

#### Solutions

Medium I: 10 m$M$ histidine (free base), 0.75 $M$ NaCl
Medium II: 10 m$M$ NaHCO$_3$, 5 m$M$ histidine (free base)
The pH of histidine (free base) dissolved in H$_2$O is about 7.5. Therefore, Medium I and II can be used directly without adjusting the pH.

*Disruption of Tissue and Isolation of Cardiac Membranes.* All operations are performed at 4°. The left ventricles and interventricular septa obtained from two dog hearts are first well-trimmed using scissors and hemostats. Endocardium, epicardium, fat, and major coronary arteries are stripped away, and the tissue is then minced into small pieces using a meat grinder. Minced tissue, 180–210 g, is weighed and portioned into six equal aliquots, and each aliquot is placed into a Beckman JA-14 polycarbonate centrifuge tube containing 120 ml of Medium I. The samples are next homogenized once for 5 sec with a Polytron PT-20 (Brinkman Instruments) set at half-maximal speed. The homogenates are centrifuged at 10,000 rpm (14,000 $g_{max}$) for 20 min in a Beckman JA-14 rotor. The supernatants are discarded, and each pellet is resuspended in 120 ml of Medium I with vigorous shaking. The samples are homogenized and centrifuged a second time as described above. The supernatants are again discarded, and each pellet is resuspended in 120 ml of Medium II with shaking. The samples are homogenized and centrifuged a third time as described above, and the supernatants are again discarded. These three cycles of brief homogenization followed by low-speed centrifugation are performed to selectively fragment the sarcoplasmic reticulum of cardiac tissue and allow it to be removed into the supernatant fractions.[1-4] This maneuver is critical for ultimate isolation of sarcolemmal vesicles free of significant sarcoplasmic reticulum contamination, and is the essence of preparation of cardiac microsomes according to Procedure II.[1-4]

Each pellet obtained from the third centrifugation described above is resuspended in 100 ml of Medium II and homogenized three times for 30 sec with the Polytron PT-20 set at half-maximal speed. This longer period of homogenization causes formation from the extracted pellets of sealed sarcolemmal vesicles that are free of sarcoplasmic reticulum. The samples are next sedimented at 10,000 rpm for 20 min in the Beckman JA-14 rotor, and the supernatants containing the partially purified sarcolemmal vesicles are then poured off into 16 Beckman JA-20 polycarbonate centrifuge tubes. The samples are centrifuged for 30 min at 19,000 rpm (44,000 $g_{max}$) in two Beckman JA-20 rotors, and the supernatants are removed by aspiration and discarded. The pellets resulting from this higher-speed centrifugation are the Procedure II microsomes, and are used directly for sucrose flotation as described below. (Further purification of these Procedure II microsomes by additional differential pelleting steps is required when our older method is used to isolate sarcolemmal vesicles. The older method employs conventional sucrose density gradient centrifugation at a later stage of the procedure.[1-4] However, these additional differential pelleting steps can be bypassed completely when the newer sucrose flotation method is utilized.)

### Isolation of Canine Cardiac Sarcolemmal Vesicles by Sucrose Flotation

*Solutions*

Medium III: 2.0 $M$ sucrose dissolved in 300 m$M$ NaCl, 50 m$M$ tetrasodium pyrophosphate, and 100 m$M$ Tris (pH 7.1)

Medium IV: 0.6 $M$ sucrose dissolved in 300 m$M$ NaCl, 50 m$M$ tetrasodium pyrophosphate, and 100 m$M$ Tris (pH 7.1)

Medium V: 0.25 $M$ sucrose, 10 m$M$ histidine (free base)

*Sucrose Flotation.* The pellets from the 19,000 rpm centrifugation step described above (Procedure II microsomes) are resuspended to a final volume of 30 ml in ice-cold deionized or distilled $H_2O$. This is done by repeated aspiration of individual pellets in a small volume of $H_2O$ through a 4-in. 14-gauge stainless steel cannula (Becton Dickinson and Co.) which is attached to a 10-ml plastic syringe. The bottom of the cannula is pressed against the bottoms of the centrifuge tubes during aspiration of individual pellets to obtain a mild shear force. Several pellets can be resuspended in the same small volume of $H_2O$ (2 to 3 ml) by moving the syringe consecutively from tube to tube. The final volume of the membrane pellet suspension is adjusted to 30 ml by placing the suspension in a 100-ml graduated cylinder, and by adding additional ice-cold $H_2O$ if necessary to bring the volume to 30 ml. Thirty milliliters of Medium III is next added to the 30 ml of membranes suspended in $H_2O$, and the cylinder is then sealed with Parafilm and inverted several times to allow uniform mixing of the membrane suspension. The total volume of suspended membranes at this point is 60 ml, and the sucrose concentration is 1.0 $M$. Ten milliliters of resuspended membranes is then added to each of six empty Ti70 polycarbonate centrifuge tubes. Seven milliliters of Medium IV (containing 0.6 $M$ sucrose) is next layered on top of the 10 ml of membrane suspension in each centrifuge tube, and approximately 7 ml of Medium V (containing 0.25 $M$ sucrose) is then layered on top of the 0.6 $M$ sucrose in each tube, which is sufficient to fill the centrifuge tubes to the tops. Aluminum caps are used to seal the tubes, and the samples are centrifuged for 54 min at 60,000 rpm (370,000 $g_{max}$) in a Beckman Ti70 rotor.

During centrifugation, the sarcolemmal vesicles in the crude membrane suspensions float toward the tops of the sucrose gradients in the six centrifuge tubes. After the run, the sarcolemmal vesicles are observed as distinct snow-white protein bands at the interfaces of the 0.25 $M$/0.6 $M$ sucrose layers. Considerable protein is also observed at the 0.6 $M$/1.0 $M$ sucrose interfaces, as well as in the 1.0 $M$ sucrose layers at the bottom of the centrifuge tubes. However, sarcolemmal vesicles are not enriched in these latter fractions. The sarcolemmal vesicles from the 0.25 $M$/0.6 $M$ sucrose interfaces are recovered by aspiration with the 10-ml syringe and the 14-gauge cannula. Sarcolemmal vesicles aspirated from each centrifuge tube are added to identical empty centrifuge tubes, and the membrane suspensions are then diluted with three to four volumes of ice-cold H$_2$O. The six samples are sedimented at 40,000 rpm for 30 min in the Ti70 rotor to pellet the purified sarcolemmal vesicles. The pellets are resuspended in a total volume of 3.5 to 4.5 ml of Medium V with use of the 14-gauge cannula, and the final sarcolemmal membrane suspension is divided into small aliquots and stored frozen at −20° for later use.

Protein content is estimated by the method of Lowry et al.[10] The protein concentration of the final sarcolemmal vesicle suspension will be approximately 5 to 7 mg/ml when vesicles are prepared and resuspended as described above. The protein yield of sarcolemmal vesicles is 13.0 ± 0.6 mg per 100 g wet weight of minced ventricular tissue (mean ±S.D. from three separate preparations). This yield is double that which was obtained by our former method.[2] The sucrose flotation method for preparation of canine cardiac sarcolemmal vesicles is quite reproducible, and has been performed independently by several different investigators.

### Purity and Percent Recovery of Canine Cardiac Sarcolemmal Vesicles

Specific activities of the sarcolemmal markers Na$^+$,K$^+$-ATPase and muscarinic–cholinergic receptors (assayed by [$^3$H]quinuclidinyl benzilate

---

[10] O. H. Lowry, N. J. Rosebrough, A. L. Farr, and R. J. Randall, *J. Biol. Chem.* **193**, 265 (1951).

TABLE I

PURIFICATION OF CANINE CARDIAC SARCOLEMMAL VESICLES BY SUCROSE FLOTATION

| | Marker activity | | | | | |
| | Na$^+$,K$^+$-ATPase[a] | | | [$^3$H]QNB binding[b] | | |
| Fraction | Specific activity ($\mu$mol P$_i$/mg/hr) | Purification (-fold) | Total activity (%) | Specific activity (fmol/mg) | Purification (-fold) | Total activity (%) |
|---|---|---|---|---|---|---|
| Homogenate[c] | 3.05 | 1.0 | 100 | 90 | 1.0 | 100 |
| Sarcolemmal vesicles | 147 | 48.2 | 3.5 | 5820 | 64.7 | 4.7 |

[a] Na$^+$,K$^+$-ATPase is reported as ouabain-sensitive activity, assayed as described.[4] Activity was measured in 50 m$M$ histidine, 4 m$M$ MgCl$_2$, 3 m$M$ ATP, 100 m$M$ NaCl, 10 m$M$ KCl, 4.5 m$M$ NaN$_3$, 0.9 m$M$ EGTA, 2.7 m$M$ phosphoenolpyruvate, and 75 $\mu$g/ml pyruvate kinase—in the presence and absence of 1 m$M$ ouabain. Latent activity was expressed by pretreating membrane fractions with optimal concentrations of sodium dodecyl sulfate,[4] 0.4 mg/ml for sarcolemmal vesicles and 0.3 mg/ml for the crude homogenate.

[b] [$^3$H]Quinuclidinyl benzilate binding, assayed as described,[3] with saturating concentrations of radioligand.

[c] A crude homogenate was prepared by adding 1 g of minced ventricular tissue to 9 ml of Medium V and then homogenizing the tissue three times for 30 sec using the Polytron.

binding) are enriched 50- to 65-fold in sarcolemmal vesicles prepared by sucrose flotation, relative to specific activities of the same markers measured in crude myocardial homogenates (Table I). Even though preparation of canine cardiac sarcolemmal vesicles by sucrose flotation doubles the yield compared to our earlier methods,[1-4] recovery of total activity of these two sarcolemmal markers is still less than 5% (Table I).

Sarcolemmal vesicles prepared by sucrose flotation are tightly sealed, as evidenced by the large stimulation produced in their ouabain-sensitive Na$^+$,K$^+$-ATPase activity, when vesicle permeability is increased by either freeze–thaw shock[11] or treatment with sodium dodecyl sulfate or alamethicin (Table II). Ca$^{2+}$,K$^+$-ATPase activity, a sarcoplasmic reticulum marker,[1-4] is very low in the sarcolemmal vesicle preparation, even when it is measured in the presence of the unmasking agent alamethicin (Table II). Several additional sarcolemmal markers have also been measured and found to be highly enriched in canine cardiac sarcolemmal vesicles prepared by sucrose flotation, including Na$^+$/Ca$^{2+}$ exchange ac-

[11] E. Van Alstyne, R. M. Burch, R. G. Knickelbein, R. T. Hungerford, E. J. Gower, J. G. Webb, S. L. Poe, and G. E. Lindenmayer, *Biochim. Biophys. Acta* **602,** 131 (1980).

TABLE II

UNMASKING OF LATENT Na⁺,K⁺-ATPase ACTIVITY IN CANINE
CARDIAC SARCOLEMMAL VESICLES

| Activity measured ($\mu$mol P$_i$/mg/hr) | Control | Unmasking condition | | |
|---|---|---|---|---|
| | | Freeze–thaw[a] | SDS[b] | Alamethicin[b] |
| Na⁺,K⁺-ATPase[c] | | | | |
| (−) Ouabain | 54.4 | 141.2 | 149.1 | 205.2 |
| (+) Ouabain | 10.6 | 7.0 | 6.2 | 14.5 |
| Difference | 43.8 | 134.2 | 142.9 | 190.7 |
| Ca²⁺,K⁺-ATPase[d] | | | | |
| (+) Ca | 42.8 | ND[e] | ND | 62.5 |
| (+) EGTA | 38.4 | — | — | 60.7 |
| Difference | 4.4 | — | — | 1.8 |

[a] Vesicles were subjected to freeze–thaw shock according to Van Alstyne et al.[11]

[b] Vesicles were pretreated with optimal concentrations of sodium dodecyl sulfate (SDS) and alamethicin as described.[4]

[c] Na⁺,K⁺-ATPase was measured as described in footnote a of Table I.

[d] Ca²⁺,K⁺-ATPase was measured in the presence of 50 m$M$ histidine, 3 m$M$ MgCl$_2$, 3 m$M$ ATP, 50 $\mu M$ CaCl$_2$, and 100 m$M$ KCl. EGTA (1 m$M$) was also included in the medium where indicated. Addition of an ATP-regenerating system or NaN$_3$ in the buffer system did not affect Ca²⁺,K⁺-ATPase activity (data not shown).

[e] ND, not determined.

tivity, Na⁺/H⁺ exchange activity, active Ca²⁺ transport, cAMP-dependent protein kinase activity, and $\beta$-adrenergic receptors (L. R. Jones, unpublished observations). The purity of sarcolemmal vesicles prepared by sucrose flotation is found to be identical to that of sarcolemmal vesicles prepared by our former methods.[1-4] Sarcolemmal vesicles prepared by sucrose flotation are predominately right-side-out in orientation (L. R. Jones, unpublished observations), and are therefore similar to our earlier preparations[1-4] in this respect as well.

Conclusions

Sarcolemmal vesicles prepared by sucrose flotation are identical to vesicles prepared by our previously reported methods,[1-4] with respect to vesicle orientation, membrane protein composition, and vesicle purity. However, preparation of sarcolemmal vesicles by sucrose flotation offers several advantages. The purification scheme is faster and easier, and gives a membrane yield (approximately 25 mg protein from two dog hearts) which is twice that formerly achieved.[1-4] The flotation method is highly reproducible, and can be followed easily by investigators who have not previously isolated sarcolemmal membranes.

In spite of the advantages inherent in isolating sarcolemmal vesicles by sucrose flotation, there are some caveats which should be mentioned when such preparations are to be used for biochemical and biophysical studies of the cardiac plasma membrane. These caveats apply to use of the older method preparations[1-4] as well. First, since the yield of purified sarcolemmal vesicles is probably less than 10%, it remains at least possible that only a specific region of the cardiac sarcolemma has actually been isolated. It is not known, for example, whether the sarcolemmal vesicles originate mainly from surface sarcolemma or transverse tubules, or whether they are derived from both parts of the plasma membrane. Second, although sarcolemmal vesicles prepared by sucrose flotation are substantially purified, they are almost certainly not purified to homogeneity. E. Van Alstyne and G. Lindenmayer have recently been successful in isolating canine cardiac sarcolemmal vesicles by a different method which utilizes Percoll density gradient centrifugation (personal communication), and vesicles isolated by this technique exhibit a higher $Na^+,K^+$-ATPase specific activity than those described here. Finally, the method described in this chapter was developed for use with canine hearts. Some modification of the preparation will probably be required if other sources of cardiac tissue are utilized.

### Acknowledgment

The author is an Established Investigator of the American Heart Association. This work was supported by grants HL-28556 and HL-06308 from the National Institutes of Health and by the Herman C. Krannert Fund.

# [9] Isolation of Canine Cardiac Sarcoplasmic Reticulum

*By* BRIAN K. CHAMBERLAIN and SIDNEY FLEISCHER

The sarcoplasmic reticulum (SR) is a specialized intracellular membrane system in muscle that regulates the myoplasmic calcium ($Ca^{2+}$) level, thereby controlling contraction and relaxation. In cardiac muscle, the SR network is less extensive than it is in skeletal muscle, and cell surface membranes (sarcolemma and transverse tubule) constitute much of the total membrane content.[1] Consequently, isolating highly purified SR from heart is more difficult than from skeletal muscle. Various adaptations of the procedure of Harigaya and Schwartz[2] have been used by many laboratories, but the resulting preparation is significantly contaminated with surface membrane vesicles.[3] Other procedures[4,5] are based upon the ability of SR to precipitate Ca$^{2+}$ oxalate inside the vesicles, thus increasing the density of the SR, but such a procedure is perhaps better suited to fractionation of subpopulations of cardiac SR[6] than it is to isolation of SR per se.

The isolation procedure described here yields canine cardiac SR of high purity, which exhibits efficient, high specific activity Ca$^{2+}$ transport and which is stable to multiple freezings and thawings and to long-term cryogenic storage.[7] These attributes have made possible the following advances: the determination of the oligomeric structure of the Ca$^{2+}$-pump protein in cardiac SR membranes using radiation inactivation analysis[8] and the reconstitution of the Ca$^{2+}$-pumping function of cardiac SR.[9] In addition, the preparation is suitable for studying Ca$^{2+}$ release from cardiac SR[10,11] and for further fractionation into cardiac SR subpopulations by loading with Ca$^{2+}$ oxalate or phosphate.[12,13]

## Preparation of SR Vesicles

### Reagents

Homogenization solution: 0.29 $M$ sucrose, 0.5 m$M$ dithiothreitol, 3 m$M$ NaN$_3$, 10 m$M$ imidazole-HCl, pH 6.9

Solution A: 0.29 $M$ sucrose, 0.65 $M$ KCl, 0.5 m$M$ dithiothreitol, 3 m$M$ NaN$_3$, 10 m$M$ imidazole-HCl, pH 6.7

[1] H. Lüllmann and T. Peters, *Clin. Exp. Pharmacol. Physiol.* **4,** 49 (1977).
[2] S. Harigaya and A. Schwartz, *Circ. Res.* **25,** 781 (1969).
[3] H. R. Besch, Jr., L. R. Jones, and A. M. Watanabe, *Circ. Res.* **39,** 586 (1976).
[4] D. O. Levitsky, M. K. Aliev, A. V. Kuzmin, T. S. Levchenko, V. N. Smirnov, and E. I. Chazov, *Biochim. Biophys. Acta* **443,** 468 (1976).
[5] L. R. Jones, H. R. Besch, Jr., J. W. Fleming, M. M. McConnaughey, and A. M. Watanabe, *J. Biol. Chem.* **254,** 530 (1979).
[6] L. R. Jones and S. E. Cala, *J. Biol. Chem.* **256,** 11809 (1981).
[7] B. K. Chamberlain, D. O. Levitsky, and S. Fleischer, *J. Biol. Chem.* **258,** 6602 (1983).
[8] B. K. Chamberlain, C. J. Berenski, C. Y. Jung, and S. Fleischer, *J. Biol. Chem.* **258,** 11997 (1983).
[9] M. Inui, B. K. Chamberlain, A. Saito, and S. Fleischer, *J. Biol. Chem.* **261,** 1794 (1986).
[10] B. K. Chamberlain, P. Volpe, and S. Fleisher, *J. Biol. Chem.* **259,** 7540 (1984).
[11] B. K. Chamberlain, P. Volpe, and S. Fleischer, *J. Biol. Chem.* **259,** 7547 (1984).
[12] B. K. Chamberlain, unpublished observations, 1983.
[13] S. Wang, M. Inui, E. Ogunbunmi, A. Saito, and S. Fleischer, *Biophys. J.* **51,** 354a (1987).

Gradients: see text

42.5% (w/v) sucrose, containing 0.65 $M$ KCl

Solution B: 0.65 $M$ KCl, 0.5 m$M$ dithiothreitol, 3 m$M$ NaN$_3$, 10 m$M$ imidazole-HCl, pH 6.7

Solution C: 0.29 $M$ sucrose, 0.2 $M$ KCl, 10 m$M$ imidazole-HCl, pH 6.7

For the sake of convenience, the pH values of solutions used in the isolation procedure are presented as the values at the preparation temperature of 25°. These solutions are then cooled to 0–4° for use in the isolation procedure.

## Procedure

Mongrel dogs of either sex (18–22 kg) are killed by intravenous injection of either 0.3 ml/kg of T-61 Euthanasia Solution (American Hoechst Corp., Somerville, NJ) or 100 mg/kg of pentobarbital. Hearts are removed immediately and immersed in 0.9% NaCl for 5–10 min. All subsequent operations (see Fig. 1) are performed at 0–4°. All centrifugation times are the time control settings and therefore include acceleration times, but not deceleration times.

Ventricles are trimmed of fat and connective tissue and sliced into small pieces, 5–10 mm along each dimension. Portions of 29–32 g are added to 5 vol (based on tissue wet weight) of homogenization solution in a 250-ml flask. The tissue is homogenized using a VirTis "45" macrohomogenizer with two perpendicular blades set together 5 mm from the bottom of the flask, for 15 sec at 21,500 rpm, followed by 35 sec at 35,800 rpm. Three homogenizations are required for a typical heart (85–100 g of trimmed ventricle).

These conditions are optimal for the particular homogenizer apparatus used to prepare the SR described in this report. However, the importance of homogenization optimization in each individual laboratory cannot be overstressed. Homogenization is perhaps the most critical stage of the preparation, and even slight changes can have profound effects on the eventual SR activity, purity, and yield. For example, just changing the blades on the homogenizer necessitates a reoptimization of the homogenization conditions (i.e., rpm and/or time). To optimize conditions, one needs only to vary the rpm and/or time in each homogenization of the tissue from one heart and (keeping the material from each homogenization separate) proceed through the preparation of the salt-washed microsomes (see below) for analysis of $Ca^{2+}$-loading and $Ca^{2+}$-dependent ATPase total and specific activities. Subsequent steps provide more-or-less constant enrichment to the SR activity present at this point. Since three different

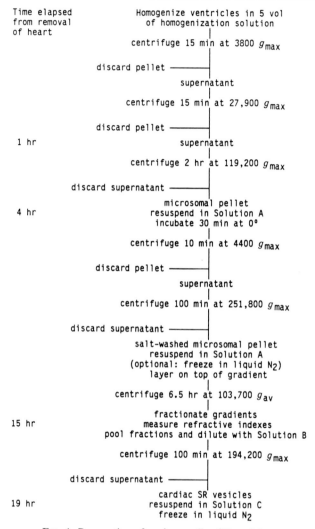

FIG. 1. Preparation of canine cardiac SR vesicles.

conditions of homogenization can be evaluated in the three homogenizations of the tissue from one heart, a satisfactory optimization should be achieved with four or fewer microsome preparations. (If, however, one wishes to adapt this procedure to isolation of cardiac SR from another species, it is advisable to reoptimize the subsequent centrifugation steps as well as the homogenization.)

The homogenates, one per polycarbonate bottle, are centrifuged 15 min at 5000 rpm (3,800 $g_{max}$) in a JA-14 rotor (Beckman Instruments, La Jolla, CA). Each supernatant is filtered through four layers of cheesecloth, adjusted to 140 ml with fresh homogenization solution, and centrifuged 15 min at 13,500 rpm (27,900 $g_{max}$) in a JA-14 rotor. After filtration through cheesecloth, the supernatants from the second centrifugation are sedimented by centrifugation for 2 hr at 32,000 rpm (119,200 $g_{max}$) in a Beckman Type 35 rotor. The supernatants from this centrifugation are removed by aspiration, and the polycarbonate bottles are set in ice for a few minutes to allow the soft pink pellet to slide off the hard clear pellet at the bottom of each tube. The soft pellets are pooled and resuspended for salt washing in solution A by hand homogenization (15 strokes) in a Potter–Elvehjem tissue grinder with a Teflon pestle. The volume is adjusted to 22 ml per VirTis homogenization.

After 30 min on ice, the suspensions are centrifuged 10 min at 6000 rpm (4400 $g_{max}$) in a Beckman JA-20 rotor to remove large aggregates. The supernatants are then centrifuged 100 min at 50,000 rpm (251,800 $g_{max}$) in a Beckman Type 60Ti rotor. (This is the maximum speed allowed for partially filled bottles.) To recover SR of high specific activity, it is essential to limit the volume in each Type 60Ti polycarbonate bottle to the minimum allowed volume of about 16 ml. After removal of the supernatants by aspiration, a minimal volume of solution A is added to each centrifuge tube, and the tubes are agitated with a vortex mixer to free the soft opaque pellets from the hard clear pellets (glycogen). The soft pellets are pooled and resuspended in solution A by hand homogenization (15 strokes) in a Potter–Elvehjem tissue grinder, and the volume is adjusted to 5.5 ml per VirTis homogenization. It is often convenient to quick-freeze these salt-washed microsomes in liquid nitrogen at this point (about 7 hr after removal of the heart) and proceed with the gradient purification of SR (required time about 12 hr) on another day. The microsomes can be stored in liquid nitrogen for at least a month.

After thawing, the microsomes are rehomogenized (8 strokes) in a Potter–Elvehjem tissue grinder. Aliquots of 5.5 ml are layered over 31-ml linear gradients of 10–26% (w/v) sucrose with 5–7% (w/v) dextran T10 (Pharmacia Fine Chemicals, Piscataway, NJ), atop a 2-ml cushion of 30% (w/v) sucrose and 7.5% (w/v) dextran T10. All gradient solutions include 0.65 $M$ KCl, 0.5 m$M$ dithiothreitol, 3 m$M$ NaN$_3$, 10 m$M$ imidazole-HCl, pH 6.7. Dextran T10, a 10,000-Da polysaccharide, is included to obtain the desired solution densities with minimal osmotic effects from the use of sucrose. Isoosmotic gradients cannot be prepared since the allowed level of dextran T10 is limited by the presence of 0.65 $M$ KCl, which is essential to reduce nonspecific aggregation. When cooled to 4° for extended periods of time, gradient solutions containing levels of dextran T10 higher than 8% (w/v) in 30% (w/v) sucrose and 0.65 $M$ KCl can either precipitate or gel.

The gradients, one tube per VirTis homogenization, are centrifuged 6.5 hr at 28,000 rpm (103,700 $g_{av}$) in a Beckman SW28 rotor. Fractions of about 1 ml are collected from the tops of the gradients by pumping a solution of 42.5% (w/v) sucrose containing 0.65 $M$ KCl into the bottom of the tubes. The refractive indexes of the fractions are measured with an Abbe 3L refractometer (Bausch & Lomb, Rochester, NY). Fractions with refractive indexes of 1.3671–1.3722 are pooled, diluted over 10 min with 0.65 vol of solution B, and sedimented by centrifugation for 100 min at 46,000 rpm (194,200 $g_{max}$) in a Beckman Type 70.1 Ti rotor. To obtain SR of high specific activity, it is essential to limit the volume in each polycarbonate tube to 4.6–5.4 ml of solution. The pellets are pooled and resuspended in Solution C by hand homogenization (15 strokes) in a Potter–Elvehjem tissue grinder. The volume is adjusted to 1.5 ml per VirTis homogenization (typically 4.5 ml per heart). Aliquots of 0.5 ml each are placed in plastic cryotubes, then quick-frozen and stored in liquid nitrogen. The average yield is about 31 mg of SR from one heart. The first time a tube is thawed for use, its contents are further split into two tubes (~1.7 mg SR each) before refreezing. In this way, the number of freeze/thaw cycles for any one tube is limited; although the preparation is stable to multiple freezings and thawings, it is generally advisable to keep such manipulations to a minimum.

Cardiac SR isolated by this procedure includes both ryanodine-sensitive and ryanodine-insensitive subfractions and can serve as starting material for preparing SR subfractions by loading with Ca²⁺ oxalate or phosphate,[12,13] using modifications of published procedures.[6] However, the proportion of ryanodine-sensitive vesicles might be lower than that in other preparations.[7] If the SR is to be further subfractionated (when there will also be further removal of contaminants), collecting a wider region of the sucrose/dextran gradient (refractive indexes 1.3655–1.3749) is recommended to increase subfraction recovery. Alternatively, it should be possible to prepare subfractions with improved Ca²⁺-transport properties by applying the Ca²⁺-loading procedure to the salt-washed microsomes, since the Ca²⁺-loading rate (1.68 ± 0.09 $\mu$mol Ca²⁺/mg protein · min, $n = 5$) and the Ca²⁺-loading efficiency (Ca²⁺/ATP = 0.97 ± 0.10, $n = 5$) of these microsomes[12] are considerably higher than the respective values for the microsome preparation previously used.[6]

### Characteristics of SR Vesicles

Although differences in assay conditions make comparisons difficult, the values for the Ca²⁺-transport rate and the Ca²⁺-loading efficiency of this preparation (Table I) are considerably higher than the values typically reported for canine cardiac SR.[6,14-17] These are the two most important criteria for evaluating maintenance of SR activity and membrane integrity during isolation. The maximum $Ca^{2+}$-loading capacity is additionally influenced by vesicle size, and the $Ca^{2+}$-dependent ATPase activity can actually be increased in less-intact preparations, since it is influenced by the "leakiness" of the membrane. Others[15] have reported isolation of cardiac SR with a $Ca^{2+}$-loading rate similar to that reported here, but the assay conditions (200 $\mu M$ $CaCl_2$, no EGTA, 10–20 m$M$ potassium oxalate, 35 $\mu$g SR/ml) are quite different from those used in this work (66.2 $\mu M$ $CaCl_2$, 30 $\mu M$ EGTA, 5 m$M$ potassium oxalate, 7 $\mu$g SR/ml). Also, the higher concentrations of $CaCl_2$ and potassium oxalate, especially in the presence of increased concentrations of SR, can produce artifacts in the $Ca^{2+}$-loading assay which result in erroneously high measurements of $Ca^{2+}$ transport.[18]

Our preparation of cardiac SR is quite stable. Incubating the SR as isolated for 24 hr at room temperature or for 48 hr at 0° has little or no effect on either $Ca^{2+}$ uptake or $Ca^{2+}$-ATPase activities. Moreover, when the SR is maintained in liquid nitrogen, it can be stored for more than a year and can be thawed and refrozen up to five times with little or no change in $Ca^{2+}$-transport properties.

Contamination by other organelles is low. As judged by ouabain-sensitive $Na^+,K^+$-ATPase activity,[7] by [$^3$H]ouabain binding,[7] and by $Ca^{2+}$-insensitive ATPase activity (Table I; 7.3% of total ATPase), contamination by surface membranes (sarcolemma and transverse tubules) is 7.3–10.7%. Succinate–cytochrome-$c$ reductase activity and cardiolipin content indicate a maximum of 1.7–2.2% contamination by mitochondrial membranes.[7] On the basis of protein content, therefore, the preparation is 87–91% pure SR. Both ryanodine-sensitive and ryanodine-insensitive subpopulations are present, and $Ca^{2+}$ transport by this preparation of cardiac SR is regulated by both cyclic AMP-dependent and calmodulin-dependent mechanisms.[7]

The $Ca^{2+}$-pump protein comprises 35–40% of the cardiac SR protein as determined by densitometry after sodium dodecyl sulfate-polyacrylamide gel electrophoresis and staining of the gel with Coomassie blue.[7] The $Ca^{2+}$-pump protein content of canine cardiac SR is thus about half that of rabbit skeletal muscle SR.[19] The lower content of $Ca^{2+}$-pump poly-

[14] C. J. LePeuch, J. Haiech, and J. G. Demaille, *Biochemistry* **18**, 5150 (1979).
[15] J. J. Feher, and F. N. Briggs, *Cell Calcium* **1**, 105 (1980).
[16] E. G. Kranias, F. Mandel, T. Wang, and A. Schwartz, *Biochemistry* **19**, 5434 (1980).
[17] G. Meissner and D. McKinley, *J. Biol. Chem.* **257**, 7704 (1982).
[18] B. K. Chamberlain, unpublished observations, 1979.
[19] G. Meissner, this series, Vol. 31, p. 238.

TABLE I
CHARACTERISTICS OF CANINE CARDIAC SR[a]

| Parameter | Value |
|---|---|
| Yield ($\mu$g protein/g muscle)[b] | 346 $\pm$ 18   (9) |
| Bound phosphorus ($\mu$g P$_i$/mg protein) | 32.6 $\pm$ 0.6 (4) |
| Ca$^{2+}$-loading[c] | |
|     Capacity ($\mu$mol Ca$^{2+}$/mg protein) | 9.45 + 0.62(7) |
|     Rate ($\mu$mol Ca$^{2+}$/mg protein $\cdot$ min) | 2.58 $\pm$ 0.10(9) |
| ATPase ($\mu$mol P$_i$/mg protein $\cdot$ min)[d] | |
|     Ca$^{2+}$-insensitive ("basal") | 0.21 $\pm$ 0.03(9) |
|     Ca$^{2+}$-stimulated | 2.65 $\pm$ 0.11(9) |
| Ca$^{2+}$-loading efficiency (Ca$^{2+}$/ATP)[e] | 0.97 $\pm$ 0.05(9) |

[a] Values are means $\pm$ SE, with the number of preparations assayed given in parentheses. The values for individual preparations were obtained by averaging the results of two to four separate determinations.

[b] Protein is determined by the procedure of O. H. Lowry, N. J. Rosebrough, A. L. Farr, and R. J. Randall [*J. Biol. Chem.* **193,** 265 (1951)] using bovine serum albumin as a standard.

[c] Ca$^{2+}$ loading is determined at 37° in a medium containing 0.1 M KCl, 4 m$M$ MgCl$_2$, 66.2 $\mu M$ $^{45}$CaCl$_2$, 30 $\mu M$ EGTA, 3.5 m$M$ Na$_2$ATP, 5 m$M$ potassium oxalate, 5 m$M$ NaN$_3$, 0.1 M sucrose, 20 m$M$ imidazole-HCl, pH 6.8. Approximately 6 $\mu M$ Ca$^{2+}$ is contamination from the other reagents. The Ca$^{2+}$ content of the assay medium was determined by atomic absorption spectrophotometry. SR vesicles (7 $\mu$g protein/ml for rate determination and 2 $\mu$g protein/ml for capacity determination) are equilibrated in the medium for 3 min prior to starting Ca$^{2+}$ uptake by addition of ATP. Reactions are stopped after 1 min by filtering aliquots through 0.22-$\mu$m filters (Millipore Corp., Bedford, MA), and $^{45}$Ca accumulated by the SR vesicles is determined from the radioactivity remaining in the filtrates.

[d] ATPase activities are determined at 37° by measuring P$_i$ according to the procedure of P. Ottolenghi [*Biochem. J.* **151,** 61 (1975)], with the modification that the reaction is stopped by adding 0.5 ml of the reaction mixture to 1 ml of ice-cold reagent IV containing 10% (w/v) trichloracetic acid instead of HCl. Total ATPase is determined under Ca$^{2+}$-loading conditions except for the omission of tracer $^{45}$Ca. Basal ATPase is determined under identical conditions except the EGTA concentration is 1 m$M$ and CaCl$_2$ is omitted. In both cases, aliquots are withdrawn at 15-sec intervals for P$_i$ determination, and ATPase values are determined by linear regression analysis of P$_i$ production over the first min-

peptide observed on polyacrylamide gels can be correlated with a reduction in the morphological characteristics attributed to $Ca^{2+}$-pump protein observed by electron microscopy.[7]

In the presence of the $Ca^{2+}$ ionophore A23187 (1 $\mu M$) to make the SR membrane freely permeable to $Ca^{2+}$, the $Ca^{2+}$-dependent ATPase activity of this preparation is increased to about 5.5 $\mu$mol $P_i$/mg protein · min,[12] or about one-third the activity of skeletal muscle SR under similar conditions.[20] This ATPase activity, free from constraints imposed by $Ca^{2+}$ gradients across the SR membrane, confirms the previous suggestions[7] that the turnover of the cardiac $Ca^{2+}$-pump is about two-thirds that of the skeletal muscle enzyme, and that the major factor in the lower $Ca^{2+}$-transport rate of cardiac SR is a lower density of $Ca^{2+}$-pump polypeptides in the cardiac SR membrane.

### Acknowledgments

Part of this work was performed during tenure of an Investigatorship from the American Heart Association, Tennessee Affiliate to B.C. The Studies were also supported by NIH Grant DK 14632 and by a grant from the Muscular Dystrophy Association of America to S.F.

[20] K. W. Anderson, R. J. Coll, and A. J. Murphy, *J. Biol. Chem.* **259**, 11487 (1984).

ute of the reaction. $Ca^{2+}$-stimulated ATPase is obtained by subtracting the basal ATPase from the total.

$^e$ Net moles of $Ca^{2+}$ accumulated per mole of $Ca^{2+}$-dependent ATP hydrolysis under identical assay conditions. This ratio approaches two in the presence of ruthenium red,[11] which reduces the $Ca^{2+}$ leak, acting to close the calcium release channels.

# [10] Junctional and Longitudinal Sarcoplasmic Reticulum of Heart Muscle

By MAKOTO INUI, SHERRY WANG, AKITSUGU SAITO
and SIDNEY FLEISCHER

Introduction[1]

In heart, muscle contraction is triggered by calcium which derives from two sources: (1) extracellular Ca$^{2+}$ enters via the plasmalemma during excitation; which then (2) induces Ca$^{2+}$ release from sarcoplasmic reticulum (SR).[2] This type of release is referred to as Ca$^{2+}$-induced Ca$^{2+}$ release.[3,4]

In fast twitch skeletal muscle, essentially all of the Ca$^{2+}$ which triggers muscle contraction derives from the sarcoplasmic reticulum subsequent to depolarization at the plasmalemma/transverse tubule. This type of release is referred to as "depolarization-induced calcium release."[3,4] In order to study the calcium release mechanism in skeletal muscle, a fraction of junctional terminal cisternae containing well-defined feet structures has been isolated.[5] The feet structures *in situ* are involved in junctional association with the terminal cisternae to form the triad junction.[5–9] The calcium release mechanism was found to be modulated by ryanodine at pharmacologically significant concentrations, and has been localized to the terminal cisternae of sarcoplasmic reticulum.[10]

In order to study the calcium release mechanism in heart, a comparable fraction of junctional terminal cisternae has been isolated.[11] This procedure is described herein.

[1] This work was supported in part by Grant NIH HL32711 to S.F. and by a Grant-in-Aid from the American Heart Association, Tennessee Affiliate to M.I. M.I. is an Investigator of the American Heart Association, Tennessee Affiliate. We thank Ms. Laura Taylor for typing this manuscript.

[2] The abbreviations used are EGTA, [ethylenebis(oxyethylenenitrilo)]tetraacetic acid; HEPES, *N*-2-hydroxyethylpiperazine-*N'*-2-ethanesulfonic acid; SDS-PAGE, sodium dodecyl sulfate-polyacrylamide gel electrophoresis; SR, sarcoplasmic reticulum.

[3] M. Endo, *Physiol. Rev.* **57**, 71 (1977).

[4] A. Fabiato, *Am. J. Physiol.* **245**, C1 (1983).

[5] A. Saito, S. Seiler, A. Chu, and S. Fleischer, *J. Cell Biol.* **99**, 875 (1984).

[6] C. Franzini-Armstrong, *Fed. Proc. Fed. Am. Soc. Exp. Biol.* **39**, 2403 (1980).

[7] C. Franzini-Armstrong and C. Nunzi, *J. Muscle Res. Cell Motil.* **4**, 233 (1983).

[8] A. V. Somlyo, *J. Cell Biol.* **80**, 743 (1979).

[9] S. Fleischer, *in* "Structure and Function of Sarcoplasmic Reticulum" (S. Fleischer and Y. Tonomura, eds.), pp. 119–145. Academic Press, Orlando, Florida, 1985.

[10] S. Fleischer, E. M. Ogunbunmi, M. C. Dixon, and E. A. M. Fleer, *Proc. Natl. Acad. Sci. U.S.A.* **82**, 7256 (1985).

[11] M. Inui, S. Wang, A. Saito, and S. Fleischer, submitted.

## Preparation of Junctional and Longitudinal SR

### Reagents and Preparations

Loading solution: 0.167 $M$ KCl, 11.1 m$M$ MgCl$_2$, 11.1 m$M$ CaCl$_2$, 11.1 m$M$ EGTA, 138.9 m$M$ potassium phosphate, pH 7.4
100 m$M$ Na$_2$ATP, pH 7.0
Gradient solution: 0.15 $M$ KCl, 5 m$M$ ATP, 5 m$M$ MgCl$_2$, 2 m$M$ CaCl$_2$, 2 m$M$ EGTA, 50 m$M$ potassium phosphate, pH 7.4
Sucrose Gradient: see text
    0.6 $M$ sucrose in Gradient solution (see above)
    0.8 $M$ sucrose in Gradient solution
    1.1 $M$ sucrose in Gradient solution
    1.6 $M$ sucrose in Gradient solution
Dilution solution: 0.3 $M$ KCl, 20 m$M$ K-HEPES, pH 7.0
Solution A: 0.3 $M$ KCl, 0.3 $M$ sucrose, 20 m$M$ K-HEPES, pH 7.0
Salt-washed microsomes are prepared from dog heart ventricles by the method of Chamberlain *et al.*[12]

### Procedures

The salt-washed microsomes (100 mg protein) are suspended in 45 ml of Loading solution. Ca$^{2+}$-loading is started by addition of 5 ml of 100 m$M$ ATP, pH 7.0. After incubation at 37° for 10 min with occasional mixing (4–5 times), the suspension was placed on ice and then immediately centrifuged for 30 min at 46,000 rpm (218,000 $g_{max}$) in a Beckman Ti 70 rotor. The pellets are resuspended in 15 ml of Gradient solution. Aliquots of 7.5 ml are layered onto a discontinuous sucrose gradient consisting of 7.5 ml each of 0.6 $M$, 0.8 $M$, 1.1 $M$, and 1.6 $M$ sucrose in Gradient solution (Fig. 1). The gradients are centrifuged for 135 min at 28,000 rpm (141,000 $g_{max}$) in a Beckman SW28 rotor. Figure 1 illustrates the typical distribution of membrane fractions on a discontinuous sucrose gradient. Fraction 1 is collected at the 0 $M$/0.6 $M$ interface, fraction 2 at the 0.6 $M$/0.8 $M$ interface, fraction 3 at the 0.8 $M$/1.1 $M$ interface, fraction 4 at the 1.1 $M$/1.6 $M$ interface. Fraction 5 is the pellet at the bottom of the gradient. The fractions are diluted with 3 volumes of Dilution solution, and then centrifuged for 30 min at 46,000 rpm (218,000 $g_{max}$) in a Beckman Ti 70 rotor. The pellets are resuspended to about 5 mg protein/ml in Solution A, and stored frozen in liquid nitrogen. Of these, fractions 2 and 3 are enriched in junctional terminal cisternae of heart SR. Fraction 5 is enriched in longitudinal tubules. Fraction 1 contains sarcolemma and/or transverse tubules, and fraction 4 is a mixture of junctional and longitudinal SR and aggregated vesicles. In a typical preparation, about 40 mg of junctional SR and

[12] B. K. Chamberlain, D. O. Levitsky, and S. Fleischer, *J. Biol. Chem.* **258**, 6602 (1983).

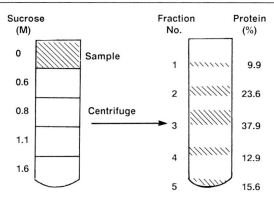

FIG. 1. Subfractionation of cardiac microsomes using a sucrose step gradient. Cardiac microsomes are preloaded with calcium phosphate and applied to the top of the gradient and centrifuged for 135 min at 28,000 rpm in a Beckman SW28 rotor. Distribution of the fractions after centrifugation is shown on the right. Fractions 2 and 3 are enriched in junctional SR. Fraction 5 is enriched in longitudinal SR.

about 10 mg of longitudinal SR were obtained from 100 g of muscle from the left ventricle.

### Characteristics of Junctional and Longitudinal SR from Heart

The subfractions of cardiac SR were isolated by calcium phosphate loading followed by sucrose density gradient centrifugation and have been characterized with regard to morphology and function.[11] Subfractionation of cardiac SR was previously reported by Jones and Cala using calcium oxalate preloading, and characterized by enhanced loading using very high concentrations of ryanodine (>200 $\mu M$).[13] In our procedure, calcium phosphate preloading instead of calcium oxalate is employed for the subfractionation of cardiac SR. An advantage of calcium phosphate loading method is that calcium remaining within the vesicles is lower than that in the vesicles prepared by calcium oxalate loading method.[11] The calcium within the vesicles could further be removed by incubating the vesicles in 0.3 $M$ KCl and 20 m$M$ HEPES, pH 7.0 for 1 hour.[11] The calcium release channels were localized to the terminal cisternae of the heart muscle SR and found to be modulated by ryanodine at pharmacologically significant (approximately nanomolar) concentrations.[11]

*Electron Microscopy.* Junctional SR (fractions 2 and 3) consists of vesicles containing some feet structures (Fig. 2B and C). *In situ,* the terminal cisternae of SR are junctionally associated with plasmalemma/ transverse tubules by way of these structures (Fig. 2A). On the other hand, longitudinal tubules of SR (fraction 5) vesicles are devoid of feet structures (Fig. 2D). These morphological characteristics resemble those

[13] L. R. Jones and S. E. Cala, *J. Biol. Chem.* **256,** 11809 (1981).

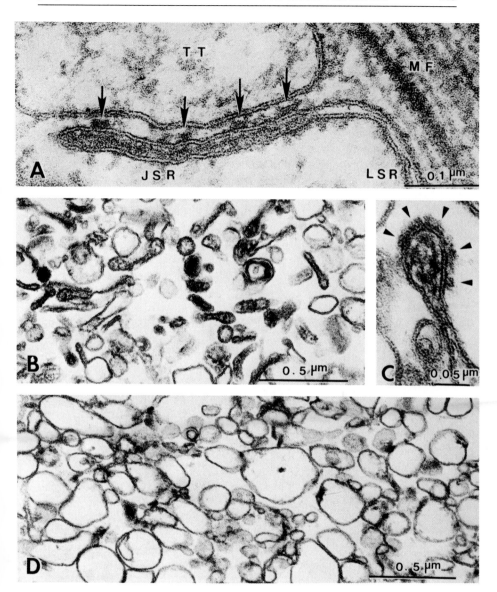

FIG. 2. Electron microscopy of cardiac membranes. (A) Thin section of cardiac tissue fixed *in situ* showing junctional SR (JSR) associated with transverse tubules (TT) by way of the feet structures (arrows). MF, myofilaments. (Magnification: 210,000×). (B and C) Isolated junctional SR (Magnification: 50,000× and 210,000×, respectively). The feet structures are indicated by arrowheads. (D) Isolated longitudinal SR (LSR) (Magnification: 50,000×).

of skeletal muscle SR,[5] but junctional SR from heart has fewer feet structures and lower $B_{max}$ for ryanodine binding (Table II).[10]

*Protein Composition.* SDS-PAGE shows that a high-molecular-weight protein with $M_r \sim 340K$ and $Ca^{2+}$-binding protein (calsequestrin) are enriched in cardiac junctional SR but not in the longitudinal tubule SR fraction. In the junctional terminal cisternae from skeletal muscle SR, a similar high-molecular-weight protein with $M_r \sim 360K$ has been observed. The isolated ryanodine receptor from skeletal muscle and heart SR has been shown to be equivalent to the feet structures[14,15] and the $Ca^{2+}$-binding protein to be compartmental contents.[9,16–18]

*Ca Transport.* Longitudinal SR shows high $Ca^{2+}$-loading rate which is not enhanced by addition of ruthenium red (7 $\mu M$) (Table I). Junctional SR has low $Ca^{2+}$-loading rate and high ATPase activity (Table I). The $Ca^{2+}$-loading rate of junctional SR is enhanced 4 to 5-fold by addition of ruthenium red, although the ATPase activity is not appreciably changed (Table I). These data indicate that the $Ca^{2+}$ release channels are localized to junctional SR but not to longitudinal SR, as has already been observed for skeletal muscle SR.[10]

*Ryanodine Binding.* [³H]Ryanodine binding activity is significantly enriched in junctional SR (Table I). Scatchard plots of total ryanodine binding in junctional SR reveal a nonlinear curve, which can be resolved into two different affinity sites with $K_d$ of 7.9 and 1087 n$M$ (Table II). In contrast with cardiac junctional SR, junctional terminal cisternae of skeletal muscle SR show only the high-affinity binding. Under identical conditions, the affinity of ryanodine binding at high-affinity sites in cardiac junctional SR is 4 to 5-fold higher than that in junctional terminal cisternae of skeletal muscle SR.[11,14,15]

*Effect of Ryanodine on $Ca^{2+}$ Release Channels.* The $Ca^{2+}$-loading rate of junctional SR is enhanced about 5-fold by ruthenium red, apparently by closing the $Ca^{2+}$ release channels. When junctional SR is preincubated with low concentrations of ryanodine (1–10 n$M$), stimulation of $Ca^{2+}$ loading by ruthenium red is blocked in a dose-dependent fashion, whereas the $Ca^{2+}$-loading rate without ruthenium red is not affected by preincubation with ryanodine. The apparent inhibition constant for ryanodine is about 7 n$M$ (Table III). When the same preincubation conditions are used for the ryanodine binding and the assay of pharmacological effect on $Ca^{2+}$

---

[14] M. Inui, A. Saito, and S. Fleischer, *J. Biol. Chem.* **262,** 1740 (1987).

[15] M. Inui, A. Saito, and S. Fleischer, *J. Biol. Chem.* **262,** 15637 (1987).

[16] The ryanodine receptor from heart has recently been identified as the $Ca^{2+}$ release channel.[17,18]

[17] F. A. Lai, K. Anderson, E. Rousseau, Q. Liu, and G. Meissner, *Biochem. Biophys. Res. Commun.* **151,** 441 (1988).

[18] L. Hymel, H. Schindler, M. Inui, and S. Flesicher, *Biochem. Biophys. Res. Commun.* **152,** 308 (1988).

TABLE I

CHARACTERISTICS OF CARDIAC SUBCELLULAR SR FRACTIONS[a]

| Fraction | Ca²⁺ phosphate loading rate ($\mu$mol Ca²⁺/mg · min) | | | Ca²⁺,Mg²⁺-ATPase activity ($\mu$mol P$_i$/mg · min) | | | [³H]Ryanodine binding[c] (pmol/mg) |
|---|---|---|---|---|---|---|---|
| | $-$RR[b] | $+$RR[b] | Ratio ($+$RR/$-$RR) | $-$RR[b] | $+$RR[b] | Ratio ($+$RR/$-$RR) | |
| Microsomes | 0.63 ± 0.02 (3) | 0.91 ± 0.06 (3) | 1.44 | 0.73 ± 0.12 (3) | 0.70 ± 0.27 (2) | 0.96 | 3.02 ± 0.41 (3) |
| LSR | 0.93 ± 0.08 (3) | 1.01 ± 0.05 (3) | 1.09 | 1.02 ± 0.36 (4) | 1.16 ± 0.50 (4) | 1.14 | 1.67 ± 0.53 (5) |
| JSR | 0.16 ± 0.02 (4) | 0.84 ± 0.16 (4) | 4.75 | 1.02 ± 0.22 (4) | 1.19 ± 0.07 (2) | 1.17 | 8.78 ± 1.87 (5)[d] |

[a] The assays were carried out as described previously.[10,11] Values are the mean ±SE from the number of separate preparations indicated in parentheses.

[b] +RR refers to addition of 7 $\mu M$ ruthenium red; $-$RR, no ruthenium red present.

[c] Ryanodine binding was measured at 210 n$M$ ryanodine in the presence of 1 $M$ KCl and 25 $\mu M$ CaCl$_2$. The binding consists of both high- and low-affinity binding.

[d] The $B_{max}$ for high-affinity binding determined by Scatchard analysis was 5.1 pmol/mg protein (58% of the total) (see Table II), so that the amount of low-affinity binding under these conditions (not $B_{max}$) is 3.68 pmol/mg (42%).

TABLE II

RYANODINE BINDING CHARACTERISTICS OF CARDIAC
JUNCTIONAL SR[a]

| Ryanodine binding | $K_d$ (n$M$) | $B_{max}$ (pmol/mg) |
|---|---|---|
| High-affinity site | 7.92 ± 1.2 (3) | 5.1 ± 0.6 (2) |
| Low-affinity site | 1087 ± 743 (3) | 32.8 ± 4.5 (2) |

[a] $K_d$ and $B_{max}$ were determined from the Scatchard analysis of ryanodine binding as described previously.[10,11] Values are the mean ±SE from the number of separate preparations indicated in parentheses.

TABLE III

MODULATION OF THE FUNCTION OF CARDIAC JUNCTIONAL
SR BY RYANODINE[a]

| Ryanodine action | $K_i$ (n$M$) | $K_m$($\mu M$) |
|---|---|---|
| Inhibition of RR stimulation of Ca²⁺ loading[b] | 7.1 ± 2.1 (3) | |
| Stimulation of Ca²⁺ loading[c] | | 1.11 ± 0.63 (3) |

[a] Values are the mean ±SE from the number of separate preparations indicated in parentheses.

[b] The effect of ryanodine on preventing ruthenium red stimulation of Ca²⁺ phosphate loading in junctional SR was examined by the type of experiment described elsewhere.[10,11] $K_i$ for ryanodine was determined from a double reciprocal plot of Δ velocity versus ryanodine concentration.[10]

[c] The effect of ryanodine on enhancing Ca²⁺ loading was examined by the type of experiment described elsewhere.[11] $K_m$ for ryanodine was determined from a double reciprocal plot of velocity versus ryanodine concentration.

loading, the $K_i$ value is essentially the same as the $K_d$ value for the high-affinity binding site (7.9 n$M$) (Tables II and III). Very high concentrations (>200 $\mu M$) of ryanodine have been reported to enhance the Ca²⁺ loading in junctional SR of heart muscle.[13] Preincubation of junctional SR with 0.5–10 $\mu M$ ryanodine stimulates the Ca²⁺-loading rate in a dose-dependent manner with the apparent $K_m$ value of 1.11 $\mu M$ (Table III). When similar conditions were used for the preincubation of the ryanodine binding assay and Ca²⁺ loading, the $K_m$ value for ryanodine was essentially the same as the $K_d$ for the low-affinity binding site (1.09 $\mu M$). These findings indicate that high-affinity ryanodine binding locks the Ca²⁺ release channels in the open state and that low-affinity binding closes the Ca²⁺ release channels of cardiac junctional SR.[11]

# [11] Regulation of Ca²⁺-Pump from Cardiac Sarcoplasmic Reticulum

*By* Michihiko Tada, Masaaki Kadoma, Makoto Inui, and Jun-ichi Fujii

## Introduction[1]

Active Ca²⁺ transport by cardiac sarcoplasmic reticulum (SR) assumes a central role in the excitation–contraction coupling of the myocardium, in that Ca²⁺-dependent ATPase (EC 3.6.1.3, ATP phosphohydrolase) of cardiac SR functions as a Ca²⁺-pump, transducing chemical energy of ATP into osmotic work during the translocation of Ca²⁺ across the membrane.[2] The translocation of Ca²⁺ is tightly coupled with the enzymatic activity of the Ca²⁺-dependent ATPase, which undergoes a complex series of reactions to form and decompose intermediate phosphoenzyme (EP).[2,3]

Ca²⁺-dependent ATPase of cardiac SR is regulated by a specific membrane protein named phospholamban,[4] which serves as a substrate for cyclic AMP (cAMP)-dependent protein kinase[5] (EC 2.7.1.37, ATP : protein phosphotransferase). Cyclic AMP-dependent phosphorylation of phospholamban, forming phosphoester phosphoproteins in serine residues of this protein,[5] results in a marked increase in the rate of Ca²⁺ transport by enhancing the turnover of the ATPase enzyme.[6,7] Under these conditions the rates of the key elementary steps, i.e., the steps at which the intermediate EP is formed and decomposed, are greatly enhanced.[8,9] Thus, phospholamban is assumed to serve as a regulator of Ca²⁺-dependent ATPase.[5] Also reported is a possibility that another protein kinase activated by Ca²⁺ and calmodulin is functional in regulating

[1] This work was supported by research grants from Ministries of Education, Science and Culture, and of Health and Welfare of Japan and by a Grant-in-Aid from the Muscular Dystrophy Association of America.

[2] M. Tada, T. Yamamoto, and Y. Tonomura, *Physiol. Rev.* **58**, 1 (1978).

[3] L. de Meis and A. L. Vianna, *Annu. Rev. Biochem.* **48**, 275 (1979).

[4] M. Tada, M. A. Kirchberger, and A. M. Katz, *J. Biol. Chem.* **250**, 2640 (1975).

[5] M. Tada and A. M. Katz, *Annu. Rev. Physiol.* **44**, 401 (1982).

[6] M. A. Kirchberger, M. Tada, and A. M. Katz, *J. Biol. Chem.* **249**, 6166 (1974).

[7] M. Tada, M. A. Kirchberger, D. I. Repke, and A. M. Katz, *J. Biol. Chem.* **249**, 6174 (1974).

[8] M. Tada, F. Ohmori, M. Yamada, and H. Abe, *J. Biol. Chem.* **254**, 319 (1979).

[9] M. Tada, M. Yamada, F. Ohmori, T. Kuzuya, M. Inui, and H. Abe, *J. Biol. Chem.* **255**, 1985 (1980).

the phospholamban-ATPase system,[10] thus suggesting the existence of a dual control system, in which both cAMP- and calmodulin-dependent phosphorylation control the rate of $Ca^{2+}$-pump by cardiac SR.

Phospholamban is a unique membrane protein from biophysical points of view because it appears to exert physiological function by a direct protein–protein interaction with the $Ca^{2+}$-pump ATPase.[5,11,12] The hydrophobic nature of this protein is thought to permit such a membrane function; however, little is known about the submolecular characteristics of this protein and its molecular interactions with the ATPase. Phospholamban is also unique from a physiological point of view[5,13] because it mediates the effects of cAMP in inducing the changes in cellular motility of myocardial cells and a number of other cells under the influence of receptor agonists, such as $\beta$-adrenergic agents, that increase intracellular cAMP. Our recent observations, utilizing the binding of antiphospholamban antibodies to microsomal proteins from many sources, indicate that phospholamban-like protein exists in microsomal membranes of other tissues, e.g., slow contracting skeletal muscle, smooth muscle, and platelets.

The function of phospholamban in the control of $Ca^{2+}$-dependent ATPase of cardiac SR may be to interpret the mechanism of the catecholamine action, at the cellular level, on the myocardial contractility, notably increased contractility (positive inotropic effect) and the abbreviation of systole (positive chronotropic effect).[5,13] Cyclic AMP-mediated increase in the rate of $Ca^{2+}$ accumulation by SR of the heart could explain abbreviation of systole because of the increased rate at which $Ca^{2+}$ would be removed from troponin. Enhanced rate of $Ca^{2+}$ accumulation could increase the amount of Ca stored within the SR by retaining some of the $Ca^{2+}$ that would otherwise be lost during diastole. This could increase the amount of $Ca^{2+}$ available for delivery to the contractile proteins in subsequent contractions, thus promoting augmentation of myocardial contractility.[5,13]

This chapter extensively surveys and summarizes the technical and procedural aspects as well as the enzymatic properties of cardiac SR $Ca^{2+}$-pump and its relationship to the putative regulator, phospholamban. Structural and functional aspects of the latter protein will be thoroughly reviewed, with emphasis on its regulatory role of the $Ca^{2+}$-pumping function by SR membranes.

[10] C. J. Le Peuch, J. Haiech, and J. G. Demaille, *Biochemistry* **18**, 5150 (1979).
[11] M. Tada, M. Yamada, M. Kadoma, M. Inui, and F. Ohmori, *Mol. Cell. Biochem.* **46**, 73 (1982).
[12] C. J. Le Peuch, D. A. M. Le Peuch, and J. G. Demaille, *Biochemistry* **19**, 3368 (1980).
[13] M. Tada and M. Inui, *J. Mol. Cell. Cardiol.* **15**, 565 (1983).

Preparation Procedures and Structural Characteristics of Cardiac SR

*Preparation Procedures*

Cardiac SR membranes are obtained by differential centrifugation of the homogenate from canine ventricular muscle.[5,14,15] SR membranes, largely condensed in the microsomal fraction,[16] are found to form resealed right-side-out vesicles with a diameter of 0.1–0.2 μm.[17]

Canine ventricular muscle is homogenized in sodium bicarbonate solution and centrifuged at 10,000 g. After the supernatant containing microsomes is sedimented at 40,000 g, the precipitate is suspended in buffer A (0.6 M KCl, 1 mM dithiothreitol, 0.3 M sucrose, and 20 mM Tris-maleate, pH 6.8); inclusion of 0.6 M KCl serves to remove contaminating actomyosin[18] as well as to stabilize the Ca$^{2+}$-pump ATPase.[19] Centrifugation at 40,000 g and resuspension in buffer A are repeated three more times and the final precipitate is suspended in buffer A. Unlike SR vesicles from skeletal muscle, cardiac SR vesicles thus obtained are less stable[20] as judged from Ca$^{2+}$ uptake and ATPase activities. The yield of microsomal proteins from canine ventricle is about 0.63 mg/g muscle, which represents approximately 10% of the SR membranes present in the muscle homogenate.[16] The resultant SR preparations, which are used commonly in most laboratories including ours, are highly enriched in SR membranes, judging from electron microscopic examination and from the marker enzyme activities such as Ca$^{2+}$ uptake, Ca$^{2+}$-dependent ATPase activity, and Ca$^{2+}$-dependent acylphosphoprotein formation.[5] The ATPase enzyme (Ca$^{2+}$-pump protein) with a molecular weight of about 100,000 accounts for up to 40% of the total protein in this SR preparation.[21] Other identified proteins are calsequestrin (1–2%)[22] and phospholamban (4–6%).[5] These preparations also contain sarcolemmal membranes, which are estimated to be up to 15%.[23] While sarcolemmal vesicles exhibit ATP-

[14] S. Harigaya and A. Schwartz, *Circ. Res.* **25**, 781 (1969).
[15] D. O. Levitsky, M. K. Aliev, A. V. Kuzmin, T. S. Levchenko, V. N. Smirnov, and E. I. Chazov, *Biochim. Biophys. Acta* **443**, 468 (1976).
[16] R. J. Solaro and F. N. Briggs, *Circ. Res.* **34**, 531 (1974).
[17] P. H. De Foor, D. Levitsky, T. Biryukova, and S. Fleischer, *Arch. Biochem. Biophys.* **200**, 196 (1980).
[18] A. Martonosi, *J. Biol. Chem.* **243**, 71 (1968).
[19] D. H. MacLennan, *J. Biol. Chem.* **245**, 4508 (1970).
[20] The cardiac SR with a high stability and a high ATPase activity is obtained by B. K. Chamberlain, D. O. Levitsky, and S. Fleischer, *J. Biol. Chem.* **258**, 6602 (1983).
[21] J. Suko and W. Hasselbach, *Eur. J. Biochem.* **64**, 123 (1976).
[22] S. E. Cala and L. R. Jones, *J. Biol. Chem.* **258**, 11932 (1983).
[23] L. R. Jones, H. R. Besch, Jr., J. W. Fleming, M. M. McConnaughey, and A. M. Watanabe, *J. Biol. Chem.* **254**, 530 (1979).

energized Ca$^{2+}$ uptake, this activity is not facilitated by oxalate, unlike Ca$^{2+}$ uptake by SR vesicles[23] (see below). The Ca$^{2+}$-dependent ATPase activity derived from slight contamination of mitochondrial membrane is inhibited by sodium azide,[24] while Ca$^{2+}$ uptake and ATPase activities of SR are not affected by this reagent.[14,16]

Centrifugation on sucrose density gradient is employed to separate SR and sarcolemmal membranes.[15,23,25] When the microsomal fraction is subjected to calcium oxalate loading in the presence of ATP prior to centrifugation on a sucrose density gradient, sarcolemmal vesicles, which are incapable of loading calcium oxalate, are concentrated in the lighter fractions and SR vesicles loaded with calcium oxalate are sedimented in the heavier fractions.[15,23] These procedures, primarily employed to isolate sarcolemmal vesicles, are not always suitable for SR preparation, since it is difficult to remove the loaded calcium oxalate that would interfere with the measurement of Ca$^{2+}$ uptake and ATPase activity of the SR vesicles. Alternative purification procedure was successfully employed by Chamberlain et al.,[25] who purified SR vesicles by linear sucrose density gradient without using oxalate loading.

### Structural Characteristics

Ca$^{2+}$-pump ATPase, representing the major protein component within cardiac SR, forms a tight complex with bilayer lipids.[2,5] Like skeletal SR, cardiac ATPase protein is asymmetrically distributed across the membrane. The ATPase of the cardiac SR is an amphipathic single polypeptide whose hydrophobic region is embedded in the lipid bilayer of the SR membrane and whose hydrophilic region is exposed to the cytoplasmic surface of the SR.[26] The structural organization of Ca$^{2+}$-ATPase protein of the SR membranes from skeletal and cardiac muscles is studied using electron microscopic techniques.[17,27–29] Negative staining of cardiac and skeletal SR shows that many small particles (about 4 nm in diameter) are present at the outer surface of the membrane, which is connected with stalks projecting from the membrane.[17] Freeze fractures of SR mem-

[24] A. M. Katz, D. I. Repke, J. E. Upshaw, and M. A. Polascik, Biochim. Biophys. Acta 205, 473 (1970).
[25] B. K. Chamberlain, D. O. Levitsky, and S. Fleischer, J. Biol. Chem. 258, 6602 (1983).
[26] M. Tada, M. Shigekawa, and Y. Nimura, in "Physiology and Pathophysiology of the Heart" (N. Sperelakis, ed.), p. 255. Nijhoff, The Hague, 1984.
[27] P. J. Pretorius, W. G. Pohl, C. S. Smithen, and G. Inesi, Circ. Res. 25, 487 (1969).
[28] D. G. Rayns, C. E. Devine, and C. L. Sutherland, J. Ultrastruct. Res. 50, 306 (1975).
[29] D. Scales and G. Inesi, Biophys. J. 16, 735 (1976).

branes indicate that globular particles (about 9 nm in diameter) are abundantly present in the cytoplasmic leaflet of the lipid bilayer.[28,29] These surface and intramembranous particles are considered to represent structural features of the ATPase proteins. The smaller 4-nm particles on the cytoplasmic surface distribute 3–4 times denser than the larger 9-nm intramembranous ones, suggesting that the ATPase exists as an oligomeric form within SR membranes.[29] However, it remains to be clarified whether such structural features would represent the functional unit of ATPase. In contrast to the cytoplasmic leaflet that contains the main body of ATPase, the luminal leaflet exhibits much less dense distribution of the protein structure. It is yet to be determined whether the latter also represents the part of the ATPase molecule that is presumably exposed to the luminal surface to accomplish $Ca^{2+}$ translocation.

The $Ca^{2+}$-dependent ATPase of cardiac SR forms acid-stable phosphoenzyme (EP), whose stability characteristics are of acyl phosphate[5]; in skeletal enzyme phosphorylation is found to occur at the $\beta$-carboxyl group of aspartic acid.[30,31] In terms of kinetic properties of $Ca^{2+}$-pump ATPase, cardiac muscle SR is less potent than its skeletal muscle counterpart. Thus, the steady-state level of EP (up to 1.3 nmol/mg protein) is approximately 3- to 4-fold less than that in skeletal SR.[32] The rates of ATP hydrolysis and $Ca^{2+}$ uptake of cardiac SR are also 3- to 4-fold lower than those of skeletal SR.[32]

It is not known whether such a difference is derived from different content of the ATPase,[32,33] or whether it is due to qualitative difference in enzymatic properties. More extensive analyses are needed to define functional and structural aspects of cardiac ATPase. The analysis of membrane crystals of ATPase protein from skeletal and cardiac SR is reported to show similar density of $Ca^{2+}$-ATPase for both types of membranes.[34] However, Chamberlain et al.[25] indicate that intramembranous particles, probably representing $Ca^{2+}$-pump ATPase, in cardiac SR are more unevenly distributed on the membrane when compared with skeletal SR.[34a] Immunologically, cardiac and skeletal ATPases do not exhibit full similarities, in that antisera against ATPase of skeletal SR show lower cross-reactivities against that of cardiac SR.[17] The findings that cardiac ATPase enzyme, but not the ATPase from fast skeletal muscle, is under a pro-

[30] F. Bastide, G. Meissner, S. Fleischer, and R. L. Post, *J. Biol. Chem.* **248**, 8385 (1973).

[31] C. Degani and P. D. Boyer, *J. Biol. Chem.* **248**, 8222 (1973).

[32] M. Shigekawa, J. M. Finegan, and A. M. Katz, *J. Biol. Chem.* **251**, 6894 (1976).

[33] M. Sumida, T. Wang, F. Mandel, J. P. Froehlich, and A. Schwartz, *J. Biol. Chem.* **253**, 8772 (1978).

[34] L. Dux and A. Martonosi, *Eur. J. Biochem.* **141**, 43 (1984).

[34a] The particle density is 70% of skeletal SR (see Ref. 89).

found control by its regulatory protein phospholamban may also indicate the critical dissimilarities of cardiac enzyme to its skeletal muscle counterpart, which is devoid of phospholamban.[5]

## Assay for Ca²⁺ Transport

### Operational Definition for Ca²⁺ Transport

The term "translocation" of $Ca^{2+}$ should be operationally defined precisely in order to comprehend the mechanism of transport across the biological membrane. When $Ca^{2+}$ is incubated with SR membranes, which form sealed, right-side-out vesicles, it is taken up by the membrane vesicles at the expense of energy supplied by ATP, either by binding to the surface of the membrane or by actual uptake into the membrane vesicle. By operational definition, $Ca^{2+}$ is considered to be translocated by the membrane when the following conditions are met: (1) $Ca^{2+}$ cannot be removed by the application, at the exterior space, of the highest possible concentration of a chelating agent such as ethylene glycol bis($\beta$-aminoethyl ether)-$N,N'$-tetraacetic acid (EGTA), which by itself does not penetrate the membrane; (2) $Ca^{2+}$ can be readily removed by the treatment of the membrane with Ca-specific ionophores (X537A or A23187[35,36]), detergents (Triton X-100[37,38]), or alkali[39] in the presence of a chelating agent. Such conditions would correspond to an actual translocation of $Ca^{2+}$ from one side of the membrane to the other, or an apparent translocation due to trapping of $Ca^{2+}$ inside the pocket formed by the carrier protein.[2]

Anions like oxalate are often included in the medium to enhance the amounts of $Ca^{2+}$ transported into vesicles. The term "oxalate-facilitated calcium uptake" is often applied to describe these conditions.[2] Oxalate is freely permeable to SR membranes and could form a calcium-oxalate complex by binding with transported $Ca^{2+}$ within the vesicular lumen, thus maintaining an intravesicular free $Ca^{2+}$ concentration and enabling more $Ca^{2+}$ to be transported.[2] In the absence of oxalate most of the $Ca^{2+}$ is bound to the membrane interior, and only a small amount exists as the ionized form, thus leaving the $Ca^{2+}$-binding capacity of the vesicles quite low.

[35] M. G. P. Vale and A. P. Carvalho, *Biochim. Biophys. Acta* **413,** 202 (1975).
[36] M. G. P. Vale, V. R. Osorio, E. Castro, and A. P. Carvalho, *Biochem. J.* **156,** 239 (1976).
[37] T. Yamamoto and Y. Tonomura, *J. Biochem.* (*Tokyo*) **62,** 558 (1967).
[38] T. Kanazawa, S. Yamada, T. Yamamoto, and Y. Tonomura, *J. Biochem.* (*Tokyo*) **70,** 95 (1971).
[39] P. F. Duggan and A. Martonosi, *J. Gen. Physiol.* **56,** 147 (1970).

## Calculation of Free $Ca^{2+}$ Concentration

Since $Ca^{2+}$-dependent ATPase has high affinity for $Ca^{2+}$, it is important to regulate $Ca^{2+}$ concentrations in the reaction medium. For example, the $Ca^{2+}$-pump of SR can be stimulated by micromolar concentrations of $Ca^{2+}$, the usual level of $Ca^{2+}$ which would exist in the deionized distilled water. Use of a $Ca^{2+}$-chelating agent, EGTA, is considered to be beneficial in controlling the free $Ca^{2+}$ concentration within the range between $10^{-9}$ and $10^{-5}$. The formation of Ca–EGTA complex is markedly influenced by the presence of adenine nucleotides like ATP and ADP, and cations like $Mg^{2+}$, $K^+$, $Na^+$, and $H^+$. A computer program is employed for calculating the free $Ca^{2+}$ concentrations in the presence of many kinds of the electrolytes and substrates.[40] This program can be operated either by entering the concentrations of free ions desired and obtaining the total concentrations of solutes (electrolytes and substrates) as the output, or by entering the total concentrations of solutes and obtaining the concentrations of free ions as the output. The widely accepted absolute association constants for Ca–EGTA complexes, obtained by Schwarzenbach, are found by Fabiato[41] to give a reliable basis for obtaining the apparent association constants under variety of conditions. Those absolute association constants of Schwarzenbach are

$$K_1 = [CaEGTA^{2-}]/[Ca^{2+}][EGTA^{4-}] = 1.00 \times 10^{11} \ M^{-1}$$
$$K_2 = [CaHEGTA^-]/[Ca^{2+}][HEGTA^{3-}] = 2.14 \times 10^5 \ M^{-1}$$

The apparent association constants of Ca–EGTA complex under a variety of electrolyte and substrate conditions are obtained according to Fabiato's suggestion.[41] In practice, when Ca–EGTA buffers giving different free $Ca^{2+}$ concentrations are desired, the concentration of $CaCl_2$ is fixed (e.g., 100, 125, or 200 $\mu M$) and EGTA concentrations are varied according to the output of computer program. Under usual circumstances, our $10\times$ stock Ca–EGTA buffer, containing $CaCl_2$ and EGTA, is added to the assay media in a one-tenth dilution.

[In presenting our previous conditions and data in this article, we recalculated free $Ca^{2+}$ concentrations by the computer program and found that they should be altered slightly. Therefore, the revised $Ca^{2+}$ concentrations are indicated in figures and reaction conditions of the present article.]

[40] A. Fabiato and F. Fabiato, *J. Physiol. (Paris)* **75**, 463 (1979).
[41] A. Fabiato, *J. Gen. Physiol.* **78**, 457 (1981).

*Assay for Oxalate-Facilitated Ca$^{2+}$ Uptake*

*Assay Medium*

Cardiac SR, 50 μg/ml
MgATP, 1 mM
KCl, 125 mM
NaN$_3$, 5 mM
Tris-oxalate, 2.5 mM
Tris-maleate, 40 mM, pH 6.8
Ca–EGTA buffer, containing 125 μM $^{45}$CaCl$_2$ (10 μCi/μmol) and 125 μM EGTA, giving calculated free Ca$^{2+}$ of 7.8 μM

*Reagents*

MgATP, 20 mM: a solution containing equimolar concentrations (20 mM) of MgCl$_2$ and ATP is adjusted to pH 6.8 with NaOH.

Tris-oxalate, 10 mM: oxalic acid is titrated with Tris to pH 6.8.

Ca–EGTA buffer: a solution containing equimolar concentrations (1.25 mM) of $^{45}$CaCl$_2$ and EGTA is adjusted to pH 6.8 by NaOH. When the free $^{45}$Ca$^{2+}$ concentration is varied, EGTA concentrations are varied, while CaCl$_2$ concentration (1.25 mM) is fixed.

NaN$_3$, 50 mM: sodium azide should be included in the assay medium to inhibit mitochondrial Ca$^{2+}$-ATPase, which is present in conventional SR preparations (see above).

*Assay Procedures.* Oxalate-facilitated Ca$^{2+}$ uptake is determined in 2 to 5 ml of a medium containing the above reagents at 25°. After preincubation for 5 min, the reaction is initiated by the addition of MgATP at a final concentration of 1 mM. At appropriate time intervals (0–10 min), a 1-ml aliquot of the reaction mixture is filtered by Millipore filter (HAWP 0.45 μm) and the amounts of $^{45}$Ca incorporated into SR vesicles are determined by measuring either (1) $^{45}$Ca radioactivity in the filtrate[7,42] or (2) the precipitate with retained $^{45}$Ca on the filter.[14,43] The latter procedure requires the washing of the filter with 1 mM EGTA in 20 mM Tris-maleate (pH 6.8) and 100 mM KCl at 0–4°. After drying, the Millipore filter is liquefied with acetone and counted for radioactivity.

Ca$^{2+}$ uptake can also be measured by a dual-wavelength spectrophotometer using the metallochromic indicators, such as arsenazo III,[44] to

[42] M. A. Kirchberger, M. Tada, D. I. Repke, and A. M. Katz, *J. Mol. Cell. Cardiol.* **4,** 673 (1972).

[43] M. Tada, M. Inui, M. Yamada, M. Kadoma, T. Kuzuya, H. Abe, and S. Kakiuchi, *J. Mol. Cell. Cardiol.* **15,** 335 (1983).

[44] A. Scarpa, this series, Vol. 56, p. 301.

monitor $Ca^{2+}$ removal from the assay solution under various conditions that are described elsewhere.[44]

### Assay for Ca²⁺ Transport at Presteady State

#### Assay Medium

Cardiac SR, 0.4 mg/ml
ATP, 10 $\mu M$
MgCl₂, 3 mM
KCl, 100 mM
NaN₃, 5 mM
Tris-maleate, 20 mM, pH 6.8
Ca–EGTA buffer, containing 100 $\mu M$ ⁴⁵CaCl₂ (25 $\mu$Ci/$\mu$mol) and 100 $\mu M$ EGTA, giving free $Ca^{2+}$ of 9.7 $\mu M$

*Assay Procedures.* The rate of ATP-supported $Ca^{2+}$ transport, in the absence of facilitating anions, could be measured by the rapid-quenching technique (Fig. 1), which enables the determination of the rapid incorporation of $Ca^{2+}$. The initial reaction of $Ca^{2+}$ transport takes place within

A  Ca²⁺-bound SR

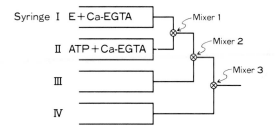

B  Ca²⁺-free SR

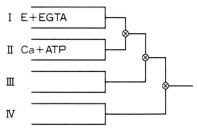

FIG. 1. Experimental design for measurements of $Ca^{2+}$ transport, EP formation, and $P_i$ liberation at the transient state. See text for further details.

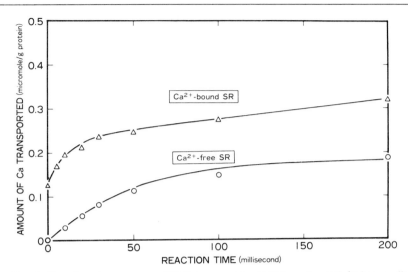

FIG. 2. Transient time courses of Ca²⁺ transport by Ca²⁺-bound and Ca²⁺-free cardiac SR. For measurements in Ca²⁺-bound SR (△), ATPase enzyme (0.4 mg/ml in final concentration) is reacted with ATP (10 $\mu M$) in the presence of 100 $\mu M$ ⁴⁵CaCl₂ (25 $\mu$Ci/$\mu$mol) and 100 $\mu M$ EGTA, which give an ionized Ca²⁺ concentration of 9.7 $\mu M$. In Ca²⁺-free SR (○), the ATPase enzyme (0.4 mg/ml) in the presence of EGTA (100 $\mu M$ in final concentration) is reacted with ATP (10 $\mu M$) containing ⁴⁵CaCl₂ (100 $\mu M$). By binding of ⁴⁵CaCl₂ and EGTA, free Ca²⁺ concentration is considered to be established instantaneously at 9.7 $\mu M$. [Reproduced with permission from Tada et al., J. Biol. Chem. 255, 1985 (1980).]

tens of milliseconds, reaching steady state within 100 msec at 20° (Fig. 2). Rapid mixing measurements are performed employing a slightly modified version of a chemical quench-flow apparatus of Froehlich et al.,[45] which is equipped with four syringes and three mixers, allowing the SR vesicles to react rapidly with substrates (Fig. 1). Syringe I contains cardiac SR (0.8 mg/ml), and syringe II contains ATP (20 $\mu M$). These are designed to mix at 20° on an equivolume basis in the first mixer, and after various times (5 to 200 msec) are quenched with a solution containing 6 m$M$ EGTA in syringe III. The standard vehicle solution for the syringes I, II, and III contains 100 m$M$ KCl, 3 m$M$ MgCl₂, 5 m$M$ NaN₃, and 20 m$M$ Tris-maleate (pH 6.8).

The initial rates of Ca²⁺ transport are markedly different when Ca²⁺-pump ATPase is at the different states [see Eq. (2) below] depending on the binding of Ca²⁺ to the ATPase. Therefore, Ca²⁺ concentration in the medium should be strictly controlled. In experiments employing Ca²⁺-

[45] J. P. Froehlich, J. V. Sullivan, and R. L. Berger, Anal. Biochem. 73, 331 (1976).

bound SR (Fig. 1A), both syringes I and II contain Ca–EGTA buffer (100 $\mu M$ $^{45}CaCl_2$ and 100 $\mu M$ EGTA) to establish state $E_1$ of ATPase before and during the reactions with substrate ATP. In contrast, Fig. 1B represents the experimental design for $Ca^{2+}$-free SR, in which $Ca^{2+}$-pump ATPase is allowed to exist as state $E_2$ before starting the reaction. Syringe I in Fig. 1B contains 200 $\mu M$ EGTA, which is mixed with syringe II containing 200 $\mu M$ $^{45}CaCl_2$. When mixed, $CaCl_2$ and EGTA instantaneously react to establish a free $Ca^{2+}$ concentration that is practically the same as Fig. 1A, allowing the ATPase enzyme to convert from state $E_2$ to $E_1$. Immediately after quenching, 2 ml of the assay solution is passed through a Millipore filter (HAWP 0.45 $\mu$m), and the precipitate with retained $^{45}Ca$ is washed with 1 m$M$ EGTA in 50 m$M$ Tris-maleate (pH 6.8) and 100 m$M$ KCl at 0–4°. After drying, the Millipore filter is liquefied with acetone and counted for radioactivity.

Figure 2 shows the time courses of $Ca^{2+}$ transport at the presteady state of the reaction. It is evident that $Ca^{2+}$ transport in $Ca^{2+}$-bound SR occurs much faster and to higher extent than that in $Ca^{2+}$-free SR, in accord with the findings for $Ca^{2+}$-dependent ATPase enzyme (see below). When the reaction is initiated at $Ca^{2+}$-bound SR (state $E_1$), the enzymes at state $E_1$ instantaneously form $E_1P$ on addition of ATP, accompanying a rapid translocation of $Ca^{2+}$ from outside to inside the membrane. When the reaction is initiated at $Ca^{2+}$-free SR (state $E_2$), $Ca^{2+}$ binding allows $E_2$ to convert into $E_1$, after which $Ca^{2+}$ translocation is attained with a simultaneous formation of $E_1P$. Since the conversion from $E_2$ to $E_1$ occurs much more slowly,[33] the rate of $Ca^{2+}$ transport under these conditions is lower than that in $Ca^{2+}$-bound SR.

Assay for $Ca^{2+}$-Dependent ATPase

*Steady-State Measurements*

*Assay Medium*

Cardiac SR, 100–150 $\mu$g/ml

[$\gamma$-$^{32}$P]ATP, 100 $\mu M$ (recommended specific radioactivities are described in the text)

$MgCl_2$, 1 m$M$

KCl, 100 m$M$

$NaN_3$, 5 m$M$

Tris-maleate, 40 m$M$, pH 6.8

Ca–EGTA buffers, containing 125 $\mu M$ $CaCl_2$ and various concentrations (1.4 m$M$–110 $\mu M$) of EGTA, giving free $Ca^{2+}$ of 0.1–20 $\mu M$

*Assay Procedures.* The rate of ATP hydrolysis is determined by measuring [$^{32}$P]P$_i$ liberation in 1 ml of assay medium at 25°. The reaction is initiated by adding [$\gamma$-$^{32}$P]ATP (10 $\mu$Ci/$\mu$mol) at a final concentration of 100 $\mu M$, and is terminated by the addition of 1 ml of 10% (w/v) trichloroacetic acid (TCA) containing 2 m$M$ ATP and 0.5 m$M$ P$_i$. After the addition of 0.2 ml of 0.63% bovine serum albumin (BSA) as carrier protein and centrifugation at 1000 g for 10 min, the amount of [$^{32}$P]phosphate in an aliquot (1 ml) of the supernatant is determined by the procedure of Martin and Doty,[46] employing the extraction mixture of *n*-butanol and benzene. To determine "basic" (Ca$^{2+}$-independent) ATPase activity, the reaction is carried out in the presence of 1 m$M$ EGTA instead of Ca–EGTA buffer. The rate of Ca$^{2+}$-dependent ATP hydrolysis (Ca$^{2+}$-dependent ATPase activity) is estimated by subtracting the "basic" ATPase activity from the rate of ATP hydrolysis in the presence of Ca$^{2+}$ ("total" ATPase activity).[2]

The rate of ATP hydrolysis can also be measured by determining the liberation of ADP. Through an appropriate coupling of pyruvate kinase with ATPase, the amount of pyruvate liberated is determined as is described below.

To simultaneously determine the formation of reaction intermediate EP and the liberation of P$_i$, the reaction, carried out at 15° in 1 ml medium, is initiated by adding 100 $\mu M$ [$\gamma$-$^{32}$P]ATP (50 $\mu$Ci/$\mu$mol). At time intervals of 5 to 30 sec, the reactions are terminated by adding TCA solutions followed by BSA as described above. While the supernatant (1-ml aliquot) obtained by centrifugation (1000 g, 10 min) is subjected to [$^{32}$P]P$_i$ determination, the pellet containing E$^{32}$P is resuspended at 0° in 3 ml of 4% perchloric acid (PCA) containing 30 m$M$ P$_i$ and 10 m$M$ PP$_i$, Centrifugation (1000 g, 10 min) and resuspension in the PCA solution are repeated three more times, then the washed final pellet is dissolved in 0.2 ml of 0.5 $N$ NaOH, followed by the addition of 1 ml H$_2$O, and counted for radioactivity.

## Transient State Measurements

The initial reaction between Ca$^{2+}$-pump ATPase and substrate ATP is a rapid process taking place within tens of milliseconds. The intermediate EP is rapidly formed, which is followed by the liberation of P$_i$ from EP (Fig. 3). It is well established that EP is the true and sole intermediate for the ATPase reaction.[2,3] In the initiation of the Ca$^{2+}$-dependent ATPase reaction, it is important to define the conditions under which the ATPase is allowed to react with substrate ATP, since the ATPase at state E$_1$ in the

[46] J. B. Martin and D. M. C. Doty, *Anal. Chem.* **21**, 965 (1949).

presence of $Ca^{2+}$ and that at state $E_2$ in the absence of $Ca^{2+}$ exhibit markedly different kinetic properties,[3,5,33] as described below [see also Eq. (2) below].

### Assay Medium

Cardiac SR, 0.4 mg/ml
$[\gamma\text{-}^{32}P]ATP$, 10 $\mu M$ (50 $\mu Ci/\mu mol$)
$MgCl_2$, 3 m$M$
KCl, 100 m$M$
$NaN_3$, 5 m$M$
Tris-maleate, 20 m$M$, pH 6.8
$Ca^{2+}$, 9.7 $\mu M$: Ca–EGTA buffer, containing 100 $\mu M$ $CaCl_2$ and 100 $\mu M$ EGTA

*Assay Procedures.* The extremely rapid formation of EP and $P_i$ can be determined by employing the chemical quench-flow device, which allows the use of time intervals as short as 4 msec.[45] It is also important to define the enzyme state of ATPase, which would be either at state $E_1$ or $E_2$, prior to its reaction with substrate ATP. Figure 1A and 1B represents the two reaction modes by which substrate is allowed to react with the ATPase enzyme (E) at state $E_1$ (condition A) and at state $E_2$ (condition B), respectively. In condition A, the enzyme at state $E_1$ (in the presence of $Ca^{2+}$) reacts with ATP to form EP under the same $Ca^{2+}$ milieu, while in condition B the enzyme at state $E_2$ (in the presence of EGTA) reacts with ATP and $Ca^{2+}$ to form $E_1P$ via conversion from state $E_2$ to state $E_1$ [see Eq. (2)].

In the ATPase reaction under condition A, syringe I contains cardiac SR (0.8 mg/ml) and Ca–EGTA buffer (100 $\mu M$ $CaCl_2$ and 100 $\mu M$ EGTA) and syringe II contains 20 $\mu M$ $[\gamma\text{-}^{32}P]ATP$ and Ca–EGTA buffer with concentrations identical to syringe I. The standard vehicle solution for syringes I and II consists of 100 m$M$ KCl, 3 m$M$ $MgCl_2$, 5 m$M$ $NaN_3$, and 20 m$M$ Tris-maleate (pH 6.8). Syringe III contains 20% TCA, 2 m$M$ ATP, and 0.5 m$M$ $P_i$. The reaction, initiated by mixing equivolume solutions of syringes I and II in the first mixer is terminated by the stopping solution in syringe III in the second mixer at appropriate time intervals (4–200 msec). A constant volume (1.8–2.0 ml) of the resultant assay solution is mixed with 1.26 mg of BSA and centrifuged at 1000 $g$ for 10 min. The supernatant and pellet are subjected to determination of $^{32}P_i$ and $E^{32}P$, respectively, by the standard procedures.

In the ATPase reaction under condition B, syringe I contains cardiac SR (0.8 mg/ml) and EGTA (200 $\mu M$) and syringe II contains 20 $\mu M$ $[\gamma\text{-}^{32}P]ATP$ and 200 $\mu M$ $CaCl_2$, with the standard vehicle solutions described above. When equivolume solutions of syringes I and II are mixed in the first mixer, $CaCl_2$ and EGTA instantaneously establish a free $Ca^{2+}$ concentration practically identical to that in condition A. The assay solution quenched by TCA at appropriate time intervals is subjected to $^{32}P_i$ and $E^{32}P$ determinations.

Employing the fourth syringe, the time course for EP decomposition is also determined by this device.[47] The EP formation, established by mixing syringes I and II, is quenched by EGTA in syringe III. The decay in EP is determined by subsequent mixing with TCA in the third mixer.

*Elementary Steps of Ca²⁺-Dependent ATPase Coupled*
  *with Ca²⁺ Transport*

During the translocation of $Ca^{2+}$ across SR membranes, $Ca^{2+}$-dependent ATPase enzyme serves as an energy transducer and a carrier for $Ca^{2+}$. $Ca^{2+}$ translocation is tightly coupled with the formation of a phosphoprotein intermediate EP, in which the terminal phosphate of ATP is incorporated into the ATPase enzyme forming an acyl phosphate.

$Ca^{2+}$-dependent ATPase of skeletal SR vesicles would assume three major roles, in a successive manner, which correspond to the recognition, translocation, and release of $Ca^{2+}$ during the process of $Ca^{2+}$ transport across the membrane.[2] The ATPase enzyme possesses $Ca^{2+}$-binding moiety at the membrane exterior that recognizes $Ca^{2+}$ among many solutes in the absence of ATP. ATP is able to energize the $Ca^{2+}$-bound ATPase to translocate $Ca^{2+}$ from outside to inside the membrane by forming EP. The translocated $Ca^{2+}$ is readily released from the enzyme into the membrane interior as EP decomposes. Among these steps, the step at which EP is decomposed is rate-determining, so that the intermediate EP is significantly accumulated.[2]

Figure 3 compares the transient time courses of EP formation and $P_i$ liberation by SR vesicles of cardiac and skeletal muscle, determined by the rapid-quenching technique.[45] The formation of EP in both types of SR takes place instantaneously (within 50 msec), which is followed by the liberation of $P_i$ after a lag phase. After the rapid initial phase, a steady state is reached, in which $P_i$ is liberated as a function of time, while EP is apparently maintained constant. The observed overshoot of EP at the initial phase, particularly seen in skeletal SR, is the matter of dispute.[2] Since this issue is beyond the scope of this article, we do not attempt to make comments.

The kinetic analyses of the overall and partial reactions of $Ca^{2+}$-dependent ATPase of both skeletal and cardiac SR are consistent with the

---

[47] J. P. Froehlich and P. F. Heller, *Biochemistry* **24,** 126 (1985).

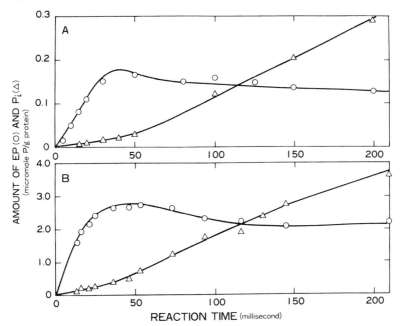

FIG. 3. Transient time courses of EP formation and $P_i$ liberation by cardiac (A) and skeletal (B) muscle SR. (A) 0.4 mg/ml cardiac SR, 10 $\mu M$ [γ-³²P]ATP (50 $\mu$Ci/$\mu$mol), Ca–EGTA buffer (100 $\mu M$ each for free Ca²⁺ of 9.7 $\mu M$), 3 m$M$ MgCl₂, 100 m$M$ KCl, and 20 m$M$ Tris-maleate, pH 6.8. (B) 0.15 mg/ml skeletal SR, 5 $\mu M$ [γ-³²P]ATP (4.5 $\mu$Ci/$\mu$mol), Ca–EGTA buffer (100 $\mu M$ CaCl₂, Ca²⁺ of 14 $\mu M$), 1 m$M$ MgCl₂, 100 m$M$ KCl, 5 m$M$ sodium oxalate, and 20 m$M$ Tris-maleate, pH 6.8. Both reactions (A) and (B) are performed under condition A. See text for details. [Reproduced with permission from Tada *et al., Mol. Cell. Biochem.* **46**, 73 (1982).]

following reaction scheme[2]:

$$E + ATP° + 2Ca° \leftrightarrow E_{ATP}^{Ca_2^o} \leftrightarrow E_P^{Ca_2^i} + ADP° \rightarrow E + ADP° + P_i° + 2Ca^i \qquad (1)$$

where i and o indicate the inside and outside of the membrane vesicles, respectively. At the exterior surface of the membrane, 2 mol of Ca²⁺ and 1 mol of ATP form a ternary complex ($E_{ATP}^{Ca_2^o}$) with the ATPase enzyme. Subsequent phosphorylation of the enzyme is coupled with the translocation of Ca²⁺ from outside to inside the membrane, thus forming $E_P^{Ca_2^i}$ [$E_1P$ in Eq. 2)]. By the operational definition described above, this form of Ca²⁺ corresponds to that occluded in the pocket of the enzyme or to that which is transported into the vesicular interior. Ca²⁺ is then released from the enzyme into the interior of the vesicular lumen [corresponding to the conversion from $E_1P$ to $E_2P$ in Eq. (2)].

From the bioenergetic points of view, $Ca^{2+}$-pump ATPase provides a unique property in which $Ca^{2+}$ translocation is coupled with EP formation. It is feasible to assume, therefore, that the formation and decomposition of EP are associated with significant conformational changes of the protein structure.[2,3] Such contention is supported by a number of findings indicating that both unphosphorylated enzyme (E) and phosphorylated enzyme (EP) undergo two distinct states,[3,5] depending upon the binding of $Ca^{2+}$. $Ca^{2+}$-bound enzyme ($E_1$) is conformationally distinct from $Ca^{2+}$-free enzyme ($E_2$), while $Ca^{2+}$-bound EP ($E_1P$) is from $Ca^{2+}$-free EP ($E_2P$). Taking these into consideration, the following reaction scheme is proposed[3,5,26]:

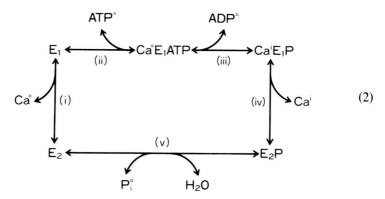

$$(2)$$

where $E_1$ and $E_2$ represent two different forms of the enzyme, which exhibit higher and lower affinities for $Ca^{2+}$, respectively; i and o indicate the inside and outside of SR membranes, respectively. $E_1P$ is the form of the intermediate having higher affinity for $Ca^{2+}$, while $E_2P$ has a lower affinity for $Ca^{2+}$. Although the amount of EP represents the sum of $E_1P$ and $E_2P$, $E_1P$ is considered to predominate under usual circumstances where $Ca^{2+}$ and $K^+$ are present at neutral pH. It is also noted that $E_1P$ is reactive to ADP, forming ATP reversibly, whereas $E_2P$ is not reactive to ADP and does not form ATP.[48,49]

At lower concentration of ATP ($<10~\mu M$), $Ca^{2+}$ induces the conversion of $E_2$ to $E_1$, while such conversion is not apparent in the presence of higher concentrations of ATP. At any rate, steps i and iv are the rate-determining steps in the overall turnover of the reaction steps. It is interesting that these are the steps at which major conformational changes of ATPase are assumed to take place.[2,3]

[48] M. Shigekawa and J. P. Dougherty, *J. Biol. Chem.* **253,** 1451 (1978).
[49] M. Shigekawa, J. P. Dougherty, and A. M. Katz, *J. Biol. Chem.* **253,** 1442 (1978).

The time courses of EP formation, seen in Fig. 3, correspond to the reactions which are initiated when the enzyme is at state $E_1$. In contrast, the initial time course of EP formation should be much lower when the reaction is initiated at state $E_2$, since EP is formed *via* the conversion of $E_2$ to $E_1$, the rate-determining step. Indeed, as seen in Fig. 7 (below), the initial rate of EP formation in $Ca^{2+}$-free SR (condition B), following the first-order kinetics, is markedly lower than that seen in $Ca^{2+}$-bound SR (condition A).

### Cyclic AMP-Dependent Phosphorylation of Phospholamban

*Assay Medium*

Cardiac SR, 1.6 mg/ml
$[\gamma$-$^{32}$P]ATP, 0.5 m$M$ (10–50 $\mu$Ci/$\mu$mol)
cAMP, 1 $\mu M$
cAMP-dependent protein kinase, 0.8 mg/ml
$MgCl_2$, 0.5 m$M$
KCl, 120 m$M$
NaF, 25 m$M$
EGTA, 2.5 m$M$
Histidine-HCl, 40 m$M$, pH 6.8

*Reagents*

cAMP-dependent protein kinase: prepared from bovine heart according to the method of Miyamoto *et al.*[50] This can be substituted by the catalytic subunit of cAMP-dependent protein kinase (50–100 $\mu$g/ml) with no addition of cAMP.

*Procedures.* cAMP-dependent phosphorylation of phospholamban in cardiac SR is performed at 25° for 1–10 min in 0.2 ml of assay medium. The reaction is initiated by the addition of ATP and is terminated and processed as described below.

*Method A.* The reaction is terminated by the addition of 2 ml of ice-cold 10% TCA and 0.1 m$M$ $P_i$ followed by 1.26 mg BSA as carrier protein. After centrifugation, the pellet is dissolved in 0.1 ml of 0.5 $N$ NaOH and is suspended in 2 ml of 10% TCA. The centrifugation in TCA and resuspension in alkali are repeated two more times. The final pellet containing $^{32}$P-bound protein is dissolved in 0.1 ml of 0.5 $N$ NaOH, followed by the addition of 1 ml $H_2O$, and counted for radioactivity.

*Method B.* The reaction is terminated by the addition of a solution (0.1

[50] E. Miyamoto, J. F. Kuo, and P. Greengard, *J. Biol. Chem.* **244**, 6395 (1969).

TABLE I
COMPARISON OF STABILITY OF TWO KINDS OF PHOSPHOPROTEINS FORMED IN
CARDIAC SR[a]

| Treatment | Amounts of phosphoprotein intermediate of ATPase | | Amounts of phosphorylated phospholamban (22,000-Da protein) | |
|---|---|---|---|---|
| | nmol P/mg protein | Percentage | nmol P/mg protein | Percentage |
| None; 0° (= control) | 0.74 | 100 | 0.67 | 100 |
| None; 90° | 0.00 | 0 | 0.62 | 92.5 |
| 0.5 N NaOH; 25° | 0.00 | 0 | 0.31 | 46.3 |
| 0.5 N NaOH; 90° | 0.00 | 0 | <0.01 | <1.0 |
| Control (0.8 M NaCl); 30° | 0.71 | 100 | 0.61 | 100 |
| 0.8 M hydroxylamine; 30° | <0.01 | <1.0 | 0.60 | 98.4 |

[a] From M. Tada, F. Ohmori, M. Yamada, and H. Abe, J. Biol. Chem. **254,** 319 (1979).

ml) containing sodium dodecyl sulfate (SDS), EDTA, and 2-mercapto-ethanol, to give final concentrations of 2%, 0.1 mM, and 1%, respectively. After standing for several minutes on ice, this mixture is incubated for approximately 10 min at 37° to solubilize the SR protein and a 0.2-ml aliquot is added to a solution (0.2 ml) containing 20 mM sodium phosphate buffer (pH 7.2), 0.1 mM EDTA, 1% 2-mercaptoethanol, 50% glycerol, and 0.005% bromphenol blue. An aliquot (50–100 μl) of this solution, containing 25 to 60 μg of SR protein, is applied to SDS-polyacrylamide gel for electrophoresis by the method of Weber and Osborn.[51] In some experiments, the assay solution in 2% SDS is boiled for 1 min prior to electrophoresis.

*Phosphorylation of 22,000-Da Protein (Phospholamban)*

A phosphoester phosphoprotein is formed when cardiac SR is incubated with cAMP-dependent protein kinase.[6] Formation of this phosphoprotein is markedly dependent on cAMP between 0.1 and 10 μM with the half-maximal activation occurring at 0.2 μM.[6] This phosphoprotein is stable in hydroxylamine and in hot acid, but is unstable in hot alkali unlike the acylphosphoprotein intermediate of ATPase, which is unstable in hydroxylamine and in alkali (Table I). Indeed, the phosphorylation occurs at the serine residues.[6,12] This phosphoprotein possesses a molecular weight

[51] K. Weber and M. Osborn, J. Biol. Chem. **244,** 4406 (1969).

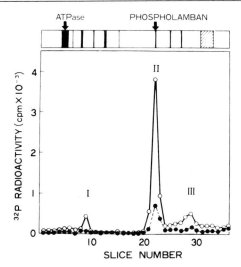

FIG. 4. cAMP-dependent protein kinase-catalyzed phosphorylation of phospholamban in cardiac SR, determined by SDS-polyacrylamide gel electrophoresis by the procedure of Weber and Osborn [*J. Biol. Chem.* **244**, 4406 (1969)]. Cardiac SR is phosphorylated by [γ-$^{32}$P]ATP and protein kinase in the presence (○) and absence (●) of 25 m$M$ NaF. Phosphoprotein formed is determined by method B, employing SDS (0.1%)-polyacrylamide (10%) gel electrophoresis. Peak II represents phosphorylation of phospholamban (22,000) and peak III its component with lower molecular weight (11,000), while peak I seen at MW 55,000 is derived from autophosphorylation of protein kinase occurring at the catalytic subunit. [Modified with permission from Tada *et al., J. Biol. Chem.* **250**, 2640 (1975).]

of 22,000 when examined by electrophoresis on SDS-polyacrylamide gel[4] (Fig. 4). Inhibitor of phosphoprotein phosphatase (NaF), augments the phosphorylation of this component, suggesting that the intrinsic phosphatase functions in the dephosphorylation of the phosphoester bond. The 22,000 MW phosphorylatable peptide is referred by Tada *et al.*[4] to as phospholamban, i.e., phosphate receptor (λαμβανειν = receive).

Existence of phospholamban phosphorylation in cardiac SR is confirmed by numerous laboratories (Table II), employing conventional preparations and preparations virtually free from sarcolemmal vesicles. The reported molecular weight for phospholamban ranges between 20,000 and 24,000 when the electrophoretic mobility of phosphoprotein is determined on SDS-polyacrylamide gel. A slightly higher molecular weight is found when electrophoresis is performed on the alkaline gel system of Laemmli[52] (see the section Structural Characteristics of Phospholamban below). Other proteins with lower molecular weights are also phosphory-

---

[52] U. K. Laemmli, *Nature (London)* **227**, 680 (1970).

TABLE II

MOLECULAR WEIGHTS OF PROTEINS PHOSPHORYLATED BY cAMP-DEPENDENT
PROTEIN KINASE IN CARDIAC SR

| Source | Molecular weight | | Ref. |
| | Phospholamban | Other phosphoprotein | |
|---|---|---|---|
| Rabbit | 20K | | a |
| Dog | 22K | 11K | b |
| Dog, cat, guinea pig, rabbit | 22K | 11K | c |
| Dog | 20K | | d |
| Dog | 20K | | e |
| Pigeon | 22K | 15K | f |
| Pig | 22K | 56K, 16K | g |
| Dog | 20K | 7K | h |
| Rat | 24K | 9K | i |
| Dog | 22K | 6K | j |
| Dog | 22K | 11K, 7K | k |
| Dog | 22K | 11K, 5.5K | l |
| Dog | 22K | 11K | m |
| Dog | 22K | 11K | n |
| Dog | 25K | 9K | o |
| Dog | 27K | 11K | p |
| Dog | 25K | 5K | q |
| Dog | 26K | 6K | r |

ᵃ P. J. LaRaia and E. Morkin, *Circ. Res.* **35,** 298 (1974).

ᵇ M. Tada, M. A. Kirchberger, and A. M. Katz, *J. Biol. Chem.* **250,** 2640 (1975).

ᶜ M. A. Kirchberger and M. Tada, *J. Biol. Chem.* **251,** 725 (1976).

ᵈ A. Schwartz, M. L. Entman, K. Kaniike, L. K. Lane, W. B. Van Winkle, and E. P. Bornet, *Biochim. Biophys. Acta* **426,** 57 (1976).

ᵉ H. L. Wray and R. R. Gray, *Biochim. Biophys. Acta* **461,** 441 (1977).

ᶠ H. Will, T. S. Levchenko, D. O. Levitsky, V. N. Smirnov, and A. Wollenberger, *Biochim. Biophys. Acta* **543,** 175 (1978).

ᵍ P. J. St. Louis and P. V. Sulakhe, *Arch. Biochem. Biophys.* **198,** 227 (1979).

ʰ L. R. Jones, H. R. Besch, Jr., J. W. Fleming, M. M. McConnaughey, and A. M. Watanabe, *J. Biol. Chem.* **254,** 530 (1979).

ⁱ J. M. J. Lamers and J. T. Stinis, *Biochim. Biophys. Acta* **624,** 443 (1980).

ʲ J. M. Bidlack and A. E. Shamoo, *Biochim. Biophys. Acta* **632,** 310 (1980).

ᵏ C. J. Le Peuch, D. A. M. Le Peuch, and J. G. Demaille, *Biochemistry* **19,** 3368 (1980).

ˡ M. A. Kirchberger and T. Antonetz, *Biochem. Biophys. Res. Commun.* **105,** 152 (1982).

ᵐ C. F. Louis, M. Maffitt, and B. Jarvis, *J. Biol. Chem.* **257,** 15182 (1982).

ⁿ M. Tada, M. Inui, M. Yamada, M. Kadoma, T. Kuzuya, H. Abe, and S. Kakiuchi, *J. Mol. Cell. Cardiol.* **15,** 335 (1983).

ᵒ A. D. Wegener and L. R. Jones, *J. Biol. Chem.* **259,** 1834 (1984).

lated. Phosphoprotein of lower molecular weight, appearing under specified conditions, is attributable to a putative monomer of phospholamban (see below).

## Properties of Phospholamban

In the membrane of cardiac SR, phospholamban serves as a substrate for cAMP-dependent protein kinase, and is considered to mediate the effect of cAMP that is augmented when myocardial cells are exposed to $\beta$-adrenergic agonists.[5] The content of phospholamban in cardiac SR is estimated to be about 4 or 6% of the total protein,[5] while that of the ATPase enzyme is up to 40%.[5] A 1:1 stoichiometry between phospholamban and the ATPase is suggested by the approximate 1:1 ratio between the amount of the two phosphoproteins formed within the membrane.[8,43] Phospholamban exhibits a proteolipid nature which is intimately associated with the SR membrane.[5] A portion of phospholamban is exposed at the outer surface of the SR vesicles, as evidenced by the observations that it can be phosphorylated by exogenous protein kinase, and that proteolytic digestion prevents its phosphorylation.[4] The failure to iodinate phospholamban probably reflects the location of the tyrosine residues within the membrane interior.[53]

Phospholamban was initially considered to represent a single polypeptide. However, the boiling of phosphorylated SR membranes in SDS results in a disappearance of the 22,000 MW phosphoprotein with concomitant appearance of the 11,000 and even lower, 5,000 to 7,000, phosphoproteins.[10,54] The possible minimum molecular weight is estimated to be approximately 5,500, based on the amino acid analysis.[12,55,56] The submolecular structure of phospholamban remains to be examined (see the section Structural Characteristics of Phospholamban below).

A phospholamban-like protein is reported to exist in sarcolemmal membranes of cardiac muscle. This protein (MW 23,000 daltons) termed calciductin,[57] also serves as a substrate for cAMP-dependent protein ki-

[53] C. F. Louis and A. M. Katz, Biochim. Biophys. Acta 494, 255 (1977).
[54] M. A. Kirchberger and T. Antonetz, Biochem. Biophys. Res. Commun. 105, 152 (1982).
[55] J. M. Bidlack, I. S. Ambudkar, and A. E. Shamoo, J. Biol. Chem. 257, 4501 (1982).
[56] M. A. Kirchberger and T. Antonetz, J. Biol. Chem. 257, 5685 (1982).
[57] M. L. Rinaldi, C. J. Le Peuch, and J. G. Demaille, FEBS Lett. 129, 277 (1981).

[p] M. Inui, M. Kadoma, and M. Tada, J. Biol. Chem. 260, 3708 (1985).
[q] L. R. Jones, H. K. B. Simmerman, W. W. Wilson, F. R. N. Gurd, and A. D. Wegener, J. Biol. Chem. 260, 7721 (1985).
[r] J. Fujii, M. Kadoma, M. Tada, H. Toda, and F. Sakiyama, Biochem. Biophys. Res. Commun. 138, 1044 (1986).

nase and is assumed to operate as a regulator of Ca channels in sarcolemma. Examining these proteins in SR and sarcolemma, Manalan and Jones[58] suggest that phospholamban primarily residing in SR may be transposed to sarcolemmal membranes. It remains to be established whether these two proteins are identical or not. It is also intriguing to examine whether phospholamban is widely present in membranes of other tissues.

### Cyclic AMP-Dependent Stimulation of Ca$^{2+}$-Pump

*Assay Medium*

*Preincubation (Phosphorylation Medium)*

Cardiac SR, 8 mg/ml
ATP, 1 m$M$
cAMP, 1 $\mu M$
cAMP-dependent protein kinase, 4 mg/ml
MgCl$_2$, 1 m$M$
KCl, 100 m$M$
Histidine-HCl, 40 m$M$, pH 6.8

*Incubation Medium for Assaying Ca$^{2+}$ Transport and*
   *Ca$^{2+}$-Dependent ATPase*

Vesicles of phosphorylated and control SR are incubated under conditions which are appropriate for steady- and presteady-state measurements of Ca$^{2+}$ transport and Ca$^{2+}$-dependent ATPase. The assay conditions are virtually similar to those described under Ca$^{2+}$ transport and Ca$^{2+}$-dependent ATPase, so that each reaction condition should be referred back to each section.

*Reagents*

cAMP-dependent protein kinase: prepared by the method of Miyamoto *et al.*[50] with slight modifications. In some experiments the catalytic subunit of cAMP-dependent protein kinase (100 $\mu$g/ml) can be substituted for the holoenzyme with no addition of cAMP.

*Procedures*

*Phosphorylation of SR Vesicles.* This process is performed as preincubation prior to assay for kinetics of Ca$^{2+}$ transport and ATPase. Cardiac

---

[58] A. S. Manalan and L. R. Jones, *J. Biol. Chem.* **257**, 10052 (1982).

SR is incubated in the phosphorylation medium with or without cAMP-dependent protein kinase and 1 $\mu M$ cAMP in a total volume of 3–5 ml. After 2 to 5 min of incubation at 25°, the mixture is cooled and applied to a column of Sephadex G-50 (1.5 × 30 cm) preequilibrated at 0–4° with 20 m$M$ Tris-maleate buffer (pH 6.8). Vesicles of phosphorylated and un-phosphorylated (control) SR are eluted by the same buffer in a void vol-ume, which is devoid of any kind of solutes; unreacted ATP and products ADP and P$_i$, which would otherwise interfere with kinetic properties of ATPase, are also removed from SR under these conditions.

The fractions containing SR vesicles (5 to 8 ml) after Sephadex G-50 chromatography are immediately assayed for Ca$^{2+}$ transport and ATPase, with an aliquot of SR solution being saved for determination of protein concentration. While SR preincubated without protein kinase serves as a control, SR preincubated with 2 m$M$ EDTA instead of MgCl$_2$ in the pres-ence of protein kinase could also serve as a control. When examined by employing [$\gamma$-$^{32}$P]ATP, the phosphoproteins formed under these condi-tions remain virtually stable throughout the procedures. Thus, the amount of phospholamban phosphorylation (0.6–1.3 $\mu$mol of phosphate/g of pro-tein) does not alter throughout, and for several hours after, column chro-matography on Sephadex G-50. Control (unphosphorylated) SR exhibits virtually no phosphorylation (<0.02 $\mu$mol of phosphate/g) under these conditions. Ca$^{2+}$-dependent ATPase and Ca$^{2+}$ transport by SR vesicles thus treated do not undergo significant alterations.

*Assay for Ca$^{2+}$ Transport and Ca$^{2+}$-Dependent ATPase.* Cardiac SR vesicles, preincubated and recollected under the above conditions, are subjected to determination of Ca$^{2+}$ transport and Ca$^{2+}$-dependent ATPase at steady and presteady conditions.

Figure 5 illustrates the relationship between phospholamban phospho-

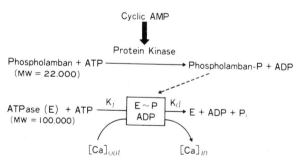

FIG. 5. Control of Ca$^{2+}$-pump ATPase of cardiac SR by cAMP-dependent phosphoryl-ation of phospholamban. [Reproduced with permission from Tada and Katz, *Annu. Rev. Physiol.* **44,** 401 (1982).]

rylation and the $Ca^{2+}$-pump ATPase system. It is important to note that phosphorylation of phospholamban regulates the elementary steps of $Ca^{2+}$-pump ATPase (see below). Under these conditions, the effect of phospholamban phosphorylation can be seen when control and phosphorylated SR are subjected to the following assay. In steady-state assay systems, the rate of $Ca^{2+}$ transport is determined by measuring the amount of $^{45}Ca$ transported into the vesicles. Similarly, the steady-state rate of ATP hydrolysis can be determined by measuring the rates of liberation of $P_i$ and ADP, while the steady-state level of EP is also determined. In presteady-state assay, the rates of formation ($k_f$) and decomposition ($k_d$) of EP, as well as transient EP levels, are determined by employing the rapid-quenching device. The presteady-state kinetics of $Ca^{2+}$ transport can also be determined by the same procedures.

Figure 6 shows effects of phospholamban phosphorylation on the steady-state kinetics of EP and $P_i$. The levels of EP and the rate of $P_i$ liberation at the steady state are determined as a function of $Ca^{2+}$ concentrations. Compared with control, phosphorylated SR exhibits a higher rate of $P_i$ liberation, whereas EP levels are significantly decreased in phosphorylated SR. It is also intriguing that $v/[EP]$, which practically

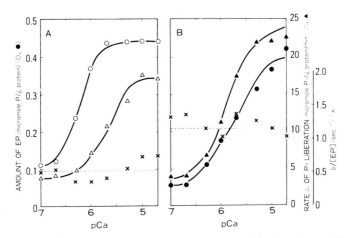

FIG. 6. Effect of phospholamban phosphorylation on the $Ca^{2+}$-dependent profiles of EP formation, the rate of $P_i$ liberation, and $v/[EP]$. Cardiac SR (2 mg/ml) is preincubated in phosphorylation medium containing ATP (1 m$M$) and cAMP-dependent protein kinase (1 mg/ml) in the presence of 2 m$M$ EDTA (control SR, panel A) and 2 m$M$ MgCl$_2$ (phosphorylated SR, panel B). The amounts of EP formed and $P_i$ liberated are determined in 0.14 mg/ml SR, 100 $\mu M$ [$\gamma$-³²P]ATP (50 $\mu$Ci/$\mu$mol), and different concentrations of free $Ca^{2+}$ under standard conditions. The ratio $v/[EP]$ at each $Ca^{2+}$ concentration is determined by dividing the value of $v$ (the rate of $P_i$ liberation) by [EP] (concentration of EP). [Reproduced with permission from Tada *et al.*, *J. Biol. Chem.* **254**, 319 (1979).]

represents the rate of EP decomposition,[2] is independent of pCa and is significantly enhanced by phosphorylation of phospholamban. These results are indicative of the possibility that the overall rate of turnover in $Ca^{2+}$-dependent enzyme is markedly enhanced through the enhancement of EP decomposition. In accord with these findings, the value of $k_d$, determined by measuring the rate of decay in EP, indicates that the rate of EP decomposition is enhanced by phospholamban phosphorylation.

Figure 7 demonstrates the transient time courses of EP formation in control and phosphorylated SR. Since the initial rate of EP formation is different when the state of the enzyme is different, the assays are performed both in $Ca^{2+}$-bound SR (state $E_1$) and in $Ca^{2+}$-free SR (state $E_2$). It is evident that the phospholamban phosphorylation in $Ca^{2+}$-bound SR does not significantly alter the initial rate of EP formation, estimated by determining the half-maximal activation $(t_{1/2})$ of EP formation. In contrast, phospholamban phosphorylation in $Ca^{2+}$-free SR profoundly enhances the initial rate of EP formation, in that $t_{1/2}$ in control SR (43 msec)

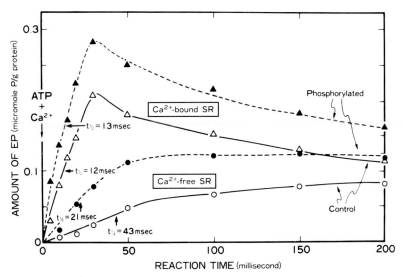

FIG. 7. Effects of phospholamban phosphorylation on the transient time courses of EP formation in $Ca^{2+}$-bound and $Ca^{2+}$-free SR. Cardiac SR (7.8 mg/ml) is preincubated in phosphorylation medium containing 1 m$M$ ATP in the absence (control SR) and presence (phosphorylated SR) of cAMP-dependent protein kinase (3.9 mg/ml). The amount of EP is determined in 0.41 mg/ml SR, 10 $\mu M$ [$\gamma$-$^{32}$P]ATP (50 $\mu$Ci/$\mu$mol), and 9.7 $\mu M$ $Ca^{2+}$ (100 $\mu M$ $CaCl_2$ and 100 $\mu M$ EGTA) under standard conditions. The half-maximal activation $(t_{1/2})$, as an index of the initial rate of EP formation, represents the time in milliseconds (msec), at which half of the maximal EP is formed. [Reproduced with permission from Tada $et$ $al.$, J. Biol. Chem. **255**, 1985 (1980).]

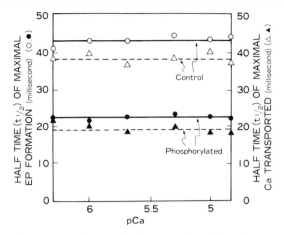

FIG. 8. The enhancement of the half-time ($t_{1/2}$) of EP formation and Ca$^{2+}$ transport induced by phospholamban phosphorylation at different pCa. [Reproduced with permission from Tada *et al.*, *Mol. Cell. Biochem.* **46**, 73 (1982).]

is markedly lowered (21 msec) in phosphorylated SR. These results are indicative of the possibility that phosphorylation of phospholamban stimulates the rate of conversion from $E_2$ to $E_1$, but not the rate of conversion from $E_1$ to $E_1P$. If the rate of conversion from $E_2$ to $E_1$ is regulated by phosphorylation of phospholamban, such an effect would be seen under a variety of conditions. We determined the value of $t_{1/2}$ of EP formation and Ca$^{2+}$ transport in phosphorylated and control SR. Interestingly, the value of $t_{1/2}$ is independent of pCa both in EP formation and Ca$^{2+}$ transport (Fig. 8). Phosphorylation of phospholamban is found to reduce $t_{1/2}$ to approximately half its control value. Thus, the average values of $t_{1/2}$ for EP formation, determined at six different Ca$^{2+}$ concentrations, are 21.7 msec for phosphorylated SR and 42.7 msec for control SR. These results are consistent with the view that phosphorylation of phospholamban augments the rates of EP formation and Ca$^{2+}$ transport by enhancing the rate of conversion from $E_2$ to $E_1$.

*Cyclic AMP Regulation of Active Ca$^{2+}$ Transport and Ca$^{2+}$-Dependent ATPase.* Active Ca$^{2+}$ transport by cardiac SR is markedly stimulated when phospholamban is previously phosphorylated by incubation of cardiac SR with cAMP-dependent protein kinase.[6,7,59] Under these conditions, the rate of oxalate-facilitated Ca$^{2+}$ uptake is more than doubled. Such an effect produced by phospholamban phosphorylation is considered to be derived from an increase in apparent affinity of the Ca$^{2+}$ transport system for Ca$^{2+}$. Thus, the rate of Ca$^{2+}$ uptake by phosphorylated SR

[59] A. M. Katz, M. Tada, and M. A. Kirchberger, *Adv. Cyclic Nucleotide Res.* **5**, 453 (1975).

TABLE III
ENZYMATIC PARAMETERS OF Ca$^{2+}$-DEPENDENT ATPase OF UNPHOSPHORYLATED
(CONTROL) AND PHOSPHORYLATED CARDIAC SR[a]

| Cardiac SR | Maximal velocity, $V_{max}$ (nmol P$_i$/mg/min) | EP decomposition | | EP formation, $t_{1/2}$[d] (msec) |
| | | $v$/[EP][b] (sec$^{-1}$) | $k_d$[c] (sec$^{-1}$) | |
| --- | --- | --- | --- | --- |
| Control | 26.3 | 0.54 | 0.55 | 42.7 |
| Phosphorylated | 54.3 | 1.14 | 1.03 | 21.7 |

[a] Modified from M. Tada, F. Ohmori, M. Yamada, and H. Abe, *J. Biol. Chem.* **254,** 319 (1979).
[b] The rate of ATP hydrolysis per unit of concentration of EP at steady state.
[c] Estimated from the rate of decay in the amount of EP after EP formation is terminated by excess EGTA.
[d] Time at which a half-maximal EP is attained when Ca$^{2+}$ is added to Ca$^{2+}$-free SR.

exhibits a half-maximal activation at 0.3 $\mu M$ Ca$^{2+}$, whereas half-maximal activation by unphosphorylated SR occurs at 1 $\mu M$ Ca$^{2+}$.[7] The stimulatory effect of phospholamban is abolished when phosphorylated phospholamban is dephosphorylated by protein phosphatase[60,61] or when phospholamban phosphorylation is inhibited by protein kinase inhibitor.[62,63]

The observed stimulation of Ca$^{2+}$ uptake is accompanied by the augmentation of ATP hydrolysis.[7] Under these conditions, the coupling stoichiometry of two is maintained between Ca$^{2+}$ taken up and ATP hydrolyzed,[7] indicating that the turnover rate of the normally coupled Ca$^{2+}$-pump is enhanced by phospholamban phosphorylation. The maximal velocity of ATP hydrolysis is markedly enhanced by phospholamban phosphorylation[8] (Table III), with no apparent alteration in $K_m$ for ATP.

The kinetic properties of Ca$^{2+}$-pump ATPase are shown to undergo a profound enhancement when phospholamban is phosphorylated by cAMP-dependent protein kinase. Since such an enhancement takes place without altering the level of EP,[8] it is feasible to assume that the enhanced $V_{max}$ of ATPase activity results from the augmented rate of EP turnover.[8,64] In fact, phospholamban phosphorylation results in a twofold increase in the ratio of $v$/[EP] (Fig. 6) and the rate of EP decay ($k_d$) (Table

[60] M. Tada, M. A. Kirchberger, and H.-C. Li, *J. Cyclic Nucleotide Res.* **1,** 329 (1975).
[61] M. A. Kirchberger and A. Raffo, *J. Cyclic Nucleotide Res.* **3,** 45 (1977).
[62] M. Tada, F. Ohmori, Y. Nimura, and H. Abe, *J. Biochem. (Tokyo)* **82,** 885 (1977).
[63] F. Ohmori, M. Tada, N. Kinoshita, H. Matsuo, H. Sakakibara, Y. Nimura, and H. Abe, in "Recent Advances in Studies on Cardiac Structure and Metabolism" (T. Kobayashi *et al.,* eds.), Vol. 11, p. 279. Univ. Park Press, Baltimore, Maryland, 1978.
[64] M. Tada, F. Ohmori, N. Kinoshita, and H. Abe, *Adv. Cyclic Nucleotide Res.* **9,** 355 (1978).

III), indicating that the decomposition of EP is enhanced by phospholamban phosphorylation. Circumferential evidence indicates that the rate at which $E_1P$ is converted to $E_2P$ [step iv, Eq. (2)] is probably accelerated by phospholamban phosphorylation,[11] the rate-determining step during EP decomposition.

It is more important to note that the rate of EP formation is also enhanced when phospholamban is phosphorylated. Employing transient kinetic analyses of EP formation, we demonstrate that the rate of conversion from $E_2$ to $E_1$ is markedly enhanced by phospholamban phosphorylation.[9] Also found is the fact that the rate of the reversed conversion ($E_1$ to $E_2$) is probably enhanced by phospholamban phosphorylation,[11] indicating that the dynamic equilibrium between $E_1$ and $E_2$ is greatly augmented when phospholamban is phosphorylated.

It is significant to point out that phospholamban appears to regulate both of the two rate-determining steps in the ATPase reaction, i.e., steps i and iv [Eq. (2)]. These are considered to represent the most important steps where conformational change of ATPase takes place. These steps are also characteristic in terms of enzyme kinetics, in which the affinities of the enzyme for the divalent cations Ca$^{2+}$ and Mg$^{2+}$ are greatly altered. Taking these into consideration, we contend that phospholamban exhibits a direct protein–protein interaction with the ATPase enzyme.[5] However, we should gain more insights into the molecular interactions between phospholamban and ATPase before such a molecular model is formulated.

### Calmodulin-Dependent Stimulation of Ca$^{2+}$-Pump

*Phosphorylation of Phospholamban by Calmodulin-Dependent Protein Kinase*

*Assay Medium*

Cardiac SR, 1–2 mg/ml
[γ-$^{32}$P]ATP, 5 mM (10–50 μCi/μmol)
Calmodulin, 0.2 μM
MgCl$_2$, 5 mM
KCl, 100 mM
Histidine-HCl, 40 mM, pH 6.8
Ca–EGTA buffers, containing 200 μM CaCl$_2$ and various concentrations (2.2 mM–50 μM) of EGTA, giving free Ca$^{2+}$ of 0.1–50 μM

*Procedures.* Calmodulin-dependent phosphorylation of phospholamban is assayed at 25° for 10 min in 0.1 ml medium. The reaction is initiated by the addition of [$\gamma$-$^{32}$P]ATP and terminated by adding 0.5 ml of 10% ice-cold TCA, followed by the addition of 1.26 mg BSA. The amount of phospholamban phosphorylation is determined either by the dilution–precipitation method (method A) or by SDS-polyacrylamide gel electrophoresis (method B). SR phosphorylation catalyzed by cAMP-dependent protein kinase is determined, under the above conditions, by incubating SR with the catalytic subunit of cAMP-dependent protein kinase (50 $\mu$g/ml). In phosphorylation catalyzed by calmodulin- and cAMP-dependent protein kinases, both the catalytic subunit of protein kinase and calmodulin are present.

## Calmodulin-Dependent Stimulation of Ca$^{2+}$ Uptake and Ca$^{2+}$-Dependent ATPase

*Phosphorylation Medium.* Prior to assay for Ca$^{2+}$ transport and ATPase, cardiac SR (2 mg/ml) is incubated at 25° in the above-mentioned medium (3 ml) in the presence of nonlabeled ATP and 2 $\mu$M Ca$^{2+}$ (200 $\mu$M CaCl$_2$ and 290 $\mu$M EGTA). Control medium contains no calmodulin and no cAMP-dependent protein kinase. After 10 min of incubation, the mixture is cooled on ice and immediately submitted to a column of Sephadex G-50 (1.5 × 30 cm) preequilibrated at 0–4° with 20 m$M$ Tris-maleate buffer (pH 6.8). The fraction containing phosphorylated SR is collected. After protein concentration is determined, an aliquot of this solution is subjected to the following assay.

### Incubation Medium for Assaying Ca$^{2+}$ Uptake and ATPase

Cardiac SR, 0.2 mg/ml: previously phosphorylated or unphosphorylated (control) in phosphorylation medium
ATP, 100 $\mu$M
MgCl$_2$, 1 m$M$
KCl, 100 m$M$
Tris-oxalate, 2.5 m$M$
Tris-maleate, 20 m$M$, pH 6.8
$^{45}$Ca–EGTA buffer, containing 500 $\mu$M $^{45}$CaCl$_2$ (10 $\mu$Ci/$\mu$mol) and various concentrations (5.5 m$M$–50.3 $\mu$M) of EGTA, giving free Ca$^{2+}$ of 0.1–20 $\mu$M
Phosphoenolpyruvate (PEP), 2.5 m$M$
Pyruvate kinase, 20 IU/ml
*Procedures.* The rate of oxalate-facilitated Ca$^{2+}$ uptake is determined by initiating the reaction by the addition of $^{45}$Ca–EGTA at 25°. At various time intervals after the start of the reaction (0.5 to 10 min), a 1-ml aliquot is taken and applied to the Millipore filter. The amount of $^{45}$Ca on the filter

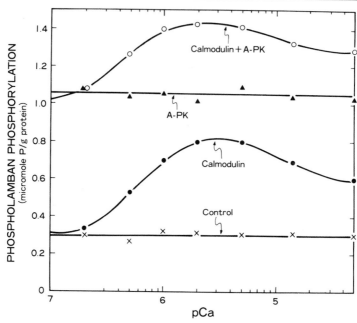

FIG. 9. Dual phosphorylation of phospholamban catalyzed by cAMP- and calmodulin-dependent protein kinases. Cardiac SR is phosphorylated by cAMP-dependent protein kinase (A-PK), calmodulin-dependent protein kinase, and both of these kinases in the presence of different Ca²⁺ concentrations. [Reproduced with permission from Tada *et al., J. Mol. Cell. Cardiol.* **15**, 335 (1983).]

is counted for radioactivity as described above (see the section Assay for Oxalate-Facilitated Ca²⁺ Uptake).

The rate of ATP hydrolysis is determined by measuring the amount of ADP liberated in the above reaction medium, in which ⁴⁵Ca is substituted by unlabeled Ca. The PEP–pyruvate kinase system included in the assay serves to maintain ATP concentration. The appropriate coupling of the ATP-regenerating system and ATPase activity produces pyruvate, the formation of which is stoichiometrically coupled with ADP formed. The amount of pyruvate liberated is determined by the colorimetric method of Reynaud *et al.*[65]

## Dual Phosphorylation of Phospholamban by Two Protein Kinases

Figure 9 shows that phospholamban in cardiac SR is phosphorylated by calmodulin-dependent protein kinase in a Ca²⁺-dependent manner.

---

[65] A. M. Reynaud, L. F. Hass, D. D. Jacobsen, and P. D. Boyer, *J. Biol. Chem.* **236**, 2277 (1961).

This phosphorylation is found to occur on top of cAMP-dependent phosphorylation, suggesting that the phosphorylation sites for the two kinases are different.[10,43,66] It is intriguing to note that calmodulin-dependent protein kinase appears to exist within SR membrane,[43,67] in contrast to cAMP-dependent protein kinase which is readily removed from SR. Stability characteristics of phosphoprotein formed by calmodulin indicate the occurrence of a phosphoester. A high-voltage electrophoretogram of phosphorylated SR indicates that this phosphorylation occurs at serine residues.[12,68] The extent of calmodulin-dependent phosphorylation (0.95 ± 0.23 nmol/mg SR protein) is almost equal to that of cAMP-dependent protein kinase (0.99 ± 0.12 nmol/mg SR protein).[43] Phosphorylation sites catalyzed by both kinases are found to exist, at least partly, in a common tryptic peptide in phospholamban.[69]

The exact number and location of phosphorylation sites in phospholamban, catalyzed by cAMP- and calmodulin-dependent protein kinases, remain to be examined. Phospholamban is reported to be phosphorylated by the third type of protein kinase (protein kinase C), resulting in phosphorylation of the phosphoester type with the amounts almost comparable to those catalyzed by the other two kinases.[69] Interestingly, part of such phosphorylation occurs within a tryptic fragment common to cAMP and calmodulin, although the three protein kinases differ in their selectivity for sites of phosphorylation.

Phosphorylation of phospholamban by calmodulin-dependent protein kinase results in significant stimulation of $Ca^{2+}$ uptake by cardiac SR.[43,70,71] The $Ca^{2+}$-dependent profile of the rate of $Ca^{2+}$ uptake indicates that the calmodulin-dependent stimulation is almost superimposable with cAMP-dependent stimulation.[43] The stimulatory effects induced by the two kinases are found to occur in an additive manner,[43] in accord with the additive phosphorylation by the two kinases.

The observed stimulation of $Ca^{2+}$ uptake by calmodulin-dependent phosphorylation accompanies the stimulation of $Ca^{2+}$-dependent ATPase activity within a range of $Ca^{2+}$ between 0.1 and 10 $\mu M$, indicating that

[66] L. M. Bilezikjian, E. G. Kranias, J. D. Potter, and A. Schwartz, *Circ. Res.* **49,** 1356 (1981).

[67] A. D. Wegener and L. R. Jones, *J. Biol. Chem.* **259,** 1834 (1984).

[68] A report suggests that a threonine residue in phospholamban is phosphorylated by calmodulin-dependent protein kinase (see Ref. 87), while the chemical nature of residues phosphorylated by protein kinase C is not well characterized (see Ref. 69).

[69] M. A. Movsesian, M. Nishikawa, and R. S. Adelstein, *J. Biol. Chem.* **259,** 8029 (1984).

[70] G. Lopaschuk, B. Richter, and S. Katz, *Biochemistry* **19,** 5603 (1980).

[71] B. A. Davis, A. Schwartz, F. J. Samaha, and E. G. Kranias, *J. Biol. Chem.* **258,** 13587 (1983).

Ca$^{2+}$-pump ATPase is directly activated by calmodulin-dependent phosphorylation of phospholamban. Under these conditions, the apparent affinity of ATPase for Ca$^{2+}$ is greatly enhanced (Table IV). When both types of phosphorylation are operational, the ATPase activity is further augmented, resulting in a further increase in the affinity of ATPase for Ca$^{2+}$ (Table IV). These findings indicate that cAMP- and calmodulin-mediated enhancement of Ca$^{2+}$ transport and Ca$^{2+}$ ATPase activity occurs in independent and additive manners.

This is in sharp contrast to the findings by Le Peuch et al.,[10] who suggest that cAMP-dependent stimulation of Ca$^{2+}$ uptake is not seen, unless there is the prior phosphorylation of phospholamban by calmodulin-dependent protein kinase. Such a misleading conclusion is derived from the ATPase assay, which is determined in a system where Ca$^{2+}$ concentration is extraordinarily high (100 $\mu M$). At 100 $\mu M$ Ca$^{2+}$, the activity of Ca$^{2+}$-pump ATPase exhibits its maximal performance, thus masking the existence of any possible influence produced by protein kinase.

## Purification of Phospholamban

Several attempts at purifying phospholamban are reported. Le Peuch et al.[12] and Kirchberger and Antonetz[56] purify phospholamban in the phosphorylated state using preparative SDS-gel electrophoresis. In these preparations, phospholamban is not fully phosphorylated by cAMP-de-

TABLE IV
EFFECTS OF cAMP- AND CALMODULIN-DEPENDENT
PHOSPHORYLATION OF PHOSPHOLAMBAN ON Ca²⁺
CONCENTRATION TO ATTAIN HALF-MAXIMAL
ATPase ACTIVITY[a]

| Conditions | $K_{Ca}{}^{b}$ ($\mu M$) |
|---|---|
| Control | $0.92 \pm 0.03$ |
| Phosphorylated | |
|   cAMP-dependent protein kinase | $0.66 \pm 0.05$ |
|   Calmodulin-dependent protein kinase | $0.64 \pm 0.03$ |
|   Both protein kinases | $0.48 \pm 0.02$ |

[a] From M. Tada, M. Inui, M. Yamada, M. Kadoma, T. Kuzuya, H. Abe, and S. Kakiuchi, *J. Mol. Cell. Cardiol.* **15**, 335 (1983).

[b] Ca$^{2+}$ concentration that attains the half-maximal ATPase activity. Averages (mean ± SE) of determination of three different preparations are shown.

pendent protein kinase after the preparation is enzymatically dephosphorylated. The phosphorylated form of phospholamban is also purified by chromatography in organic solvents,[72,73] employing the method to extract lipids and lipoproteins.[74] Although physicochemical characteristics of these preparations are documented, they fail to report that these preparations retain properties to be catalyzed by cAMP-dependent protein kinase. A harsh treatment, such as the extraction with organic solvents or SDS employed in these reports, may alter the inherent properties of phospholamban. Phospholamban is reported to be purified in the unphosphorylated state by gel filtration in the presence of an extremely low concentration of deoxycholate (0.04% or 0.29 $\mu$g/mg of SR protein).[55] However, a much higher concentration of deoxycholate is required to extract phospholamban (see below).

We recently developed procedures for purification of phospholamban from cardiac SR that enable us to obtain a highly purified and phosphorylatable preparation.[75] Employing fractionation by gel-permeation chromatography in the presence of nonionic detergent octaethylene glycol $n$-dodecyl ether (C$_{12}$E$_8$), phospholamban is purified to near homogeneity. The method for this procedure is detailed below.

*Materials and Solutions*

*Buffers for Purifying Phospholamban*

Buffer A: 0.6 $M$ KCl, 0.3 $M$ sucrose, 20 m$M$ Tris-HCl, pH 7.5, 1 m$M$ dithiothreitol

Buffer B: 2 $M$ KI, 0.2% C$_{12}$E$_8$, 20 m$M$ Tris-HCl, pH 7.5, 5 m$M$ 2-mercaptoethanol

Buffer C: 1 $M$ KI, 0.1% C$_{12}$E$_8$, 20 m$M$ Tris-HCl, pH 7.5, 5 m$M$ 2-mercaptoethanol

Buffer D: 50 m$M$ KCl, 0.1% C$_{12}$E$_8$, 20 m$M$ Tris-HCl, pH 7.5, 5 m$M$ 2-mercaptoethanol

Buffer E: 50 m$M$ KCl, 20 m$M$ Tris-HCl, pH 7.5

Stock C$_{12}$E$_8$ solution (20%, w/v): 25 g of C$_{12}$E$_8$ obtained from Nikko Chemicals, Tokyo, is liquefied by warming (about 40°) and is diluted into warmed H$_2$O to get a final volume of 125 ml stock solution

[72] J.-P. Capony, M. L. Rinaldi, F. Guilleux, and J. G. Demaille, *Biochim. Biophys. Acta* **728,** 83 (1983).
[73] J. H. Collins, E. G. Kranias, A. S. Reeves, L. M. Bilezikjian, and A. Schwartz, *Biochem. Biophys. Res. Commun.* **99,** 796 (1981).
[74] J. Folch and M. Lees, *J. Biol. Chem.* **191,** 807 (1951).
[75] M. Inui, M. Kadoma, and M. Tada, *J. Biol. Chem.* **260,** 3708 (1985).

*Column Chromatography*

TSK-G3000SW column (60 cm × 21.5 mm)
CM-Sepharose CL-6B (2 ml resin bed)

*Medium for Phosphorylating Phospholamban*

SR proteins or purified phospholamban, 0.1–1 mg/ml
[$\gamma$-$^{32}$P]ATP (or unlabeled ATP), 0.1 m$M$ (10–50 $\mu$Ci/$\mu$mol)
Catalytic subunit of cAMP-dependent protein kinase, 10 $\mu$g/ml
MgCl$_2$, 1 m$M$
KCl, 100 m$M$
EGTA, 0.1 m$M$
Tris-HCl, 20 m$M$, pH 7.5

## Methods

*Detection of Phospholamban during Purification Steps.* SR proteins are analyzed by SDS-polyacrylamide gel electrophoresis using the buffer system of Laemmli[52] in 12.5 or 15% polyacrylamide gel and gradient gels of 10–25% polyacrylamide, employing a minislab electrophoresis apparatus (gel size, 50 mm in length and 1 mm in thickness). SDS-polyacrylamide gel electrophoresis is also carried out according to Weber and Osborn[51] on 15% polyacrylamide gels. Proteins are stained by Coomassie Brilliant Blue. The protein fractions, phosphorylated by the catalytic subunit of cAMP-dependent protein kinase under standard conditions, are subjected to electrophoresis on SDS-polyacrylamide. While the quantitative determination of protein at each purification step is made by the method of Lowry *et al.*,[76] the dye absorption method of Schaffner and Weissmann[77] is also employed.

To detect $^{32}$P-bound protein, the reaction (0.2 ml) is started at 0° by adding ATP and is terminated after 10 min by adding 2 ml of ice-cold TCA containing 0.1 m$M$ P$_i$. After Millipore filtration and washing by 25 ml of TCA solution, the filter is air-dried and counted for radioactivity. An aliquot of the phosphorylation medium is also subjected to electrophoresis followed by $^{32}$P autoradiography or protein staining.

*Procedures for Purification of Phospholamban*

*Step 1: Solubilization and Fractionation of Cardiac SR.* Cardiac SR (100 mg) at a concentration of 10 mg/ml in buffer A is solubilized with

---

[76] O. H. Lowry, N. J. Rosebrough, A. L. Farr, and R. J. Randall, *J. Biol. Chem.* **193**, 265 (1951).

[77] W. Schaffner and C. Weissmann, *Anal. Biochem.* **56**, 502 (1973).

deoxycholate (DOC) (1 mg/mg of protein) at 4° for 20 min. Insoluble materials are removed by centrifugation at 150,000 $g$ for 20 min. The solubilized proteins are fractionated with ammonium sulfate to the final concentration of 25% saturation. The slurry is stirred at 4° for 20 min and centrifuged at 150,000 $g$ for 20 min. The resulting pellet is dissolved in 1 ml of buffer B.

*Step 2: Gel-Permeation High-Performance Liquid Chromatography.* The sample dissolved in buffer B is applied to high-performance liquid chromatography on a TSK-G3000SW column, equilibrated with buffer C. The sample is eluted with buffer C at a flow rate of 3 ml/min. Monitoring at 280 nm, 3-ml fractions are collected. The phospholamban-enriched fractions (about 20 ml), determined by SDS-polyacrylamide gel electrophoresis are pooled and dialyzed overnight against 4 liters of buffer D at 4°.

*Step 3: Ion-Exchange Chromatography (CM-Sepharose CL-6B).* The dialyzed sample is applied to a CM-Sepharose CL-6B preequilibrated with buffer D. After washing with 10 bed volumes of buffer D, proteins

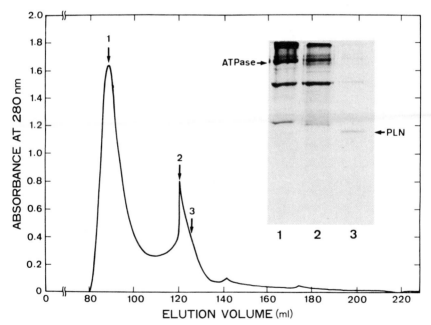

FIG. 10. Resolution of phospholamban from other SR components by gel-permeation chromatography in C$_{12}$E$_8$ and KI. The fractions indicated by the arrows are analyzed with 12.5% Laemmli gel electrophoresis (inset: lanes 1–3). ATPase, SR Ca$^{2+}$-dependent ATPase; PLN, phospholamban. [Reproduced with permission from Inui *et al., J. Biol. Chem.* **260,** 3708 (1985).]

are eluted with a linear gradient of 0.05 to 1 $M$ KCl in buffer D; phospholamban is eluted at approximately 0.4 $M$. Fractions containing phospholamban are combined and dialyzed against 2 liters of buffer E. The dialyzed samples, containing purified phospholamban, are condensed up to 100 $\mu$g/ml by the membrane filtration system, and stored frozen at −70° in buffer E.

Figure 10 shows the protein elution profile from a gel-permeation column of high-performance liquid chromatography (Step 2) in the presence of $C_{12}E_8$ and KI, when the ammonium sulfate pellet dissolved in $C_{12}E_8$ and KI is submitted to the column. Phospholamban is well resolved from other SR components. $Ca^{2+}$-dependent ATPase is eluted in fractions from the void volume peak to the second peak. Phospholamban is obtained in fractions at the shoulder of the second peak. The inclusion of a high concentration (1 $M$) of chaotropic agent KI is required, since the substitution of KI by KCl gives poor resolution of phospholamban from SR components.

The selected fractions, combined and dialyzed, are then applied to a CM-Sepharose CL-6B column (Step 3). Elution with a linear KCl gradient

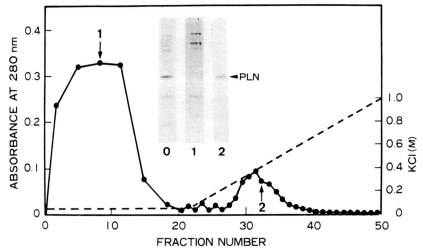

FIG. 11. Elution of phospholamban by KCl gradient in ion-exchange chromatography. The phospholamban-enriched fractions from Step 2 are applied to CM-Sepharose CL-6B column. After washing the column with buffer D, a linear KCl gradient is applied to dissociate phospholamban from the column. The fractions indicated by the arrows are analyzed by 12.5% Laemmli gel electrophoresis (inset, lanes 1 and 2); lane 0 represents a gel electrophoretogram of the phospholamban-enriched fraction, submitted to the ion-exchange chromatography column. [Reproduced with permission from Inui *et al.*, *J. Biol. Chem.* **260,** 3708 (1985).]

allows phospholamban to be fractionated at a KCl concentration of approximately 0.4 *M* (Fig. 11). An SDS-polyacrylamide gel electrophoretogram of this fraction shows that phospholamban is purified to near homogeneity (more than 98% pure) (Fig. 12). The yield of phospholamban and its phosphorylation sites at each step of purification are summarized in Table V. Approximately 0.18 mg of phospholamban is purified from 100 mg of canine cardiac SR. Amounts of cAMP-dependent phosphorylation are estimated by incubating with the [$\gamma$-$^{32}$P]ATP and the catalytic subunit of cAMP-dependent protein kinase. Purified phospholamban incorporates

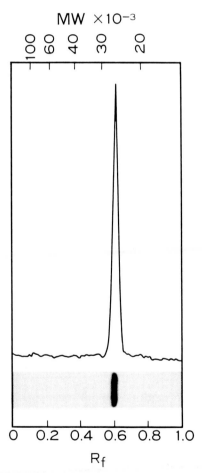

FIG. 12. SDS-polyacrylamide gel electrophoresis of purified phospholamban. Purified phospholamban, after condensation, is electrophoresed on 12.5% Laemmli polyacrylamide gel and scanned by densitometer. [Reproduced with permission from Inui *et al., J. Biol. Chem.* **260,** 3708 (1985).]

TABLE V
PURIFICATION OF PHOSPHOLAMBAN FROM CARDIAC SR[a]

| | Total protein (mg) | cAMP-dependent phosphorylation | | | |
| | | Protein $^{32}$P (nmol P/mg) | Total $^{32}$P (nmol P) | Recovery (%) | Purity (−fold) |
| Purification step | | | | | |
|---|---|---|---|---|---|
| Cardiac SR | 100 | 1.24 | 124 | 100 | 1 |
| Deoxycholate extract | 68.5 | 1.70 | 116 | 93.5 | 1.37 |
| Ammonium sulfate pellet | 27.5 | 2.29 | 63.0 | 50.8 | 1.85 |
| Gel-permeation chromotography (phospholamban-enriched fraction) | 2.05 | 20.7 | 42.4 | 34.2 | 16.7 |
| CM-Sepharose CL-6B chromotography (purified phospholamban) | 0.18 | 41.7 | 7.5 | 6.0 | 33.6 |

[a] From M. Inui, M. Kadoma, and M. Tada, *J. Biol. Chem.* **260,** 3708 (1985).

about 42 nmol of phosphate/mg protein, in contrast to the original SR vesicles which incorporate about 1.2 nmol of phosphate/mg of SR protein. These findings indicate a 34-fold purification with overall recovery of 6% from cardiac SR. When the protein concentration is determined by dye absorption method,[77] the phosphate incorporated in purified phospholamban amounts to approximately 138 nmol/mg or 3.73 mol/mol of the holoprotein (27,000 Da).

## Antiphospholamban Antibody and Immunochemical Studies

### Immunochemical Identification of Phospholamban

#### Reagents

Antiphospholamban antisera (rabbit): 10 $\mu$l antisera are dissolved in 1 ml of BSA buffer

FITC (fluorescein isothiocyanate)-conjugated antirabbit IgG (goat): 10 $\mu$l antisera is dissolved in 1 ml of BSA buffer

Buffer A: 0.9% NaCl, 10 m$M$ Tris-HCl, pH 7.5

BSA buffer: 3% bovine serum albumin in buffer A

*Procedures.* Immunization against purified phospholamban is performed by the method of Vaitukaitis.[78] About 10 $\mu$g of purified phospholamban (2 ml) is emulsified with an equal volume of Freund's complete

[78] J. L. Vaitukaitis, this series, Vol. 73, p. 46.

adjuvant. An aliquot (30 to 50 $\mu$l) of the emulsion is injected into rabbits at each of approximately 100 sites. Antisera are collected after 3 weeks and checked by the Ouchterlony double diffusion test. Immunoblotting procedures are carried out by the method of Towbin *et al.*[79] Protein samples (15 $\mu$g SR protein and 1 $\mu$g of purified phospholamban) are submitted to each gel lane for one-dimensional SDS-polyacrylamide slab gel electrophoresis by the method of Laemmli.[52] Western blotting is subsequently performed to transfer protein vertically from SDS-gel to nitrocellulose sheets. After nitrocellulose sheets are equilibrated with BSA buffer for 1 hr at room temperature, they are incubated with a solution containing antisera (10 $\mu$l/ml of BSA buffer) overnight. Resulting nitrocellulose sheets, washed with buffer A, are reacted with FITC-conjugated antirabbit IgG (10 $\mu$l/ml BSA buffer) for 3 hr at room temperature. The sheets are washed with buffer A and inspected under long-wave UV light.

*Immunostaining of Purified Phospholamban and Cardiac and Skeletal SR*

The obtained antisera are found to bind to a single band of 27,000 Da, corresponding to phospholamban, when tested against either cardiac SR or purified phospholamban (Fig. 13B, lanes 1 and 5). When cardiac SR and purified phospholamban are boiled in SDS prior to Laemmli slab gel electrophoresis, the antisera are specifically bound to the 11,000-Da band (Fig. 13B) in accord with the appearance of 11,000-Da protein band and the complete disappearance of the 27,000-Da protein band (Fig. 13A). No reactivity is detected in fast skeletal muscle SR by immunofluorescent staining of Western blots (Fig. 13B, lanes 3 and 4). Applying the immunofluorescent technique to microsomal fractions from other tissues, we have demonstrated that antiphospholamban antisera bind to a 27,000-Da protein in membranes of slow skeletal muscle and vascular and visceral smooth muscles.

Structural Characteristics of Phospholamban

The molecular weight of phospholamban is originally reported to be 22,000,[4] based on electrophoretic mobility of ³²P-labeled phospholamban on the Weber and Osborn gel system. A number of reports indicate that the molecular weight, determined by similar procedures, is in accord with the original report (Table II). However, higher values are obtained,[67,80] particularly when electrophoresis is performed in the Laemmli system.

[79] H. Towbin, T. Staehelin, and J. Gordon, *Proc. Natl. Acad. Sci. U.S.A.* **76**, 4350 (1979).
[80] Y. Iwasa and M. M. Hosey, *J. Biol. Chem.* **258**, 4571 (1983).

**A**          **B**

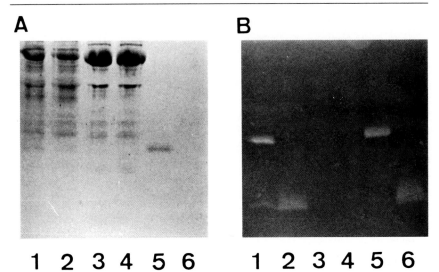

**1  2  3  4  5  6          1  2  3  4  5  6**

FIG. 13. Identification of phospholamban by immunofluorescent staining of Western blots in cardiac and fast skeletal muscle SR. (A) Cardiac (lanes 1 and 2) and fast skeletal muscle (lanes 3 and 4) SR and the purified phospholamban (lanes 5 and 6) are applied to 12.5% Laemmli gel electrophoresis. Prior to electrophoresis, samples are incubated at 25° for 10 min (lanes 1, 3, and 5) or boiled for 1 min (lanes 2, 4, and 6) in 2% SDS, 10% glycerol, 5% 2-mercaptoethanol, and 50 m$M$ Tris-HCl, pH 6.8. The gel is stained by Coomassie Blue. (B) Protein samples after electrophoresis are blotted to the nitrocellulose membrane and treated with antiphospholamban antisera, followed by staining with antirabbit IgG (goat) conjugated with fluorescein isothiocyanate. [Reproduced with permission from Inui *et al.*, *J. Biol. Chem.* **260**, 3708 (1985).]

Employing purified phospholamban in the unphosphorylated form, we have demonstrated that the apparent molecular weight of phospholamban varies with varying the gel system for electrophoresis.[75] In the Weber and Osborn neutral gel system, the molecular weight of phospholamban is 22,000 as originally reported, while it is 27,000 in the Laemmli alkaline gel system (Fig. 14). The similar shift of electrophoretic mobility on SDS gel is reported in another SR protein, calsequestrin, in which the molecular weight is 44,000 in neutral gel system while it is 55,000 in the alkaline system.[81]

The electrophoretic mobilities of phospholamban in SDS-polyacrylamide gel in both the Weber and Osborn and Laemmli systems significantly change under specific conditions, in that the boiling in 2% SDS prior to electrophoresis results in a total conversion into the lower-molecular-weight component (Fig. 14). Also observed is the phosphorylation-

---

[81] K. P. Campbell, D. H. MacLennan, A. O. Jorgensen, and M. C. Mintzer, *J. Biol. Chem.* **258**, 1197 (1983).

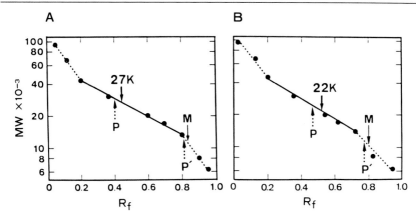

FIG. 14. Estimation of the molecular weight of phospholamban by 0.1% SDS–15% poly-acrylamide gel electrophoresis according to the methods of Laemmli (A) and Weber and Osborn (B). 27K and 22K represent the apparent molecular weight on the Laemmli (A) or Weber and Osborn (B) gel systems, respectively. M refers to the mobility of phospholamban on the gel when the sample is boiled for 1 min prior to electrophoresis, corresponding to a molecular weight of 11,000 in A and 10,500 in B. P and P′ show the mobilities of phospho-lamban phosphorylated by cAMP-dependent protein kinase. [Reproduced with permission from Inui et al., J. Biol. Chem. **260**, 3708 (1985).]

induced increase in the apparent molecular weight, which is seen in both nonboiled and boiled phospholamban (Fig. 14). Thus, in the Laemmli system (Fig. 15), phospholamban of 27,000 Da is converted into a compo-nent of 11,000 Da after boiling, while phosphorylated phospholamban of 29,000 Da is converted into a component of 13,000 Da after boiling (Figs. 14A and 15). Essentially similar mobility changes are observed in the Weber and Osborn system (Fig. 14B).

Phospholamban is first considered to represent a single polypeptide based on its electrophoretic mobility on SDS-polyacrylamide gel. Le Peuch et al.[12] contend that phospholamban represents a dimer composed of monomers of identical molecular weight. When SR is boiled in SDS prior to electrophoresis, phospholamban is totally converted into the lower-molecular-weight component.[54,82] This conversion is partially inhib-ited by a high concentration of Mg$^{2+}$; freezing of the sample boiled in SDS results in the reconversion into its original molecular weight.[82] However, these investigators[12,54,82] report a different molecular weight for the lower-molecular-weight component; the value ranges between 5,500 and 11,000. These results suggest the possibility that phospholamban consists of an oligomer.

[82] C. F. Louis, M. Maffitt, and B. Jarvis, J. Biol. Chem. **257**, 15182 (1982).

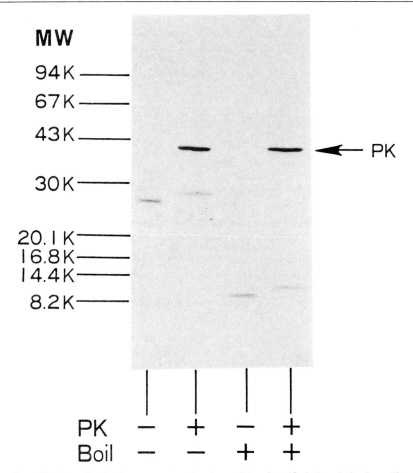

FIG. 15. Alteration in the apparent molecular weight of purified phospholamban. Phospholamban in the unphosphorylated (PK −) and the phosphorylated (PK +) states are incubated at 25° for 10 min (Boil −) or boiled for 1 min (Boil +) in 2% SDS, 10% glycerol, and 5% 2-mercaptoethanol and submitted to Laemmli 10–25% polyacrylamide gel electrophoresis. Phospholamban phosphorylation is performed using unlabeled ATP and the catalytic subunit of protein kinase (PK). The gel is stained with Coomassie Blue. [Reproduced with permission from Inui *et al.*, *J. Biol. Chem.* **260,** 3708 (1985).]

The existence of an oligomeric assembly is confirmed in purified preparations of phospholamban.[83,84] High-resolution SDS-polyacrylamide gel

[83] L. R. Jones, H. K. B. Simmerman, W. W. Wilson, F. R. N. Gurd, and A. D. Wegener, *J. Biol. Chem.* **260,** 7721 (1985).
[84] J. Fujii, M. Kadoma, M. Tada, H. Toda, and F. Sakiyama, *Biochem. Biophys. Res. Commun.* **138,** 1044 (1986).

electrophoresis using several lower-molecular-weight calibration standards indicates that the lowest-molecular-weight component of phospholamban is about 6000.[84] The value is in good agreement with the calculated molecular weight from the amino acid sequence of the phospholamban monomer deduced from the cDNA base sequence.[85] The maximal amount of phosphate incorporated by cAMP-dependent protein kinase is estimated to be 138 nmol/mg protein or 3.73 mol/27,000-Da phospholamban based on protein determination by the dye absorption method.[75] A similar value is observed by Wegener and Jones.[67]

It is intriguing from biophysical points of view how the electrophoretic mobility of phospholamban is altered by phosphorylation (Fig. 15). The conversion of 27,000 Da to 29,000 Da appear to occur in a stepwise fashion, in that at least four intermediary components can be identified.[67] These findings may be associated with the complex submolecular features of phospholamban. The possibility that phospholamban consists of a pentamer is proposed based on electrophoretic findings that graded phosphorylation results in five distinctly different conformations of phospholamban[67,83,86] and that three intermediary electrophoretic bands appear between the 6,000- and 26,000-Da components when phospholamban is treated with SDS at 50–70°.[84] The holoprotein may consist of homologous monomers since peptide mapping of tryptic fragments of phospholamban indicates the existence of identical phosphopeptide among five different conformations induced by graded phosphorylation.[86] Assuming that the 27,000 to 29,000-Da moiety of phospholamban represents an oligomeric functional unit, it is of utmost importance to determine the minimal chemical unit of phospholamban.

Amino acid sequencing of purified phospholamban indicates that the monomeric unit of phospholamban consists of 52 amino acids with a calculated molecular weight of 6080.[85] Amino-terminal methionine is acetylated. Sequence analysis of phospholamban-derived peptides fragmented by cyanogen bromide or by $N$-tosyl-L-phenylalanylchloromethyl ketone-treated trypsin determines the partial amino acid sequence initiated with $N^{\alpha}$-acetylated methionine followed by 44 amino acid residues intervened by two unidentified residues.[84] The remaining carboxy-terminal residues are not determined probably due to high hydrophobicity. A canine cardiac cDNA library is screened by hybridization with a mixture of synthetic oligodeoxyribonucleotide probes composed of all possible complementary sequences predicted from a partial amino acid sequence from glutamic acid-19 to alanine-24.[85] The resulting amino acid sequence

[85] J. Fujii, A. Ueno, K. Kitano, S. Tanaka, M. Kadoma, and M. Tada, *J. Clin. Invest.* **79**, 301 (1987).

[86] T. Imagawa, T. Watanabe, and T. Nakamura, *J. Biochem. (Tokyo)* **99**, 41 (1986).

```
                                           5'-AGAAAACTTTCTAACTAAACAC  -159

CGATAAGACTTCATACAACTCACAATACTTTATATTGTAATCATCACAAGAGCCAAGGCTACCTAAAAGAAGAGAGTGG   -80

TTGAGCTCACATTTGGCCGCCAGCTTTTTACCTTTCTCTTCACCATTTAAAACTTGAGACTTCCTGCTTTCCTGGGGTC    -1
```

```
1                            10                            20
Met Asp Lys Val Gln Tyr Leu Thr Arg Ser Ala Ile Arg Arg Ala Ser Thr Ile Glu Met
ATG GAT AAA GTC CAA TAC CTC ACT CGC TCT GCT ATT AGA AGA GCT TCA ACC ATT GAA ATG   60

21                           30                            40
Pro Gln Gln Ala Arg Gln Asn Leu Gln Asn Leu Phe Ile Asn Phe Cys Leu Ile Leu Ile
CCT CAA CAA GCA CGT CAA AAT CTT CAG AAC CTA TTT ATA AAT TTC TGT CTC ATT TTA ATA   120

41                           50      52
Cys Leu Leu Leu Ile Cys Ile Ile Val Met Leu Leu End
TGT CTC TTG TTG ATC TGC ATC ATT GTG ATG CTT CTC TGA AGTTCTGCTGCAATCTCCAGTGATGCA   187
```

```
ACTTGTCACCATCAACTTAATATCTGCCATCCCATGAAGAGGGGAAAATAATACTATATAACAGACCACTTCTAAGTAG   266

AAGATTTTACTTGTGAAAAGGTCAAGATTCAGAACAAAAGAAATTATTAACAAATGTCTTCATCTGTGGGATTTTGTAA   345

ACATGAAAAGAGCTTTATTTTCAAAAATTAACTTCAAAATGACTATAGGTGCGCATAATGTAATTGCTGAATTCCTCAA   424

CAAAGCTTGTAAAAGTTTCTATGCCAAATTTTTTCTGAGGGTAAAGTAGGAGTTTAGTTTTAAAACTGCTCTGCTAACC   503

AGTTCACTTCACATATAAAGCATTAGCTTCACTATTTGAGCTAAATATTTATATTGTACTGTAAATGCCTATGTAATGT   582

TTATTAAGATTTTTCAAGTCTCCGCTAAGTACGAAAATAATCATCCA⎡AATGAA⎤GTCATCATTTGAAATAGC-3'   652
```

FIG. 16. Amino acid sequence of phospholamban monomer deduced from complementary DNA. Nucleotide residues are numbered in the 5' to 3' direction, beginning with the first residue of ATG triplet encoding the initiator methionine. Nucleotides on the 5' side of residue 1 are indicated by negative numbers. A poly(A) tail on the 3'-end is not shown. The predicted amino acid sequence of phospholamban is displayed above the nucleotide sequence with its residue number beginning with the initiator methionine. A box indicates the presumptive polyadenylation signal. The in-frame stop codon preceding the initiator codon is underlined. [Reproduced with permission from Fujii et al., J. Clin. Invest. **79**, 301 (1987).]

deduced from the cDNA sequence shows that the phospholamban monomer consists of 52 amino acid residues with a molecular weight of 6080 (Fig. 16). The amino-terminal sequence of phospholamban, exhibiting 45 residues, corresponds precisely to the cDNA-deduced amino acid sequence.[85] The encoded protein starts with the first methionine determined for protein sequencing, indicating that phospholamban is translated without a signal peptide. The hydropathy profile of phospholamban indicates that the protein is divided into two domains; one with a phosphorylatable and hydrophilic portion at the amino-terminal region (positions 1–30) and the other with a hydrophobic portion at the carboxy-terminal region (positions 31–52).[85] The partial amino acid sequence of phospholamban is also determined by Simmerman et al.,[87] who identify a serine and a threonine

[87] H. K. B. Simmeman, J. H. Collins, J. L. Theibert, A. D. Wegener, and L. R. Jones, J. Biol. Chem. **261**, 13333 (1986).

residue as phosphorylation sites for cAMP- and calmodulin-dependent protein kinases, respectively. These residues correspond to serine-16 and threonine-17 in our sequence (Fig. 16). It remains to be seen whether the site for molecular assembly resides in the cytoplasmic amino-terminal domain or in membranous carboxy-terminal domain. The cytoplasmic moiety having serine-16 and threonine-17 is probably responsible for making a direct interaction with the site in Ca$^{2+}$-pump ATPase. A phospholamban monoclonal antibody, which blocks phosphorylation and dephosphorylation of phospholamban, stimulates Ca$^{2+}$-pump activity, suggesting the importance of phosphorylation sites for the phospholamban-Ca$^{2+}$-pump interaction.[88] These results indicate the possibility that conformational change of phospholamban, induced by phosphorylation, is required to induce the control over Ca$^{2+}$-pump.

Secondary structure prediction of the protein suggested that domain I consists of an $\alpha$-helix. In view of an unusual behavior of phospholamban molecule, it is intriguing to ask whether this helix breaks into two portions, possibly at around proline-21, allowing side chains of each to express hydrophobic interactions. Although alternative molecular models are possible, such an assumption may explain phosphorylation-induced structural changes of the phospholamban molecule, leading to profound functional consequences. Figure 17 illustrates the pentameric model of phospholamban, in which each of the phosphorylation site faces the cytoplasmic milieu with proline-21 serving as a hinge. This model would probably explain the conformational change of phospholamban when protein kinase-catalyzed phosphorylation alters the charge distribution. It is intriguing to assume that the hydrophobic residues inside the pocket formed by the pentameric assembly could exhibit direct protein–protein interaction with the certain key residues of Ca$^{2+}$-pump ATPase. This model would also assume that the protein kinase-catalyzed phosphorylation serves to relieve the inhibitory action on the ATPase, supporting the view that phospholamban functions as the suppressor of Ca$^{2+}$-pump ATPase.[5,89]

### Concluding Remarks

Elucidation of the phospholamban-ATPase system provides the first evidence to support the view that the intracellular effects of cAMP and Ca$^{2+}$ are tightly linked.[5] Such mechanisms would explain $\beta$-adrenergic actions of catecholamines on the myocardial cells,[5,13] in which an increase of intracellular cAMP activates protein kinase to augment Ca$^{2+}$-pump

[88] T. Suzuki and J. H. Wang, *J. Biol. Chem.* **261**, 7018 (1986).

[89] M. Inui, B. K. Chamberlain, A. Saito, and S. Fleischer, *J. Biol. Chem.* **261**, 1794 (1986).

Phosphorylated State          Unphosphorylated State

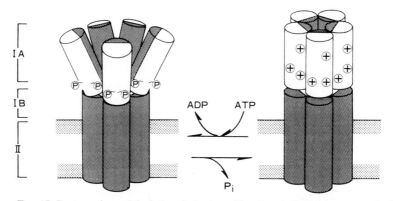

FIG. 17. Pentameric model of phospholamban. The phospholamban monomer is divided into two domains. The amino-terminal domain from methionine-1 to asparagine-31 (domain I) is composed largely of hydrophilic amino acids, whereas the other domain, representing 22 amino acids from leucine-31 to leucine-52 (domain II), is highly hydrophobic. Domain I breaks into two portions (domains IA and IB), possibly at around proline-21. When each monomer is assembled into a pentamer, the hydrophilic residues, including phosphorylatable serine-16 and threonine-17, face the cytoplasmic milieu. The positive charge of the arginine residue in unphosphorylated phospholamban is indicated by $\oplus$, whereas the phosphate group covalently bound to the side chain of serine-16 and/or threonine-17 is indicated by an encircled P. The cross-hatched area indicates the hydrophobic region of the $\alpha$-helix. In domain IA, the helical wheel analysis indicates that the hydrophobic residues face the core of pentamer.

ATPase through phosphorylation of phospholamban. The *in vivo* evidence supporting these intracellular mechanisms was obtained by a couple of groups,[80,90,91] who documented that the addition of isoproterenol to isolated heart[90,91] or sliced heart tissue[80] perfused with [³²P]Pᵢ results in increased ³²P incorporation into phospholamban *in situ*, with the simultaneous increase in the rates of contraction and relaxation.[91] Interestingly, cholinergic agonists are found to antagonize the isoproterenol-induced augmentation of phospholamban phosphorylation.[80,91] Calmodulin inhibitor (fluphenazine) significantly reduces *in vivo* phosphorylation of phospholamban,[90] although the physiological relevance of such an effect is not entirely clarified.

[90] C. J. Le Peuch, J.-C. Guilleux, and J. G. Demaille, *FEBS Lett.* **114**, 165 (1980).
[91] J. P. Lindemann, L. R. Jones, D. R. Hathaway, B. G. Henry, and A. M. Watanabe, *J. Biol. Chem.* **258**, 464 (1983).

There is other evidence that is consistent with the mechanism by which $Ca^{2+}$ fluxes of SR are controlled by the cAMP-phospholamban system. Employing a skinned cardiac cell, which exhibits cycles of phasic contractions upon addition of $Ca^{2+}$, Fabiato and Fabiato[92] demonstrated that a brief preincubation with cAMP results in an increased amplitude of contraction and faster rates of tension development and relaxation. More direct evidence was obtained by Allen and Blinks[93] who measured intracellular $Ca^{2+}$ using aequorin, a $Ca^{2+}$-sensitive bioluminescent protein, and found that isoproterenol augments the initial rate of $Ca^{2+}$ release from SR during the early phase of contraction, with simultaneous enhancement in the rate of reduction of $Ca^{2+}$ at the onset of relaxation.

Substantial evidence indicates that a protein similar to phospholamban is operational in other muscular tissues. Microsomes isolated from slow-contracting skeletal muscle contain phospholamban that, when phosphorylated by cAMP-dependent protein kinase, is associated with stimulated $Ca^{2+}$ uptake[94]; neither a phosphorylatable 22,000-Da protein nor $Ca^{2+}$ uptake stimulation is seen in fast-contracting skeletal muscle.[94] These findings are in accord with immunochemical data from attempts to detect phospholamban, which found that antiphospholamban antisera bind to the 27,000-Da component of SR from slow skeletal muscle, but that no reaction is detected in fast skeletal muscle SR.[75,95] Existence of a similar mechanism possibly exerted by phospholamban is also considered in microsomes from platelets[96] and visceral[97] and vascular[98] smooth muscles. In platelet microsomes a phospholamban-like protein is found to function as a regulator of $Ca^{2+}$ uptake.[99] $Ca^{2+}$ uptake by smooth muscle microsomes is stimulated by cAMP-dependent protein kinase,[97,98] which phosphorylates a phospholamban-like protein.

Cyclic AMP-induced augmentation of a Ca-selective ion channel in cardiac sarcolemma is proposed by Sperelakis and Schneider[100] to be mediated by cAMP-dependent phosphorylation of a membrane protein. This assumption is followed by Rinaldi et al.[57] who content that phospho-

[92] A Fabiato and F. Fabiato, Nature (London) 253, 556 (1975).
[93] D. G. Allen and J. R. Blinks, Nature (London) 273, 509 (1978).
[94] M. A. Kirchberger and M. Tada, J. Biol. Chem. 251, 725 (1976).
[95] Cardiac and slow muscles express the same phospholamban gene. See J. Fujii, J. Lytton, M. Tada, and D. H. MacLennan, FEBS Lett. 227, 51 (1988).
[96] R. Käser-Glanzmann, M. Jakabova, J. N. George, and E. F. Lüscher, Biochim. Biophys. Acta 466, 429 (1977).
[97] M. Kimura, I. Kimura, and S. Kobayashi, Biochem. Pharmacol. 26, 994 (1977).
[98] E. Suematsu, M. Hirata, and H. Kuriyama, Biochim. Biophys. Acta 773, 83 (1984).
[99] R. Käser-Glanzmann, E. Gerber, and E. F. Lüscher, Biochim. Biophys. Acta 558, 344 (1979).
[100] N. Sperelakis and J. A. Schneider, Am. J. Cardiol. 37, 1079 (1976).

rylation of 23,000-Da protein may be related to Ca channel activities, while such phosphorylation in sarcolemma is suggested by Lamers and Stinis[101] to be related to Ca$^{2+}$-pump ATPase of sarcolemma. After purifying such a sarcolemmal protein, termed as calciductin, Caponey et al.[72] note that this protein of proteolipid in nature resembles phospholamban isolated from cardiac SR. More extensive characterization of calciductin and phospholamban was performed by Manalan and Jones,[58] who documented that the former exhibits properties essentially similar to phospholamban. They also suggest that sarcolemmal calciductin is probably derived from phospholamban primarily residing in SR membranes. Whether the two proteins are identical should await further examination by more direct chemical analysis.

Phospholamban is proposed to exert its actions by regulating the cation-mediated conformational changes of Ca$^{2+}$-pump ATPase.[5] This assumption, primarily derived from kinetic analyses of ATPase enzyme, should be proved by more substantial evidence that directly relates molecular characteristics of phospholamban to the structure and function of Ca$^{2+}$-pump ATPase. In reconstituted cardiac SR, Inui et al.[89] indicate that Ca$^{2+}$-pump in such vesicles is fully activated so that it is not under control by phospholamban phosphorylation. These observations are suggestive of the possibility that phospholamban in a dephosphorylated state acts as a suppressor of Ca$^{2+}$-pump, with its phosphorylation triggering the reversal of suppression. More insights into the molecular interactions between Ca$^{2+}$-pump ATPase and its regulatory component phospholamban should be gained before a more precise molecular model is formulated.

[101] J. M. J. Lamers and J. T. Stinis, Biochim. Biophys. Acta **624,** 443 (1980).

## [12] Kinetic and Equilibrium Characterization of an Energy-Transducing Enzyme and Its Partial Reactions

By GIUSEPPE INESI, MARK KURZMACK, and DAVID LEWIS

### Introduction

In its simplest formulation, the catalytic transformation of a substrate (S) to a product (P) by an enzyme (E) may be represented as shown in Scheme 1.

On the other hand, coupling of enzyme catalysis to active transport of a ligand (L) requires a minimal sequence of partial reactions. Step 1 in Scheme 2 yields activated enzyme (E') through binding of a specific ligand

$$E + S \rightleftharpoons ES$$
$$ES \rightleftharpoons EP$$
$$EP \rightleftharpoons E + P$$

SCHEME 1

(L). The substrate (S) is then utilized through binding (step 2) and formation of an intermediate species with a different orientation of the L site (out → in), and a lower affinity for L (L*E-I). The ligand then dissociates against a concentration gradient and with a vectorial direction (out to in), and finally the intermediate species yields product (P) which is dissociated to allow further cycling of the enzyme (E). This is, in essence, the mechanism on which all reaction schemes for transport ATPases are based.

$$L_{out} + E \rightleftharpoons LE'$$
$$LE' + S \rightleftharpoons LE'S$$
$$LE'S \rightleftharpoons L*E-I$$
$$L*E-I \rightleftharpoons L_{in} + *E-I$$
$$*E-I \rightleftharpoons EP$$
$$EP \rightleftharpoons E + P$$

SCHEME 2

To investigate such a system the experimental accessibility of the concentrations of S, P, and L before binding to, and after disassociation from E, and of sequential reactions related to interactions of L, S, and P with the enzyme is desirable. The sarcoplasmic reticulum $Ca^{2+}$-ATPase is ideal for demonstration of experimental methods that are useful in characterization of coupling mechanisms in enzyme catalysis and active transport (for review see Ref. 1). Such a characterization may be carried out by equilibrium and/or kinetic experimentation, taking advantage of specific features of each partial reaction. Furthermore, the number of well-defined experimental parameters (together with the widespread availability of microcomputers) allows tailored analysis of postulated mechanisms by simulation of the experimental data. We will give here a description of some of these methods, mostly based on work performed in our laboratory over the past several years.

The Experimental System

The most commonly used experimental system consists of sarcoplasmic reticulum (SR) vesicles obtained by differential centrifugation from skeletal muscle homogenates. Two populations of SR vesicles can be

[1] G. Inesi, *Annu. Rev. Physiol.* **47**, 573 (1985).

separated based on their density.[2,3] In addition to the Ca$^{2+}$-pump ATPase, the heavy vesicles deriving from cisternal and/or junctional SR have a heterogeneous protein composition including junctional processes, calsequestrin, and a channel for passive diffusion of Ca$^{2+}$.[4] The light vesicles deriving from longitudinal SR have a protein composition which is mostly limited to the Ca$^{2+}$-pump ATPase (Fig. 1).[5,6] Such a selective protein composition confers structural and functional specificity to the light SR vesicles. The ATPase molecules are homogeneously oriented in all the light vesicles, with their catalytic portion protruding from the outer surface of the membrane, and a nonpolar portion intruding into the membrane bilayer. Thereby, a mechanism for ATP utilization on the outer surface and active transport of Ca$^{2+}$ into the lumen of the vesicles is provided. The light vesicles lack channels for passive leak of Ca$^{2+}$. Therefore, the Ca$^{2+}$ gradient formed as a consequence of active transport is maintained for an experimentally useful time. Meanwhile, the presence of channels for anions and monovalent cations allows passive fluxes of electrolytes to counterbalance charge displacements that may be produced by active transport.[7] For these reasons the light SR vesicles are an ideal system for studies of ATPase activity and coupled Ca$^{2+}$ transport.

Characterization of the partial reactions of the ATPase cycle sometimes requires ready accessibility of the lumen of the vesicles to experimental manipulations of the medium. To this end, it is possible to use leaky vesicles which are obtained by addition of appropriate ionophores,[8] or by treatment with detergents. For other purposes, such as determination of site stoichiometry or reconstitution experiments, "purified" ATPase may be obtained by subjecting the SR vesicles to selective solubilizations and extractions.[2,9,10]

We will refer to light vesicles, leaky vesicles, and purified ATPase as the systems used throughout the experimentation described in this chapter.

### Transport and ATPase Activity

Addition of ATP to SR vesicles preincubated in the presence of Ca$^{2+}$ in a suitable reaction medium is followed by Ca$^{2+}$ uptake by the vesi-

[2] G. Meissner, G. Conner, and S. Fleischer, *Biochim. Biophys. Acta* **298**, 246 (1973).
[3] G. Meissner, *Biochim. Biophys. Acta* **389**, 51 (1975).
[4] H. Yamamoto and M. Kasai, *J. Biochem. (Tokyo)* **92**, 485 (1982).
[5] S. Eletr and G. Inesi, *Biochim. Biophys. Acta* **282**, 174 (1972).
[6] H. Morii, H. Takisawa, and T. Yamamoto, *J. Biol. Chem.* **260**, 11536 (1985).
[7] T. Kometani and M. Kasai, *J. Membr. Biol.* **56**, 159 (1980).
[8] A. Scarpa, J. Baldassare, and G. Inesi, *J. Gen. Physiol.* **60**, 735 (1972).
[9] D. MacLennan, *J. Biol. Chem.* **245**, 4508 (1970).
[10] H. Barrabin, H. Scofano, and G. Inesi, *Biochemistry* **23**, 1542 (1984).

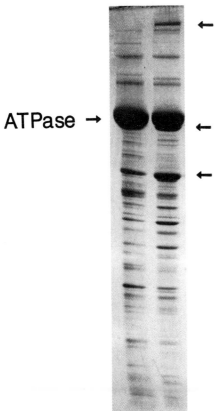

ATPase →     ← Junctional Protein

← Phosphorylase

← Calsequestrin

**1    2**

FIG. 1. SDS-PAGE of light (1) and heavy (2) vesicles. The two preparations were obtained in our laboratory from rabbit muscle, according to the methods of Eletr and Inesi,[5] and Morii et al.,[6] respectively. The lower ATPase content, the higher content of calsequestrin and phosphorylase, and the presence of junctional proteins are distinctive features of the heavy vesicles.

cles.[11,12] This can be demonstrated by the use of metallochromic indicators[13] undergoing light absorption changes when the $Ca^{2+}$ concentration in the medium decreases (Fig. 2). Of various indicators used for this purpose, murexide has the lowest sensitivity but can be used at a relatively

[11] W. Hasselbach and W. Makinose, *Biochem. Z.* **333,** 518 (1961).
[12] S. Ebashi and F. Lipman, *J. Cell Biol.* **14,** 389 (1962).
[13] G. Inesi and A. Scarpa, *Biochemistry* **11,** 356 (1972).

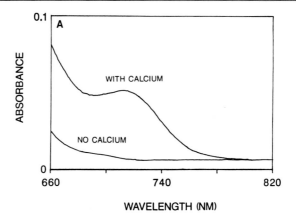

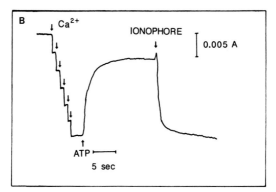

FIG. 2. Use of metallochromic indicators for measurements of $Ca^{2+}$ transients. (A) Light absorption spectra of antipyrylazo III in the presence or absence of calcium; it is apparent from these spectra that a maximal calcium signal can be obtained by differential absorption measurements, setting the reference wavelength at 790 nm, and the signal wavelength at 720 nm. (B) Differential wavelength (720 versus 790 nm) light absorption changes by successive addition of 5 $\mu M$ CaCl$_2$, and then 0.5 m$M$ ATP to a suspension of SR vesicles (0.3 mg protein/ml) in the presence of 50 $\mu M$ antipyrylazo, 20 m$M$ 3-($N$-morpholino)propane-sulfonic acid (MOPS) (pH 7), 100 m$M$ KCl, and 2 m$M$ MgCl$_2$. The ATP solution was buffered with $Ca^{2+}$ so that no signal was obtained by adding ATP to the reaction mixture in the absence of SR. The calcium accumulated by the SR vesicles is released upon addition of a divalent cation ionophore. The measurements were obtained with an Aminco DW2 spectrophotometer. Calcium, ATP, and ionophore additions were made with a microsyringe pushing the needle through black tape used to mask the opening of the cuvette. The reaction mixture was subjected to continuous stirring and constant temperature (25°).

high $Ca^{2+}$ concentration range (5–100 $\mu M$); tetramethylmurexide is twice as sensitive as murexide, but it may penetrate the lumen of the vesicles; antipyrylazo III has a good sensitivity and a useful range between 1 and 20 $\mu M$ $Ca^{2+}$; arsenazo III is the most sensitive, but has a narrow range and may inhibit the transport system in some cases.

Metallochromic indicators may be sensitive to $Mg^{2+}$, $H^+$, and ionic strength, in addition to $Ca^{2+}$. Therefore a careful study of their light absorption spectra in the presence and in the absence of relevant cations is required. The specificity and the sensitivity to changes of $Ca^{2+}$ concentration is maximized by the careful choice of two selective wavelengths for differential light absorption measurements in kinetic experiments (Fig. 2). A new generation of highly sensitive $Ca^{2+}$ indicators with fluorescent properties is also available.[14] In all cases, it is necessary to ascertain that the response of the indicator is linear within the $Ca^{2+}$ concentration range pertinent to the experiment. The time resolution of these measurements can be increased by the use of stopped flow instrumentation.[13]

Calcium uptake can be also measured by following the distribution of radioactive calcium tracer. For this purpose, SR vesicles are incubated with ATP in the presence of the required cofactors, the reaction is interrupted at serial times by rapid filtration (Millipore HAWP, 0.45- or 0.65-$\mu$m pore size). The filters retaining the loaded vesicles are then washed with a cold lanthanum solution, and used for determination of radioactive calcium. A better time resolution (Fig. 3) is obtained by using the rapid filtration apparatus distributed by Cosmologic (Pullman, Washington). In this case, the unloaded vesicles are placed on the filter, and then the loading solution (containing radioactive calcium and ATP) is passed through the filter for controlled time intervals.

An even greater time resolution of the initial phase of calcium uptake is obtained by quenching the reaction at serial times with lanthanum or EGTA[15] with the aid of a quench-flow apparatus. The quenched reaction mixture is then passed through a Millipore filter. Reaction times as short as 3 msec (Fig. 14) can be obtained by this method.

It is possible to demonstrate that following addition of ATP at 25°, $Ca^{2+}$ uptake reaches an asymptote within approximately 30 sec. At this time, addition of the divalent cation ionophore produces rapid efflux of the $Ca^{2+}$ taken up by the vesicles following the addition of ATP (Fig. 2). This demonstrates that the ATP-dependent $Ca^{2+}$ uptake results in the establishment of a $Ca^{2+}$ gradient which can be readily collapsed by increasing the passive permeability of the SR membrane to $Ca^{2+}$.

[14] G. Grymkiewicz, M. Poenie, and R. Y. Tsien, *J. Biol. Chem.* **260**, 3440 (1985).
[15] M. Chiesi and G. Inesi, *J. Biol. Chem.* **254**, 10370 (1979).

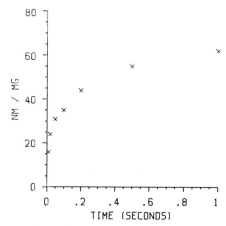

FIG. 3. Measurement of ATP-dependent calcium uptake by radioactive tracer and rapid filtration (25°). SR vesicles (80 $\mu$g/ml) were preincubated in 20 m$M$ MOPS (pH 6.8), 80 m$M$ KCl, 2 m$M$ MgCl$_2$, and 50 $\mu$$M$ $^{45}$CaCl$_2$. An aliquot (1 ml = 80 $\mu$g protein) of the preincubation mixture was passed through a Millipore filter (0.65 $\mu$m), which retained the SR vesicles. An aliquot of an identical medium (containing 0.5 m$M$ ATP) was then passed for variable time intervals through the filter. Velocity and times of flow were controlled with a rapid filtration apparatus (Cosmologic, Pullman, WA). Finally, the filters were washed with 3 ml of a medium containing 20 m$M$ MOPS (pH 6.8), 5 m$M$ MgCl$_2$, and 1 m$M$ LaCl$_3$ and then collected for determination of radioactivity.

The inhibition of transport activity observed at the asymptote is caused by a rise of the intravesicular Ca²⁺ concentration above the intrinsic capacity of the pump. In this case, the "capacity" of the pump is a combination of the concentration gradient, the passive permeability of the membrane, and the degree of saturation of the low-affinity calcium sites. Such an inhibition can be delayed by the use of phosphate or oxalate, which diffuse freely across the SR membrane and form insoluble calcium complexes when the Ca²⁺ concentration inside the vesicles increases as a consequence of active transport. Thereby a large rise of the intravesicular Ca²⁺ concentration is prevented, and the activity of the calcium pump is prolonged for several minutes. It was first demonstrated with the aid of this expedient that calcium transport is coupled with ATP hydrolysis, and the Ca/ATP ratio is 2 in optimal conditions.[16] In these experiments, calcium transport is monitored with radioactive tracer by counting the radioactivity in the medium and/or the vesicles that are separated by filtration at serial time intervals. In parallel experiments, ATP hydrolysis is monitored by measuring production of P$_i$ in acid-quenched samples. A colori-

[16] W. Hasselbach, *Prog. Biophys. Biophys. Chem.* **14,** 167 (1964).

metric method is adequate for this purpose. We find that the molybdo-vanadate reaction[17,18] is very convenient for $P_i$ determination, requiring only a single addition for quenching and color development.

The molybdovanate method requires the following preparations:

1. Ammonium molybdate

   10 g of $(NH_4)Mo_7O_{24} \cdot 4H_2O$ and 1 ml of ammonia (specific gravity 0.90) are dissolved in $H_2O$ to make 100 ml.

2. Ammonium metavanadate

   0.235 g of $NH_4VO_3$ is dissolved in 40 ml of hot (100°) $H_2O$ and cooled under tap water immediately after the vanadate goes into solution. 0.61 ml of concentrated (specific gravity 1.42) nitric acid which has been diluted with 1.4 ml of $H_2O$ is added to the vanadate solution and the entire mixture is brought to 100 ml with $H_2O$.

3. 20% SDS solution

   20 g of SDS is dissolved in $H_2O$ to make 100 ml. (Note: Bio-Rad SDS is satisfactory. Some SDS sources contain significant $P_i$ contamination and should not be used.)

4. Mixed reagent

   The above solutions are mixed together with 37 ml of concentrated nitric acid (specific gravity 1.42) and made up to 1 liter with $H_2O$. This solution will keep for 1 month, and should be discarded when a significant amount of precipitate is seen on the bottom of the bottle. Storage in repipets is not advised unless the solution is in daily use, as solution will precipitate inside the repipet, clogging it. This precipitate can be removed by washing with concentrated NaOH.

A typical reaction mixture may contain: 80 m$M$ KCl, 20 m$M$ MOPS (pH 6.8), 2.0 m$M$ MgCl$_2$, 0.1 m$M$ CaCl$_2$, 0.1 m$M$ EGTA, 0.05 mg SR/ml, and possibly a divalent cation ionophore to render the SR vesicles leaky if constant rates of activity are wanted. At serial times, quenching and color development are obtained by adding aliquots (1 ml) of the reaction mixture to an equal volume of the mixed reagent. The optical density should be read at 350 nm wavelength. Dilution of the sample with $H_2O$ may be necessary to reduce the OD to the working range of the spectrophotometer being used. A standard $P_i$ curve is obtained by adding 0.05, 0.1, 0.15, 0.2, and 0.25 ml of 1 m$M$ $P_i$ to a series of test tubes, bringing the volume to 1 ml with reaction medium (no ATP), and developing the color with 1 ml of mixed reagent.

[17] J. Lecocq and G. Inesi, *Anal. Biochem.* **15,** 160 (1966).
[18] T. Lin and M. Morales, *Anal. Biochem.* **77,** 10 (1977).

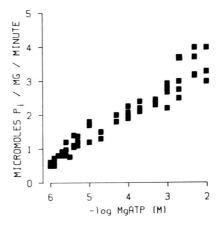

FIG. 4. Substrate concentration dependence of SR ATPase. The reaction was run in the presence of saturating $Ca^{2+}$ (150 $\mu M$) and an ATP-regenerating system (2 m$M$ phosphoenolpyruvate and 10 pyruvate kinase units/ml). SR vesicles rendered leaky by addition of the ionophore A23187 (10 $\mu M$) were used to obtain constant velocities of hydrolysis (steady state). (From Kosk-Kosicka et al.[20])

Experiments with oxalate or phosphate as precipitating anions, entail some uncertainty as to whether the inhibition by intravesicular $Ca^{2+}$ is in fact *totally* relieved, and whether some precipitation occurs outside, in addition to inside the vesicles, when the concentration of anions is high. Furthermore, analysis of these experiments requires calculation of solubility products, thereby adding some complexity.

Steady-state parameters of ATPase kinetics can be obtained quite nicely with leaky vesicles. In this case the rise of intravesicular $Ca^{2+}$ is prevented by passive leak. Therefore, ATPase activity proceeds at constant rates for a long time, and can be followed by colorimetric determinations of $P_i$ production at serial times. It is then possible to determine the dependence of steady-state enzyme activity on substrate, $Ca^{2+}$, $Mg^{2+}$, pH, temperature, etc.

Most preparations of light and leaky vesicles catalyze low rates of ATP hydrolysis (basic ATPase) in the absence of $Ca^{2+}$ and in the presence of high (m$M$) ATP (Hasselbach[16]). This basic ATPase is not found in purified ATPase, and is likely due to a contaminating enzyme. The activity of all preparations is sharply increased by the addition of $Ca^{2+}$ in the micromolar range.[19] This $Ca^{2+}$-dependent activity has a complex ATP concentration dependence, with a first rise in the 1 to 100 $\mu M$ range, and a

[19] A. Weber, R. Herz, and I. Reiss, *Biochem. Z.* **345,** 329 (1966).

further rise at higher ATP concentrations (Fig. 4).[20] $Mg^{2+}$, in the milli-molar range is also a requirement, and $K^+$ in the 100 mM range is an activator of the $Ca^{2+}$-dependent activity.[21] The pH optimum is near neutrality.[22] The temperature dependence yields an activation energy of 25–30 kcal mol$^{-1}$ within the 5 to 20° range, and 15–18 kcal mol$^{-1}$ within the 20 to 37° range.[23]

## Calcium Binding to SR ATPase

Since $Ca^{2+}$ is both the specific ATPase activator and the transported species, it is expected to bind to the enzyme even in the absence of substrate (see Scheme 2). We find that the best method for measuring calcium binding in the absence of ATP is equilibration of SR ATPase with a medium containing radioactive calcium tracer, followed by molecular sieve chromatography. The excess radioactivity (over the base line) eluting with the protein peak (Fig. 5)[24] represents the calcium bound, while the radioactivity trough eluting after the protein peak is related to the calcium taken up from the volume of medium originally equilibrated with the enzyme and then delayed by the molecular sieve column. Determinations of radioactivity and protein give accurate values for calcium bound per unit weight of protein.

It should be pointed out that experiments on high-affinity binding require $Ca^{2+}$ concentrations lower than 10 $\mu M$. In practice, such low concentrations can only be obtained with the aid of Ca–EGTA buffers. The actual concentration of free $Ca^{2+}$ is calculated from total calcium and total EGTA on the basis of the affinity constant, which is $9.33 \times 10^{10}$ for the [Ca–EGTA]/[$Ca^{2+}$][EGTA$^{4-}$] equilibrium,[25] and the constants for the pH dependence of the distribution of EGTA ionization species which are

$$\frac{[\text{HEGTA}^{3-}]}{[\text{H}^+][\text{EGTA}^{4-}]} = 3.8 \times 10^9$$

$$\frac{[\text{HEGTA}^{2-}]}{[\text{H}^+][\text{HEGTA}^{3-}]} = 9.12 \times 10^8$$

$$\frac{[\text{HEGTA}^-]}{[\text{H}^+][\text{HEGTA}^{2-}]} = 575$$

$$\frac{[\text{HEGTA}]}{[\text{H}^+][\text{HEGTA}^-]} = 126$$

[20] D. Kosk-Kosicka, M. Kurzmack, and G. Inesi, *Biochemistry* **22**, 2559 (1983).
[21] A. Martonosi and R. Feretos, *J. Biol. Chem.* **239**, 648 (1964).
[22] G. Inesi and T. Hill, *Biophys. J.* **44**, 271 (1983).
[23] G. Inesi, M. Millman, and S. Eletr, *J. Mol. Biol.* **81**, 483 (1973).
[24] G. Inesi, M. Kurzmack, C. Coan, and D. Lewis, *J. Biol. Chem.* **255**, 3025 (1980).
[25] G. Schwartzenbach, H. Senn, and G. Anderegg, *Helv. Chim. Acta* **40**, 1886 (1957).

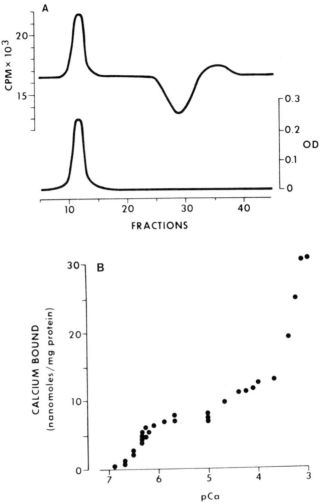

FIG. 5. Measurement of calcium binding at equilibrium. (A) Elution of radioactive cal-cium (upper curve) and sarcoplasmic reticulum protein (lower curve) from a chromatogra-phy column for determination of calcium binding. (B) Calcium binding to sarcoplasmic reticulum vesicles includes specific (high affinity) and nonspecific sites. (From Inesi *et al.*[24])

according to Blinks *et al.*[26] The influence of electrolytes such as $Mg^{2+}$ and ATP, when present, must also be taken into consideration. These calculations are facilitated by computer programs such as that originally described by Fabiato and Fabiato[27] (see also this volume [31]).

[26] J. Blinks, W. Wier, P. Hess, and F. Prendergast, *Prog. Biophys. Mol. Biol.* **40,** 1 (1982).
[27] A. Fabiato and F. Fabiato, *J. Physiol. (Paris)* **75,** 463 (1979).

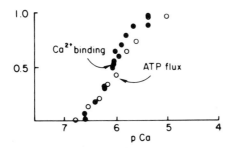

Fig. 6. Fractional values of maximal calcium binding and ATPase activation, as functions of $Ca^{2+}$ concentration. Calcium binding (●) was determined at equilibrium as for Fig. 4. Steady-state ATPase activity [ATP flux (○)] sustained by leaky vesicles was estimated by measurements of $P_i$ production, as for Fig. 3. (From Inesi et al.[28])

Measurements of calcium binding to light SR vesicles at equilibrium, in the absence of ATP and over a wide range of $Ca^{2+}$ concentrations, uncover a large number of low-affinity sites, and a few high-affinity sites saturating within the micromolar range (Fig. 5). The large number of low-affinity sites observed *in the absence of ATP* are most likely nonspecific. The high-affinity sites, on the other hand, are involved in enzyme activation (Fig. 6)[28] and transport. High-affinity calcium binding is retained by purified ATPase with a stoichiometry of two calcium ions per 110,000 MW ATPase chain.[29]

Another indication of calcium binding is a change in intrinsic fluorescence of the SR ATPase, as first noted by Dupont.[30] The fluorescence signal is particularly useful in kinetic experiments, and lends itself to stopped-flow measurements (Fig. 7).[31] An advantageous feature of the fluorescence signal is its specificity inasmuch as it is only obtained following calcium occupancy of high-affinity sites involved in enzyme activation.

Having obtained equilibrium and kinetic data on calcium binding, it is important to consider what useful mechanistic inference can be construed from them, and to devise an analysis to challenge the postulated mechanism with simulations of the experimental data. For instance, we note that the equilibrium binding isotherms are cooperative (Figs. 6 and 7) and, therefore, we expect interaction and participation of at least two domains in the binding mechanism. In fact, we know that the specific, high-affinity

[28] G. Inesi, M. Kurzmack, and D. Lewis, in "Structure and Function of Sarcoplasmic Reticulum" (S. Fleischer and Y. Tonomura, eds.), p. 191. Academic Press, Orlando, Florida, 1985.
[29] H. Scofano, H. Barrabin, G. Inesi, and J. Cohen, Biochim. Biophys. Acta 819, 93 (1985).
[30] Y. Dupont, Biochem. Biophys. Res. Commun. 71, 544 (1976).
[31] F. Fernandez-Belda, M. Kurzmack, and G. Inesi, J. Biol. Chem. 259, 9687 (1984).

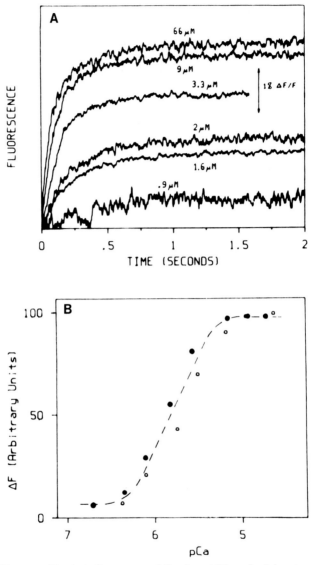

FIG. 7. Changes of intrinsic fluorescence following addition of calcium to sarcoplasmic reticulum ATPase in the absence of ATP. (A) Transients obtained in stopped-flow experiments. (B) Fluorescence changes following equilibration with various Ca²⁺ concentrations. The excitation wavelength was 290 nm and emission was passed through a 0-54 Corning cutoff filter. (From Fernandez-Belda et al.[31])

binding entails two calcium ions for each 110,000 MW ATPase chain. Finally, we take note of the relatively slow kinetics of intrinsic fluorescence change upon calcium binding, suggesting a conformational change that may be instrumental in the cooperative interaction between the binding domains, as well as in enzyme activation. We consider then, for example, a sequential binding mechanism[24] for the binding of calcium ion to SR ATPase (E), whereby binding of the first calcium induces a conformational change which in turn uncovers a second binding site of higher affinity.

The conformational change in Step 2 of Scheme 3 was postulated to explain the slow fluorescent changes measured upon the addition or re-

Step 1: $E + Ca^{2+} \underset{k_{-1}}{\overset{k_1}{\rightleftharpoons}} ECa$     $k_1 = 4.25 \times 10^7 \, M^{-1} \, sec^{-1}, \, k_{-1} = 450 \, sec^{-1}$

Step 2: $ECa \underset{k_{-2}}{\overset{k_2}{\rightleftharpoons}} E'Ca$     $k_2 = 15 \, sec^{-1}, \, k_{-2} = 33 \, sec^{-1}$

Step 3: $E'Ca + Ca^{2+} \underset{k_{-3}}{\overset{k_3}{\rightleftharpoons}} E'Ca_2$     $k_3 = 1 \times 10^8 \, M^{-1} \, sec^{-1}, \, k_{-3} = 16 \, sec^{-1}$

SCHEME 3

moval of calcium and still be consistent with the cooperative mechanism observed with binding studies. It is then assumed that $E'Ca$ and $E'Ca_2$ are species with higher intrinsic fluorescence than E and ECa.

It is noteworthy that a combination of equilibrium binding studies (to measure the apparent affinity and demonstrate the cooperative character of the isotherms), as well as rapid kinetic fluorescent studies (to infer the rates in Step 2) was needed to postulate a three-step mechanism rather than a simple two-step mechanism.

From the postulated reaction mechanism, an equation may be written to define the flux for each species:

$$\frac{d[E]}{dt} = -k_1[E][Ca^{2+}] + k_{-1}[ECa]$$

$$\frac{d[ECa]}{dt} = k_1[E][Ca^{2+}] - (k_{-1} + k_2)[ECa] + k_{-2}[E'Ca]$$

$$\frac{d[E'Ca]}{dt} = k_2[ECa] - (k_{-2} + k_3[Ca^{2+}])[E'Ca] + k_{-3}[E'Ca_2]$$

$$\frac{d[E'Ca_2]}{dt} = k_3[E'Ca][Ca^{2+}] - k_{-3}[E'Ca_2]$$

In addition, another equation may be written for mass conservation in the system:

$$E_{total} = [E] + [ECa] + [E'Ca] + [E'Ca_2]$$

At equilibrium, the net flux for each species is zero, and the equations may be solved for the concentrations of each enzyme species:

$$[E] = \frac{k_{-1}k_{-2}k_{-3}[E'Ca_2]}{k_1k_2k_3[Ca^{2+}]^2}$$

$$[ECa] = \frac{k_{-2}k_{-3}[E'Ca_2]}{k_2k_3[Ca^{2+}]}$$

$$[E'Ca] = \frac{k_{-3}[E'Ca_2]}{k_3[Ca^{2+}]}$$

$$[E'Ca_2] = \frac{[E_t]k_1k_2k_3[Ca^{2+}]^2}{k_1k_2k_3[Ca^{2+}]^2 + k_{-1}k_{-2}k_{-3} + k_1k_{-2}k_{-3}[Ca^{2+}] + k_1k_2k_{-3}[Ca^{2+}]}$$

The results may then be compared to the experimental data obtained at equilibrium (Fig. 8).

Before equilibrium is reached, however, the net flux of each species will change with time. In order to solve for the concentrations of the enzyme species during the transient state, we must first start with the enzyme in a known state at time = 0. For instance, all the enzyme may be forced into the E form by adding EGTA to chelate calcium. By then adding calcium and controlling the free calcium concentration, we can simulate the reequilibration of the enzyme. This is done by calculating the instantaneous fluxes at time = 0, and numerically integrating the equations over a small time interval (for instance, 0.05 msec) in order to predict the change in the species concentration. By successive iterations, it is possible to obtain the full transient state kinetic simulation, as well as the final (equilibrium) levels (Fig. 8).

## Enzyme Phosphorylation with ATP and P$_i$ Production

Addition of [$\gamma$-$^{32}$P]ATP to SR ATPase preincubated with Ca$^{2+}$ is rapidly followed by a rapid burst of [$^{32}$P]phosphoryl transfer from ATP to an aspartyl residue of the ATPase catalytic site, [32-35] and then by steady-state [$^{32}$P]P$_i$ production.

[32] T. Yamamoto and Y. Tonomura, *J. Biochem.* (*Tokyo*) **62,** 558 (1967).
[33] M. Makinose, *Eur. J. Biochem.* **10,** 74 (1969).
[34] F. Bastide, G. Meissner, S. Fleischer, and R. L. Post, *J. Biol. Chem.* **248,** 8385 (1973).
[35] C. Degani and P. Boyer, *J. Biol. Chem.* **248,** 8222 (1973).

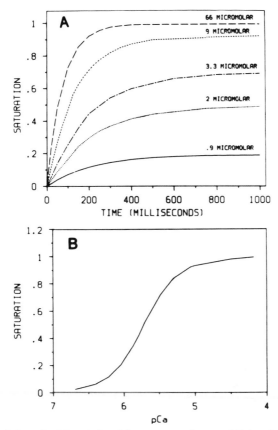

FIG. 8. Simulation of calcium-induced fluorescence changes: (A) transients; (B) equilibria. The simulated curves were obtained assuming the binding mechanism described in the text. Note the satisfactory agreement with the experimental measurements in Fig. 7.

The [$^{32}$P]phosphoenzyme (EP) is acid stable and, therefore, it can be measured in acid-quenched samples. To make these measurements, the reaction of the SR ATPase with [$\gamma$-$^{32}$P]ATP is quenched with 0.125 $M$ perchloric acid (PCA) and 2 m$M$ P$_i$, and approximately 1 mg of the quenched protein is centrifuged down at low speed and resuspended in 0.125 $M$ PCA and 2 m$M$ P$_i$. This washing procedure is repeated five times to eliminate nonreacted [$\gamma$-$^{32}$P]ATP, and the final sediment is dissolved in 0.25 ml of a solution containing 0.1 $N$ NaOH, 2% Na$_2$CO$_3$, 2% SDS, and 5 m$M$ P$_i$. The solubilized sample is then diluted with 1 ml of water before determination of radioactive phosphorus (scintillation counting) and protein (Folin method). This method yields accurate results, but requires approximately 1 mg protein per sample, and is time-consuming. Alternatively, a volume of quenched sample corresponding to 50 $\mu$g protein is passed through a 0.45-$\mu$m HAWP Millipore filter. The filters, placed on a filtration apparatus, are rinsed with water immediately before filtering the quenched sample. After passing the samples through, the filters are washed six times with 5 ml of 0.125 $M$ PCA and 2 m$M$ P$_i$. Finally, the washed filters (with the protein adsorbed onto them) are placed in scintillation vials, dissolved in 1 ml dimethylformamide, and the radioactivity is determined by scintillation counting after the addition of a suitable scintillation fluid. It should be pointed out that each Millipore filter absorbs only a maximum of 80 $\mu$g quenched protein. Excess protein goes through the filter and is lost. This method is fast and convenient. However, the amount of protein corresponding to the radioactivity found in the filter is not determined directly in the final sample, but simply assumed on the basis of the protein concentration in the original reaction mixture.

The amount of [$^{32}$P]P$_i$ produced during the incubation of the SR ATPase with the [$\gamma$-$^{32}$P]ATP can be measured in the same experiment in which EP is determined. To this end, the first supernatant following centrifugation of the quenched reaction mixture is collected. The [$\gamma$-$^{32}$P]ATP is first extracted by vortex mixing 0.5 ml of the sample (first supernatant) with 0.5 ml of 0.1% (w/v) Norit (prewashed with acid). The mixture is passed through a Millipore filter which is then washed twice with 1 ml 5% TCA. The filtrate and the two washings are collected, and 10 $\mu$l of 100 m$M$ (nonradioactive) P$_i$ carrier, 0.3 ml acetone, and 1.5 ml 5% ammonium molybdate in 2.5 $N$ H$_2$SO$_4$ is added in sequence while vortex mixing. The final concentration of acid in the sample should be 0.8–1.2 $N$. Finally 2.0 ml of a 1 : 1 isobutanol–benzene solution is added, and the phosphomolybdate complex is extracted by vortex mixing for 30–40 sec. Following a centrifugation at low speed for 5 min, a fraction of the organic phase is pipetted out and the radioactive phosphorus determined by scintillation counting. Appropriate blanks for the radioactivity obtained from samples in which [$\gamma$-$^{32}$P]ATP is added to prequenched protein, as well as controls for the extraction efficiency starting with standard amounts of [$^{32}$P]P$_i$, must be obtained. Finally the amount of [$^{32}$P]P$_i$ produced per protein unit weight, specifically by the ATPase reaction, is calculated by taking into account all the dilution factors, as well as blank and yield corrections.

The methods described above are quite sensitive and can be used to determine accurately the EP and P$_i$ resulting from a single enzyme cycle in the presteady state following addition of ATP. Since maximal levels of EP are reached in less than 1 sec, kinetic resolution of EP formation requires the use of rapid mixing devices in order to start the reaction by addition of ATP and then quench it by the addition of acid in the millisec-

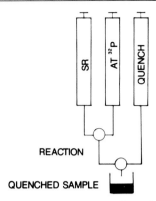

FIG. 9. Diagram of a quench-flow mixer.

ond time scale. In their simplest configuration these devices consist of three syringes and two mixing chambers (Fig. 9). The first two syringes push SR ATPase and ATP into the first mixing chamber. The reaction mixture continues to flow through a delay line into a second mixing chamber where it is quenched by acid pushed by the third syringe. The quenched reaction mixture is then collected for determination of EP and $P_i$. The reaction time is determined by the length of the delay line between the first and the second mixing chambers and by the flow velocity.

Devices of this type can be built in the laboratory. On the other hand, excellent apparatuses have been made available commercially. In our laboratory we have used for several years a Dionex multimixer in which the syringes are pushed by gas pressure, and the flow velocity is monitored electromagnetically. The instrument permits reaction times varying from a minimum of 16 msec to a maximum of 10 sec through a combination of continuous flow and electronically controlled aging modes.

We have also used extensively the highly precise Froehlich–Berger mixer (Commonwealth Technology, Alexandria, VA) in which the syringes are pushed by a stepping motor, and reaction times as short as 3 msec can be obtained. Both devices have diverging flow pathways and solenoids for purging the old reaction mixture from the instrument dead volumes, before a new sample is collected. Rapid mixing devices are also manufactured by other companies such as High-Tech Scientific Limited (Salisbury, England).

Kinetic resolution of EP and $P_i$ production, following addition of ATP to SR ATPase preincubated with calcium, is shown in Fig. 10. It is clear that the EP is formed before $P_i$ production begins, as expected if EP is an "intermediate" and $P_i$ the "final product" of the ATPase reaction. In fact, the initial velocity of EP formation at saturating ATP concentrations is too fast to be resolved. Therefore, initial velocities must be measured at

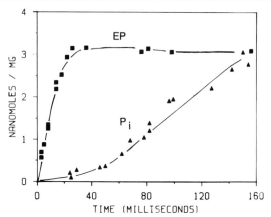

FIG. 10. Kinetic resolution of phosphoenzyme formation (EP) and $P_i$ production follow-ing addition of ATP to SR vesicles preincubated with calcium. Reaction mixture: 20 m$M$ MOPS (pH 6.8), 2 m$M$ MgCl$_2$, 80 m$M$ KCl, 100 $\mu M$ CaCl$_2$, 0.3 mg SR protein/ml, and 10 $\mu M$ ATP. Mixing was carried out with a Froehlich–Berger multimixer.

increasing concentrations of ATP within the nonsaturating range,[36,37] and the resulting values extrapolated to infinite ATP (Fig. 11) to estimate the maximal velocity of phosphorylation. Considering the reaction sequence shown in Scheme 4, the rate constant $k_2$ can be obtained by dividing the

$$Ca_2 \cdot E + ATP \underset{k_{-1}}{\overset{k_1}{\rightleftharpoons}} Ca_2 \cdot E \cdot ATP \underset{k_{-2}}{\overset{k_2}{\rightleftharpoons}} Ca_2 \cdot E\text{-}P \cdot ADP$$

SCHEME 4

maximal velocity at infinite ATP concentration by the stoichiometry of the active sites (see below for determination of catalytic site stoichiome-try). For instance, a maximal velocity of 375–450 nmol EP formed per milligram protein per second, divided by 3 nmol of catalytic sites per milligram protein, yields a turnover number of 125–150 per second, which is equivalent to the first-order rate constant $k_2$ in Scheme 4. The reverse rate constant $k_{-2}$ can be estimated from experiments on ATP synthesis (see below) to be approximately 450 sec$^{-1}$. The experiment shown in Fig. 10 permits also an estimate of the $K_m$ from linear extrapolation of the reciprocal velocities of EP formation to a value of $2 \times 10^{-5}$ $M$. Consider-ing that EP hydrolysis is negligible soon after addition of ATP (Fig. 10), and the reversal of the phosphorylation reaction is negligible (owing to rapid ADP dissociation) at very low ADP concentrations, it can be as-

[36] J. Froehlich and E. Taylor, *J. Biol. Chem.* **250**, 2013 (1975).
[37] S. Verjovski-Almeida, M. Kurzmack, and G. Inesi, *Biochemistry* **17**, 5006 (1978).

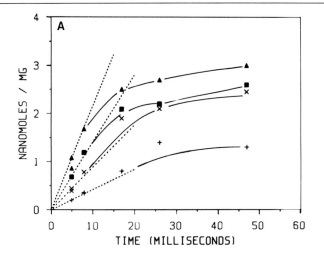

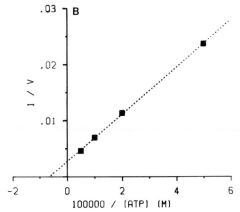

FIG. 11. Determination of the rate constant for transfer of ATP terminal phosphate to SR ATPase. (A) Various concentrations of [γ-³²P]ATP were added to SR vesicles (50 μg/ml) in the presence of 20 m$M$ MOPS (pH 6.8), 80 m$M$ KCl, 2 m$M$ MgCl$_2$, and 100 μ$M$ CaCl$_2$. Mixing was carried out with a Froehlich–Berger multimixer. (+, 2 μ$M$; ×, 5 μ$M$; ■, 10 μ$M$; ▲, 20 μ$M$) (B) The initial velocities of phosphorylation are plotted as a function of ATP concentration.

sumed that a $K_m$ value (*limited to enzyme phosphorylation with ATP*) approximates:

$$K_m = (k_{-1} + k_2)/k_1$$

Since the off rate constant ($k_{-1}$) for ATP dissociation has been estimated to be 20–30 sec$^{-1}$, based on experiments on ATP synthesis (see below), the

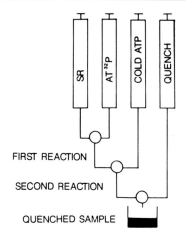

FIG. 12. Diagram of a quench-flow mixer in the four-syringe configuration.

ATP association rate constant $(k_1)$ will be $(20 \ \text{sec}^{-1} + 125 \ \text{sec}^{-1})/2 \times 10^{-5} M$ $= 7 \times 10^6 \ \text{sec}^{-1} M^{-1}$. In conclusion, the set of reactions in Scheme 4 are characterized with the following set of equilibrium and kinetic constants: $K_1 = 3.5 \times 10^5 M$, $k_1 = 7.0 \times 10^6 \ \text{sec}^{-1} M^{-1}$, $k_{-1} = 20 \ \text{sec}^{-1}$; $K_2 = 0.33$, $k_2 = 150 \ \text{sec}^{-1}$, $k_{-2} = 450 \ \text{sec}^{-1}$.

Another type of experiment has been devised to study the breakdown of EP in the forward direction. In this experiment, radioactive EP is first made by a brief incubation of calcium-activated ATPase with $[\gamma\text{-}^{32}P]ATP$, and then the reaction mixture is diluted with an excess of nonradioactive ATP to prevent further formation of radioactive EP. Acid quenching is then carried out at serial times and the breakdown of radioactive EP is monitored. This can be accomplished with mixing devices in a four-syringe configuration (Fig. 12). It turns out that the kinetics of EP breakdown are not monophasic, and their interpretation requires analysis that may be better studied in the original literature.[38] In general, however, it can be concluded from these experiments that the kinetics of EP breakdown are sufficiently fast to be consistent with the overall enzyme turnover, thereby establishing that the EP is in fact a kinetically competent enzyme intermediate.

## Functional Characterization of the Phosphoenzyme

The kinetically complex breakdown of EP made with ATP in the presence of $Ca^{2+}$ suggests that a transition related to vectorial release of $Ca^{2+}$

[38] Y. Nakamura, M. Kurzmack, and G. Inesi, J. Biol. Chem. **261**, 3090 (1986).

against a concentration gradient occurs before hydrolytic cleavage of $P_i$. It is then appropriate to mention here some of the experimental methods which have been used for functional characterization of EP. First, one would think that the affinity of the specific calcium sites on the ATPase must be reduced as a consequence of enzyme phosphorylation, to account for the release of the divalent cation against a concentration gradient. Ideally, such a reduction in affinity should be shown by a displacement of the binding isotherms to a higher $Ca^{2+}$ concentration range. However, it is experimentally difficult to detect the specific binding of a few nanomoles of calcium in the presence of more than a hundred nanomoles of nonspecific calcium binding occurring at millimolar calcium concentrations (see Fig. 5). This difficulty has been overcome by measuring spectroscopic parameters that are sensitive *only* to calcium occupancy of *specific* sites. It can be shown by this expedient (Fig. 13)[39] that the apparent calcium binding affinity is reduced three orders of magnitude (approximately, from $10^6$ to $10^3$ $M^{-1}$) as a consequence of enzyme phosphorylation. This is in agreement with the $Ca^{2+}$ concentration (millimolar range) dependence of ATPase inhibition and its reversal, which may be attributed to calcium occupancy of the specific sites of the phosphoenzyme.

Another feature of functional interest is that, in addition to a reduction in affinity, the specific calcium sites undergo a change in orientation as a consequence of enzyme phosphorylation.[40] This can be demonstrated in rapid mixing experiments analogous to those described for characterization of EP formation. In this case, however, radioactive calcium tracer is used instead of radioactive ATP. Furthermore, the reaction is quenched with EGTA or lanthanum,[15] rather than with acid. These agents limit leak of accumulated calcium from the SR vesicles, while preventing further calcium occupancy of the specific sites and recycling of the enzyme. It is then possible to demonstrate by parallel experiments with acid and lanthanum (or EGTA) quenching procedures, that formation of EP is rapidly followed by a burst (Fig. 14) of calcium "internalization," i.e., the calcium bound to specific sites cannot be displaced anymore by lanthanum following enzyme phosphorylation.

It is noteworthy that addition of ATP analogs that are not utilized as substrates for enzyme phosphorylation is not accompanied by calcium internalization. The stoichiometric ratio of the EP and calcium burst is 2 under optimal conditions. Following the phosphorylation and calcium burst, steady-state $P_i$ production and calcium uptake begin in parallel,

[39] C. Coan, S. Verjovski-Almeida, and G. Inesi, *J. Biol. Chem.* **254**, 2968 (1979).
[40] M. Kurzmack, S. Verjovski-Almeida, and G. Inesi, *Biochem. Biophys. Res. Commun.* **78**, 772 (1977).

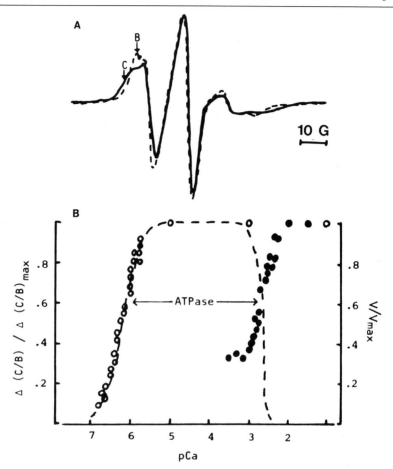

FIG. 13. (A) EPR spectra of spin-labeled SR ATPase in the absence (dotted line) and the presence (solid line) of Ca²⁺ and nucleotide. In the lower figure the change in the spectral parameter C/B (○ and ●) and ATPase activity (broken line) are plotted as a function of pCa, in the presence of ATP (●) or AMP-PNP (○). Previous to experimentation the SR vesicles were treated with Triton X-100 or sonication to prevent net accumulation of Ca on addition of ATP. These measurements demonstrated a shift in the Ca²⁺ affinity of the SR ATPase following enzyme phosphorylation with ATP. (From Coan *et al.*[39])

after a kinetic delay which is evidently related to Ca²⁺ release inside the vesicles and precedes hydrolytic cleavage of the EP.

### Hydrolytic Cleavage of the Phosphoenzyme and Its Reversal

Before undergoing hydrolytic cleavage, the phosphoenzyme formed by phosphoryl transfer from ATP to E · Ca₂ must undergo a slow transition related to Ca²⁺ release into the lumen of the vesicles. This transition

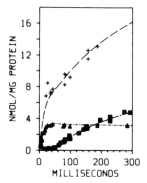

FIG. 14. Early phase of phosphoenzyme formation (▲), calcium uptake (+), and $P_i$ production (■). ATP (10 $\mu M$) was added to SR vesicles preincubated with calcium at 25°. For phosphoenzyme and $P_i$ measurements [$\gamma$-$^{32}$P]ATP and acid quenching were used. For calcium uptake measurements, $^{45}$Ca and lanthanum quenching were used. (From Fernandez-Belda *et al.*[31])

imposes a rate-limiting block on phosphoenzyme decomposition, thereby masking the intrinsic kinetics of the hydrolytic reaction. Therefore, the hydrolytic reaction can be studied more conveniently if the EP species directly undergoing hydrolytic cleavage is first formed by phosphorylation of SR ATPase with $P_i$[41–43] through reversal of the hydrolytic reaction shown in Scheme 5. This reaction is very interesting from the methodological point of view, because it can be studied by both equilibrium and kinetic experiments. In these experiments the enzyme (E) is in fact a reagent rather than just a catalyst. Phosphorylation of SR ATPase with $P_i$ requires addition of $Mg^{2+}$ and removal of $Ca^{2+}$ with EGTA.[43] It is favored by low pH[44,45] and high (25° as opposed to 5°) temperature.[46]

$$E\text{-}P + HOH \rightleftharpoons E \cdot P_i \rightleftharpoons E + P_i$$

SCHEME 5

Equilibrium experiments are carried out by incubating SR ATPase with [$^{32}$P]$P_i$ for a few seconds, followed by acid quenching and determination of [$^{32}$P]phosphoenzyme as described for the EP obtained with ATP. It is important to purify the radioactive $P_i$ stock within a week before the

[41] M. Makinose, *FEBS Lett.* **25**, 113 (1972).
[42] S. Yamada, M. Sumida, and Y. Tonomura, *J. Biochem.* (*Tokyo*) **72**, 1537 (1972).
[43] H. Masuda and L. deMeis, *Biochemistry* **12**, 4581 (1973).
[44] L. deMeis and R. Tume, *Biochemistry* **16**, 4455 (1977).
[45] G. Inesi, D. Lewis, and A. Murphy, *J. Biol. Chem.* **259**, 996 (1984).
[46] M. Sumida and Y. Tonomura, *J. Biochem.* (*Tokyo*) **75**, 283 (1974).

experiments in order to avoid a high background. For this purpose we usually start from 5 mCi of $^{32}P_i$, add nonradioactive $P_i$ to yield a 5 m$M$ concentration in 1 ml volume, and 0.2 ml of concentrated HCl to yield approximately a 2 $N$ concentration. This mixture is then extracted by vortexing with 2 ml 1 : 1 isobutanol–benzene (IBB), centrifuged for 3 min at 2000 $g$, and the IBB aspirated off and discarded. This extraction is repeated twice, and then 0.2 ml acetone and 0.6 ml of 0.06 $M$ ammonium molybdate in 0.02 $N$ HCl are added. [The latter solution is prepared once a month by dissolving 7.42 g of $(NH_4)_6Mo_7O_{24} \cdot 4H_2O$ in less than 100 ml of water, adding 0.17 ml HCl, and then bringing to 100 ml volume. The solution is filtered and stored in the refrigerator.] The molybdate–phosphate complex is then extracted by vortexing and, when separated, the organic phase is collected into a tube containing 2 ml 1 $N$ HCl. The remaining aqueous phase is vortexed again following the addition of 0.1 ml acetone and 0.5 ml IBB; the organic phase is collected and added to the previously collected organic phase. This HCl and IBB mixture is vortexed, and the organic phase transferred into a conical centrifuge tube to which 0.7 ml of a 5 $N$ $NH_4OH$–1 $N$ $NH_4Cl$ solution are added. [The latter solution is made by dissolving 2.41 g of $NH_4Cl$ in 30 ml of water, and adding 15 ml of concentrated (15 $M$) $NH_4OH$.] Following a thorough vortexing, the organic phase is discarded. 0.5 ml of a "precipitating" mixture (this mixture is made by dissolving 5.5 g of $MgCl_2 \cdot 6H_2O$ + 10 g $NH_4Cl$ in 60 ml water, adding 10 ml of concentrated (15 $M$) $NH_4OH$, and bringing up the volume to 100 ml) is then added, and the mixture is incubated in ice for 30 min. During this time a white precipitate appears, which is centrifuged down and washed twice with 2 ml 5 $N$ $NH_4OH$ (cold). The final precipitate is dissolved in a few drops (just enough to dissolve the precipitate) of 1 $N$ HCl, and then some water is added to make a convenient stock solution of radioactive phosphate.

Under certain conditions, such as acid pH and/or in the presence of organic solvents,[45,47] phosphorylation of SR ATPase with $P_i$ is very much favored and the resulting $EP/E \cdot P_i$ ratios are greater than 10. Therefore, in the presence of saturating $P_i$,[10] the equilibrium level of EP is an approximate measure of the total number of phosphorylation (i.e., catalytic) sites in a preparation of SR vesicles or purified enzyme. If denaturation is avoided during the preparation, the number of phosphorylation sites is found to be nearly equal to the number of 110,000 MW chains (i.e., one catalytic site per chain).

[47] L. deMeis, O. Martins, and E. Alves, *Biochemistry* **19**, 4252 (1980).

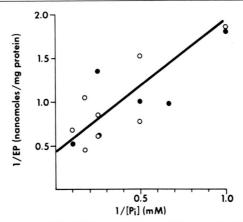

FIG. 15. Phosphorylation of SR ATPase with $P_i$ at equilibrium. SR ATPase was incubated with various concentrations of $[^{32}P]P_i$ in the presence of 20 m$M$ MOPS (pH 6.8), 80 m$M$ KCl, 10 m$M$ MgCl$_2$, and 0.5 mM EGTA. Following 5 min incubation at 25°, the reaction was acid quenched and the phosphoenzyme measured. (From Inesi *et al.*[48])

When SR vesicles are equilibrated with various concentrations of $P_i$ in the absence of $Ca^{2+}$, at neutral pH and in the presence of $Mg^{2+}$ and $K^+$ concentrations commonly used in studies of ATPase activity, the maximal EP level is found to correspond to approximately half the catalytic sites present in any given preparation (Fig. 15).[48] Therefore, $K_{eq} = EP/E \cdot P_i = 1$. Under these conditions, the $P_i$ concentration yielding half-maximal EP is 3 m$M$.

If the $[^{32}P]EP$ obtained by equilibration of SR ATPase with $[^{32}P]P_i$ is suddenly diluted with a large excess of nonradioactive $P_i$, the $[^{32}P]EP$ undergoes an exponential decay (Fig. 16) as nonradioactive EP is formed. Thereby a rate constant for the EP hydrolytic cleavage is obtained. In media identical to those used in ATPase studies, the rate constant of EP hydrolytic cleavage is approximately 60 sec$^{-1}$. Since under these conditions $K_{eq} = 1$, the reverse rate constant (i.e., enzyme phosphorylation) must be also approximately 60 sec$^{-1}$.

Experiments and analysis on the phosphorylation reaction of SR ATPase with $P_i$ have been carried out in great detail.[43,45,49,50]

[48] G. Inesi, M. Kurzmack, D. Kosk-Kosicka, D. Lewis, H. Scofano, and H. Guimaraes-Motta, *Z. Naturforsch. C* **37C,** 685 (1982).
[49] C. Punzengruber, R. Prager, N. Kolassa, F. Winkler, and J. Suko, *Eur. J. Biochem.* **92,** 349 (1978).
[50] J. Lacapere, M. Gingold, P. Champeil, and F. Guillain, *J. Biol. Chem.* **256,** 2302 (1981).

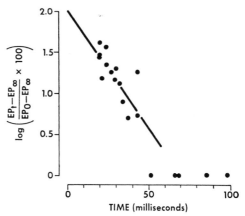

FIG. 16. Hydrolytic cleavage of phosphoenzyme obtained with $P_i$. [³²P]Phosphoenzyme was first obtained by preincubation of SR vesicles with [³²P]$P_i$, 10 m$M$ MgCl₂, and 0.5 m$M$ EGTA, at pH 6.0 and 25°. The reaction mixture was then diluted with a 12-fold excess of nonradioactive $P_i$, and the medium composition changed to 80 m$M$ KCl, 10 m$M$ MgCl₂, and pH 6.8. Acid quenching was carried out at serial times for determination of residual [³²P]phosphoenzyme. Rapid mixing was obtained with a Dionex multimixer. (From Inesi et al.[48])

## ATP Synthesis

It was originally found by Makinose and Hasselbach[51] that the calcium pump of SR can be reversed, and that efflux of calcium from loaded vesicles is accompanied by synthesis of ATP. As long as the Ca²⁺ concentration remains high inside the vesicles, and low (below $10^{-7}$ $M$) outside the vesicles, repeated cycles of enzyme phosphorylation by $P_i$ and phosphoryl transfer to ADP occur simultaneously with Ca²⁺ efflux from the vesicles. It was also demonstrated that if SR ATPase (reassembled in leaky vesicles) is phosphorylated with $P_i$ in the absence of Ca²⁺, addition of millimolar Ca²⁺ and ADP is followed by a single cycle of phosphoryl transfer from the EP to ADP.[44,52] It is apparent that calcium binding to the transport sites (in the low-affinity and inward-oriented state) is required to generate the phosphorylation potential for ATP synthesis.

Experiments on ATP synthesis include formation of [³²P]EP (with acetyl [³²P]phosphate in the forward, or with [³²P]$P_i$ in the reverse direction of the ATPase cycle), and then addition of ADP and other cofactors. Quenching with 0.125 $N$ (final concentration) perchloric acid is carried out

[51] M. Makinose and W. Hasselbach, FEBS Lett. **12,** 271 (1971).
[52] A. Knowles and W. Racker, J. Biol. Chem. **250,** 1949 (1975).

at serial times following the addition of ADP, and the quenched samples are centrifuged for 5 min at 5000 $g$. In these experiments it is important to measure the kinetics of both EP disappearance and ATP formation. Determination of [$^{32}$P]EP is performed with the sediment as described in a previous section. For determination of [$\gamma$-$^{32}$P]ATP, the supernatant is neutralized and, after the addition of nonradioactive ATP carrier, an aliquot is passed through an anion-exchange HPLC column (e.g., Partisil 10 SAX). Differential elution of nucleotides is obtained with a $P_i$ gradient, and the newly formed ATP is estimated by measurements of radioactivity. If high concentrations of [$^{32}$P]$P_i$ are used in the reaction mixture, it is advisable to remove it after quenching and centrifugation, and before HPLC. For this purpose, 0.4 ml of concentrated HCl, 0.75 ml of 60 m$M$ ammonium molybdate in 10 m$M$ HCl, and 0.6 ml acetone, are added to 3.0 ml of the quenched sample supernatant. Extraction is then carried out with 8 ml of 1 : 1 isobutanol : benzene.

An informative experiment on the kinetics of ATP synthesis can be performed by first loading SR vesicles with calcium, using acetyl [$^{32}$P]phosphate as a substrate. When the vesicles are loaded with calcium and steady-state levels of [$^{32}$P]EP are obtained, addition of ADP produces stoichiometric transformation of [$^{32}$P]EP into [$\gamma$-$^{32}$P]ATP. As shown by Pickart and Jencks,[53] this transformation occurs with two discrete kinetic components (Fig. 17): a fast component (450 sec$^{-1}$) due to the actual phosphoryl transfer (reaction 1 in Scheme 6 with $K_{eq} \cong 3$), and a slower component (35 sec$^{-1}$) due to slow dissociation of the newly formed ATP from the enzyme (reaction 2 in Scheme 6) and reequilibration of the phosphorylation reaction:

$$Ca_2 \cdot E \sim P \cdot ADP \overset{1}{\rightleftharpoons} Ca_2 \cdot E \cdot ATP \overset{2}{\rightleftharpoons} Ca_2 \cdot E + ATP$$

SCHEME 6

It should be understood that the ATP measured after acid quenching corresponds to the total amount of ATP formed at any time, including ATP and $Ca_2 \cdot E \cdot ATP$.

It is of interest that if [$^{32}$P]EP is obtained by incubating SR ATPase (reassembled in leaky vesicles) with [$^{32}$P]$P_i$ in the absence of $Ca^{2+}$, addition of $Ca^{2+}$ and ADP yields some [$\gamma$-$^{32}$P]ATP.[44,52] In this case, the kinetic pathway most likely includes a calcium-induced, EP transition to a form that is able to phosphorylate ADP, as in Scheme 7.

$$*E\text{-}P + 2Ca^{2+} + ADP \overset{1}{\rightleftharpoons} Ca_2 \cdot E \sim P \cdot ADP \overset{2}{\rightleftharpoons} Ca_2 \cdot E \cdot ATP \overset{3}{\rightleftharpoons} Ca^{2+} \cdot E + ATP$$

SCHEME 7

[53] C. Pickart and W. Jencks, *J. Biol. Chem.* **257**, 5319 (1982).

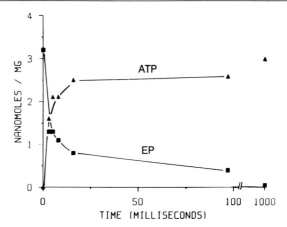

Fig. 17. Phosphoenzyme decay and ATP formation following addition of ADP to SR vesicles filled with calcium using acetyl [³²P]phosphate as a substrate. SR vesicles were first incubated for 1 min in the presence of 50 mM MOPS (pH 6.8), 80 mM KCl, 10 mM MgCl₂, 1 mM EGTA, 1.05 mM CaCl₂, and 2 mM acetyl [³²P]phosphate. Formation of ATP was started by addition of 3 mM ADP and 20 mM EGTA. Quenching at serial times was obtained with Froehlich–Berger mixer. The reaction was carried out at 25°. (From Fernandez-Belda and Inesi.[54])

Formation of ATP in these conditions is likely to be a transient phenomenon related to the pathway for ligand-induced reequilibration of the system.[54]

### Exchange Reactions

Although the catalytic cycle of SR ATPase includes several intermediate steps that can be measured *directly,* a number of exchange reactions yield useful although indirect information on the catalytic mechanism.

### *ATP ⇌ ADP Exchange*

Scheme 8 entails the exchange of the adenosine moiety between ADP and ATP and requires formation of a phosphorylated enzyme intermediate, as well as equilibration of free and bound nucleotides:

$$ATP + E \cdot Ca_2 \rightleftharpoons E{\sim}P \cdot Ca + ADP$$
$$[^{14}C]ATP \qquad\qquad\qquad [^{14}C]ADP$$

SCHEME 8

[54] F. Fernandez-Belda and G. Inesi, *Biochemistry* **25,** 8083 (1986).

Thereby, under conditions permitting enzyme phosphorylation by nonradioactive ATP and its reversal, addition of ADP, radioactively labeled in its adenosine moiety, yields radioactive ATP by exchange of radioactive ADP.[12,55] This exchange can be useful to assess the rate of reversal of enzyme phosphorylation with ATP under certain conditions, i.e., various $Ca^{2+}$ concentrations, empty versus filled vesicles, etc. Measurement of ATP $\rightleftharpoons$ ADP exchange requires acid quenching at serial times, neutralization of the quenched sample's supernatant, and separation of the nucleotides by anion-exchange HPLC as described for ATP synthesis.

## $ATP \rightleftharpoons P_i$ Exchange

This reaction entails exchange of the ATP terminal phosphate with medium $P_i$, and requires formation of a phosphorylated enzyme intermediate *as well as* hydrolytic cleavage of the phosphorylated intermediate, with equilibration between bound and free $P_i$ and ATP (Scheme 9).

$$ATP + E \cdot Ca_2 \rightleftharpoons (ADP \cdot E\text{-}P \cdot Ca_2) \rightleftharpoons ADP + 2Ca_2 + E + P_i$$
$$[\gamma\text{-}^{32}P]ATP \qquad\qquad\qquad\qquad\qquad [^{32}P]P_i$$

SCHEME 9

Thereby, under conditions permitting enzyme phosphorylation with nonradioactive ATP and hydrolytic cleavage of the EP, the presence of radioactive $P_i$ yields radioactive ATP. This exchange is useful to assess the rates of reversal of the entire cycle. For instance, it was originally demonstrated by ATP $\rightleftharpoons$ $P_i$ exchange that reversal is greatly increased when net dissociation of $Ca^{2+}$ from the EP is reduced by increasing the $Ca^{2+}$ concentration to the millimolar level.[56]

Measurement of ATP $\rightleftharpoons$ $P_i$ exchange requires the use of purified $[^{32}P]P_i$, acid quenching at serial times, and chromatographic separation of nucleotides as described for ATP synthesis.

## $P_i \rightleftharpoons HOH$ Exchange

This reaction entails exchange of oxygen between water and $P_i$. During formation and hydrolysis of EP (assuming attack on the phosphorus atom), there is an obligatory incorporation of one oxygen atom from water into the $P_i$ released from the EP in Scheme 10. If before being released, enzyme-bound $P_i$ reacts again with the enzyme to form EP,

[55] G. Inesi and J. Almendares, *Arch. Biochem. Biophys.* **126**, 733 (1968).
[56] L. deMeis and M. Carvalho, *J. Biol. Chem.* **251**, 1413 (1976).

hydrolysis of this species can yield a $P_i$ product with an additional oxygen atom(s) from water. The extent of this process is dependent on the relative rate constants of EP formation and $P_i$ release from $E \cdot P_i$, the rotational freedom of enzyme-bound $P_i$, and the exchange of $H_2O$ at the site with the medium.[57]

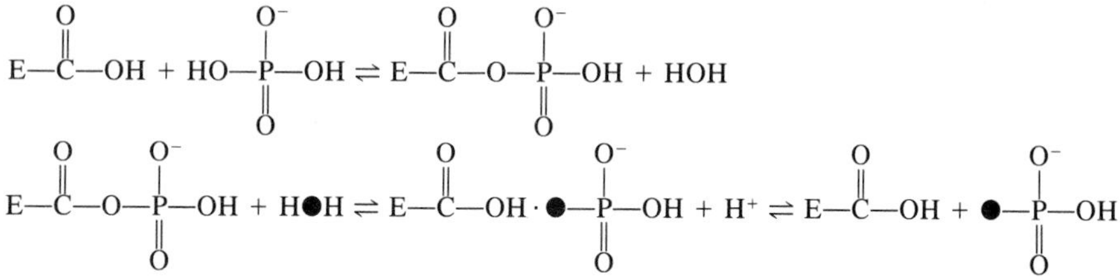

$$
\begin{array}{c}
\text{O} \qquad\qquad \text{O}^- \qquad\qquad \text{O} \qquad\qquad \text{O}^- \\
\parallel \qquad\qquad \parallel \qquad\qquad \parallel \qquad\qquad \parallel \\
\text{E—C—OH + HO—P—OH} \rightleftharpoons \text{E—C—O—P—OH + HOH} \\
\qquad\qquad\qquad \parallel \qquad\qquad\qquad\qquad \parallel \\
\qquad\qquad\qquad \text{O} \qquad\qquad\qquad\qquad \text{O}
\end{array}
$$

$$
\begin{array}{c}
\text{O} \quad \text{O}^- \qquad\qquad\qquad \text{O} \qquad \text{O}^- \qquad\qquad\qquad \text{O} \qquad \text{O}^- \\
\parallel \quad \parallel \qquad\qquad\qquad \parallel \qquad \parallel \qquad\qquad\qquad \parallel \qquad \parallel \\
\text{E—C—O—P—OH + H}{\bullet}\text{H} \rightleftharpoons \text{E—C—OH} \cdot {\bullet}\text{—P—OH + H}^+ \rightleftharpoons \text{E—C—OH + }{\bullet}\text{—P—OH} \\
\qquad\quad \parallel \qquad\qquad\qquad\qquad\qquad\qquad\qquad \parallel \qquad\qquad\qquad\qquad\qquad\qquad\qquad \parallel \\
\qquad\quad \text{O} \qquad\qquad\qquad\qquad\qquad\qquad\qquad \text{O} \qquad\qquad\qquad\qquad\qquad\qquad\qquad \text{O}
\end{array}
$$

SCHEME 10

The reaction is defined as "medium" exchange when $[^{16}O]P_i$ is used as the substrate for enzyme phosphorylation, and the exchange occurs with $[^{18}O]HOH$. The reaction is defined as "intermediate" exchange when $[\gamma\text{-}^{18}O]ATP$ is used as the substrate for enzyme phosphorylation, and then the exchange occurs between the $[^{18}O]P_i$ derived from this EP and $[^{16}O]HOH$.

The distribution of isotopic species in the acid-quenched samples is assessed by extraction of $P_i$ and analysis by mass spectroscopy for determination of species containing zero to four $^{18}O$ atoms per $P_i$. The partition coefficient ($P_c = k_2/k_2 + k_{-1}$) is close to 1 when the $P_i$ bound to the enzyme undergoes many reversals of step 2 before being released, and loses all its $^{18}O$ by exchange with water; this indicates that $k_{-1}$ is much lower than $k_2$. The partition coefficient is close to zero when only one oxygen atom is exchanged per productive encounter of $P_i$ with the enzyme; this indicates that $k_{-1}$ is much higher than $k_2$. This analysis, of course, assumes that the bound $P_i$ undergoes fast rotation within the binding site on the enzyme, exposing all its oxygen atoms to rapid exchange with water within one phosphorylation cycle, and that the catalytic water exchanges very rapidly with the medium.

Studies of $P_i \rightleftharpoons HOH$ exchange in SR ATPase have demonstrated that release of bound $P_i$ is considerably slower than its hydrolytic cleavage from the phosphoenzyme.[58,59] An advantage of these studies is that the pertinent rate constants can be estimated in the presence of all substrates and cofactors permitting the entire enzyme cycle to occur in the steady

[57] D. Hackney and P. Boyer, *Proc. Natl. Acad. Sci. USA* **75**, 3133 (1978).
[58] T. Kanazawa and P. Boyer, *J. Biol. Chem.* **248**, 3163 (1973).
[59] D. McIntosh and P. Boyer, *Biochemistry* **22**, 2867 (1983).

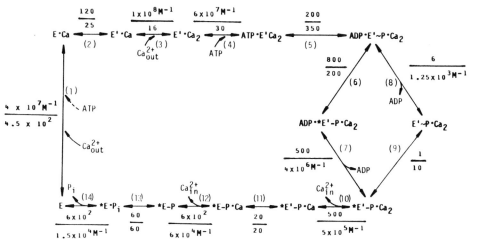

FIG. 18. Reaction scheme for the catalytic cycle and coupled calcium transport in SR ATPase. The rate constants are expressed in sec⁻¹ time units, with the top values representing the forward constants in the clockwise direction of the cycle. (From Nakamura et al.[38])

state. Thereby, the effects of these ligands on the pertinent rate constants can be tested.

### Analysis of Experimental Data

It is common to obtain an immediate evaluation of experimental results by submitting them to well-known graphic manipulations such as the Hill and Scatchard plots for binding data, or Lineweaver–Burke and similar plots for steady-state enzyme catalysis. These plots, however, are based on simple and fixed assumptions, and do not allow testing of different and more complex mechanisms that may be suggested by the behavior of the system. In our laboratory, we prefer a direct display of the experimental data, attempting to fit them with equations derived from a postulated reaction scheme. An example of this type of analysis is given in the section on calcium binding, as it relates to the equilibrium and kinetic behavior of a partial reaction studied under conditions (i.e., absence of ATP) separating it from the entire ATPase cycle. Similar analysis can be performed with regards to phosphorylation of the enzyme with $P_i$, in the absence of $Ca^{2+}$ and ATP.[45,49]

A more complex analysis is required when the behavior of the entire ATPase cycle needs to be simulated. For instance, based on the experimental behavior of the enzyme under appropriate conditions, we have recently considered a scheme for the entire ATPase and transport cycle (Fig. 18). This scheme includes a stepwise enzyme activation by $2Ca^{2+}$ (as suggested by the analysis of this reaction), and a branched pathway for release of ADP and transition of the phosphoenzyme to the inward-oriented and low-affinity state. It should be emphasized that this scheme *does not provide a unique solution* of the enzyme mechanism, but simply an example of the analysis that can be carried out to *test if any proposed mechanism fits a set of experimental data and is thereby possible*. We will concentrate here on how the analysis is developed rather than the justification for each step.[38]

We may first write an equation derived from the postulated reaction mechanism to define the flux to each species, based on the forward ($k$) and reverse ($r$) rate constants shown in the reaction scheme.

$$\frac{d[E]}{dt} = -k_1[Ca_{out}^{2+}][E] + r_1[ECa] + k_{14}[{}^{*}EP_i] - r_{14}[E][P_i]$$

$$\frac{d[ECa]}{dt} = k_1[Ca_{out}^{2+}][E] + r_2[E'Ca] - [ECa](r_1 + k_2)$$

$$\frac{d[E'Ca]}{dt} = k_2[ECa] + r_3[E'Ca_2] - [E'Ca](r_2 + k_3[Ca_{out}^{2+}])$$

$$\frac{d[E'Ca_2]}{dt} = k_3[E'Ca][Ca_{out}^{2+}] + r_4[E'Ca_2ATP] - [E'Ca_2](r_3 + k_4[ATP])$$

$$\frac{d[E'Ca_2ATP]}{dt} = k_4[E'Ca_2][ATP] + r_5[ADPE'{\sim}PCa_2]$$
$$- [E'Ca_2ATP](r_4 + k_5)$$

$$\frac{d[ADPE'{\sim}PCa_2]}{dt} = k_5[E'Ca_2ATP] + r_6[ADP^{*}E'\text{-}PCa_2]$$
$$+ r_8[E'{\sim}PCa_2][ADP]$$
$$- [ADPE'{\sim}PCa_2](r_5 + k_6 + k_8)$$

$$\frac{d[ADP^{*}E'\text{-}PCa_2]}{dt} = k_6[ADPE'{\sim}PCa_2] + r_7[{}^{*}E'\text{-}PCa_2][ADP]$$
$$- [ADP^{*}E'\text{-}PCa_2](r_6 + k_7)$$

$$\frac{d[{}^{*}E'\text{-}PCa_2]}{dt} = k_7[ADP^{*}E'\text{-}PCa_2] + k_9[E'{\sim}PCa_2] + r_{10}[{}^{*}E'\text{-}PCa][Ca_{in}^{2+}]$$
$$- [{}^{*}E'\text{-}PCa_2](r_7[ADP] + r_9 + k_{10})$$

$$\frac{d[E'{\sim}PCa_2]}{dt} = k_8[ADPE'{\sim}PCa_2] + r_9[{}^{*}E'\text{-}PCa_2]$$
$$- [E'{\sim}PCa_2](r_8[ADP] + k_9)$$

$$\frac{d[*E'\text{-}PCa]}{dt} = k_{10}[*E'\text{-}PCa_2] + r_{11}[*E\text{-}PCa] - [*E'\text{-}PCa](r_{10}[Ca_{in}^{2+}] + k_{11})$$

$$\frac{d[*E\text{-}PCa]}{dt} = k_{11}[*E'\text{-}PCa] + r_{12}[*E\text{-}P][Ca_{in}^{2+}] - [*E\text{-}PCa](r_{11} + k_{12})$$

$$\frac{d[*E\text{-}P]}{dt} = k_{12}[*E\text{-}PCa] + r_{13}[*EP_i] - [*E\text{-}P](r_{12}[Ca_{in}^{2+}] + k_{13})$$

$$\frac{d[*EP_i]}{dt} = k_{13}[*E\text{-}P] + r_{14}[E][P_i] - [*EP_i](r_{13} + k_{14})$$

There are many mathematical methods available for solving differential equations. We have chosen to use a Runge–Kutta method[60] because it does not require the calculation of prior values to solve the first iteration. Each step is divided into four substeps, the equations are evaluated for each of the substeps, and then a weighted average is used to calculate the change in each species for the desired time change. Since the concentration of each species is dependent on both the previous species and the following species, it is necessary to recalculate the concentrations until a self-consistent solution is calculated for each particular time. Three iterations at each time have proven to be sufficient.

Choosing the correct step size is very important. Too small a step size can lead to excessive roundoff error and long simulation times because of unnecessary iterations. Too large a step size can lead to instability in the calculations. In general, the optimal step size is determined by the fastest rate in the reaction cycle. In the scheme given above, the fastest rate is usually one of the second-order constants. At a reasonable [Ca$_{out}$] or ATP concentration of 10 to 100 $\mu M$, the rates will be on the order of $10^2$ to $10^5$ per second.

We have found that an initial step size of 0.02 to 0.05 msec is sufficiently small to maintain stability and still remain within the portion of the exponential change in the species concentration which can be approximated as a linear function. As the enzyme species reequilibrate in the slower first-order portions of the cycle, it is feasible to increase the step size gradually to 0.25 msec. Since a change of step size from 0.025 to 0.25 msec will decrease computational time by a factor of ten, it is obviously advantageous to increment the step size as soon as possible. It is important to note that higher substrate concentrations or faster reaction rates in a cycle will require smaller step sizes and longer computational time.

There are two major sources of computational error in the simulations. These are from the roundoff error due to the limited accuracy with which

[60] R. LaFara, "Computer Methods for Science and Engineering," p. 237. Hayden, Rochelle Park, New Jersey, 1973.

the computer handles numbers and the error due to the Runge–Kutta approximation of the changes in the species concentrations. We reduce the effect of roundoff error by using a 14-digit precision binary-coded decimal (BCD) version of BASIC (MEGABASIC from American Planning Corp., Alexandria, Va). The BCD coding prevents errors from the conversion from base 10 to binary. Since the number of sites is fixed, it is possible to calculate the error after each set of iterations by summing all the species and comparing to the total number of sites. In general, it is reasonable to strive for accuracy in the species concentrations to 0.01 nmol/mg protein. Roundoff errors are generally on the order of $10^{-12}$ nmol/mg, while actual errors in calculation from the approximation can usually be tolerated up to $10^{-4}$ before the calculations become unstable. We have arbitrarily chosen to decrease the step size if the error is greater than $2 \times 10^{-5}$ and increase the step size if the error is less than $5 \times 10^{-7}$. To prevent wasted calculations, the step size is evaluated only at intervals of one twenty-fifth of the total simulation time. We still must deal with the buildup of error after each iteration cycle. To do this, the species with the largest concentration is adjusted by the error after each iteration cycle. The adjustment is generally less than $10^{-5}$ on a concentration on the order of 1 nmol/mg. This prevents a buildup of error after a large number of iterations.

An example of the numerical solution for the concentrations of intermediate species considered in scheme, and of the steady-state substrate flux at fixed ATP, ADP, $Ca^{2+}_{out}$, and $Ca^{2+}_{in}$ concentrations is given in Table I. The calculated transients in the concentrations of intermediate species upon a sudden change of ATP from 10 $\mu M$ to zero are also shown. These simulations can be compared to experimental data, and are very helpful in testing whether a postulated reaction scheme is altogether possible.[38]

Although the methods of analysis described in this chapter are able to deal with detailed reaction mechanisms, including several steps and cooperative interactions, they are strictly empirical and include approximations related to the lack of statistical treatment. It should be pointed out that analysis based on fundamental molecular considerations has also been applied to equilibrium and steady-state phenomena of the SR ATPase.[61,62] The reader interested in these methods is referred to their appropriate presentations.[63,64]

[61] T. Hill and G. Inesi, *Proc. Natl. Acad. Sci. USA* **79**, 3978 (1982).
[62] G. Inesi and T. Hill, *Biophys. J.* **44**, 271 (1983).
[63] T. Hill, "Statistical Thermodynamics," p. 1. Addison-Wesley, Reading, Massachusetts, 1960.
[64] T. Hill, "Cooperative Theory in Biochemistry: Steady State and Equilibrium System," p. 1. Springer-Verlag, Berlin and New York, 1985.

TABLE I

FORMATION AND DECAY OF PHOSPHORYLATED INTERMEDIATE[a]

| | Time (milliseconds) | | | | | | |
|---|---|---|---|---|---|---|---|
| Species | 0 | 8 | 20 | 40 | 80 | 200 | 1000 |
| **Formation** | | | | | | | |
| E | 5.00 | 1.98 | 1.08 | 0.51 | 0.50 | 0.52 | 0.45 |
| E·Ca | 0.00 | 1.53 | 0.84 | 0.37 | 0.35 | 0.37 | 0.32 |
| E'·Ca | 0.00 | 0.19 | 0.10 | 0.05 | 0.04 | 0.04 | 0.04 |
| E'·Ca$_2$ | 0.00 | 0.04 | 0.02 | 0.01 | 0.01 | 0.01 | 0.01 |
| ATP·E'·Ca$_2$ | 0.00 | 0.82 | 1.00 | 0.53 | 0.32 | 0.35 | 0.30 |
| ADP·E'~P·Ca$_2$ | 0.00 | 0.15 | 0.22 | 0.12 | 0.07 | 0.07 | 0.07 |
| ADP·*E'-P·Ca$_2$ | 0.00 | 0.13 | 0.25 | 0.15 | 0.08 | 0.09 | 0.07 |
| *E'-PCa$_2$ | 0.00 | 0.08 | 0.25 | 0.17 | 0.10 | 0.11 | 0.09 |
| E'~P·Ca$_2$ | 0.00 | 0.00 | 0.04 | 0.11 | 0.17 | 0.32 | 0.91 |
| *E'-P·Ca | 0.00 | 0.08 | 1.08 | 2.41 | 2.37 | 2.20 | 1.93 |
| *E-P·Ca | 0.00 | 0.00 | 0.03 | 0.08 | 0.08 | 0.07 | 0.06 |
| *E-P | 0.00 | 0.00 | 0.08 | 0.47 | 0.83 | 0.78 | 0.69 |
| *E·P$_i$ | 0.00 | 0.00 | 0.01 | 0.04 | 0.08 | 0.07 | 0.06 |
| Total EP | 0.00 | 0.44 | 1.96 | 3.50 | 3.70 | 3.64 | 3.82 |

| | Time (milliseconds) | | | | | | |
|---|---|---|---|---|---|---|---|
| | 0 | 40 | 80 | 120 | 200 | 400 | 600 | 1000 |
| **Decay** | | | | | | | |
| E | 0.10 | 0.39 | 0.29 | 0.17 | 0.06 | 0.03 | 0.03 | 0.02 |
| E·Ca | 0.00 | 0.28 | 0.21 | 0.12 | 0.05 | 0.02 | 0.02 | 0.02 |
| E'·Ca | 0.00 | 0.05 | 0.06 | 0.06 | 0.06 | 0.06 | 0.07 | 0.07 |
| E'·Ca$_2$ | 0.00 | 0.97 | 2.12 | 2.84 | 3.46 | 3.82 | 4.02 | 4.30 |
| ATP·E'·Ca$_2$ | 0.32 | 0.00 | 0.00 | 0.00 | 0.00 | 0.00 | 0.00 | 0.00 |
| ADP·E'~P·Ca$_2$ | 0.07 | 0.00 | 0.00 | 0.00 | 0.00 | 0.00 | 0.00 | 0.00 |
| ADP·*E'-P·Ca$_2$ | 0.08 | 0.00 | 0.00 | 0.00 | 0.00 | 0.00 | 0.00 | 0.00 |
| *E'-PCa$_2$ | 0.10 | 0.02 | 0.01 | 0.01 | 0.00 | 0.00 | 0.00 | 0.00 |
| E'~P·Ca$_2$ | 1.43 | 1.39 | 1.34 | 1.29 | 1.20 | 0.98 | 0.81 | 0.55 |
| *E'-P·Ca | 2.02 | 1.23 | 0.61 | 0.32 | 0.12 | 0.05 | 0.04 | 0.03 |
| *E-P·Ca | 0.07 | 0.04 | 0.02 | 0.01 | 0.00 | 0.00 | 0.00 | 0.00 |
| *E-P | 0.74 | 0.58 | 0.32 | 0.16 | 0.05 | 0.02 | 0.02 | 0.01 |
| *E·P$_i$ | 0.07 | 0.05 | 0.03 | 0.02 | 0.00 | 0.00 | 0.00 | 0.00 |
| Total EP | 4.51 | 3.26 | 2.30 | 1.79 | 1.37 | 1.06 | 0.87 | 0.59 |

[a] Example of computer-simulated, time-dependent changes of various intermediate species of the SR ATPase cycle. The cycle and rate constants are those displayed in Fig. 18. The formation reaction is started by the addition of ATP to enzyme saturated with calcium. Phosphoenzyme decay is initiated by sudden reduction of the ATP concentration to zero.

Although this chapter is mostly devoted to methods, it should be pointed out that equilibrium and kinetic characterization of an energy-transducing enzyme and its partial reactions unveils the *sequence of events that must take place to satisfy the mechanism of energy transduction*. Furthermore, the values obtained for the microscopic and overall equilibria permit accounts of energy expenditure and utilization as discussed elsewhere.[1,65–67]

*Note:* Several of the computer programs described in the review, including those for calculation of free calcium concentration, nonlinear regression fitting of the Hill equation, and simulation of equilibrium and transient calcium binding, are available on disk for PC compatibles from Dr. G. Inesi, Department of Biochemistry, University of Maryland, 660 West Redwood, Baltimore, Maryland 21201.

### Acknowledgments

The development of methods and performance of the experiments described in this chapter were supported by the National Institutes of Health (HL 27867) and the Muscular Dystrophy Association of America. The authors are indebted to Drs. L. deMeis, R. Nakamoto, and S. Verjovski-Almeida for their association in the laboratory and their help in the acquisition of these methods.

[65] T. Hill and E. Eisenberg, *Q. Rev. Biophys.* **14**, 1 (1981).
[66] C. Tanford, *Annu. Rev. Biochem.* **52**, 379 (1983).
[67] C. Pickart and W. Jencks, *J. Biol. Chem.* **259**, 1629 (1984).

## [13] Approaches to Studying the Mechanisms of ATP Synthesis in Sarcoplasmic Reticulum

### By Leopoldo de Meis

After the proposal of the chemiosmotic theory by Mitchell[1] it has been recognized that different membrane-bound enzymes are able to use the energy derived from ionic gradients for the synthesis of ATP. These include the Ca$^{2+}$-ATPase of sarcoplasmic reticulum, the F$_1$-ATPase of mitochondria and chloroplasts, and the Na$^+$,K$^+$-ATPase of plasma membrane. In these systems the process of energy transduction is fully reversible. The enzyme can use the energy derived from the hydrolyis of ATP to build up a concentration gradient of ions across the membrane and, in the reverse process, use the energy derived from the gradient to synthesize ATP. The mechanism by which the energy derived from the gradients is used by membrane-bound enzymes to catalyze the synthesis of ATP is

[1] P. Mitchell, *Eur. J. Biochem.* **95**, 1 (1979).

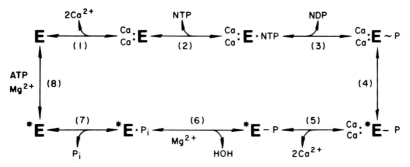

FIG. 1. Steps involved in the process of ATP hydrolysis and $Ca^{2+}$ transport. See text for details.

still far from understood. Among the different enzymes studied, the $Ca^{2+}$-ATPase of the sarcoplasmic reticulum provides the most knowledge about the mechanism of energy transduction. Experimental conditions allow us to move the different steps of the catalytic cycle of the enzyme in the direction of ATP synthesis. Thus, ATP can be attained after a single catalytic cycle in the absence of a transmembrane $Ca^{2+}$ gradient. The net synthesis of ATP can be promoted by a variety of perturbations, including $Ca^{2+}$, pH, temperature, and water activity. The experiments indicate that during the catalytic cycle different forms of energy are interconverted by the $Ca^{2+}$-dependent ATPase (for review see Refs. 2–5).

Catalytic Cycle of $Ca^{2+}$-ATPase

The reaction sequence shown in Fig. 1 describes the steps involved in the process of ATP hydrolysis and $Ca^{2+}$ transport.[6] This sequence includes two distinct functional states of the enzyme, E and *E. The $Ca^{2+}$ binding site in the E form faces the outer surface of the vesicle and has an apparent $K_m$ for $Ca^{2+}$ in the range of 0.2–2 $\mu M$ at pH 7.0 (high affinity). In the *E form the $Ca^{2+}$-binding site faces the inner surface of the vesicles and has an apparent $K_m$ for $Ca^{2+}$ in the range of 1–3 m$M$ at pH 7.0 (low affinity). The E form is phosphorylated by nucleoside triphosphate (NTP) but not by $P_i$, while the form *E is phosphorylated by $P_i$ but not by NTP. Reaction 8 can be the rate-limiting step, depending on the NTP used.

[2] L. de Meis and A. L. Vianna, *Annu. Rev. Biochem.* **48**, 275 (1979).

[3] L. de Meis, "The Sarcoplasmic Reticulum, Transport and Energy Transduction" ( E. E. Bittar, ed.), Transport in the Life Sciences, Vol. 2. Wiley. New York, 1981.

[4] W. Hasselbach, *Annu. Rev. Physiol.* **45**, 325 (1983).

[5] C. Tanford, *Crit. Rev. Biochem.* **17**, 123 (1984).

[6] M. G. C. Carvalho, D. O. Souza, and L. de Meis, *J. Biol. Chem.* **251**, 3629 (1976).

Conversion of *E into E occurs very slowly in the absence of NTP and increasing concentrations of NTP activate the rate of interconversion, ITP being much less effective than ATP. The cycle proceeds from reaction 1 to reaction 8 when the enzyme catalyzes the hydrolysis of ATP and in the reverse direction when ATP is synthesized. In either case, three different forms of phosphoenzyme appear in sequence. Among these forms, only Ca$_2 \cdot$ E~P is able to transfer its phosphate to ADP (reaction 3). The rate constant for the hydrolyis of *E-P (reactions 6 and 7) is much higher than that for the hydrolysis of Ca$_2 \cdot$ *E-P and Ca$_2 \cdot$E~P. This means that, during the catalytic cycle, virtually the only phosphoenzyme form hydrolyzed is *E-P (reactions 6 and 7). This chapter presents a brief account of the intermediary reactions involved in the process of ATP synthesis. After the description of the reaction, some of the methods used are described.

### Phosphorylation by P$_i$

The synthesis of ATP is initiated by the phosphorylation of an aspartyl residue located in the catalytic site of the enzyme by P$_i$, forming an acyl phosphate residue. The part of the enzyme phosphorylated by P$_i$ faces the outer surface of the vesicle.[7] When the reversal of the Ca$^{2+}$ pump was first described, it was thought that the energy derived from the Ca$^{2+}$ gradient was used to drive this reaction.[8–10] This conclusion was based on the fact that, in an aqueous medium, the $\Delta G$ of hydrolysis of an acyl phosphate residue is about $-8$ kcal/mol. Later it was shown that leaky vesicles or soluble Ca$^{2+}$-ATPase can be spontaneously phosphorylated by P$_i$, forming an acyl phosphate residue.[11–13] This finding shows that the energy derived from the Ca$^{2+}$ gradient is not necessary for the first step of the reversal of the Ca$^{2+}$-pump.

The apparent $K_m$ for P$_i$ varies significantly with the pH of the medium and according to whether leaky vesicles or vesicles previously loaded with Ca$^{2+}$ are used (Table I). The pH dependence is abolished and there is a drastic reduction of the apparent $K_m$ for P$_i$ when the water activity of the medium is decreased by the addition of organic solvents to the medium (Table I). These effects can be attained with different organic solvents such as dimethyl sulfoxide, ethylene glycol, and glycerol.[14] The true sub-

[7] L. de Meis and M. G. C. Carvalho, *J. Biol. Chem.* **251**, 1413 (1976).

[8] M. Makinose, *FEBS Lett.* **25**, 113 (1972).

[9] S. Yamada, M. Sumida, and Y. Tonomura, *J. Biochem. (Tokyo)* **72**, 1537 (1972).

[10] T. Kanazawa and P. D. Boyer, *J. Biol. Chem.* **248**, 3163 (1973).

[11] H. Masuda and L. de Meis, *Biochemistry* **12**, 4581 (1973).

[12] L. de Meis and H. Masuda, *Biochemistry* **13**, 2057 (1974).

[13] T. Kanazawa, *J. Biol. Chem.* **250**, 113 (1975).

[14] L. de Meis, O. B. Martins, and E. W. Alves, *Biochemistry* **19**, 4252 (1980).

TABLE I
APPARENT $K_m$ FOR $P_i$ AT DIFFERENT pH VALUES[a]

| Conditions | Apparent $K_m$ $(M)$ | | |
|---|---|---|---|
| | pH 6.0 | pH 7.0 | pH 8.0 |
| Vesicles loaded with calcium phosphate | $7 \times 10^{-4}$ | $2 \times 10^{-3}$ | — |
| Leaky vesicles | $2 \times 10^{-3}$ | $10^{-2}$ | $>>10^{-2}$ |
| Leaky vesicles in 40% (v/v) dimethyl sulfoxide | $7 \times 10^{-6}$ | $7 \times 10^{-6}$ | $2 \times 10^{-6}$ |

[a] Data shown are from Refs. 12, 14, and 22.

strate of the enzyme is $H_2PO_4^-$.[15,16] Magnesium is required for the phosphorylation of the ATPase in both the presence and absence of a transmembrane gradient. There is evidence that $Mg^{2+}$ and $P_i$ bind in random sequence to the enzyme, and the binding of one ionic species facilitates the binding of the other.[16,17] In optimal conditions, practically all enzyme units are phosphorylated by $P_i$[14,18] with the formation of 7 to 8 $\mu$mol of phosphoenzyme per gram of enzyme protein (MW 110,000 to 120,000). Both in the presence and absence of a $Ca^{2+}$ gradient, phosphorylation by $P_i$ is inhibited when $Ca^{2+}$ is added to the medium. This inhibition is promoted by the binding of $Ca^{2+}$ to a high-affinity site located on the portion of enzyme which faces the outer surface of the membrane.[3,11,13] Half-maximal inhibition is attained in the presence of $10^{-5}$ $M$ $Ca^{2+}$ at pH 6.0 and in the presence of $10^{-6}$ $M$ $Ca^{2+}$ at pH 7.0. In the presence of high concentration of organic solvents (e.g., 40% v/v dimethyl sulfoxide), inhibition is only observed when the $Ca^{2+}$ concentration is raised above $10^{-4}$ $M$ regardless of the pH value of the medium.[19] The inhibition by $Ca^{2+}$ is overcome when a low concentration of ATP (0.05 or 0.2 m$M$) or 0.1 to 5 m$M$ ITP or GTP is added to the assay medium. In these conditions, part of the enzyme unit is phosphorylated by the nucleotide and part is phosphorylated by $P_i$.[12] During hydrolyis of the nucleotide the enzyme cycles between two different forms. One of them binds $Ca^{2+}$ with high affinity and is phosphorylated only by the nucleotide and the other form has a very low affinity for $Ca^{2+}$ and is phosphorylated only by $P_i$ (E and *E in Fig. 1). Phosphorylation by $P_i$ is progressively impaired when the temperature of the assay medium is decreased from 20 to 0°. This effect is abolished when the water activity of the medium is decreased with organic solvents.[13,14]

[15] F. U. Beil, D. Chak, and W. Hasselbach, *Eur. J. Biochem.* **81**, 151 (1977).
[16] G. Inesi, D. Lewis, and A. J. Murphy, *J. Biol. Chem.* **259**, 996 (1984).
[17] J. Suko, B. Plank, P. Preis, N. Kolassa, G. Hellmann, and W. Conca, *Eur. J. Biochem.* **119**, 225 (1981).
[18] H. Barrabin, H. M. Scofano, and G. Inesi, *Biochemistry* **23**, 1548 (1984).
[19] L. de Meis and G. Inesi, *Biochemistry* **24**, 922 (1985).

[$^{32}$P]P$_i$ Purification

Regardless of the source of the isotope, [$^{32}$P]P$_i$ ([$^{32}$P]orthophosphate) solutions usually contain radioactive contaminants which represent a frequent source of error in measurements of phosphoenzyme formation from P$_i$. The chemical nature of the contaminants is not known. With purification, the level of contaminants decreases, but they reappear on standing. This observation suggests that radiolysis may promote the formation of odd radioactive phospho compounds. The radioactive contaminants were first reported by Meyerhof and Green[20] who proposed that they were a mixture of radioactive metaphosphate and phosphite. In the purification method described below [$^{32}$P]P$_i$ is extracted as a phosphomolybdate with n-butyl acetate, reextracted into aqueous phase with ammonium hydroxide solution, and precipitated as the MgNH$_4$$^{32}$PO$_4$ salt.[21,22] After this procedure, the amount of radioactive contaminant represents less than 0.001% of the total radioactivity. It is recommended to repeat the purification when the [$^{32}$P]P$_i$ is stored for more than 2 weeks at room temperature. The solution of [$^{32}$P]P$_i$ is acidified with HCl to a final concentration of 1.3 $M$ and extracted with 5 volumes of n-butyl acetate. The organic phase is discarded, and to each milliliter of the water phase 0.25 ml of a 60 m$M$ ammonium molybdate solution in 0.01 $M$ HCl is added. The mixture is then extracted with 2 ml n-butyl acetate. After phase separation the yellow organic phase is removed and set aside. After addition of 0.1 ml of a 20 m$M$ KH$_2$PO$_4$ solution, the aqueous phase is extracted again with 2 ml n-butyl acetate. The organic phase is pooled with the organic phase obtained in the preceding extraction. The 4 ml n-butyl acetate sample is vigorously stirred with 50 ml 1 $M$ HCl. After phase separation, the organic phase is transferred to a 15-ml conical test tube and mixed with 1 ml of a solution containing 5 $M$ NH$_4$OH and 1 $M$ NH$_4$Cl. The mixture is vigorously stirred on a Vortex mixer for 1 min. After phase separation, the organic phase is discarded and the aqueous phase is cooled in ice and then mixed with 0.8 ml of an ice-cold solution containing 0.27 $M$ MgCl$_2$, 1.9 $M$ NH$_4$Cl, and 1.5 $M$ NH$_4$OH. After 30 min in ice, the mixture is centrifuged at 600 $g$ for 3 min in a clinical centrifuge and the white MgNH$_4$$^{32}$PO$_4$ precipitate is washed three times with 5-ml samples of ice-cold 5 $M$ NH$_4$OH solution. The precipitate is solubilized by the successive addition of 0.1 ml aliquots of 1 $M$ HCl. The [$^{32}$P]P$_i$ is stored in diluted HCl, about 0.1 $M$. For the different extraction described above, the n-butyl acetate can be replaced by an isobutanol–benzene mixture (v/v) previously saturated with water.

[20] O. Meyerhof and H. Green, *J. Biol. Chem.* **183,** 377 (1950).
[21] P. D. Boyer and D. M. Bryan, this series, Vol. 10, p. 60.
[22] L. de Meis, *J. Biol. Chem.* **259,** 6090 (1984).

Measurement of Phosphoenzyme

Two different methods can be used. One of them involves the use of Millipore filters. It has the advantage of being rapid and consuming small amounts of enzyme. However, it is less precise than the second method in which the denatured phosphoprotein is centrifuged and washed repeatedly. Regardless of the method used, the final concentration of sarcoplasmic reticulum vesicles in the assay medium should not exceed 0.4 mg protein per milliliter. At protein concentrations higher than 0.5 mg per milliliter the yield of phosphoprotein decreases. This is probably related to the formation of vesicle aggregates which impede the access of $[^{32}P]P_i$ to some of the vesicles. For the Millipore method, a sample of the assay mixture containing 0.050 to 0.100 mg vesicle protein is quenched with 5 volumes of ice-cold 0.20 $M$ HClO$_4$ solution containing 5 m$M$ NaH$_2$PO$_4$. The sample is filtered through a Millipore filter (pore size 0.45 $\mu$m). The filter is washed 10 to 15 times with 20-ml samples of 0.125 HClO$_4$ solution containing 4 m$M$ P$_i$. The filter is counted in a liquid scintillation counter. When desired, the filter can be dissolved in 1 ml dimethylformamide before the addition of the scintillation mixture. In each experimental condition, a control should be performed where the vesicles are quenched with the acid solution before the addition of the reaction mixture containing $[^{32}P]P_i$. The amount of phosphoprotein formed is calculated assuming that all the protein was retained by the filter, and taking into account the specific activity of $[^{32}P]P_i$ in the assay medium and the difference of counts between the sample and control. The amount of protein filtered in each assay should not exceed the range 0.05–0.10 mg. Larger amounts of protein obstruct the filter and impair the washings with the acid solution.

For the centrifugation method the amount of vesicles needed in each assay is in the range of 0.7 to 1.0 mg protein per milliliter. The assay medium, usually 2 to 3 ml, is quenched with 2 vol of ice-cold 0.25 $M$ HClO$_4$ plus 4 m$M$ P$_i$ carrier. The precipitate attained with HClO$_4$ is easier to wash than the precipitate attained with trichloroacetic acid (TCA). The denatured phosphoprotein is sedimented in the cold by centrifugation at 600 to 800 $g$ for 10 min. The supernatant is discarded, a small piece of ice is introduced in the test tube and this is vigorously stirred on a Vortex for 30 to 40 sec. The dispersed precipitate is resuspended in 4 ml ice-cold 0.125 $M$ HClO$_4$ containing 4 m$M$ P$_i$ carrier and centrifuged at 600 to 800 $g$ for 10 min. The phosphoprotein is washed four times, as described above. The washed pellet is dissolved in 0.25 ml of a solution containing 0.2 $M$ Na$_2$CO$_3$, 0.1 $N$ NaOH, 70 m$M$ sodium dodecyl sulfate, and 5 m$M$ P$_i$. An aliquot of this solution is used for liquid scintillation counting and another aliquot is used for determination of protein concentration by the Folin method,[23] standardized with bovine albumin. In order to measure unspecific binding of [$^{32}$P]P$_i$, controls should be performed in which the vesicles are quenched with $HClO_4$ before the addition of the assay medium.

### Leaky Vesicles and Soluble Ca²⁺-ATPase

These are used to measure phosphorylation by [$^{32}$P]P$_i$ in the absence of a transmembrane $Ca^{2+}$ gradient. The best preparations of leaky vesicles are those attained following the purification procedure described by Mac-Lennan[24,25] and by Meissner et al.[26] Alternatively, sarcoplasmic reticulum vesicles can be rendered leaky as described by Duggan and Martonosi.[27] The vesicles (10 to 15 mg protein per milliliter) are suspended in a medium containing 2 m$M$ ethylene glycol bis($\beta$-aminoethyl ether)-$N,N'$-tetraacetic acid (EGTA) and 100 m$M$ Tris adjusted to pH 9.0. After 20 min at room temperature, the pH is readjusted to 6.5 by the addition of small aliquots of a 5 $M$ maleic acid solution. After this treatment the ATPase activity is maintained but the vesicles are no longer able to accumulate $Ca^{2+}$.

Usually, phosphorylation by P$_i$ is measured in media without $Ca^{2+}$ (excess EGTA). Following solubilization with different nonionic detergents, the $Ca^{2+}$-ATPase is inactivated in less than 5 min when $Ca^{2+}$ is omitted from the assay medium.[13] This inactivation is not observed with the use of membrane-bound ATPase of leaky vesicles. Inactivation of the soluble $Ca^{2+}$-ATPase can be prevented by the addition of organic solvents to the assay medium. In the presence of 30% glycerol (v/v) and excess EGTA, the $Ca^{2+}$-ATPase solubilized with Triton X-100 remains fully active for several hours.[28]

### Ca²⁺-Loaded Vesicles

These are used to measure phosphorylation by P$_i$ in the presence of a transmembrane $Ca^{2+}$ gradient. The vesicles, 0.2 to 0.7 mg protein per milliliter, are loaded actively with either calcium phosphate or calcium oxalate by incubation at 30° for 5 to 10 min in a medium containing 20 m$M$

[23] O. H. Lowry, N. Y. Rosebrough, A. L. Farr, and R. J. Randall, *J. Biol. Chem.* **193**, 266 (1951).
[24] D. H. MacLennan, *J. Biol. Chem.* **245**, 4508 (1970).
[25] D. H. MacLennan, P. Seeman, G. H. Iles, and C. C. Yip, *J. Biol. Chem.* **246**, 2702 (1971).
[26] G. Meissner, G. E. Conner, and S. Fleisher, *Biochim. Biophys. Acta* **298**, 246 (1973).
[27] P. F. Duggan and A. Martonosi, *J. Gen. Physiol.* **56**, 147 (1970).
[28] O. B. Martins and L. de Meis, *J. Biol. Chem.* **260**, 6776 (1985).

Tris-maleate buffer (pH 6.5), 5 m$M$ MgCl$_2$, 1 m$M$ $^{45}$CaCl$_2$, 0.9 m$M$ EGTA, 1 m$M$ ITP, and either 20 m$M$ P$_i$ (pH 6.5) or 4 m$M$ potassium oxalate. The mixture is centrifuged at either 40,000 $g$ for 30 min or at 88,000 $g$ for 15 min. The supernatant is poured and aliquots are counted to estimate the amount of calcium taken up by the vesicles. The pellet is kept in ice. The vesicles should be resuspended in the desired media a few seconds before utilization.

When Ca$^{2+}$-loaded vesicles are used, the phosphoenzyme formed readily transfer its phosphate to ADP, leading to the synthesis of ATP. The apparent $K_m$ of the phosphoenzyme for ADP is in the range of 0.01 to 0.06 m$M$, whereas that for IDP is in the range of 1.0 to 2.0 m$M$.[6,29,30] Therefore, for loading the vesicles, ITP is used instead of ATP in the phosphoenzyme assay in order to minimize the effect of nucleoside diphosphate contamination which might be derived from the loading mixture. Oxalate and phosphate increase the Ca$^{2+}$ storage capacity of the vesicles by providing a sink for the entering calcium, which precipitates inside the sarcoplasmic reticulum vesicles as calcium oxalate or calcium phosphate.[31,32] Once these precipitates are formed, the free Ca$^{2+}$ concentration inside the vesicles remains constant as determined by the solubility product. When loaded with calcium phosphate, the free Ca$^{2+}$ concentration inside the vesicles is estimated to be in the range of 10 to 20 m$M$, while in vesicles loaded with calcium oxalate it is in the range of 0.2 to 0.5 m$M$. When incubated in media containing excess EGTA, the dimension of the Ca$^{2+}$ concentration gradient formed across the vesicles membrane depends on the ionic calcium concentration in the lumen of the vesicles. Thus, the Ca$^{2+}$ gradient in calcium phosphate-loaded vesicles is larger than that formed in vesicles loaded with calcium oxalate.

## ATP $\rightleftharpoons$ P$_i$ Exchange Reaction

In 1972, Makinose[33] observed that, when a Ca$^{2+}$ gradient is formed across the vesicles membrane, the Ca$^{2+}$-ATPase can catalyze simultaneously the hydrolysis of ATP and the synthesis of [$\gamma$-$^{32}$P]ATP from ADP and [$^{32}$P]P$_i$. As a result, there is an exchange between P$_i$ and the $\gamma$-phosphate of ATP. The ATP $\rightleftharpoons$ P$_i$ exchange reaction is observed when vesicles loaded with calcium phosphate (Ca$^{2+}$ concentration inside the

[29] L. de Meis, *J. Biol. Chem.* **251,** 2055 (1976).
[30] W. Hasselbach, *Biochim. Biophys. Acta* **515,** 23 (1978).
[31] W. Hasselbach and M. Makinose, *Biochem. Z.* **333,** 518 (1961).
[32] L. de Meis, W. Hasselbach, and R. D. Machado, *J. Cell Biol.* **62,** 505 (1974).
[33] M. Makinose, *FEBS Lett.* **12,** 269 (1971).

vesicles $\sim 10$ m$M$) are incubated in a medium containing a low $Ca^{2+}$ concentration ($\sim 10^{-6}$ $M$), $MgCl_2$, $P_i$, ATP, and ADP. In these conditions, the ATP $\rightleftharpoons P_i$ exchange is abolished when the membrane of the vesicles is rendered leaky by various procedures that do not affect the transport ATPase. These data infer that when the steady state between $Ca^{2+}$ influx and $Ca^{2+}$ efflux is reached, the energy derived from the hydrolysis of ATP is used to maintain the $Ca^{2+}$ gradient and the energy derived from the gradient is used for the synthesis of ATP.[33,34] Shortly after the Makinose report,[33] it was found that the $Ca^{2+}$-ATPase can catalyze a rapid ATP $\rightleftharpoons P_i$ exchange in the absence of a transmembrane $Ca^{2+}$ gradient.[34] Leaky vesicles or soluble $Ca^{2+}$-ATPase catalyze only the hydrolysis of ATP when incubated in media containing a $Ca^{2+}$ concentration sufficient to saturate only the enzyme high-affinity $Ca^{2+}$-binding site ($Ca^{2+}$, $10^{-6}$ to $10^{-4}$ $M$). As the $Ca^{2+}$ concentration in the medium is raised to the range needed to saturate the $Ca^{2+}$ binding of low affinity (0.5 to 10 m$M$), there is a progressive inhibition of the ATPase activity and activation of the synthesis of ATP. The $Ca^{2+}$ concentration required for half-maximal activation of ATP $\rightleftharpoons P_i$ exchange varies with the pH of the assay medium.[35] At pH 6.0 saturation is not reached even in the presence of 20 m$M$ $Ca^{2+}$. At pH 7.0 and 8.0, half-maximal rates of ATP $\rightleftharpoons P_i$ exchange are attained in the presence of 2.0 and 0.3 m$M$ $Ca^{2+}$, respectively. At pH 7.0 the $Ca^{2+}$-ATPase is phosphorylated by $[^{32}P]P_i$ in the presence of either 0.1 or 4 m$M$ $Ca^{2+}$. However, formation of $[\gamma\text{-}^{32}P]$ATP is observed only in presence of 4 m$M$ $Ca^{2+}$.[6,36] These data show that the osmotic energy derived from the $Ca^{2+}$ gradient is not required for the activation of the ATP $\rightleftharpoons P_i$ exchange reaction and that the binding of $Ca^{2+}$ to a site of low affinity allows the transfer of phosphate from the phosphoenzyme to ADP. This $Ca^{2+}$-binding site should be located in the inner surface of the vesicles, since the ATP $\rightleftharpoons P_i$ exchange in leaky vesicles or soluble ATPase is in the same range as those found inside the vesicles when a $Ca^{2+}$ gradient sufficient to activate ATP $\rightleftharpoons P_i$ exchange is formed. There is no net synthesis of ATP during the exchange reaction.

In the different experimental conditions tested, the rate of ATP hydrolysis was found to be always faster than the rate of ATP synthesis. Therefore, the data obtained with leaky vesicles and soluble $Ca^{2+}$-ATPase indicate that the system must be able to conserve some of the energy released from ATP hydrolysis in a form that permits synthesis of ATP. The mechanism of energy conservation is related to the binding of $Ca^{2+}$ to

[34] L. de Meis and M. G. C. Carvalho, *Biochemistry* **13**, 5032 (1974).
[35] S. Verjovski-Almeida and L. de Meis, *Biochemistry* **16**, 329 (1977).
[36] L. de Meis and M. M. Sorenson, *Biochemistry* **14**, 2739 (1975).

the low-affinity site of the enzyme. This triggers the conversion of the phosphoenzyme from low energy or ADP-insensitive into high energy or ADP-sensitive (reactions 5 and 4 backwards in Fig. 1). The $Ca^{2+}$-ATPase can also catalyze an ITP $\rightleftharpoons$ $P_i$ exchange, both in the presence and absence of a transmembrane $Ca^{2+}$ gradient.[6,37] Both in presence and absence of a $Ca^{2+}$ concentration gradient, the apparent $K_m$ for $P_i$ is in the range of 2 to 4 mM and the ratio between the hydrolysis and synthesis of nucleotide varies between 2 and 6 when either 0.05 to 0.2 mM ATP or 2 to 5 mM ITP is used. In the presence of a high ATP concentration (2 to 5 mM), the apparent $K_m$ for $P_i$ and the ratio between the velocities of hydrolysis and synthesis of ATP are higher in leaky vesicles than in $Ca^{2+}$-loaded vesicles. High ratios between hydrolysis and synthesis of ATP are also attained with the soluble $Ca^{2+}$-ATPase regardless of the ATP concentration used. This ratio decreases to values similar to those attained with intact vesicles ($Ca^{2+}$ gradient) when the water activity of the assay medium is decreased with organic solvents.[28]

In the sequence of Fig. 1 the NTP $\rightleftharpoons$ $P_i$ exchange is represented by reaction 2 to 7 flowing forward (NTP hydrolyis) and backward (NTP synthesis). The ratio of hydrolysis to synthesis approaches unity when the catalytic cycle is confined between reactions 2 and 7, as it is in the presence of ITP or a low ATP concentration. The ratio of hydrolysis to synthesis is higher than one to the extent that some of the enzyme units *E are converted to $Ca_2E$ instead of being driven in the reverse direction through phosphorylation by $P_i$. Therefore, the faster the forward rate of reactions 8 and 1, the faster the rate of hydrolysis and the slower the rate of synthesis.

## Measurements of ATP $\rightleftharpoons$ $P_i$ Exchange

This is determined by measuring the incorporation of $[^{32}P]P_i$ into $[\gamma\text{-}^{32}P]ATP$, the excess of $[^{32}P]P_i$ being extracted from the assay medium as phosphomolybdate with an isobutanol–benzene mixture.[21,34,38] After quenching the reaction with either $HClO_4$ or TCA, the sample is centrifuged at 600 to 800 g for 10 min in order to remove the denatured protein. A 0.4-ml aliquot of the supernatant is mixed with 0.3 ml acetone and 0.4 ml of a 2.5 N $H_2SO_4$ solution containing 5% (w/v) ammonium molybdate, followed by 2 ml of an isobutanol–benzene mixture (v/v) previously saturated by water. The tube is vigorously stirred on a Vortex

[37] B. Plank, G. Hellmann, C. Punzengruber, and J. Suko, *Biochim. Biophys. Acta* **550,** 259 (1979).
[38] M. Avron, *Biochim. Biophys. Acta* **40,** 257 (1960).

mixer for 30 sec. After phase separation, the yellow isobutanol–benzene layer is removed, and 0.020 ml of 20 m$M$ KH$_2$PO$_4$ and 0.3 ml acetone are added to the aqueous layer, followed by extraction with 2 ml of the isobutanol–benzene mixture. The addition of carrier P$_i$ and acetone and the extraction with isobutanol–benzene are repeated four times, after which a sample of the aqueous phase is taken for liquid scintillation counting of $^{32}$P. For the extractions, the isobutanol–benzene mixture can be replaced by $n$-butyl acetate. The phosphomolybdate complex is poorly soluble in water. Depending on the P$_i$ concentration in the medium, a yellow precipitate may be formed when the ammonium molybdate solution is added. The precipitate is readily dissolved following the addition of acetone. This improves the extraction with isobutanol–benzene. Practically all the acetone added is extracted together with the phosphomolybdate complex into the organic layer. There is practically no ATP hydrolysis during the extraction. In control experiments using a 1 m$M$ [γ-$^{32}$P]ATP solution and no added [$^{32}$P]P$_i$, it is found that less than 3% of the total ATP is hydrolyzed 30 min after the addition of the ammonium molybdate solution. A frequent source of error is derived from the phosphorolysis reaction.[39,40] The β-γ-phosphoanhydride bond of either ATP or ITP can be attacked by inorganic phosphate leading to the formation of ADP and pyrophosphate.

$$ATP + [^{32}P]P_i \rightarrow ADP + [^{32}P]PP_i$$

The phosphorolysis reaction occurs spontaneously in the absence of enzyme, i.e., it is not catalyzed by the Ca$^{2+}$-ATPase. The rate of phosphorolysis is very slow in the presence of Mg$^{2+}$ but it increases severalfold when Ca$^{2+}$ (1 to 10 m$M$) or organic solvents is included in the assay medium. Thus, after extraction with isobutanol–benzene a small part of the radioactivity found in the aqueous phase may be derived from radioactive pyrophosphate. The amount of radioactive pyrophosphate formed can be estimated in controls where the enzyme is quenched with HClO$_4$ before the addition of the reaction mixture containing [$^{32}$P]P$_i$.

## ATP or ITP Hydrolysis

The ATPase and ITPase activities vary depending on the concentration of P$_i$, ADP, and IDP in the assay medium. In order to measure the ratio between the velocities of hydrolysis and synthesis of either ATP or ITP, the nucleoside triphosphatase activity must be measured in the same

[39] J. B. Lowenstein, *Biochem. J.* **70**, 222 (1958).
[40] A. Vieyra, J. R. Meyer-Fernandes, and O. B. H. Gama, *Arch. Biochem. Biophys.* **238**, 574 (1985).

conditions as those used for measuring the synthesis of ATP and ITP, i.e., in the presence of nucleoside diphosphate and with a $P_i$ concentration varying between 1 and 10 m$M$. This can be done with the use of [$\gamma$-$^{32}$P]ATP or [$\gamma$-$^{32}$P]ITP. The amount of [$^{32}$P]$P_i$ produced can be measured either by removing the [$\gamma$-$^{32}$P]nucleoside triphosphate with charcoal[41] or by measuring the [$^{32}$P]$P_i$ produced with ammonium molybdate and $n$-butyl acetate.[34] For the charcoal method, a 0.5-ml sample of the assay medium is quenched with 0.1 ml of ice-cold 0.6 $M$ HClO$_4$, followed by the addition of 1 ml of a suspension of acid-washed charcoal (250 mg per milliliter) in 0.1 $N$ HCl. The mixture is stirred on a Vortex for 20 sec and centrifuged for 5 min at 800 $g$ in a clinical centrifuge. The [$^{32}$P]$P_i$ content of a sample of the supernatant is determined by liquid scintillation counting. In the second method, the reaction is quenched and centrifuged as described for the ATP $\rightleftharpoons$ $P_i$ exchange reaction. A 0.4-ml aliquot of the supernatant is mixed with 0.3 ml acetone and 0.4 ml of a 2.5 $N$ H$_2$SO$_4$ solution containing 5% (w/v) ammonium molibdate followed by 1 ml $n$-butyl acetate. The tube is vigorously stirred on a Vortex mixer for 30 sec. After phase separation, a sample of the $n$-butyl acetate layer is taken for liquid scintillation counting of $^{32}$P.

## Net Synthesis of ATP

When vesicles previously loaded with Ca$^{2+}$ are incubated in a medium containing EGTA, Ca$^{2+}$ flows out of the vesicles at a slow rate due to the low Ca$^{2+}$ permeability of the membrane. The Ca$^{2+}$ efflux is sharply increased when ADP, $P_i$, and Mg$^{2+}$ are added to the incubation medium. The increment of Ca$^{2+}$ efflux is not observed when one of these reactants is omitted from the medium.[30,42] The fast efflux of Ca$^{2+}$ is coupled with the synthesis of ATP.[30,43] For every two calcium ions released from the vesicles, one molecule of ATP is synthesized. The vesicles derived from the sarcoplasmic reticulum are able to accumulate Sr$^{2+}$ at the expense of ATP hydrolysis. However, the catalytic cycle of the ATPase is not reversed when a Sr$^{2+}$ gradient is formed across the vesicles membrane, i.e., there is no synthesis of ATP and the efflux of Sr$^{2+}$ is not enhanced when vesicles loaded with Sr$^{2+}$ are incubated in a medium containing excess EGTA, ADP, $P_i$ and Mg$^{2+}$.[44]

In vesicles loaded with Ca$^{2+}$ the synthesis of ATP is initiated by the

[41] C. Grubmeyer and H. S. Penefsky, *J. Biol. Chem.* **256**, 3718 (1981).
[42] B. Barlogie, W. Hasselbach, and M. Makinose, *FEBS Lett.* **12**, 267 (1971).
[43] M. Makinose and W. Hasselbach, *FEBS Lett.* **12**, 271 (1971).
[44] H. Guimarães-Motta, M. P. Sande-Lemos, and L. de Meis, *J. Biol. Chem.* **259**, 8699 (1984).

phosphorylation of the enzyme by P$_i$. This is not accompanied by an increment of the Ca$^{2+}$ efflux. The addition of ADP to the medium leads to an enhancement of the rate of Ca$^{2+}$ efflux and synthesis of ATP.[8] Therefore, the fast release of Ca$^{2+}$ is triggered by the transfer of the phosphate from the phosphoenzyme to ADP. The phosphoenzyme formed during reversal of the Ca$^{2+}$-pump can transfer its phosphate to either ADP or IDP, leading to the formation of ATP or ITP. The apparent $K_m$ of the phosphoenzyme for ADP is in the range of 0.01 to 0.06 m$M$, whereas that for IDP is in the range of 1 to 2 m$M$. The ADP dependence for ATP synthesis follows simple Michaelis–Menten kinetics.[6,29,30] The reversal of the Ca$^{2+}$-pump is inhibited by the end products of the overall reaction, that is, by ATP and by increases of the Ca$^{2+}$ concentration in the assay medium. In both cases, the phosphorylation of the enzyme by P$_i$ is impaired. When a large excess of EGTA is present in the medium, the inhibition promoted by ATP is overcome by raising the P$_i$ concentration in the medium, indicating a competitive process.[11,29] Inhibition by Ca$^{2+}$ is due to the binding of this cation to the high-affinity site of the enzyme which faces the outer surface of the vesicles. This impairs phosphorylation of the enzyme by P$_i$ regardless of the magnitude of the Ca$^{2+}$ gradient formed across the vesicles membrane.[29]

Under appropriate conditions, arsenate uncouples the fast efflux of Ca$^{2+}$ and the synthesis of ATP. Although the fast release of Ca$^{2+}$ is maintained, both the enzyme phosphorylation by P$_i$ and the synthesis of ATP are impaired. This effect of arsenate is reversible.[45]

Leaky vesicles prepared from purified Ca$^{2+}$ ATPase can synthesize a small amount of ATP in the absence of a Ca$^{2+}$ gradient.[46,47] This is achieved by a two-step procedure where, initially, the enzyme is phosphorylated by P$_i$ in the absence of Ca$^{2+}$. In the second step, ADP and a high Ca$^{2+}$ concentration ($\sim$10 m$M$ at pH 7.0) are added to the medium (Ca$^{2+}$ jump). The amount of ATP synthesized by this procedure is proportional to and never exceeds the number of enzyme sites phosphorylated by P$_i$. In other words, the Ca$^{2+}$ jump permits the enzyme to complete only one catalytic cycle, after which the system reaches equilibrium and no more ATP is synthesized. In the first step, Ca$^{2+}$ must be omitted from the medium in order to form the "low-energy" or "ADP-insensitive" phosphoenzyme. This is represented by reactions 7 and 6 in the sequence shown in Fig. 1. In the second step, the binding of Ca$^{2+}$ to the low-affinity site of the enzyme allows the conversion of the phosphoenzyme from a

[45] W. Hasselbach, M. Makinose, and A. Migala, *FEBS Lett.* **20**, 311 (1972).
[46] A. F. Knowles and E. Racker, *J. Biol. Chem.* **250**, 1943 (1975).
[47] L. de Meis and R. K. Tume, *Biochemistry* **16**, 4455 (1977).

low-energy into a high-energy or ADP-sensitive form, which transfers its phosphate to ADP, leading to the synthesis of ATP. This is represented by reactions 5, 4, 3, and 2 in Fig. 1. The $Ca^{2+}$ concentrations added in the second step of the $Ca^{2+}$-jump procedure are sufficient to saturate both $Ca^{2+}$-binding sites of high and low affinity. Synthesis of ATP is observed because the rate at which the phosphoenzyme transfers its phosphate to ADP (reactions 5, 4, 3, and 2 backward) is about two orders of magnitude faster than the rate of hydrolysis of the ADP-insensitive phosphoenzyme and formation of the enzymatic complex $Ca_2E$ (reactions 6, 7, 8, and 1 forward).[47–50] The binding of $Ca^{2+}$ to the high- and low-affinity binding sites of the enzyme can be altered by varying either the pH, temperature, or the water activity of the medium with organic solvents. Thus, synthesis of ATP can be attained with leaky vesicles in conditions where the $Ca^{2+}$ concentration is maintained constant and the apparent affinity of the $Ca^{2+}$-binding sites is varied by a sudden change of either the pH, temperature, or water activity of the medium.[14,47,51]

### Measurements of ATP Synthesis and $Ca^{2+}$ Efflux

In order to measure simultaneously the rate of $Ca^{2+}$ efflux and of ATP synthesis, vesicles loaded with either [$^{45}Ca$]calcium oxalate or calcium phosphate are used. These are prepared as described for the ATP $\rightleftharpoons P_i$ exchange reaction. The reaction is started by the addition of vesicles to a medium containing 15 m$M$ EGTA and the desired concentrations of $Mg^{2+}$, $P_i$, and ADP. To stop the reaction, the vesicles are removed by filtration with Millipore filters (average pore size 0.45 $\mu$m). An aliquot of the filtrate is taken for liquid scintillation counting of the $^{45}Ca$ released by the vesicles. The ATP synthesized is assayed enzymatically using hexokinase (EC 2.7.1.1) and glucose-6-phosphate dehydrogenase (EC 1.1.1.49). To 0.6 ml of the filtrate is added 0.2 ml of a mixture containing 250 m$M$ Tris-HCl buffer (pH 8.5), 100 m$M$ EGTA (pH 8.0), 100 m$M$ MgCl$_2$, 80 m$M$ glucose, and 1 unit each of hexokinase and glucose-6-phosphate dehydrogenase per milliliter. The reaction is started by adding 0.025 ml of a 10 m$M$ solution of NADP. After exactly 10 min incubation at room temperature, the optical extinction at 340 nm is compared to that of a control solution identical with the assay medium but which has not been incubated with vesicles. In order to measure the passive efflux of $Ca^{2+}$ not

[48] B. Rauch, D. Chak, and W. Hasselbach, *Z. Naturforsch. C.* **32C,** 828 (1977).
[49] R. M. Chaloub, H. Guimarães-Motta, S. Verjovski-Almeida, L. de Meis, and G. Inesi, *J. Biol. Chem.* **254,** 9464 (1979).
[50] H. Guimarães-Motta and L. de Meis, *Arch. Biochem. Biophys.* **203,** 395 (1980).
[51] L. de Meis and G. Inesi, *J. Biol. Chem.* **257,** 1289 (1982).

coupled with ATP synthesis, $P_i$ or ADP should be omitted from the assay medium.

Synthesis of ATP can also be measured with the use of $[^{32}P]P_i$, as described for the ATP $\rightleftharpoons P_i$ exchange reaction. In these experiments, the denatured protein can be used to measure the level of phosphoenzyme formed from $[^{32}P]P_i$.

### Synthesis of ATP in the Absence of a Ca²⁺ Gradient

When leaky vesicles are used, the amount of ATP synthesized is small and can not be measured enzymatically as described above. In these experiments synthesis is assayed measuring the formation of $[\gamma\text{-}^{32}P]ATP$ from ADP and $[^{32}P]P_i$ as described for the ATP $\rightleftharpoons P_i$ exchange reaction. In some experiments it is worthwhile to confirm that the radioactive material remaining in the water phase is indeed $[\gamma\text{-}^{32}P]ATP$. In this case, the excess of ammonium molybdate remaining in the water phase after extraction of $[^{32}P]P_i$ should be removed. The reaction is quenched with HCl to a final concentration of 1.3 $M$. The mixture is centrifuged at 600 $g$ for 10 min in order to remove the denatured protein. To each milliliter of the supernatant, 0.25 ml of a 60 m$M$ ammonium molybdate solution in 0.01 $M$ HCl and 0.3 ml acetone are added. The mixture is then extracted with 2 ml $n$-butyl acetate. After phase separation, the yellow organic layer is removed, 0.020 ml of 20 m$M$ KH$_2$PO$_4$ and 0.3 ml acetone are added to the aqueous layer followed by extraction with 2 ml of $n$-butyl acetate. This is repeated five times. Then, an amount of $P_i$ equivalent to that of ammonium molybdate remaining in the solution is added. This is calculated taking into account the amount of $P_i$ used in the assay medium and the amount of carrier $P_i$ used in the extractions. The mixture is extracted with 2 ml $n$-butyl acetate. A small sample of the aqueous phase is taken for liquid scintillation counting of $^{32}P$ and the remainder is used to confirm that the radioactive material is $[\gamma\text{-}^{32}P]ATP$.

There are several methods to identify ATP.[46,47,52,53] A simple method is to characterize the synthesized $[\gamma\text{-}^{32}P]ATP$ both kinetically and by autoradiography of thin-layer chromatograms.[22] A known amount of nonradioactive ATP is added to the aqueous phase and the pH of the mixture is adjusted to 6.5 with a 3 $M$ KOH solution, followed by the addition of CaCl$_2$, MgCl$_2$ and Ca²⁺-ATPase to final concentrations of 0.1 m$M$, 10 m$M$, and 0.2 mg protein per milliliter, respectively. After different incubation intervals at 30°, 0.3 ml samples of the mixture are quenched with

[52] R. P. Magnusson, A. R. Portis, Jr., and R. E. McCarty, *Anal. Biochem.* **72**, 653 (1976).
[53] S. M. Penningroth, K. Olehnik, and A. Cheung, *J. Biol. Chem.* **255**, 9545 (1980).

0.1 ml of a 2.5 $N$ $H_2SO_4$ solution containing 5% (w/v) ammonium molybdate and extracted with 1 ml of isobutanol–benzene mixture. After phase separation, a sample of the isobutanol–benzene layer is taken for liquid scintillation counting of $^{32}P$ and another sample is used to measure total $P_i$ by the method of Ernster et al.[54] Two other samples are quenched with 0.1 ml 1 $N$ HCl, one before the addition of the $Ca^{2+}$-ATPase and the other after most of the carrier ATP is hydrolyzed by the enzyme. These samples are used to identify [$\gamma$-$^{32}P$]ATP by autoradiography of ascending thin-layer chromatograms performed as described by Penningroth et al.[53] using a 0.75 $M$ $KH_2PO_4$ solution adjusted to pH 3.4 with phosphoric acid. In this system the $R_f$ values of pyrophosphate, ATP, and ADP are 0.14, 0.21, and 0.39, respectively. If all the radioactive material of the water phase is indeed ATP then, in the $P_i$ measurements, the radioactive material is hydrolyzed at the same rate as the carrier ATP and, in the chromatograms, the radioactive material has the same $R_f$ as ATP before the addition of the ATPase, and moves near the front together with $P_i$ after cleavage by the ATPase.

### ATP Contamination

For the synthesis of ATP in the absence of $Ca^{2+}$ gradient, care must be taken to ensure that no ATP contamination is introduced in the system. When ADP and $Ca^{2+}$ are added to the phosphorylating medium ($Ca^{2+}$ jump), the final mixture has a $Ca^{2+}$ concentration sufficient for the enzyme to catalyze an ATP $\rightleftharpoons$ $P_i$ exchange reaction. Therefore, if a significant amount of ATP is introduced into the system, the appearance of [$\gamma$-$^{32}P$]ATP in the water phase after isobutanol–benzene extraction might not necessarily indicate that there is net synthesis of ATP.[47] This source of error is not relevant when synthesis of ATP is measured with $Ca^{2+}$-loaded vesicles in the presence of excess EGTA. The enzyme does not catalyze an ATP $\rightleftharpoons$ $P_i$ exchange when $Ca^{2+}$ is removed from the assay medium. ATP contamination might arise from either addition of commercial ADP preparations or indirectly from contamination of the ATPase with adenylate kinase, which will form ATP and AMP from added ADP. Commercial ADP preparations contain 1 to 6 mol of ATP for each 100 mol of ADP. This contaminant is difficult to identify in high-pressure liquid chromatography systems but can be assayed using hexokinase and glucose-6-phosphate dehydrogenase as described above. Advantage can be taken of this enzymatic method to decrease the ATP contamination. A solution containing 40 m$M$ ADP, 50 m$M$ glucose, 40 m$M$ $MgCl_2$ is incu-

[54] L. Ernster, R. Zetterstrom, and O. Lindberg, Acta Chem. Scand. **4**, 942 (1950).

bated with 10 units per milliliter hexokinase for 1 hour at 30°. The reaction is stopped by heating the incubation medium for 30 sec in boiling water, followed by rapid cooling in ice. The denatured enzyme is removed by centrifugation and the clear supernatant stored at −5°.

The sarcoplasmic reticulum vesicles are contaminated with adenylate kinase. Leaky vesicles attained by the purification method described by MacLennan[24,25] are practically free of adenylate kinase. Therefore, purified Ca²⁺-ATPase should be used to measure synthesis of ATP in Ca²⁺-jump experiments.

## [14] Fluorimetric Detection and Significance of Conformational Changes in Ca²⁺-ATPase

By Y. DUPONT, F. GUILLAIN, and J. J. LACAPERE

### Introduction

The reaction cycle of sarcoplasmic reticulum Ca²⁺-ATPase and the methods used to investigate its fine structure are described in [12, 13] of this volume. This chapter deals more particularly with an important aspect of the cycle: the detection and measurement of conformational changes of Ca²⁺-ATPase protein induced by substrate binding and Ca²⁺ transport. As in the case of other transport ATPases (Na⁺, K⁺ and H⁺-ATPase) we know very little about the structure of Ca²⁺-ATPase. It follows that the present view of its functioning is mostly derived from conventional enzymology.

The simplest scheme used by the vast majority of scientists to study calcium transport in sarcoplasmic reticulum involves an alternate-site mechanism in which the transport sites are alternately exposed to each side of the membrane, with a simultaneous change in the binding constant of the ions transported and the reactivity of the intermediate phosphorylated.

According to this concept, the calcium ions remain bound to the calcium-binding sites during the transport process and the sites move physically from one side of the membrane to the other. Such a process obviously involves an important conformational change in the protein. This principle constitutes the basis of the $E \rightleftharpoons {}^*E$ scheme for Ca²⁺-ATPase[1] and of the closely related $E_1 \rightleftharpoons E_2$ scheme for the Na⁺, K⁺-ATPase.

[1] L. de Meis, in "The Sarcoplasmic Reticulum" (E. E. Bittar, ed.). Wiley, New York, 1981.

This simple and very mechanistic view of Ca$^{2+}$-ATPase functioning was recently questioned for several reasons. First, in order to account for the fluorescence and kinetic data reported during the last few years, some authors proposed more complicated schemes including a larger number of conformational states.[2-4] Despite the different implications of the conformations envisaged by these authors, the idea of a simple two-state enzyme will probably be abandoned.

In another hypothetical scheme[5] it was proposed that the conformational changes undergone by Ca$^{2+}$-ATPase were simply related to substrate binding and/or enzyme activation, and had nothing to do with the transport itself, which must occur within any major change in the structure of the enzyme.

Whichever mechanism finally proves to be correct, the investigation of structural changes in Ca$^{2+}$-ATPase is clearly a subject of major interest since it might help to solve the most challenging problem regarding Ca$^{2+}$-ATPase.

The first attempts to detect conformational changes in Ca$^{2+}$-ATPase were undertaken in the late 1960s. Using optical techniques such as light scattering and optical rotatory dispersion (ORD),[6,7] various groups tried to detect structural changes associated with the transport process; others used extrinsic dyes or electron paramagnetic resonance (EPR) spin labels.[8] In most cases the changes detected were connected with calcium accumulation and not with the conformation of the enzyme. Other attempts at detection used X-ray diffraction, a technique which yielded useful information about the insertion of Ca$^{2+}$-ATPase in the bilayer[9] and also indicated that an important conformational transition occurred when the Ca$^{2+}$-ATPase SH groups were modified. The significance, however, of all these investigations is still poorly understood, largely because the knowledge of the reaction cycle is insufficient.

It was not until 1976, after significant progress in the enzymology of Ca$^{2+}$-ATPase, that better experimental protocols were worked out for the detection of its conformational transitions. The most realistic scheme, published in 1976 by deMeis and colleagues,[1] showed that the simplest

[2] Y. Dupont, *Biochim. Biophys. Acta* **688**, 75 (1982).

[3] Y. Dupont, *in* "Structure and Function of Sarcoplasmic Reticulum" (Y. Tonomura and S. Fleischer, eds.). Academic Press, Orlando, Florida, 1985.

[4] F. Fernandez-Belda, M. Kurzmack, and G. Inesi, *J. Biol. Chem.* **259**, 9687 (1984).

[5] Y. Dupont, *FEBS Lett.* **161**, 14 (1983).

[6] W. F. H. M. Mommaerts, *Proc. Natl. Acad. Sci. U.S.A.* **58**, 2476 (1967).

[7] T. Onishi and T. Terasaki, *J. Biochem.* (*Tokyo*) **61**, 812 (1967).

[8] J. M. Vanderkoi and A. Martonosi, *Biochemistry* **16**, 1262 (1977).

[9] Y. Dupont and W. Hasselbach, *Nature* (*London*) **246**, 41 (1973).

way of detecting any structural change in this enzyme was to monitor at equilibrium, in the absence of turnover, the changes induced by the binding of the substrates, calcium, ATP, or phosphate. Conversely, during catalysis, all the conformations are present and their concentrations are regulated in a very complex manner by all the substrates. Therefore explanation of conformational transitions of an enzyme during its activity involves extremely complex problems, and can only be undertaken after basic knowledge of the main steps of the reaction cycle has been acquired.

Various approaches began to yield interesting results. For example, calcium or ATP binding to Ca$^{2+}$-ATPase was found to alter both the EPR spectra,[10] SH group reactivity,[11,12] and intrinsic Ca$^{2+}$-ATPase fluorescence.[13] The fluorescence approach has proved the most efficient and productive, probably because of the high sensitivity and time resolution of the fluorescence technique.

Methodology of Intrinsic Fluorescence Measurement

In the partial reactions studied so far, the changes in intrinsic fluorescence induced by the binding of specific ligands never exceeded a few percent. This led us to adjust the fluorescence measurements to the reaction studied, in order to make the signal-to-noise ratio optimal. The main improvements are briefly described below.

1. In order to make the best choice of wavelengths, it is necessary to now the emission spectra of the different species present. Because these spectra are only slightly different, an accumulation device is necessary to permit the averaging, normalization, and subtraction of spectra. Figure 1a and b shows the changes in fluorescence induced by Ca$^{2+}$ and Mg$^{2+}$ ions and recorded by this procedure. The difference spectra (Fig. 1c and d) facilitate the choice of emission wavelength for the study of a given transition.

2. To obtain accurate dose–response curves at given excitation and emission wavelengths, the signal-to-noise ratio of the set-up must be as high as possible. For this purpose, the commercial spectrofluorimeters are often quite unsuitable. A good arrangement is a UV source of medium power like a 75- or 150-W xenon or xenon(Hg) arc lamp (high power usually involves instability), connected to a very well-regulated power

[10] P. Champeil, F. Bastide, C. Taupin, and C. M. Gary-Bobo, *FEBS Lett.* **63**, 270 (1976).
[11] N. Ikemoto, *J. Biol. Chem.* **253**, 8027 (1978).
[12] A. J. Murphy, *J. Biol. Chem.* **253**, 385 (1978).
[13] Y. Dupont, *Biochem. Biophys. Res. Commun.* **71**, 544 (1976).

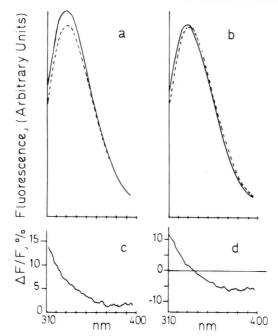

FIG. 1. Emission fluorescence spectra of the *E, E · $Ca_2$, and E?Mg states of sarcoplasmic reticulum ATPase and relative difference spectra for the *E → E · $Ca_2$ and *E → E?Mg transitions. Sarcoplasmic reticulum vesicles (50 µg/ml) were suspended in 2.5 ml of 150 m$M$ MOPS-Tris, pH 7.2, 2 m$M$ EGTA, at 20°. Calcium (a,c) and magnesium (b,d) induced transitions were induced by 1 m$M$ $CaCl_2$ and 100 m$M$ $MgCl_2$, respectively, conditions which ensured full transition in each case. Observation conditions: $\lambda_{ex}$, = 290 nm; bandwidth, 4 nm for excitation and emission.

supply. The excitation wavelength can be selected by a monochromator which does not allow any white light into the excitation wavelength, and the emission wavelength can be advantageously selected by an interference filter. The cell holder is also a very important part of the apparatus. It must be very well thermostatted because of the temperature sensitivity of fluorescence: a rise of 1° in the temperature corresponds to a 1.2% decrease in fluorescence. Furthermore, solutions to be analyzed must be well stirred and the cell holder must permit small volumes to be added without opening the fluorescence cuvette.

3. The above improvements in the fluorescence measurement technique were adapted to a stopped-flow apparatus for kinetic studies. For such studies the most important requirements are (a) the possibility of accumulating several stopped-flow traces, and (b) good temperature regulation and homogeneity. The first problem has been very easily solved

using commercially available equipment, but the second remains because, with most of the current commercial stopped-flow instruments, temperature artifacts produce fluorescence changes of higher amplitude than those induced by the substrate (these artifacts are mainly due to differences in the temperature of the drive syringes and the observation chamber).

4. The last technical difficulty is inherent in the turbidity of vesicular systems. For the ligands to be effective, some of them must be present at high concentrations which, as a side effect, tend to induce changes in swelling and vesicle aggregation, and this in turn alters light scattering and optical density. An example is shown in Fig. 2, in which it can be seen that addition of 1 m$M$ calcium induces very little change in light transmission. In contrast, addition of 10 m$M$ magnesium causes a large, fast increase in transmission, followed by a slow return to the initial level. These changes in transmission were first attributed to a sudden outflow of water caused by the osmotic shock, followed by a slow combined inflow of solute and water. This type of experiment underlines the importance of the choice of excitation and emission wavelengths, and shows that when these wavelengths become very close together, the use of a very sharp cut-off filter is imperative, to avoid the presence of scattered light in the emission channel. When all these conditions are met, the fluorescence signal is not affected by light scattering (Fig. 2).

5. Finally, it should be noted that fluorescence experiments in themselves are not sufficient to permit the unambiguous attribution of a fluorescence level to a given enzymatic species. Other techniques are therefore needed to enable distinctions to be made between several interpretations or to acquire important information like phosphorylation stoichiometry.[14] The quench-flow technique,[15] the oxygen-exchange technique,[14] and the rapid filtration technique[3,16] have proved essential complements for the interpretation of the significance of fluorescence signals.

## Application of Fluorescence Measurements

### Calcium Binding

Binding of calcium ions to the high-affinity sites of Ca²⁺-ATPase was first found to change its intrinsic fluorescence in experiments reported in 1976.[13] These experiments were subsequently reproduced[17] and the

[14] F. Guillain, P. Champeil, and P. Boyer, *Biochemistry* **23**, 4754 (1984).

[15] F. Guillain, P. Champeil, J. J. Lacapere, and M. P. Gingold, *J. Biol. Chem.* **256**, 6140 (1981).

[16] Y. Dupont, *Anal. Biochem.* **142**, 504 (1984).

[17] F. Guillain, M. P. Gingold, S. Buschlen, and P. Champeil, *J. Biol. Chem.* **255**, 2072 (1980).

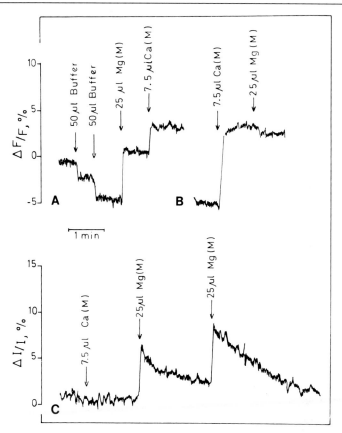

FIG. 2. Fluorescence and light scattering changes upon addition of $CaCl_2$ and $MgCl_2$ to sarcoplasmic reticulum vesicles. Experiments were carried out in the same pH 7.0 medium as that reported in Fig. 1. (A) Two additions of buffer, corresponding to 2% dilutions, were made to calibrate the fluorescence signal. Subsequent additions leading to final concentrations of 10 m$M$ $Mg^{2+}$ and 1 m$M$ free $Ca^{2+}$, respectively, induced the $*E \rightarrow E?Mg$ and $E?Mg \rightarrow E \cdot Ca_2$ transitions. (B) Illustration of the $*E \rightarrow E \cdot Ca_2$ transition and shows the nonadditivity of the two transitions. (C) Shows the change in light scattering induced by 10 m$M$ $MgCl_2$, whereas a preliminary addition of 3 m$M$ $CaCl_2$ had no effect on the scattered light.

results thoroughly analyzed in a large number of studies. A similar change in fluorescence was also observed with solubilized Ca²⁺-ATPase. Calcium binding to sarcoplasmic reticulum vesicles enhanced the fluorescence and induced a blue shift of the emission spectrum,[18] and it has been shown that the tryptophan residues responsible for the fluorescence change after cal-

[18] F. Guillain, M. P. Gingold, and P. Champeil, *J. Biol. Chem.* **257,** 7366 (1982).

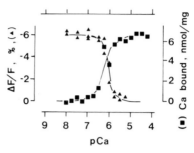

FIG. 3. Comparison of ⁴⁵Ca binding to sarcoplasmic reticulum vesicles with the pCa dependence of intrinsic fluorescence. ⁴⁵Ca experiments were conducted from a stock suspension of sarcoplasmic reticulum vesicles (100 μg/ml) in 150 m$M$ MOPS-Tris, 20 m$M$ MgCl₂, and 50 μ$M$ CaCl₂, pH 7.0, at 20°. The suspension was divided into 5-ml aliquots and the free Ca²⁺ adjusted by addition of EGTA. Each sample was filtered through an HA 0.45-μm Millipore filter, the filters were dried and radioactivity was estimated by scintillation counting. Fluorescence experiments were conducted with sarcoplasmic reticulum vesicles (50 μg/ml) suspended in the same medium as for the filtration experiments, except that the CaCl₂ concentration was 100 μ$M$. Free Ca²⁺ concentrations were adjusted as above. λ_{ex} = 290 nm. Emitted light was analyzed through an MTO 324 filter.

cium addition are located in a hydrophobic environment and are no doubt deeply buried in the intramembranous part of the molecule.

The dependence of the fluorescence change at pH 7.0 as a function of the calcium concentration indicates a dissociation constant for calcium of about 0.3 μ$M$, a value very similar to that found by direct titration of the high-affinity binding sites with ⁴⁵Ca (Fig. 3). It is now well established that one molecule of Ca²⁺-ATPase is able to bind two calcium ions in a cooperative and extremely pH-dependent manner. From the steepness of the binding curve in Fig. 3, this cooperativity is evident (10 to 90% saturation, with only a 10-fold change in the calcium concentration).

The E ⇌ *E scheme[1] does not include any proposals concerning the role of the interaction between the two binding sites for the two Ca²⁺ ions, which are implicitly taken to be equivalent. However, the results of various fluorescence experiments indicate that this is not the case, and also revealed that magnesium ions have an important role in the differentiation of the two calcium binding sites. Guillain et al.[18,19] showed that magnesium is a potent competitor for calcium binding to the high-affinity sites, and studies at equilibrium tend to favor a reaction mechanism in which magnesium only competes with one of the two calcium sites. Stopped-flow studies indicate that the effect of magnesium is more intriguing. Figure 4 represents two stopped-flow traces of the fluorescence signal

[19] P. Champeil, M. P. Gingold, F. Guillain, and G. Inesi, J. Biol. Chem. **258**, 4453 (1983).

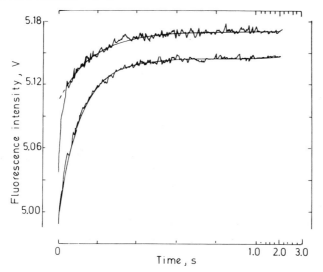

FIG. 4. Stopped-flow recording of the fluorescence change induced by calcium binding to sarcoplasmic reticulum vesicles in the presence or absence of magnesium. The enzyme syringe contained 150 m$M$ MOPS-Tris, 2 m$M$ EGTA, 100 $\mu$g/ml sarcoplasmic reticulum vesicles, and either no magnesium (lower trace) or 20 m$M$ MgCl$_2$ (upper trace). The substrate syringe contained the same medium plus 6 m$M$ CaCl$_2$ (pCa 3 after mixing) instead of sarcoplasmic reticulum vesicles. Spectroscopic conditions as in Fig. 3. Both traces represent the average signals accumulated after 10 experiments.

induced by calcium binding. In the absence of magnesium, the signal rises as a single exponential, whereas in its presence, the trace is unmistakably biphasic. The simplest and most rational explanation for these results was given by Dupont[2] and Champeil et al.,[19] who analyzed the concentration dependence of the fast and slow traces. They proposed that the initial fast phase corresponds to the binding of calcium to a rapidly accessible site of low affinity (100 $\mu M$), and that this binding induces a slow transconformation, thus unmasking a second calcium-binding site with a higher affinity (0.1 $\mu M$), the rate constant of the first step being modulated by magnesium. The existence of this biphasic calcium binding and of the nonequivalence of the two calcium-binding sites was confirmed by direct time-resolved studies of $^{45}$Ca binding, first at low temperature,[2] and then with the rapid filtration technique.[3,16]

Scheme 1 implies that one calcium binding site is accessible in the basal *E state, and that the binding of one calcium ion to this site triggers the *E $\rightleftharpoons$ E transition. As this result disagrees with the basic proposal of the *E $\rightleftharpoons$ E scheme, an alternative scheme was proposed (Scheme 2) which retains the idea of a basal *E state with no externally accessible calcium binding site and an E state with two high-affinity sites, but in-

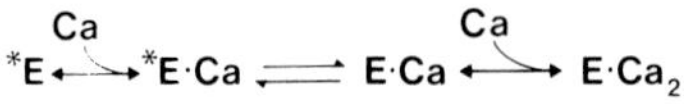

$$\text{*E} \xrightarrow{\text{Ca}} \text{*E·Ca} \rightleftharpoons \text{E·Ca} \xleftarrow{\text{Ca}} \text{E·Ca}_2$$

SCHEME 1

cludes a third intermediate with only one externally accessible calcium binding site. According to Scheme 2, the first *E ⇌ **E transconformation is regulated by magnesium. In favor of the latter scheme are experiments showing that vanadate poisoning, which stabilizes the lowest fluorescence level (*E), completely inhibits calcium binding at concentrations below 100 $\mu M$.[5]

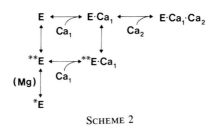

SCHEME 2

*ATP Binding*

In the absence of calcium, MgATP binds to sarcoplasmic reticulum ATPase. A dissociation constant of 3 $\mu M$ has been reproducibly reported in direct binding measurements with radioactive ATP and filtration.

Under similar experimental conditions (i.e., in the absence of calcium), MgATP binding induces a large increase in fluorescence,[20] which however, is smaller than that induced by calcium (3 to 4%). The change in fluorescence caused by ATP indicates a dissociation constant equal to the constant measured with labeled nucleotides (Fig. 5). At present, knowledge regarding the conformational change induced by ATP is far less advanced than knowledge of the change induced by other substrates like calcium or phosphate. Important work remains to be done in order to understand the mechanisms governing ATP binding and phosphate transfer. However, several unpublished attempts to elucidate the time course of ATP binding by following the intrinsic fluorescence change induced by ATP have been carried out. These experiments have not been successful, since the rate constant of the transition is extremely fast, much faster than that observed with calcium. Such experiments would be very helpful in establishing whether or not calcium binding affects the affinity of Ca²⁺-ATPase for MgATP.

[20] Y. Dupont, N. Bennett, and J. J. Lacapere, *Ann. N.Y. Acad. Sci.* **402**, 569 (1982).

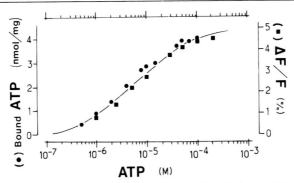

FIG. 5. Comparison of [$^{14}$C]ATP binding to sarcoplasmic reticulum vesicles with the ATP dependence of intrinsic fluorescence. Filtrations and fluorescence experimental procedures were similar to Fig. 3 protocols, but the following medium was used: 200 m$M$ MOPS-Tris, 4 m$M$ EGTA, and 2 m$M$ MgCl$_2$, pH 7.2, at 20°. $\lambda_{ex}$ = 290 nm, using a 330-nm interference filter.

## $Ca^{2+}$-ATPase Phosphorylation by ATP and $P_i$

In the presence of calcium, ATP binding to the enzyme leads to very fast formation of a phosphorylated intermediate.[1] Recording of the intrinsic fluorescence of $Ca^{2+}$-ATPase under the same conditions revealed that ATP rapidly reduces the fluorescence from the E · Ca$_2$ level to an intermediate level higher than the fluorescence of the basal *E state.[4,20] The amplitude of this drop saturates at very low concentrations of ATP (almost equal to the phosphoenzyme concentration), and remains in this state as long as ATP is present, but returns to the high E · Ca$_2$ level when ATP is completely hydrolyzed.[20] The rate constant of this drop in fluorescence and the titration of its amplitude indicate that the intermediate fluorescence level observed after addition of ATP in the presence of calcium corresponds to that of the phosphorylated enzyme. This idea has been confirmed recently by Fernandez-Belda et al.[4] On the other hand, it was reported in 1973 that, under appropriate conditions, sarcoplasmic reticulum ATPase can be directly phosphorylated by inorganic phosphate in the absence of calcium. Punzengruber et al.[21] proposed the following reaction (Scheme 3).

A large number of kinetic studies used Scheme 3 in order to measure its various equilibrium constants, essentially the $K_5$ constant, which contains important information on the thermodynamics of the phosphory-

[21] C. Punzengruber, R. Prager, N. Kolassa, F. Winkler, and J. Suko, *Eur. J. Biochem.* **92**, 349 (1978).

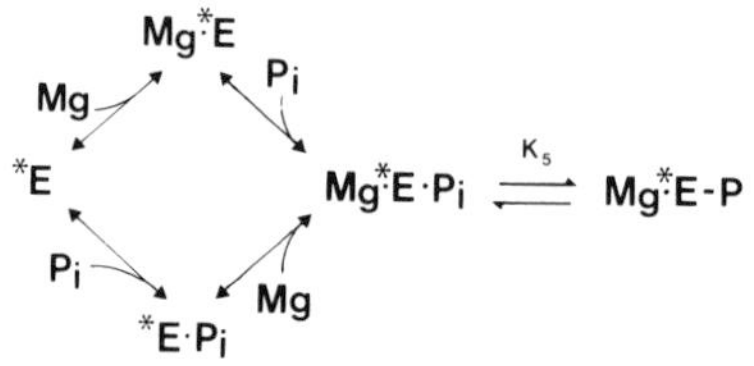

$$
\begin{array}{ccc}
 & \text{Mg}^*\text{E} & \\
\text{Mg}\nearrow & & \searrow \text{P}_i \\
^*\text{E} & & \text{Mg}^*\text{E}\cdot\text{P}_i \xrightleftharpoons{K_s} \text{Mg}^*\text{E-P} \\
\text{P}_i\searrow & & \nearrow \text{Mg} \\
 & ^*\text{E}\cdot\text{P}_i &
\end{array}
$$

SCHEME 3

lation reaction. Most of the initial studies agreed in assigning this constant a value very close to unity which, at that time, appeared satisfactory to the authors because it was difficult to imagine the existence of a thermodynamically favorable phosphorylation reaction by P$_i$ and also because this value explained the incomplete phosphorylation reported for the active site (only 3 to 5 nmol/mg).

In 1980, Lacapere et al.[22] showed that phosphorylation by P$_i$ significantly enhanced Ca$^{2+}$-ATPase fluorescence, and also that the amplitude of the changes induced by various concentrations of P$_i$ and magnesium was in agreement with the above scheme. In addition, the intrinsic fluorescence signal offered a very convenient way of testing all aspects of the reaction mechanism under conditions very difficult to obtain by chemical-quench techniques (i.e., with high P$_i$ concentrations), and thus of evaluating the rate and equilibrium constants more accurately. One very crucial result was obtained in this work, showing that the rate constant of the conformational transition induced by phosphorylation is an increasing function of the P$_i$ concentration and does not saturate at P$_i$ concentrations as high as 50 m$M$. The main conclusion drawn from these studies was that the equilibrium constant $K_5$ is far larger than unity, indicating that the phosphorylation reaction is almost completely displaced towards the formation of the covalent intermediate *E-P. This conclusion was later confirmed by $^{18}$O measurements[14] and reevaluation of the chemical-quench experiments.[23]

In addition to the measurement of the $K_5$ equilibrium constant, the fluorescence experiments just mentioned indicate that P$_i$ phosphorylation is probably linked to a conformational transition of Ca$^{2+}$-ATPase. It is highly probable that this conformational change is related to the mechanism which makes phosphorylation by inorganic phosphate a favorable reaction. Reports by de Meis et al.[24] and Dupont[25] favor the idea that

[22] J. J. Lacapere, M. P. Gingold, P. Champeil, and F. Guillain, J. Biol. Chem. **256**, 2302 (1981).

[23] G. Inesi and T. Watanabe, Ann. N.Y. Acad. Sci. **402**, 501 (1982).

[24] L. de Meis, O. B. Martins, and E. W. Alves, Biochemistry **19**, 4252 (1980).

[25] Y. Dupont and R. Pougeois, FEBS Lett. **156**, 93 (1983).

conformational transition is associated with dehydration of the $Ca^{2+}$-ATPase active site. Release of a large number of water molecules during phosphorylation might explain the considerable change in entropy during that reaction.

*Magnesium-Induced Fluorescence Changes*

To account for the multiple aspects of the modulation of $Ca^{2+}$-ATPase activity by magnesium ions, Makinose and Boll[26] proposed that magnesium enters the cycle either by direct binding to $Ca^{2+}$-ATPase or by complexing with ATP and thus changing the substrate to MgATP. This suggestion was derived from the study of the ATP–ADP exchange reaction as a function of the magnesium ion concentration. Recently, Guillain *et al.*[18] showed that the binding of magnesium in the absence of calcium induces a blue shift of 2 nm in the emission spectrum of the intrinsic fluorescence of $Ca^{2+}$-ATPase (Fig. 1).

The existence of this spectral change allowed the equilibrium and kinetics of magnesium binding to be studied in the absence of any other ligand. Under these conditions, the binding curves indicate a single type of binding site with pH-dependent affinity ($K_d = 5$ m$M$ at pH 7.0 and 1 m$M$ at pH 8.0).

As discussed above, it seems very likely that magnesium ions compete with calcium ions for binding to one of the two calcium binding sites. This idea has been further documented by kinetic studies showing that the rate of magnesium binding is very similar to that of calcium binding in the absence of magnesium (4 $sec^{-1}$ at pH 7.0)[19] and by the results of Marcsek *et al.*,[27] who showed that micromolar concentrations of calcium and millimolar concentrations of magnesium induce the same changes in the tryptic proteolysis of $Ca^{2+}$-ATPase.

Additional Changes

The more precise scheme used so far for $Ca^{2+}$-ATPase is the E $\rightleftharpoons$ *E scheme, proposed for the first time in 1976 by deMeis *et al.*[1] A simplified form is shown in Scheme 4. The most important contribution of the fluorescence technique reviewed here has been to permit accurate measurement of the conformational transitions and therefore verification of Scheme 4.

[26] M. Makinose and W. Boll, *in* "Function and Molecular Aspects of Biomembrane Transport" (E. Quagliariello, F. Palmieri, and S. Papa, eds.). Elsevier/North-Holland, Amsterdam, 1979.

[27] Z. Marcsek, *Biophys. J.* **41**, 18A (1983).

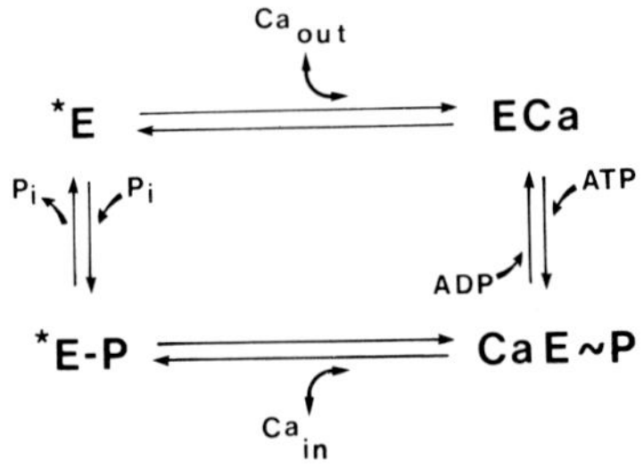

Ca out
*E
ECa
Pi
Pi
ATP
ADP
*E-P
CaE~P
Ca in


SCHEME 4

The results of this technique do not support this simple two-state scheme for the following reasons:

1. The detailed sequence of calcium binding implies that the *E ⇌ E transition is biphasic and that one-half the calcium sites are always rapidly accessible from the cytoplasmic side of the membrane. Fast binding to these sites induces a slow transition revealing a second class of calcium sites.

2. The E-P ⇌ *E-P transition is not associated with reversal of the conformational change observed during the *E ⇌ E transition.

3. There is a need to introduce a third **E state, stabilized by either ATP binding or phosphoenzyme formation, independently of the energy contents of the phosphoenzyme formed.

The notion which now emerges from this survey of $Ca^{2+}$-ATPase fluorescence studies is the existence of three distinct conformational states, as described in Scheme 5. The newly introduced third state **E is observable under transport conditions (presence of ATP or phosphorylated enzyme) or perhaps transiently, during calcium binding.

ATP
ADP
E·Ca2
Ca out
Ca out
Ca in
**E·Ca
**E·ATP
**E Ca2 ~P
**E-P
Ca out
ATP
Pi
H2O
*E


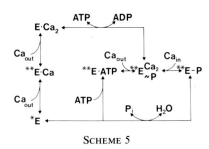

SCHEME 5

It has been proposed that this three-state system is compatible with the classical E ⇌ *E two-state scheme, by assuming that the active ATPase is a dimer and that, during turnover, the accumulated species **E corresponds to a hybrid intermediate $E \cdot *E$.[2,3] Whatever the final solution, it is clear that the simple $E \rightleftharpoons *E$ scheme should now be modified in such a way as to account for the entire body of experiments made possible by the use of $Ca^{2+}$-ATPase intrinsic fluorescence.

The final point for discussion concerns the state of the calcium ions during the transport sequence. During transport, the calcium bound to the ATPase evolves into three species, in the following sequence: high-affinity binding, $ATP \cdot E\text{-}Ca_2$; occluded form, $P\text{-}E \cdot \langle Ca_2 \rangle$; and low-affinity binding, $P\text{-}*E \cdot\cdot Ca_2$. The mechanics of this transport process is at present completely unknown. Are both calcium ions always bound to the same site during the transport? Is it these calcium sites which, by a change in their orientation, release the calcium ions into the lumen of the sarcoplasmic reticulum by a shuttle mechanism? The fact that the transport occurs at a fixed calcium/E-P coupling ratio makes this type of process a very popular concept. An alternative idea is that the transported calcium ions are first bound to the high-affinity sites and then occluded at another location when the enzyme becomes phosphorylated, after which they are submitted to a force which drives the ions across the membrane via low-affinity binding sites. There seems to be an important distinction between these two models, in terms of the role of the conformational changes in $Ca^{2+}$-ATPase.

If the alternative hypothesis is correct, these changes are not the mechanical events responsible for calcium transport but, as with most other enzymes, are simply a way for the enzyme to modify the shape of the active site around the substrate, and to optimize the enzyme–substrate interaction. In this case, the vectorial nature of the transport process should be a direct consequence of the geometry of the active site and/or of the channel formed by the hydrophobic part of $Ca^{2+}$-ATPase located in the bilayer.

Two observations support this idea: first, the E-P to *E-P transition, which is the heart of the energy transduction process, is not associated with an important conformational change in $Ca^{2+}$-ATPase[3,20] and second, the changes in active-site polarity, which is thought to play an important part in the transport process, are also not related to the intrinsic fluorescence changes of $Ca^{2+}$-ATPase.[5,25]

## [15] Synthesis of ATP from $Ca^{2+}$ Gradient by Sarcoplasmic Reticulum $Ca^{2+}$ Transport ATPase

*By* E. Fassold and Wilhelm Hasselbach

### Introduction

Sarcoplasmic reticulum vesicles isolated from rabbit skeletal muscle[1,2] consist of tightly sealed membranes which only allow a slow passive back-diffusion of $Ca^{2+}$ taken up[3] previously by a pathway not yet identified. An accelerated $Ca^{2+}$ release and a concomitant ATP synthesis were observed by Makinose and Hasselbach[4] and Barlogie et al.[5] from vesicles loaded with $Ca^{2+}$ after the additon of ADP and ethylene glycol bis($\beta$-aminoethyl ether)-$N,N'$-tetraacetic acid (EGTA) in the presence of $P_i$. The driving force for ATP formation was the $Ca^{2+}$ gradient existing between the vesicular volume and the external medium. Conditions were chosen in such a way that nearly all $Ca^{2+}$ that was released could contribute to ATP synthesis. These authors obtained a ratio of 2 mol $Ca^{2+}$ released per mol ATP synthesized, the same ratio with which the forward reaction—the ATP-driven $Ca^{2+}$ uptake—proceeds. Subsequent studies confirmed the reversibility of the transport reaction[6,7] via a phosphorylated intermediate,[8,9] even if isolated ATPase[10] or leaky instead of closed vesicles[11] were phosphorylated and the ATP synthesis was initiated by high medium concentrations of $Ca^{2+}$ and ADP, a design which allows only one backward cycle of the reaction.

### Principle of Enzyme Action

The reaction scheme, as illustrated in Fig. 1, includes the main partial steps and reaction intermediates occurring during the course of enzyme

[1] W. Hasselbach and M. Makinose, *Biochem. Z.* **339**, 94 (1963).
[2] L. de Meis and W. Hasselbach, *J. Biol. Chem.* **246**, 4759 (1971).
[3] J. J. Feher and F. N. Briggs, *J. Biol. Chem.* **257**, 10191 (1982).
[4] M. Makinose and W. Hasselbach, *FEBS Lett.* **12**, 271 (1971).
[5] B. Barlogie, W. Hasselbach, and M. Makinose, *FEBS Lett.* **12**, 267 (1971).
[6] R. Panet and Z. Selinger, *Biochim. Biophys. Acta* **255**, 34 (1972).
[7] D. W. Deamer and R. J. Baskin, *Arch. Biochem. Biophys.* **153**, 47 (1972).
[8] M. Makinose, *Int. Congr. Biophys., 2nd, Vienna* p. 276 (1966).
[9] T. Yamamoto and Y. Tonomura, *J. Biochem. (Tokyo)* **62**, 558 (1967).
[10] A. F. Knowles and E. Racker, *J. Biol. Chem.* **250**, 1949 (1975)
[11] L. de Meis and R. K. Tume, *Biochemistry* **16**, 4455 (1977).

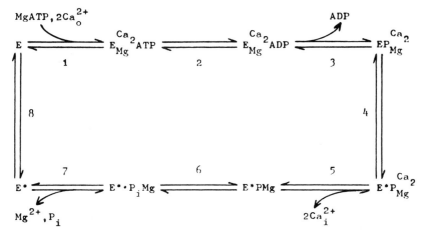

FIG. 1. Reaction scheme of the sarcoplasmic reticulum $Ca^{2+},Mg^{2+}$-ATPase.

action.[12,13] E and E* represent the functional states of the enzyme with high-affinity $Ca^{2+}$ binding sites located on the exterior and low-affinity $Ca^{2+}$ binding sites located on the interior surface of the membrane. It is demonstrated that the phosphoenzyme is the intermediate state between the events on the external and the internal surface. A steadily renewed formation of the $Ca^{2+}$ bound form $E*P_{Mg}^{Ca_2}$ is required for a continuing $Ca^{2+}$ release-driven synthesis of ATP, which proceeds via reversal of the partial steps 4, 3, 2, and 1 as long as the intravesicular $Ca^{2+}$ concentration is sufficient to saturate the low-affinity binding sites ($K_d = 1$–10 m$M$) and the extravesicular $Ca^{2+}$ is kept below 0.1 $\mu M$. This is achieved by addition of a nonpenetrating chelating agent.

The mere chelation of external free $Ca^{2+}$ causes the vesicles to release $Ca^{2+}$ slowly on a diffusional pathway. On the simultaneous or subsequent addition of ADP an accelerated continuous $Ca^{2+}$ outward flow commences and lasts until the $Ca_i/Ca_o$ ratio drops to low values, depending on the concentrations of the reactants.

### Considerations in Preparation of the Experiment

Loading of the vesicles is usually performed in the presence of acetyl phosphate (AcP) as energy-yielding substrate and inorganic phosphate ($P_i$) serving as precipitating anion and substrate for the backward reaction. $Ca^{2+}$ loads obtained by slow passive diffusion of millimolar concen-

[12] M. Makinose, *FEBS Lett.* **37**, 140 (1973).
[13] M. G. C. Carvalho, D. O. G. deSouza, and L. de Meis, *J. Biol. Chem.* **251**, 3629 (1976).

trations (<100 nmol/mg protein) or by active transport in the absence of a precipitating anion (100–200 nmol/mg protein) set limits to the duration of the two fast-occurring processes and to the accuracy of the results obtained. AcP is split about 10 times more slowly by the enzyme than ATP,[14] but ATP as energy donator is naturally not expedient as it obscures its subsequent measurement as the product. In the presence of phosphate, $Ca^{2+}$ loads up to 4000 nmol/mg can be achieved. It is advisable to stay well beyond this value, for a luminal calcium phosphate concentration surmounting 1000 nmol/mg causes a deviation from the $Ca^{2+}$ release/ATP synthesis ratio of 2[7]. This might be a consequence of the externalization of precipitate which, at first amorphous, converts to a crystalline state on standing.[15] For the same reason the usage of oxalate as $Ca^{2+}$-precipitating anion is not expedient.[16] as a higher $Ca^{2+}$ load is of no advantage due to the lower solubility product of the precipitate. Also, the rate of $Ca^{2+}$ release that can be achieved after loading with calcium oxalate and subsequent addition of phosphate, ADP, and EGTA is several times lower than that after calcium phosphate loading.[5,17]

The Lineweaver–Burk plot of the rates of ATP synthesis versus cotransported phosphate as used in our experiments has yielded an apparent $K_m$ in the range of 1.5 m$M$. The flattening slope in the higher $P_i$ concentration range suggests, however, that the true $K_m$ for the back-reaction is even lower.

To establish a sufficiently high $Ca^{2+}$ gradient excess EGTA is added to the medium which reduces the external free $Ca^{2+}$ below 0.1 $\mu M$. The resulting slow passive $Ca^{2+}$ release rate cannot simply be subtracted from rates obtained when ADP is present, in addition, for two reasons: due to the different release rates, EGTA-induced release still proceeds after EGTA/ADP-induced release and concurring ATP synthesis have stopped; also, it cannot be excluded that part of the EGTA-induced release occurs via the pump itself. This pathway might be closed when ADP is present. If the time courses the $Ca^{2+}$ release and ATP formation are expressed in percentages of total they correspond very well.

The concentration of ADP was chosen according to the affinity constant given by Beil et al.[18]

In order to prevent an accumulation of formed ATP which would drive the reaction in the forward direction its $\gamma$-$P_i$ is trapped by the glucose/

[14] W. Hasselbach and J. Suko, Biochem. Soc. Spec. Publ. No. 4, 159 (1974).
[15] L. Raeymaekers, B. Agostini, and W. Hasselbach, Histochemistry 70, 139 (1981).
[16] J. J. Feher and G. B. Lipford, Biochim. Biophys. Acta 818, 373 (1985).
[17] H. Masuda and L. de Meis, Biochim. Biophys. Acta 332, 313 (1974).
[18] F. U. Beil, D. v. Chak, and W. Hasselbach, Eur. J. Biochem. 81, 151 (1977).

hexokinase system.[4] The activity of the adenylate kinase contamination was inhibited by P¹, P⁵-diadenosine 5'pentaphosphate.

The ionic strength[19] and the $Mg^{2+}$ concentration[20] have to be considered too, as increasing ionic strength decelerates the reaction rate and $Mg^{2+}$ reduces the concentration of the substrate, free ADP.

The pH optimum for $Ca^{2+}$ release and ATP synthesis established in control experiments is in the range of 6.8, which is in agreement with an earlier report by de Meis[21] and depends on the $P_i$ concentration. $Ca^{2+}$ uptake proceeds with its fastest rate at pH 7.2 of the three proton concentrations tested (pH 6.4, 6.8, and 7.2).

## $Ca^{2+}$ Uptake and $Ca^{2+}$ Release

### Reagents

The reaction mixture for $Ca^{2+}$ uptake consists of 5 m$M$ $MgSO_4$, 40 m$M$ KCl, 30 m$M$ 3-($N$-morpholino)propanesulfonic acid (MOPS), pH 6.8, 100 m$M$ glucose, 100 $\mu M$ $CaCl_2$ (200 cpm $^{45}Ca^{2+}$/nmol $CaCl_2$), 2, 4, or 8 m$M$ $P_i$, pH 6.8, 0.5 m$M$ acetyl phosphate, and 0.2 mg vesicles/ml reaction medium.

The mixtures to induce $Ca^{2+}$ release consist of 2 m$M$ EGTA, pH 6.8, 100 $\mu M$ P¹,P⁵-diadenosine 5'-pentaphosphate, 0.05 mg hexokinase/ml without (EGTA-induced $Ca^{2+}$ release, mixture A) or with 1 m$M$ ADP (EGTA/ADP-induced $Ca^{2+}$ release, mixture B). The concentrations as given are achieved after addition to the reaction medium.

### Procedure

The $Ca^{2+}$ uptake reaction is performed in 5 ml total volume under constant stirring at 20–22°. Before adding the vesicles, an aliquot of the protein-free reaction medium (0.5 ml minus 9/10 of the total vesicular volume to be added) is removed and filtered (zero control). Either Millipore filters (HA 0.45) or Schleicher and Schuell BA 85 membrane filters, 0.45-$\mu$m pore diameter, are used which have been prerinsed with water. To remove $Ca^{2+}$ adhering to the filter, it is washed with 3 ml 100 m$M$ sucrose and removed for liquid scintillation counting.

At 15 and 19 min after the addition of protein, 0.5-ml aliquots are filtered to determine $Ca^{2+}$ uptake. The resulting uptake curve is extrapo-

[19] S. Yamada and N. Ikemoto, *J. Biol. Chem.* **255**, 3108 (1980).

[20] M. Makinose and W. Boll, in "Cation Fluxes Across Biomembranes" (Y. Mukohota and L. Packer, eds.), p. 89. Academic Press, New York, 1979.

[21] L. de Meis, *J. Biol. Chem.* **251**, 2055 (1976).

lated to 20 min, at which point $Ca^{2+}$ release is induced by addition of 0.0665 ml mixture A or 0.1015 ml mixture B. After 15, 50, 100 sec, and 5 min, 0.5-ml aliquots are filtered as above. The dilution of the protein concentration is considered when calculating the $Ca^{2+}$ release rates. Fifty microliters of the 10 m$M$ $^{45}CaCl_2$ stock solution (500 nmol) is counted as the radioactivity standard; a direct determination from the assay media (0.1 ml) is performed for comparison.

## $Ca^{2+}$ Uptake and ATP Synthesis

### Reagents

The reaction mixture for $Ca^{2+}$ uptake is identical with that used in the previous section with the exception that $^{40}Ca^{2+}$ and [$^{32}$P]P$_i$ (about 1000 cpm/nmol P$_i$) are used. Mixture A consisted of the same components as described above.

### Procedure

$Ca^{2+}$ uptake is allowed to continue for nearly 20 min. Subsequently mixture A (0.095 ml) is added under stirring and immediately a 1-ml aliquot is combined with 1.5 ml ice-cold 20% trichloroacetic acid as a zero time control. Then 0.04 ml 100 m$M$ ADP is added and 15, 50, 100 sec, and 5 min later 1-ml aliquots each are acid-quenched as above. The samples are centrifuged to remove the protein. The ATP synthesized is determined in 2 ml of the supernatant according to Avron.[22]

### Characteristics of the Reactions

Figure 2A illustrates that up to 460 of 500 nmol $Ca^{2+}$/mg protein present can be taken up during 20 min, depending on the P$_i$ concentration. The EGTA-induced $Ca^{2+}$ release rate is rather high with 80–100 nmol/mg · min, the reason for this is not clear. EGTA/ADP-induced $Ca^{2+}$ release is nearly complete after 1 min (with an initial rate of about 450–700 nmol/mg · min depending on the phosphate present) and is in good agreement with the time course of ATP synthesis (Fig. 2B). Previously Makinose and Hasselbach reported[4] that no more ATP is synthesized after 2 min under comparable conditions. Figure 3A and B illustrates the time courses expressed as percentages of total (5 min values). From the comparison of the figures, it can easily be deduced that the $Ca^{2+}$ release/ATP synthesis ratio is close to 2 with 2, 4, or 8 m$M$ P$_i$ present.

[22] M. Avron, *Biochim. Biophys. Acta* **40**, 257 (1960).

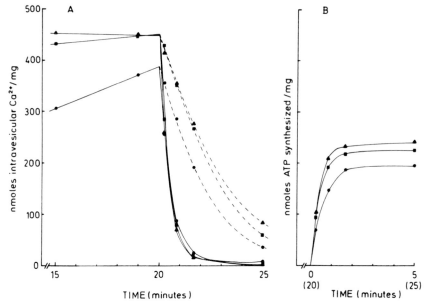

Fig. 2. Time courses of Ca²⁺ release and Ca²⁺ release-driven ATP synthesis. (A) Ca²⁺ uptake was performed in the presence of 2 (●), 4 (■), or 8 (▲) m$M$ P$_i$. At 20 min Ca²⁺ release was initiated by EGTA (- - -) or EGTA/ADP (——). (B) After 20 min of Ca²⁺ uptake ATP synthesis was initiated by subsequent addition of EGTA and ADP; symbols as above. The composition of the reaction media and the reaction procedure are described in the text.

The $K_m$ of P$_i$ for ATP synthesis (and Ca²⁺ release) of 1.5 m$M$ or less (not shown) is lower than the one reported by Masuda and de Meis.[17] Using calcium oxalate-loaded vesicles, they found a $K_m$ of 4.7 m$M$. This is probably due to a competition between P$_i$ and oxalate for binding sites.

In the absence of a Ca²⁺ gradient (equally low concentrations of Ca$_i^{2+}$ and Ca$_o^{2+}$) or in the presence of a Ca ionophore, no ATP synthesis occurs. This is also evident from the time courses of Ca²⁺ release and ATP synthesis in Fig. 2A and B.

Other Approaches

Makinose[23] proved the reversibility of the ATPase reaction by demonstrating simultaneously occurring ATP hydrolysis and synthesis when a steady state had been reached after ATP-driven calcium phosphate loading of closed vesicles. This ATP ⇌ P$_i$ exchange is accompanied by an in-

²³ M. Makinose, *FEBS Lett.* **12**, 269 (1971).

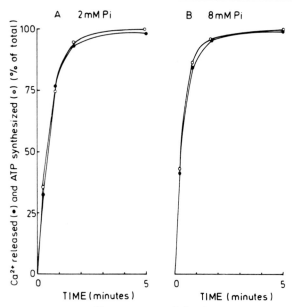

FIG. 3. Time courses expressed in percentages. Ca$^{2+}$ released and ATP synthesized after 5 min (see Fig. 2A and B) represent 100%.

and outflow of Ca$^{2+}$. It can also be obtained with leaky vesicles if the Ca$^{2+}$ concentration is in the millimolar range[24] and can be stimulated to a large degree by dimethyl sulfoxide (DMSO).[25]

Knowles and Racker[10] were the first to show that isolated ATPase that has been phosphorylated at pH 6.3 is able to transfer a large proportion of the P$_i$ residues of the previously formed phosphoenzyme to ADP if high Ca$^{2+}$ is added simultaneously. ATP synthesis can also be observed with the use of leaky vesicles.[11] It should be stressed that this one-cycle reaction sequence only occurs if the low-affinity Ca$^{2+}$-binding sites of the luminal section of the ATPase molecule are freely accessible to external Ca$^{2+}$. (For details, see [13], this volume.)

### Effect of Arsenate

Arsenate has been proved to be a potent initiator of Ca$^{2+}$ release[26] (Fig. 4). Even though an arsenylated intermediate is apparently formed

[24] L. de Meis and M. G. C. Carvalho, *Biochemistry* **13**, 5032 (1974).
[25] L. de Meis and G. Inesi, *FEBS Lett.* **185**, 135 (1985).
[26] W. Hasselbach, M. Makinose, and A. Migala, *FEBS Lett.* **20**, 311 (1972).

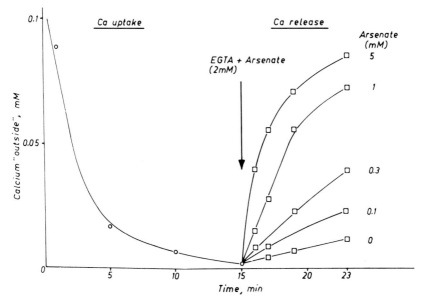

FIG. 4. Arsenate-induced Ca²⁺ release from Ca²⁺-loaded vesicles.[26] The vesicles were loaded with calcium phosphate in the presence of 5 m$M$ MgCl$_2$, 2 m$M$ acetyl phosphate, 20 m$M$ P$_i$, 0.1 m$M$ CaCl$_2$, and 0.1 mg vesicular protein/ml. Ca²⁺ release was initiated by the addition of arsenate and 2 m$M$ EGTA.

which contains occluded Ca²⁺ during the course of the reaction,[27] the resulting Ca²⁺ release is specifically uncoupled from ATP formation. Ca²⁺ leaves the enzyme, while the arsenylated intermediate is hydrolyzed. To investigate the effect of arsenate, Mg²⁺ should be present, but ADP in concentrations above 10 $\mu M$, as well as ATP, should be absent. In the presence of phosphate, arsenate binding is inhibited. In contrast to AcP, ATP prevents the effect of arsenate, indicating that nucleotides exert additional effects.

Concluding Remarks

    The sarcoplasmic reticulum native vesicles which are easily accessible in large amounts represent a comparatively simple membrane system and therefore provide a good model for studying osmochemical energy inter-conversion. To evaluate the described reactions quantitatively, the membrane potential can be completely disregarded, as the membranes are

[27] U. Pick and S. Bassilian, *Eur. J. Biochem.* **131**, 393 (1983).

highly permeable to small cations.[28] This does not exclude the possibility that the elementary Ca$^{2+}$ translocation step occurs electrogenically.[29,29a] The superiority of sarcoplasmic microsomes is further demonstrated if ATP production by human red blood cell inside-out vesicles is compared[30] under similar conditions. The yield of 14.9 pmol ATP formed/mg protein · min on EGTA/ADP plus P$_i$-induced Ca$^{2+}$ release from a transmembrane Ca$^{2+}$ gradient mirrors the low concentration of the Ca$^{2+}$,Mg$^{2+}$-ATPase in the plasma membrane.

Mitchell's chemiosmotic hypothesis[31] that ATP is generated by a transmembrane electrochemical potential difference-driven proton efflux is not contradicted in principle (considering protons to be substituted by other cations[32]). A proton gradient produced *in vitro* using chloroplast membranes,[33,34] however, yields only small amounts of ATP during the short incubation time possible. Likewise ATP synthesis is restricted when the Na$^+$,K$^+$-ATPase of the plasma membrane is employed, due to the inborn properties of the system.[35]

The driving force in sarcoplasmic reticulum vesicular membranes has been recognized to be the distinct Ca$^{2+}$ affinity of the enzyme in its E or E* conformational state, respectively, which leads to the Ca$^{2+}$ gradient-dependent continuous ATP synthesis.

[28] G. Meissner, *Mol. Cell. Biochem.* **55**, 65 (1983).
[29] P. Zimniak and E. Racker, *J. Biol. Chem.* **253**, 4631 (1978).
[29a] K. Hartung, E. Grell, W. Hasselbach, and E. Bamberg, *Biochim. Biophys. Acta* **900**, 209 (1987).
[30] A. Wüthrich, H. J. Schatzmann, and P. Romero, *Experientia* **35**, 1589 (1979).
[31] P. Mitchell, *Biol. Rev. Cambridge Philos. Soc.* **41**, 445 (1966).
[32] P. Mitchell, *FEBS Lett.* **33**, 267 (1973).
[33] A. T. Jagendorf and E. Uribe, *Proc. Natl. Acad. Sci. U.S.A.* **55**, 170 (1966).
[34] P. Gräber, U. Junesch, and G. H. Schatz, *Ber. Bunsenges, Phys. Chem.* **88**, 599 (1984).
[35] I. M. Glynn and V. L. Lew, *J. Physiol. (London)* **207**, 393 (1970).

# [16] Occluded Ca$^{2+}$

*By* HARUHIKO TAKISAWA and MADOKA MAKINOSE

Ion-transport enzymes undergo a sequence of reaction steps in which the respective ions and phosphates interact with the enzyme mutually. One enzyme state in the main reaction cycle has been defined as occluded with respect to the ion to be translocated, i.e., the ions bound to the enzyme in this state are slowly exchanged with free ions in the medium. In order to measure the occluded ions bound to the enzyme, free ions are removed from assays as rapidly as possible. Several procedures have been applied to studies of the sodium–potassium[1] and the calcium pump.[2–4]

The centrifuge column procedure[5] is employed to identify calcium occlusion.[4] The method allows one to measure not only occluded calcium but also the state of phosphorylation of the enzyme. The latter is determined in a separate assay under the same experimental conditions.

### Assay for Measuring Calcium Binding

One-milliliter disposable columns are prepared as described by Penefsky.[5] Sephadex G-50 fine is swollen at 0° in a solution containing 0.4 $M$ KCl, 0.3 $M$ sucrose, and 50 m$M$ Tris-[$N$-tris(hydroxymethyl)methyl-2-aminoethanesulfonic acid] (TES) buffer, pH 8.0. Volume and concentration of the Sephadex suspension are selected to give a 1-ml bed volume after the column stops dripping spontaneously. Subsequently the column is suspended in a test tube (15 × 105 mm) and centrifuged at 150 $g$ for 2 min at 5°.

Solubilized Ca$^{2+}$-ATPase (3 mg/ml)[4] is incubated with $^{45}$CaCl$_2$ (0.1–1.0 m$M$) in a solution containing 0.4 $M$ KCl, 0.3 $M$ sucrose, 3.5 mg/ml deoxycholate, MgCl$_2$ (80 $\mu$M–5 m$M$), and 50 m$M$ Tris–TES, pH 8.0, at 0° for 2 min. An aliquot (50 $\mu$l) is transferred to the top of the column after it is suspended in a new test tube (trapping tube) and immediately centrifuged at 150 $g$ for 2 min at 5°. The volume applied to the column is obtained as effluent.

Radioactive calcium and the protein[6] are determined in aliquots of 10–20 $\mu$l of the effluent. The radioactivity in the effluent is normalized against the active site concentration of the enzyme in the effluent. The amount of the active site is estimated from the maximal amount of phosphoprotein (EP) which is determined in a separate assay containing 3 m$M$ CaCl$_2$, 1 m$M$ MgCl$_2$, and 100 $\mu$M [$\gamma$-$^{32}$P]ATP.

Table I shows the typical results obtained with solubilized ATPase enzyme. Since one active site of the enzyme maintains at least 1 mol of high-affinity Ca$^{2+}$-binding site,[2] the results in Table I indicate that in the absence of ATP, the enzyme loses most of its bound Ca$^{2+}$ when centrifuged through the column. Ca$^{2+}$ binding is not affected when 200 $\mu$M

[1] I. M. Glynn, Y. Hara, and D. E. Richards, *J. Physiol.* (*London*) **351**, 531 (1984).
[2] H. Takisawa and M. Makinose, *Nature* (*London*) **290**, 271 (1981).
[3] Y. Nakamura and Y. Tonomura, *J. Biochem.* (*Tokyo*) **91**, 449 (1982).
[4] H. Takisawa and M. Makinose, *J. Biol. Chem.* **258**, 2986 (1983).
[5] H. S. Penefsky, this series, Vol. 56, p. 527.
[6] O. H. Lowry, N. J. Rosebrough, A. L. Farr, and R. J. Randall, *J. Biol. Chem.* **193**, 265 (1951).

TABLE I
RETENTION OF BOUND CALCIUM ON THE PHOSPHORYLATED SARCOPLASMIC
TRANSPORT ENZYME[a]

| Addition | Effluent contents | | Moles bound Ca²⁺/mole phosphorylable site |
| | Calcium (nmol) | Phosphorylable site (nmol) | |
| --- | --- | --- | --- |
| None | 0.099 ± 0.005 | 0.306 ± 0.004 | 0.324 ± 0.015 |
| +100 $\mu M$ ATP | 0.371 ± 0.011 | 0.259 ± 0.019 | 1.437 ± 0.076 |
| +200 $\mu M$ AMPPCP | 0.091 ± 0.007 | 0.281 ± 0.007 | 0.322 ± 0.024 |

[a] The solubilized Ca²⁺-ATPase (3 mg protein/ml) is incubated with 1 m$M$ ⁴⁵CaCl₂ in the presence or absence of 100 $\mu M$ ATP or 200 $\mu M$ AMPPCP in a solution containing 0.4 $M$ KCl, 0.3 $M$ sucrose, 3.5 mg/ml deoxycholate, 80 $\mu M$ MgCl₂, and 50 m$M$ Tris–TES, pH 8.0, at 0° for 2 min. The amount of Ca²⁺ bound to the ATPase and that of protein in the effluent are measured as described in the text. The active site concentration is estimated as 5.1 nmol/ mg protein in the other assay. Values are the means of four determinations ± SEM. In the control experiment without the ATPase protein, the Ca²⁺ content of the effluent both in the presence and absence of ATP is smaller than 0.002 nmol.

adenylyl($\beta,\gamma$-methylene) diphosphonate (AMPPCP) is added to the incubation mixture. However, the addition of 100 $\mu M$ ATP causes a nearly fivefold increase in the amount of bound Ca²⁺. Increasing Ca²⁺ binding is assigned to Ca²⁺ occlusion that is related to enzyme phosphorylation.

### Assay for Measuring Phosphorylation

The amount of EP is measured by applying the same procedure as described for Ca²⁺ binding except that [$\gamma$-³²P]ATP and nonradioactive Ca²⁺ are used instead of nonradioactive ATP and radioactively labeled calcium. In the presence of excess EGTA, which prevents enzyme phosphorylation, no [$\gamma$-³²P]ATP can be detected in the effluent. In the presence of Ca²⁺, EP is formed and can be precipitated in 0.5 ml of 4% trichloroacetic acid (TCA), 0.1 m$M$ ATP, and 0.1 m$M$ P$_i$ placed at the bottom of the trapping tube. The radioactivity in the denatured protein[7] and in an aliquot of the supernatant is determined. More than 95% of the radioactivity is contained in the protein fraction.

For the measurement of ADP-sensitive EP, 50 $\mu$l of the assay described above is centrifuged (150 $g$ for 2 min) in a trapping tube filled with

[7] M. Makinose, *Eur. J. Biochem.* **10**, 74 (1969).

0.5 ml of 1 m$M$ ADP, 2 m$M$ EGTA, 0.1 m$M$ ATP, 3.5 mg/ml deoxycholate, 0.3 $M$ sucrose, 0.4 $M$ KCl, and 50 m$M$ Tris–TES, pH 8.0, at 0°. The main part (~90%) of the phosphate residue of ADP-sensitive EP in the effluent is transferred to ATP immediately and the rest (~10%) is converted into ADP-insensitive EP.[4] The presence of ATP and the chelation of Ca$^{2+}$ by EGTA minimize the hydrolysis of the radioactive ATP formed. Meanwhile, a small part of the ADP-insensitive phosphoprotein is hydrolyzed in E and P$_i$. After further incubation for 30 sec in an ice bath the reaction is stopped by the addition of 80 $\mu$l of 50% TCA. The denatured protein is removed by centrifugation at 2000 rpm for 10 min at 0°. In 100-$\mu$l samples radioactive inorganic phosphate[8] and total radioactivity in the supernatant are determined. The calculations are based on a total volume of 630 $\mu$l, since the fluctuation in the effluent volume (about 50 $\mu$l) is relatively small compared to the total volume of the incubation buffer and TCA (580 $\mu$l). The protein precipitates are used for measuring the residual ADP-insensitive phosphoprotein. The amount of [$\gamma$-$^{32}$P]ATP, which is descended from the main part of the ADP-sensitive phosphoprotein, is obtained from the difference between the total radioactivity in the supernatant and the radioactivity of inorganic phosphate. The initial amount of ADP-sensitive phosphoprotein per active site of the enzyme in the effluent can be obtained as follows: 1.1 times the mole of [$^{32}$P]ATP formed in effluent/mole of active site in effluent. Factor 1.1 is to compensate the conversion of ADP-sensitive EP to ADP-insensitive EP during the incubation time.

Since Ca$^{2+}$ binding of the ATPase is not affected by the existence of ATP in the medium (see Table I, 200 $\mu M$ AMPPCP), the amount of Ca$^{2+}$ bound to EP (EPCa) can be estimated as follows:

$$\text{EPCa} = \text{ECa}_t(+\text{ATP}) - \text{ECa}_t(-\text{ATP}) \times (\text{E}_t - \text{EP})/\text{E}_t$$

ECa$_t$(+ATP) and ECa$_t$(−ATP) represent the amounts of Ca bound to the enzyme after the incubation with and without ATP, respectively. (E$_t$ − EP)/E$_t$ represents the relative amount of unphosphorylated enzyme in the presence of ATP.

Figure 1 shows the dependence of ADP-sensitive phosphoprotein and Ca$^{2+}$ bound to EP on added MgCl$_2$. Both the amount of ADP-sensitive phosphoprotein and of Ca$^{2+}$ bound to phosphoprotein decrease concomitantly with increasing MgCl$_2$ added. The ratio of about 2 between the latter and the former is observed throughout all measurements. The results thus indicate that 2 mol of Ca$^{2+}$ are occluded in 1 mol of ADP-sensitive phosphoprotein.

[8] H. Takisawa and Y. Tonomura, *J. Biochem. (Tokyo)* **86**, 425 (1979).

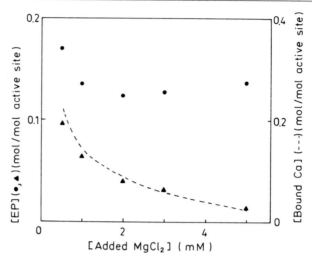

FIG. 1. Dependence of phosphorylated intermediates and $Ca^{2+}$ bound to EP on added $MgCl_2$ concentrations. The solubilized $Ca^{2+}$-ATPase (3 mg/ml) is incubated with 0.1 m$M$ $CaCl_2$ in the presence of 100 $\mu M$ ATP in a solution containing 0.4 $M$ KCl, 0.3 $M$ sucrose, 3.5 mg/ml deoxycholate, various concentrations of $MgCl_2$, and 50 m$M$ Tris–TES, pH 8.0, at 0° for 2 min. The amounts of total EP (●) and ADP-sensitive EP (▲) are measured as described in the text. Broken line shows the $Ca^{2+}$ bound to EP.

## Comments on the Centrifuge Column Procedure

The centrifuge column procedure is the most suitable method to identify occluded $Ca^{2+}$-binding site and the corresponding phosphoprotein intermediates. The procedure is rather simple and the sensitivity is high. For extended use of this method, the following precautions should be observed. First, when membranous vesicular preparations are used, the membranes should be completely permeable to the transported solute so that only the bound and not the trapped solute is measured. Second, the rate at which the intermediate decomposes should be relatively slow so that no significant decomposition takes place during the column centrifugation. Finally, it should be mentioned that the recovery of membrane protein is poor compared to soluble protein. The recovery of soluble protein is reported to be higher than 95%.[5]

In the case of sarcoplasmic $Ca^{2+}$-transport ATPase, the recovery of protein in the effluent is only 30–40% and fluctuates. Taking into account the problem of the stability using the Lowry method, a particularly careful execution of the protein measurement is recommended. The deviation of results in Table I arises mainly from this point.

In the case of untreated SR vesicles, almost no protein is recovered by the procedure described here. In order to recover vesicular protein, it is necessary to use Sephadex G-50 (coarse) and the time required for centrifugation should be kept to less than 20 sec in order to avoid contamination of free ligand in the effluent.

## [17] Modified Membrane Filtration Methods for Ligand Binding on ATP-Driven Pumps during ATP Hydrolysis

*By* MOTONORI YAMAGUCHI and TAKAHIDE WATANABE

The study of ligand binding on the ATP-driven pumps during ATP hydrolysis is important to clarify the molecular mechanism of active transport of cation. Direct measurement of the ligand binding on the pumps has been done using equilibrium dialysis or ultracentrifugation. These are inappropriate for measurement of the ligand binding during ATP hydrolysis because they require long periods for measurement. A conventional membrane filtration is a relatively rapid and sensitive method for detecting the ligand binding in the course of enzyme turnover as long as two conditions are met. (1) Dissociation is sufficiently slow that the bound ligand is not lost when the enzyme is collected on a filter. (2) Affinity of the enzyme for ligand is satisfactorily high for accurate measurement. These conditions are not always met for the ligand binding on the pumps. Therefore, we have devised new membrane filtration methods which enable us to measure the ligand binding on the ATP-driven pumps during ATP hydrolysis.[1,2]

### Monovalent Cation Binding to the $Na^+,K^+$-ATPase during ATP Hydrolysis

The amount of monovalent cations bound to the $Na^+,K^+$-ATPase during ATP hydrolysis is measured using the double-membrane filtration method. This method, as shown in Fig. 1, uses a set of two membrane filters (upper for a molecular sieve, lower for trapping the filtrate) and the double-labeling technique (cation tracer and D-[$^3$H]glucose).[1,3] The

[1] M. Yamaguchi and Y. Tonomura, *J. Biochem. (Tokyo)* **86,** 509 (1979).
[2] T. Watanabe, D. Lewis, R. Nakamoto, M. Kurzmack, C. Fronticelli, and G. Inesi, *Biochemistry* **20,** 6617 (1981).
[3] M. Yamaguchi and Y. Tonomura, *J. Biochem. (Tokyo)* **88,** 1387 (1980).

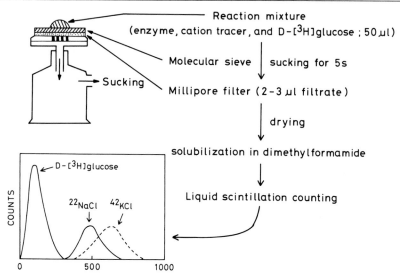

FIG. 1. Procedures for measurement of the amounts of cation binding to the enzyme. For explanation, see the text.

Na$^+$,K$^+$-ATPase partially purified from the outer medulla of porcine kidney is used in these experiments.[4] To measure $^{22}$Na$^+$, $^{42}$K$^+$, or $^{86}$Rb$^+$ (a K$^+$ congener) binding during ATP hydrolysis, the reaction is started by adding 10 $\mu$l of 0.5 m$M$ Tris-ATP to 40 $\mu$l of the reaction mixture to give final concentrations of 5–18 mg/ml enzyme, 0.8 m$M$ $^{22}$NaCl, 0.24 m$M$ $^{42}$KCl, or 0.042 m$M$ $^{86}$RbCl, 2 m$M$ MgCl$_2$, 0.5 m$M$ EDTA, 50 m$M$ D-[$^3$H]glucose, 75 m$M$ choline-Cl at pH 7.5 and 0°. The reaction is carried out in a cellulose nitrate tube (Beckman Instrument Inc., Palo Alto, CA), since monovalent cations are nonspecifically absorbed by a glass tube.

At appropriate intervals after the start of reaction, the reaction mixture is poured onto a set of two membrane filters which is then sucked under vacuum for 5 sec at cold room temperature of 3°. A small volume of the reaction mixture (2–3 $\mu$l) can be trapped in the lower filter by this suction, even when a high concentration of the enzyme (5–18 mg/ml) is used. For efficient suction, an acrylic plate, about 2 mm thick, and 20 mm in diameter, with 4–5 pores (1 mm diameter) in its center is used on the medium grade sintered glass filter (Fig. 1). D-[$^3$H]Glucose is used to determine the volume of the filtrate as described later. Choline-Cl is usually added to the reaction mixture to reduce nonspecific binding of monova-

[4] L. K. Lane, J. H. Copenhaver, Jr., G. E. Lindenmayer, and A. Schwartz, *J. Biol. Chem.* **248**, 7179 (1973).

lent cations to the enzyme by increasing the ionic strength of the reaction mixture.[5] The $Na^+,K^+$-ATPase reaction is unaffected by D-[$^3$H]glucose and choline-Cl under the condition used here. Furthermore, the equilibrium of the binding reaction of monovalent cations to the enzyme may not be significantly altered by the filtration, since the volume of the filtrate is less than 6% of the reaction mixture as mentioned above.

No significant interaction of cation tracers or D-[$^3$H]glucose with the upper filter is observed, and no protein leak through the upper filter is detected. These should be the criteria for selection of the upper filter. The 0.4-$\mu$m Nucleopore filter (Nucleopore Co., Pleasanton, CA), GS 0.22-$\mu$m Millipore filter (Nihon Millipore Ltd., Tokyo), and Amicon Diaflo XM100A (Amicon Co. Ltd., Lexington, MA) are available for the upper filter (25 mm diameter). The 0.45–3.0 $\mu$m Millipore filter (Nihon Millipore Ltd., Tokyo) is used for the lower filter (10 mm diameter). The lower Millipore filter is dried for 2 hr, then solubilized with 0.5 ml of dimethylformamide for 1 hr. The solution (0.4 ml) is used for counting radioactivities with 12 ml of xylene or toluene scintillation cocktail. The radioactivities of cation tracer and $^3$H are counted separately by liquid scintillation counter programmed for double-label analysis (Beckmann LS-9000 with the dual-label DPM calculation routine of the data-reduction system).

The radioactivities of cation tracer and D-[$^3$H]glucose can be measured separately in the same filtrate because the energies of $\beta$-radiation emitted by $^{22}Na^+$, $^{42}K^+$, and $^{86}Rb^+$ are high, while that by D-[$^3$H]glucose is low (Fig. 1). In the case of $^{86}Rb^+$, 17.8% of its radioactivity is in the range of the radioactivity of $^3$H, and the amount of D-[$^3$H]glucose is obtained by subtracting the counts of $^{86}Rb^+$ from the total counts in the range of energy of $^3$H. The amount of cation bound to the ATPase protein is calculated as follows: if the concentration of added cation in the reaction mixture is $A$ m$M$, the ratio of radioactivities ($^{22}N^+/^3H$) of the filtrate in the presence of the ATPase is $x$, and the ratio in the absence of the ATPase is $y$, the concentration of bound cation is $A(1 - x/y)$ m$M$. The amount of bound cation (nmol/mg protein) is calculated from this value by correcting for the amount of the ATPase protein used. The ratio of radioactivities can be measured accurately when the radioactivity of D-[$^3$H]glucose used is 10 times higher than that of cation tracer.

Nonradioactive $Na^+$ and $K^+$ contaminating the reaction mixture cause the decrease in the specific radioactivity of $^{22}Na^+$ and $^{42}K^+$. Reagents of special grade should be used to prevent it and the concentration of free $Na^+$ and $K^+$ must be calculated by correcting for the decrease in the

[5] K. Kaniike, G. E. Lindenmayer, E. T. Wollick, L. K. Lane, and A. Schwartz, *J. Biol. Chem.* **251**, 4794 (1976).

specific radioactivity of $^{22}$Na$^+$ and $^{42}$K$^+$.[6] The standard deviations of the amounts of bound cations are dependent on the purity of the enzyme preparation and the concentrations of the ATPase protein and monovalent cations. They are usually 5–10% for Na$^+$ and K$^+$ binding when 0.8 m$M$ NaCl or 0.24 m$M$ KCl is used. In the case of Rb$^+$ binding, the deviation is less than 5% in the presence of 0.1 m$M$ RbCl.

There are two important aspects to the reaction mechanism of the Na$^+$,K$^+$-ATPase reaction. One aspect is that the Na$^+$,K$^+$-ATPase can take at least two different conformations, a Na-bound form, E$_1$Na, and a K$^+$-bound form, E$_2$K. The other aspect is that the affinities of the ATPase for Na$^+$ and K$^+$ change with the conformational change accompanying the phosphorylation cycle of the ATPase reaction. The amounts of Na$^+$ and K$^+$ bound to the ATPase must be measured at various enzymatic states to know which step in the ATPase reaction actually induces the affinity change. Na$^+$ bound to the enzyme in the presence of Mg$^{2+}$ is dissociated rapidly by the formation of phosphoenzyme, then rebinds to the enzyme gradually as the amount of phosphoenzyme is reduced. On the other hand, as shown in Fig. 2, on addition of ATP, about 2 mol of Rb$^+$ bind rapidly per mole of active site, with the formation of phosphoenzyme. Similar results are also observed for the binding of K$^+$. Under these conditions, almost all phosphoenzyme formed at the steady state is ADP-insensitive and K$^+$-sensitive, that is, E$_2$P. Thus, the formation of E$_2$P induces marked changes in affinities of the Na$^+$,K$^+$-ATPase for monovalent cations. It should be noted that 5 min after the addition of ATP, E$_2$P disappears almost completely while the enzyme still binds Rb$^+$ (Fig. 2). This may indicate that the enzyme occludes K$^+$ just after the dephosphorylation, as suggested previously by Post et al.[7]

### Ca$^{2+}$ Binding to the SR Ca$^{2+}$-ATPase during ATP Hydrolysis

The double-membrane filtration method is applicable for measuring the Ca$^{2+}$ binding during ATP hydrolysis, which provides us with useful information about the SR Ca$^{2+}$-ATPase reaction.[8,9] This method allows the relatively fast time course of Ca$^{2+}$ dissociation from the enzyme to be followed accurately. However, it is not available for measurement of a very small amount of the enzyme-bound Ca$^{2+}$. Therefore a single-membrane filtration method with the double-labeling technique has been devised.[2] The SR Ca$^{2+}$-ATPase purified from rabbit skeletal muscle by the

[6] H. Matsui and H. Homareda, *J. Biochem. (Tokyo)* **92**, 193 (1982).

[7] R. L. Post, C. Hegyvary, and S. Kume, *J. Biol. Chem.* **247**, 6530 (1972).

[8] Y. Nakamura and Y. Tonomura, *J. Biochem. (Tokyo)* **91**, 449 (1982).

[9] M. Shigekawa, S. Wakabayashi, and H. Nakamura, *J. Biol. Chem.* **258**, 8698 (1983).

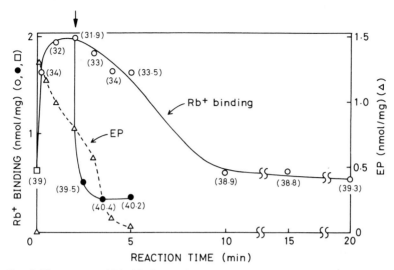

FIG. 2. Time course of Rb$^+$ binding to the enzyme after the start of the ATPase reaction. The amount of Rb$^+$ binding was measured after the ATPase reaction was started by adding 10 $\mu$l of 0.5 m$M$ ATP (○) to 40 $\mu$l of the enzyme solution (5 mg/ml) in the presence of 100 m$M$ NaCl, 42 $\mu M$ $^{86}$RbCl, 2 m$M$ MgCl$_2$, 0.5 m$M$ EDTA, 50 m$M$ D-[$^3$H]glucose, 75 m$M$ choline-Cl, 12.5 m$M$ imidazole-HCl, at pH 7.5 and 0°. The amount of Rb$^+$ binding at zero time was determined by adding H$_2$O (□) instead of ATP. When 10 m$M$ KCl was added to the reaction mixture 2 min after the start of the ATPase reaction (↓), the amount of Rb$^+$ binding decreased markedly (●). The numbers in parentheses indicate the concentrations of free Rb$^+$ ions ($\mu M$). The phosphorylation reaction was carried out using [$\gamma$-$^{32}$P]ATP under the same conditions (△).

method of MacLennan[10] is used in this experiment. Ca$^{2+}$ binding to the SR Ca$^{2+}$-ATPase protein is assayed in 10 m$M$ D-[$^3$H]glucose and 50 $\mu M$ $^{45}$CaCl$_2$ under various conditions. At serial time intervals following addition of ATP, two 0.1-ml aliquots of the reaction mixture are placed in duplicate on a 0.45-$\mu$m Millipore filter (13 mm diameter), prewashed with nonradioactive medium. The filtration time is 5 sec. The filter is then counted to determine the amount of total Ca$^{2+}$ ($^{45}$Ca$^+$ tracer) and the volume of medium (D-[$^3$H]glucose) retained on the filter. $^{45}$Ca$^{2+}$ and $^3$H in the filter are counted separately by differential settings on a liquid scintillation counter as mentioned above.

The amount of Ca$^{2+}$ bound to the enzyme is calculated as follows: if the ratio of the radioactivities ($^{45}$Ca$^{2+}$/$^3$H) in the reaction mixture is $x$ and the radioactivities of $^{45}$Ca$^{2+}$ and $^3$H retained on the filter are $y$ and $z$, respectively, the radioactivity of Ca$^{2+}$ bound to the ATPase is $y - zx$. The

[10] D. MacLennan, *J. Biol. Chem.* **245**, 4508 (1970).

amount of bound $Ca^{2+}$ (nmol/mg protein) is calculated from this value by correcting for the specific radioactivity of $^{45}Ca^{2+}$ and the amount of the ATPase protein used. For these calculations, the concentration of free $Ca^{2+}$ in the trapped volume is assumed to be equal to that of the reaction mixture, because the amount of $^{45}Ca^{2+}$ bound to the SR ATPase is small compared to the concentration of $^{45}Ca^{2+}$ in the reaction mixture (<10%) (Fig. 3).

The reaction mechanism of the SR $Ca^{2+}$-ATPase seems to be similar to that of the $Na^+,K^+$-ATPase. The affinity of the ATPase for $Ca^{2+}$ changes with the conformational change accompanying the phosphorylation cycle of the ATPase reaction. Many important problems concerning the affinity change remain unsolved, and direct measurement of $Ca^{2+}$ binding during ATP hydrolysis is an immediate goal. As shown in Fig. 3, $Ca^{2+}$ bound to the enzyme is dissociated rapidly with the formation of phosphoenzyme, and $Ca^{2+}$ binds again on phosphoenzyme decomposition in the presence of 30% dimethyl sulfoxide. Under this condition, ADP-insensitive EP, that is, $E_2P$ is accumulated.[11] Thus, it is clear that $Ca^{2+}$ dissociation and rebinding are due to change in affinity of the $Ca^{2+}$ binding site on the ATPase, related to the formation of $E_2P$ and its decomposition.

### ATP Binding on the ATP-Driven Pumps and Other ATPases during ATP Hydrolysis

The amount of ATP bound to the SR $Ca^{2+}$-ATPase during ATP hydrolysis is measured by the double-membrane filtration method, using $[\alpha$-$^{32}P]ATP$ as a substrate.[12] The reaction is started by mixing 20 $\mu l$ of the enzyme solution with 80 $\mu l$ of the $[\alpha$-$^{32}P]ATP$ solution containing 20 m$M$ creatine phosphate, 2 mg/ml creatine kinase, 1 m$M$ $CaCl_2$, 20 m$M$ $MgCl_2$, 100 m$M$ D-$[^3H]glucose$ at pH 8.8 and 0°. Excess amounts of creatine kinase and creatine phosphate are added as an ATP-regenerating system which removes ADP from the reaction mixture during ATP hydrolysis. At various intervals after the start of reaction, 30-$\mu l$ portions of the reaction mixture are withdrawn and applied to a set of two membrane filters (upper, Amicon Diaflo XM100A; lower, 0.8-$\mu m$ Millipore filter; 10 mm each in diameter), and suction is applied for 5 sec. The amount of bound ATP is calculated as described above.

The double-membrane filtration method has also been applied to the measurement the ATP, ADP, or $P_i$ binding to the $Na^+,K^+$-ATPase,[13]

[11] M. Shigekawa and A. A. Akowitz, *J. Biol. Chem.* **254**, 4726 (1979).
[12] Y. Nakamura and Y. Tonomura, *J. Bioenerg. Biomembr.* **14**, 307 (1982).
[13] M. Yamaguchi and Y. Tonomura, *J. Biochem.* (*Tokyo*) **88**, 1377 (1980).

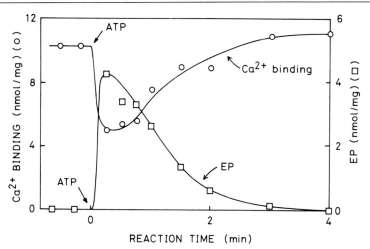

FIG. 3. $Ca^{2+}$ dissociation and the formation of phosphorylated intermediate following addition of ATP. $Ca^{2+}$ binding ($\bigcirc$) was measured in the presence of 0.5 mg/ml enzyme, 50 $\mu M$ $^{45}CaCl_2$, 5 m$M$ $MgCl_2$, 80 m$M$ KCl, 10 m$M$ D-[$^3$H]glucose, 30% dimethyl sulfoxide, and 20 m$M$ MES at pH 6.1 and 25°. ATP (200 $\mu M$) was added at zero time ($\downarrow$). Phosphorylation of the enzyme was measured with the aid of [$\gamma$-$^{32}$P]ATP ($\square$), in conditions identical with those used for $Ca^{2+}$ measurements.

mitochondrial $F_1$-ATPase,[14,15] myosin ATPase,[16] and dynein ATPase,[17] using [$\alpha$-$^{32}$P]- or [$\gamma$-$^{32}$P]ATP, [$\alpha$-$^{32}$P]ADP, or [$^{32}$P]P$_i$ as substrates. The single-membrane filtration method with the double-labeling technique has also been devised to measure the Adopp(NH)P, unhydrolyzable ATP analog, binding to the Na$^+$,K$^+$-ATPase, using [$^3$H]Adopp(NH)P and [$^{14}$C]sucrose.[18] Their calculation of bound nucleotide takes into consideration the reduction of free ligand concentration due to the binding to the enzyme in the reaction mixture.

## Comments

The double-membrane filtration method is appropriate for the measurement of the ligand binding to the ATP-driven pumps during ATP hydrolysis as described so far. This method is faster than equilibrium

[14] I. Matsuoka, T. Watanabe, and Y. Tonomura, *J. Biochem.* (*Tokyo*) **90**, 967 (1981).
[15] J. Sakamoto, *J. Biochem.* (*Tokyo*) **96**, 475 (1984).
[16] K. Furukawa, A. Inoue, and Y. Tonomura, *J. Biochem.* (*Tokyo*) **89**, 1283 (1981).
[17] S. Terashita, T. Kato, H. Sato, and Y. Tonomura, *J. Biochem.* (*Tokyo*) **93**, 1575 (1983).
[18] F. M. A. H. Schuurmans Stekhoven, H. G. P. Swarts, J.J. H. H. M. de Pont, and S. L. Bonting, *Biochim. Biophys. Acta* **649**, 533 (1981).

dialysis, ultracentrifugation, or the centrifuge–Sephadex column method, and exhibits relatively good time resolution in following the ligand dissociation and rebinding during ATP hydrolysis. The equilibrium of the binding reaction of ligands to the enzyme may not be altered significantly by the filtration as mentioned above. However, the standard deviations of the amounts of bound cations are usually 5–10% for Na$^+$ and K$^+$ binding to the Na$^+$,K$^+$-ATPase in the presence of 0.8 m$M$ NaCl or 0.24 m$M$ KCl, although they are dependent on the purity of the enzyme preparation and the concentration of the ATPase protein used in the experiment. The single-membrane filtration method with the double-labeling technique enables us to detect a small amount of the enzyme-bound ligand, but it may involve a shift in the equilibrium of the enzyme and ligand during filtration. Thus, both methods are useful for measurement of the ligand binding during ATP hydrolysis as long as they are used properly, and they should be refined further.

### Acknowledgments

Dedicated to the memory of Prof. Yuji Tonomura. We would like to thank Prof. Giuseppe Inesi, University of Maryland School of Medicine, and Dr. Akio Inoue, Faculty of Science, Osaka University, for their valuable advice during this study. We are also grateful to Prof. Hideo Matsui, Kyorin University School of Medicine, for his valuable comments.

## [18] Active Cation Pumping of Na$^+$,K$^+$-ATPase and Sarcoplasmic Reticulum Ca$^{2+}$-ATPase Induced by an Electric Field

*By* TIAN YOW TSONG

The peptide unit of a protein molecule is an electric dipole of about 3.5 Debye.[1,2] Thus, commonly occurring secondary structures of proteins possess electric moment(s). Most enzymes are also highly charged at neutral pH. Charged groups and structure units with electric moments are susceptible to various types of electrical perturbation. Because of the limited electric field one can apply to an aqueous solution without causing

[1] W. G. J. Hol, P. T. van Duijnen, and H. J. C. Berendsen, *Nature* (*London*) **273**, 443 (1978).
[2] A. Wada, *Adv. Biophys.* **9**, 1 (1976).

the breakdown of water, these perturbations are usually small.[3,4] However, if these molecules are embedded in the bilayer of a membrane vesicle which is suspended in an aqueous medium, the effect of the electric field experienced by these molecules is greatly enhanced. The amplification factor is roughly 1.5 $R/d$, where $R$ and $d$ are, respectively, the outer radius of the vesicle and the thickness of the cell membrane.[5] For an integral protein in human erythrocyte membranes, this amplification factor is about 1000. Membrane proteins under an intense electric field may undergo a conformational change. Their affinity to ions may also be altered. We have recently found that $Na^+,K^+$-ATPase of human erythrocytes and $Ca^{2+}$-ATPase of sarcoplasmic reticulum (SR) from rabbit can be induced by external electric fields to pump potassium and calcium ions against their respective concentration gradients.[6-9] This article will describe experimental procedures and summarize experimental findings. Extensive reviews are available concerning other effects of electric fields on cell functions and the activity of membrane-bound ATPases.[5,10-12]

## Instrument Arrangement

### Basic Principles of Electric Stimulation

Before selecting a particular instrument for experiment, several factors have to be considered. The most obvious one is the range of transmembrane potential needed to induce transport reaction. This value is not known in most cases, and an estimate has to be made. In our case, we selected induced transmembrane potential of about 10 mV to start our experiment. Second, one needs information on size, or size distribution of membrane vesicles, in order to estimate what range of applied field to use. In a simple case where the membrane vesicles or cells are spherical, the

[3] M. Eigen and L. de Maeyer, *Tech. Org. Chem.* **8**, 845 (1963).
[4] C. F. Bernasconi, "Relaxation Kinetics." Academic Press, New York, 1976.
[5] T. Y. Tsong, *Biosci. Rep.* **3**, 487 (1983).
[6] J. Teissie and T. Y. Tsong, *J. Membr. Biol.* **55**, 133 (1980).
[7] E. H. Serpersu and T. Y. Tsong, *J. Membr. Biol.* **74**, 191 (1983).
[8] E. H. Serpersu and T. Y. Tsong, *J. Biol. Chem.* **259**, 7155 (1984).
[9] E. H. Supersu, B. E. Knox, and T. Y. Tsong, *Fed. Proc. Fed. Am. Soc. Exp. Biol.* **42**, 1933 (1983).
[10] K. S. Cole, "Membranes, Ions and Impulses." Univ. of California Press, Berkeley, 1968.
[11] P. Lauger, R. Benz, G. Stark, E. Bamberg, P. C. Jordan, A. Fahr, and W. Brock, *Q. Rev. Biophys.* **14**, 513 (1981).
[12] U. Zimmermann, *Biochim. Biophys. Acta* **694**, 227 (1982).

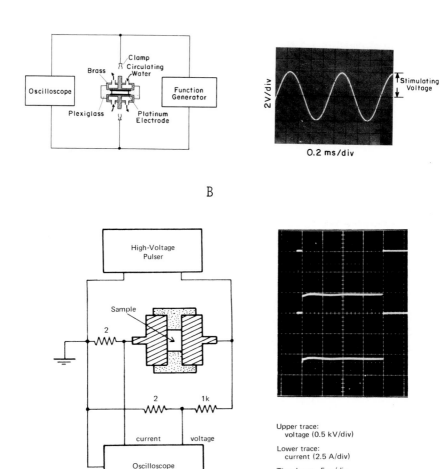

FIG. 1. Instruments for voltage stimulation of cation transport. (A) A function generator (Heath/Zenith SG-1271) is used to provide an ac field for the stimulation of Rb⁺ uptake by human erythrocytes. A typical waveform recorded with a Tektronix 7704A oscilloscope is shown. The stimulation chamber is composed of two hollow brass blocks through which cooling water circulates; a Plexiglas spacer of 3-mm thickness with a 7.5-mm center hole that holds cell suspension; and two flat premium disks that serve as electrodes. The sample is introduced through a V-spaced slit that opens a passage to the center hole of the Plexiglas spacer. (B) A high-voltage generator (in this case a Cober 605P) is used to generate kilovolt electric pulses for the induction of Ca²⁺ uptake by SR vesicles. The circuit shown here allows direct monitoring of the voltage and the current across the two electrodes. An

induced transmembrane potential ($\Delta\psi$) as a function of applied field strength may be calculated using the following relation.[5,12-15]

$$\Delta\psi_m = 1.5\ RE$$

In the equation, the maximum transmembrane membrane potential, $\Delta\psi_m$, is expressed in volts, the radius of vesicle, $R$, in centimeters, and the applied field strength, $E$, in volts/centimeter. The induced membrane potential for cells of different shape has been discussed previously.[15] For all practical purposes, $R$ in the equation may be replaced with the major axis of the cell.

*Experimental Setup*

Two different types of apparatus used in our experiments are mentioned here. The first type has been used for the stimulation of the $Na^+$, $K^+$-ATPase in human erythrocytes. The radius of human erythrocytes is about 4 $\mu$m. To induce a transmembrane potential of 10 mV, one needs an electric field strength of about 17 V/cm. This is easy to achieve by using an inexpensive function generator. We use a Heath/Zenith Model SG-1271 function generator. The arrangement and the waveform used in our experiment are illustrated in Fig. 1A. Usually, alternating current (ac field) is used to avoid electrode polarization when electric stimulation needs to last for longer than a few milliseconds. Two thin platinum sheets, about 2-cm diameter, coated with platinum black[16] are used as electrodes. The two electrodes sit on the back of two hollow brass blocks which are connected to the function generator. A Plexiglas sheet with a center opening separates the two electrodes and serves as the sample chamber. The thickness of the Plexiglas sheet is 3 mm, and the central opening has a diameter of 0.5 to 1.5 cm. Two capillary passages opened to the center hole allow filling and drawing of sample. The temperature of the stimulation chamber is controlled by circulating cooling water through the brass blocks, and is checked whenever necessary by a microthermistor probe inserted into the sample solution. Ionic transport in stimulated and controlled samples are monitored by a radioactivity assay as will be described

[13] K. Kinosita, Jr. and T. Y. Tsong, *Proc. Natl. Acad. Sci. U.S.A.* **74**, 1923 (1977).

[14] E. Neumann and K. Rosencheck, *J. Membr. Biol.* **14**, 194 (1973).

[15] K. Kinosita, Jr. and T. Y. Tsong, *Biochim. Biophys. Acta* **554**, 479 (1979).

[16] A. M. Feltham and M. Spiro, *Chem. Rev.* **71**, 177 (1971).

example is given in the oscillograph. In this arrangement, current in amperes is equal to one-half the reading of oscilloscope current signal in volts, and field strength is equal to one-five-hundredth the reading of voltage signal in volts divided by the distance between the two electrodes in centimeters. The stimulation chamber is identical to that explained in (A).

later. In our experiments, clean and uncontaminated platinum electrodes were found to be essential to see the voltage-induced ionic movement.

Another instrument used in our experiment is the Cober Model 605P high-voltage generator (Cober Electronics, 102 Hamilton Ave., Stamford, CT 06904). Instruments from other manufacturers with similar capability should be equally applicable. This setup (Fig. 1B) is capable of delivering about 2 kV, and is suitable for membrane vesicles with small size. The sarcoplasmic reticulum vesicles we use have an average diameter of about 90 nm. To achieve a transmembrane potential of 10 mV it requires an electric field strength of 1.5 kV/cm or greater, and one needs a more powerful generator such as a Cober 605P. The stimulation chamber is the same as the one used above for human erythrocytes. This setup is applicable for stimulation using direct current (dc), although each electric pulse should be limited to less than 1-msec duration to avoid severe heating.[3,13] Repetitive pulses with an interval between each pulse of about 30 sec are allowed. We also use reversed polarity for each electric pulse to avoid electrode polarization.

## Electrogenic Transport of Rb Ion by Na⁺,K⁺-ATPase

Several years ago, we observed that an electric pulse of a few kilovolts/centimeter and of duration in microseconds can puncture erythrocyte membranes so as to render them nearly freely permeable to Na, K, and Rb ions, and other small molecules.[13,15,17–19] In an attempt to locate the specific site of pore formation, we have determined that about 20–30% of these pores occurred at the Na⁺,K⁺-ATPase site.[6] An experiment was then designed to reversibly open and close this enzyme by a low-amplitude ac field,[20] and to investigate the effect of electric fields on its ion-transport activity.[7,8] In the latter experiments, radioactive tracers were used to monitor the cation movements and the response of these movements to cardiac glycosides.

### Ion Uptake and Efflux Measurements

Washed human red blood cells (RBC) are suspended in an isotonic mixture of NaCl, KCl, or RbCl, sucrose, and 1 m$M$ MgCl$_2$, 10 m$M$ Tris buffer at pH 7.4, with or without 50 $\mu M$ ouabain. Radioactive tracer, either ²²Na or ⁸⁶Rb, is added at time zero, and the sample is divided into

[17] K. Kinosita, Jr. and T. Y. Tsong, *Nature (London)* **268,** 438 (1977).
[18] K. Kinosita, Jr. and T. Y. Tsong, *Nature (London)* **272,** 258 (1978).
[19] E. H. Serpersu, K. Kinosita, Jr., and T. Y. Tsong, *Biochim. Biophys. Acta* **812,** 779 (1985).
[20] J. Teissie and T. Y. Tsong, *J. Physiol. (Paris)* **77,** 1043 (1981).

two halves. One half (designated S, i.e., stimulated) is stimulated with an ac field of defined voltage and frequency, using the instrument shown in Fig. 1A, and the other half (NS, i.e., nonstimulated) is incubated at the same temperature as the stimulated sample. The ouabain-containing sample is treated similarly (OS, ouabain-treated and voltage-stimulated; ONS, ouabain-treated, not stimulated). At intervals, aliquots from each sample are drawn and ionic movement in the time interval is determined by radioactivity counting. Ionic composition of the cytoplasm of stimulated and controlled cells is determined by flame photometry. When it is necessary to change the ionic composition of the cytoplasm, we either use the high-voltage perforation and resealing method of Kinosita and Tsong[13] or simply incubate the red cells in a medium overnight at 4°, and determine ionic composition afterward by flame photometry. Ionic uptake in a voltage-stimulation experiment is expressed in attomoles ($10^{-18}$) per cell per hour, which is equivalent to 0.011 mmol per liter cells per hour.[8]

### Active Rb⁺ Uptake Induced by the Electric Field

The range of electric field tested is 1 to 50 V/cm. This induces membrane potentials of 0.6 to 30 mV. Uptake and efflux of $Na^+$, $K^+$, and $Rb^+$ are monitored as a function of time for S, NS, OS, and ONS samples. The basal activity is taken as the value of NS minus ONS and the net stimulated activity is the value of S minus NS. Experiments are done at 4°, at which basal activity of $Na^+,K^+$-ATPase is negligible, and at 26°, at which $Na^+,K^+$-ATPase activity can be measured. Typical results are given in Table I and Refs 7 and 8 have established that $Rb^+$ and $K^+$ uptake is stimulated by the electric field and this stimulated uptake is against the concentration gradient of $K^+$ and $Rb^+$. With an ac field of 20 V/cm at 1 kHz, the stimulated $Rb^+$ uptake is 22.2 attomoles (amol)/RBC · hr at 26°, roughly one-half the basal $Rb^+$ pumping activity of the $Na^+,K^+$-ATPase, which is 39.8 amol/RBC · hr. The voltage-induced uptake of $Rb^+$ or $K^+$ is completely inhibited by ouabain, a potent inhibitor of $Na^+,K^+$-ATPase. Our recent experiments show that the Na pump of the enzyme can also be activated by an ac field of 20 V/cm, although the frequency for the activation is 1 MHz or higher.[21] The maximal stimulated $Na^+$ efflux is approximately 30 amol/RBC · hr at 4°.

### Stimulated Rb⁺ Uptake Mediated by Na⁺, K⁺-ATPase

Another interesting result of these experiments is that the voltage-stimulated $Rb^+$ or $K^+$ uptake shows an optimum around a field strength of 20 V/cm, and a frequency optimum of 1 kHz when 20 V/cm ac field is used

---

[21] T. Y. Tsong, D.-S. Liu, F. Chauvin, and R. D. Astumian, *Biophys. J.* **53**, 623a (1988).

TABLE I
VOLTAGE-STIMULATED Rb⁺ UPTAKE BY HUMAN ERYTHROCYTES[a]

| | Basal activity | | | | Net stimulated activity | | | |
|---|---|---|---|---|---|---|---|---|
| | Influx | | Efflux | | Influx | | Efflux | |
| Conditions | Rb⁺ | Na⁺ | Rb⁺ | Na⁺ | Rb⁺ | Na⁺ | Rb⁺ | Na⁺ |
| 1. Cellular concentration: 5.1 m$M$ Na⁺, 57 m$M$ K, 28 m$M$ Rb⁺. External concentration: 2.5 m$M$ Na⁺, 12.5 m$M$ Rb⁺, 1 m$M$ Mg²⁺, sucrose, etc., at 4° | 2.1 (3.7) | 0.2 (0.4) | 5.3 (8.0) | 3.9 (5.5) | 14.8 (4.2) | 0 (0.8) | 5.2 (3.0) | −5.8 (1.3) |
| 2. Cellular concentration 7.0 m$M$ Na⁺, 98 m$M$ K⁺. External concentration: 2.5 m$M$ Na⁺, 12.5 m$M$ Rb⁺, 1 m$M$ Mg²⁺, sucrose, etc. at 26° | 39.8 (2.7) | 3.7 (6.5) | 40.0 (13.9) | 62.5 (9.3) | 22.2 (2.8) | 1.0 (3.7) | 9.0 (13.0) | −2.8 (13.9) |
| 3. ATP depleted to 0.017 m$M$. Other conditions as in 2 | 19.9 (1.9) | ND | ND | ND | 22.8 (0.6) | ND | ND | ND |
| 4. ATP restored to 0.18 m$M$. Other conditions as in 2 | 34.0 (1.9) | ND | ND | ND | 20.3 (1.5) | ND | ND | ND |
| 5. RBC with higher basal ATPase activity. Conditions as in 2 | 79.9 (4.6) | ND | ND | 120.5 (10.0) | 39.7 (4.5) | ND | ND | ND |

[a] Human erythrocytes were loaded with either Rb⁺ or Na⁺ and radioactive tracer by incubation at 4° overnight. After washing, the ionic composition of cytoplasm was determined by flame photometry and radioactive counting. An ac field of 20 V/cm at 1.0 kHz was used in these experiments. Sample was exposed to the electric field for 1 hr in the absence (S) and presence of 50 $\mu M$ ouabain (OS). In the control sample, the electric field was not applied with (ONS) and without (NS) ouabain. Values are given in atto-moles ($10^{-18}$) per red blood cell (RBC) per hour, which is equivalent to 0.0108 mmol per liter cells per hour. Value for the basal activity equals NS minus ONS, and value for the net stimulated activity equals S minus NS. Normal cellular ATP concentration was 0.6–1.0 m$M$. Standard deviation from 6 to 18 measurements is given in parentheses. ND, not determined.

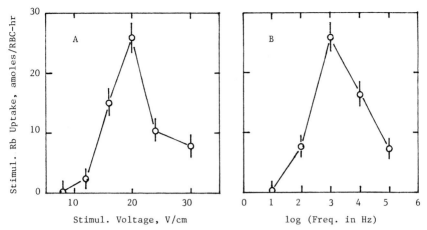

FIG. 2. An ac-stimulated Rb⁺ uptake by human erythrocytes. (A) Dependence on the electric field strength: The frequency of the ac field was 1 kHz. After 1 hr stimulation by the ac field of different field strength at 4°, the net Rb⁺ uptake as defined in the text was measured and plotted. See text for details. The ionic composition of the medium was 12.5 m$M$ RbCl, 2.5 m$M$ NaCl, 243 m$M$ sucrose, 10 m$M$ Tris at pH 7.4. (B) Dependence on the frequency: An ac field of 20 V/cm with varied frequency was used to stimulate the ouabain-sensitive Rb⁺ uptake. Conditions of experiment and the definition of the net uptake are the same as for (A). (Data were adapted from Serpersu and Tsong.[7])

(Fig. 2). These results rule out Joule heating as a source of the observed effect (Joule heating is less than 0.5° in all cases). That the uptake is mediated by Na⁺,K⁺-ATPase is supported by the fact that it is completely inhibited by inhibitors of the enzyme, ouabain (50% inhibition at 0.35 $\mu M$), oligomycin (8.0 $\mu M$), and ouabagenin (3 $\mu M$). The uptake, on the other hand, is not sensitive to 4,4′-bis(isothiocyano)-2,2′-stilbene sulfonate (inhibitor of band 3 protein-mediated Li/Na exchange), phloretin (inhibitor of Li/Na exchange), and vanadate (inhibitor of the Na pump), in micromolar concentrations.[22] The uptake is dependent on the external K⁺ concentration, with a $K_m$ of 1.7 m$M$, and on the internal Na⁺ concentration, with a $K_m$ of 5.5 m$M$, consistent with the activity of the Na⁺,K⁺-ATPase. An individual with twice the basal Na⁺,K⁺-ATPase activity also showed twice the voltage-stimulated activity.[8]

It is now known that electric fields can induce ATP synthesis in chloroplast and mitochondrial ATPases.[23–27] This is unlikely to occur in our

[22] E. H. Serpersu and T. Y. Tsong, *Fed. Proc. Fed. Am. Soc. Exp. Biol.* **43**, 2053 (1984).

[23] H. T. Witt, E. Schlodder, and P. Graber, *FEBS Lett.* **69**, 272 (1976).

[24] M. Rogner, K. Ohno, T. Hamamoto, N. Sone, and Y. Kagawa, *Biochem. Biophys. Res. Commun.* **91**, 362 (1979).

experiments with erythrocytes because the range of electric field we use is low and the induced membrane potential of 10 mV is less than one order of magnitude of that of the potential required for the synthesis of ATP. We have carefully monitored ATP concentration in the cells exposed to the ac field and found it to be quite constant in the course of the voltage experiment. The voltage-stimulated $Rb^+$ uptake is not sensitive to the cellular ATP level in the range 10 to 1000 $\mu M$.

The above observations support the notion that the $Na^+,K^+$-ATPase in the red cells is activated by the externally applied electric field. The optimum field strength of 20 V/cm can induce a transmembrane potential of 12 mV, and this potential could trigger an enzyme conformational change. However, the fact that the two pumps of the enzyme are activated by ac fields of different frequencies suggests that the Na pump and the K pump can be functionally uncoupled. When 1 kHz ac is used the K pump is active and the Na pump is idle; and the opposite is true when 1 MHz ac field is used. *In vivo,* the two pumps must work in parallel. Whether the phosphorylation of the enzyme *in vivo* plays a role in functionally coupling the two pumps remains unknown.

## Voltage-Induced Ca²⁺ Uptake by Sarcoplasmic Reticulum Vesicles

To examine whether other ion transport systems may also be susceptible to voltage stimulation we have also studied $Ca^{2+}$-ATPase activity of erythrocytes. However, our attempts to stimulate $Ca^{2+}$ uptake in human erythrocytes were not successful (unpublished results), possibly because the $Ca^{2+}$-ATPase in the red cell is not electrogenic. Attention was then turned to the sarcoplasmic reticulum vesicles of rabbit muscle. So far, we have limited data on this system. The R1 fraction of MacLennan[28] was used in our experiment. The mean diameter of the SR vesicles was 90 nm. We have used the Cober high-voltage generator described in Fig. 1B to generate transmembrane potentials in the range 5–50 mV. The sample chamber was made of 3-mm-thick Plexiglas, with a circular opening of 5-mm diameter. Approximately 0.2 ml of SR suspension is processed at one time. Several pulses are applied to each sample, with roughly 30-sec intervals between two pulses to allow for cooling due to Joule heating. The initial temperature of the sample is 7°, and each pulse raises the

[25] C. Vinkler, R. Korenstein, and D. L. Farkas, *FEBS Lett.* **145,** 235 (1982).

[26] J. Teissie, B. E. Knox, T. Y. Tsong, and J. Wehrle, *Proc. Natl. Acad. Sci. U.S.A.* **78,** 7473 (1981).

[27] B. E. Knox and T. Y. Tsong, *J. Biol. Chem.* **259,** 4757 (1984).

[28] D. H. MacLennan, *J. Biol. Chem.* **245,** 4508 (1970).

temperature of the sample by 10–15°. A pulse duration of 280 $\mu$sec and up to 9 kV/cm field strength are used in our experiment.

The suspension contains 0.25 $M$ sucrose, 0.5 m$M$ MgCl$_2$, 5 m$M$ KCl, 1 mg/ml bovine serum albumin (BSA), 1 m$M$ histidine, 0.1 m$M$ CaCl$_2$ and $^{45}$Ca$^{2+}$ tracer, and 10 m$M$ HEPES at pH 7.0. ATP concentration in the medium is negligible. After 10 pulses, SR vesicles are quickly spun down, washed twice with cold medium and prepared for radioactive counting. Control samples follow each step except the voltage-pulse step. To ensure that voltage-stimulated uptake of Ca$^{2+}$ requires Ca$^{2+}$-ATPase activity, as a double control we use an irreversible inhibitor of the enzyme, CrATP, to block ATPase-dependent Ca$^{2+}$ transport. A typical result is shown in Fig. 3A. Here calcium uptake, expressed in picomoles per milligram protein per pulse, is plotted for the voltage-treated samples with and without 0.15 m$M$ of CrATP. The corresponding controls, not exposed to voltage pulses, are also given. Figure 3B shows that SR vesicles are not severely damaged by the voltage treatment, at least for the first several pulses applied. Beyond that, vesicles seem to deteriorate and CrATP-sensitive Ca$^{2+}$ uptake does not increase further.

The rate of the stimulated Ca$^{2+}$ uptake is approximately 35 pmol/mg · pulse, with an 8-kV/cm, 280-$\mu$sec pulse. This is equivalent to 0.13 $\mu$mol/mg · sec. The normal ATP-dependent Ca$^{2+}$-pumping activity of the vesicles was 0.5 $\mu$mol/mg · sec.

Summary

To explain these results, we have proposed that the applied electric field induces a shift in the enzyme conformational equilibria through coulombic interaction. When the electric field is removed, or when it reverses its sign, the system returns to its initial state, and in the process of returning the free energy absorbed from the electric field is released. This released energy is used to drive active transport of an ion or to synthesize ATP. Models based on the concept of the "electroconformational coupling" have been analyzed and shown to work as predicted.[29-32]

Most cells maintain a steady-state transmembrane potential. This is an energy-consuming process, and as such, it must serve certain functions or purposes. Experiments discussed here would suggest that activation or

[29] T. Y. Tsong and R. D. Astumian, *Bioelectrochem. Bioenerg.* **15**, 457 (1986).
[30] H. Westerhoff, T. Y. Tsong, P. B. Chock, Y.-D. Chen, and R. D. Astumian, *Proc. Natl. Acad. Sci. U.S.A.* **83**, 4734 (1986).
[31] R. D. Astumian, P. B. Chock, T. Y. Tsong, T.-D. Chen, and H. Westerhoff, *Proc. Natl. Acad. Sci. U.S.A.* **84**, 434 (1987).
[32] T. Y. Tsong and R. D. Astumian, *Ann. Rev. Physiol.* **50**, 273–290 (1988).

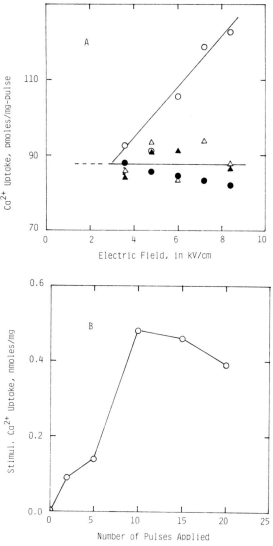

FIG. 3. Voltage-induced Ca²⁺ uptake by SR vesicles. The high-voltage generator shown in Fig. 1B was used to drive Ca²⁺ uptake by rabbit SR vesicles. The duration of each electric pulse was 280 μsec. (A) Ca²⁺ uptake is plotted for voltage-treated (○), voltage-treated but in the presence of 0.15 mM CrATP (●), control sample (△), and control sample in the presence of CrATP (▲). The control samples were not voltage treated. (B) Net stimulated Ca²⁺ uptake of a sample treated with 8 kV/cm 280 μsec pulses is shown. After more than ten pulses, the sample apparently degraded and the Ca²⁺ uptake was no longer stimulated by the electric pulses. See text for experimental details. Data were obtained by Serpersu and Knox.[9]

regulation of ion-transport system is one of these functions. Other experiments have shown that electric fields can stimulate ATP synthesis, cell proliferation, membrane fusion, DNA and RNA biosynthesis, etc. (for review see, e.g., Refs. 5 and 12). Apparently, the activation of ATPases is the most fundamental because this class of enzymes can transduce energy from one form to another, including transduction of electric energy to chemical bond energy of ATP and chemical potential energy of cations.

### Acknowledgments

Drs. K. Kinosita, Jr., B. E. Knox, J. Teissie, E. H. Serpersu, R. D. Astumian, F. Chauvin, and D.-S. Lin contributed to the work discussed here. This work has been supported by NIH Grant GM28795 and NSF Grant DCB86-11836.

## [19] Chemical Derivatization of Ca²⁺-Pump Protein from Skeletal Muscle with N-Substituted Maleimides and 5-(2-Iodoacetamidoethyl)aminonaphthalene 1-Sulfonate

*By* Masao Kawakita, Kimiko Yasuoka-Yabe, Kimiko Saito-Nakatsuka, Akiko Baba, and Tetsuro Yamashita

It has been difficult to achieve selective modification of a particular SH group which is well-defined in terms of either its location or functional role, since a large number of SH groups are present in Ca²⁺,Mg²⁺-ATPase of the sarcoplasmic reticulum (12–17 reactive SH groups in an ATPase polypeptide).[1-3] In spite of the fact that the ATPase molecule has long been known to contain SH groups essential for the functioning of the transport system, and that their number has been suggested to be rather small,[1,4,5] they have not yet been identified. SH groups were also utilized occasionally as a target for covalent attachment of various probes to monitor conformational changes of transport ATPase during its catalytic cycle.[6-9] In many cases, however, information gained from those studies

[1] A. J. Murphy, *Biochemistry* **15**, 4492 (1976).

[2] D. A. Thorley-Lawson and N. M. Green, *Biochem. J.* **167**, 739 (1977).

[3] N. Ikemoto, J. F. Morgan, and S. Yamada, *J. Biol. Chem.* **253**, 8027 (1978).

[4] W. Hasselbach and K. Seraydarian, *Biochem. Z.* **345**, 159 (1966).

[5] H. Yoshida and Y. Tonomura, *J. Biochem. (Tokyo)* **79**, 649 (1976).

[6] P. Champeil, F. Bastide, C. Taupin, and C. M. Gary-Bobo, *FEBS Lett.* **63**, 270 (1976).

[7] C. R. Coan and G. Inesi, *J. Biol. Chem.* **252**, 3044 (1977).

[8] C. Coan, S. Verjovski-Almeida, and G. Inesi, *J. Biol. Chem.* **254**, 2968 (1979).

[9] F. Guillain, P. Champeil, J.-J. Lacàpere, and M. P. Gingold, *J. Biol. Chem.* **256**, 6140 (1981).

has been rather limited by the fact that procedures for site-specific labeling have been lacking and thus the attachment sites of the probes were not sufficiently well-characterized.

In this article, simple procedures for selective derivatization of a limited number of specific SH groups with N-ethylmaleimide (MalNEt) and related fluorescent and paramagnetic probes as well as a procedure for labeling with 5-(2-iodoacetamidoethyl)aminonaphthalene 1-sulfonate (IAEDANS), a fluorescent derivative of iodoacetamide, are described.

### Factors Affecting the Reactivity of SH Groups toward Maleimide Compounds

The brief description which follows is intended to provide readers with background information pertinent to the understanding of the rationale of the procedures described below.

Among various factors which affect the specificity of derivatization, pH of the reaction mixture and the reagent used for the derivatization are critically important. Thus at pH 8.0, 16–17 SH groups per ATPase polypeptide reacted rather uniformly with SH reagents such as 5,5'-dithiobis(2-nitrobenzoic acid), although a certain degree of distinction among them can be noticeable.[1,10] On the other hand, at pH 7.0 SH groups were much less reactive and the difference in reactivity among them was more clearly discernible.[11,12] N-Substituted maleimides and iodoacetamide prefer different SH groups on the Ca$^{2+}$-pump protein. Thus, limited tryptic cleavage[2,13] of labeled SR membranes revealed that MalNEt-reactive SH groups are located on the A$_1$ fragment, which is derived from the middle portion of the ATPase polypeptide,[14] whereas iodoacetamide and IAEDANS specifically modify the B-fragment portion, the C-terminal half of the ATPase molecule, at the same pH and temperature.[15]

Four SH groups per ATPase polypeptide as listed in Table I are modified with MalNEt at pH 7.0 and 30° in the presence of 50 $\mu M$ Ca$^{2+}$.[11] They differ in reactivity from each other, the order of reactivity being SH$_N$ > SH$_D$ > SH$_{N'}$ > SH$_F$, and this provides the basis for a selective modification. SH$_N$, the most reactive one, and SH$_{N'}$, the third most reactive under this particular set of conditions are not essential for the functioning of the Ca$^{2+}$-transport system. On the other hand, integrity of SH$_D$ and SH$_F$ is

[10] A. J. Murphy, *J. Biol. Chem.* **253**, 385 (1978).

[11] M. Kawakita, K. Yasuoka, and Y. Kaziro, *J. Biochem. (Tokyo)* **87**, 609 (1980).

[12] S. Yamada and N. Ikemoto, *J. Biol. Chem.* **253**, 6801 (1978).

[13] D. A. Thorley-Lawson and N. M. Green, *Eur. J. Biochem.* **40**, 403 (1973).

[14] K. Saito, Y. Imamura, and M. Kawakita, *J. Biochem. (Tokyo)* **95**, 1297 (1984).

[15] A. Baba, T. Nakamura, and M. Kawakita, *J. Biochem. (Tokyo)* **100**, 1137 (1986).

TABLE I
SH GROUPS OF $Ca^{2+}Mg^{2+}$-ATPase REACTIVE WITH MalNEt AT pH 7.0 AND
THEIR CHARACTERISTICS

| Reactivity | | Number (residue/mole) | Function | Remarks |
|---|---|---|---|---|
| $SH_N$ | High | 1 | | |
| $SH_D$ | | 1 | E-P decomposition | Less reactive at low $Ca^{2+}$ |
| $SH_{N'}$ | | 1 | | Partially protected by AMP-P(NH)P |
| $SH_F$ | Low | 1 | E-P formation | Protected by AMP-P(NH)P |

required for the system to work properly. Modification of the former leads to an impairment of phosphoenzyme (E-P) decomposition and results in inactivation of ATPase and $Ca^{2+}$ transport, while the latter is required for E-P formation. SR membranes become heavily aggregated in parallel with $SH_F$ modification.

In the presence of AMP-P(NH)P, $SH_F$ is specifically protected from modification with MalNEt. Although $SH_{N'}$ also reacted more slowly in the presence of the nucleotide than in its absence, it is eventually modified after prolonged incubation with MalNEt. Reactivity of $SH_N$ and $SH_D$ is not affected by AMP-P(NH)P but appears to be altered to some extent by $Ca^{2+}$. Thus in the absence of $Ca^{2+}$ $SH_D$ is modified slower than in its presence, as judged by slower inactivation of the $Ca^{2+}$-transporting activity, despite the fact that more SH groups are actually reacted with MalNEt under this condition.[11]

$SH_N$, $SH_D$, and the IAEDANS-reactive SH group are particularly suitable for the purpose of site-specific derivatization, since the reactivity of each group is very distinct from the others under carefully chosen conditions[16,17] (details are described below). Recently, we have identified the location of $SH_N$[18] and the IAEDANS-reactive SH[19] with reference to the reported amino acid sequence of the ATPase molecule.[20] A unique SH group (Cys-674) has been identified as the IAEDANS binding site, while

[16] K. Yasuoka-Yabe and M. Kawakita, J. Biochem. (Tokyo) 94, 665 (1983).
[17] K. Yasuoka-Yabe, A. Tsuji, and M. Kawakita, J. Biochem. (Tokyo) 94, 677 (1983).
[18] K. Saito-Nakatsuka, T. Yamashita, I. Kubota, and M. Kawakita, J. Biochem. (Tokyo) 101, 365 (1987).
[19] T. Yamashita and M. Kawakita, J. Biochem. (Tokyo) 101, 377 (1987).
[20] C. J. Brandl, N. M. Green, B. Korczak, and D. H. MacLennan, Cell (Cambridge, Mass.) 44, 597 (1986).

SH$_N$ has turned out to consist of two nearby SH groups (Cys-344 and Cys-364) which show half-of-the-site reactivity.

Another method of site-specific derivatization using fluorescein isothiocyanate (FITC) has recently been reported.[21,22]

Labeling Reagents

A series of maleimide derivatives (Table II) as well as MalNEt can be utilized in A$_1$ fragment-directed site-specific labeling. It can be seen that more hydrophobic groups, serving as fluorescent or paramagnetic probes, are attached to the maleimide nitrogen of these compounds in place of the ethyl group of MalNEt. Reactivity of these probes toward the SH groups of Ca$^{2+}$-transport ATPase was increased in general as a result of respective substitution. Nevertheless, the difference in relative susceptibility of the four SH groups was found to be conserved.

If labeling specifically directed at the B fragment is desired, IAEDANS is a reagent of choice. In contrast to maleimide derivatives whose target site specificities are qualitatively similar to each other, two other fluorescent iodoacetamide derivatives tested, 5-iodoacetamido-fluorescein and 5-iodoacetamidoeosin, did not bind specifically to the B-tryptic fragment.

MalNEt and IAEDANS can be dissolved in H$_2$O, but other probes cannot and must be dissolved in 1,2-dimethoxyethane and added to reaction mixtures for labeling. The concentration of 1,2-dimethoxyethane in the reaction mixture should be kept below 0.2% to avoid nonspecific inactivation of the Ca$^{2+}$-pump protein. Fluorescent labels, BIPM,[23] ANM,[24] and DACM[25] are commercially available from Wako Pure Chemical Industries (Osaka, Japan) and IAEDANS from Aldrich (Milwaukee, WI). A nitroxide spin label, MSL, is available from Syva (Palo Alto, CA). The concentration of the labeling reagents is estimated from the ultraviolet absorption in water (for MalNEt and IAEDANS) or in ethanol (for other labels) by using molar absorption coefficients at appropriate wave-

[21] U. Pick, *Eur. J. Biochem.* **121,** 187 (1981).
[22] C. Mitchinson, A. F. Wilderspin, B. J. Trinaman, and N. M. Green, *FEBS Lett.* **146,** 87 (1982).
[23] Y. Kanaoka, M. Machida, K. Ando, and T. Sekine, *Biochim. Biophys. Acta* **207,** 269 (1970).
[24] Y. Kanaoka, M. Machida, M. I. Machida, and T. Sekine, *Biochim. Biophys. Acta* **317,** 563 (1973).
[25] M. Machida, N. Ushijima, T. Takahashi, and Y. Kanaoka, *Chem. Pharm. Bull.* **24,** 1417 (1976).

TABLE II
FLUORESCENT AND PARAMAGNETIC N-SUBSTITUTED MALEIMIDE COMPOUNDS EMPLOYED IN
SPECIFIC DERIVATIZATION

| Name | Abbreviation | Structure |
|------|-------------|-----------|
| N-[p-(2-Benzimidazolyl) phenyl]maleimide | BIPM | |
| N-(1-Anilinonaphthyl-4)- maleimide | ANM | |
| N-(7-Dimethylamino-4- methyl-3-coumarynyl) maleimide | DACM | |
| 4-Maleimido-2,2,6,6- tetramethylpiperidino oxyl | MSL | |

lengths as follows: MalNEt, 302 nm, 0.62 × 10$^3$ cm$^{-1}M^{-1}$; BIPM, 315 nm, 30.0 × 10$^3$ cm$^{-1}M^{-1}$; ANM, 351.5 nm, 13.2 × 10$^3$ cm$^{-1}M^{-1}$; DACM, 381 nm, 22.6 × 10$^3$ cm$^{-1}M^{-1}$; MSL, 210 nm, 8.2 × 10$^3$ cm$^{-1}M^{-1}$; IAEDANS, 337 nm, 6.1 × 10$^3$ cm$^{-1}M^{-1}$.

## Labeling of SH$_N$

### Monitoring the Course of the Modification Reaction

The extent of SH$_N$ labeling has to be assessed by measuring the amount of label covalently attached to ATPase. It is equally important to make sure that Ca$^{2+}$-transporting activity remains unchanged during the course of the reaction. The amount of MalNEt bound to SR membranes can be easily determined by using [³H]MalNEt and by measuring acid-insoluble radioactivity after the reaction has been stopped by excess 2-mercaptoethanol. The amount of DACM and ANM introduced into SR membranes is determined by measuring the ultraviolet absorption at appropriate wavelengths after solubilization with sodium dodecyl sulfate (SDS). To do this, samples are acid-precipitated with 5% trichloroacetic acid after the reaction has been stopped. The denatured protein is washed once with 0.6% trichloroacetic acid and twice with ethyl ether by centrifugation, and then dissolved in 2.5% SDS. The amount of DACM bound to SR can be estimated from the value of absorption at 380 nm, the isosbestic point of succinimide and succinamic acid forms of the adduct, using 19.8 × 10³ cm⁻¹M⁻¹ as a molar absorption coefficient. A molar absorption coefficient of 13.2 × 10³ cm⁻¹M⁻¹ at 351.5 nm is used for ANM. BIPM binding is difficult to assess because of partial overlapping of its absorption spectrum with proteins, but may be assumed to be similar to that of DACM and ANM in view of the closely similar reactivity of these reagents, as judged from their effect on activities of SR ATPase. Binding of MSL can be estimated during the reaction by following the decrease in electron spin resonance (ESR) signal from unbound spin label.[7,8]

### Pretreatment to Block Highly Reactive SH Groups on Non-ATPase Proteins of SR Membranes[18]

Several minor components of SR membranes contain SH groups which are much more reactive than $SH_N$.[26] These have to be blocked with nonradioactive MalNEt prior to the reaction with radioactive MalNEt or various spectroscopic probes in order that $SH_N$ can be labeled as specifically as possible. This is attained simply as follows: SR membranes are suspended in a solution containing 40 m$M$ $N$-tris(hydroxymethyl)methyl-2-aminoethanesulfonic acid (TES)-NaOH buffer (pH 7.0), 0.1 $M$ KCl, 5 m$M$ $MgCl_2$, and 50 $\mu M$ MalNEt at a concentration of 10 mg protein/ml. The suspension was kept at 0° for 30 min and then was ready for subsequent steps of specific labeling.

By using radioactive MalNEt in this step we can see that 1.0–1.5 nmol of MalNEt usually binds to 1 mg of SR protein. SDS-polyacrylamide gel electrophoresis followed by autoradiography mainly revealed three radiolabeled components of apparent molecular weights of 50,000, 32,000, and 30,000, respectively, as well as a few others to which much weaker radioactivity was incorporated. The ATPase molecule was virtually unreactive toward MalNEt under this condition.

[26] C. Hidalgo and D. D. Thomas, *Biochem. Biophys. Res. Commun.* **78,** 1175 (1977).

*Labeling of $SH_N$ with $[^{14}C]MalNEt$*[16,18]

After the pretreatment as described above, the suspension was diluted with solution A [40 m$M$ TES–NaOH buffer (pH 7.0), 0.1 $M$ KCl, and 5 m$M$ MgCl$_2$] to a protein concentration of 3 mg/ml, and [$^{14}C$]MalNEt of an appropriate specific radioactivity is added to give a final concentration of 50 $\mu M$. The suspension is incubated for 45 min at 30° and then the reaction is stopped by adding 0.1 vol of 1 $M$ 2-mercaptoethanol. The labeled membranes were washed once by centrifugation at 160,000 $g$ for 30 min in solution B [40 m$M$ Tris-maleate buffer (pH 6.5), 0.1 $M$ KCl, 5 m$M$ MgCl$_2$, and 0.3 $M$ sucrose], resuspended in the same solution at a concentration of 5–10 mg/ml, and stored frozen at −80°.

By this treatment, 6–7 nmol of [$^{14}C$]MalNEt usually binds to 1 mg of SR protein. Assuming a value of 70% for the ATPase content of SR membrane, 0.9–1.0 mol SH per mole ATPase is labeled under this condition. At least 90% of both $Ca^{2+}$-transport and ATPase activities should be retained after labeling $SH_N$ by this procedure.

*Labeling with Fluorescent and Paramagnetic Probes*[17]

Slight modifications of the above procedure are necessary to attain sufficiently selective labeling of $SH_N$ with the spectroscopic labels listed in Table II because of their increased reactivity toward SH groups. Use of these labeling reagents at lower concentrations is recommended, because the reagent-to-ATPase ratio rather than the concentration of the reagent itself was found to be critical in determining the extent of the modification reaction. This is in contrast to the case with MalNEt where the extent of the reaction can be controlled by mainly adjusting the concentration of the reagent. Thus, the fluorescent probes, DACM, ANM, and BIPM, are used at 27 $\mu M$ and a spin label, MSL, at 25 $\mu M$ in the otherwise-the-same reaction mixture given above. The reagent-to-ATPase ratio is roughly 1.2–1.3 on a molar basis under these conditions. Reactions are stopped at 40 min.

Labeling with fluorescent labels can be conveniently stopped by adding 2-mercaptoethanol to give a final concentration of 0.1 $M$, and the labeled membranes are recovered and stored in exactly the same way as described above. However, this obviously is not applicable for MSL, since nitroxide radical would be reduced by 2-mercaptoethanol. For this reason, the reaction of spin labeling was terminated by 5-fold dilution with cold solution C containing 40 m$M$ Tris-maleate buffer (pH 7.0), 0.1 $M$ KCl, and 0.3 $M$ sucrose. The labeled membranes are then washed twice by centrifugation at 160,000 $g$ for 30 min in solution C, resuspended in this solution at a concentration of 50–75 mg/ml, and stored at −80°.

Labeling of SH$_D$ with Fluorescent and Paramagnetic Probes[17]

Selective labeling of SH$_D$ with various spectroscopic probes is achieved in a two-step reaction. In this procedure, SH$_N$ as well as highly reactive non-ATPase-SH groups are first blocked with MalNEt. An appropriate labeling reagent is then added to this SH$_N$-blocked SR in the second step, resulting in a selective attachment of the probe to SH$_D$.

SH$_D$ and SH$_N$ were originally assumed to be distinct sulfhydryl groups because of the markedly different consequences of their modification. Quite recently, however, Cys-344 and Cys-364 were assigned to both SH$_N$ and SH$_D$.[18,27] These Cys residues are most likely located very close to each other and the modification of either one, which does not affect the enzyme activities, would greatly diminish the reactivity of the other on the same polypeptide chain. When both Cys-344 and Cys-364 on one ATPase peptide are modified, E$_1$-P to E$_2$-P conversion does not take place, and a long-lived E$_1$-P can be accumulated on addition of ATP to the SH$_D$-modified ATPase. SH$_D$ modification may thus be useful in conformational studies of this important reaction intermediate.

*First-Step Reaction (Blocking of SH$_N$)*

SR membranes (3.0 mg protein/ml) are suspended in solution A containing 70 $\mu M$ MalNEt and incubated at 30° for 35 ± 5 min (see below). The reaction with MalNEt was stopped by 6-fold dilution with solution A. The diluted suspension is ready for the second-step reaction.

The absolute reactivity of SH groups toward maleimides usually differs to some extent with different batches of SR membranes. For this reason, it is advisable to carry out a pilot experiment prior to a large-scale labeling to determine an incubation period just appropriate for a particular batch. The time course of the blocking of SH$_N$ can be monitored by measuring the binding of radioactive MalNEt as well as Ca$^{2+}$-transporting activity (a typical example is shown in Fig. 1). An incubation period which causes 7–8 nmol of MalNEt to bind to 1 mg of SR protein while retaining about 80% of the original Ca$^{2+}$-transport activity would be satisfactory.

A higher concentration of MalNEt as compared to that for SH$_N$ labeling is used here. As a result, some 20% loss of Ca$^{2+}$-transport activity is usually observed at the end of this step. Obviously, this can be avoided by using a lower concentration of MalNEt, but we prefer complete blocking of SH$_N$ even at the expense of blocking a small fraction of SH$_D$. In this

[27] M. Kawakita and T. Yamashita, *J. Biochem. (Tokyo)* **102**, 103 (1987).

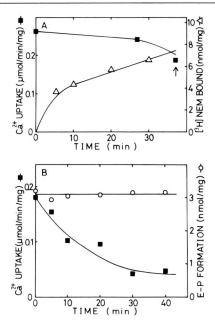

FIG. 1. Labeling of SH$_D$ with DACM in the two-step reaction. (A) Time course of blocking of SH$_N$. At the time indicated by an arrow, the sample was diluted 6-fold with solution A. (B) Time course of labeling at SH$_D$. See text for details of the reaction.

way, the site of fluorescence- or spin-labeling during the second-step reaction can be confined to SH$_D$.

*Second-Step Reaction (Labeling of SH$_D$)*

AMP-P(NH)P is added to the reaction mixture diluted at the end of the first step to give a final concentration of 1 m$M$. The second-step reaction is then started by addition of any one of the fluorescent probe listed in Table II at 27 $\mu M$ (final) or MSL at 0.3 m$M$ (final). AMP-P(NH)P is added to protect SH$_{N'}$ and SH$_F$ from modification. The reaction is allowed to proceed at 30° and terminated at 25 ± 5 min as described in the section on SH$_N$ labeling. Again, the use of 2-mercaptoethanol is avoided when MSL is used. The labeled membranes are recovered and stored as described above.

Ca$^{2+}$-transport activity would be rapidly lost during the reaction, while E-P forming activity should remain unchanged. The length of incubation period may be changed based on the results of a pilot experiment such as the one shown in Fig. 1. The amount of protein-bound dye is usually in the range of 1.5–2.0 mol/mol ATPase. This suggests that the selectivity of the modification with these probes may not be as strict as MalNEt.

Lowering the protein concentration to 0.5 mg/ml, with the concentration of the reagent being kept sufficiently low, was found effective in achieving the selective labeling of $SH_D$. This is particularly true for the fluorescent labels whose reactivity is much higher than MalNEt.

## Labeling with IAEDANS[15,19]

### Pretreatment with MalNEt

The pretreatment with MalNEt at 0° as described above in the section on labeling of $SH_N$ is also recommended in this case. This enables us to block highly reactive SH groups on non-ATPase proteins of SR membranes, leaving the IAEDANS-reactive sites completely intact. Exactly the same procedure as above may be followed.

### Reaction with IAEDANS

After the pretreatment with MalNEt, the reaction mixture is diluted with solution A to a protein concentration of 3 mg/ml and IAEDANS is added to give a final concentration of 50 $\mu M$. The reaction mixture is incubated for 4–6 hr at 30° in the dark to avoid photosensitized side reactions, chilled on ice, and centrifuged for 30 min at 160,000 g to recover the labeled membranes. The membranes are washed once by centrifugation in solution B, resuspended in the same solution at a concentration of 5–10 mg/ml, and stored at −80°.

### Monitoring the Course of Modification Reaction

Reactivity of the SH groups of ATPase towards IAEDANS is somewhat different quantitatively for each preparation, although qualitatively always the same SH group is modified first. For this reason it is advisable to carry out a pilot reaction prior to every preparative work to determine the length of the modification reaction period appropriate for the particular lot of SR membranes.

To estimate the amount of fluorescent label bound to the SR membranes, the membranes are separated from unreacted dye by centrifugal gel filtration[28] as follows: A 1-ml Sephadex G-25 (coarse) column in a plastic syringe equilibrated with solution A is centrifuged at 200 g for 2

---

[28] H. S. Penefsky, *J. Biol. Chem.* **252**, 2891 (1977).

min to drain external liquid. A 30-$\mu$l aliquot of the reaction mixture for labeling is placed on top of the column and effluent containing SR membranes is recovered by centrifugation (1100 rpm, 200 $g$, 2 min). The column is washed by another 30-$\mu$l aliquot of the above solution by centrifugation. Recovery of SR in the combined effluent is about 80% without any trace of contamination by unreacted labeling reagent. SR membranes in the effluent are solubilized by adding one-fourth the volume of 10% SDS and the absorbance is measured at pH 8.0 at 337 nm.

# [20] Use of Detergents to Solubilize the Ca$^{2+}$-Pump Protein as Monomers and Defined Oligomers

By JESPER V. MØLLER, MARC LE MAIRE, and JENS P. ANDERSEN

## Introduction

A central question in the study of the structural basis of transport protein function concerns the size of the transporting unit: Does each copy (monomeric polypeptide or protomer) possess all the attributes of transport, including channels for passage of the transported compound? Or does transport require the concerted action of an oligomeric structure? It is difficult to approach this question in a direct way since there is no direct method by which the self-associated state can be estimated in the membrane-bound form of the protein. In addition, an investigation of this kind requires comparisons between functional properties of the protein in various self-associated states in the membrane, and no established procedures as yet exist for this purpose.

The use of detergents to solubilize membrane proteins provides an approach to the problem which has the advantage that it is possible by analytical ultracentrifugation to define the aggregational state and to relate it to functional properties. These advantages are being realized in the case of the sarcoplasmic reticulum Ca$^{2+}$-ATPase, where it is possible to prepare active Ca$^{2+}$-ATPase in monomeric and oligomeric forms with the aid of nonionic detergents.[1-11] In addition, detergents are useful for characterization of the gross conformation of membrane proteins.[12,13] For this

[1] M. le Maire, J. V. Møller, and C. Tanford, *Biochemistry* **15**, 2356 (1976).

[2] M. le Maire, K. E. Jørgensen, H. Røigaard-Petersen, and J. V. Møller, *Biochemistry* **15**, 5805 (1976).

[3] W. L. Dean and C. Tanford, *J. Biol. Chem.* **252**, 3551 (1977).

[4] W. L. Dean and C. Tanford, *Biochemistry* **17**, 1683 (1978).

purpose we also describe the use of deoxycholate to prepare monomeric Ca$^{2+}$-ATPase solubilized in a stable, although inactive state, resembling the native form.

## Methods and Materials

The principle of the procedures for preparation of detergent-solubilized Ca$^{2+}$-ATPase described below is to solubilize the protein and lipid component of purified Ca$^{2+}$-ATPase membranes by an efficiently solubilizing detergent, and to separate mixed micelles and variously sized forms of the Ca$^{2+}$-ATPase by gel filtration in the presence of detergent.

### Ca$^{2+}$-ATPase Preparation

Sarcoplasmic reticulum membranes are prepared from rabbit skeletal muscle either by zonal centrifugation[14] or by differential centrifugation.[15] The sarcoplasmic reticulum membranes are extracted with a low concentration of deoxycholate[14] to remove extrinsic proteins. The purified Ca$^{2+}$-ATPase membranes are suspended in 1 m$M$ HEPES (pH 7.5) and 0.3 $M$ sucrose and stored at $-70°$ at a protein concentration of 12–15 mg/ml. The purity of the preparation with respect to Ca$^{2+}$-ATPase, as judged by sodium dodecyl sulfate-polyacrylamide gel electrophoresis, is 90% and phospholipid content is around 0.45 mg/mg protein. Enzyme activity (measured in the presence of 5 m$M$ MgATP, 0.1 m$M$ Ca$^{2+}$, 1 m$M$ Mg$^{2+}$, pH 7.5, 20°) is 4–5 $\mu$mol/mg/min, and maximal phosphorylation is 4–5 nmol/mg.

### Detergents

Octaethylene glycol dodecyl monoether (C$_{12}$E$_8$) is obtained in homogeneous form from Nikkol Chemicals, through the Kouyoh Trading Com-

[5] M. le Maire, K. E. Lind, K. E. Jørgensen, H. Røigaard, and J. V. Møller, *J. Biol. Chem.* **253,** 5051 (1978).

[6] K. E. Jørgensen, K. E. Lind, H. Røigaard-Petersen, and J. V. Møller, *Biochem. J.* **169** (1978).

[7] J. V. Møller, K. E. Lind, and J. P. Andersen, *J. Biol. Chem.* **255,** 1912 (1980).

[8] M. le Maire, J. V. Møller, and A. Tardieu, *J. Mol. Biol.* **150,** 273 (1981).

[9] A. J. Murphy, M. Pepitone, and S. Highsmith, *J. Biol. Chem.* **257,** 3551 (1982).

[10] J. L. Silva and S. Verjovski-Almeida, *Biochemistry* **22,** 707 (1983).

[11] H. Lüdi and W. Hasselbach, *J. Chromatogr.* **297,** 111 (1984).

[12] C. Tanford, Y. Nozaki, J. A. Reynolds, and S. Makino, *Biochemistry* **13,** 2369 (1974).

[13] J. V. Møller, M. le Maire, and J. P. Andersen, in "Progress in Protein–Lipid Interactions" (A. Watts and J. J. H. H. M. de Pont, eds.), Vol. 2. In press, 1987.

[14] G. Meissner, G. E. Conner, and S. Fleischer, *Biochim. Biophys. Acta* **298,** 246 (1973).

[15] L. de Meis and W. Hasselbach, *J. Biol. Chem.* **246,** 4759 (1971).

pany (4-1, 2-chome, Iwamoto-cho, Chiyoda-Ku, Tokyo, Japan). Tween 80 is from Serva (Heidelberg, FDR), and [14]C-labeled $C_{12}E_8$ and Tween 80 are from Centre Energie Atomique (Saclay, France). Deoxycholic acid, purchased from Merck AG (Darmstadt, FDR), is treated with charcoal to remove colored impurities and recrystallized from ethanol. [*carboxyl*-[14]C]Deoxycholic acid is obtained from The Radiochemical Centre (Amersham, England).

## Chromatographic Materials

For ordinary gel filtration, the following agarose gels have been used: Sepharose CL-6B (Pharmacia, Uppsala, Sweden) or BioGel A 1.5 m (Bio-Rad, Richmond, CA, with 8% cross-linking). Both gels provide equivalent separation of variously sized Ca²⁺-ATPase forms, but the latter gel material provides better resolution between monomeric Ca²⁺-ATPase and mixed micelles of lipid and $C_{12}E_8$. High-performance liquid chromatography (HPLC) is performed on 1 × 60 cm columns of TSK 3000 SW or TSK 4000 SW, manufactured by Toyo Soda Manufacturing Co. (Tokyo, Japan).

## Enzyme Activity

ATP hydrolysis is followed by monitoring NADH oxidation with a spectrophotometer, equipped with a recorder (e.g., Unicam SP1800 or Acta, Beckman). The procedure is as follows: Transfer 2.5 ml medium, containing 0.1 mM Ca²⁺, 0.1 M KCl, 0.01 M N-tris(hydroxymethyl)-methyl-2-aminoethanesulfonic acid TES (pH 7.5), and detergent at the appropriate concentration (usually equal to that in the column eluant) to a cuvette with an optical path of 1 cm, thermostatted at 20° in the cuvette compartment of the spectrophotometer. Make, in sequence, the following additions to the cuvette: 0.15 ml MgATP (0.1 M, adjusted to pH 7.5 with Tris base), 0.03 ml NADH (15 mM), 0.05 ml magnesium phosphoenolpyruvate (50 mM MgCl₂, and 50 mM phosphoenolpyruvate), 0.03 ml lactate dehydrogenase (from rabbit muscle, crystalline suspension in 3.2 M ammonium sulfate, Boehringer, Mannheim, FDR), and 0.05 ml pyruvate kinase [Sigma, grade III (lyophilized), solubilized in 0.01 M TES (pH 7.5) and 0.1 M KCl at a protein concentration of 5 mg/ml]. Following the addition of pyruvate kinase, there is a drop in absorbance at 340 nm, due to the presence of contaminating ADP. When the absorption has become constant, add the Ca²⁺-ATPase sample and measure the decrease in absorption as a function of time. A slight presteady-state curvature (concave) is observed initially, but after 1 min the slope is constant for membranous Ca²⁺-ATPase. For detergent-solubilized Ca²⁺-ATPase, slightly concave slopes are often observed throughout the time course as the result of inactivation during enzymatic turnover of the enzyme.[7,16] In this case, activity is best measured from the tangent of the curve after 1 min. The amount of $Ca^{2+}$-ATPase added should aim at an absorption decrease in the range of 0.05–0.15 per minute.

## Estimation of Size and Aggregational State

The size and homogeneity of the detergent-solubilized preparations may be checked either by analytical ultracentrifugation or by calibration of the columns with water-soluble standard proteins. Sedimentation coefficients and homogeneity may be determined from the rate of movement and shape of the boundary profile at high speeds (e.g., 44,000 rpm in the Beckman Model E analytical ultracentifuge, or the MSE Centriscan equipped with photoelectric scanner). Complete characterization requires sedimentation equilibrium runs at different rotor speeds for determination of molecular weight. As an alternative, molecular weight may be calculated from an estimate of the Stokes radius, based on the elution position of the detergent-solubilized $Ca^{2+}$-ATPase from the column.[13] However, it should be noted that this is a less precise method, complicated by a systematic error, since detergent-solubilized membrane proteins elute prior to the water-soluble standard proteins of the same Stokes radius.[17] Except for serum albumin the elution of standard proteins is not affected by the detergents used.

## Other Procedures

Protein concentrations are measured by the method of Lowry et al.[18] or by the biuret reaction.[19] Phosphorylation capacity is measured by ³²P incorporation into trichloric acid-precipitated protein[20] after exposure to $0.1 \, mM \, Ca^{2+}$, $0.1 \, mM \, MgATP$, containing [γ-³²P]ATP, at 0°. Phospholipid content in detergent-solubilized $Ca^{2+}$-ATPase is measured from phosphate content after wet ashing.[21] Detergent binding is measured by the method of Hummel and Dreyer[22] after equilibration of the column with radioactive detergent.

[16] D. W. Martin, J. A. Reynolds, and C. Tanford, *Proc. Natl. Acad. Sci. U.S.A.* **81**, 6623 (1984).

[17] M. le Maire, E. Rivas, and J. V. Møller, *Anal. Biochem.* **106**, 12 (1980).

[18] C. H. Lowry, N. J. Rosebrough, A. L. Farr, and R. J. Randall, *J. Biol. Chem.* **193**, 265 (1951).

[19] J. Goa, *Scand. J. Clin. Lab. Invest.* **5**, 218 (1953).

[20] J. P. Andersen, K. Lassen, and J. V. Møller, *J. Biol. Chem.* **260**, 371 (1985).

[21] G. R. Bartlett, *J. Biol. Chem.* **234**, 466 (1959).

[22] J. P. Hummel and W. J. Dreyer, *Biochim. Biophys. Acta* **63**, 530 (1962).

Procedures for Preparation of Detergent-Solubilized $Ca^{2+}$-ATPase

## Oligomeric $Ca^{2+}$-ATPase Solubilized in Tween 80

A Sepharose Cl-6B column (1.5 × 90 cm) is prepared and equilibrated with degassed buffer, containing 0.1 mg/ml Tween 80, 0.01 $M$ TES (pH 7.5), 0.1 $M$ KCl, 0.1 m$M$ $Ca^{2+}$, and 1 m$M$ azide (as a bacterial preservative). The procedure is as follows.

After thawing the $Ca^{2+}$-ATPase membranes (3–6 mg protein), add a stock solution of $C_{12}E_8$ (40 mg/ml) to obtain a weight ratio of detergent to protein of 3.6 : 1. Then solubilize 3–4 mg dithiothreitol in the slightly turbid sample and centrifuge for 10 min at 500–1000 $g$ to remove insoluble residues that otherwise might be retained on the top of the gel. Apply the supernatant to the Sepharose column and elute with column equilibrium buffer. Collect fractions of 1.5 ml for 20 hr at a rate of 6 ml per hour. Measure absorption at 280 nm, enzyme activity, and protein concentration.

The result of a typical experiment is shown in Fig. 1. Solubilized $Ca^{2+}$-ATPase elutes as a major and a minor peak between the void volume (indicated by insolubilized $Ca^{2+}$-ATPase) and the internal volume (indicated by dithiothreitol). Mixed micelles of lipid and $C_{12}E_8$ elute a little later. Enzyme activity is only associated with the major peak of the solubilized $Ca^{2+}$-ATPase which has approximately the same elution position as $\beta$-galactosidase. The protein molecular weight as measured by sedimentation equilibrium in the analytical ultracentrifuge corresponds to a trimeric–tetrameric aggregational state of the $Ca^{2+}$-ATPase polypeptide on the basis of a molecular weight of the $Ca^{2+}$-ATPase polypeptide chain of 110,000 and is fairly constant across the peak.[1,5] The minor and enzymatically inactive peak corresponds to a mixture of monomer and dimer. Approximately 0.15 g phospholipid is retained per gram ATPase (corresponding to 20–25 mol phospholipid), which apparently is essential for retention of enzyme activity. Binding of Tween 80 is 0.25 g per gram protein, whereas little $C_{12}E_8$ remains associated with the trimer–tetramer (0.01–0.03 g per gram protein). Thus the procedure is an example of exchange of one detergent with another by gel filtration. The rationale for this approach is that the enzyme is more rapidly inactivated by $C_{12}E_8$ than by Tween 80, while Tween 80 in contrast to $C_{12}E_8$ cannot be used for solubilization of the $Ca^{2+}$-ATPase membranes. In fact, the half-life of the Tween 80 preparation is the same as that of membranous $Ca^{2+}$-ATPase when stored under the same conditions at 0°.[1] Some inactivation occurs during passage through the column, resulting in a decrease of phosphorylation capacity to about 40–75% of that obtained before detergent solubili-

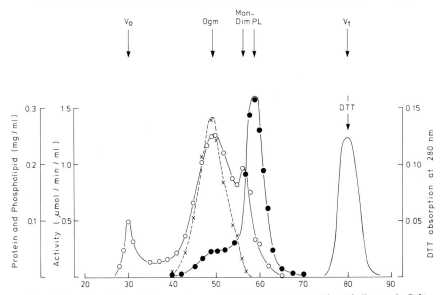

FIG. 1. Representative gel chromatogram obtained in the preparation of oligomeric Ca$^{2+}$-ATPase. Vesicles of Ca$^{2+}$-ATPase (3 mg) were solubilized by 10.8 mg C$_{12}$E$_8$ and chromatographed on a 1 × 90 cm Sepharose CL-6B column in the presence of 0.1 mg Tween 80/ml, 0.01 M TES buffer (pH 7.5), 0.1 M KCl, 0.1 mM Ca$^{2+}$, and 1 mM azide. $V_0$ and $V_t$, void volume and total volume, respectively; Ogm, trimeric–tetrameric Ca$^{2+}$-ATPase; PL, phospholipid; DDT, dithiothreitol in the applied sample, eluting at the total volume. O—O, protein; ●—● phospholipid; ×--×, ATP hydrolysis activity.

zation. The self-associated state of the preparation is also stable, i.e., it is not affected by dilution or storage.[5]

The Tween 80 preparation and similar oligomeric forms have been used in the study of various functional aspects of the enzyme.[23–25] Tween 80, in contrast to C$_{12}$E$_8$, has no perturbing effects on enzyme activity. The formation of an oligomeric peak in a trimeric–tetrameric aggregational state with the appearance shown in Fig. 1 stringently depends on a proper balance between the protein concentration, C$_{12}$E$_8$ concentration, and column material. If, for instance, BioGel A 1.5 m is used, a lower ratio of C$_{12}$E$_8$ to protein (2 : 1) is required to obtain an equivalent result. The reason for this different behavior appears to be that with BioGel A 1.5 m, the mixed micelles are better separated from the solubilized Ca$^{2+}$-

[23] Y. Dupont and M. le Maire, *FEBS Lett.* **115**, 247 (1980).
[24] H. Takisawa and Y. Tonomura, *J. Biochem. (Tokyo)* **86**, 425 (1979).
[25] T. Yamamoto, R. E. Yantorno, and Y. Tonomura, *J. Biochem. (Tokyo)* **95**, 1783 (1984).

ATPase. Apparently, for the same ratio of $C_{12}E_8$ to protein, this results in lower activity and phospholipid content. On the other hand, if too little $C_{12}E_8$ is added, a broad peak of solubilized $Ca^{2+}$-ATPase, extending to the void volume, is obtained.[5]

A question which is often raised in connection with the Tween 80 preparation is why the monomer/dimer peak is inactive, considering the evidence for solubilization of enzymatically active $Ca^{2+}$-ATPase in monomeric form by $C_{12}E_8$.[4,5,7,26] We suggest that oligomer formation during chromatography reflects a tendency for enzymatically active monomer of $Ca^{2+}$-ATPase to (re)aggregate when exposed to a nondissociating detergent like Tween 80. By contrast, $Ca^{2+}$-ATPase, *inactivated* in the monomeric or dimeric state, does not appear to have the same tendency for self-association. Stability of aggregational state after detergent inactivation is also a feature which is observed after solubilization with deoxycholate, as discussed in the next section.

### Preparation of Monomeric $Ca^{2+}$-ATPase

*Solubilization with Deoxycholate.* A Sepharose CL-6B column (1.5 × 90 cm) is equilibrated with 5 mM deoxycholate, solubilized in 0.03 M Tris buffer (pH 8.3), 1 mM ethylene glycol bis($\beta$-amino ethyl ether)-$N,N'$-tetraacetic acid (EDTA), and 1 mM azide. Solubilization is performed as follows: Add slowly, while stirring, 0.14 ml of 205 mM deoxycholate stock solution, solubilized in elution buffer, to 8 mg ATPase protein so that the detergent-to-protein weight ratio becomes 1.5 : 1. After centrifugation at 500–1000 g for 10 min, apply the solubilized sample to the column.

Figure 2 shows that the $Ca^{2+}$-ATPase elutes as a narrow peak well before the mixed micelles of phospholipid and deoxycholate. The predominant species is monomeric with a molecular weight of 115,000. However, aggregated (mainly dimeric) protein is present at the leading edge. When fractions are pooled corresponding to the hatched area shown on the chromatogram, the level of aggregated $Ca^{2+}$-ATPase is below that which can be detected by analytical ultracentrifugation. (By contrast, attempts to prepare monomeric $Ca^{2+}$-ATPase with nonionic detergents by gel filtration on agarose columns have always resulted in the admixture of appreciable amounts of dimeric $Ca^{2+}$-ATPase.)

$Ca^{2+}$-ATPase treated according to the protocol described is totally delipidated and enzymatically inactive. Binding of deoxycholate is 0.2 g per gram protein. To retain activity after solubilization with deoxycholate requires sucrose, a high ionic strength (e.g., 0.4 M KCl), phospholipid,

---

[26] D. W. Martin, *Biochemistry* **22,** 2276 (1983).

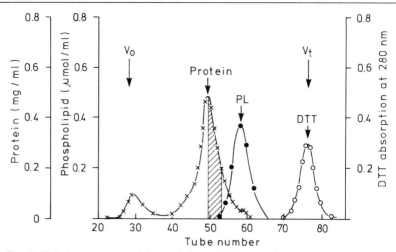

FIG. 2. Gel chromatogram of deoxycholate-solubilized $Ca^{2+}$-ATPase. Vesicles of $Ca^{2+}$-ATPase (6 mg) were solubilized by deoxycholate at a weight ratio of 1.5 : 1, and chromatographed on a Sepharose CL-6B column in the presence of 5 m$M$ deoxycholate, 0.03 $M$ Tris (pH 8.3), 1 m$M$ EDTA, and 1 m$M$ azide. The shaded area indicates fractions that may be used as a pure monomeric preparation of $Ca^{2+}$-ATPase. $V_0$, void volume; $V_t$, total volume; PL, phospholipid (●—●); DDT, dithiothreitol (○—○); protein (×—×). (From le Maire et al.[8])

and $Ca^{2+}$ to saturate the high-affinity sites of the $Ca^{2+}$-ATPase.[6] The shape and secondary structure of the two forms appears to be similar as judged by sedimentation analysis,[2] circular dichroism,[2] and low-angle X-ray scattering profile.[8] The unique feature of the gel filtration preparation is that it is structurally stable, with no tendency for either reversible or slow irreversible aggregation upon storage, even at high protein concentrations (5–10 mg/ml). This makes the preparation well suited for characterization of the gross conformation of the protein. Detailed structural investigation by analytical ultracentrifugation[2] or low-angle X-ray scattering[8] and freeze-fracture[27] has shown the monomer to have an elongated shape with an uneven distribution of mass between the two poles, presumably consisting of a large hydrophilic head and a hydrophobic tail which, in the native state, corresponds to the part which is inserted in the membrane.[28]

*Solubilization with $C_{12}E_8$ and HPLC Chromatography.* An attractive feature of $C_{12}E_8$, as compared to deoxycholate, is the possibility of pre-

[27] M. le Maire, J. V. Møller, and T. Gulik-Krzywicki, *Biochim. Biophys. Acta* **643**, 115 (1981).

[28] J. V. Møller, J. P. Andersen, and M. le Maire, *Mol. Cell. Biochem.* **42**, 83 (1982).

paring monomeric Ca$^{2+}$-ATPase with retention of enzyme activity. This requires rapid and efficient separation between the various forms of the Ca$^{2+}$-ATPase. We have only been able to achieve satisfactory results by HPLC. Furthermore, the use of an HPLC setup has the advantage that a whole series of chromatograms can be obtained from the same sample within 1 day. For example, the effect on aggregational state of solubilization conditions, protein concentration, and detergent concentration in the column can be systematically studied in this way.

A procedure for the preparation of enzymatically active Ca$^{2+}$-ATPase is as follows: Equilibrate the gel filtration column (TSK G-3000 SW or TSK G-4000 SW) with 20 m$M$ TES (pH 7.0), 0.1 m$M$ Ca$^{2+}$, and 10 m$M$ Mg$^{2+}$, and the desired concentration of C$_{12}$E$_8$. Solubilize 2 mg Ca$^{2+}$-ATPase by addition of 5 mg C$_{12}$E$_8$ (stock solution 100 mg C$_{12}$E$_8$/ml) and add the column buffer (without detergent) to a total volume of 0.5 ml. Remove unsolubilized material by centrifugation in a Beckman Airfuge at 130,000 $g$ for 25 min. Inject a 100-$\mu$l sample into the HPLC system and follow the elution pattern on a recorder attached to an ultraviolet absorption spectrophotometer or Uvicord (LKB, Bromma, Sweden), monitoring absorption at 226 or 280 nm.

Representative examples of results obtained by this procedure are shown in Fig. 3. At 2 mg C$_{12}$E$_8$/ml (Fig. 3, chromatogram 1) the protein elutes as a sharp peak of active Ca$^{2+}$-ATPase, preceded by smaller peaks of aggregated and inactive protein. Analytical ultracentrifugation indicates that the sharp peak represents monomeric Ca$^{2+}$-ATPase. Both enzyme activity (6 $\mu$mol/mg/min) and phosphorylation capacity (5–5.5 nmol/mg/min) are higher than for membranous Ca$^{2+}$-ATPase, suggesting separation of inactive from active Ca$^{2+}$-ATPase by centrifugation and gel filtration. Less than 0.02 mg phospholipid per milligram protein is associated with the monomer. If the sample is left for several hours before chromatography, an increased amount of inactive aggregates is formed at the expense of the monomeric peak, which remains enzymatically active. Thus HPLC is also an interesting method for separation of active and inactive forms of the Ca$^{2+}$-ATPase.

The elution pattern is unaltered by a decrease in the column eluant concentration of C$_{12}$E$_8$ to 0.5 mg C$_{12}$E$_8$/ml. As the C$_{12}$E$_8$ concentration is further reduced, a shoulder peak of Ca$^{2+}$-ATPase emerges, as shown in Fig. 3, chromatogram 2, at 0.075 mg C$_{12}$E$_8$/ml. At the critical micellar concentration (0.05 mg C$_{12}$E$_8$, Fig. 3, chromatogram 3), a whole spectrum of enzymatically active Ca$^{2+}$-ATPase peaks is observed, extending to the void volume. These represent reversibly associated states of Ca$^{2+}$-ATPase (in contrast to the Tween 80 preparation), since monomeric Ca$^{2+}$-ATPase is reformed by rechromatography. Analytical ultracentrifugation

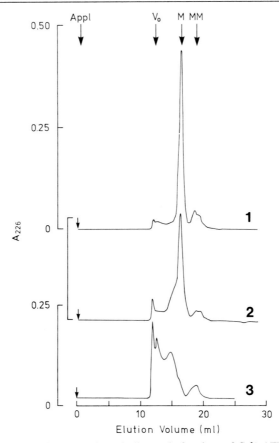

FIG. 3. Separation of monomeric and oligomeric fractions of $Ca^{2+}$-ATPase by HPLC. $Ca^{2+}$-ATPase (4 mg/ml) was solubilized with $C_{12}E_8$ (10 mg/ml), and 100 $\mu$l of the mixture was applied to a TSK 4000 SW column (1 × 60 cm), operated at 0.8 ml/min. The sample was eluted with 20 m$M$ TES (pH 7.0), 0.1 $M$ KCl, 0.1 m$M$ $Ca^{2+}$, 10 m$M$ $Mg^{2+}$, and various concentrations of $C_{12}E_8$. Chromatogram 1, 1 mg/ml; chromatogram 2, 0.075 mg/ml; chromatogram 3, 0.050 mg/ml. Applic., application of sample. $V_0$, void volume; $M$, monomer; $MM$, mixed micelles of lipid and $C_{12}E_8$. Chromatograms were registered with an Uvicord using a 226-nm interference filter.

evidence indicates that it is very likely that the shoulder peak associated with monomeric $Ca^{2+}$-ATPase (Fig. 3, chromatogram 3) represents dimeric $Ca^{2+}$-ATPase. Similar elution patterns as those shown in Fig. 3 are observed by conventional gel filtration, but the resolution of the chromagrams in terms of individual components is inferior to that obtained by HPLC.

# [21] Analysis of Two-Dimensional Crystals of $Ca^{2+}$-ATPase in Sarcoplasmic Reticulum

*By* K. A. TAYLOR, L. DUX, S. VARGA, H. P. TING-BEALL, and A. MARTONOSI

## Introduction

The $Ca^{2+}$-transport ATPase of sarcoplasmic reticulum (SR) is an intrinsic membrane protein with a molecular weight of 109,000.[1] It is asymmetrically distributed in the SR membrane[2,3] with much of its mass exposed on the cytoplasmic surface, where it can be visualized by negative staining in the form of 40-Å-diameter surface particles.[4,5] The intramembranous portion of the $Ca^{2+}$-ATPase is revealed by freeze-etching as particles of about 75 Å diameter that are more numerous in the cytoplasmic than in the luminal fracture face of the membrane.[2,6,7]

During ATP-dependent $Ca^{2+}$ transport, the $Ca^{2+}$-ATPase alternates between two distinct conformations, $E_1$ and $E_2$.[1] The $Ca^{2+}$ transport is initiated by interaction of the $Ca^{2+}$-ATPase with $Ca^{2+}$ and ATP in the $E_1$ conformation. Phosphorylation of the enzyme by ATP is followed by conversion from the $E_1$ into the $E_2$ enzyme form. The $Ca^{2+}$ transport is completed by the release of $Ca^{2+}$ from the enzyme in the $E_2$ conformation, and the subsequent hydrolysis of the phosphoenzyme intermediate. The $E_1$ conformation is stabilized by saturation of the high-affinity $Ca^{2+}$ binding site of the enzyme with $Ca^{2+}$ or lanthanides. Removal of $Ca^{2+}$ from the enzyme with ethylene glycol bis($\beta$-aminoethyl ether)-$N,N,N'N'$-tetraacetic acid (EGTA) and binding of vanadate stabilize the $E_2$ conformation.

The $Ca^{2+}$-ATPase has been successfully crystallized in the native SR membranes in both conformations[8-11] and in detergent solubilized SR in the presence of 20 m$M$ $Ca^{2+}$ at pH 6.0.[12-14] The two crystal forms pro-

---

[1] A. Martonosi and T. J. Beeler, *in* "Handbook of Physiology—Skeletal Muscle" (L. Peachey and R. H. Adrian, eds.), p. 417. Am. Physiol. Soc., Washington, D.C., 1983.

[2] D. W. Deamer and R. J. Baskin, *J. Cell Biol.* **42**, 296 (1969).

[3] A. Saito, C. Wang, and S. Fleischer, *J. Cell Biol.* **79**, 601 (1978).

[4] A. Martonosi, *Biochim. Biophys. Acta* **150**, 694 (1968).

[5] N. Ikemoto, F. A. Sreter, A. Nakamura, and J. Gergely, *J. Ultrastruct. Res.* **23**, 216 (1968).

[6] R. L. Jilka, A. N. Martonosi, and T. W. Tillack, *J. Biol. Chem.* **250**, 7511 (1975).

[7] C. Peracchia, L. Dux, and A. N. Martonosi, *J. Muscle Res. Cell Motil.* **5**, 431 (1984).

[8] L. Dux and A. Martonosi, *J. Biol. Chem.* **258**, 2599 (1983).

duced in native membranes are quite different from each other, suggesting that the transition between the $E_1$ and $E_2$ conformations involves a relatively large change in the structure of the enzyme that affects significantly the ATPase–ATPase interactions.

## Crystallization of Ca²⁺-ATPase in Native SR Membranes by Vanadate and EGTA

Sarcoplasmic reticulum vesicles prepared as described[15] are suspended in a medium of $0.1 M$ KCl, $10 mM$ imidazole (pH 7.4), $5 mM$ $MgCl_2$, $0.5 mM$ EGTA, and $5 mM$ $Na_3VO_4$ at a final protein concentration of 1 mg/ml, and allowed to crystallize for 1–3 days at a temperature of 2° (in ice). Crystalline arrays of Ca²⁺-ATPase molecules develop on the surface of 60–70% of the vesicles in SR preparations obtained from fast-twitch skeletal muscle, and on less than 10% of the vesicles in preparations of slow-twitch and cardiac muscle.[16] Vesicles with extensive crystalline arrays usually acquire an elongated tubular shape (Fig. 1). The vanadate-induced Ca²⁺-ATPase crystals consist of helical chains of Ca²⁺-ATPase dimers. Occasionally the crystals unwind (Fig. 2), revealing isolated dimer chains that adhere to the support film. Decavanadate is more effective than monovanadate in inducing rapid crystallization of the Ca²⁺-ATPase.[17] The vanadate-induced crystalline arrays are disrupted by Ca²⁺, ATP, and negative membrane potential (inside), presumably by shifting the equilibrium of the Ca²⁺-ATPase in favor of the $E_1$ form.[18,19]

## Crystallization of the Ca²⁺-ATPase in Native SR Membranes by Ca²⁺ and Lanthanides

Sarcoplasmic reticulum vesicles prepared from rabbit skeletal muscle are diluted in a medium of $0.1 M$ KCl, $10 mM$ imidazole (pH 7.4), and

[9] L. Dux, K. A. Taylor, H. P. Ting-Beall, and A. Martonosi, *J. Biol. Chem.* **260,** 11730 (1985).

[10] K. A. Taylor, L. Dux, and A. Martonosi, *J. Mol. Biol.* **174,** 193 (1984).

[11] K. A. Taylor, L. Dux, and A. Martonosi, *J. Mol. Biol.* **187,** 417 (1986).

[12] L. Dux, S. Pikula, N. Mullner, and A. Martonosi, *J. Biol. Chem.* **262,** 6439 (1987).

[13] S. Pikula, N. Mullner, L. Dux, and A. Martonosi, *J. Biol. Chem.* **263,** 5277 (1988).

[14] K. A. Taylor, N. Mullner, S. Pikula, L. Dux, C. Peracchia, S. Varga, and A. Martonosi, *J. Biol. Chem.* **263,** 5287 (1988).

[15] H. Nakamura, R. L. Jilka, R. Boland, and A. N. Martonosi, *J. Biol. Chem.* **251,** 5414 (1976).

[16] L. Dux, and A. Martonosi, *Eur. J. Biochem.* **141,** 43 (1984).

[17] S. Varga, P. Csermely, and A. Martonosi, *Eur. J. Biochem.* **148,** 119 (1985).

[18] L. Dux and A. Martonosi, *J. Biol. Chem.* **258,** 11896 (1983).

[19] L. Dux and A. Martonosi, *J. Biol. Chem.* **258,** 11903 (1983).

$5 \, \text{m}M$ $MgCl_2$ to a protein concentration of 1–2 mg/ml. For crystallization, $CaCl_2$ ($100 \, \mu M$) or lanthanide ions ($1$–$8 \, \mu M$) were added to the SR suspension.[9] After incubation at 2° for 5–48 hr crystalline tubules are observed on the surface of 10–20% of the vesicles in SR preparations obtained from fast-twitch rabbit skeletal muscle (Figs. 3–5). The crystalline arrays induced by $Ca^{2+}$ or lanthanides are presumed to reflect the $E_1$ conformation of the $Ca^{2+}$-ATPase. They are usually less well organized than the vanadate-induced crystals, and occur on fewer vesicles.

The optimum concentration for lanthanide-induced crystallization (5–$8 \, \mu M$) is just sufficient to saturate the high-affinity binding site of the $Ca^{2+}$-ATPase, and at slightly higher lanthanide concentration the frequency of crystalline vesicles sharply declines. On addition of slight excess of EGTA, the $Ca^{2+}$ and lanthanide-induced crystals rapidly disappear.

## Crystallization of $Ca^{2+}$-ATPase in Detergent-Solubilized SR

Crystals of $Ca^{2+}$-ATPase molecules develop in detergent-solubilized SR during incubation for several weeks at 2°, under nitrogen, in a medium of $0.1 \, M$ KCl, $10 \, \text{m}M$ K-MOPS (pH 6.0), $3 \, \text{m}M$ $MgCl_2$, $20 \, \text{m}M$ $CaCl_2$, 20% glycerol, $3 \, \text{m}M$ $NaN_3$, $5 \, \text{m}M$ dithiothreitol, 25 IU/ml Trasylol, $2 \, \mu g/ml$ 1,6-di-*tert*-butyl-*p*-cresol, 2 mg/ml protein, and 2–4 mg detergent/mg protein.[12,13] These crystals differ from those produced in intact SR vesicles in that they consist of extensive two-dimensional sheets that aggregate into stacked lamellar arrays.[14]

Under these same conditions, but in the absence of detergents, $Ca^{2+}$ at $20 \, \text{m}M$ concentration promoted the formation of stable two-dimensional crystalline tubules in intact SR vesicles.[12] These crystals have greater stability than the $E_1$ crystals obtained earlier,[9] and may be more suitable for detailed structural analysis by electron microscopy.

## Observation of Crystalline Arrays by Negative Staining Electron Microscopy

Vesicle suspensions containing the $Ca^{2+}$-ATPase crystals are most easily observed in electron microscopy (EM) by the negative staining method. The mechanism of preservation by negative stain involves replacement of aqueous regions of the sample with a nonvolatile salt of high atomic number. The image that is obtained is an indirect representation of the structure, since what is actually observed is the stain. Protein and the lipid bilayer are identified as regions of stain exclusion. Because deposition of stain is confined to the aqueous regions, no information on the protein shape in the hydrophobic regions of the membrane is obtained.

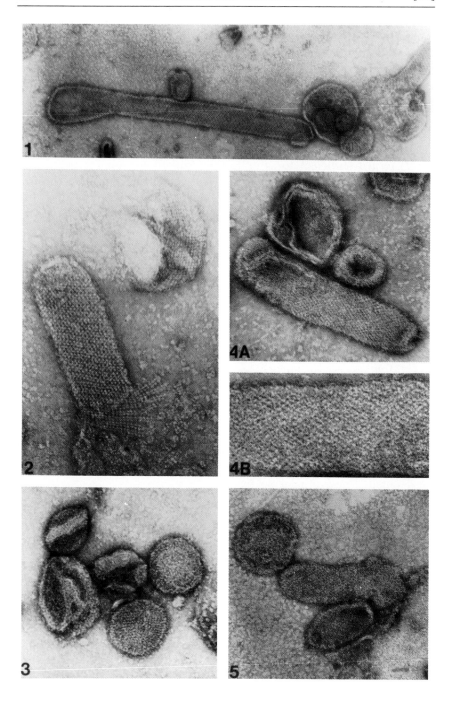

As in most EM sample preparations, many factors contribute to the quality of the specimen and often changes in one parameter can be compensated by alterations in another. Vesicles can be deposited on carbon-coated Parlodion films, or carbon films, and stained with 1–2% uranyl acetate or uranyl formate (pH 4.3). Staining with 1% potassium-phosphotungstate (pH 7.0) has given less satisfactory images.

Negatively stained preparations of crystalline arrays of $Ca^{2+}$-ATPase in native SR membranes appear in the electron microscope as elongated tubular shapes (see Figs. 6–8). Two types of images are frequently observed. Tubules of large diameter form the largest population and usually appear as long double-layered sheets. Smaller diameter tubules usually retain a cylindrical profile with a stain-filled core. Occasionally both types of morphology can be seen in the same tubule, indicating nonuniform flattening onto the support film of the EM specimen grid.

The vanadate-induced $E_2$ crystals display doublet rows of densities running at an angle 57.6° ± 2.1° to the long axis of the sheets (Fig. 6A). The doublet rows arise from chains of $Ca^{2+}$-ATPase dimers. Unpaired rows run at an angle of about 20° to the tubule axis. The superposition of the structural features arising from the two layers frequently shows a diamondlike arrangement. In addition, an interrupted line of density can

---

FIG. 1. Vanadate-induced $Ca^{2+}$-ATPase crystals in sarcoplasmic reticulum. Sarcoplasmic reticulum vesicles were incubated in $0.1 M$ KCl, $10 mM$ imidazole (pH 7.4), $5 mM$ $MgCl_2$, $0.5 mM$ EGTA, and $5 mM$ $Na_3VO_4$ for 48 hr at 2°. The diagonal lattice arises from superimposition of images from the front and rear surfaces of the collapsed cylinder. Magnification: 70,000×.

FIG. 2. Separation of $Ca^{2+}$-ATPase "chains" in a disrupted vanadate-treated sarcoplasmic reticulum vesicle. Sarcoplasmic reticulum vesicles were treated with $Na_3VO_4$ for 48 hr, as described in Fig. 1. At the end of the crystalline tubule, $Ca^{2+}$-ATPase chains are seen unwinding from crystalline arrays, which consist of rows of $Ca^{2+}$-ATPase dimers. The width of the flattened crystalline tubule is about 1000 Å. Magnification: 139,000×.

FIG. 3. $Ca^{2+}$-ATPase crystals induced by $Ca^{2+}$. Sarcoplasmic reticulum vesicles were incubated with Chelex chelating resin (100 mg wet resin per milligram protein) at 2° overnight. The Chelex particles were removed by centrifugation at low speed and the crystallization was induced in a medium of $0.1 M$ KCl, $10 mM$ imidazole (pH 8.0), $5 mM$ $MgCl_2$, and $0.1 mM$ $CaCl_2$ by incubation for 16 hr at 2°. Protein concentration: 1 mg/ml. Negative staining with 1% uranyl acetate. Magnification: 136,000×.

FIG. 4. $Ca^{2+}$-ATPase crystals induced by $GdCl_3$. (A) Sarcoplasmic reticulum vesicles were incubated in $0.1 M$ KCl, $10 mM$ imidazole (pH 7.2), $5.0 mM$ $MgCl_2$, and $10^{-6} M$ $GdCl_3$ at 2° for 48 hr, and samples were taken for negative staining with 1% uranyl acetate. Magnification: 136,000×. (B) A portion of the vesicle seen in A at higher magnification (272,000×).

FIG. 5. $Ca^{2+}$-ATPase crystals induced by $LaCl_3$. Sarcoplasmic reticulum vesicles were incubated in $0.1 M$ KCl, $10 mM$ imidazole (pH 7.2), $5.0 mM$ $MgCl_2$ in the presence of $10^{-6} M$ $LaCl_3$ at 2° for 5 days. Samples were stained with 1% uranyl acetate. Magnification: 136,000×.

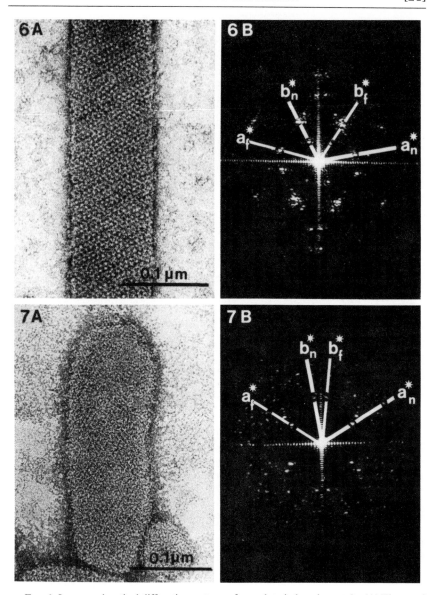

FIG. 6. Image and optical diffraction pattern of vanadate-induced crystals. (A) Flattened tubules show superimposed images of the crystalline arrays on both sides. Doublet rows of stain excluding densities running at an angle of $57.6 \pm 2.1°$ to the tubule axis. Magnification: $222,000\times$. The optical diffraction pattern (B) separates the contributions from the top and bottom sides. a* and b* represent the two reciprocal lattice vectors.

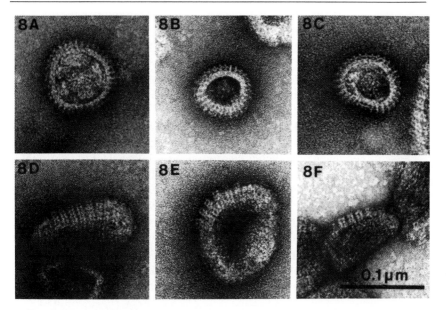

FIG. 8. Decavanadate-induced crystals viewed down the *a* and *b* axes. Panels A–C are projected images down the *b* axis and illustrate the connecting lobe feature of the structure. Panels D–F are projected images down the *a* axis of the unit cell, which is equivalent to sighting down the axis of the dimer ribbons. Magnification: 230,000×.

be seen running parallel to the tubule axis and separated, in projection, from the tubule surface by a narrow line of negative stain.

Optical diffraction analysis provides information on the unit cell dimensions. Both sides of the tubule contribute to the diffraction pattern (Fig. 6B). For the vanadate-induced $Ca^{2+}$-ATPase crystals, the unit cell of the side of the tubule stabilized by the carbon support film has dimensions of $a = 65.9$ Å, $b = 114.4$ Å, and $\gamma = 77.9°$.[10] The *a* dimension of the unit cell for the side of the tubule not stabilized by the carbon support film is usually shrunk by up to 5% in images of negatively stained tubules.

All preparations of the vanadate-induced crystals show views corresponding to projections down the two axes of the unit cell within the membrane plane (Fig. 8). Projections down the *a* axis amount to views

FIG. 7. Image and optical diffraction pattern of praseodymium-induced crystals. (A) Crystallization was induced with $8 \mu M$ $PrCl_3$. The doublet tracks so prominent in vanadate-induced crystals are not evident in crystals induced with lanthanides. This results in an approximate halving of the *b* axis of the unit cell. Magnification: 222,000×. (B) The image of the superimposed top and bottom lattices of the flattened cylinder gives rise to two separate diffraction patterns.

down the axis of the dimer chains and appear as inverted U shapes. The profile of the membrane bilayer is not, however, clearly visualized in this orientation. Projections down the $b$ axis are nearly perpendicular to the dimer chains. In this orientation, Ca$^{2+}$-ATPase molecules extend normal to the plane of the bilayer and are connected through a lobe that is clearly seen in computer-reconstructed images. This view often shows the bilayer profile clearly.

Crystalline arrays of Ca$^{2+}$-ATPase induced by Ca$^{2+}$ and by lanthanides also show the tubular morphology.[9] The crystals consist of chains of ATPase molecules wound helically around the tubule axis, but the double-stranded ribbons and stain-filled grooves between them are not seen in the E$_1$ crystals (Fig. 7A). The E$_1$ crystals are not as well preserved in negative stain as the E$_2$ crystals. Tubular membranes often show crystalline arrays on only a small patch of the membrane area.

Analysis of diffraction patterns of the E$_1$ crystals (Fig. 7B) gives unit cell dimensions in the range of $a = 61.7$–$71.9$ Å, $b = 50.4$–$54.4$ Å, and $\gamma = 103$–$114°$.[9] The $a$ axis of the unit cell is oriented at an angle of $70$–$80°$ to the tubule axis. The unit cell dimensions show that the crystals are made up of monomers of Ca$^{2+}$-ATPase.

Two distinct patterns are observed in negatively stained crystals from detergent solubilized SR.[14] In one view, layers of densities are seen that repeat at $130$–$170$ Å. These represent side views of stacked lamellar arrays of ATPase molecules. We assume that the cores of the lamellae contain a lipid-detergent phase into which the hydrophobic tail portions of the ATPase molecules are inserted symmetrically on both sides. Variable size of lamellae within the stacks suggests that they arise through aggregation of two-dimensional crystalline sheets.

In the second view, the projected image normal to the plane of the lamella shows ordered arrays of stain excluding particles with periodicities of $60$–$70$ Å. The particles are presumed to represent views of the cytoplasmic domains of Ca$^{2+}$-ATPase molecules. Under the solvent conditions in which the crystals are formed and are stable, images of the negatively stained crystals viewed from this orientation always show superlattice periodicities or moire fringes that arise from a slight rotational misalignment between the successive stacks of two-dimensional sheets.

### Observation of Unstained Crystalline Arrays of Ca$^{2+}$-ATPase

Structural information of protein–lipid distribution within the lipid bilayer requires examination of specimens that are not preserved with heavy metal stains. The most common preparation methods used to achieve this purpose are glucose embedding and rapid freezing of the sample in a thin aqueous film. For two-dimensional crystalline arrays, such as bacteriorhodopsin, that are highly ordered and insensitive to the solute concentrations that occur during drying, glucose embedding is an excellent method.[20] However, when the solution chemistry is important, analysis of frozen aqueous suspensions is the method of choice. Several techniques exist for preparing these specimens.[21,22] Rapid freezing is an important criterion.

Contrast in unstained specimens at low resolution is provided through the inherent difference in electron density between protein and lipid and between protein and the aqueous solution. The latter can be altered through the addition of solutes. There are two critical requirements for electron microscopy of frozen specimens: a stable cold specimen stage that can reach at least $-160°$ and a familiarity with low-dose procedures. Frozen hydrated suspensions are extremely sensitive to electron irradiation and frequently bubble when viewed directly in the electron beam, thereby destroying the structure.[23,24]

Unstained specimens behave as weak phase objects in the electron microscope and thus the contrast is entirely phase contrast. Underfocusing by large amounts (i.e., 2–5 $\mu$m) is a necessary requirement for visualizing any structural detail on the micrograph itself. A surprising amount of detail can be seen in highly underfocused images of the unstained $Ca^{2+}$-ATPase crystals frozen in vitreous ice films (Fig. 9). Protein, which scatters electrons more strongly than lipid or water, appears dark in the underfocused image, exactly the opposite of the situation with negative stain.

As in negative stain, views that are essentially projections down the *a* and *b* axes can be found in images of unstained $Ca^{2+}$-ATPase crystals (Fig. 10). These are very similar to the negatively stained structures except that in the images of frozen hydrated membranes some detail also can be seen within the membrane bilayer (Fig. 10A).

### Observation of Crystalline Arrays by Freeze-Fracture-Etch Electron Microscopy

Two-dimensional crystalline arrays of $Ca^{2+}$-ATPase can also be observed by freeze-fracture-etch electron microscopy. This technique is one

[20] P. N. T. Unwin and R. Henderson, *J. Mol. Biol.* **94**, 425 (1975).
[21] M. Adrian, J. Dubochet, J. Lepault, and A. W. McDowall, *Nature (London)* **308**, 32 (1984).
[22] K. A. Taylor and R. M. Glaeser, *J. Ultrastruct. Res.* **55**, 448 (1976).
[23] R. M. Glaeser and K. A. Taylor, *J. Microsc. (Oxford)* **112**, 127 (1978).
[24] Y. Talmon, M. Adrian, and J. Dubochet, *J. Microsc. (Oxford)* **141**, 375 (1986).

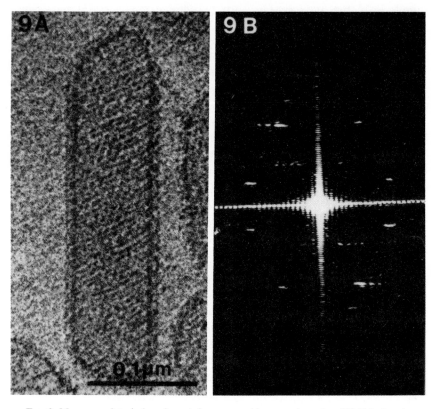

FIG. 9. Monovanadate-induced crystals preserved in amorphous ice. (A) Tubules shown here have not been flattened by drying and illustrate the diffraction pattern typical of helical cylinders. Protein is represented by the darker features of the image. Magnification: 300,000×. (B) Diffraction pattern.

method for producing a direct high-resolution image of both large areas of membrane surfaces and internal fracture faces. Furthermore, complementary replica preparations provide information on depth of protein penetration into the lipid bilayer.

In vanadate-induced $E_2$ crystals of $Ca^{2+}$-ATPase, regular arrays of oblique parallel ridges with spacings of ∼105–110 Å appear on the concave P-fracture (protoplasmic) faces and complementary grooves or furrows on the convex E-fracture (luminal) faces[7,25] (Fig. 11A and B and Ref. 7). Occasionally the ridges seem to break up into 60-Å particles repeating at every 55 Å. Lanthanide- and $Ca^{2+}$-induced $E_1$ crystalline lattices also

[25] H. P. Ting-Beall, F. M. Burgess, L. Dux, and A. Martonosi, *J. Muscle Res. Cell Motil.* **8**, 252 (1987).

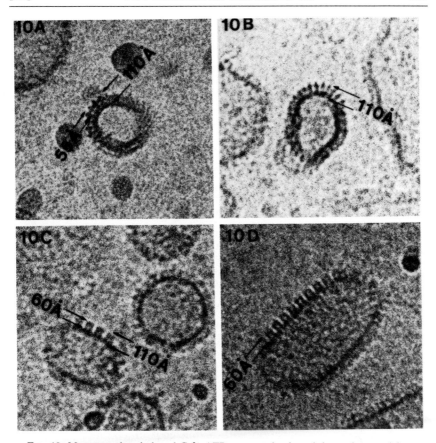

FIG. 10. Monovanadate-induced Ca$^{2+}$-ATPase crystals viewed down the $a$ and $b$ axes. Views similar to those shown in Fig. 8 can be obtained of unstained membranes in ice. Views down the $b$ axis (A and B) illustrate the lobe feature of the structure and views down the $a$ axis (C and D) show a bridge that connects Ca$^{2+}$-ATPase monomers to form the dimer. Magnification: 364,000×.

appear oblique in freeze-fracturing; however, the particles of 60-Å diameter are more individually spaced at about 60–70 Å along both axes. Complementary pits are also visible on the convex E faces.[9]

Etching the specimens after fracturing exposes the true protoplasmic surface. Deep etching and rotary shadowing reveal oblique crests on the protoplasmic surface consisting of dimeric particles of 85 × 55 Å in size, in which each monomer can be resolved frequently into two structural domains.[7] The difference between the particle size of 60 Å in the studies of Dux et al.[9] and Ting-Beall et al.[25] and the earlier reported 85 Å could be due to differences in the amount of metal replicated.

Conventional ways of replica cleaning and mounting onto grids are

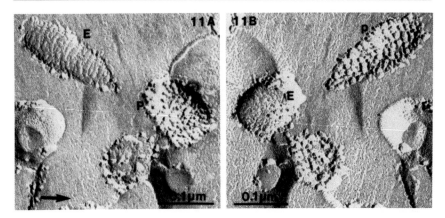

FIG. 11. (A and B) Freeze-fracture complementary pair of vanadate-induced $Ca^{2+}$-ATPase crystals in sarcoplasmic reticulum. Sarcoplasmic reticulum vesicles were incubated in $0.1\,M$ KCl, $10\,mM$ imidazole (pH 7.4), $5\,mM$ $MgCl_2$, $0.5\,mM$ EGTA, $5\,mM$ decavanadate for 24 hr, and fixed in 2% glutaraldehyde overnight. The fixed vesicles were ultrarapidly frozen in liquid propane without any cryoprotectant [M. J. Costello, *Scanning Electron Microscopy II*, 361 (1980)] and fractured in Cryofract 250 (Reichert-Jung, Paris, France) at $-183°$ and $2.5 \times 10^{-9}$ Torr, and replicated immediately with Pt/C at a 20° angle followed by C at a 90° angle. The thickness of Pt and C was monitored to be 11 Å and 110 Å, respectively. Concave P-fracture faces (P) show oblique parallel ridges and complementary furrows are seen on the convex E-fracture faces (E). The arrow indicates the direction of shadowing. Magnification: 150,000×.

well established,[26] but the procedures for double replicas are more complicated. Special care must be taken to prevent the contamination and decoration prior to replication with a thin coat of platinum. A special procedure for cleaning the replica and mounting of grids has been discussed in detail.[27] Briefly, replicas are stabilized by collodion[28] and by 50-mesh gold grids during the cleaning process. Finally the replicas adsorbed onto the gold grids are picked up on thin Formvar films spread over platinum wire loops. These Formvar films also help to stabilize the replicas in the electron beam.

Best results of complementary preparations are obtained when fracturing and replicating are performed at very high vacuum and at very low temperature. For example, under standard vacuum conditions ($10^{-6}$ Torr) and at a temperature of $-100°$, freshly fractured membrane faces can be contaminated with water vapor within a second.[29] However, lowering the

[26] J. H. M. Willison and A. J. Rowe, *in* "Practical Methods in Electron Microscopy" (A. G. Glarret, ed.), Vol. 8, p. 164–165. North-Holland Publ., Amsterdam, 1980.

[27] R. D. Fetter and M. J. Costello, *J. Microsc. (Oxford)* **141**, 277 (1986).

[28] J. R. Sommer and R. A. Waugh, *Am. J. Pathol.* **82**, 192 (1976).

[29] H. Gross, O. Kuebler, E. Bas, and H. Moor, *J. Cell Biol.* **79**, 646 (1978).

vacuum to $10^{-9}$ Torr reduces contamination by a factor of $10^3$. Gross *et al.*[30] have shown that fracturing and replicating at $-260°$ result in better structural preservation and substantially reduce granularity of Pt/C and thus provide better structural details.

Analysis of Crystalline Structure by Two-Dimensional
    Image Reconstruction

Structural analysis of the crystalline arrays begins with optical diffraction to survey images and is completed with a three-dimensional reconstruction to the limit of the periodic resolution. For crystals, the conventional approach utilizes Fourier methods, much as in X-ray crystallography. The methodology requires, in addition to an electron microscope, a densitometer to digitize the electron micrographs and a fairly large computer to handle the data analysis. Some software packages are available for this analysis,[31,32] otherwise a large amount of computer programming is necessary if a complete system is to be developed from scratch.

Briefly, the procedure involves selection from a large population of images of a few that exhibit the best diffraction pattern. These are digitized using a raster that is at least twice as fine as the highest resolution periodic feature in the image. Fourier transformation of the digitized image yields the structure factors, both amplitude and phase, of the crystal. The space group is determined from both the unit cell dimension and the symmetry elements present.[33]

Symmetry elements are deduced from systematic absences and by phase origin refinement using structure factors for the in-plane projection, a process that amounts to shifting the origin to different positions within the unit cell and comparing the recalculated phases with values that are restricted according to symmetry rules for the various two-sided plane groups. In the case of two-dimensional crystalline arrays of protein molecules, there are only 17 possible two-sided plane groups,[34] 15 of which require an interaxis angle of either 90 or 120°. Only two of these, $p1$ and $p2$, can accommodate a lattice in which the interaxis angle is not 90 or 120°. The distinction between $p1$ and $p2$ depends upon the presence of twofold rotation axes normal to the membrane plane. When the phase

[30] H. Gross, T. Muller, I. Wildhaber, H. Winkler, and H. Moor, *Proc. 42nd Annu. Meet. Electron Microsc. Soc. Am.*, p. 12 (1984).

[31] J. Frank, B. Shimkin, and H. Dowse, *Ultramicroscopy* **6**, 343 (1981).

[32] M. Van Heel and W. Keegstra, *Ultramicroscopy* **7**, 113 (1981).

[33] L. A. Amos, R. Henderson, and P. N. T. Unwin, *Prog. Biophys. Molec. Biol.* **39**, 183 (1982).

[34] W. T. Holser, *Z. Kristallogr.* **110**, 266 (1958).

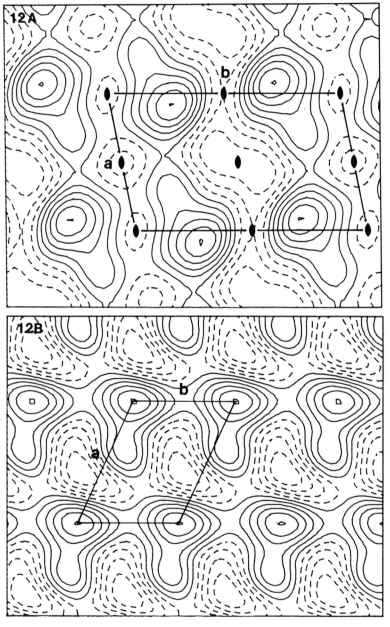

FIG. 12. Projection maps of vanadate- and praseodymium-induced crystals. (A) Monovanadate-induced crystals. From K. A. Taylor, L. Dux, and A. Martonosi, *J. Mol. Biol.* **174,** 193 (1984). (B) Praseodymium-induced crystals. From L. Dux, K. A. Taylor,

origin is located on the twofold, the phases from the in-plane projection are either 0 or 180°. The $p2$ space group for the vanadate-induced crystals is supported by this analysis. The same analysis carried out on the lanthanide crystals failed to demonstrate the presence of twofold rotation symmetry in projection. In this case, the space group is $p1$. Absolute proof of the space group cannot be achieved, however, without a three-dimensional reconstruction.

The analysis of the in-plane projection of the vanadate-[10] and praseodymium[9]-induced crystalline arrays yielded pear-shaped profiles for the $Ca^{2+}$-ATPase monomer (Fig. 12). In both cases, the long axis of the monomer was oriented in a very similar fashion relative to the tubule axis. The major difference between the two crystalline forms appears to be the presence of dimers in the vanadate-induced crystals. The subtle differences seen in the shapes of the $Ca^{2+}$-ATPase monomers at this stage do not appear to be significant.

Computed diffraction patterns from images of crystals from detergent-solubilized sarcoplasmic reticulum indicate that the unit cell dimensions are 164.2 ± 2.2 and 55.5 ± 1.5 Å with the included angle of 90°.[14] The diffraction pattern shows a set of systematic absences with observed spots obeying the selection rule $h + k = 2n$ indicative of a lattice centered on the $c$ face of the crystal. Only two of the two-sided plane groups are $c$ centered, $c12$ and $c222$. Both of these would show diffraction patterns with millimeter symmetry (two sets of orthogonal mirror lines in the $hk0$ zone).

The area of the unit cell in these crystals is 9113 Å.[2] This compares with 7371 Å[2] for the vanadate-induced crystals,[10] which contain two ATPase molecules per unit cell, and with 3070–3560 Å[2] for lanthanum-induced crystals,[9] which contain one ATPase molecule per repeating unit. Both of these are produced in native SR membranes. This area is large enough to give a reasonable but somewhat dense packing to the four ATPase molecules required for the $c12$ plane group, but it is too small for the eight molecules that would be required for two-sided plane group $c222$, if there is to be any space for a lipid-detergent phase. The 1840 Å[2] per molecule is less than that observed with the two crystal forms induced in native membranes and probably arises from the molecular packing, which symmetrically distributes the bulky cytoplasmic domains on both sides of the lipid-detergent phase. It is compatible, however, with the

---

H. P. Ting-Beall, and A. Martonosi, *J. Biol. Chem.* **260,** 11730 (1985). Although the two crystal forms are quite different in their oligomeric association, the $Ca^{2+}$-ATPase monomers have a similar orientation relative to the long axis of the tubule in both cases. Map scale: 0.55 mm/Å.

cross-sectional area of 1256 Å$^2$ for the transmembrane domains of individual ATPase molecules measured from three-dimensional reconstructions obtained from images of frozen, hydrated, vanadate-induced crystalline membranes.[35]

The thickness of the sheets from detergent-solubilized SR is much greater than that observed in images and three-dimensional reconstructions of frozen, hydrated, vanadate-induced crystalline tubules. The latter, formed in native sarcoplasmic reticulum membranes, have a thickness of 100–110 Å and have all ATPase molecules extending out the cytoplasmic side of the bilayer.[35] The observed thickness of the sheets is reasonably compatible with the crystal dimension of ATPase molecules in the native membranes, if it is assumed that molecules extend out both sides of the 40 Å thick lipid-detergent phase.

The two-sided plane group $c$12 requires that ATPase dimers be present in the crystals. However, unlike the ATPase dimers observed in vanadate-induced crystalline tubules, which are related by a 2-fold rotation axis normal to the membrane plane, the dimers in the sheets are related by a 2-fold rotation axis within the membrane plane.

### Image Reconstruction in Three Dimensions

There are two approaches for the three-dimensional reconstruction analysis. Because the native tubules are essentially helical cylinders, Fourier–Bessel reconstruction is a logical approach[36] provided that extensive flattening can be avoided. The Fourier–Bessel reconstruction method has both advantages and disadvantages. Depending upon the helical parameters, a single view is often sufficient to determine the structure. The method requires preservation of the cylindrical profile of the structure. Its primary drawback is that reconstructions tend to be noisy due to the fact that the transform data are spread out in continuous layer lines rather than being concentrated into discrete spots. Procedures for averaging particles of identical helical symmetries[37] are useful in reducing the noise level.

The alternative approach is the tilt view reconstruction method first carried out by Henderson and Unwin.[38] An excellent review of the subject of tilt view three-dimensional reconstruction has been published recently,[33] so only those aspects of the problem relevant to the Ca²⁺-ATPase will be discussed here. For this reconstruction method, structure

[35] K. A. Taylor, M-H. Ho, and A. Martonosi, *Ann. N.Y. Acad. Sci.* **483,** 31 (1986).
[36] D. J. DeRosier and P. B. Moore, *J. Mol. Biol.* **52,** 355 (1970).
[37] L. A. Amos and A. Klug, *J. Mol. Biol.* **99,** 51 (1975).
[38] R. Henderson and P. N. T. Unwin, *Nature (London)* **257,** 28 (1975).

factor data obtained from micrographs of tilted membranes are combined using a computer to determine the three-dimensional structure. The advantage of this approach is a high signal-to-noise ratio, due to the large number of identically oriented unit cells. It requires uniformly flat areas but has the disadvantage that some of the transform data, notably the data extending through the origin, are not accessible by tilting. Both Fourier–Bessel and tilt view methods have been applied to the crystalline tubules of sarcoplasmic reticulum $Ca^{2+}$-ATPase.[11,35,39]

A problem that occurs in dealing with images of double-layered crystals is overlapping diffraction spots. Such data cannot be used in the reconstruction process because they come from both sides of the double-layered membrane. The slightly different helical parameters of the various tubules help to ensure that the same diffraction spots do not always overlap in different crystals. A second problem that arises with processing images of the flattened $Ca^{2+}$-ATPase crystals in negative stain is the presence of some distortion in the lattice on drying. The effect of this on the computer-processed image can be reduced by processing small areas, since the disorder introduced is essentially one of long range. A second approach is to compensate for the disorder by correlation averaging techniques.[40–42] This method has been applied to the vanadate-induced crystals with the result that the resolution in the reconstruction did not increase, although there was a slight improvement in the reconstructed image.[43]

The three-dimensional reconstruction of the vanadate-induced crystals preserved in uranyl acetate stain yielded an image of the cytoplasmic region of the molecule (Fig. 13).[11] The profile of the molecule is pear-shaped as in the projection but the three-dimensional density map has the additional information that ATPase–ATPase interactions are of two types and occur at different levels above the membrane surface. Molecules are linked to form dimers by a bridge that is located about 42 Å above the bilayer surface. The lobe also seen in the projection maps is located about 28 Å above the membrane surface. The reconstruction of negatively stained membranes was featureless on the luminal surface, suggesting that little if any of the protein projects into the lumen.

The three-dimensional reconstruction of the unstained, vanadate-induced $Ca^{2+}$-ATPase tubule has density features contained within a cylindrical shell between 193 and 303 Å radius and shows both the cytoplasmic

[39] L. Castellani, P. Hardwicke, and P. Vibert, *J. Mol. Biol.* **185**, 579 (1985).
[40] J. Frank, *in* "Computer Processing of Electron Microscope Images" (P. W. Hawkes, ed.), Vol. 13, p. 187. Springer-Verlag, Berlin and New York, 1980.
[41] J. Frank, *Optik* **63**, 67 (1982).
[42] R. H. Crepeau and E. K. Frahm, *Ultramicroscopy* **6**, 7 (1981).
[43] E. L. Buhle, B. E. Knox, E. Serpersu, and U. Aebi, *J. Ultrastruct. Res.* **85**, 186 (1983).

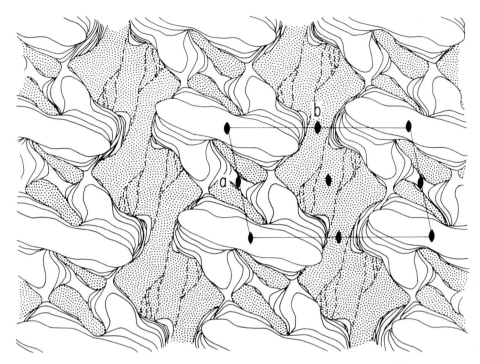

FIG. 13. Three-dimensional reconstruction of vanadate-induced crystals. This structure was derived from near and far sides of seven separate tilt series. The surface contour is shown with hidden lines removed. The reconstruction is viewed from the cytoplasmic side. Ca²⁺-ATPase monomers are connected by a bridge to form the dimers. Dimers are connected to form ribbons through the projecting lobe. The bridge axis is approximately parallel to the tubule axis. Stippling has been added to clarify the cytoplasmic surface of the molecule and to give an indication of the surface of the lipid bilayer [K. A. Taylor, L. Dux, and A. Martonosi, *J. Mol. Biol.* **187**, 417 (1986)]. Scale: 0.44 mm/Å.

and intramembranous regions of the protein.[35] The overall thickness of the membrane is in reasonable agreement with values derived from other techniques.[44] The most striking feature of the cytoplasmic surface of the reconstruction is the deep helical groove between dimer chains, while on the luminal surface, a deep groove extends down the center of the dimer chains. The cytoplasmic region of each Ca²⁺-ATPase molecule contains domains that bind molecules to form the dimer and that link dimers to form the dimer chains. The single intradimer connection forms a bridge centered at a radius of 270 Å with its long axis oriented approximately

[44] J. K. Blasie, L. G. Herbette, D. Pascolini, V. Skita, D. H. Pierce, and A. Scarpa, *Biophys. J.* **48**, 9 (1985).

parallel with the tubule axis. Dimers are connected into chains primarily through the lobe domain located at a radius of 253 Å.

Within the lipid bilayer region of the density map, the major feature is a somewhat cylindrical domain about 40 Å in diameter that extends through the bilayer at a slight angle, thereby placing $Ca^{2+}$-ATPase molecules within the dimer further apart on the luminal side than on the cytoplasmic side. A single connection near the cytoplasmic surface of the assumed position of the lipid bilayer at a radius of about 230 Å is oriented along steep left-handed helical tracks and links the dimer chains together to form the surface lattice.

Our knowledge of the structure of the $Ca^{2+}$-ATPase has advanced considerably through production and analysis of these crystalline arrays. Further efforts to produce more extensive two-dimensional sheets of the enzyme suitable for high-resolution analysis can be expected to advance considerably the understanding of the structural basis of ion transport.

### Acknowledgments

This work was supported by Grants GM30598, AM26545, GM27804 from the NIH, PCM 84-03679 from the NSF, a research gift from R.J.R./Nabisco, Inc., and grants from the Muscular Dystrophy Association. Portions of this work were done during the tenure of an Established Investigatorship of the American Heart Association to K. Taylor.

## [22] cDNA Cloning of Sarcoplasmic Reticulum Proteins

*By* CHRISTOPHER J. BRANDL, LARRY FLIEGEL, and DAVID H. MACLENNAN

Recombinant DNA technology is a powerful tool for obtaining information on primary structures and developmental alterations in expression of specific proteins. This is especially true for intractable membrane proteins. cDNAs encoding slow-twitch/cardiac[1,2] and fast-twitch[2,3] forms of the $Ca^{2+}$-ATPase and a fast-twitch form of calsequestrin[4] have been cloned and studies have been initiated on the structure and developmental pattern of these proteins of the sarcoplasmic reticulum. In this chapter,

[1] D. H. MacLennan, C. J. Brandl, B. Korczak, and N. M. Green, *Nature (London)* **316,** 696 (1985).

[2] C. J. Brandl, S. de Leon, D. R. Martin, and D. H. MacLennan, *J. Biol. Chem.* **262,** 3768 (1987).

[3] C. J. Brandl, N. M. Green, B. Korczak, and D. H. MacLennan, *Cell (Cambridge, Mass.)* **44,** 597 (1986).

[4] L. Fliegel, M. Ohnishi, M. R. Carpenter, V. K. Khanna, R. A. F. Reithmeier, and D. H. MacLennan, *Proc. Natl. Acad. Sci. U.S.A.* **84,** 1167 (1987).

we describe optimal conditions for the rapid construction of muscle cDNA libraries containing full-length transcripts. Although many approaches and vector systems are available, we have chosen to use the basic strategy for cloning in the vector pBR322.[5] Because both calsequestrin[4] and the Ca$^{2+}$-ATPase[6,7] were partially sequenced, we were able to use synthetic oligonucleotide probes to screen for specific cDNAs. Conditions for synthetic oligonucleotide screening and priming of cDNA synthesis are also described.

### General Considerations

Since RNase is a ubiquitous molecule, extensive precautions must be taken in the preparation and handling of materials and reagents used for isolation of RNA and for first-strand cDNA synthesis. Where possible, sterile plasticware is used and when this is not suitable, the use of oven-baked glassware (180°, 2 hr) is recommended. To minimize handling, reagents for solutions are transferred directly or with a baked spatula to preweighed sterile tubes. After reweighing, the appropriate amount of sterile distilled H$_2$O is added. If the pH of a solution is to be adjusted, the required volume of acid or base is predetermined by titrating an aliquot of the solution used only for this purpose. Solutions that are heat-stable are then autoclaved.

For ease in handling, all reactions involving cDNA are performed in 1.5-ml disposable tubes that can be centrifuged in a desk-top microcentrifuge at about 12,000 to 15,000 $g$ (in this chapter "centrifugation" refers to this form of centrifugation). For particular steps where losses of cDNA on tube walls can be substantial, tubes are siliconized prior to use.[8] After all phenol/chloroform extractions, the organic phase is back-extracted with a half-volume of Tris–EDTA (TE) to optimize the yield of cDNA (see below). Centrifugations at 4–8° are preferable, although not essential.

Some of the general procedures involving handling of nucleic acids are not described in detail. The reader is referred to Volumes 65, 68, and 152 in this series[9,10,10a] and the comprehensive laboratory manual of Maniatis et al.[11]

[5] L. Villa-Komaroff, A. Efstratiadis, S. Broome, P. Lomedico, R. Tizard, S. P. Naker, W. L. Chick, and W. Gilbert, *Proc. Natl. Acad. Sci. U.S.A.* **75,** 3727 (1978).

[6] G. Allen, B. J. Trinnaman, and N. M. Green, *Biochem. J.* **187,** 591 (1980).

[7] N. M. Green and E. J. Toms, *Biochem. J.* **231,** 425 (1985).

[8] R. Roychoudhury and R. Wu, this series, Vol. 65, pp. 43–62.

[9] L. Grossman and K. Moldave, this series, Vol. 65, 1980.

[10] R. Wu, this series, Vol. 68, 1979.

[10a] S. L. Berger and A. R. Kimmel, this series, Vol. 152, 1987.

[11] T. Maniatis, E. F. Fritsch, and J. Sambrook, "Molecular Cloning, a Laboratory Manual." Cold Spring Harbor Lab., Cold Spring Harbor, New York.

Isolation of mRNA

*Reagents*

Chloroform/isoamyl alcohol (4 : 1)

20 m$M$ ethylene glycol bis($\beta$-aminoethyl ether)-$N,N'$-tetracetic acid (EDTA), pH 7.5

TE: 10 m$M$ Tris-HCl, 1 m$M$ EDTA, pH 7.5

In a previous volume of this series, we have described our protocols for isolation of poly(A)$^+$ RNA from muscle tissue[12] and the reader is referred to this chapter. We have used these procedures to prepare RNA for cDNA synthesis with the following modifications. Buffers are prepared in autoclaved water without the addition of diethyl pyrocarbonate since trace amounts of this chemical can react with RNA making it unsuitable for full-length cDNA synthesis. Deoxycholate and sodium dodecyl sulfate (SDS) are not used in any steps of the preparation. Following guanidine-HCl extraction, the RNA is dissolved in approximately 2 ml of 20 m$M$ EDTA, pH 7.5, heated at 55° for 3 min, chilled on ice, and extracted with an equal volume of chloroform/isoamyl alcohol (4 : 1). The upper aqueous phase is ethanol precipitated as described previously. The RNA is prepared for oligo(dT) chromatography after a second ethanol precipitation by dissolving in TE and adjusting to 10 $A_{260}$ units/ml.

First-Strand cDNA Synthesis

*Reagents*

Moloney murine leukemia virus (M-MLV) reverse transcriptase (BRL)

200 m$M$ dithiothreitol

10× reverse transcription (RT) buffer: 500 m$M$ Tris-HCl, pH 7.8 (at 37°), 750 m$M$ KCl, 30 m$M$ MgCl$_2$ (must be prepared just prior to use from concentrated stocks)

Oligo(dT)$_{12-18}$, 10 mg/ml

Deoxynucleotide triphosphates (dNTPs): dATP, dGTP, dCTP, dTTP are purchased in a form neutralized to pH 7.0 (Pharmacia) and are combined in a single solution containing each dNTP at 25 m$M$.

[$\alpha$-$^{32}$P]dCTP, SA 3000 Ci/mmol (Amersham)

500 m$M$ EDTA, pH 7.5

Phenol/chloroform/isoamyl alcohol: equal volumes of water-saturated phenol, pH 7.5, and chloroform/isoamyl alcohol [24/1 (v/v)]

---

[12] D. H. MacLennan and S. de Leon, this series, Vol. 96, pp. 570–579.

TE: 10 mM Tris-HCl, 1 mM EDTA, pH 7.5
4 M ammonium acetate
1 M Tris-HCl, pH 7.5
Diethyl ether

*Procedure.* Eight to twelve micrograms of RNA is pelleted by centrifugation from ethanol suspension and washed once with ice-cold 80% ethanol. The ethanol is carefully removed by a pipet, the tube is recentrifuged, and any residue ethanol is removed. It is not necessary to vacuum dry this pellet. The RNA is dissolved in 30 $\mu$l of freshly diluted 5 mM Tris-HCl, pH 7.5, and heated for 3 min at 68°. The RNA is placed directly in a 37° bath and sufficient $H_2O$ to make a final reaction volume of 100 $\mu$l, 10 $\mu$l of 10× RT buffer, 2 $\mu$l of 25 mM dNTPs, 10 $\mu$l of 10 mg/ml oligo(dT), 5 $\mu$l of 200 mM dithiothreitol, and 2000 units of reverse transcriptase are added rapidly in that order. For later calculation of cDNA synthesis, 10 $\mu$l of the reaction mixture is removed immediately and added to 0.5 $\mu$l of [³²P]dCTP in a separate tube. By performing the labeled reaction separately, radiodecay will not affect the stability of the cDNA. The reaction is continued for 1 hr before being stopped by the addition of 4 $\mu$l of 0.5 M EDTA. Six microliters of the labeled side reaction is then returned to the reaction mix to enable the cDNA to be followed in subsequent steps. The reaction mix is extracted twice with an equal volume of phenol/chloroform and three times with about 1 ml of diethyl ether to remove residual phenol from the aqueous phase. One volume of 4 M ammonium acetate and 2.5 volumes of 100% ethanol are added and precipitation of the nucleic acids is facilitated by placing the reaction tube in a dry-ice ethanol bath for a minimum of 0.5 hr.

Aliquots of the radioactive side reaction are used to calculate total synthesis either by trichloroacetic acid (TCA) precipitation of cDNAs and collection of the precipitate on a filter, or by binding the soluble DNA to a diethylaminoethyl (DEAE) filter, washing off background counts with 0.5 M $Na_2HPO_4$, and counting of radioactivity incorporated.[11]

## Second-Strand Synthesis

*Reagents*

10× second-strand buffer: 200 mM Tris-HCl, pH 7.5, 50 mM $MgCl_2$,
  1 M KCl
100 mM $(NH_4)_2SO_4$
1.5 mM $\beta$-NAD
500 $\mu$g/ml bovine serum albumin (BSA)
400 $\mu$M dNTPs
DNA polymerase I
*Escherichia coli* DNA ligase

RNase H
Klenow fragment DNA polymerase I

*Procedure.* Because of endogenous RNase H and DNA polymerase activities in the M-MLV reverse transcriptase, RNA-primed second-strand DNA synthesis occurs during the first-strand reaction. The measurement of first-strand DNA synthesis thus actually represents both first- and second-strand synthesis and, in a successful reaction, should be 20 to 40% by weight of the RNA template. Second-strand synthesis is completed by strand replacement.[13,14] A 100-$\mu$l reaction volume is required for each microgram of cDNA. The cDNA suspension from the first-strand synthesis reaction is warmed to room temperature to allow free unincorporated nucleotides to dissolve. The cDNA suspension is then centrifuged for 10 min. The cDNA pellet is washed twice with 80% ethanol at room temperature and dissolved in 20 m$M$ Tris-HCl, pH 7.5, 5 m$M$ MgCl$_2$, 10 m$M$ (NH$_4$)$_2$SO$_4$, 100 m$M$ KCl, 0.15 m$M$ $\beta$-NAD, 50 $\mu$g/ml BSA, and 40 $\mu M$ dNTPs. It is convenient to add salts other than (NH$_4$)$_2$SO$_4$ from a combined 10× stock and the other reagents individually from separate stocks. $\beta$-NAD solutions should be freshly prepared. *E. coli* RNase H (9 U/ml), DNA polymerase I (230 U/ml), and *E. coli* DNA ligase (10 U/ml) are added and the reaction is incubated for 60 min at 12° followed by 60 min at 22°. To ensure blunt-ended fragments, aliquots of the Klenow fragment of DNA polymerase I (50 U/ml) and RNase H (9 U/ml) are added and the reaction continued for 20 min at 22°. EDTA is then added to 20 m$M$ and the reaction mixture is extracted with phenol/chloroform.

### Size Fractionation of cDNA

When cloning a large cDNA of relatively low abundance, it is beneficial to size fractionate the cDNA accurately by agarose gel electrophoresis. Size fractionation can also be performed, though less efficiently, by Sepharose 4B chromatography. It is essential to perform one of these steps in order to remove contaminants from the cDNA before tailing.

*Separation by Agarose Gel Electrophoresis*[14a]

*Materials and Reagents*

Agarose: standard low $M_r$
DNA loading buffer: 25% Ficoll 400, 25 m$M$ EDTA, pH 8.0, 1% SDS, 0.05% bromphenol blue

[13] H. Okayama and P. Berg, *Mol. Cell. Biol.* **3**, 280 (1983).
[14] U. Gubler and B. Hoffman, *Gene* **25**, 263 (1983).
[14a] We are grateful to Dr. Gary Shull for introducing us to this technique.

10× TBE: 890 m$M$ Tris-borate, 890 m$M$ boric acid, 2 m$M$ EDTA
10 mg/ml ethidium bromide
Elution buffer: 50 m$M$ arginine (free base), 1.0 $M$ NaCl
NA45 paper (Schleicher and Schuell): washed 10 min in 10 m$M$
    EDTA, pH 7.6, 5 min in 0.5 $M$ NaOH, and several times in sterile
    distilled H$_2$O.
H$_2$O-saturated $n$-butanol

*Procedure*. The nucleic acids are ethanol precipitated in the presence
of 2 $M$ ammonium acetate to concentrate the cDNA after phenol/chloro-
form extraction (see above) and after centrifugation, the pellet is dissolved in 20 $\mu$l of TE. Five microliters of DNA loading buffer is added and
the solution heated to 65° for 3 min. The cDNA is separated electropho-
retically at 4 V/cm on a horizontal 0.8% agarose gel containing TBE and
0.1 $\mu$g/ml ethidium bromide. The time of electrophoresis depends only on
the resolution required, and this can be monitored by transfer of the gel to
a UV light box to visualize the size markers which are separated in parallel with the cDNA. Care should be taken to avoid exposure of the cDNA
to UV light by shielding the portion of the gel containing cDNA. A strip of
NA45 paper, treated with 10 m$M$ EDTA, pH 7.6, and 0.5 $M$ NaOH as
described by the manufacturer, is wedged into a slit in the gel in front of
the desired size classes of cDNA. Positioning of the paper is facilitated by
placing it along a razor blade and gently inserting the razor into the gel at
the desired position. The gel is resubmerged in buffer and electrophoresis
is continued until the desired size class of cDNA becomes trapped on the
paper. This process can be followed by visualizing the cDNA under a
hand-held UV lamp. The paper may become saturated with DNA, in
which case DNA is seen on the distal side of the paper. A new piece of
paper should then be inserted.

The cDNA is eluted from the paper by submerging the NA45 paper
strip in 100 $\mu$l of a solution of 50 m$M$ arginine (free base), 1.0 $M$ NaCl at
68° for 1 hr. At this stage, all carrier nucleic acid has been removed from
the cDNA and yields will be greatly reduced unless siliconized tubes are
used. The elution is repeated until 80% or more of the radioactivity has
been displaced from the filter. Ethidium bromide is removed from the
cDNA by extracting three times with an equal volume of water-saturated
$n$-butanol. Particles of gel, if present, can be removed by brief centrifugation. The cDNA is precipitated overnight at $-20°$ after addition of 2.5
volumes of 100% ethanol, pelleted by centrifugation for 40 min, and suspended in 100 $\mu$l of TE. From this stage onward, it is essential to increase
the centrifugation times in order to pellet all of the cDNA. The cDNA is
extracted twice with phenol/chloroform and once with chloroform before
being ethanol precipitated in the presence of 2 $M$ ammonium acetate.

Again, the times for precipitation and centrifugation are extended to ensure complete recovery.

## Sepharose 4B Chromatography

### Materials and Reagents

NET: 100 mM NaCl, 1 mM EDTA, 10 mM Tris-HCl, pH 7.5
1 mg/ml tRNA
Sepharose 4B
Glass column: 20-cm length, 1-cm diameter

*Procedure.* A 20-cm column with a 1-cm diameter is packed with Sepharose 4B, suspended in NET. Nonspecific nucleotide binding sites are saturated by equilibrating the column with tRNA and the column is then washed thoroughly with NET. cDNA is applied directly to the column and 15-drop fractions are collected. Elution of cDNA is followed by Cerenkov counting of the eluted fractions. Desired fractions are collected and concentrated by vacuum desiccation so that ethanol precipitation can be carried out in a single, siliconized tube. Preparation for tailing is the same from this point for both agarose-purified and Sepharose-purified cDNAs.

## Tailing of cDNA

### Reagents

500 $\mu M$ dCTP
Terminal deoxynucleotide transferase
5× tailing buffer (supplied by manufacturer): 500 mM potassium cacodylate, pH 7.2, 10 mM $CoCl_2$, 1 mM dithiothreitol (DTT)
[$\alpha$-$^{32}$P]CTP, SA 3000 Ci/mmol

*Procedure.* Conditions for deoxycytidine (dC) tailing of the cDNA can be tested by trials on blunt-ended DNA fragments of similar size. The tailing reaction, however, is sensitive to impurities in DNA. Thus, while basic conditions can be established on trial fragments it is still necessary to monitor the addition of radioactively labeled nucleotides to the cDNA. A labeled trial is performed with a sample of the cDNA and, if successful, the remainder is tailed in an unlabeled reaction. The cDNA is pelleted from ethanol suspension by centrifugation (40 min), washed three times with 80% ethanol, dried under vacuum, and suspended in $H_2O$ at a concentration such that 0.1 pmol of cDNA can be tailed in a 20 $\mu$l reaction. The cDNA solution is briefly centrifuged to remove any suspended particles that may have been collected, transferred to a fresh tube, and heated to 45° for 1 min. Immediately thereafter, 4 $\mu$l 5× tailing buffer, 2 $\mu$l 500

$\mu M$ dCTP, 1 $\mu$l [³²P]dCTP, 5 U terminal deoxynucleotidyltransferase, and H$_2$O to 20 $\mu$l are added and the tube is transferred to 37°. Aliquots are taken at regular intervals up to 25 min to monitor incorporation of labeled nucleotide. The reaction time ultimately selected is that in which 20 residues are added to each 3′ end of the cDNA. The reaction volume is scaled proportionally, depending on the amount of cDNA. To stop the reaction, EDTA is added to 10 m$M$ and the reaction mixture is phenol/chloroform extracted, chloroform extracted, and ethanol precipitated.

The most likely cause of an unsuccessful trailing reaction is contamination of the cDNA. Often, after reextraction with phenol/chloroform and ethanol precipitation the cDNA can be tailed. In some cases we observed no apparent incorporation of labeled dCTP into the cDNA, yet the cDNA has been successfully transformed into *E. coli*. This may have resulted from tailing of only a small subpopulation of cDNA molecules or from the presence of a reaction inhibitor only in the radioactively labeled reaction mixture. Thus if tailing cannot be monitored even after a second attempt, it may be worthwhile to test the cDNA in a transformation assay before assuming that tailing has been unsuccessful.

### Annealing of cDNA and Vector

*Reagents*

Cut and dG-tailed pBR322
10× annealing buffer: 1 $M$ NaCl, 0.1 $M$ Tris-HCl, pH 7.8, 1 m$M$ EDTA

*Procedure.* Tailed plasmid vectors that give a background level of transformation of less than 1% can be purchased from several manufacturers. Such low background levels are difficult to obtain in the laboratory. Tailed cDNA is pelleted by centrifugation (40 min), washed with 80% ethanol, vacuum dried, and dissolved with an equimolar amount of cut and tailed pBR322 in 100 m$M$ NaCl, 10 m$M$ Tris-HCl, pH 7.8, 0.1 m$M$ EDTA at a total DNA concentration of 1 $\mu$g/ml. The tube is submerged in a 65° bath which is then turned off and allowed to cool slowly overnight to 30°.

### Transformation of cDNA : Vector

It is essential to obtain a high transformation efficiency and, depending on the source of competent cells, variations of 10-fold or more can be observed in transformation trials. *E. coli* cells made competent by the procedure of Hanahan[15] should be used. Several suppliers produce competent cells that give a high transformation efficiency ($10^8$ per microgram of pBR322). The protocols outlined by the manufacturer are followed with 5 to 10 $\mu$l of annealed cDNA per 100 $\mu$l of transformation reaction. Aliquots of transformed cells are plated on LB agar plates with tetracycline (15 $\mu$g/ml) from a trial transformation to estimate the volume required for optimal plating density. The remaining cDNA is transformed immediately or stored at 4°.

Colony size is very variable. Numerous small colonies become visible after approximately 20 hr at 37°. These colonies do contain cDNAs but there is no correlation between rate of growth and insert size.

### Screening cDNA Libraries with Synthetic Oligonucleotide Probes

*General Considerations*

When at least part of a protein sequence is known it is possible to use synthetic oligonucleotide probes to screen for cDNAs encoding the protein. Purified synthetic oligonucleotides in the 14–30 mer range are available commercially. The longer the oligonucleotide and the fewer redundancies, the more specific it will be for the desired DNA species. Sequences containing amino acids with 1 or 2 codons should be chosen since this will provide the lowest redundancy. The base in the wobble position of the last codon can be omitted, reducing the redundancy of the probe. It is not desirable to omit some less frequently used codons[16] from probes, since even a single base mismatch can drastically affect the stability of the oligonucleotide–cDNA complex.[17] The substitution of deoxyinosine at ambiguous codon positions[18] is useful in highly degenerate probes since deoxyinosine will base pair with A, C, and U, and thus a smaller mixture of synthetic oligonucleotides is necessary. Synthetic oligonucleotides complementary to the mRNA are most useful since they can be used to probe Northerns of mRNA and to be used in primer extension experiments (described below).

We have successfully isolated ATPase clones with a mixture of 4

---

[15] D. Hanahan, *J. Mol. Biol.* **166**, 557 (1983).

[16] R. Granstham, C. Gautier, M. Gouy, M. Jacobson, and R. Mercier, *Nucleic Acids Res.* **9**, r43 (1981).

[17] R. B. Wallace, J. Shaffer, R. F. Murphy, and J. Bonner, *Nucleic Acids Res.* **6**, 3543 (1979).

[18] E. Ohtsuka, S. Matsuki, M. Ikehara, Y. Takahashi, and K. Matsubara, *J. Biol. Chem.* **260**, 2605 (1985).

oligonucleotides 14 bases long[1,3] and calsequestrin clones with a mixture of 4 oligonucleotides 17 bases long.[4]

## Labeling of the Synthetic Oligonucleotide

### Materials and Reagents

0.5 $M$ Tris-HCl, pH 9.0
100 m$M$ MgCl$_2$
100 m$M$ DTT
10 m$M$ EDTA
10 m$M$ KH$_2$PO$_4$
1 $M$ KH$_2$PO$_4$
100 m$M$ EDTA
tRNA: 1 mg/ml
T$_4$ polynucleotide kinase
Adenosine 5'-triphosphate [$\gamma$-$^{32}$P]ATP, carrier-free

Synthetic oligonucleotides are labeled at 5' free hydroxyl groups by T$_4$ polynucleotide kinase catalyzed transfer of [$^{32}$P] from the terminal phosphate of [$\gamma$-$^{32}$P]ATP to the 5' hydroxyl of the oligonucleotide. Free 5' ends are usually obtained by treatment of crude fractions with ammonium hydroxide and most suppliers base-treat and purify their oligonucleotides prior to sale. Crude ATP is available at considerably reduced prices and has been found to be satisfactory for probe labeling. Although several procedures are available,[11,19] the following procedure generates probes with specific activities greater than 10$^9$ cpm/$\mu$g of 17-mer. A 100 $\mu$l reaction contains the following reagents added from sterile stocks: 50 m$M$ Tris-HCl, pH 9.0, 10 m$M$ MgCl$_2$, 10 m$M$ DTT, 0.2 m$M$ EDTA, 10–20 U polynucleotide kinase, 0.5 $\mu$g probe, and 0.8–1 mCi of carrier-free [$\gamma$-$^{32}$P]ATP. The reaction is continued for 1 hr at 37° and is terminated by the addition of 20 $\mu$l of 100 m$M$ EDTA plus 10 $\mu$l of 1 $M$ KH$_2$PO$_4$ to inhibit contaminating phosphatases. The reaction mixture is heated to 65° for 3 min to inactivate the kinase and 1 $\mu$l of tRNA (1 mg/ml) is added to reduce loss of the probe through nonspecific binding during subsequent purification steps.

## Purification of the Labeled Probe

### Materials and Reagents

50% glycerol in distilled water
Elution buffer: 10 m$M$ Tris-HCl, pH 8.0, 1 m$M$ EDTA, 0.1 SDS%, 0.2 $M$ NaCl, 0.5 $M$ Na$_2$HPO$_4$

---

[19] G. Chaconas and J. H. Van De Sande, this series, Vol. 65, pp. 75–85.

10 mg/ml salmon sperm DNA
Sephadex G-25 (Fine)
Whatman DE81 filter paper (2.4-cm disks)
*Procedure.* Separation of the probe from [γ-$^{32}$P]ATP is achieved by passage of the sample through a Sephadex G-25 (Fine) column in 10 m$M$ Tris-HCl, pH 8.0, 1 m$M$ EDTA, 0.1% SDS, and 0.2 $M$ NaCl. The column is made in plastic 10-ml disposable pipets containing 1.25–1.5 g of Sephadex G-25 (Fine) and is equilibrated with tRNA, deoxynucleotides, or salmon sperm DNA to reduce nonspecific binding of probe to the column. The tip of the column contains siliconized glass wool. A half-volume of 50% glycerol is added to the sample before it is applied to the column to allow it to flow under the elution buffer. The column is eluted from a reservoir 30–40 cm above the column and 300-$\mu$l fractions are collected in Eppendorf tubes inserted inside 17 × 100 mm test tubes. This procedure is carried out behind a Plexiglas radiation shield. The fractions are Cerenkov counted to find the peaks of radioactivity of probe and ATP. The probe peak appears first but may overlap slightly with free ATP. To identify the appropriate oligonucleotide-containing fractions, 5 $\mu$l samples are removed and spotted onto DE81 filter papers. After drying, the filters are washed with 0.5 $M$ Na$_2$HPO$_4$ and bound radioactivity is determined by scintillation counting.[11] The column must be washed thoroughly after use and can be used a maximum of three times due to increasing contamination with radioactivity. Starting with 0.5 $\mu$g of probe, at least 5 × 10$^7$ cpm of labeled probe can be obtained routinely.

## Hybridization

### Materials and Reagents

Wash solution: 50 m$M$ Tris-HCl, pH 8.0, 1 $M$ NaCl, 0.1% SDS, 1 m$M$ EDTA
20× SSCP: 2.4 $M$ NaCl, 0.3 $M$ Sodium citrate, 0.26 $M$ KH$_2$PO$_4$, 0.02 $M$ EDTA, pH 7.2
100× Denhardt's solution[11]
10 mg/ml polyadenylic acid
Millipore HA, 0.45-$\mu$m nitrocellulose filters, 85-mm diameter
Nalgene straight-side, wide-mouth jars: 10-cm diameter, 12-cm high, with a 10-cm diameter screw top

*Procedure.* Transfer of cDNA colonies to autoclaved Millipore nitrocellulose filters has been described by Grunstein and Hogness.[20] After transfer and fixation, the filters are baked for 2 hr under vacuum at 80° and transferred to Nalgene straight-side, wide-mouth jars. All subsequent ma-

[20] M. Grunstein and D. S. Hogness, *Proc. Natl. Acad. Sci. U.S.A.* **72**, 3961 (1975).

nipulations are carried out in these jars. The filters are washed with agitation for at least 2 hr at 42° in wash solution to remove cellular debris and prehybridized in $5\times$ SSCP, $10\times$ Denhardt's, and 100 $\mu$g/ml poly(A) for 6 to 14 hr at room temperature. We prefer the use of poly(A) as a nonspecific nucleotide sequence rather than salmon sperm DNA because it is possible that short probes could bind to homologous regions on the salmon sperm DNA. The prehybridization solution is replaced with a fresh solution of the same composition to which is added freshly labeled probe at a concentration of $1.5 \times 10^6$ cpm/ml of hybridization solution. The filters are shaken gently overnight at room temperature. Up to 30 filters can be screened simultaneously in 100 ml, and a minimum of 50 ml is usually necessary to cover even one or two filters. After hybridization, the filters are rinsed several times with $5\times$ SSCP at room temperature and washed twice with $2\times$ SSCP at 37° for 30 min, with gentle shaking. At this point radioactivity in the filters can be checked with a Geiger counter. If there is a low background of bound radioactivity, they are dried at room temperature for exposure on Kodak X-Omatic X-ray film. If the filters retain large, easily detectable amounts of bound radioactivity spread evenly over the filter they are washed again at 37° in $1\times$ SSCP. It is possible to let the filters dry, expose them to X-ray film, and then continue washing at higher stringencies if the background is high or results unclear. The next wash should be in $0.5\times$ SSCP at 37°.

Apparent positive clones are picked from the original plate with sterilized toothpicks, and streaked in a small grid for a second screen. Colonies from these plates are screened with the same probe in the same manner and positive clones are streaked or diluted to obtain single positive colonies in a third screen.

### Primer Extension

#### General Considerations

When a particular protein is present in low abundance in cellular mRNA or the particular synthetic oligonucleotide probe cross-hybridizes with other nondesirable cDNA clones, it may be desirable to enrich the amount of the specific cDNA by primer extension. In this technique a synthetic oligonucleotide, synthesized on the basis of known protein sequence, is used to prime the synthesis of the first cDNA strand. Second-strand synthesis and construction of the library are completed as described earlier. The result should be a small, easy to screen library enriched in the desired cDNA. Positive clones will be devoid of the 3′ ends of the sequence but such cDNAs can be used for ready screening of libraries made with an oligo(dT) primer. In our experiments with calsequestrin,[4] a 17-mer probe (a mixture of 16 combinations) near the 3' end of the coding region was used as the primer and the library was screened with a second oligonucleotide 5' to the first. The 3' end was obtained from clones isolated after rescreening a full-length library.

### Materials and Reagents

15 $\mu$g poly(A)$^+$ RNA
TE: 10 m$M$ Tris-HCl, 1 m$M$ EDTA, pH 7.5
1 $M$ KCl
2× reverse transcriptase buffer: 120 m$M$ Tris-HCl, 80 m$M$ KCl, 20 m$M$ MgCl$_2$, 60 m$M$ DTT, dATP, dGTP, dCTP, dTTP 2 m$M$ for each deoxynucleotide triphosphate, pH 8.3, at 37° Avian myeloblastosis virus (AMV) reverse transcriptase (Boehringer-Mannheim)
[$\alpha$-$^{32}$P]CTP, 3000 Ci/mmol
Stop buffer: 2% SDS, 10 m$M$ EDTA, pH 7.0
Synthetic oligonucleotide primer (50 ng/$\mu$l)
0.25 $M$ EDTA, pH 8.0

*Procedure.* Primer extension has been described in detail by Nathans and Hogness.[21] The bulk of the reaction is unlabeled but a side reaction is performed in the presence of [$\alpha$-$^{32}$P]dCTP (see above). At least 15 $\mu$g of polyadenylated RNA is pelleted out of ethanol suspension, washed twice with ice-cold 80% ethanol, and dried *in vacuo*. The following steps are carried out as follows: Add 10 $\mu$l of TE and 1 $\mu$l of the synthetic oligonucleotide primer (50 ng/$\mu$l) to an Eppendorf tube; heat to 75° for 1 min; add 3 $\mu$l of 1 $M$ KCl and incubate for 15 min at 30°; add 45 $\mu$l of 1× reverse transcriptase buffer containing 20 units of reverse transcriptase; remove 1 $\mu$l immediately to a tube containing 10 $\mu$Ci of [$\alpha$-$^{32}$P]dCTP; remove 2 $\mu$l from this reaction mix to 8 $\mu$l stop solution and a further 2 $\mu$l at the end of the reaction to monitor the progress of incorporation of radioactivity.[11] First-strand synthesis, with and without label, is carried out at 30° for 10 min and at 37° for 60 min. After incubation the unlabeled reaction is terminated by the addition of 10 $\mu$l of 0.25 $M$ EDTA, pH 8.0, and 6 $\mu$l of the labeled reaction is returned to the unlabeled reaction so that further monitoring can be carried out.

The amount of cDNA synthesized will be much lower than in reactions containing oligo(dT) because of the smaller amount of priming even though nonspecific priming occurs. Care must be taken to ensure that the small amount of cDNA synthesized is not lost during subsequent handling steps since only 10–40 ng of first strand is synthesized. All

[21] J. Nathans and D. S. Hogness, *Cell (Cambridge, Mass.)* **34,** 807 (1983).

ethanol precipitations should be of sufficient time to ensure complete recoveries and all Eppendorf tubes should be siliconized prior to use to prevent binding of the cDNA to the tubes. Recoveries of cDNA are monitored by Cerenkov counting in order to be able to add the appropriate amount of vector in the subsequent annealing reaction. A different synthetic oligonucleotide, 5′ to the first, should be used to screen the library because the original one might pick up too many false positives primed nonspecifically. Synthesis of the second strand and of cDNA cloning and screening is carried out as described earlier, with the exception that the reaction mixture is treated with 50 ng RNase for 5 min at room temperature after second-strand synthesis is complete. This facilitates removal of excess RNA, which is desirable since ribonucleotides may inhibit terminal transferase activity.[22]

### Acknowledgments

CJB was a predoctoral fellow of the Medical Research Council of Canada and LF was a postdoctoral fellow of the Ontario Heart and Stroke Foundation.

[22] R. B. Bhalla, M. K. Schwartz, and M. J. Modak, *Biochem. Biophys. Res. Commun.* **76,** 1056 (1977).

## [23] Reconstitution of Skeletal Muscle Sarcoplasmic Reticulum Membranes: Strategies for Varying the Lipid/Protein Ratio

*By* LIN HYMEL and SIDNEY FLEISCHER

Dissociation and reconstitution of membranes is a powerful approach to assess the function of specific polypeptides and to manipulate the individual protein and lipid components in membranes for biophysical studies. We describe reconstitution methodology for the preparation of functionally active reconstituted SR membrane vesicles with high protein content comparable to that in the original SR.[1,2] The basic dissociation

[1] Supported in part by grants from NIH DK 14632 and from the Muscular Dystrophy Association.
[2] Abbreviations: CPP, calcium-pump protein; DOC, deoxycholic acid; EDTA, ethylenediaminetetraacetic acid; EGTA, ethylene glycol bis($\beta$-aminoethyl ether)-$N,N'$-tetraacetic acid; HEPES, $N$-2-hydroxyethylpiperazine-$N'$-2-ethanesulfonic acid; SR, sarcoplasmic reticulum.

and reconstitution procedure is similar to that described previously by Meissner and Fleischer.[3,4] The modifications described here are particularly useful for preparing vesicles of defined lipid content with up to twofold higher or lower lipid content than the native vesicles, and for incorporating exogenous phospholipids.

The nature of the dissociation and reconstitution process for SR using deoxycholate has recently been elucidated.[5] Titration of SR with DOC first involves the selective extraction of bilayer phospholipid to yield a "limit membrane" consisting of 48 mol of phospholipid per mole of CPP (compared with 115 for the original SR membrane) and, second, the solubilization of the CPP and associated phospholipid from the limit membrane. Membrane assembly is essentially the reversal of dissociation. Raising the temperature results in reformation of a limit membrane. As detergent is removed, phospholipid gradually adds to the limit membrane, the phospholipid content increases, and vesicles are formed.

## I. Reagents and Materials

Sarcoplasmic reticulum vesicles are prepared from rabbit skeletal muscle as previously described.[6] They are stored frozen at $-70°$ in solution B at more than 15 mg protein/ml (see below) and thawed immediately before use. We have used the unfractionated fast-twitch SR preparation which contains both light and heavy SR.

Sarcoplasmic reticulum phospholipid suspension, in dissociation buffer (90 to 144 $\mu$g lipid phosphorus/ml) is adjusted to the same deoxycholate concentration as selected for SR dissociation (see Section V, Method 3). SR phospholipids are prepared by extraction according to the method of Folch as modified by Rouser and Fleischer.[7] The lipids are stored under nitrogen at $-20°$ in chloroform/methanol (2/1) for up to several months. To prepare the suspension, the required amount of lipid solution is evaporated *in vacuo* and/or under a stream of nitrogen and resuspended in dissociation buffer containing deoxycholate using glass beads and a vortex mixer. Finally, the mixture is sonicated for 10 min under nitrogen in a bath sonicator (Laboratory Supplies Company, Hicksville, NY). The suspension can be stored at $-20°$, but should be resonicated just prior to use.

[3] G. Meissner and S. Fleischer, this series, Vol. 32, p. 475.
[4] G. Meissner and S. Fleischer, *J. Biol. Chem.* **249,** 302 (1974).
[5] L. Hymel, A. Maurer, and S. Fleischer, in preparation.
[6] G. Meissner, G. E. Conner, and S. Fleischer, *Biochim. Biophys. Acta* **298,** 246 (1973).
[7] G. Rouser and S. Fleischer, this series, Vol. 10, p. 385.

Solution A is used for dissociation: 0.5 $M$ sucrose, 0.8 $M$ KCl, 3 m$M$ MgCl$_2$, 2 m$M$ EDTA, 0.2 m$M$ CaCl$_2$, and 20 m$M$ Tris, pH 7.9 at 0°.

Solution B is used for storage of SR vesicles: 0.3 $M$ sucrose, 0.1 $M$ KCl, and 5 m$M$ HEPES, pH 7.1 at 0°.

Dissociation buffer is a mixture of equal parts solutions A and B [0.4 $M$ sucrose, 0.45 $M$ KCl, 1 m$M$ EDTA, 1.5 m$M$ MgCl$_2$, 0.1 $M$ CaCl$_2$, 10 m$M$ Tris-HCl, pH 7.9 at 0°.

Dialysis buffer consists of 0.25 $M$ sucrose, 0.4 $M$ KCl, 1.5 m$M$ MgCl$_2$, 1 m$M$ EDTA, 0.1 m$M$ CaCl$_2$, and 5 m$M$ HEPES, pH 7.25 at 20°.

Deoxycholate (DOC) stock solution (10%): Deoxycholic acid is decolorized and recrystallized as described.[3] The purified acid is dissolved in water by slow titration with KOH to a pH of 8.0 while monitoring with a pH meter. The concentration of the stock solution is determined enzymatically using the 3$\alpha$-hydroxysteroid dehydrogenase assay described by Turley and Dietschy.[8] Hydroxysteroid dehydrogenase (EC 1.1.1.50) is obtained from Worthington (Freehold, NJ). This assay typically indicates that 10–15% of the weight of the crystals is due to residual solvent (ethanol), even after extensive air drying. Concentrations of deoxycholate are based on the formula weight of deoxycholic acid rather than potassium deoxycholate. The stock solution is stored frozen at −20°.

Dialysis tubing (12,000 MW cutoff, obtained from A. H. Thomas, Co., Philadelphia, PA): The tubing, approximately 10-cm lengths, is pretreated by soaking in several changes of deionized water followed by storage at 4° in 10 m$M$ Na$_2$EDTA, 10 m$M$ Tris-Cl, pH 7.9. The tubings are thoroughly rinsed in deionized water and soaked in dialysis buffer immediately prior to use.

Assay medium for calcium loading and ATPase activity: 0.3 $M$ sucrose, 0.1 $M$ KCl, 5 m$M$ potassium oxalate, 10 m$M$ HEPES, pH 7.0, 5.5 m$M$ MgCl$_2$, 5 m$M$ Na$_2$ATP, 60 $\mu M$ EGTA, 100 $\mu M$ CaCl$_2$.

## II. Basic Reconstitution Protocol

1. Thaw SR and resuspend in dissociation buffer at 6 mg protein/ml using the appropriate amounts of solutions A and B.

2. Add DOC to obtain limited to extensive solubilization of the SR (2.2 to 3.0 mg/ml, see Section V Method 2). Incubate 10 min on ice.

3. Centrifuge to remove membrane fragments (40,000 rpm for 75 min, 0°, 75Ti rotor). Decant supernatant containing solubilized SR (see Section III).

4. Warm the solubilized SR from 0–4 to 25° (see Section IV).

---

[8] S. D. Turley and J. M. Dietschy, *J. Lipid Res.* **19**, 924 (1978).

5. (Optional) Supplement with phospholipid suspension in deoxycholate. The same concentration of DOC as for solubilization is used to increase the lipid content of the reconstituted membrane (see Section V Method 3).

6. Dialyze against 500 volumes dialysis buffer at room temperature for 0–24 hr (see Section IV for steps 6–9).

7. Dilute with equal volume fresh dialysis buffer. Sediment vesicles (45,000 rpm for 75 min, 20°, 75Ti rotor).

8. Wash pellet in 25 times the original dialysis volume using dialysis buffer. Centrifuge as in step 7 above (45,000 rpm for 90 min, 20°, 60Ti rotor).

9. Resuspend in solution B (~0.5 ml per milliliter of original dialysis volume).

The methodology is described in detail below as are a number of options which can be applied to prepare membranes of defined composition.

III. Dissociation Procedure (See Section II, steps 1–3)

Dissociation is carried out in the cold at 0–4°. A suspension of SR vesicles at 6 mg protein/ml in dissociation buffer is used. This is prepared by diluting the SR to 12 mg protein/ml with solution B, and mixing with an equal volume of solution A sucrose. Dissociation of SR is achieved by adding the required amount of DOC stock solution (see Fig. 1) by carefully layering onto the top of the suspension, and inverting the tube several times to obtain good mixing. Slow titration of DOC with continued mixing yields similar results. The mixture is kept on ice for 10 min to ensure equilibration, and is then centrifuged as described below to sediment membrane fragments.

A noticeable reduction in the turbidity of the suspension occurs after mixing the SR with DOC which can be correlated with the amount of the added detergent. The extent of solubilization is a critical parameter which is subject to cumulative small errors in pipetting, protein concentration, and the concentration of the DOC stock solution. The measurement of turbidity has been used as a reliable calibration method to achieve the desired extent of solubilization (Fig. 1).

Two important factors in obtaining the desired lipid/protein ratio in the reconstituted membranes are the extent of dissociation of the membrane and the time of dialysis (see below). In practice, the degree of dissociation can be reproducibly controlled by adding slightly less DOC than calculated, measuring the turbidity, and titrating the remaining DOC in small aliquots to obtain the appropriate turbidity.

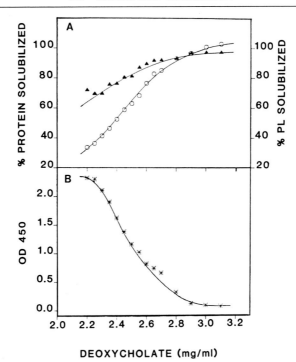

FIG. 1. Correlation of solubilization of SR membrane vesicles with decrease in turbidity as a function of deoxycholate (DOC) concentration. (A) SR was solubilized with different concentrations of DOC. (B) Before centrifugation to remove insoluble material, the turbidity of each suspension (optical density at 450 nm) was measured at 0°. After centrifugation, the percentages of solubilized protein (○)[8b] and phospholipid (PL) (▲)[7] were determined from the amounts remaining in the supernatant. The detergent first preferentially extracts phospholipid. Only the portion of the DOC concentration curve which begins to solubilize membrane protein is shown. Note that the decrease in turbidity parallels better the solubilization of protein than phospholipid.

The amount of solubilized CPP remaining in the supernatant is determined in part by the effectiveness of centrifugation. Our standard condition, which was used to obtain the calibration curve in Fig. 1, is to centrifuge 2.4-ml aliquots in a Beckman 75Ti rotor for 75 min at 40,000 rpm (2°). If larger volumes are desired, the speed of centrifugation is increased to sediment a 20S particle according to the clearing factor ($k$) equation (replacing $k$ with $tS$):

$$20S = 20 \times 10^{-13}\,\text{sec} = [\ln(r_{max}/r_{min})/\omega^2 t](1/3600)$$

[8b] D. H. Lowry, N. J. Rosebrough, A. C. Farr, and R. J. Randall, *J. Biol. Chem.* **193**, 265 (1951).

where $S$ is the sedimentation coefficient in dissociation buffer at 2°, $\omega$ is the angular velocity (radians/sec), $t$ is the time in hours, and $r_{max}$ and $r_{min}$ are the maximum and minimum radii of the sample in the rotor.[9] Under our centrifugation conditions, the supernatant (solubilized SR) is devoid of membranous material as characterized by electron microscopy and gel filtration.[5]

The sucrose concentration used in the dissociation buffer also determines the recovery.[10] Reducing the sucrose concentration in this buffer increases the amount of CPP which sediments, and thereby reduces the amount remaining in the supernatant. Such changes also bear on the functional characteristics of the reconstituted preparation.[5]

IV. Reconstitution Procedure (See Section II, steps 4–9)

Reconstitution of SR vesicles from solubilized SR occurs optimally at room temperature, whereas little or no $Ca^{2+}$ transport activity is recovered when the procedure is carried out at 0°.[3] The solubilized SR is first incubated in a 25° water bath for 10 min before transferring it into dialysis bags. When a limited solubilization has been performed (less than about 75% of the protein solubilized), the sample becomes visibly turbid during warming, reflecting the initial reassembly of membrane.[5,11]

After warming, the samples (usually 2.0 ml) are transferred into dialysis bags and sealed using dialysis tubing clamps (Fisher Scientific, Pittsburgh, PA), leaving an air bubble approximately the diameter of the tubing. The bags are then loaded onto a rocker dialyzer,[12] a device which rocks the tubular bags at 4 cycles per minute, so that the two ends take turns rocking up and down with respect to each other, enabling mixing of the contents by the air bubble. The sample is dialyzed against 500 volumes of dialysis buffer at room temperature. After a given time (see below) the sample is retrieved, diluted with 1 volume fresh dialysis buffer, and centrifuged at 45,000 rpm for 75 min in a 75Ti rotor at 20°. The pellet is washed by resuspending in 25 volumes of fresh dialysis buffer (based on the initial dialysis volume) and centrifuged at 45,000 rpm for 90 min in a 60Ti rotor at 20°. These two centrifugation steps are designed to remove residual DOC; solubilized components which have not been incorporated into the membrane are also removed. The final pellet is resuspended in about 0.5 ml of solution B per milliliter sample dialyzed, frozen in liquid nitrogen, and stored at −70°.

[9] Beckman Rotor Manual, Spinco Div., Beckman Instruments, Palo Alto, California, 1981.
[10] D. I. Repke, J. C. Spivak, and A. M. Katz, *J. Biol. Chem.* **251**, 3169 (1976).
[11] L. Hymel, Ph.D. Thesis, Vanderbilt Univ., 1982.
[12] S. Fleischer and B. Fleischer, this series, Vol. 10, p. 406.

## V. Controlling the Lipid/Protein Ratio of Reconstituted SR Membrane Vesicles

### Method 1. Dialysis Time Is Varied

As DOC is slowly removed by dialysis, phospholipid is gradually transferred from the mixed DOC/phospholipid micelles to the newly formed reconstituted SR membrane.[5] This process continues until the maximal phospholipid/CPP ratio is reached, reflecting the composition of the solubilized SR. Under our conditions, about 8 hr are required to obtain maximal lipid content. Residual DOC is then further removed by washing.

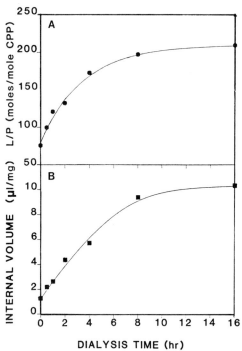

FIG. 2. Correlation of lipid content (●) and internal volume (■) with dialysis time. Solubilized SR was prepared using limited solubilization (2.35 mg DOC/ml) resulting in 43% solubilization of protein. After centrifugation, the supernatant was warmed to 25° for 10 min, then divided into 2-ml aliquots which were dialyzed for various times against 1 liter dialysis buffer per aliquot. Afterward the samples were washed to remove residual detergent as described in the text. (A) The lipid/protein (L/P) ratio and (B) the dextran-impermeable water space (measured as the internal vesicle volume), were measured as described by Rottenberg (see Eq. 7 of Ref. 13).

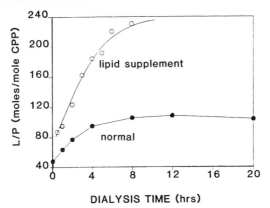

DIALYSIS TIME (hrs)

FIG. 3. The lipid/protein (L/P) ratio of reconstituted SR vesicles can be controlled by dialysis time and lipid supplementation. Extensively solubilized SR (2.80 mg DOC/ml, 83% protein solubilization) was dialyzed in 2-ml aliquots for various times at 23° ("normal" sample). A similar sample was supplemented with approximately 1 equivalent of SR phospholipid suspension (an equal volume of SR phospholipid at 144 $\mu$g P/ml with 2.80 mg DOC/ml in dissociation buffer) and likewise dialyzed in 2-ml aliquots for various times. After dialysis the samples were washed to remove residual DOC and their lipid/protein ratios determined.

The time-dependent increase in lipid/protein ratio with dialysis time is illustrated in Fig. 2 for limited solubilization (up to 55% SR protein is solubilized). The increase in vesicle lipid content is accompanied by an increase in vesicle volume (and diameter). A comparable experiment is shown in Fig. 3 for more extensively solubilized SR (85% or more SR protein is solubilized). The pattern of lipid readdition is similar, although the curve plateaus at a lower lipid/protein ratio than in Fig. 2, reflecting the lower ratio of solubilized lipid to protein in the more extensively solubilized SR (see below).

*Method 2. The Extent of Solubilization Is Varied*

The extent of solubilization of the original SR also determines the lipid/protein ratio in the solubilized extract and hence in the reconstituted membrane.[5,11] A more limited solubilization results in a higher ratio of solubilized lipid to CPP, and a correspondingly higher lipid/protein ratio in the reconstituted vesicles (compare Fig. 2 for limited solubilization with Fig. 3 for extensive solubilization). The maximal lipid/protein ratios are approximately proportional to the starting ratios in the solubilized SR.

---

[13] H. Rottenberg, this series, Vol. 55, p. 547.

By varying both the extent of solubilization and the dialysis time, reconstituted vesicles with any desired lipid content from about 48 mol phospholipid/mol CPP (the ratio of the *limit membrane*) to 250 mol phospholipid/mol CPP (obtained with very limited solubilization and 8 hr or more dialysis).[5]

## Method 3. Lipid Supplementation

The third approach is to supplement the solubilized SR with a phospholipid suspension containing detergent. The supplemented phospholipid can be extracted from SR; other phospholipid samples containing various types of probes or radioactivity can be used.

We have previously described a procedure for lipid supplementation in which a fixed dialysis time was used and the amount of added exogenous phospholipid was varied.[14] We now describe a more reproducible variant of this method (cf. Fig. 3). The phospholipid and DOC concentrations of the lipid suspension for supplementation are set to similar values as in the solubilized SR using the calibration curves in Fig. 1. This method ensures that the equilibria between DOC, phospholipid, and the solubilized CPP remain close to those in the solubilized extract and do not vary with the amount of added phospholipid.

An example of how to implement this method is now provided (consult Fig. 1). If we select conditions to solubilize a percentage of the protein, e.g., ~55% protein (Fig. 1A), this requires that 2.4 mg DOC/ml be added (Fig. 1B) to the 6.0 mg SR protein which contains a total amount of 144 μg P (phospholipid). By correlating Fig. 1A and B, it can be seen that this condition solubilizes about 75% of the phospholipid (108 μg P). Supplementation with, for example, 1 and 2 Eq of phospholipid can be obtained by adding, per milliliter of solubilized SR, 1 and 2 ml of 108 μg P (phospholipid) containing 2.4 mg DOC/ml.

Following lipid supplementation, the samples are dialyzed as usual, either for a fixed period of time (with the amount of added lipid varied to achieve the desired lipid content) or by varying the time of dialysis. Longer dialysis times can also be combined with lipid supplementation to achieve higher lipid/protein ratios than those obtainable with strictly the endogenous lipid in the solubilized SR. It should be noted that, for reconstituted SR above 250 mol lipid/mol CPP, the sucrose should be removed near the end of dialysis and during washing, otherwise the buoyancy precludes sedimentation.

---

[14] C.-T. Wang, A. Saito, and S. Fleischer, *J. Biol. Chem.* **254**, 9209 (1979).

VI. Additional Comments

A variety of procedures can be used to remove the detergent concentration to achieve reconstitution. Dialysis is the procedure selected for this methodology. The rocker dialyzer is an effective means to obtain efficient and reproducible mixing during reconstitution.[12] Simpler mixing can be adequate when the dialysis time is prolonged. The reconstituted SR membrane vesicles of different lipid/protein ratio are relatively uniform as judged by sucrose density gradient centrifugation (Fig. 4). The

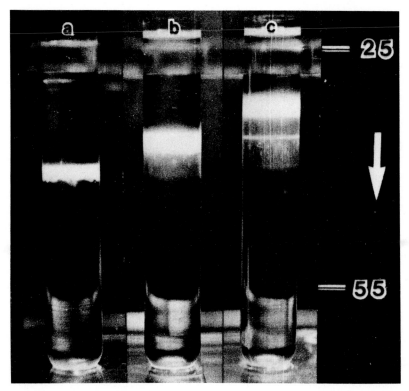

FIG. 4. Homogeneity of reconstituted SR preparations. Solubilized SR (2.80 mg DOC/ml, 89% protein solubilization) was dialyzed for 0, 1, and 4 hr (a, b, and c, respectively), then diluted with an equal volume (2 ml) of dialysis buffer, sedimented, and resuspended in 1 ml each of dialysis buffer. The samples were then layered onto a linear sucrose gradient composed of 25 to 55% (w/w) sucrose in dialysis buffer, and centrifuged for 18 hr at 38,000 rpm in a Beckman SW 41 rotor at 20°. The bands were collected and their lipid/protein ratio and isopycnic density in the gradients were determined: (a) 59 mol PL/mol CPP, 40.4% sucrose; (b) 70 mol PL/mol CPP, 38.2% sucrose; and (c) 88 mol PL/mol CPP, 35.1% sucrose.

recovery of CPP increases with the extent of dialysis. At prolonged dialysis time, 8 hr or longer, it is 60–70% based on the amount of CPP solubilized.[11]

*Factors Which Influence Functional Characteristics of
Reconstituted SR*

Important factors which determine SR function are the extent of SR solubilized in the dissociation procedure and the lipid/protein ratio of the reconstituted SR membrane vesicles. The Ca²⁺-loading and Ca²⁺-ATPase

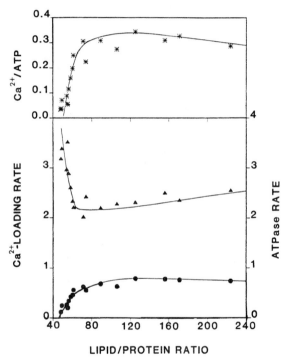

LIPID/PROTEIN RATIO

FIG. 5. Transport characteristics of reconstituted SR as a function of lipid/protein ratio. The lipid/protein ratio of reconstituted SR vesicles was varied by using different amounts of DOC for dissociation (see Fig. 1), thereby altering the ratio in the solubilized SR. Dialysis was for 2 hr. The calcium loading (●) is optimal when the lipid/protein ratio exceeds 65 and is negligible when the ratio approaches that of the limit membrane, 48 mol PL/mol CPP. The Ca²⁺-ATPase (▲) is highest at the low lipid/protein ratio and decreases to a minimum when optimal pumping is achieved. Calcium loading was determined using ⁴⁵Ca and Millipore filtration, and Ca²⁺-ATPase activity was measured as the release of inorganic phosphate.[15] Both activities were measured at 25° in assay medium. The initial rates were calculated by linear regression analysis of values obtained at 20, 40, 60, and 80 sec after starting the reaction with ATP.

TABLE I
FUNCTIONAL CHARACTERISTICS OF RECONSTITUTED
SARCOPLASMIC RETICULUM[a]

| Dissociation condition (% protein solubilized) | Calcium-loading rate[b] | ATPase rate[b] | $Ca^{2+}$/ATP |
|---|---|---|---|
| Limited solubilization (55% of protein) | 0.75–0.80 | 2.0–2.2 | 0.34–0.40 |
| Extensive solubilization (≥85% of protein) | 0.3–0.6 | 3.5–4.0 | 0.08–0.17 |
| Control SR[c] | 1.6 | 2.5 | 0.64 |
| Original SR | 2.0–2.5 | 1.9–2.5 | 0.9–1.1[d] |

[a] Values represent a range of lipid/protein ratios in the reconstituted membranes (~100 to 150 mol phospholipid/mol CPP) which give optimal activity ($n = 5$) (see Fig. 5).

[b] All rates in units of $\mu$mol/min · mg protein at 25° (see Fig. 5).

[c] Control SR was subjected to the same manipulations as for the preparation of reconstituted vesicles.

[d] These assay conditions (low free $[Mg^{2+}]$) do not give optimal $Ca^{2+}$/ATP stoichiometry. Raising the free $[Mg^{2+}]$ to 5 m$M$ increases the $Ca^{2+}$/ATP ratio to 1.75 for native SR vesicles, but does not affect the reconstituted SR. The increase in $Ca^{2+}$/ATP ratio may be due to the closing of $Ca^{2+}$ release channels in the terminal cisternae.[16]

rates for a variety of samples over the range from 50 to 230 mol PL/mol CPP are summarized in Fig. 5.[5] Below a lipid/protein ratio of about 70 in the reconstituted membrane, there is a sharp decline in $Ca^{2+}$ loading paralleled by a reciprocal increase in ATPase activity. Above this lipid content, the functional properties remain essentially unchanged over the range investigated. A likely explanation for this sharp transition is that vesicles with less than 70 mol phospholipid/mol CPP do not contain sufficient bilayer lipid to form tightly sealed membrane vesicles for the accumulation of $Ca^{2+}$ ions.

Reconstituted vesicles obtained using limited solubilization have a better calcium-pumping rate and efficiency ($Ca^{2+}$/ATP coupling ratio) than those obtained from more extensively solubilized SR (Table I).[5] Since the extent of solubilization is so critical, a series of samples for comparison are best produced from the same solubilized extract.

The amount of DOC remaining after reconstitution was found to be proportional to the final lipid content of the reconstituted vesicles

[15] E. S. Baginski and P. P. Foa, *Clin. Chem. Acta* **15,** 155 (1967).

[16] A. Chu, P. Volpe, B. Costello, and S. Fleischer, *Biochemistry* **25,** 8315 (1986).

(~0.07 mol/mol phospholipid). Additional washing was found to further reduce the DOC content, but did not affect function.[11]

We have also examined the effect of the steps in the reconstitution procedure on native SR vesicles, leaving out the detergent (see Section II, Basic Reconstitution Protocol). There is a 20–25% decrease in Ca$^{2+}$-loading rate and Ca$^{2+}$-pumping efficiency through steps 1 to 4 and 6; there is essentially no further loss thereafter (steps 7–9). This loss of activity due to handling undoubtedly contributes to decreased function in the reconstituted SR. Another factor relevant to decreased pumping function is that the reconstitution results in the bidirectional alignment of the calcium pump in the reconstituted membrane.[17,18] Therefore, about half of the pumps are silent with regard to Ca$^{2+}$ pumping, although the Ca$^{2+}$-ATPase is still active in leaky vesicles (see Table I).

[17] S. Fleischer, in "Structure and Function of Sarcoplasmic Reticulum" (S. Fleischer and Y. Tonomura, eds.), pp. 119–145. Academic Press, New York, 1985.
[18] A. Saito, C.-T. Wang, and S. Fleischer, J. Cell Biol. 79, 601 (1978).

# [24] Reconstitution of Calcium Pumping of Cardiac Sarcoplasmic Reticulum

By Makoto Inui and Sidney Fleischer

## Introduction[1]

The β-adrenergic action of catecholamines on heart muscle is, in part, explained by modulation of Ca$^{2+}$ fluxes by way of the action of their intracellular second messengers such as cAMP and Ca$^{2+}$ itself.[2–4] At the level of cardiac sarcoplasmic reticulum (SR), the Ca$^{2+}$-pump which transports Ca$^{2+}$ from the cytoplasm into the lumen of the SR can be modulated by cAMP and Ca$^{2+}$-calmodulin.[5] A number of findings indicate that phos-

[1] These studies were supported by NIH Grant HL 32711 to S.F. and a postdoctoral fellowship from the Muscular Dystrophy Association of America to MI.
[2] A. M. Katz, Adv. Cyclic Nucleotide Res. 11, 303 (1979).
[3] M. Tada and M. Inui, J. Mol. Cell. Cardiol. 15, 565 (1983).
[4] R. W. Tsien, Adv. Cyclic Nucleotide Res. 8, 363 (1977).
[5] M. Tada and A. M. Katz, Annu. Rev. Physiol. 44, 401 (1982).

phorylation of the ~27,000-Da membrane protein, phospholamban,[5] is paralleled by the modulation of the $Ca^{2+}$-pump [a decrease in $K_m$ for $Ca^{2+}$ $(K_{Ca})$]. Phosphorylation of phospholamban is catalyzed by cAMP-dependent,[6,7] and $Ca^{2+}$-calmodulin-dependent protein kinases[8] and protein kinase C.[9] We have studied the molecular mechanism of the modulation of $Ca^{2+}$-pump by protein kinase-catalyzed phosphorylation using the dissociation and reconstitution approach.[10] This approach has been extensively utilized for the $Ca^{2+}$-pump of skeletal muscle SR[11] but has not been achieved for heart muscle SR. We describe the reconstitution of cardiac SR transport function, which has provided new insight into the nature of the modulation of $Ca^{2+}$ transport.[10]

## Solubilization and Reconstitution of Cardiac SR Vesicles

### Reagents and Preparations

Triton X-100 (octylphenoxypolyethoxyethanol) (Sigma, St. Louis, MO)

Bio-Beads SM-2 (polystyrenedivinylbenzene) (Bio-Rad, Richmond, CA): Bio-Beads are washed with methanol and water before use

Solution A: 0.3 $M$ sucrose, 0.5 $M$ KCl, 20 m$M$ imidazole-HCl, pH 7.0

Solution B: 0.6 $M$ sucrose, 1 $M$ KCl, 40 m$M$ imidazole-HCl, pH 7.0

$CaCl_2$ solution: 100 m$M$ $CaCl_2$

Solution C: 0.1 $M$ sucrose, 0.1 $M$ KCl, 20 m$M$ imidazole-HCl, pH 6.8

Highly purified and stable cardiac SR is prepared from dog heart ventricles using a combination of differential and density gradient centrifugation according to Chamberlain et al.[12] The final pellets are resuspended in solution A at a protein concentration of 15–20 mg/ml. The SR is quick-frozen and stored in liquid nitrogen.

*Reconstitution Procedure.* This procedure is outlined in Fig. 1. The cardiac SR is thawed at room temperature. Solubilization of cardiac SR membranes is initiated by adding an aliquot of Triton X-100 solution (10% w/v) to give 6 mg/ml of Triton X-100 and 7.5 mg/ml of cardiac SR (a detergent-to-protein weight ratio of 0.8) in a medium containing 0.5 $M$ KCl, 5 m$M$ $CaCl_2$, 0.3 $M$ sucrose, and 20 m$M$ imidazole-HCl (each given

---

[6] M. A. Kirchberger, M. Tada, and A. M. Katz, *J. Biol. Chem.* **249**, 6166 (1974).

[7] M. Tada, M. A. Kirchberger, and A. M. Katz, *J. Biol. Chem.* **250**, 2640 (1975).

[8] C. J. LePeuch, J. Haiech, and J. G. Demaille, *Biochemistry* **18**, 5150 (1979).

[9] M. A. Mousesian, M. Nishikawa, and R. S. Adelstein, *J. Biol. Chem.* **259**, 8029 (1984).

[10] M. Inui, B. K. Chamberlain, A. Saito, and S. Fleischer, *J. Biol. Chem.* **261**, 1794 (1986).

[11] G. Meissner and S. Fleischer, *J. Biol. Chem.* **249**, 302 (1974).

[12] B. K. Chamberlain, D. O. Levitsky, and S. Fleischer, *J. Biol. Chem.* **258**, 6602 (1983).

## Solubilization of Cardiac SR Vesicles

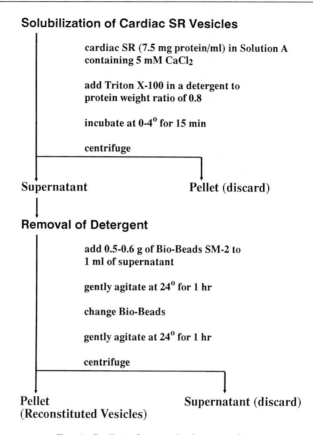

cardiac SR (7.5 mg protein/ml) in Solution A containing 5 mM CaCl₂

add Triton X-100 in a detergent to protein weight ratio of 0.8

incubate at 0-4° for 15 min

centrifuge

Supernatant                    Pellet (discard)

## Removal of Detergent

add 0.5-0.6 g of Bio-Beads SM-2 to 1 ml of supernatant

gently agitate at 24° for 1 hr

change Bio-Beads

gently agitate at 24° for 1 hr

centrifuge

Pellet                         Supernatant (discard)
(Reconstituted Vesicles)

FIG. 1. Outline of reconstitution procedure.

as final concentration). For example, when the solubilization and reconstitution is carried out in a final volume of 1 ml using 15 mg protein/ml of cardiac SR preparation, 0.5 ml of cardiac SR, 0.25 ml of solution B, 0.05 ml of CaCl₂ solution, and 0.14 ml of the water are mixed and kept on ice, and then 0.06 ml of 100 mg/ml Triton X-100 is added to the sample. The sample is maintained at 0–4° for 15 min and then centrifuged for 20 min at 30 psi in a Beckmann Airfuge A-100 rotor. For a larger scale preparation (more than 1 ml), the sample is centrifuged for 60 min at 50,000 rpm in a Beckman 70.1Ti rotor. The supernatant, solubilized cardiac SR, is collected. One milliliter of the supernatant is transferred into a test tube (inner diameter 10 mm) containing 0.5–0.6 g (wet weight) of Bio-Beads SM-2. The mixture is gently agitated at 24° for 2 hr using a wrist-action tube shaker. After 1 hr of incubation, the Bio-Beads are changed. The sample is recovered from the Bio-Beads using a 0.25-ml syringe with a 0.006-inch inner diameter needle, and transferred into a new tube containing 0.5–0.6 g of Bio-Beads SM-2. At the end of incubation, the sample is recovered from the Bio-Beads as for the change of Bio-Beads. The reformed vesicles are sedimented in a Beckmann Airfuge A-100 rotor at 30 psi for 15 min or in a Beckmann 70.1Ti rotor at 50,000 rpm for 60 min. The pellet is resuspended in solution C and can be quick-frozen and stored at −80°.

*Notes on Methodology.* The solubilization and reconstitution procedures for cardiac SR are fast and simple. The entire procedure can be accomplished within 3 hr. Two hours incubation with Bio-Beads SM-2 at 24° with a change in Bio-Beads is sufficient to remove Triton X-100 from the solubilized sample, since additional changes of Bio-Beads have no effect on membrane morphology and $Ca^{2+}$-pumping function. Prolonged incubation with Bio-Beads leads to decreased $Ca^{2+}$-pumping of the reconstituted vesicles.

The concentration of detergent used for solubilization greatly affects $Ca^{2+}$-pumping function of the reconstituted vesicles. Concentrations above a weight ratio of detergent-to-protein of 0.8 result in a lower $Ca^{2+}$-loading rate and pumping efficiency. The ATPase activity is higher at concentrations below a weight ratio of 0.6. Thus, the most efficient $Ca^{2+}$-pumping function is obtained at a Triton X-100 concentration range of between 4.5 and 6 mg/ml in the presence of 7.5 mg/ml cardiac SR (detergent-to-protein weight ratio of 0.6–0.8).[10]

The composition of the buffer during solubilization and reconstitution also determines the nature of the reconstitution of $Ca^{2+}$-pumping function.[10] A concentration of $CaCl_2$ between 5 to 10 m$M$ gives optimal $Ca^{2+}$-loading rate and efficiency. $MgCl_2$ has an inhibitory effect on the reconstitution of $Ca^{2+}$-pumping function, and, therefore, is not included. A KCl concentration of 0.5 $M$ is optimal; below or above this concentration, the $Ca^{2+}$-loading rate and the efficiency of $Ca^{2+}$-pumping declined. pH 7.0 has been selected for reconstitution since the range between pH 6.8 and 7.2 is optimal for $Ca^{2+}$-pumping function.

### Characteristics of Reconstituted Cardiac SR Vesicles

Functional membrane vesicles capable of ATP-dependent oxalate-facilitated $Ca^{2+}$ transport are obtained by the method described above. The $Ca^{2+}$-loading rate and $Ca^{2+}$-pumping efficiency of the reconstituted vesicles is 70% that of the original cardiac SR (Table I). The reconstitution of $Ca^{2+}$ transport and $Ca^{2+}$-pumping efficiency is comparable to the best reported for skeletal muscle SR.[11] The reconstituted vesicles have, in

TABLE I
CHARACTERISTICS OF ORIGINAL AND RECONSTITUTED
CARDIAC SR[a]

| Characteristic | Original SR | Reconstituted SR |
|---|---|---|
| Lipid content ($\mu$mol P/mg) | 1.06 + 0.05 (2) | 1.14 + 0.05 (3) |
| Particle density[b] (particles/$\mu^2$) | | |
| Concave (P) face | 3919 + 188 [5] | 2824 + 299 [5] |
| Convex (E) face | 720 + 228 [5] | 2795 + 331 [5] |
| Ca$^{2+}$-loading rate[c] ($\mu$mol Ca/mg · min) | 2.31 + 0.31 (4) | 1.65 + 0.31 (5) |
| ATPase activity[d] ($\mu$mol P$_i$/mg · min) | | |
| Ca$^{2+}$-insensitive | 0.25 + 0.05 (4) | 0.08 + 0.03 (5) |
| Ca$^{2+}$-activated | 2.45 + 0.36 (4) | 2.39 + 0.25 (5) |
| Ca$^{2+}$-loading efficiency | 0.95 + 0.13 (4) | 0.69 + 0.09 (5) |

[a] Values are mean ±SD with the number of experiments given in parentheses.

[b] The intramembrane particles observed at the hydrophobic fracture face by freeze-fracture electron microscopy were counted from photographs at a magnification of 120,000. The number of photographs is given in brackets.

[c] Ca$^{2+}$-loading rate was determined at 37° in a medium containing 0.1 $M$ KCl, 4 m$M$ MgCl$_2$, 66.2 $\mu M$ $^{45}$CaCl$_2$, 30 $\mu M$ EGTA, 3.5 m$M$ Na$_2$ATP, 5 m$M$ potassium oxalate, 5 m$M$ NaN$_3$, 0.1 $M$ sucrose, 20 m$M$ imidazole-HCl, pH 6.8. SR vesicles (7–10 $\mu$g/ml) are equilibrated in the medium for 3 min prior to starting Ca$^{2+}$-uptake by adding ATP. Reactions are stopped after 30 and 60 sec by filtering aliquots through 0.22-$\mu$m pore size Millipore filters (type GS), and $^{45}$Ca accumulated by the vesicles is measured.

[d] ATPase activity is determined at 37° by measuring P$_i$ release as described previously.[12]

the main, a similar protein composition to that of the original vesicles, as analyzed by sodium dodecyl sulfate (SDS)-polyacrylamide gel electrophoresis. However, the reconstituted vesicles are practically devoid of the Ca$^{2+}$-binding protein (calsequestrin) with $M_r$ equivalent to about 57,000, and three other bands (340K, 180K, 20K) are decreased.[10] The phospholipid content is also similar (Table I). The particle density for the reconstituted SR membranes observed by freeze-fracture electron microscopy is in the same range as that for original SR (Table I), although the distribution of particles is no longer asymmetric, indicating that the Ca$^{2+}$-pump protein is bidirectionally oriented in the reconstituted membrane vesicles.[10] These characteristics of the reconstituted cardiac SR membrane vesicles are essentially the same as those reported in the reconstitution of skeletal muscle SR.[11,13,14]

In cardiac SR, phospholamban is phosphorylated by adding catalytic subunit of cAMP-dependent protein kinase or via an intrinsic $Ca^{2+}$-calmodulin-dependent protein kinase.[5] Phospholamban in the reconstituted vesicles is also phosphorylated by the catalytic subunit of cAMP-dependent protein kinase to about the same extent as that in the original vesicles (Table II). The latter finding could be explained by either (1) phospholamban is preferentially oriented to the outer surface of the membrane even though the $Ca^{2+}$-pump protein is bidirectionally oriented, or (2) phospholamban is not properly rebound to the $Ca^{2+}$-pump so that the extent of phosphorylation may be different. From this perspective, phospholamban appears to be uncoupled from $Ca^{2+}$-pump protein in the reconstituted vesicles. By contrast, no significant amount of $Ca^{2+}$-calmodulin-dependent phosphorylation is found in the reconstituted vesicles, suggesting that the intrinsic $Ca^{2+}$-calmodulin-dependent protein kinase has been inactivated or lost (Table II).

In the original cardiac SR, it is known that a significant stimulation of $Ca^{2+}$ loading is observed when phospholamban is phosphorylated, by way of a decrease in the $Ca^{2+}$ concentration for the half-maximal $Ca^{2+}$-loading rate ($K_{Ca}$).[7] When the original cardiac SR is pretreated with catalytic subunit of cAMP-dependent protein kinase, $K_{Ca}$ for $Ca^{2+}$-pumping decreases from 1.35 to 0.75 $\mu M$ (Table II). On the other hand, the $K_{Ca}$ of reconstituted vesicles is appreciably lower than that of phosphorylated original cardiac SR (Table II) and is the same (0.47–0.50 $\mu M$) for nonphosphorylated and cAMP-dependent kinase-phosphorylated membranes. Thus, the $Ca^{2+}$-pump of the reconstituted vesicles is fully activated (decreased $K_{Ca}$), and phosphorylation of phospholamban does not modulate $Ca^{2+}$-pumping.

These observations from reconstituted cardiac SR vesicles provide new insight into the mode of regulation of the $Ca^{2+}$-pump by phospholamban. The $K_{Ca}$ of the $Ca^{2+}$-pump in cardiac SR is known to be higher than that in skeletal muscle SR.[15] The higher value of $K_{Ca}$ in cardiac SR was considered to be due to the suppression of the $Ca^{2+}$-pump by phospholamban[15,16] because skeletal muscle SR does not have phospholamban.[17] In the study of reconstituted cardiac SR vesicles, we find that the $Ca^{2+}$-pump of cardiac SR has a low $K_{Ca}$ value that is comparable to that of skeletal

[13] C.-T. Wang, A. Saito, and S. Fleischer, *J. Biol. Chem.* **249**, 302 (1974).
[14] L. Hymel and S. Fleischer, *Fed. Proc.* **42**, 2215a (1983), and this volume [23].
[15] M. J. Hicks, M. Skigekawa, and A. M. Katz, *Circ. Res.* **44**, 384 (1979).
[16] A. M. Katz, *Trends Pharmacol. Sci.* **1**, 434 (1980).
[17] M. A. Kirchberger and M. Tada, *J. Biol. Chem.* **251**, 725 (1976).

TABLE II

PROTEIN KINASE-CATALYZED PHOSPHORYLATION AND MODULATION OF
Ca²⁺ TRANSPORT IN ORIGINAL AND RECONSTITUTED CARDIAC SR[a]

| Sarcoplasmic reticulum | Phosphorylation[b] (nmol P/mg protein) | $K_{Ca}$[c] ($\mu M$) |
|---|---|---|
| Original | | |
|   Control | 0 | 1.35 ± 0.08 (4) |
|   cAMP-dependent protein kinase[d] | 1.73 ± 0.16 (4) | 0.75 ± 0.12 (4) |
|   CaM-dependent protein kinase[e] | 1.28 ± 0.07 (3) | — |
| Reconstituted | | |
|   Control | 0 | 0.47 ± 0.03 (4) |
|   cAMP-dependent protein kinase[d] | 1.80 ± 0.13 (4) | 0.50 ± 0.02 (4) |
|   CaM-dependent protein kinase[e] | 0.17 ± 0.05 (5) | — |

[a] Values are the mean ±SD from the number of separate experiments indicated in parentheses.

[b] Phosphorylation by phospholamban was measured at 25° with [γ-³²P]ATP (10⁵ cpm/nmol) in a medium containing 0.1 $M$ KCl, 0.1 $M$ sucrose, 2 m$M$ MgCl₂, 20 m$M$ imidazole-HCl, pH 6.8, 2 mg of protein/ml of SR, with 0.1 mg/ml catalytic subunit of cAMP-dependent protein kinase and/or by an intrinsic Ca²⁺-calmodulin-dependent protein kinase in the presence of 0.2 m$M$ CaCl₂, 0.2 m$M$ EGTA, and 0.1 mg/ml calmodulin. The reaction was started by addition of 2 m$M$ [γ-³²P]ATP (final concentration) and terminated after 5 min by adding 10 volumes of 10% ice-cold trichloroacetic acid containing 0.1 m$M$ KH₂PO₄ (stop solution). The sample was washed once with 0.5 $N$ NaOH and four times with stop solution, using sedimentation in the clinical centrifuge to recover the pellet. Radioactivity was then measured.

[c] The sample was preincubated under the same conditions as phosphorylation described above but with unlabeled ATP. An aliquot of the sample (5 μg) was added to the Ca²⁺-loading assay medium (0.75 ml) containing 0.1 $M$ KCl, 4 m$M$ MgCl₂, 0.2 m$M$ ⁴⁵CaCl₂, various amounts of EGTA (0.1–2 m$M$), 5 m$M$ NaN₃, 0.1 $M$ sucrose, 3.5 m$M$ ATP, 5 m$M$ oxalate, 20 m$M$ imidazole-HCl, pH 6.8. Ca²⁺ loading was measured at 25° by filtration through 0.22-μm pore size Millipore filters (type GS). $K_{Ca}$ was determined from the Ca²⁺ concentration profile of Ca²⁺-loading rate.

[d] The catalytic subunit of cAMP-dependent protein kinase was added to both original and reconstituted SR.

[e] Only Ca²⁺ and calmodulin (CaM) have been added. The Ca²⁺-calmodulin-dependent protein kinase is intrinsic to the cardiac SR but appears not to be present or has been inactivated in the reconstituted cardiac SR.

muscle under conditions where phospholamban is uncoupled from the Ca²⁺-pump. Therefore, phospholamban appears to act as a suppressor of the Ca²⁺-pump (elevates $K_{Ca}$) in the native cardiac SR. In heart SR, phosphorylation of phospholamban serves to reverse the suppression, resulting in the lowering of the $K_{Ca}$.

# [25] Purification and Crystallization of Calcium-Binding Protein from Skeletal Muscle Sarcoplasmic Reticulum

By ANDREAS MAURER, MASASHI TANAKA, TAKAYUKI OZAWA, and SIDNEY FLEISCHER

## Introduction

The calcium-binding protein of skeletal muscle sarcoplasmic reticulum (also referred to as calsequestrin) is the major compartmental protein and is localized within the terminal cisternae (heavy SR) rather than longitudinal cisternae (light SR).[1,2] Purification of CBP from fast skeletal muscle has been described by Meissner et al.,[3] by MacLennan and Wong,[4] and by Ikemoto,[5] and, more recently, CBP has been purified from heart muscle.[6] The molecular weight of CBP is about 40,000.[7] The CBP has a large capacity to bind $Ca^{2+}$, about 50 $Ca^{2+}$ per CBP molecule.[3-7] Thus, the CBP appears to serve as a reservoir for the storage of $Ca^{2+}$ in the terminal cisternae. The CBP appears to be tightly associated with the junctional face membrane of the terminal cisternae,[2,8] and it may serve an additional role in the calcium release process. Saito et al. observed crystalline arrays within the lumen of isolated terminal cisternae.[2] This was the clue that CBP could be crystallized. The purification and crystallization of CBP with high yield and purity has been achieved[9] and is described.

## Purification and Crystallization of CBP

The CBP is released from the SR compartment with the use of a small amount of detergent, sufficient to make the membrane leaky. Divalent

[1] Abbreviations used are $C_{12}E_8$, Octaethylene glycol mono-$n$-dodecyl ether; HEPES, $N$-2-hydroxyethylpiperazine-$N'$-2-ethanesulfonic acid; CBP, calcium-binding protein; CPP, calcium-pump protein; SR, sarcoplasmic reticulum; SDS-PAGE, sodium dodecyl sulfate-polyacrylamide gel electrophoresis.

[2] A. Saito, S. Seiler, A. Chu, and S. Fleischer, J. Cell Biol. **99**, 875 (1984).

[3] G. Meissner, G. E. Conner, and S. Fleischer, Biochim. Biophys. Acta **298**, 246 (1973).

[4] D. H. MacLennan and P. T. S. Wong, Proc. Natl. Acad. Sci. U.S.A. **68**, 1231 (1971).

[5] N. Ikemoto, G. M. Bhatnager, and J. Gergely, Biochem. Biophys. Res. Commun. **44**, 1510 (1971).

[6] S. E. Cala and L. R. Jones, J. Biol. Chem. **258**, 11932 (1983).

[7] B. Cozens and R. A. F. Reithmeier, J. Biol. Chem. **259**, 6248 (1984).

[8] B. R. Costello, A. Saito, A. Chu, A. Maurer, and S. Fleischer, J. Cell Biol. **99**, 399a (1984).

[9] A. Maurer, M. Tanaka, T. Ozawa, and S. Fleischer, Proc. Natl. Acad. Sci. U.S.A. **82**, 4031 (1985).

cations are then used to induce selective crystallization. The washed crystals are essentially pure CBP.

### Isolation of SR and Extraction of Proteins

#### Reagents and Preparations

Octaethylene glycol mono-*n*-dodecyl ether (C$_{12}$E$_8$) (Nikko Chemical, Tokyo, Japan)

Sucrose, ultrapure grade (Schwartz-Mann, Spring Valley, NY) *N*-2-hydroxyethylpiperazine-*N*-ethanesulfonic acid (HEPES) (Calbiochem-Behring Corp., La Jolla, CA)

All solutions were prepared in deionized water.

0.3 *M* sucrose, 0.1 *M* KCl, 5 m*M* K-HEPES, pH 7.4 (SR storage medium)

0.3 *M* sucrose, 0.6 *M* KCl, 5 m*M* K-HEPES, pH 7.4

0.3 *M* sucrose, 5 m*M* K-HEPES, pH 7.4

0.6 *M* sucrose, 100 m*M* Tris-HCl, pH 7.8

10% solution of C$_{12}$E$_8$ (v/v) in water

Sarcoplasmic reticulum is prepared from fast-twitch skeletal muscle of female New Zealand white rabbits using a combination of differential and isopycnic zonal ultracentrifugation according to Meissner *et al.*[10] SR vesicles are obtained from the 32–39% sucrose regions of the linear sucrose gradient. The purified SR is then washed in 0.3 *M* sucrose, 0.6 *M* KCl, and 5 m*M* K-HEPES buffer, pH 7.4, and sedimented at 150,000 *g* for 60 min in a fixed-angle rotor. The salt-treated SR are suspended in the above buffer but without KCl, then resedimented and resuspended in SR storage medium

*Procedure.* The SR is adjusted to a protein concentration of 12 mg/ml, and a KCl concentration of 80 to 90 m*M* by a combination of SR storage medium and this medium without KCl. To one part of SR suspension is added one part of 0.6 *M* sucrose, 0.1 *M* Tris-HCl, pH 7.8, and C$_{12}$E$_8$ solution (10% v/v), 15 $\mu$l per ml, to give a final concentration of 0.15%. After 10 min at 0°, the solution is centrifuged at 150,000 $g_{max}$ for 90 min in a Beckman 70Ti (40,000 rpm) or 45Ti (35,000 rpm) fixed-angle rotor. The supernatant, enriched in CBP protein, is collected.

### Crystallization of CBP

#### Reagents

Crystal resuspension buffer: 0.4 *M* sucrose, 1 m*M* CaCl$_2$, 0.15% (v/v) C$_{12}$E$_8$, 43 m*M* KCl, 0.05 *M* Tris-HCl, pH 7.8, 100 m*M* CaCl$_2$

[10] G. Meissner, this series, Vol. 31, pp. 238–246.

*Procedure.* For each milliliter of supernatant, 10 $\mu$l of $CaCl_2$ (100 m$M$) is added to a final concentration of 1 m$M$ $CaCl_2$ and maintained in the coldroom ($\sim4°$). After 120 hr at 0–4°, most of the CBP crystallizes out of solution and the supernatant is carefully removed and replaced with the same volume of crystal resuspension buffer. The crystals are again allowed to settle out at 0–4° for 24 hr. The CBP is largely in crystalline form.

Seeding with CBP crystals speeds up the process so that once seed crystals are available, crystal formation can be complete in 24–48 hr.

## Purification and Crystallization of CBP from Microsomal Fractions

Microsomes are prepared from the supernatant of the first blendate of skeletal muscle (S1) by the method of Saito *et al.*[2] and washed as described for SR above (see Isolation of SR and Extraction of Proteins). The CBP is extracted in the same way as described for purified SR (see above). $CaCl_2$ is added to the extract to a final concentration of 1 m$M$. The crystals are centrifuged after 24 hr at 0–4° for 10 min at 5000 $g$. The supernatant is discarded and the crystals resuspended in crystal resuspension buffer to the same volume and allowed to settle for another 24 hr. The supernatant is removed by aspiration and the wash repeated one more time as before.

It should be noted that when microsomes are used instead of SR, the time for crystallization is limited to 24–28 hr rather than 120 hr to minimize coprecipitation. The extra washes are useful to remove contaminants. Seeding with crystals after addition of the $CaCl_2$ is recommended to speed up the rate of crystal formation.

## Solubilization and Recrystallization of CBP Crystals

### Reagents

0.4 $M$ KCl, 50 m$M$ Tris-HCl, pH 7.8
0.12 $M$ KCl
1 m$M$ $CaCl_2$, 50 m$M$ Tris-HCl, pH 7.8

The crystals can readily be dissolved by resuspension in 0.4 $M$ KCl, 50 m$M$ Tris-HCl, pH 7.8. Recrystallization can then be obtained by dialysis versus 0.05 $M$ KCl, 1 m$M$ $CaCl_2$, 50 m$M$ Tris-HCl, pH 7.8.

### Notes on Methodology

When the purified SR preparation is used as the starting material, the detergent extract is clear and colorless, and no turbidity is observed when it is kept at 4° for several days. By contrast, the extract from the microsomal fraction is turbid, and the turbidity increases with time.

The concentration of $C_{12}E_8$ is an important factor in extraction. A low detergent-to-protein ratio is used which is just sufficient to release the CBP. In this way, the leaky membrane vesicles can readily be sedimented.

Besides $Ca^{2+}$, only two other cations of alkaline earth elements, $Mg^{2+}$ and $Sr^{2+}$, induce the CBP to crystallize at 0.5 to 2 m$M$ concentrations. The amount of CBP precipitated increases with greater divalent ion concentrations, suggesting that the equilibrium between soluble and crystalline form of CBP is determined by the concentration of divalent cations. At concentrations higher than 3 m$M$ divalent cation, the CBP does not crystallize, but precipitates out as droplets. Combinations of these three divalent cations also lead to crystals of CBP, so long as the total concentration of the cations is less than 2 to 3 m$M$. Other divalent cations, e.g., $Ba^{2+}$, $Mn^{2+}$, $Cu^{2+}$, $Zn^{2+}$, $Cd^{2+}$, $Sn^{2+}$, and $Hg^{2+}$, does not give rise to crystals. Instead, these cations precipitate not only CBP but also other proteins in the $C_{12}E_8$ extract, resulting in little or no purification of CBP.

### Properties of the CBP Crystals

The purification and crystallization of CBP from rabbit skeletal muscle SR is summarized in Table I.[9,11] CBP represented approximately 6% of the total protein in this SR preparation. $C_{12}E_8$ extracted 15% of the total SR protein and 53% of $Ca^{2+}$-binding capacity from SR. By measuring specific $Ca^{2+}$-binding by equilibrium dialysis in the presence of 1 m$M$ $Ca^{2+}$ and 100 m$M$ KCl, the purification at this step was about fourfold. By addition of $Ca^{2+}$ 29% of the original $Ca^{2+}$-binding capacity was harvested in a crystalline form as CBP. The purified CBP bound 903 nmol $Ca^{2+}$/mg protein. This represented a 17-fold increase in specific $Ca^{2+}$-binding capacity compared with the original SR.

Treatment of purified SR with low concentrations of $C_{12}E_8$, under the conditions described, leads to the extraction of mainly CBP and $M_{55}$ and a number of minor constituents (Fig. 1). The CBP (lane 2) separated into two bands, both staining blue with Stains-all.[13] They represent two slightly different forms of CBP as reported by MacLennan.[14] The residue contains mainly CPP and a small amount of CBP and $M_{55}$ (lane 3). The calcium content in the extract and residue is 116 and 24 nmol/mg protein,

[11] M. Tanaka, T. Ozawa, A. Maurer, J. D. Cortese, and S. Fleischer, *Arch. Biochem. Biophys.* **251**, 369 (1986).

[12] U. K. Laemmli, *Nature (London)* **227**, 680 (1970).

[13] K. P. Campbell, D. H. MacLennan, and A. O. Jorgensen, *J. Biol. Chem.* **258**, 11267 (1983).

[14] D. H. MacLennan, *J. Biol. Chem.* **249**, 980 (1974).

TABLE I
PURIFICATION AND CRYSTALLIZATION OF CALCIUM-BINDING PROTEIN FROM RABBIT
SKELETAL SARCOPLASMIC RETICULUM[a]

| Fraction | Protein mg | % | $Ca^{2+}$-binding capacity nmol | % | Specific $Ca^{2+}$-binding (nmol/mg) | Purification (-fold) |
|---|---|---|---|---|---|---|
| SR | 128 | 100 | 8282 | 100 | 65 (54)[b] | 1.0 |
| Residue after extraction with $C_{12}E_8$ | 109 | 84.7 | 5036 | 61 | 46 | 0.9 |
| Extract with $C_{12}E_8$ | 19.9 | 15.5 | 4358 | 53 | 219 | 4.1 |
| Crystalline CBP | 2.7 | 2.1 | 2438 | 29 | 903 | 17 |
| Supernatant after sedimentation of the crystals | 15.5 | 12.1 | 3007 | 36 | 194 | 3.6 |

[a] Data from a representative experiment are given. Calcium-binding capacity was measured by equilibrium dialysis as described previously[9] with some modifications.[11] The "binding" assay consisted of dialyzing 1.5 ml of the sample (the purified and resolubilized CBP and from each step of the purification) against 1 liter of 5 mM K-HEPES buffer, pH 7.4, containing 1 mM $CaCl_2$ and 0.1 M KCl with stirring at 4° for 46 hr. Aliquots of samples were taken for protein determination and for atomic absorption spectrophotometry in duplicate or triplicate.

[b] Since there are 5.7 nmol of CPP/mg protein in normal SR, which has two $Ca^{2+}$ binding sites per monomer, the binding referable to CPP (11.4 nmol/mg protein) was subtracted from the specific $Ca^{2+}$ binding of SR. This corrected value (in parentheses) was used to estimate the specific calcium binding attributable to CBP in SR and was used to calculate the purification in the last column.

respectively (not equilibrium dialysis values), indicating that most of the CBP has been extracted. The crystalline CBP is essentially pure (lanes 5 and 7).

The CBP crystals are needle-shaped, up to 500 μm long and 50 μm wide, as observed by light microscopy (see Fig. 2). The crystals stain red with ruthenium red or deep blue with the cationic carbocyanine dye, Stains-all, indicating the acidic nature of the CBP. Essentially no phospholipid is associated with the crystalline CBP as revealed by phosphorus analysis.

The crystals are highly ordered as revealed by electron microscopy (Fig. 3). The spacings *a* and *b* are 11 nm each and the angle between the two axes is about 90°. The diffraction pattern obtained by laser light diffraction using an optical bench[15] exhibits intense reflections out to the

[15] A. Klug and J. E. Berger, *J. Mol. Biol.* **10,** 565 (1964).

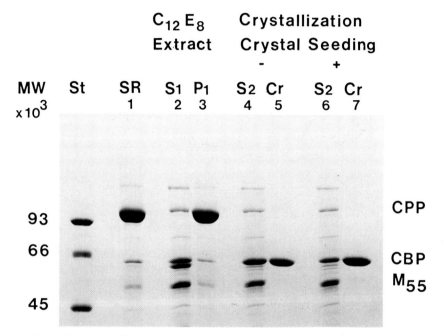

FIG. 1. Protein profile of fractions in the purification and crystallization of CBP. SDS-PAGE was performed according to Laemmli[12] with an acrylamide concentration of 7.5%. The gels were stained with Coomassie brilliant blue. Approximately 7 µg protein were applied per lane. Phosphorylase *b* (92.5K), bovine serum albumin (66K), and ovalbumin (45K) served as molecular weight standards. The starting material, SR (lane 1), the extract of SR with 0.15% $C_{12}E_8$ (S1, lane 2), and the residue (P1, lane 3) are compared with the crystals (Cr) of CBP crystallized in the presence of 1 m$M$ CaCl₂ and 0.1 $M$ KCl (lanes 5 and 7) and the supernatants (S2) from the crystals (lanes 4 and 6). A small amount of seed crystals of CBP was added to the $C_{12}E_8$ extract together with the 1 m$M$ CaCl₂ in order to facilitate the crystallization (lanes 6 and 7).

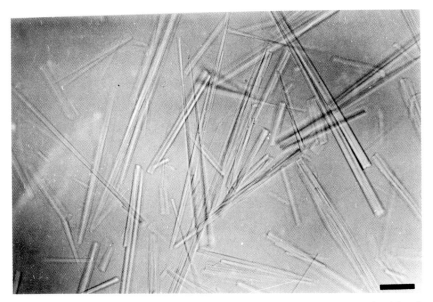

FIG. 2. Light micrograph of crystals of CBP. The crystals of CBP obtained by the addition of 1 m$M$ CaCl$_2$ to the C$_{12}$E$_8$ extract of SR were observed under polarized light. Bar indicates 50 $\mu$m.

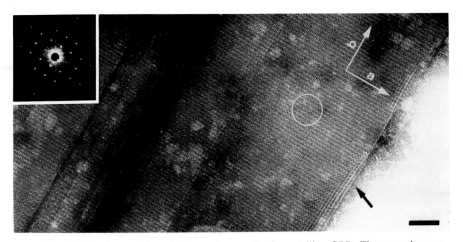

FIG. 3. Negative-staining electron micrograph of crystalline CBP. The crystals were absorbed onto carbon films, washed twice with distilled water, and stained with uranyl acetate (1%, w/v). The specimens were examined in a JEOL 100S electron microscope. On the edge of the crystal (black arrow) a single layer of the crystal is visible. The crystal was induced by the addition of 1 m$M$ MgCl$_2$. The inset in the upper left corner represents the optical diffraction pattern, which was obtained by examination of the area of the image enclosed by the circle in a light-optical diffractometer.[15] The magnification is approximately 63,000. (The bar represents 100 nm.)

fourth order. Thin-section electron microscopy and freeze-drying give complementary information.[9]

X-ray diffraction analysis confirms that the CBP is crystalline. A powder pattern shows diffraction beyond 3 Å Bragg spacing. The crystals are stable in the X-ray beam for over 22 hr. Thus, the crystals of CBP induced with divalent cations are well-ordered so that this type of crystallization, or a modification thereof, may be of value for structural analysis of the protein.

## [26] Purification of Calmodulin and Preparation of Immobilized Calmodulin

By Thomas J. Lukas and D. Martin Watterson

Calmodulin is an acidic, low-molecular-weight, calcium-modulated protein which has been isolated from a diverse number of sources and characterized in terms of its amino acid sequence and biochemical activities.[1] The protein is highly conserved[2,3] in both structure and function, although there are quantitative differences in some activator activities among calmodulins. As a consequence, it may be important that functional studies of calmodulin-binding proteins be carried out with calmodulins from the appropriate biological sources. Because calmodulin is a member of a family of closely related calcium-binding proteins that can coexist in the same cell, purification procedures designed to give proteins of high purity are also necessary for many studies.

Purification procedures for calmodulins from different sources have appeared in preceding volumes of this series and in the literature. Recent advances in high-performance liquid chromatography (HPLC) and protein sequence analysis, however, have allowed a more detailed characterization of calmodulins. The results of these studies indicate that certain procedures used previously may not give calmodulins free of chemical modifications. Therefore, in this chapter we discuss some of the problems encountered during calmodulin purifications and, as an example, describe

---

[1] C. B. Klee and T. C. Vanaman, *Adv. Protein Chem.* **35**, 213 (1981).

[2] D. M. Watterson, W. H. Burgess, T. J. Lukas, D. Iverson, D. R. Marshak, M. Schleicher, B. W. Erickson, K.-F. Fok, and L. J. Van Eldik, *Adv. Cyclic Nucleotide Protein Phosphorylation Res.* **16**, 205 (1984).

[3] T. J. Lukas, D. Iverson, M. Schleicher, and D. M. Watterson, *Plant Physiol.* **75**, 788 (1984).

METHODS IN ENZYMOLOGY, VOL. 157

a simple isolation of calmodulin from chicken gizzard. We also briefly describe the preparation of immobilized calmodulin for the isolation of calmodulin-binding proteins and mention some of the conditions and considerations used in previous studies of calmodulin-binding proteins from membranes.

In practical terms, the decision of exactly what purification protocol to use and how much characterization is necessary should be based on the ultimate goal of the studies. For example, if a large amount of biological tissue must be processed to purify a relatively low-abundance calmodulin-binding protein, then the question of chemical homogeneity and activity of the immobilized calmodulin is a reasonable concern. However, if one is screening a large number of samples simply for the ability to be activated by calmodulin, this may be less of a concern. In the purification of calmodulin, the protocol may require fine tuning if calmodulin isotypes or closely related proteins are present in the same tissue. Similarly, if amino acid sequence analysis or analysis of posttranslational modifications is the goal of the studies, then the requirements of the protocol are different from those of an experiment designed simply to produce a calmodulin standard for gel electrophoresis.

## Calmodulin Purifications

### General Considerations

Calmodulin, although a relatively heat-stable protein, is quite labile to oxidation of methionine residues and deamidation of certain asparagine residues. Some calmodulin purification procedures utilize heat treatment steps that may inactivate proteases and act as a purification step by removing unwanted proteins, but these treatments could result in the oxidation of methionine residues and catalyze deamidation of glutamine or asparagine residues. Invariably some oxidation and deamidation may occur. The former may be partially reversed by using reducing agents such as thiols, but the latter is an irreversible process. It has been observed[2] that heat- or alkali-treated calmodulin is a better substrate for carboxymethyltransferase enzymes than native calmodulin. These enzymes have been shown to specifically methylate the carboxyl groups of isoaspartyl residues of peptides which have been deamidated at Asn-Gly sequences.[4,5] The mechanism of deamidation has been proposed to proceed by way of a cyclic aspartic acid imide intermediate which then hydrolyzes

[4] E. D. Murray, Jr. and S. Clarke, *J. Biol. Chem.* **259,** 10722 (1984).
[5] D. W. Aswad, *J. Biol. Chem.* **259,** 10714 (1984).

to normal and isopeptide linkages.[6] As shown in Fig. 1,[2,3,7–10] the amino acid sequence of bovine brain calmodulin[2] contains two such Asn-Gly sequences within the calcium-binding loops of the second and third domains. We have found that calmodulin peptides, treated briefly with 0.1 $M$ NaOH and then acidified, have aspartic acids at residues 60 and 97 upon subsequent amino acid sequence analyses. The untreated peptides have asparagine residues at these positions. The ratio of isoaspartyl to normal peptide bonds in the deaminated peptides was not determined. To what extent this reaction occurs during calmodulin purifications or *in vivo,* and how effector properties of calmodulin may be altered requires further investigation. However, these results indicate that purification procedures should be designed to minimize the risk of these modifications. Specifically, exposure of calmodulin to high pH or elevated temperatures should be avoided in purification protocols.

*Materials*

> Chicken gizzard, 1 kg, frozen (Pel-Freez)
> Buffer H: 0.05 $M$ Tris-HCl, pH 7.5, containing 0.001 $M$ ethylene glycol bis($\beta$-aminoethyl ether)-$N,N'$-tetraacetic acid (EGTA) and 0.001 $M$ 2-mercaptoethanol
> Buffer B: 0.01 $M$ Tris-HCl, pH 8.0, containing 0.001 $M$ EGTA, 0.001 $M$ 2-mercaptoethanol, and 0.1 $M$ NaCl
> Buffer C: 0.05 Tris-HCl, pH 7.5, containing 0.001 $M$ CaCl$_2$ and 0.001 $M$ 2-mercaptoethanol
> Buffer D: buffer C containing 0.001 $M$ EGTA in place of CaCl$_2$
> DEAE-cellulose (Whatman DE-52, preswollen) equilibrated in buffer B)
> Phenyl-Sepharose CL-4B (Pharmacia)
> 50% sulfuric acid, purified ammonium sulfate (e.g., grade I from Sigma), and cheesecloth

*Purification of Calmodulin*

Chicken gizzard is a readily available and inexpensive source from which a relatively homogeneous preparation of calmodulin can be ob-

---

[6] B. A. Johnson and D. W. Aswad, *Biochemistry* **24**, 2581 (1985).

[7] G. A. Jamieson, Jr., D. D. Bronsen, F. H. Shachat, and T. C. Vanaman, *Ann. N.Y. Acad. Sci.* **356**, 36 (1980).

[8] D. R. Marshak, M. Clarke, D. M. Roberts, and D. M. Watterson, *Biochemistry* **23**, 2891 (1984).

[9] M. Yazawa, K. Yaga, H. Toda, K. Kondo, K. Narita, R. Yamazaki, K. Sobue, S. Kakiuchi, S. Nagao, and Y. Nozawa, *Biochem. Biophys. Res. Commun.* **99**, 1051 (1981).

[10] T. J. Lukas, M. E. Wiggins, and D. M. Watterson, *Plant Physiol.* **78**, 477 (1985).

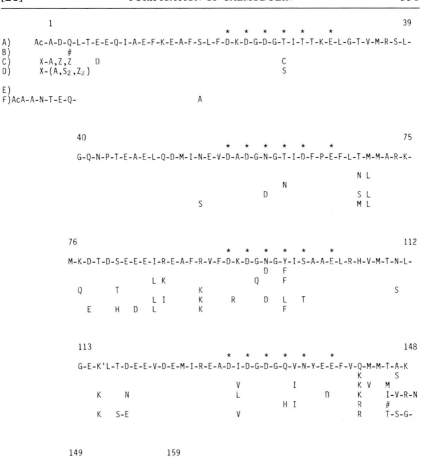

FIG. 1. Amino acid sequences of calmodulins from (A) vertebrate,[2] (B) invertebrate,[7] (C) higher plant,[3] (D) slime mold,[8] (E) protozoan,[9] and (F) green algae.[10] The bovine brain sequence is shown in (A) and sequence differences in the other calmodulins appear in the lines below. Amino acid residues which are tentative ligands for $Ca^{2+}$ in the binding loops have an asterisk above them. Standard single letter codes are used except K′ is $N^{\varepsilon}$-trimethyllysine, and # is an amino acid gap introduced for the purpose of alignment.

tained in good yield with a minimal number of steps. The procedure outlined here essentially uses five steps: differential centrifugation, ammonium sulfate precipitation, isoelectric precipitation, ion-exchange chromatography, and affinity-based adsorption chromatography. Further purification is usually done by HPLC using silica-based resins. The use of phenyl-Sepharose[11] or immobilized calmodulin antagonist[12] has been described previously and will not be reiterated here. Homogeneity of calmodulin preparations is usually assessed by amino acid analysis, UV absorption spectra using known molar extinction coefficients, and polyacrylamide gel electrophoresis in the presence and absence of sodium dodecyl sulfate.

The frozen gizzards are thawed overnight at 4°. The tissue is homogenized in two volumes (2 liters) of buffer H with three 30-sec bursts of a Waring blender. The homogenate is then centrifuged in polycarbonate bottles at 4° in a Sorvall GS-3 rotor at 8000 rpm for 30–60 min. The supernatant is poured through two layers of cheesecloth and kept at 4°. The pellet is then rehomogenized with 1 volume of buffer H, centrifuged, and filtered through cheesecloth. The exact details of the homogenization will vary with the tissue. For example, some precautions must be taken to prevent modification by secondary metabolites when working with some plant tissues. This is often prevented by the inclusion of polyvinylpolypyrrolidone in the homogenization step. Similarly, the rehomogenization step is not used for most plant tissues in order to reduce the risk of modification. Rehomogenization can also be eliminated with brain tissue and still allow a respectable final yield, although the yield of calmodulin will be lower than that obtained with rehomogenization.

The combined supernatants (about 3 liters) are brought to 60% saturation with solid ammonium sulfate (390 g/liter) with stirring at 4°. It is important that the ammonium sulfate not be added in lumps or too rapidly. The procedure is essentially a "salting out" of proteins and should be done as such while attempting to minimize the formation of foam. After the salt has dissolved, the solution is allowed to stir for a minimum of 30 min. The mixture is centrifuged, as described above, for at least 60 min.

The supernatant is poured through cheesecloth into a beaker and the pellet discarded, if it is not to be used for the purification of calmodulin-binding proteins. The solution is then brought to pH 4.1 with 50% (v/v) sulfuric acid. The pH is adjusted carefully with constant stirring at 4°. After stirring for 60 min, the mixture is centrifuged, as described above, for 60 min. The supernatant is carefully decanted and discarded. The pH 4 pellet is then resuspended with the aid of a rubber policeman in a few milliliters of deionized water. One-molar Tris is added dropwise to bring

[11] R. Gopalakrishna and W. B. Anderson, *Biochem. Biophys. Res. Commun.* **104**, 830 (1982).

[12] D. M. Watterson, D. B. Iverson, and L. J. Van Eldik, *Biochemistry* **19**, 5762 (1980).

the pH to about 7.5. The resuspended mixture is then dialyzed at 4° versus 4 liters of buffer B. The dialysis is changed several times until the solution inside the dialysis bag has a conductivity and pH approximating that of the column buffer used in the diethylaminoethyl (DEAE) chromatography step. Any particulate matter is removed by centrifugation prior to loading the sample onto the DEAE-cellulose column. Clearly, a number of variations of the theme can be done, depending on the preferences of the laboratory. The important point is to work rapidly and at 4° in the early part of the protocol, and to have a sample at this point that is equilibrated with the column buffer and does not contain any particulate matter.

Although we have used both DEAE-Sephadex and DE-52, the description given below is for the DE-52 resin. The ionic strength of the column buffer needs to be slightly lower if DE-52 is used rather than DEAE-Sephadex. In addition, the conditions need to be fine tuned for the particular calmodulin that is being purified. For example, *Chlamydomonas* calmodulin has a more positive net charge than vertebrate calmodulin and will elute from the DEAE-cellulose at a lower ionic strength than vertebrate calmodulin. The dialyzed sample is loaded onto a column (usually about 4 × 20 cm) of DE-52. After the sample is loaded onto the column, the column is washed with buffer B until the effluent of the column has an absorption at 280 nm that approximates that of a column blank or the buffer. This is usually a minimum of 3–4 column volumes of buffer B. The adsorbed calmodulin is then eluted with a linear salt gradient consisting of 1 liter of buffer B and 1 liter of buffer B containing 0.5 $M$ NaCl. A flow rate of approximately 100 ml per hour is used and fractions of 10 ml or less are collected. After the procedure is established in the laboratory, the elution position of calmodulin can be estimated from the conductivity of the fractions (for DE-52 chromatography this is in the range of 12–14 mmhos). The actual location is determined by polyacrylamide gel analysis of fractions. We routinely use small 12.5% (w/v) or 15.0% (w/v) polyacrylamide slab gels. These can be run with or without sodium dodecyl sulfate in the gel. Regardless of the gel recipe used, we look for a protein band that comigrates with a calmodulin standard and that elutes from the column in the appropriate part of the gradient.

The decision of which fractions to pool at this step in the protocol is a critical one that depends on the ultimate goal of the experiment and the tissue being studied. The decision is usually one of sacrificing yield for purity. If there is a problem with removing a contaminating protein later in the preparation, especially one that is similar to calmodulin in physical properties and affinity ligand binding properties, the DEAE-cellulose chromatography step is usually where cautious pooling of fractions will be helpful later in the preparation. In brain tissue, calmodulin-enriched and

S100-enriched fractions can be made at this point to allow ease of fractionation later, as well as the preparation of two calcium-modulated proteins from one homogenate. In this regard, DEAE-Sephadex A-50 is better than DE-52 for fractionating S100 and calmodulin. This separation is very pH-dependent in the range of pH 7.2–8.0.

Fractions containing calmodulin are pooled, reduced in volume, and equilibrated in the buffer for the affinity-based adsorption chromatography step (buffer C). The reduction in volume can be accomplished by an isoelectric point precipitation followed by resuspension in the column buffer, or the sample can be dialyzed at 4° against 20 liters of 0.01 $M$ ammonium bicarbonate, lyophilized to dryness, and resuspended in the column buffer. Regardless of the method used, the sample should have a conductivity and pH equal to that of the column buffer and the sample should be clear of particulate matter.

The sample is applied to a column (2 × 10 cm) of phenyl-Sepharose that has been previously equilibrated with buffer C at 4°. Alternatively, immobilized calmodulin antagonists such as phenothiazine or W7-Sepharose or drug–silica conjugates (see below) could be used at this step. For the phenyl-Sepharose chromatography, the sample is applied to the column and allowed to enter the resin bed, but the resin is not allowed to dry out. The effluent flow is shut off for several minutes (for convenience we use about 30 min), the column is washed with several column volumes of buffer until the absorbance of the effluent at 280 nm is down to baseline, followed by a similar wash with buffer C containing 0.5 $M$ NaCl. The column is then washed again with five column volumes of buffer C followed by elution of the calmodulin with buffer D. The sample is then dialyzed against two changes of 0.05 or 0.01 $M$ ammonium bicarbonate (20 liters each) at 4°, then lyophilized to dryness. At this point in the protocol the yield of calmodulin from 1 kg of chicken gizzards should be in the range of 40–60 mg.

### Analysis of Calmodulin by High-Performance Liquid Chromatography

To further purify calmodulins after the "affinity" chromatography step we have used reversed-phase HPLC routinely for removal of chromophoric impurities, calmodulin forms resulting from oxidation and other modifications, and any trace contaminating proteins detected upon analysis of the sample. A chromatogram of chicken gizzard calmodulin run on a preparative octadecyl silica column is shown in Fig. 2A. Smaller (0.46 × 25 cm) columns can also be used for this purpose with similar results. When used with a neutral phosphate–EGTA buffer (Fig. 2), calmodulin elutes at 30–32% (v/v) acetonitrile. Sensitivity of the system with the preparative column and a 214-nm detector is less than 0.25 μg of calmodulin. The performance of the system is illustrated in Fig. 2B with samples of oxidized vertebrate calmodulin and an aged sample of spinach calmodulin (Fig. 2C). As shown, the HPLC can resolve modified protein from

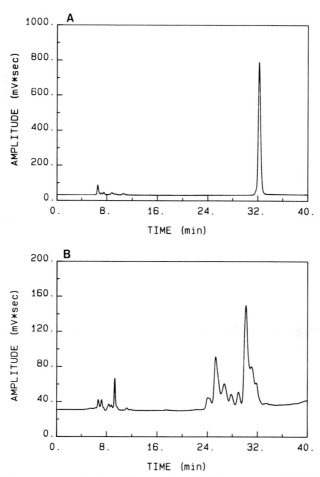

FIG. 2. Chromatograms of calmodulin samples. (A) Chromatogram of chicken gizzard calmodulin purified as described in the text. Solvent A was 5.0 m$M$ NaH$_2$PO$_4$, pH 6.25, containing 0.10 m$M$ EGTA. Solvent B was acetonitrile. The HPLC column is an RP-P (Synchrom, Linden, IN; a 6-μm particle size, 300 Å pore size, octadecylsilanyl silica) with dimensions 1.0 cm × 25 cm run at a flow rate of 2.0 ml/min. The gradient program used for elution of calmodulin is initial conditions, 20% B; 10 min, 20% B; 15 min, 25% B; 20 min, 25% B; 24 min, 32% B; 36 min, 32% B. (B) A sample of peroxide-treated bovine brain calmodulin. (C) A sample of spinach calmodulin which had been stored in water for several months at −20°. The $y$ axis on the graphs is the detector output at 214 nm. (*Figure continues.*)

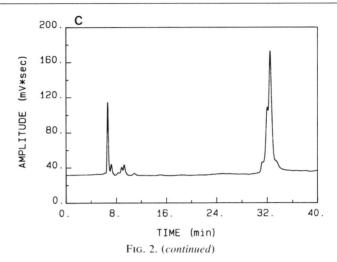

TIME (min)

FIG. 2. (*continued*)

the native protein. Under the current conditions, however, calmodulins which have been deamidated cannot be separated very readily from the native protein.

### Preparation and Use of Immobilized Calmodulin

*General Considerations*

In light of the discussion presented earlier, the preparation of immobilized calmodulin should be carried out under conditions which minimize any additional residue transformations other than those involved in the formation of the covalent bond with the resin. The preparation of calmodulin-Sepharose described here is similar to that originally published.[13] Commercially available activated Sepharose has been used routinely as well as activation of the resin at the time of coupling.[13] The preparation of the calmodulin-Sepharose conjugate can be done in small batches or in a large batch as described below. Most of the considerations in the preparation of immobilized calmodulin are those common to the use of immobilized proteins (see Volume 104 of this series).

*Materials*

8 g CNBr-activated Sepharose (Pharmacia)
30–35 mg of vertebrate calmodulin (lyophilized powder)

¹³ D. M. Watterson and T. C. Vanaman, *Biochem. Biophys. Res. Commun.* **73**, 40 (1976).

0.001 $M$ HCl
0.2 $M$ sodium bicarbonate solution (2 liters)
Ethanolamine (Aldrich)

*Preparation of Immobilized Calmodulin*

Approximately 8 g of CNBr-activated Sepharose is swollen in 75 ml of 0.001 $M$ HCl. The swollen volume will be about 30 ml. The resin is transferred to a 600-ml sintered glass funnel and washed with 200 ml of 0.001 $M$ HCl, 500 ml of deionized water, and 200 ml of 0.2 $M$ sodium bicarbonate solution. The resin is then transferred to a beaker as a 1 : 1 (v/v) slurry in the sodium bicarbonate solution. Calmodulin, 30–35 mg (by weight) is dissolved in 3 ml of 0.2 $M$ sodium bicarbonate solution and added with mixing to the CNBr-Sepharose slurry. The mixture is then swirled on a shaker for 3 hr at room temperature. After the reaction period, the slurry is washed with 0.2 $M$ sodium bicarbonate to remove the unreacted calmodulin. The wash may be saved and calmodulin recovered. To the slurry, 40 ml of 1 $M$ ethanolamine adjusted to pH 8.0 is added and the mixture swirled on a shaker at room temperature for 30 min. The resin is then transferred to a funnel and washed with 100 ml of deionized water, 250 ml of a solution containing 0.1 $M$ sodium acetate, 1 $M$ NaCl, pH 4.0, 50 ml of deionized water, 250 ml of a buffer containing 0.01 $M$ Tris, 0.001 $M$ MgCl$_2$, 0.001 $M$ 2-mercaptoethanol, and 0.002 $M$ EGTA, pH 7.5. The calmodulin resin is then washed with the same buffer containing 0.002 $M$ CaCl$_2$ instead of EGTA and stored as a 1 : 1 (v/v) slurry at 4°. The amount of calmodulin coupled to Sepharose can be determined by hydrolysis of an aliquot of the slurry and amino acid analysis. In the experiment described here, 0.25 to 0.45 mg of calmodulin was covalently attached per milliliter of equilibrated resin. More importantly, analysis of the activity of the protein by means of 3'5'-cAMP phosphodiesterase[14] or myosin light-chain kinase[15] assay indicates that more than 50% of the bound calmodulin is capable of activating these enzymes.

We also have investigated the use of silica supports for the immobilization of calmodulin antagonists and calmodulin. The primary advantages of silica columns are the ability to use a wide variety of solvents and elution conditions normally incompatible with agarose supports, and faster load, wash, and elution times. The coupling of drugs such as W-7 [*N*-(6-aminohexyl-5-chloro-1-naphthalenesulfonamide hydrochloride] or CAPP (2-chloro-10-aminopropylphenothiazine) to derivatized silica yields

---

[14] D. M. Watterson, D. B. Iverson, and L. J. Van Eldik, *J. Biochem. Biophys. Methods* **2**, 139 (1980).
[15] D. K. Blumenthal and J. T. Stull, *Biochemistry* **19**, 5608 (1980).

a support which binds acidic proteins of the calmodulin family. The overall properties of columns prepared with these supports are quite similar to reversed-phase HPLC columns.[16] The immobilized drug columns have allowed a facile separation of calmodulin and S100 proteins.[16]

We have also coupled calmodulin to a silica support and tested the properties of the immobilized protein. The technique used for coupling was reductive alkylation of an aldehyde-functionalized silica.[16] Calmodulin (2 mg) was coupled to 1 g of support in 2 ml of a 0.1 $M$ phosphate buffer, pH 7.0, in the presence of 0.1 $M$ sodium cyanoborohydride overnight at room temperature. Ethanolamine was then added to react with the residual aldehyde groups on the resin. This procedure may give the support a cationic backbone since the reduced Schiff's base is now a secondary amine. In the preparation of the immobilized drug supports, the reducing agent was lithium borohydride which reduces the unreacted aldehydes to alcohols. From amino acid analysis of the hydrolyzed support, there were 38 $\mu$g of calmodulin bound per gram of silica. This level of incorporation is about one-tenth that obtained with CNBr-activated Sepharose. Preliminary experiments with the immobilized calmodulin–silica suggested that it selectively bound proteins from a crude bovine brain fraction containing calmodulin-binding proteins. The low capacity of the column, however, precluded more definitive experiments with this support. At least one HPLC manufacturer has recently made available a silica support containing an epoxide functional group which can be derivatized with drugs and peptide or protein ligands. This support offers the same advantages as the aldehyde–silica described earlier and may be capable of greater density of ligand coupled. Clearly, this is an area of development that could be a tremendous asset in future attempts to purify calmodulin-binding proteins from membrane fractions.

*Use of Calmodulin-Sepharose*

Calmodulin-Sepharose has been used to purify a number of calmodulin-dependent enzymes. These include myosin light-chain kinase,[15,17] calmodulin-dependent protein kinases from brain and liver,[18,19] plant NAD kinase,[20] calmodulin-dependent phosphatase,[21] and 3′,5′-cyclic nu-

[16] D. R. Marshak, T. J. Lukas, C. M. Cohen, and D. M. Watterson, *in* "Calmodulin Antagonists and Cellular Physiology" (H. Hidaka and D. J. Hartshorne, eds.), pp. 495–510. Academic Press, Orlando, Florida, 1985.

[17] R. S. Adelstein and C. B. Klee, this series, Vol. 85, pp. 298–308.

[18] J. Kuret and H. Schulman, *Biochemistry* **23**, 5495 (1984).

[19] M. E. Payne, C. M. Schworer, and T. R. Soderling, *J. Biol. Chem.* **258**, 2376 (1983).

[20] P. Dieter and D. Marme, *Cell Calcium* **1**, 279 (1980).

[21] E. A. Tallant and W. Y. Cheung, *Arch. Biochem. Biophys.* **232**, 269 (1984).

cleotide phosphodiesterase.[13] In most cases, the calmodulin affinity step produces a highly purified but not necessarily homogeneous enzyme. The use of immobilized calmodulin requires that the endogenous calmodulin in the extract be removed before the proteins of interest will bind to the column. This task can be achieved by dissociation of the calmodulin–enzyme complex and chromatography on supports which allow separation of the binding protein and calmodulin. This is often done by adsorbing calmodulin to DEAE-cellulose in the presence of a chelator with subsequent gradient elution of the column. To extend the life of the immobilized calmodulin column and take full advantage of the affinity nature of this stage of a purification protocol, it should be used in the latter stages of the purification process. Consideration should also be given to the resolution of calmodulin-binding proteins from each other before utilizing the calmodulin-Sepharose chromatography step.

Although calmodulin-Sepharose chromatography is usually done in aqueous buffers, there are reports of the successful use of detergents for calmodulin-Sepharose chromatography. For example, the purification of membrane-associated calmodulin-binding proteins such as brain adenylate cyclase[22] and a $Ca^{2+}$-dependent ATPase from erythrocytes[23] were done in the presence of nonionic detergent. Westcott and co-workers[22] used Lubrol PX, while Niggli *et al.*[23] used Triton X-100 in the calmodulin-Sepharose chromatography step. Obviously, the use of these detergents may not preserve the functional activity of other calmodulin-binding proteins, necessitating an analysis of the effect of other detergents or organic solvents for solubilization of these proteins and subsequent calmodulin-Sepharose chromatography.

## Acknowledgments

We thank our current and former colleagues in the laboratory for their contributions and Ms. Janis Elsner for assistance in the preparation of this manuscript. Supported in part by NIH Grant GM 30861 and NSF Grant DMB 8405374.

[22] K. R. Westcott, D. C. LaPorte, and D. R. Storm, *Proc. Natl. Acad. Sci. U.S.A.* **76**, 204 (1979).
[23] V. Niggli, J. T. Penniston, and E. Carafoli, *J. Biol. Chem.* **254**, 9955 (1979).

# [27] Purification, Reconstitution, and Regulation of Plasma Membrane Ca$^{2+}$-Pumps

By JOHN T. PENNISTON, ADELAIDA G. FILOTEO,
CAROL S. MCDONOUGH, and ERNESTO CARAFOLI

The Ca$^{2+}$-pumping ATPase of cell plasma membranes is a ubiquitous and critical element in the control of intracellular Ca$^{2+}$. The concentration of free Ca$^{2+}$ within cells is kept very low (about 0.1 $\mu M$) by this and other Ca$^{2+}$-pumps, exchangers, and channels. The low Ca$^{2+}$ concentration allows the free Ca$^{2+}$ concentration to change by a large factor with the movement of small total amounts of Ca$^{2+}$. Thus, control of intracellular processes is accomplished with a small number of transporting systems per cell. This is favorable for the cell's function, but it makes these systems difficult to study, because of their low concentration.

The first studies on purified ion pumps were carried out in specialized situations in which the pump was present in high concentration. In such cases, the application of the standard techniques of enzymology adapted for use in membranes were successful.[1,2] However, the pump which moves Ca$^{2+}$ across the plasma membrane usually constitutes only 0.1% of the total membrane protein. Attempts to purify this pump by standard techniques were not successful; such a purification became a real possibility only with the discovery that calmodulin was a regulator of the ATPase.[3,4] It regulates the ATPase by interacting directly with it. Thus, when covalently coupled to Sepharose, calmodulin is a very desirable material for purification of the ATPase by affinity chromatography.[5] It not only binds the ATPase tightly and specifically in the presence of Ca$^{2+}$, but it also releases it easily when free Ca$^{2+}$ is reduced to low levels by addition of ethylenediaminetetraacetic acid (EDTA) or ethylene glycol bis($\beta$-aminoethyl ether)-$N,N'$-tetraacetic acid (EGTA). The methods described here were introduced by our laboratories to purify this ATPase from human erythrocytes; by use of calmodulin-Sepharose we were able to obtain an active enzyme which retained the properties of the native en-

---

[1] P. L. Jorgensen, this series, Vol. 32, pp. 277–290.
[2] D. H. MacLennan, this series, Vol. 32, pp. 291–302.
[3] H. W. Jarrett and J. T. Penniston, *Biochem. Biophys. Res. Commun.* **77**, 1210 (1977).
[4] R. M. Gopinath and F. F. Vincenzi, *Biochem. Biophys. Res. Commun.* **77**, 1203 (1977).
[5] R. K. Sharma, W. A. Taylor, and J. A. Wang, this series, Vol. 102, pp. 210–219.

zyme.[6,7] Calmodulin-Sepharose chromatography is also effective when used on extracts of the plasma membranes of nerve and muscle cells. Since the $Ca^{2+}$-pump is approximately as abundant in the erythrocyte membrane as in other plasma membranes, and pure erythrocyte membranes can be made easily in large quantities, erythrocytes are the system of choice for initial studies.

The plasma membrane $Ca^{2+}$-ATPase of erythrocytes resembles the other ion-transporting ATPases of the $E_1$-$E_2$ type in its formation of a high-energy phosphorylated intermediate and in the general mechanism by which it operates. However, it differs in size, modes of regulation, and physiological function. The topic of plasma membrane $Ca^{2+}$-pumps has recently been reviewed.[8,9]

Prior to preparation of the ATPase, both the membranes and calmodulin-Sepharose must be prepared. All operations must be performed at 0–5°, using cold solutions, refrigerated centrifuges, ice buckets, and cold rooms as appropriate.

Preparation of Erythrocyte Membranes

The method of preparation of the membranes does not appear to influence the product greatly; membranes prepared by methods similar to those described here have also yielded good preparations of ATPase.[6,10,11] It is important to make the membranes free of calmodulin by having EDTA or EGTA present during the washing steps and by washing the membranes thoroughly. Either stored or fresh blood may be used; in either case, the erythrocytes are centrifuged at low speed (5 min at 4000 g), the plasma poured off, and the light-colored layer on the top of the erythrocytes removed by aspiration. The cells are then washed two more times in the same way, using isotonic phosphate buffer to suspend them. The isotonic buffer is prepared by taking stock solutions of $0.155\ M$ $NaH_2PO_4$ and $0.103\ M$ $Na_2HPO_4$; 864 volumes of the $Na_2HPO_4$ solution are then mixed with 136 vol of the $NaH_2PO_4$ solution. The final pH is adjusted to $7.40 \pm 0.05$ at 0–5° by addition of concentrated NaOH or phosphoric acid solutions. The membranes may be prepared using normal

[6] V. Niggli, J. T. Penniston, and E. Carafoli, J. Biol. Chem. **254**, 9955 (1979).

[7] E. Graf, A. K. Verma, J. P. Gorski, G. Lopaschuk, V. Niggli, M. Zurini, E. Carafoli, and J. T. Penniston, Biochemistry **21**, 4511 (1982).

[8] J. T. Penniston, in "Calcium and Cell Function" (W. Y. Cheung, ed.), Vol. 4, pp. 99–149. Academic Press, New York, 1983.

[9] E. Carafoli and M. Zurini, Biochim. Biophys. Acta **683**, 279 (1982).

[10] K. Gietzen, M. Tejcka, and H. U. Wolf, Biochem. J. **188**, 81 (1980).

[11] D. R. Nelson and D. J. Hanahan, Arch. Biochem. Biophys. **236**, 720 (1985).

centrifugal techniques[12] but the processing of large amounts of membranes is more rapid when a continuous flow filter system such as the Millipore Pellicon cassette system is used (Millipore Co., Bedford, MA). This system also reduces the need for centrifuges. A description of the use of this system to make erythrocyte membranes has been published.[13] The cassette system is used with 5 square feet of HVLP000C5, a membrane with 0.5-$\mu$m pore size.

The filter system is placed in a cold room or refrigerator the day before the preparation, so that it is at 4° before use. The tubing in the pump is rotated or replaced regularly to lessen wear and prevent breakage during the run. Before use, the cassette is flushed with 2 liters of cold water, followed by 2 liters of hypotonic phosphate buffer [the hypotonic buffer is prepared by mixing 2 volumes of the isotonic buffer, described above, with 29 vol of cold water; the solution is made 0.1 m$M$ in phenylmethylsulfonyl fluoride (PMSF) and 1 m$M$ in benzamidine; the final pH is adjusted to 7.40 ± 0.05 at 0–5°]. About 1 liter of washed packed erythrocytes is measured into a large graduated cylinder. These packed erythrocytes are then poured, with stirring, into 1 liter of hypotonic buffer per 100 ml of packed erythrocytes. Stirring is continued for about 5 min before washing is started, and is continued during the washing procedure. To keep the volume of solution to a manageable size, the lysis is usually performed in two or three batches, using a 4-liter plastic beaker. This beaker is not only deeply imbedded in ice, but ice cubes (made by freezing hypotonic buffer) are placed in the hemolysate. The hemolysate is continually stirred by means of a heavy-duty magnetic stirrer. The tubing is coiled into two or three loops after it exits from the pump and these loops are also buried in ice, since the pumping action warms the solution. These precautions must be taken in order to keep the solutions as cold as possible. If they are taken, the procedure can be performed on an ordinary laboratory bench.

Initially, the membranes are not only washed, but concentrated: the cassette system is set up in the "recirculating mode," which involves the discard of all of the filtered material to the drain and the recirculation of the retained material back into the vessel containing the hemolysate. The manufacturer's instructions show photographs of the apparatus set up in the recirculating and constant-volume modes. The temperature of the solution as it flows back into the hemolysate should be measured, and should not be above 4°. A small amount of back pressure is applied to the retentate side of the system to facilitate filtration (about 5 psi of pressure

[12] D. J. Hanahan and J. E. Ekholm, this series, Vol. 31, pp. 168–172.
[13] T. L. Roseberry, J. F. Chen, M. M. L. Lee, T. A. Moulton, and P. Onigman, *J. Biochem. Biophys. Methods* **4**, 39 (1981).

is used). The total volume of the retained hemolysate is gradually reduced to 0.5–0.8 liter. At this point, the concentrated hemolysate is transferred to a 1-liter filter flask (with sidearm) which is hooked up in the constant-volume mode. In this mode, the vessel containing the hemolysate is still deeply embedded in ice, but now the vessel is stoppered and the sidearm is connected by a piece of tubing to 28 m$M$ Tris-HCl, 0.1 m$M$ PMSF, 1 m$M$ benzamidine, 1 m$M$ EDTA-Tris, pH 7.4, in a beaker also embedded in ice. The stoppering of the flask (with the tubings being passed through the stopper using glass tubing) causes a vacuum which automatically draws enough buffer into the flask to replace that which escapes as filtrate. If this were not done, the hemolysate would soon become unmanageably thick. The membranes are then washed with 8 liters of this medium, with continuing use of back pressure at the retentate side of the filter system. Finally, the membranes are washed with 2 liters of 0.1 m$M$ PMSF, 1 m$M$ benzamidine, 28 m$M$ Tris-HCl, pH 7.4, to remove EDTA. At the end of this wash, the tubing is removed from the "make-up" buffer and the volume of the retained solution is reduced to 500–600 ml by continued filtration. The supply tubing is lifted from the membrane suspension and the membranes pumped out of the filter system back into the filter flask. Then the membranes remaining in the filter system are flushed out by pumping in 100–200 ml of 28 m$M$ Tris-HCl, pH 7.4, directly into the cassette, with the retentate being collected in the filter flask. During this flushing, the cassette is rocked back and forth to allow complete removal of the membranes. This is done twice, with the total volume kept below 1000 ml. The membrane suspension is concentrated further by centrifuging for 20 min at 100,000 $g$. The membranes may be stored in liquid nitrogen indefinitely until needed.

In order to maintain efficiency of the cassette, the system should be cleaned *immediately* after use, following the manufacturer's instructions. Isotonic NaCl is used, followed by 0.1 $M$ sodium hydroxide and large volumes of deionized water. Every 3–4 runs an additional wash using 0.1% Triton X-100 is done to remove stubborn protein build-up in the membrane; 30–40 liters deionized water is used to completely remove the detergent. If the system is not to be used for more than 1 week, 0.02% sodium azide in water is pumped in and the system stored with this solution in it.

## Preparation of Calmodulin from Bovine Brain

The preparation of calmodulin-Sepharose columns requires large amounts of calmodulin. For this reason, it is more economical to prepare calmodulin from bovine brain than it is to buy it, at least at 1988 prices.

Any pure preparation of calmodulin can be used; we find that the following procedure produces about 50 mg of calmodulin from 1 kg of bovine brain relatively quickly and easily.

One kilogram of brain is cut into small chunks, thawed in isotonic saline solution, and rinsed free of hemoglobin and membranous materials. The brain is then homogenized for 90 sec with about 1 liter of cold 50 m$M$ Tris-HCl, pH 7.0, at room temperature, containing 1 m$M$ EDTA and 0.43 m$M$ PMSF, using a Waring blender. The temperature is kept below 10°. This homogenate is then centrifuged at 20,000 $g$ for 30 min and the supernatant saved. The pellet is rehomogenized in 1 liter of the same medium and centrifuged again the same way. The second supernatant is pooled with the first, while the pellet is discarded.

Diethylaminoethyl (DEAE)-cellulose previously equilibrated in 50 m$M$ Tris-HCl, pH 7.0, 50 m$M$ NaCl is stirred into the supernatant and allowed to stand for about 30 min. This is done using 250 g dry weight of DEAE-cellulose/kg brain; the dry weight of DEAE-cellulose can be estimated from the weight of a wet cake with excess water pulled off by vacuum in a Büchner funnel; such a cake weighs 425 g for each 100 g dry weight. The cellulose is filtered in a Büchner funnel using vacuum and washed extensively with the same buffer until free from hemoglobin. This requires 3–5 liters added in 1.5-liter portions; always pull the cake dry after each addition of buffer. Resuspend the DEAE-cellulose in 1 liter of 50 m$M$ Tris-HCl containing 150 m$M$ NaCl, pull dry, and repeat; this step removes loosely bound impurities. Calmodulin can now be eluted from the DEAE-cellulose by resuspending in 1.2–1.4 liters of the same buffer containing 600 m$M$ NaCl, pulling dry, and saving the eluate.

The pH of the 600 m$M$ NaCl eluate is adjusted to 7.0 with NH$_4$OH and ammonium sulfate added to 50% saturation (313 g added to one liter of eluate). Ammonium sulfate powder is added in small quantities, over a period of 15–20 min, to ensure solution before each addition. The pH of the solution is checked again, adjusted to 7.0, and the solution allowed to stir for 1 hr. The whitish precipitate is removed by spinning at 20,000 $g$ for 30 min at 4°. The supernatant is adjusted to pH 4.0 with a minimum volume of H$_2$SO$_4$ in 50% (NH$_4$)$_2$SO$_4$, and allowed to stir for 1 hr. The precipitate containing calmodulin is collected by centrifuging as above.

The precipitate is dissolved in 20 m$M$ $N$-tris(hydroxymethyl)methyl-2-aminoethanesulfonic acid–triethanolamine (TES–TEA), pH 7.4, 300 m$M$ NaCl, 1 m$M$ 2-mercaptoethanol buffer (buffer A). The solution is made 1.0 m$M$ in CaCO$_3$ and centrifuged at 100,000 $g$ for 30 min at 4°. The supernatant is then applied to a 2-chloro-10-(3-aminopropyl)phenothiazinyl(CAPP)-Sepharose column which has been previously equilibrated in the same buffer. Half of this supernatant is applied to a 30-ml column containing approximately 100 mg CAPP with a 35-ml plastic syringe serving as column; the flow rate is 1.5 ml/min and absorbance is monitored at 280 nm. The column is then washed with 30 ml buffer A containing 1 m$M$ Ca$^{2+}$ followed by buffer A containing 50 $\mu M$ Ca$^{2+}$ until the optical density at 280 nm returns to the baseline level (about 120 ml).

Calmodulin is eluted with 2 m$M$ EGTA in buffer A. The pooled fractions containing calmodulin are immediately made 2.0 m$M$ in CaCO$_3$. The solution is dialyzed extensively against H$_2$O, then lyophilized and stored at −20°. The column is washed with 50 ml 2 $M$ guanidine-HCl, then washed with the 1 m$M$ Ca$^{2+}$ buffer (containing 0.02% NaN$_3$ for long storage) until the next use.

The calmodulin may be coupled to Sepharose 4B by a standard method such as that described in a previous volume of this series.[5]

## Isolation of Ca$^{2+}$-ATPase

This procedure can be done on a scale suited to the needs of the investigator; we will describe the preparation utilizing the membranes derived from 4 liters of packed erythrocytes. This will yield 2–4 mg of purified Ca$^{2+}$-ATPase if human erythrocytes are used; porcine erythrocytes can give five times as much purified ATPase. All of the procedures described here can also be applied to porcine erythrocytes. The stored membranes from four preparations of membranes (each preparation will yield about 350 ml of packed membranes) are thawed, mixed, and the protein concentration determined using the method of Lowry *et al.* as modified.[14] The membrane suspension is made 0.3 $M$ in KCl and centrifuged at 100,000 $g$ for 20 min at 4°. These frozen and thawed membranes pack much more tightly than do fresh membranes, so that the final volume of the pellet will be in the vicinity of 500 ml. The supernatant is discarded and enough buffer B is added so that the final membrane protein concentration is 8 mg/ml. Concentrated solutions of CaCO$_3$ and MgCl$_2$ are added to make the suspension 1 m$M$ in MgCl$_2$ and 0.1 m$M$ in CaCO$_3$.

The composition of buffer B is as follows: 10 m$M$ TES–TEA, pH 7.4, 300 m$M$ KCl, 300 m$M$ sucrose, 2 m$M$ dithiothreitol, 1 m$M$ benzamidine, 0.1 m$M$ PMSF, and 0.1% phosphatidylcholine (Sigma P8640). The phosphatidylcholine is prepared as a suspension in water by stirring the lipid in water overnight, then sonicating intermittently with a Branson sonicator (standard tip) using twelve 5-sec bursts. The beaker containing the lipid is embedded in ice during the sonication.

Other lipids can be used in the preparation; the lipid recommended

---

[14] A. Bensadoun and D. Weinstein, *Anal. Biochem.* **70**, 241 (1976).

here is chosen because it is inexpensive, and contains enough acidic lipid (as an impurity) to keep the enzyme active yet still responsive to calmodulin. Omission of the lipid during the preparation produces an inactive enzyme, and also causes a variable yield of purified protein. The sucrose may be omitted, but the resulting enzyme may be unstable to freezing and thawing.

The extraction of $Ca^{2+}$-ATPase from the membranes is done in batches, so that the enzyme is not exposed to Triton X-100 for longer than 1 hr before the enzyme is bound to the calmodulin-Sepharose column. Therefore, the size of the batch should be determined by the size of the column and the flow rate to be used. Enough of a concentrated solution of Triton X-100 in water is mixed with a portion of the membrane suspension to bring the final Triton concentration to 0.4%. The mixture is allowed to stand on ice for 10–13 min, after which it is centrifuged at 100,000 $g$ for 20 min at 4°.

The supernatant from the centrifugation is immediately applied to a calmodulin-Sepharose affinity column previously equilibrated in buffer B containing 1 m$M$ MgCl$_2$, 50 $\mu M$ CaCO$_3$, and 0.05% Triton X-100. The column used is about 75 ml in volume (5 cm inside diameter by 4 cm high) and contains 150 mg of calmodulin. The capacity of calmodulin-Sepharose columns is less than calculated, presumably because much of the calmodulin bound to the column is inaccessible to the ATPase. The flow rate of the column is 5 ml/min, and about 2.5 hr are required to load the supernatant. After all of the supernatant has been passed into the column, the column is washed with buffer B containing 1 m$M$ MgCl$_2$, 20 $\mu M$ CaCO$_3$, and 0.05% Triton X-100. A minimum of 10 column volumes should be passed through the column. In cases where freedom of the enzyme from impurities is especially important, a more prolonged wash may be used; this gives a somewhat purer enzyme, although probably at the cost of high specific activity. In preparing the enzyme for protein structure work, we use an overnight wash of 4 liters. A final wash with buffer B containing MgCl$_2$, 1 $\mu M$ CaCO$_3$, and 0.05% Triton X-100 is useful to eliminate Mg$^{2+}$ and most of the Ca$^{2+}$ before elution of the enzyme; we use 500 ml of this washing medium.

The $Ca^{2+}$-ATPase is finally released from the column with 1 m$M$ EDTA and 0.05% Triton X-100 in buffer B. Elution of the enzyme should be monitored either by UV absorbance at 280 nm or by assay of enzyme activity, using the coupled enzyme assay. The enzyme usually elutes after about 20 ml of the elution medium has been introduced into the column and its elution is complete after about 120 ml. The appropriate fractions from the column are pooled, the protein concentration determined,[14] and the enzyme stored in liquid nitrogen. Sodium dodecyl sulfate (SDS)-polyacrylamide gel electrophoresis of the pooled fractions containing Ca$^{2+}$-dependent ATPase activity shows that the enzyme contains a single 138-K polypeptide chain. The measures described here can give a pure preparation such as is shown in Fig. 1. Overnight washing of the column and strict maintenance of low temperatures are necessary to obtain this result. Sometimes, the band at 138K is accompanied by a 124-K band. Further work has shown this to be a proteolytic product of the ATPase, probably split off by endogeneous membrane-bound proteases. A band of $M_r$ between 250K and 300K is also frequently seen in SDS gels. It probably corresponds to the dimer of the purified ATPase.[15]

Purified enzyme stored in liquid nitrogen keeps at least 3–4 months; longer storage is possible, but enzyme activity may be reduced or absent after extremely long storage. The usual concentration of enzyme obtained is 30–70 $\mu$g/ml. The specific activity of the enzyme may vary substantially from preparation to preparation. This may be in part due to the instability of the enzyme in high concentrations of Triton, but it also may be related to the variability of the activity in different batches of the original membranes. This variability has been commented on by many investigators,[16] and may be due to isozymes or regulatory effects on the enzyme. Low enzyme activity does not appear to be associated with the presence of protein impurities.

### Reconstitution of the ATPase into Liposomes

This may be done rather simply by mixing the enzyme with lipid dissolved in Triton, and removing the Triton using a column of polystyrene beads (Bio-Beads SM-2, Bio-Rad Corp., Richmond, CA). The properties of the enzyme depend strongly on the type of lipid used. Pure phosphatidylcholine produces a relatively inactive enzyme which is strongly activated by calmodulin; use of phosphatidylserine produces highly active enzyme which is not activated by calmodulin.[15] Mixtures of phosphatidylcholine with small amounts of phosphatidylinositol 4,5-bisphosphate (PIP$_2$) also produce an activated enzyme. PIP$_2$ is broken down in response to the action of agonists which mobilize Ca$^{2+}$; variations in its concentration may provide a physiological basis for regulation of this enzyme.[17]

The column is prepared by taking about 15 ml of Bio-Beads in a beaker; the beads are washed with about 1 liter of water and then equili-

[15] V. Niggli, E. S. Adunyah, J. T. Penniston, and E. Carafoli, *J. Biol. Chem.* **256,** 395 (1981).
[16] M. Reinila, E. MacDonald, N. Salem, M. Linnoila, and E. G. Trams, *Anal. Biochem.* **124,** 19 (1982).
[17] D. Choquette, G. Hakim, A. G. Filoteo, G. A. Plishker, J. R. Bostwick, and J. T. Penniston, *Biochem. Biophys. Res. Commun.* **125,** 908 (1984).

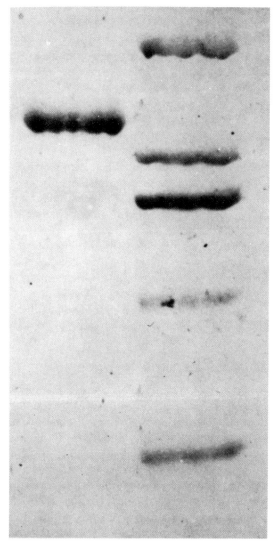

FIG. 1. A Coomassie Blue-stained electrophoresis pattern showing erythrocyte $Ca^{2+}$-ATPase purified from the methods described. (Reproduced by permission.[7])

brated with buffer containing 20 m$M$ HEPES, pH 7.4, 50 $\mu M$ MgCl$_2$, 250 m$M$ KCl. The equilibrated beads are poured into a column of 0.5 to 0.75-cm diameter, which is kept at 4°.

The lipid suspension is prepared by dissolving 15 mg of the desired lipid(s) in chloroform. The chloroform is then removed by flowing dry nitrogen into the flask while swirling in a manner which produces a film of lipid covering the lower half of the flask. After drying, the nitrogen is flowed through the flask for 10 min more, to remove residual chloroform.

To the flask is added 150 $\mu$l 20% Triton X-100, 150 $\mu$l water, 172.5 $\mu$l 3 $M$ KCl, 5 $\mu$l 1 $M$ HEPES, pH 7.4, 4 $\mu$l 500 m$M$ NaCl. The flask is flooded once again with dry nitrogen and sealed. The solution is swirled over the dried lipids until they are fully hydrated and lipid is no longer evident on the sides of the flask. Then the suspension is clarified by placing the flask in a low-powdered bath sonicator containing ice and water and operating it for three 5-min periods with 2-min breaks between the sonication periods. The turbidity of the sonicated lipids should be reduced dramatically after sonication. The lipid suspension should be kept on ice as much as possible.

To 15 mg of the lipid in suspension, 1 ml of the pump protein as eluted from the calmodulin–Sepharose column is added. The protein concentration of this mixture should be 30–100 $\mu$g/ml; the figure of 3–10 $\mu$g/ml quoted previously[17] was a typographical error. The mixture is well mixed by swirling and is immediately applied to the Bio-Bead column. The column is eluted with the buffer used to equilibrate the column, at a flow rate of 6–7 ml/hr; fractions are collected every 5 min. The fractions containing the liposomes will be easily detectable by eye, by their turbidity; liposomes usually come off at fractions 10–13. The three most concentrated fractions are pooled; if more than three fractions look approximately equally concentrated, the earlier ones (which will contain the smaller liposomes) are chosen. Bio-Beads SM-2 can be reused after regeneration following the manufacturer's recommendations.

Another method of reconstitution of the ATPase, cholate dialysis,[18] gives vesicles which are very impermeable to Ca$^{2+}$. This impermeability to Ca$^{2+}$ is demonstrated by the fact that the splitting of ATP slows down very rapidly (due to accumulation of intraliposomal Ca$^{2+}$) when the ATPase is measured in the absence of a divalent cation ionophore. Addition of divalent cation ionophore causes resumption of rapid ATP splitting activity.[19] Reconstitution is carried out as follows: about 25 mg of phospholipid is dispersed in 100 $\mu$l of 400 m$M$ sodium cholate, pH 7.0, and 400 $\mu$l of a buffer containing 120 m$M$ KCl, 20 m$M$ HEPES-NaOH, 1.2 m$M$ MgCl, and 60 $\mu M$ CaCl$_2$. This mixture is sonicated to clarity at 0° (2–3 min), and the cholate-dispersed lipid is then added to 100 $\mu$l of the purified ATPase (use the ATPase-containing column eluate directly). After mixing on a vortex mixer, the mixture is dialyzed at 4° for 16 hr against 1000 vol of 130 m$M$ KCl, 20 m$M$ HEPES-NaOH, pH 7.2, 7 m$M$ dithiothreitol, and

[18] Y. Kagawa and E. Racker, *J. Biol. Chem.* **246**, 5477 (1971).

[19] A. Clark and E. Carafoli *Cell Calcium* **4**, 83 (1983).

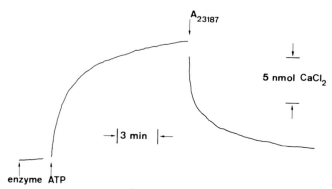

FIG. 2. $Ca^{2+}$ transport by purified $Ca^{2+}$-ATPase reconstituted into asolectin liposomes. $Ca^{2+}$ transport was measured using arsenazo III and the enzyme was reconstituted into liposomes by the cholate dialysis method. (Reproduced by permission.[15])

50 $\mu M$ $MgCl_2$. The tightest liposomes can be obtained with the crude lipid mixture "asolectin"; in this case stimulation by calmodulin is not seen, due to the presence of the acidic phospholipid cardiolipin in asolectin.

### Assays

The ATPase activity of this enzyme can be measured either in the "solubilized" state or after reconstitution into liposomes. A coupled enzyme assay[15] may be carried out spectrophotometrically, and is good for measuring initial rates, and for experiments which require points to be taken at many different concentrations of a single variable, such as free $Ca^{2+}$. In such an experiment, a number of points of different concentrations can be measured on a single sample. Addition of the divalent cation ionophore A23187 allows recycling of $Ca^{2+}$ across the bilayer and makes splitting of ATP linear for a longer time. When large numbers of individual samples are unavoidable, any of the numerous assays based on release of inorganic phosphate from ATP can be utilized.[20]

$Ca^{2+}$ transport can be measured in reconstituted liposomes using a $Ca^{2+}$ electrode, arsenazo III, or uptake of $^{45}Ca^{2+}$ (Fig. 2).[15]

### Other Plasma Membranes

Variations on this procedure allow the purification of similar ATPases from plasma membranes of rat brain,[20] dog heart,[21] pig smooth muscle,[22]

[20] G. Hakim, T. Itano, A. K. Verma, and J. T. Penniston, *Biochem. J.* **207**, 225 (1982).
[21] P. Caroni and E. Carafoli, *J. Biol. Chem.* **256**, 3263 (1981).
[22] F. Wuytack, G. deSchutter, and R. Casteels, *Biochem. J.* **198**, 265 (1981).

and skeletal muscle.[23] Plasma membranes from some other cell types have Ca²⁺-ATPases that do not appear to be calmodulin-regulated, and, therefore, are not purifiable by techniques of the sort utilized here. These tissues include liver[24] and corpus luteum.[25]

## Other Modes of Regulation

In addition to the regulation of lipid mentioned above, Ca²⁺-ATPase can also be activated by calmodulin,[3,4] by proteolysis,[26] and by phosphorylation by means of a cAMP-dependent protein kinase.[23,27] All of these mechanisms may provide means for physiologically regulating the pump.

## Acknowledgment

Supported in part by NIH grant GM 28835 and by Swiss National Funds Grant 3.189-0.82.

[23] M. Michalak, K. Famulski, and E. Carafoli, *J. Biol. Chem.* **259**, 15540 (1984).
[24] K.-M. Chan and K. D. Junger, *J. Biol. Chem.* **258**, 4404 (1983).
[25] A. K. Verma and J. T. Penniston, *J. Biol. Chem.* **256**, 1269 (1981).
[26] M. Zurini, J. Krebs, J. T. Penniston, and E. Carafoli, *J. Biol. Chem.* **259**, 618 (1984).
[27] P. Caroni and E. Carafoli, *J. Biol. Chem.* **256**, 9371 (1981)

# [28] Isolation and Reconstitution of Ca²⁺-Pump from Human and Porcine Platelets

## By WILLIAM L. DEAN

There is considerable evidence showing that Ca²⁺ plays an important role in the control of platelet function.[1,2] ATP-dependent Ca²⁺ transport has been observed in internal platelet membranes by several groups,[3-5]

[1] Abbreviations used: BHT, 2,6-di-*tert*-butyl-*p*-cresol; EDTA, ethylenediaminetetraacetic acid; EGTA, ethylene glycol bis($\beta$-aminoethyl ether)-$N,N'$-tetraacetic acid; HEPES, $N$-2-hydroxyethylpiperazine-$N'$-2-ethanesulfonic acid; PMSF, phenylmethylsulfonyl fluoride; SDS, sodium dodecyl sulfate; TES, $N$-tris(hydroxymethyl)methyl-2-aminoethanesulfonic acid; TIU, trypsin inhibitor units as defined by Sigma Chemical Co.; Tris, tris(hydroxymethyl)aminomethane; SDS, sodium dodecyl sulfate.

[2] T. C. Detwiler, I. F. Charo, and R. D. Feinman, *Thromb. Haemostasis* **40**, 207 (1978).
[3] S. Menashi, F. S. Authi, F. Carey, and N. Crawford, *Biochem. J.* **222**, 413 (1984).
[4] D. O. Levitsky, V. A. Loginov, A. V. Lebedev, T. S. Levchenko, and V. L. Leytin, *FEBS Lett.* **171**, 89 (1984).
[5] J. Enouf, R. Bredoux, B. Boizard, J.-L. Wautier, H. Chap, J. Thomas, M. De Metz, and S. Levy-Toledano, *Biochem. Biophys. Res. Commun.* **123**, 50 (1984).

and this activity is clearly an important aspect of platelet $Ca^{2+}$ metabolism. We have recently demonstrated that a $Ca^{2+}$-ATPase (EC 3.6.1.3, ATP phosphohydrolase) in platelet membranes is immunochemically[6] and functionally[7] related to the sarcoplasmic reticulum $Ca^{2+}$-pump from skeletal muscle.

## Methods

### Preparation of Crude Human and Porcine Platelet Membranes

#### Reagents

Anticoagulant: 0.03 $M$ citric acid, 0.085 $M$ sodium citrate, 0.11 $M$ glucose

Buffer A: 40 m$M$ glucose, 24 m$M$ NaPO₄, 4 $\mu M$ PMSF, 0.001 TIU/ml aprotinin (Sigma Chemical Co.), pH 6.8

Buffer B: 0.15 $M$ NaCl, 3 m$M$ EDTA, 4 $\mu M$ PMSF, 0.001 TIU/ml aprotinin, pH 7.0

Buffer C: 0.15 $M$ NaCl, 4 $\mu M$ PMSF, 0.001 TIU/ml aprotinin

Buffer D: 30 m$M$ KCl, 5 m$M$ MgCl₂, 10 m$M$ sodium oxalate, 20 m$M$ Tris, 0.3 $M$ sucrose, 1 m$M$ MgATP, 40 $\mu M$ PMSF, 1 m$M$ BHT, 10 $\mu$g/ml pepstatin A, 1 mg/ml chicken egg white trypsin inhibitor (Sigma Chemical Co. type II-O), 0.01 TIU/ml aprotinin, pH 7.0

Storage buffer: 0.01 $M$ TES, 0.1 $M$ KCl, 5 m$M$ MgCl₂, 1 m$M$ EGTA, 15% (v/v) glycerol, 1 m$M$ dithiothreitol, 40 $\mu M$ PMSF, 1 m$M$ BHT, 1 mg/ml trypsin inhibitor, 10 $\mu$g/ml pepstatin A, 0.01 TIU/ml aprotinin, pH 7.5

*Procedure.* Outdated platelet concentrates are obtained from the American Red Cross for isolation of crude human membranes. Each bag contains approximately 50 ml of platelet concentrate and usually 10–20 bags are processed in a single preparation. Platelets can be used up to 5 days after the stated expiration date, but longer storage generally results in significantly lower $Ca^{2+}$-ATPase activity. For fresh porcine platelets, 8 liters of blood are collected directly into a 10-liter container with 1.44 liters of citrate anticoagulant.[8] After mixing, the blood is centrifuged at 177 $g$ for 15 min to remove erythrocytes and then at 4400 $g$ for 10 min to pellet the platelets. The supernatant from this step is saved and 800 ml of the supernatant is added back to the platelet pellets for resuspension. This represents approximately the same concentration of platelets as for human platelets supplied by the American Red Cross.

[6] W. L. Dean and D. M. Sullivan, *J. Biol. Chem.* **257,** 14390 (1982).

[7] W. L. Dean, *J. Biol. Chem.* **259,** 7343 (1984).

[8] R. H. Aster and J. H. Jandl, *J. Clin. Invest.* **48,** 843 (1964).

Preparation of crude membranes is the same for human and porcine platelets and is a modification[6] of the procedure of Kaser-Glanzmann *et al.*[9] Platelet concentrates are diluted with an equal volume of buffer A. This platelet volume before addition of buffer A ($V_i$) is used to calculate the volume of buffers to be added at subsequent steps. The protease inhibitors PMSF, aprotinin, trypsin inhibitor, and pepstatin A are added to the buffers immediately prior to use. Concentrated (100-fold) stock solutions are stored at $-20°$ except for aprotinin, which is stored at $4°$. Contaminating erythrocytes are removed by a 1-min centrifugation step at 160 $g$. Platelets are then pelleted by centrifugation at 3000 $g$ and resuspended in 0.5$V_i$ of buffer B. The pellets are again centrifuged at 160 $g$ for 10 min to remove erythrocytes and then at 3000 $g$ for 5 min. The pellet is resuspended in 0.4$V_i$ of buffer C and pelleted by centrifugation at 3000 $g$ for 10 min, and again in buffer C at 0.2$V_i$ by centrifugation at 3000 $g$ for 10 min. The washed platelets are resuspended in 0.05$V_i$ of buffer D (MgATP is added immediately prior to use along with the protease inhibitors) and transferred to 45-ml plastic centrifuge tubes. The platelets are then sonicated three times for 15 sec at a setting of 6 on a Branson Cell Disruptor. Care must be taken to keep the temperature below $10°$. The sonicated platelets are then centrifuged at 13,000 $g$ for 20 min and the supernatant is removed and centrifuged for 30 min at 150,000 $g$. The platelet membranes are resuspended in storage buffer, frozen in liquid nitrogen, and stored at $-80°$. The Ca$^{2+}$-ATPase activity is stable for more than 2 months. For 10 consecutive crude human platelet membrane preparations, the Ca$^{2+}$-ATPase activity was $58 \pm 8$ nmol ATP hydrolyzed/min/mg while that for 12 porcine preparations was $26 \pm 4$. In a typical membrane preparation from 10 bags of outdated human platelets or from 8 liters of porcine blood, 30 mg of crude membrane protein is obtained.

*Assay of Ca$^{2+}$-ATPase Activity and Ca$^{2+}$ Transport*

*Reagents*

10-fold concentrated assay buffer: 100 m$M$ TES, 1 $M$ KCl, 50 m$M$ MgCl$_2$, pH 7.5
0.1 $M$ MgATP
4.2 m$M$ phosphoenolpyruvate
10 m$M$ NADH
Pyruvate kinase (750 IU/ml) and lactate dehydrogenase (1800 IU/ml) in 60% (v/v) glycerol

---

[9] R. Kaser-Glanzmann, M. Jakabova, J. N. George, and E. F. Luscher, *Biochim. Biophys. Acta* **466**, 429 (1977).

1 m$M$ arsenazo III, treated with Chelex 100 (Bio-Rad)[10] to remove $Ca^{2+}$, the extinction coefficient is 2.8 × 10⁻⁴ $M^{-1}$ in 5 m$M$ $Ca^{2+}$

10 m$M$ CaCl₂

10 m$M$ EGTA, pH 7.5

1 $M$ KPO₄, pH 7.5

*Procedure.* $Ca^{2+}$-ATPase activity is measured using a coupled ATPase assay where ATP hydrolysis is linked to NADH oxidation.[11] The assay mixture is prepared by diluting concentrated assay buffer 1 : 10 in a final volume of 1 ml containing 0.42 m$M$ phosphoenolpyruvate, 5 m$M$ MgATP, 0.2 m$M$ NADH, 7.5 IU of pyruvate kinase, and 18 IU of lactate dehydrogenase. After preincubation at 30° for 2 min, the rate of NADH oxidation in the presence of 1 m$M$ EGTA is subtracted from the rate in the presence of 0.1 m$M$ CaCl₂, and the resulting rate is converted to ATP hydrolysis using the extinction coefficient for NADH of 6.22 m$M^{-1}$ (1 mol ATP hydrolyzed/mol NADH oxidized). The reaction is started by addition of the membrane sample.

$Ca^{2+}$ transport is measured with the $Ca^{2+}$-specific dye arsenazo III. The assay mixture is prepared by diluting concentrated assay buffer 1 : 10 in a final volume of 1 ml containing 5 m$M$ MgATP, 50 m$M$ KPO₄, and 20 $\mu M$ arsenazo III. The assay is performed in a double-beam spectrophotometer using 20 $\mu M$ arsenazo III in assay buffer as a reference. The response of arsenazo III at 654 nm to 6–10 nmol additions of CaCl₂ (final $[Ca^{2+}]$ is 60 $\mu M$) is recorded prior to addition of the platelet membranes. The reaction is started by addition of the platelet membrane sample after preincubation at 30° for 2 min. The rate of $Ca^{2+}$ uptake is determined from the initial calibration.

Protein concentration is determined by the method of Lowry *et al.*[12] using bovine serum albumin as a standard unless detergent is present. In this case, the samples are first precipitated with trichloroacetic acid after addition of deoxycholate as described by Bensadoun and Weinstein.[13]

*Purification of Human and Porcine Platelet Membranes*

*Reagents*

0.01 $M$ TES, 0.1 $M$ KCl, pH 7.5

80% glycerol (w/v) in 0.01 $M$ TES buffer containing 0.1 $M$ KCl, concentration determined by refractive index

[10] N. C. Kendrick, R. W. Ratzlaff, and M. P. Blaustein, *Anal. Biochem.* **83**, 433 (1977).

[11] W. L. Dean and C. Tanford, *J. Biol. Chem.* **252**, 3551 (1977).

[12] O. H. Lowry, N. J. Rosebrough, A. L. Farr, and R. J. Randall, *J. Biol. Chem.* **193**, 265 (1951).

[13] A. Bensadoun and D. Weinstein, *Anal. Biochem.* **70**, 241 (1976).

0.1 *M* MgATP
0.1 *M* dithiothreitol
4 m*M* PMSF
10 TIU/ml aprotinin

*Procedure.* Discontinuous gradients are prepared in 1.5 × 10.2 mm polycarbonate tubes compatible with the Beckman SW-28 swinging bucket rotor. Just prior to centrifugation, glycerol solutions are made from the 80% (w/v) stock solution by dilution with 0.01 *M* TES buffer, pH 7.5, containing 0.1 *M* KCl. Ten microliters each of the stock solutions of MgATP, dithiothreitol, PMSF, and aprotinin are added per milliliter of glycerol solution (referred to as gradient buffer). For human platelets, 2.5 ml each of 24, 37, 47, and 56% (w/v) glycerol are added to the centrifuge tube and 2 ml of crude platelet membranes is layered on top. For porcine platelets, the procedure is the same except that the four glycerol concentrations are 26, 44, 56, and 73% (w/v). The gradients are centrifuged for 3 hr at 26,000 rpm in the SW-28 rotor. For human platelets the membranes at the interfaces between 24 and 37% and 37 and 47% glycerol are collected with a Pasteur pipette, diluted with an equal volume of water, and pelleted by centrifugation at 150,000 *g* for 30 min. For porcine membranes the material at the interface between 44 and 56% glycerol is collected and treated as for human platelet membranes. Pellets are resuspended in one-tenth the original membrane volume of gradient buffer containing 20% (v/v) glycerol and the specified protease inhibitors. Preparation of purified membranes usually yields a 2- or 3-fold purification with 10–30% recovery of activity. Thus from 30 mg of crude membrane protein, 2–3 mg of purified membrane protein is usually obtained (see Table I).

*Solubilization, Purification, and Reconstitution of the Ca²⁺-ATPase*

*Reagents*

100 mg/ml brain lipid extract (Sigma Chemical Co., type I) in 2 : 1 chloroform–methanol (prepared by dissolving 200 mg/ml lipid in chloroform–methanol and removing insoluble material by centrifugation. Concentration is determined by a dry weight measurement)
100 mg/ml octylglucoside
100 mg/ml Triton X-100
Sepharose 4B (Pharmacia)
Hydroxylapatite (Bio-Rad HTP)
Affi-Gel Blue (Bio-Rad)

*Human Platelet Membranes.* We have developed two procedures for purification of the human platelet Ca²⁺-ATPase. The first, which employs octylglucoside as the detergent,[7] is most suitable when reconstitution of

Ca$^{2+}$-pumping activity is the goal. The second procedure, which employs Triton X-100 as the detergent, is more rapid and is the method of choice when reconstitution is not a goal.

For the octylglucoside procedure, purified membranes are treated with octylglucoside below the critical micelle concentration to remove soluble proteins trapped within the membrane vesicles. This procedure is a modification of the method of Meissner et al.[14] The membranes are diluted to yield a final protein concentration of 2.5 mg/ml in 10 m$M$ Tris at pH 8.0 containing 0.5 $M$ KCl, 0.3 $M$ sucrose, 1.5 m$M$ MgCl$_2$, 1 m$M$ EDTA, 20 $\mu M$ CaCl$_2$, 2.5 m$M$ dithiothreitol, and 0.5 mg/ml octylglucoside. After 10 min at 4°, the membranes are pelleted by centrifugation at 150,000 $g$ for 20 min and are resuspended in one-half the initial volume of the same solution. The membranes are again pelleted and resuspended in one-half the initial volume of 5 m$M$ HEPES at pH 7.75 containing 0.3 $M$ sucrose, 0.5 $M$ KCl, 0.5 m$M$ MgCl$_2$, and 10 $\mu M$ CaCl$_2$. The membranes are pelleted as before and resuspended in one-fourth the initial volume of gradient buffer containing 40% glycerol, 10 $\mu$g/ml pepstatin A, and 1 mg/ml trypsin inhibitor.

The octylglucoside-treated membranes are then solubilized by the addition of 3 g of octylglucoside/g protein in the presence of 1 g brain extract/g protein at a protein concentration of 3 mg/ml [membranes are diluted with gradient buffer containing 20% glycerol (v/v), 1 mg/ml trypsin inhibitor, and 10 $\mu$g/ml pepstatin A]. The appropriate amount of brain lipid, dissolved in chloroform–methanol, is dried with a stream of nitrogen in a test tube, and the lipid is dissolved in the octylglucoside solution before addition to the membranes. After 10 min at 4°, the unsolubilized protein is removed by centrifugation at 50,000 $g$ for 15 min. The solubilized protein (0.5–1.0 ml containing 1–2 mg protein) is applied to a Sepharose 4B column (0.9 × 28 cm) equilibrated at 4° with gradient buffer containing 40% glycerol, 6 mg/ml octylglucoside, and 1 mg/ml brain lipid extract. The column is eluted at a flow rate of 8 ml/hr, and Ca$^{2+}$-ATPase activity is determined. The active fractions are pooled and then added to an equal volume of hydroxylapatite equilibrated with the same buffer used for Sepharose 4B chromatography. After intermittent mixing for 15 min at room temperature, the supernatant is removed and this is the final purified preparation.

The purified Ca$^{2+}$-ATPase can be reconstituted into lipid vesicles by dialysis. Reconstitution of Ca$^{2+}$-pumping activity is achieved by dialysis overnight against two changes of 0.05 $M$ KPO$_4$ at pH 7.5 containing 20%

[14] G. Meissner, G. E. Conner, and S. Fleischer, *Biochim. Biophys. Acta* **298**, 246 (1973).

glycerol, 1 m$M$ dithiothreitol, 40 $\mu M$ PMSF, and 0.01 TIU/ml aprotinin at 4°.

For the Triton X-100 procedure, purified membranes are solubilized exactly as described for the octylglucoside procedure except that Triton X-100 (10 mg/ml final concentration) is used instead of octylglucoside and the pH of the solution is 7.0. The solubilized membranes (0.5 ml) are loaded onto a 1-ml Affi-Gel Blue column equilibrated with 10 m$M$ TES buffer, pH 7.0, containing 20% (v/v) glycerol, 10 mg/ml Triton X-100, 3.3 mg/ml brain lipid extract, 5 m$M$ MgCl$_2$, 0.5 m$M$ CaCl$_2$, 1 m$M$ dithiothreitol, 0.3 TIU/ml aprotinin, 10 $\mu$g/ml pepstatin A, and 40 $\mu M$ PMSF. The column is washed with 2 ml of equilibration buffer and the Ca$^{2+}$-ATPase is eluted with an additional 12 ml of equilibration buffer.[15] The fractions containing Ca$^{2+}$-ATPase activity are pooled and stored at −20°.

*Porcine Platelet Membranes.* The Ca$^{2+}$-ATPase in purified porcine platelet membranes can be partially purified by a similar method to that used for human membranes. However, as a result of the lability of the porcine enzyme, extensive purification has not been achieved. The octylglucoside wash step is omitted and membranes instead are pelleted and resuspended three times at a final concentration of 3 mg protein/ml in 10 m$M$ TES, pH 7.5, containing 10 m$M$ MgATP, 1 m$M$ dithiothreitol, 1 m$M$ EGTA, and the protease inhibitors used in the storage buffer at the same concentrations (this is termed the low-salt wash). The washed membranes are solubilized with a 3 : 1 weight ratio of octylglucoside to protein at a protein concentration of 3 mg/ml as described for human membranes. Solubilized membranes are further purified by treatment with hydroxylapatite as described for the human Ca$^{2+}$-ATPase.

Purification of the human and porcine Ca$^{2+}$-ATPases using octylglucoside, and the human enzyme using Triton X-100 is summarized in Table I. Both procedures result in purification of the human Ca$^{2+}$-ATPase to an activity of over 1 $\mu$mol ATP hydrolyzed/min/mg protein and yield approximately 100 $\mu$g of purified ATPase from 30 mg of crude membrane protein. The procedure employing Triton X-100 and Affi-Gel blue is more rapid and yields better recovery of the enzyme. Scans of SDS-polyacrylamide gels showing the protein composition of fractions from three steps during the octylglucoside procedure are shown in Fig. 1.[16] The purified Ca$^{2+}$-ATPase obtained by both procedures always exhibits two polypeptides in the 100,000 molecular weight range and small amounts of

---

[15] A. Gafni and P. D. Boyer, *Biochemistry* **23**, 4362 (1984).
[16] K. Weber and M. J. Osborn, *J. Biol. Chem.* **244**, 4406 (1969).

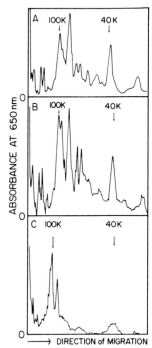

FIG. 1. SDS-polyacrylamide gel electrophoresis of purified platelet membranes and platelet $Ca^{2+}$-ATPase preparations. Samples were precipitated with 7% (w/v) trichloroacetic acid and were solubilized with 1% SDS. Electrophoresis was carried out on 7.5% polyacrylamide gels according to the procedure of Weber and Osborn.[16] After staining with Coomassie blue, gels were scanned at 650 nm. (A) 50 $\mu$g of purified human platelet membranes. (B) 20 $\mu$g of platelet $Ca^{2+}$-ATPase obtained after Sepharose 4B chromatography. (C) 10 $\mu$g of purified platelet $Ca^{2+}$-ATPase obtained after hydroxylapatite adsorption (specific activity of 1152 nmol ATP hydrolyzed/min/mg). Arrows indicating molecular weight were determined by calibration of gels with standard proteins. (From Dean.[7])

actin (40,000). Both of the major 100,000 polypeptides bind antibodies raised against rabbit skeletal muscle $Ca^{2+}$-ATPase from sarcoplasmic reticulum.[7]

Purification of the $Ca^{2+}$-ATPase from porcine platelet membranes is much more difficult because of the lability of the enzyme. However, 7-fold purification can be obtained using the octylglucoside procedure if the Sepharose 4B and octylglucoside wash steps are omitted (Table I). Approximately 450 $\mu$g of partially purified $Ca^{2+}$-ATPase are obtained from 30 mg of crude membrane protein.

$Ca^{2+}$ transport by crude and purified membranes measured during the first 20 sec of the reaction yields a maximum pump stoichiometry of 2.0 as

TABLE I

PURIFICATION OF Ca²⁺ATPase ACTIVITY FROM HUMAN AND PORCINE
PLATELET MEMBRANES

| Source/step | Specific activity (nmol ATP hydrolyzed/ min/mg) | Recovery (% of initial activity) | Purification (ratio of specific activities) |
|---|---|---|---|
| Human platelet/octylglucoside | | | |
| Crude membranes | 78 | 100 | 1 |
| Purified membranes | 194 | 20 | 2.5 |
| Octylglucoside wash | 298 | 18 | 3.8 |
| Sepharose 4B | 684 | 9 | 8.8 |
| Hydroxylapatite | 1152 | 7 | 14.8 |
| Human platelet/Triton X-100 | | | |
| Crude membranes | 65 | 100 | 1 |
| Purified membranes | 195 | 30 | 3 |
| Triton solubilized | 170 | 26 | 2.6 |
| Affi-Gel Blue | 1500 | 10 | 23.0 |
| Porcine platelets/octylglucoside | | | |
| Crude membranes | 26 | 100 | 1 |
| Purified membranes | 83 | 13 | 11 |
| Low-salt wash | 129 | 11 | 4.9 |
| Hydroxylapatite | 190 | 11 | 7.2 |

shown in Table II. Ca²⁺ continues to be transported for over an hour with maximum uptake of 0.1–0.3 $\mu$mol Ca²⁺/mg protein, but the pump stoichiometry decreases to as low as 0.1 Ca²⁺/ATP. Purified membranes exhibit a 2- to 3-fold increase in the initial rate of Ca²⁺ transport/mg protein and also exhibit continued Ca²⁺ uptake for longer than 1 hr.

Reconstitution of the human platelet Ca²⁺-ATPase purified by the octylglucoside procedure yields Ca²⁺-ATPase activities that are reduced by

TABLE II

Ca²⁺-PUMP STOICHIOMETRY FOR CRUDE AND PURIFIED HUMAN PLATELET MEMBRANES
AND RECONSTITUTED HUMAN Ca²⁺-ATPase[a]

| Source | ATP hydrolysis (nmol/min/mg) | Calcium transport (nmol/min/mg) | Stoichiometry Ca²⁺/ATP |
|---|---|---|---|
| Crude membranes | 83 ± 2 | 156 ± 14.4 | 1.9 ± 0.1 |
| Purified membranes | 182 ± 19 | 328 ± 106 | 1.8 ± 0.4 |
| Reconstituted calcium pump | 850 ± 100 | 1780 ± 200 | 2.1 ± 0.2 |

[a] Values are mean ± SD for three preparations of each source.

approximately 50% during the overnight dialysis. Initial Ca$^{2+}$ transport rates of 1.5–2.0 times the rate of ATP hydrolysis are obtained with this reconstituted preparation, as shown in Table II.

### Acknowledgments

This work was supported by USPHS grant HL 28397. The skilful assistance of Paul Eichenberger, Charlene Lordon, and Samuel Evans Adunyah is gratefully acknowledged.

# [29] Purification of Phospholamban from Canine Cardiac Sarcoplasmic Reticulum Vesicles by Use of Sulfhydryl Group Affinity Chromatography

By LARRY R. JONES, ADAM D. WEGENER, and HEATHER K. B. SIMMERMAN

Phospholamban is an integral membrane protein localized to cardiac sarcoplasmic reticulum which regulates the activity of the sarcoplasmic reticulum Ca$^{2+}$-pump in response to phosphorylation. Phospholamban in isolated sarcoplasmic reticulum vesicles is a substrate for cAMP-dependent protein kinase,[1] Ca$^{2+}$/calmodulin-dependent protein kinase,[2,3] and protein kinase C.[4,5] Phospholamban is also readily phosphorylated in sarcoplasmic reticulum in intact hearts in response to β-adrenergic stimulation.[6] Although much circumstantial evidence exists suggesting that phosphorylated phospholamban is physically associated with and modulates the activity of the sarcoplasmic reticulum Ca$^{2+}$-pump,[1] it has so far not been possible to reconstitute both proteins into liposomes and demonstrate such a functional interaction directly. This type of approach seems necessary to prove that phospholamban is indeed capable of forming a regulatory complex with the Ca$^{2+}$-pump, and is also required for detailed understanding of the molecular mechanism of stimulation of Ca$^{2+}$ trans-

[1] M. Tada and A. M. Katz, *Annu. Rev. Physiol.* **44**, 401 (1982).
[2] C. J. LePeuch, J. Haiech, and J. G. Demaille, *Biochemistry* **18**, 5150 (1979).
[3] L. R. Jones, S. W. Maddock, and D. R. Hathaway, *Biochim. Biophys. Acta* **641**, 242 (1981).
[4] Y. Iwasa and M. M. Hosey, *J. Biol. Chem.* **259**, 534 (1984).
[5] M. A. Movsesian, M. Nishikawa, and R. S. Adelstein, *J. Biol. Chem.* **259**, 8029 (1984).
[6] J. P. Lindemann, L. R. Jones, D. R. Hathaway, B. G. Henry, and A. M. Watanabe, *J. Biol. Chem.* **258**, 464 (1983).

port. A prerequisite for studying the protein in an isolated, reconstituted system is a method for preparing phospholamban from myocardium in reasonable yield and purity. Recently, we have been successful in purifying phospholamban to homogeneity from canine cardiac sarcoplasmic reticulum vesicles after extraction of the protein from the membranes with use of the anionic detergent sodium cholate.[7,8] Central to the isolation scheme is the use of sulfhydryl group affinity chromatography at the final stage of purification. The purpose of this report is to describe our isolation method in detail. Hopefully, the availability of a highly purified phospholamban preparation will be useful in future studies directed to probe more deeply into the molecular structure and function of the protein.

## Preparation of Cardiac Membrane Vesicles Enriched in Sarcoplasmic Reticulum

### Solutions

Medium I: 10 m$M$ NaHCO$_3$
Medium II: 0.6 $M$ KCl, 30 m$M$ histidine (pH 7.0)
Medium III: 0.25 $M$ sucrose, 30 m$M$ histidine (pH 7.4)

*Procedure.* Phospholamban is isolated from canine cardiac sarcoplasmic reticulum vesicles prepared as described previously.[9] All operations are performed at 4°. 180–200 g of canine left ventricle is divided into six equal portions, and each portion is homogenized three times for 30 sec in 120 ml of medium I with use of a Polytron PT-20 set at half-maximal speed. The samples are then centrifuged at 8,700 rpm for 20 min in a Beckman JA-14 rotor. The supernatants are collected and recentrifuged at 10,000 rpm for 20 min. The supernatants from this second centrifugation are next sedimented at 19,000 rpm (45,000 $g_{max}$) for 30 min in Beckman JA-20 rotors to yield pellets of crude sarcoplasmic reticulum vesicles. Contractile proteins are extracted from the crude vesicles by resuspending them in medium II and subjecting them to an identical centrifugation. The extracted pellets, enriched in cardiac sarcoplasmic reticulum membranes,[9] are resuspended in medium III to a protein concentration of approximately 10–12 mg/ml and stored frozen at −20°. Protein is determined by the method of Lowry *et al.*[10]

[7] A. D. Wegener and L. R. Jones, *J. Biol. Chem.* **259**, 1834 (1984).
[8] L. R. Jones, H. K. B. Simmerman, W. W. Wilson, F. R. N. Gurd, and A. D. Wegener, *J. Biol. Chem.* **260**, 7721 (1985).
[9] L. R. Jones and S. E. Cala, *J. Biol. Chem.* **256**, 11809 (1981).
[10] O. H. Lowry, N. J. Rosebrough, A. L. Farr, and R. J. Randall, *J. Biol. Chem.* **193**, 265 (1951).

Cholate Extraction of Phospholamban from Cardiac Sarcoplasmic
   Reticulum Vesicles and Preparation of the Reconstituted
   Particulate Fraction

Generally, eight to ten sarcoplasmic reticulum preparations are
thawed and pooled for isolation of phospholamban. However, the procedure described below can be scaled-down for use with one preparation
provided that all volumes are reduced proportionately. In the typical
isolation described here, nine preparations were pooled for isolation of
3.34 mg of pure phospholamban.

### Cholate Extraction

Nine preparations of cardiac sarcoplasmic reticulum vesicles are
thawed and poured into a beaker. The volume of the pooled membrane
vesicle suspensions is 156.6 ml and the final protein concentration is 11.1
mg/ml, as determined by the method of Lowry et al.[10] The suspension is
stirred continuously at room temperature with use of a stir bar and other
additions are then made as follows:
   Membrane vesicles (11.1 mg/ml), 156.6 ml
   0.25 M dithiothreitol, 6.94 ml
   4 M NaCl, 17.36 ml
   Medium III, 160.0 ml
   20% sodium cholate, 6.94 ml
   Total volume 347.84 ml
After addition of the various solutions, the final cholate concentration
used to extract phospholamban from membranes becomes 0.4%. Cholate
extraction occurs at a final protein concentration of 5 mg/ml, and the final
concentrations of dithiothreitol and NaCl are 5 mM and 200 mM, respectively. The pH is approximately 7.5. After stirring the membrane suspension in the presence of cholate for 10 min at room temperature, it is
poured into centrifuge tubes and sedimented at 105,000 $g_{max}$ in Beckman
fixed-angle rotors. 344 ml of supernatant is collected, which contains
solubilized phospholamban.

*Reconstituted Particulate Fraction.* Solubilized phospholamban in the
cholate extract is next batch adsorbed to calcium oxalate. 344 ml of cholate extract is poured into a beaker at room temperature and stirred continuously. 43.3 ml of 1 M CaCl$_2$ is added to the cholate extract, followed
by rapid addition of 43.3 ml of 1 M oxalic acid (adjusted to pH 7.0 with
Tris base). A white calcium oxalate precipitate forms immediately (final
concentration approximately 100 mM). The suspension is stirred for 30
min at room temperature, divided into two aliquots, and then sedimented

for 30 min at 8000 rpm with use of a Beckman JA-14 rotor. Two calcium oxalate pellets are thus obtained.

The calcium oxalate pellets, containing adsorbed phospholamban, are next dissolved with use of ethylene glycol bis($\beta$-aminoethyl ether)-$N,N,N',N'$-tetraacetic acid (EGTA). The two pellets are resuspended in 228 ml of 500 m$M$ EGTA (adjusted to pH 7.5 with Tris base), 11.4 ml of 5 $M$ NaOH, and 198 ml of $H_2O$. The pellets are dispersed and dissolved with use of the Polytron PT-20, and/or by repeated aspiration through a 14-gauge cannula. At this point the solution containing the resuspended pellets should be mostly clarified; however, a small amount of background turbidity is normal. If required, the pH is readjusted to 7.5 with use of additional NaOH.

The suspension containing the dissolved calcium oxalate pellets is next subjected to dialysis at 4° to yield the "reconstituted particulate fraction," which is enriched approximately 4- to 5-fold in phospholamban.[8] 437 ml of suspension is dialyzed overnight against 15 liters of 10 m$M$ histidine, 5 m$M$ MgCl$_2$, and 1 m$M$ EGTA (pH 7.5). An additional 24 hr of dialysis is then performed in the same medium without EGTA. The material inside the dialysis bags becomes increasingly cloudy during dialysis, and after the second period of dialysis, the contents of the dialysis bags are emptied into a beaker at room temperature. 900 ml of dialysate is collected, which contains aggregated material comprising the reconstituted particulate fraction. Peripheral proteins (including cardiac calsequestrin) are extracted from the reconstituted particulate fraction by addition of solid NaCl with stirring,[8] or in the method described here, by alkaline treatment with $Na_2CO_3$. 6.76 g of solid sodium carbonate is added to the 900 ml of reconstituted particulate fraction (final $Na_2CO_3$ concentration, 0.1 $M$), and the pH of the suspension is adjusted to 11.3 by addition of NaOH. The suspension is stirred for 30 min at room temperature, and is then sedimented at 105,000 $g_{max}$ for 30 min in Beckman fixed-angle rotors. The pellets obtained are resuspended to a total volume of 24.7 ml in medium III to yield the reconstituted particulate fraction (salt-extracted). This fraction can be stored frozen indefinitely at −20°.

Treatment of the reconstituted particulate fraction with $Na_2CO_3$ as described above inactivates the intrinsic $Ca^{2+}$/calmodulin-dependent protein kinase phosphorylating phospholamban in this fraction, but gives slightly higher yields of phospholamban when the protein is subsequently purified by sulfhydryl group affinity chromatography (unpublished observations). To retain an active $Ca^{2+}$/calmodulin-dependent protein kinase in the reconstituted particulate fraction, extraction should be with NaCl[7,8] instead of $Na_2CO_3$.

Purification of Phospholamban from the Reconstituted Particulate
Fraction by Sulfhydryl Group Chromatography

*Solutions*

Zwittergent 3-16: 10% in H$_2$O (w/v)
Zwittergent 3-14: 10% in H$_2$O (w/v)
Buffer A: 0.1% Zwittergent 3-14, 0.5 *M* NaCl, 10 m*M* 3-(*N*-morpholino)propanesulfonic acid (MOPS) (pH 7.2)
Buffer B: 0.1% Zwittergent 3-14, 10 m*M* MOPS (pH 7.2)
Buffer C: 0.1% Zwittergent 3-14, 10 m*M* MOPS, 20 m*M* dithiothreitol (pH 7.2)
Buffer D: 0.25% Zwittergent 3-14, 10 m*M* MOPS, 20 m*M* dithiothreitol (pH 7.2)

*Procedure.* 23.8 ml of Na$_2$CO$_3$-extracted reconstituted particulate fraction (protein concentration 4.2 mg/ml according to the Amido Black method[11]) is thawed and added to a centrifuge tube. 2.38 ml of 10% Zwittergent 3-16 (purchased from Calbiochem-Behring) is then added with vortexing, and after 20 min at room temperature, 1.58 ml of 1 *M* CaCl$_2$ is added, causing a white precipitate to form. The sample is then sedimented at 40,000 rpm for 30 min in a Beckman 70Ti fixed-angle rotor. The pellet obtained is resuspended in 18.2 ml of medium III at room temperature by repeated aspiration through a 14-gauge cannula. 3.6 ml of 10% Zwittergent 3-14 (purchased from Calbiochem-Behring) is next added with vortexing to solubilize phospholamban, and after 20 min at room temperature, 1.2 ml of 1 *M* CaCl$_2$ is added with additional vortexing. The suspension is then sedimented for 30 min at 40,000 rpm in a Beckman 70Ti rotor. The supernatant obtained is applied directly to a column of *p*-hydroxymercuribenzoate agarose, as described below, for purification of phospholamban to homogeneity by sulfhydryl group affinity chromatography.

Approximately 23 ml of the supernatant containing phospholamban solubilized in Zwittergent 3-14 is applied to 12 ml of *p*-hydroxymercuribenzoate agarose (purchased from Sigma Chemical Co., contains a 6-carbon spacer), which is preequilibrated with buffer B in a small column. The column is then washed four consecutive times with 12 ml of buffer A (fractions 1–4), and four consecutive times with 12 ml of buffer B (fractions 5–8). Purified phospholamban is then eluted from the column with six consecutive 12-ml washes of buffer C (fractions 9–14), followed by three consecutive 12-ml washes of buffer D (fractions 15–17). Wash fractions 9–17 are pooled and concentrated to a volume of 2.22 ml with

---

[11] W. Schaffner and C. Weissman, *Anal. Biochem.* **56**, 502 (1973).

use of an Amicon PM10 membrane. The protein concentration of purified phospholamban in this concentrate is 1.27 mg/ml, as determined by the Amido Black method.[11] After collecting 2.22 ml of the original Amicon concentrate, the PM10 membrane can be rinsed with one additional milliliter of 0.1% Zwittergent 3-14, 200 m$M$ NaCl, and 10 m$M$ MOPS (pH 7.0) to recover an additional 0.52 mg of pure phospholamban. The purified phospholamban suspensions are stored frozen at −20° and are stable indefinitely.

## Quantitation of Phospholamban by Phosphorylation with [γ-³²P]ATP and the Catalytic Subunit of cAMP-Dependent Protein Kinase

To monitor the purity and recovery of phospholamban in different fractions during its isolation, the protein is phosphorylated by [γ-³²P]ATP and the catalytic subunit of cAMP-dependent protein kinase. To one 500-U bottle of the catalytic subunit of cAMP-dependent protein kinase (obtained from Sigma Chemical Co.) is added 500 μl of buffer containing 100 m$M$ piperazine-$N,N'$-bis(2-ethanesulfonic acid)/Tris, 25 m$M$ MgCl$_2$, 20 m$M$ EGTA, 1 m$M$ dithiothreitol, and 0.6% Triton X-100 (pH 6.8). 35 μl of catalytic subunit buffer is then added to small reaction tubes followed by addition of 5 μl of different phospholamban fractions, and after 10 min at 30°, reactions are initiated by adding 10 μl of [γ-³²P]ATP to a final concentration of 0.4 m$M$. Reactions are terminated at 10 min by adding 20 μl of sodium dodecyl sulfate (SDS)-stop solution, and samples are then subjected to SDS-polyacrylamide gel electrophoresis.[7,8] Radioactive phospholamban is localized in the dried gels by autoradiography, and the protein in the different fractions is quantitated by excising it and subjecting it to liquid scintillation counting.[7,8]

Table I shows results obtained for purification of phospholamban by the method described here. 3.34 mg of phospholamban were isolated from approximately 1.26 g of cardiac membrane protein. The percent recovery (18%) and purification (-fold) of phospholamban (68.8-fold) are similar to that recently reported.[8] The level of ³²P incorporated into phospholamban (approximately 200 nmol/mg protein) is consistent with a molecular weight of 5000 for the individual subunits, assuming that cAMP-dependent protein kinase phosphorylates one site per monomer.[7] The high purity of phospholamban prepared by this method is confirmed by subjecting the protein to SDS-polyacrylamide gel electrophoresis followed by protein staining with Coomassie blue (Fig. 1). A single major protein band of apparent $M_r$ 25,000 is observed in the SDS gel for control phospholamban (form 5), which is dissociated into four distinct higher mobility forms

TABLE I
PURIFICATION OF CANINE CARDIAC PHOSPHOLAMBAN[a]

| Fraction | Total protein (mg) | Specific activity (nmol $P_i$/mg) | Total activity (nmol $P_i$) | Yield (%) | Purification (-fold) |
|---|---|---|---|---|---|
| Membrane vesicles | 1262 | 2.95 | 3723 | 100 | 1.0 |
| Reconstituted particulate | 104 | 12.7 | 1320 | 35 | |
| Purified phospholamban | 3.34 | 203 | 678 | 18 | 68.8 |

[a] Phospholamban is purified from canine cardiac sarcoplasmic reticulum vesicles (membrane vesicles) as described in the text. Purified phospholamban is recovered in fractions 9–17 eluted from the p-hydroxymercuribenzoate agarose column. Protein contents are determined by the Amido Black method.[11] Phospholamban is quantitated in the different fractions by phosphorylation with the catalytic subunit of cAMP-dependent protein kinase, followed by SDS-polyacrylamide gel electrophoresis and autoradiography (see text).

(forms 1–4) by boiling the protein in SDS prior to electrophoresis (Fig. 1). These results suggest that phospholamban may be a pentamer of five identical monomers, which has been proposed previously.[7,8]

## Amino Acid Composition of Purified Phospholamban

Phospholamban samples are hydrolyzed with 6 N HCl in evacuated tubes for 22, 48, and 72 hr at 110°, and the resulting amino acids are analyzed on a Model 121 Beckman amino acid analyzer by the method of Spackman et al.[12] Approximately 100 µg of phospholamban is hydrolyzed for each determination, with detection of amino acids in the eluate by reaction with ninhydrin. Values obtained for serine and threonine are extrapolated to zero time. Tryptophan is determined by hydrolysis in 4 N methanesulfonic acid,[13] and cysteine as cysteic acid after performic acid oxidation.[14]

Results of amino acid analysis of purified phospholamban are shown in Table II. Values reported from other laboratories are also shown for comparison. Phospholamban prepared by our method contains a relatively high proportion of cysteine, consistent with the ability to isolate the protein by sulfhydryl group affinity chromatography. The observation that phospholamban contains cysteine is in apparent conflict with results reported earlier by others, who have claimed that the protein does not

[12] D. H. Spackman, S. Moore, and N. H. Stein, *Anal. Chem.* **30**, 1190 (1958).
[13] R. J. Simpson, M. R. Neuberger, and T.-Y. Liu, *J. Biol. Chem.* **251**, 1936 (1976).
[14] S. Moore, *J. Biol. Chem.* **238**, 235 (1963).

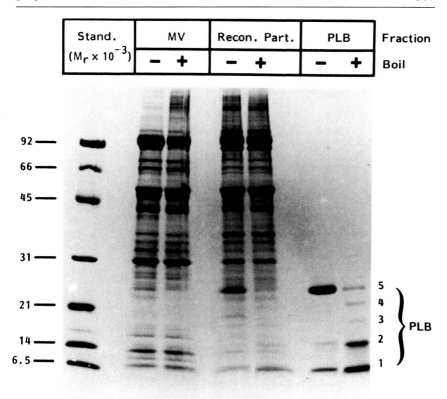

FIG. 1. SDS-polyacrylamide gel electrophoresis of membrane vesicles (MV), reconstituted particulate fraction (Recon. Part.), and purified phospholamban (PLB). MV (57 μg), Recon. Part. (57 μg), and PLB (10 μg) fractions were subjected to electrophoresis in a 10–20% gradient gel, and the gel was then stained with Coomassie Blue. − and + indicate whether samples were boiled in SDS prior to electrophoresis. (From Jones et al.[8])

contain this amino acid.[15,16] Other differences are noted between the amino acid composition of phospholamban isolated by our method and the compositions determined previously in other laboratories,[15-17] particularly with regard to levels of glycine, methionine, isoleucine, leucine, histidine, and arginine (Table II). The amino acid composition determined by Collins et al.,[18] however, is similar to that obtained here (Table II).

[15] C. J. LePeuch, D. A. LePeuch, and J. G. Demaille, this series, Vol. 102, p. 261.

[16] J. M. Bidlack, I. S. Ambudakar, and A. E. Shamoo, J. Biol. Chem. **257**, 4501 (1982).

[17] M. A. Kirchberger and T. Antonetz, J. Biol. Chem. **257**, 5685 (1982).

[18] J. H. Collins, E. G. Kranias, A. S. Reeves, L. B. Bilezikjian, and A. Schwartz, Biochem. Biophys. Res. Commun. **99**, 796 (1982).

TABLE II

AMINO ACID COMPOSITION OF PURIFIED PHOSPHOLAMBAN[a]

| Amino acid | Present study | Collins et al.[18] | LePeuch et al.[15] | Kirchberger and Antonetz[17] | Bidlack et al.[16] |
|---|---|---|---|---|---|
| Asx | 8.8 | 8.9 | 11.1 | 8.8 | 10 |
| Thr | 4.4 | 4.8 | 5.5 | 4.8 | 5.5 |
| Ser | 4.0 | 5.8 | 5.5 | 10.3 | 9.5 |
| Glx | 13.3 | 16.7 | 11.4 | 12.7 | 12.2 |
| Pro | 2.5 | 2.3 | 5.5 | 2.5 | 3.3 |
| Gly | 1.6 | 1.8 | 10.6 | 12.3 | 11.7 |
| Ala | 7.6 | 8.6 | 11.0 | 6.8 | 8.3 |
| Val | 3.4 | 5.1 | 6.5 | 5.0 | 4.4 |
| Met | 4.9 | 5.9 | 1.7 | 1.6 | 2.1 |
| Ile | 10.0 | 9.8 | 4.9 | 5.9 | 2.2 |
| Leu | 14.3 | 13.3 | 9.5 | 9.5 | 6.7 |
| Tyr | 1.5 | 2.3 | 2.1 | 3.8 | 2 |
| Phe | 3.9 | 3.7 | 3.4 | 4.2 | 10.6 |
| Lys | 4.3 | 2.5 | 4.9 | 4.8 | 6.1 |
| His | 0.6 | <0.1 | 2.0 | 1.7 | 3.3 |
| Arg | 8.8 | 8.6 | 4.4 | 4.6 | 2.8 |
| Cys | 5.9 | ND | 0 | 0.9 | 0 |
| Trp | 0.20 | ND | ND | ND | ND |

[a] The amino acid composition of purified phospholamban was determined as described in the text. For comparison, values reported in other laboratories are also given. ND, not determined. From Jones et al.[8]

Phospholamban prepared by our method also differs in several other major respects from the proteins that have been characterized in other laboratories.[15,16] Notable is its very basic isoelectric point, and its poor extractability into acidified organic solvents.[8] Thus, phospholamban prepared by our method does not seem to be an acidic proteolipid, as others have claimed.[15–17] It is possible that different proteins may have been isolated in different laboratories.

Conclusions

The method described here for isolation of cardiac phospholamban is relatively simple in that it relies on differential extraction of the protein from membranes by detergents, and requires only a single chromatography step at the final stage of purification. Thus, the method should be readily adaptable for use in other laboratories that work routinely with membrane proteins. Phospholamban prepared by our method[7,8] appears to be identical to the protein subsequently isolated by Inui et al.,[19] but is apparently different from the proteins previously characterized by Le-Peuch et al.,[15] Bidlack et al.,[16] and Kirchberger and Antonetz.[17] Studies are now under way to sequence the protein, as well as to reconstitute it into lipid membranes to gain further information on its structure and function and possible interaction with the $Ca^{2+}$-pump in cardiac sarcoplasmic reticulum vesicles.

### Acknowledgments

We gratefully acknowledge the excellent secretarial assistance of Anna Wells. Larry R. Jones is an Established Investigator of the American Heart Association. This work was supported by Grants HL-28556 and HL-06308 from the National Institutes of Health and by the Herman C. Krannert Fund.

[19] M. Inui, M. Kadoma, and M. Tada, J. Biol. Chem. 260, 3708 (1985).

## [30] Modification of Phospholipid Environment in Sarcoplasmic Reticulum Using Nonspecific Phospholipid Transfer Protein

### By Joël Lunardi, Paul DeFoor, and Sidney Fleischer

Introduction[1]

The calcium-pump membrane of sarcoplasmic reticulum (SR)[2] from fast twitch skeletal muscle is highly specialized for calcium transport. The calcium-pump protein accounts for 90% of the protein of the calcium-pump membrane of highly purified light SR, which derives from the longitudinal tubules of SR.[3,4] Phospholipid comprises 35% of the mass of the membrane of which more than 90% is phospholipid.[5,6] The $Ca^{2+}$-pump

[1] These studies were supported in part by grants to SF from NIH DK 14632 and the Muscular Dystrophy Association of America and a Biomedical Research Support Grant from the NIH, administered by Vanderbilt University.

[2] Abbreviations used: GABA, γ-aminobutyric acid; HEPES, N-2-hydroxyethylpiperazine-N'-2-ethanesulfonic acid; P, phosphorus; PC, phosphatidylcholine; PE, phosphatidylethanolamine; PI, phosphatidylinositol; PS, phosphatidylserine; SM, sphingomyelin; SR, sarcoplasmic reticulum.

[3] S. Fleischer, in "Structure and Function of Sarcoplasmic Reticulum" (S. Fleischer and Y. Tonomura, eds.), pp. 119–145. Academic Press, Orlando, Florida, 1985.

[4] A. Chu, A. Saito, and S. Fleischer, Arch. Biochem. Biophys. 258, 13 (1987).

[5] G. Meissner and S. Fleischer, Biochim. Biophys. Acta 241, 356 (1971).

[6] G. Meissner and S. Fleischer, Biochim. Biophys. Acta 255, 19 (1972).

protein is oriented transmembrane with a major portion extending out from the cytoplasmic face.[7] The membrane lipids appear to be required for both hydrolysis of ATP and Ca$^{2+}$-pumping activity.[6,8]

Approaches to study the role of phospholipids in membrane function have depended either on lipid depletion with organic solvents, phospholipases, or with detergents followed by reconstitution with added lipids,[6,8–11] or detergent-aided exchange,[12,13] or by chemical modifications of the phospholipids in the membrane.[14,15] However, these methods have the severe limitation that they are perturbing to membrane organization. In order to investigate the relationship between the lipid composition and the calcium-pumping activity, the phospholipid content of the SR membrane has been modified[16] with the use of the nonspecific lipid transfer protein from bovine liver.[17,18] This exchange protein catalyzes exchange between lipid and membrane vesicles of a variety of polar lipids including PC, PE, sphingomyelin, phosphatidic acid, phosphatidylglycerol, PI, PS, cholesterol, and some glycolipids.

The phospholipid exchange procedure consists of incubating the SR membrane vesicles with the transfer protein in the presence of phospholipid vesicles of defined phospholipid composition. The exchange of phospholipid takes place with retention of the unidirectional orientation of the calcium-pump protein in the membrane.[7,16]

Procedures

*Isolation of Transfer Protein*

The nonspecific phospholipid transfer protein is purified from beef liver essentially as described by Zilversmit and co-workers.[17,18] The trans-

---

[7] L. Herbette, P. H. DeFoor, S. Fleischer, D. Pascolini, A. Scarpa, and J. K. Blasie, *Biochim. Biophys. Acta* **817**, 103 (1985).

[8] A. Martonosi, *Curr. Top. Membr. Transp.* **3**, 83 (1972).

[9] S. Fleischer and B. Fleischer, this series, Vol. 10, pp. 406–433.

[10] H. Nakamura, R. L. Jilka, R. Boland, and A. N. Martonosi, *J. Biol. Chem.* **251**, 5414 (1976).

[11] C. Hidalgo, D. D. Thomas, and N. Ikemoto, *J. Biol. Chem.* **253**, 6879 (1978).

[12] G. B. Warren, P. A. Toom, N. J. M. Birdsall, A. G. Lee, and J. C. Metcalf, *Proc. Natl. Acad. Sci. U.S.A.* **71**, 622 (1974).

[13] A. Johannsson, C. A. Keightley, G. A. Smith, C. D. Richards, T. R. Hesketh, and J. C. Metcalf, *J. Biol. Chem.* **256**, 1643 (1981).

[14] C. Hidalgo, D. A. Petrucci, and C. Vergara, *J. Biol. Chem.* **257**, 208 (1982).

[15] J. M. East and A. G. Lee, *Biochemistry* **21**, 4144 (1982).

[16] J. Lunardi, P. DeFoor, and S. Fleischer, in preparation.

[17] R. C. Crain and D. B. Zilversmit, *Biochemistry* **19**, 1433 (1980).

[18] B. Bloj and D. B. Zilversmit, this series, Vol. 98, p. 574.

fer protein, prepared as described but omitting the octylagarose column chromatography, is adequate for exchange of phospholipid in the membrane. The nonspecific phospholipid transfer activity is eluted from the CM-52 cellulose column (5 × 10 cm) with 25 mM NaHPO$_4$, 45 mM NaCl, 5 mM 2-mercaptoethanol, 3 mM NaN$_3$, pH 8.0.[18] The eluted fractions were heated at 90° for 5 min., then cooled at 4° and centrifuged at 10,000 g for 10 min. The supernatant was collected and the pH adjusted to 7.4, and glycerol added to a final concentration of 10% (v/v). The solution was concentrated at 4° in an Amicon ultrafiltration cell with a PM10 membrane to final protein concentration of 2–4 mg/ml. The concentrated solution was then dialyzed extensively against 50 mM HEPES, 5 mM 2-mercaptoethanol, 3 mM NaN$_3$, 10% glycerol, pH 7.4. Aliquots (1–2 ml) of the solution were rapidly frozen and kept in liquid nitrogen without loss of activity for at least 2 months.

*Preparation of Phospholipid Unilamellar Vesicles (Liposomes)*

*Source of Lipids*

Egg yolk PC (type V-E, Sigma Chemical Company, St. Louis, MO)
Soybean PE (Natterman Chemical Company, Cologne, FDR)
L-α-dioleyl-PC, L-α-dielaidoyl-PC, L-α-dielaidoyl-PE (Avanti Polar Lipids Inc., Birmingham, AL)
[*carboxyl*-14C]Triolein (99 mCi/mmol, New England Nuclear, Boston, MA) was used after purification by thin-layer chromatography on silica gel plates in hexane–ethyl ether–acetic acid (60 : 40 : 1, v/v/v).
1-Palmitoyl-2-oleyl-*sn*-[9,10-3H]glycero-3-phosphocholine was synthesized and purified in the laboratory.[19]

Preparation of donor liposomes for phospholipid transfer activity. The phospholipid was evaporated from chloroform/methanol (2/1) under a stream of nitrogen and then further dried under vacuum in a desiccator for 30 min. The lipid was dispersed by sonication under nitrogen atmosphere in 0.3 M sucrose, 5 mM HEPES, pH 7.4. When PE liposomes are prepared, phosphatidic acid is added in a 1% molar ratio to PE, to ensure good dispersion of the phospholipid.

For assay of the nonspecific phospholipid transfer activity, phospholipids were dispersed at 1–2 mg/ml and then sonicated to clarity in a bath sonicator (Laboratory Supplies Company, Inc., Hicksville, NY) at 25–30° under a nitrogen atmosphere. For phospholipid transfer experiments,

[19] H.-J. Eibl, J. O. McIntyre, E. A. M. Fleer, and S. Fleischer, this series, Vol. 98, pp. 614–623.

when higher amounts of phospholipids were used, the phospholipids were dispersed in buffer at 5–10 mg and vortexed for 10–20 min. After hydration for 30 min, the mixture of phospholipids was sonicated using a Branson Sonicator (Model W-350 sonifier) with a 0.5-in. tipped horn. All these steps are carried out under nitrogen atmosphere and at a controlled temperature of 25–30°. After probe sonification, the liposomes were centrifuged for 30 min at 100,000 g at 20° to remove titanium fragments and undispersed phospholipids.

*Assay of Phospholipid Transfer Activity*

The transfer of phospholipid to SR membranes is measured by radioactivity or chemical analysis. A trace of radiolabeled triolein is included in the donor liposomes as a nonexchangeable marker.[18] Liposomes (see Preparation of Phospholipid Unilamellar Vesicles above) containing dioleyl-PC (80 μg of phosphorus), [³H]PC and a trace amount of [¹⁴C]triolein are prepared, with known [³H]PC specific activity and ³H/¹⁴C ratio. The donor liposomes are added to a mixture containing SR membranes (40 μg of phosphorus, prepared as described in Section D below) and nonspecific transfer protein (50–100 μg) in 0.3 M sucrose, 5 mM HEPES, pH 7.4 (preincubated at 30° for 10 min) in a final volume of 1 ml. After 10 or 20 min, the reaction is terminated by dilution with ice-cold buffer. The mixture is then placed on the top of a 1.5 ml 20% sucrose layer and the sample is centrifuged at 45,000 rpm in a Beckman 75Ti rotor. The pellet containing SR is rinsed with cold buffer, solubilized in 1 ml protosol (New England Nuclear), and then assayed for [³H]PC and [¹⁴C]triolein by scintillation counting. The counts of [³H]PC transferred to the pellet are corrected for sticking of liposomes by subtracting the counts of [¹⁴C]triolein in the pellet multiplied by the ³H/¹⁴C ratio of the initial liposomes. The specific activity of the nonspecific transfer protein prepared as described in the section Isolation of Transfer Protein (above) is about 40–50 units/mg of protein where 1 unit is the nanomoles of PC transferred to the SR membrane/minute under these conditions.

*Preparation of Sarcoplasmic Reticulum Vesicles*

The sarcoplasmic reticulum (SR) vesicles from rabbit skeletal muscle were isolated by zonal centrifugation as previously described by Meissner *et al.*[20] with modification to eliminate the use of salt.[21] The purified vesi-

[20] G. Meissner, G. E. Conner, and S. Fleischer, *Biochim. Biophys. Acta* **298**, 246 (1973).
[21] G. Meissner, *Biochim. Biophys. Acta* **389**, 51 (1974).

cles are suspended in a solution containing 0.3 $M$ sucrose, 100 m$M$ KCl, 5 m$M$ HEPES, pH 7.4, at 0° and stored at −80°.

### Phospholipid Exchange Procedure

#### Materials

Sarcoplasmic reticulum vesicles in 0.3 $M$ sucrose, 100 m$M$ KCl, 5 m$M$ HEPES, pH 7.4

Egg PC, dioleoyl- or dielaidoyl-PC, soybean PE, or dielaidoyl-PE liposomes in 0.3 $M$ sucrose, 5 m$M$ HEPES, pH 7.4 (10–20 mg phospholipids/ml) containing [$^3$H]PC and a trace of [$^{14}$C]triolein when exchange is determined by isotope determination

Nonspecific transfer protein (2–4 mg protein/ml) in 50 m$M$ HEPES, 3 m$M$ 2-mercaptoethanol, 3 m$M$ NaN$_3$, pH 7.4

20% sucrose, 5 m$M$ HEPES, pH 7.4

0.3 $M$ sucrose, 100 m$M$ KCl, 5 m$M$ HEPES, pH 7.4

Bovine plasma albumin (0.66%), crystalline (devoid of fatty acids), e.g., Metrix Division, Armour Pharmaceutical Co., Chicago, IL 60690

*Procedure.* In a typical transfer experiment, 20 mg of SR membranes (440 $\mu$g of P) in 1 ml of 0.3 $M$ sucrose, 100 m$M$ KCl, 5 m$M$ HEPES, pH 7.4, are mixed with a 10-fold excess of donor phospholipid to be exchanged, 15 ml of the phospholipid liposomes (4400 $\mu$g of lipid P) in 0.3 $M$ sucrose, 5 m$M$ HEPES, 0.66% bovine plasma albumin, then 2.5 ml of nonspecific transfer protein (300 activity units) in 50 m$M$ HEPES, 3 m$M$ 2-mercaptoethanol, 3 m$M$ NaN$_3$, pH 7.4, is added. The sucrose concentration is adjusted to 0.3 $M$ and the volume adjusted to 20 ml. The bovine serum albumin is added to minimize sticking of liposomes to the SR vesicles. After 2 hr incubation at 20°, four aliquots of 5 ml of the mixture are layered on the top of 2.5 ml of 20% sucrose, 5 m$M$ HEPES, pH 7.4, and the tubes are centrifuged in a Beckman Ti50 rotor for 45 min at 100,000 $g_{av}$ (40,000 rpm). The surface of the pellets is rinsed with 0.3 $M$ sucrose, 100 m$M$ KCl, 5 m$M$ HEPES, pH 7.4, and resuspended in the same buffer at 10 mg of protein/ml.

*Second Lipid Transfer.* A second lipid transfer can be carried out to see whether reversal of the modification in membrane phospholipid composition can restore the original membrane function. This second transfer is performed directly after the first and differs only in the type of phospholipid which is used.

It is important to note that the activity of the bovine liver transfer enzyme decreases linearly with ionic strength. The transfer activity of 0.2 $M$ NaCl is only one-fourth that at 0.05 $M$.[17,18]

*Membrane Analysis*

The resuspended SR vesicles are assayed for phosphorus as a measure of phospholipid,[22] and protein[23] using serum albumin as standard. The protein analysis procedure is modified when 2-mercaptoethanol is present.[24] The [¹⁴C]triolein content is used to correct the phospholipid content of the samples for the presence of cosedimented or adsorbed liposomes. Under our conditions, the contribution of the cosedimented transfer protein used to the final membrane protein content is not significant.

*Analysis of Phospholipid Composition.* Lipid extracts were prepared using 20 vol of chloroform/methanol (2/1, v/v) and back-extracting to remove nonlipid materials.[22] The phospholipid composition of the lipid extracts of normal and exchanged SR vesicles is determined by two-dimensional thin-layer chromatography and phosphorus analysis as described by Rouser and Fleischer.[22] Precoated silica gel H plates (Analtech, Newark, DE), a chloroform–methanol–28% NH₄OH (65 : 25 : 5, v/v/v) solvent system I, and a chloroform–acetone–methanol–acetic acid–water (5 : 2 : 1 : 1 : 0.5, v/v/v/v/v) solvent system II are used. Lipids are visualized by charring, i.e., spraying with concentrated sulfuric acid–30% formaldehyde (97 : 3, v/v) followed by heating in a vented oven at 180° for 20 min. The phospholipids are identified by their relative migration and the phospholipid composition is determined by measuring the phosphorus content of the spots. The fatty acid composition of normal and exchanged sarcoplasmic reticulum membranes is determined by gas–liquid chromatography. The fatty acids are transesterified to their methyl esters using a BF₃–methanol (14% w/v) esterification kit (Supelco Inc., Bellefonte, PA) and analyzed by gas liquid chromatography, e.g., using a 6-foot column, 1/8-in. diameter, containing 10% SP 2330 on 100/200 mesh Chromosorb W AW as support (Supelco Inc. Bellefonte, PA). The analysis made use of a 170–200° temperature program.

*Calculation of Phospholipid Transfer*

Some sticking of liposomes to the SR membrane occurs during phospholipid transfer that must be subtracted in order to determine the phospholipid content specific to the membrane. This is achieved by including a trace amount of [¹⁴C]triolein in the donor liposomes to be incubated with the SR membranes. Triolein is considered to be a nontransferable

[22] G. Rouser and S. Fleischer, this series, Vol. 10, p. 385.
[23] O. H. Lowry, N. J. Rosebrough, and R. J. Randall, *J. Biol. Chem.* **103**, 205 (1951).
[24] E. Ross and G. Schatz, *Anal. Biochem.* **54**, 304 (1973).

marker.[18] The presence of [$^{14}$C]triolein in the recovered membrane pellet, in conjunction with the known ratio of dpm of [$^3$H]phospholipid/ [$^{14}$C]triolein in the initial liposomes, allows estimation of the amount of each of these components in the pellet due to liposome sticking rather than molecular transfer into the membrane bilayer. The correction for liposomes sticking is given below, where phospholipids are labeled with $^3$H and triolein with $^{14}$C.

$$[\text{dpm } ^3\text{H}]_{\text{total}} = [\text{dpm } ^3\text{H}]_{\text{transferred}} + [\text{dpm } ^3\text{H}]_{\text{sticking}} \qquad (1)$$

$$[\text{dpm } ^3\text{H}]_{\text{transferred}} = [\text{dpm } ^3\text{H}]_{\text{total}} - [\text{dpm } ^3\text{H}]_{\text{sticking}} \qquad (2)$$

$$[\text{dpm } ^3\text{H}]_{\text{sticking}} = [^3\text{H}/^{14}\text{C}]_{\text{liposomes}} \times \text{dpm } ^{14}\text{C in the pellet} \qquad (3)$$

$[^3\text{H}/^{14}\text{C}]_{\text{liposomes}}$ represents the initial ratio of dpm $[^3\text{H}]$PC/$[^{14}\text{C}]$triolein in the liposomes. The amount of phospholipid transferred to the membrane [Eq. (2)] is readily calculated using the specific activity (dpm/$\mu$g lipid phosphorus) of the $^3$H-labeled phospholipids in the donor liposomes. A sample calculation is provided on page 693.

It should be noted that exchange values based on isotope determinations gave an overestimation of exchange when 1-palmitoleyl-2-oleoyl-[9,10 $^3$H]PC was used as a probe with L-$\alpha$-dielaidoyl-PC and L-$\alpha$-dioleoyl-PE liposomes. This overestimation is the result of preferential transfer of the molecular species of labeled PC compared with the two other molecular species of PC. It is known that the nature of the polar head, and/or the aliphatic chains of the phospholipid determine the rate of transfer catalyzed by the nonspecific phospholipid transfer protein.[25] An additional factor could be a nonhomogeneous distribution of the labeled probe in the liposome bilayer.[26] When dielaidoyl-PC and dioleoyl-PE liposomes were used, the extent of the phospholipid exchange was measured by fatty acid determination. Fatty acid composition of the SR membranes was determined as described above and the amount of transfer was evaluated by comparison of the fatty acids percentages before and after exchange. The fatty acid composition of the membranes after exchange must also be corrected for sticking of liposomes to the SR membranes. The amount of sticking of phospholipid is evaluated as the difference between the total amount of phospholipid present in the SR membrane before and after the exchange. Since the phospholipid present in the liposomes consist of a single type of fatty acid, the correction in the fatty acid composition for the sticking to the SR membranes can readily be made.

When exchange was determined by both isotope determination and fatty acid analysis, in the case of exchange performed using L-$\alpha$-dioleoyl-PC liposomes as phospholipid donor, a good correlation was found be-

[25] J. Zborowski and R. A. Demel, *Biochim. Biophys. Acta* **688**, 381 (1982).
[26] H. Sandermann, Jr., *Biochim. Biophys. Acta* **515**, 209 (1978).

tween exchange parameters calculated from data given by the two methods.

### Application: Correlation of Phospholipid Composition with Function

The use of a nonspecific phospholipid transfer protein is an effective and gentle way to modify the phospholipid composition in the SR membrane. With this methodology, the presence of two different classes of phospholipids in the SR membranes is observed and interpreted in terms of membrane sidedness (Fig. 1). The first pool represents 40–45% of the total phospholipids present in the outer monolayer of the membrane. The second class of phospholipid exchanged at a much slower rate and is interpreted in terms of transmembrane migration, also referred to as flip-flop. The half-time of transmembrane migration varied with the type of lipid: dioleoyl-PC, 15 hr; dielaidoyl-PC, 28 hr; and soybean PE, 20 hr.

When exchange is performed with PC liposomes as phospholipid donor, the phospholipid class composition of the outer monolayer can only be moderately modified since PC accounts for 70% of the phospholipid present in the outer layer.[7] Using defined molecular species of PC liposomes as phospholipid donor and the nonspecific phospholipid transfer

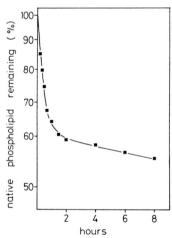

FIG. 1. Exchange of phospholipids between sealed sarcoplasmic reticulum vesicles and dielaidoylphosphatidylcholine donor liposomes. Sealed sarcoplasmic reticulum vesicles (900 μg P) were incubated with donor liposomes (11,500 μg P) and 6 mg of exchange protein in 50 ml of 0.3 M sucrose, 5 mM HEPES, pH 7.4, at 30°. At different time points, aliquots of the mixture (5 ml) were centrifuged for 45 min at 120,000 $g_{av}$ through a 20% sucrose, 5 mM HEPES layer. Pellets were carefully rinsed and resuspended in 0.3 M sucrose, 100 mM KCl, 5 mM HEPES, pH 7.4. Phospholipids were then extracted and the amount of phospholipids exchanged determined by fatty acid analysis.

TABLE I
CALCIUM-PUMPING ACTIVITY OF SR VESICLES AS A FUNCTION OF THE PE/PC
RATIO IN MEMBRANE[a]

| Incubation time (min) | PE/PC ratio | $Ca^{2+}$-pumping activity (%) | $Ca^{2+}$-loading ($\mu$mol $Ca^{2+}$/mg) |
|---|---|---|---|
| 0 | 0.20 | 100 | 3.75 |
| 15 | 0.45 | 90 | 3.63 |
| 30 | 0.66 | 85 | 3.70 |
| 60 | 1.04 | 76 | 3.80 |
| 120 | 1.46 | 68 | 3.65 |
| 180 | 1.78 | 60 | 3.55 |
| 240 | 1.98 | 55 | 3.58 |

[a] SR membranes (35 mg of protein, 23.7 $\mu$mol of phospholipids) were incubated at 30° in 36.5 ml of 0.3 M sucrose, 5 mM HEPES, 0.5% bovine serum albumin (BSA), pH 7.4, with 7.8 mg of nonspecific transfer protein (440 units) and soybean PE liposomes (149 $\mu$mol of phospholipids). At indicated times, 6.5 ml of the mixture was layered over 20% sucrose, 5 mM HEPES, pH 7.4, and sedimented by centrifugation at 100,000 g for 45 min. The pellet was resuspended in 0.5 ml of 0.3 M sucrose, 100 mM KCl, 5 mM HEPES, pH 7.4. The protein content, phospholipid composition and $Ca^{2+}$-loading activity were then measured.[16] Under these conditions, $Ca^{2+}$-loading activity of the control SR, corresponding to 100%, was 1.75 $\mu$mol $Ca^{2+}$/min · mg.

protein to modify the phospholipid composition of the membrane, up to 80–85% of the phospholipids in the outer layer were replaced. We find that $Ca^{2+}$-stimulated ATPase and $Ca^{2+}$-pumping activities are not significantly affected by exchange of dioleoyl or dielaidoyl PC in the outer monolayer of the SR membrane.

When PE liposomes were used as phospholipid donor for exchange, the initial PE/PC ratio of 0.3 was increased to a value of 2 (Table I).[16] The increase in PE content of the SR membrane is correlated with an inhibition of the $Ca^{2+}$-pumping activity while the $Ca^{2+}$-stimulated ATPase activity is almost unaffected. This inhibition was referable to modification of the PE/PC ratio in the membrane, since a second back-exchange carried out with PC liposomes as phospholipid donor restored the initial PE/PC ratio and restored the $Ca^{2+}$-pumping activity of the SR vesicles.

The nonspecific phospholipid transfer protein has been used to modify the cholesterol content of synaptosomal plasma membrane. Lowering of the cholesterol/phospholipid ratio caused a marked reduction in GABA uptake. Transport function was reconstituted by a second exchange which restored the original cholesterol content.[28,29]

[27] G. Meissner and S. Fleischer, *J. Biol. Chem.* **249**, 302 (1974).
[28] P. North and S. Fleischer, this series, Vol. 98, 56.
[29] P. North and S. Fleischer, *J. Biol. Chem.* **258**, 1242 (1983).

# [31] Computer Programs for Calculating Total from Specified Free or Free from Specified Total Ionic Concentrations in Aqueous Solutions Containing Multiple Metals and Ligands[1]

*By* ALEXANDRE FABIATO

## Introduction and Availability of the Programs

Many experiments in biology, biochemistry, biophysics, and physiology require aqueous solutions with concentrations of free cations far below those obtainable by simply adding the corresponding salts to the solutions. Then metal buffers made up of the metallic ion (or cation) and the appropriate ligand (or anion) must be used. Similarly, living cells contain ligands, such as cation-binding proteins, that cause the total to differ from the free concentrations of metals. Thus, these experimental solutions or the intracellular milieu contain ligands and metals which present between them equilibria of the type:

$$M + L \rightleftarrows ML$$

where M is the free metal, L is the free ligand, and ML is the complex. Such an equilibrium is characterized by a stability constant (or binding constant or association constant) $K_{ML}$ defined as:

$$K_{ML} = [ML]/[M][L]$$

where [ML], [M], and [L] are the concentrations of the complex, the free metal, and the free ligand, respectively.

Ligands also bind hydrogen ions so that instead of a single complex between metal and ligand, several complexes form, corresponding to various degrees of protonation. Each has its own stability constant. Thus, apparent stability constants ($K_{app}$) have to be calculated to characterize the multiple equilibria among one metal, one ligand, and the hydrogen ion at a given pH value.

Additional complexity arises when two or more metals in the solution bind to a given ligand or when more than one ligand binds to a given metal. Then equations for multiple equilibria must be solved to calculate the total concentrations necessary to obtain specified free metal concentrations.

[1] Supported by Grant No. HL19138 from the National Heart, Lung, and Blood Institute.

Experimental solutions must often have not only specified free concentrations of metals but also a specified ionic strength. The ionic strength ($\Gamma/2$) is defined by:

$$\Gamma/2 = \Sigma C_i z_i^2/2$$

where $C_i$ is the concentration and $z_i$ is the charge of the ion. The calculation of the ionic strength must include the ionic species involved in all the equilibria among metals, ligands, and hydrogen ions. Thus, complex calculations are needed for computing not only the ionic composition of the solutions but also their ionic strength. The ionic strength is not always the most relevant expression of the ionic composition of the solutions. For instance, the sensitivity of the myofilaments of skinned cardiac cells (i.e., single cardiac cells from which the sarcolemma has been removed by microdissection) to calcium is more dependent on the nature and total concentration of the major monovalent cation than on the ionic strength.[2]

Many experiments require the specification of the free ionic concentrations. Then the program has to be operated to calculate the total concentrations of metals and ligands necessary to obtain the specified free ionic concentrations and, if desired, either the specified ionic strength or the specified total concentration of the major monovalent cation. In contrast, when the experiments have already been done, one knows the total concentrations of metals and ligands that have been used and the program has to be operated in the reverse mode, that is, to calculate the free ionic concentrations resulting from specified total concentrations of metals and ligands. This is also the case when one wants to compute the free ionic concentrations in an intracellular milieu where the total concentrations of ligands and metals are known.

A major problem in interpreting the results of many types of experiments is the uncertainty about the stability constants between metals and ligands. To evaluate the consequences of this uncertainty, it is useful to first do the computation with one set of apparent stability constants and then redo the computation with a different set of apparent stability constants. Thus, in this case, one should first specify the free ionic concentrations and calculate the total concentrations with a first set of apparent stability constants and then calculate from these total concentrations the free ionic concentrations with a different set of apparent stability constants. This is the most general case for the computer program, which has to do both a computation of total ionic concentrations from specified free concentrations and then the reverse: computation of free ionic concentrations from specified total concentrations. This is what is accomplished by

[2] A. Fabiato and F. Fabiato, *J. Physiol. (Paris)* **75**, 463 (1979).

the program called "SPECS" on which this article is focused. Using parts of this program permits the two other types of computations: simple calculations of total ionic concentrations from specified free concentrations ("computation from free to total concentrations") and of free concentrations from specified total concentrations ("computation from total to free concentrations").

Previously,[2] we had reported much simpler programs for a programmable calculator which was very popular at that time, the Texas Instruments TI-59. These programs accomplished the computation from free to total and from total to free concentrations. The most original feature of these programs was the use of successive approximations to obtain free from specified total concentrations by iteration, a problem which cannot be solved explicitly when the number of interacting metals and ligands becomes too large. This method was also used for calculating total from specified free concentrations in order to give more general flexibility to the program than the approach that we had used previously, which consisted of specifically solving equations following the method derived from Reuben *et al.*[3]

These calculator programs have been used by many other investigators. Unfortunately, the TI-59 calculator is no longer manufactured. In addition, the calculations were limited by the memory space and the number of program steps allotted for the computation. Finally, the computations, especially those from total to free concentrations, were slow (up to 30 min for the most complex problems). While some investigators asked me to do the calculations for them, an increasing number preferred a program that they could use on their computers, especially since the microcomputers became increasingly popular. The primary aim of this article is to meet this request. In addition, since all my articles published during the past ten years relied heavily on these computations, I thought it was important to provide the readers of these articles with the information necessary to redo my computations.

The main thrust of the present article is to describe a FORTRAN program named "SPECS" in such a way that it can be readily used by the largest number of investigators possible.[4] This is accomplished (1) by publishing the complete source code of this program to serve those investigators proficient in computer programming who want to adapt the pro-

[3] J. P. Reuben, P. W. Brandt, M. Berman, and H. Grundfest, *J. Gen. Physiol.* **57**, 385 (1971).
[4] I am much indebted to my former technician Mr. J. Campbell for his invaluable help in writing the original version of this program and to the Division of Academic Computing of Virginia Commonwealth University, Medical College of Virginia campus, for its help in the compilation of this program with the Microsoft FORTRAN.

gram to their own computer (this is possible because no feature particular to any computer or software has been incorporated in the program); (2) by distributing two diskettes containing the compiled version of this program directly usable on the microcomputers for which such a program has been by far the most frequently requested of me, the International Business Machines series of microcomputers: IBM-PC, IBM-XT, and IBM-AT (as well as all the IBM compatible microcomputers from many trademarks); (3) by giving detailed instructions on how to input the data and examples which cover practically all the particular cases in the use of these programs that have been encountered in ten years of experience with not only my own problems, but those posed by the research of many other investigators for whom I did computations. Flow charts, rationale, and equations are deleted as they are identical to those published in detail for the calculator programs.[2]

For running on the microcomputers IBM-PC, -XT, or -AT, the program requires a math coprocessor (8087 for IBM-PC or -XT, 80287 for IBM-AT).

SPECS and the ancillary programs are available at no charge[5] to those requesting them.[6] All the programs are on two 360-kilobyte diskettes. The first diskette contains the compiled version of SPECS. This leaves enough space for the users to copy the files to a bootable diskette containing their own version of the DOS system and the useful part of a word-processor software, such as WordStar (in the nondocument mode, i.e., without high bit flags), which can be used to enter the data (the three files WS.COM, WSMSGS.OVR, and WSOVLY1.OVR suffice for the limited edition required here). Finally, this diskette contains a batch file to call for the compressed mode of the most frequently requested printers. The output of the computer assumes a 132-column dot-matrix printer. If an 80-column printer is used the compressed mode should be employed. This is easily done by push-buttons with some recent printers. With older models a series of character strings must be entered. That can be done through a batch file. For the convenience of the users of any 80-column Epson or IBM printer, a batch file called "SPECSE" is included. It contains the appropriate series of character strings in a file called "EPSOCOMP.EXE". Thus, the users of any 80-column Epson or IBM printer should call the program through "SPECSE" instead of "SPECS".

The other diskette contains (1) the source code for SPECS called

[5] If this is possible, reimbursement of the cost of two diskettes and the shipping would be appreciated.

[6] A. Fabiato, Department of Physiology, Medical College of Virginia, Richmond, VA 23298-0551, U.S.A.

SPECS.FOR as it is given in the next section of this article (as previously indicated, this source code contains no feature particular to any computer, thus it would require a minor edition before it can be compiled with the Microsoft FORTRAN compiler), (2) the input data files for the four examples studied in this article, and (3) two programs for preliminary calculations.

The data files for the examples are ASCII text files. They can be edited using WordStar in the nondocument mode or any ASCII word processor or text editor. WordStar is mentioned only because it was the text editor most frequently used by those who requested from me programs immediately usable with detailed step-by-step instructions.

The first program for preliminary calculations called "ABSTOAPP" achieves the same goals as program No. 1 of the three previously published calculator programs.[2] This interactive program, compiled with Turbo PASCAL, asks whether the user wants to calculate apparent stability constants from absolute stability constants or the ionic strength contributed by the pH buffer. Then the program interactively gives instructions on how to enter the data. The output gives either the apparent stability constants and the square root of the mean squared charge (rms for "root mean squared") for the ligands and complexes or the ionic strength contributed by the pH buffer.

A major cause of error for nonspecialist users in computing free or total ionic concentrations in solutions containing multiple metals and ligands is the use of incorrect stability constants. Accordingly, another program on the second diskette gives the stability constants for the complexes between the most frequently used metals and ligands and also the correction of these stability constants for temperature. This program called "STACONS" has been compiled with Microsoft BASIC Compiler. It is interactive but only asks the user to specify the temperature and pH. Then a list of the absolute stability constants at the selected temperature is given followed by a list of the apparent stability constants at the specified temperature and pH. The output also lists the articles in which the reasons for the selection of these stability constants and references to the literature are given. I plan to update the list of stability constants and send an updated diskette No. 2 to the users of the original programs.

SPECS can calculate the equilibria for a total of up to 20 metals and ligands, provided that no more than 9 have undefined total concentrations, and with up to 30 stability constants. This is sufficient for all the problems submitted to me during the past ten years, and much more powerful than the programs for the Texas Instruments calculator, which were limited to 4 ligands and 5 metals (3 divalent and 2 monovalent).

## Source Code for the SPECS Program

```
      DIMENSION VAL(30),RECORD(601,21),VARY(30,2),
     1LOCCOM(10,4),LABL(10,10),IHEAD1(10,19),IHEAD2(22),
     2START(30),CMVAL(30),INF(10),LRLABL(5,10),
     3PRINT(10),KFMT(11),NFMT(10),IFMT(8),IHEAD(20),LFMT(10),LFMT1(11)
      COMMON /LREG/ LNREG(5,10),REGRNG(5,2)
      COMMON /TLFR/ TOTAL(20),FREE(20),NAME(30,2),DECIDE(30),METCL,IONS
      COMMON /TFCALC/ STBLTY(30),IALPHA(20,30),MET(10,10),LCL(10),
     1CMCHRG(30),CHARGE(20),NLNM,ADDIN,LRNSTR,LMIS,PH,IONSTR,
     2NC,LV1,LMCNST,LTL2FR,IEBIND(10)
      COMMON /OUT/ NFMT1(10),IFMT1(8),IOUT(10),MAX,NOIONS,
     1BUFFER,CMNAME(30,2)
      REAL METCL,IONSTR
      INTEGER DECIDE,CMNAME
      LOGICAL LR,LSTORE,LRNSTR,ICOMP,LVO,LV1,LTL2FR,NOIONS
      LOGICAL IERR
      DATA ICNST/'CNST'/,NCOMP/'NCO'/,NIONS/'NIO'/,NLR/'NLR'/,
     1NCMPLX/'NCOM'/,IPLUS1/' ' + '/,IM/' M'/,IEG/'EGTA'/,IED/'EDTA'/,
     2IAN/'A- '/,MORE/'MORE'/,MOR/'MOR'/
      DATA IG1/'GO '/,IC/'CM '/,IG/'GO '/,IPS/'PS '/,IT/'TL '/,
     1IMG/'MG '/,IMG1/' MG'/,IBL/' '/,IS/'IS '/,IST/'STOR'/,
     2KAPP/2HK'/,IPAR1/' (TL'/,IPAR2/')'/    '/,IPLUS/'+ TL'/,IB/' '/,
     3LETP/'P '/,ICL/'CL '/,IDASH/'-'/,IDIS/'D:IS'/,LETBP/' P'/,
     4IOH/'OH '/,ICA/'CA '/,IWRT/'WRT'/,ILR/'LR '/,IS1/'I.S.'/
      DATA LFMT/'(1X,',4H'TL ,'A- =',' 2 *',4H (',,'N',4H('TL,4H ',A,
     1'2,A3',')'))/
      DATA LFMT1/4H('+',',',',M','(10X',',),','M',4HX,'+,' TL ',
     14H',A2,4H,'A-,2H')/
      DATA NFMT(1)/'(1X,',/,NFMT(2)/4HI3,',',NFMT(3)/') TO'/,
     1NFMT(4)/4HTAL',/,NFMT(5)/',3X,',/,NFMT(7)/'(1PE'/,NFMT(8)/'10.3'/,
     2NFMT(9)/',2X)'/,NFMT(10)/')'/
      DATA IFMT(1)/'(6X,',/,IFMT(2)/4H' PFR/,IFMT(3)/4HEE '/,
     1IFMT(4)/',3X,'/,IFMT(6)/'(F7.'/,IFMT(7)/'3,5X'/,IFMT(8)/'))'/
      DATA KFMT/'(1X,',4HI3,',4H)',1,'X,1P','G10.','3,',,'N',,'(1X,',
     1'1PE1','0.3)',')'/
      DATA INF/'1','2','3','4','5','6','7','8','9','10'/
      DO 1 I=1,10
      NFMT1(I)=NFMT(I)
      DO 5 I=1,8
      IFMT1(I)=IFMT(I)
      ICOMP=.FALSE.
      LSTORE=.FALSE.
      LR=.FALSE.
      LRNSTR=.FALSE.
      LVO=.TRUE.
      LV1=.TRUE.
      LTL2FR=.TRUE.
      NOIONS=.FALSE.
      IERR=.FALSE.
      NSTORE=0
      NREC=0
      NLINES=0
      LMIS=0
      LMCNST=0
      IONIS=0
      PH=0.
      IONSTR=0.
```

```
       BUFFER=0.
       IBIG=0
       K=0
       KBIND=0
       DO 10 I=1,20
       FREE(I)=0.
       TOTAL(I)=0.
       CHARGE(I)=0.
       DO 10 J=1,30
       IALPHA(I,J)=0
10     CONTINUE
       DO 20 I=1,30
       CMCHRG(I)=0.
       STBLTY(I)=0.
       VAL(I)=0.
       START(I)=0.
       VARY(I,1)=0.
       VARY(I,2)=0.
20     CONTINUE
       DO 30 I=1,10
       LCL(I)=0
       LOCCOM(I,1)=0
       LOCCOM(I,2)=0
       LOCCOM(I,3)=0
       LOCCOM(I,4)=0
       IOUT(I)=0
       IEBIND(I)=0
       DO 30 J=1,10
       MET(I,J)=0
30     CONTINUE
       DO 40 I=1,5
       DO 40 J=1,10
       LNREG(I,J)=0
       LRLABL(I,J)=IBL
40     CONTINUE
       NC=0
50     NC=NC+1
       READ(5,5000) (CMNAME(NC,J),J=1,2),CMCHRG(NC),STBLTY(NC)
       IF(CMNAME(NC,1).NE.IG1) GOTO 50
       NC=NC-1
       NLNM=0
60     NLNM=NLNM+1
       READ(5,5010) ILAB,NAME(NLNM,1),CHARGE(NLNM),VAL(NLNM),
      1(VARY(NLNM,J),J=1,2)
       IF(ILAB.EQ.IG) GOTO 110
       IF(ILAB.EQ.IC) GOTO 70
       IF(ILAB.EQ.IPS) GOTO 80
       IF(ILAB.EQ.IT) GOTO 90
       WRITE(6,6000)
       STOP
70     ICOMP=.TRUE.
       DECIDE(NLNM)=-1
       GOTO 100
80     DECIDE(NLNM)=0
       GOTO 100
90     DECIDE(NLNM)=1
100    START(NLNM)=VAL(NLNM)
       GOTO 60
110    NLNM=NLNM-1
       ICM=NLNM+1
       DO 115 I=1,NLNM
       IF(DECIDE(I).EQ.1) GOTO 115
       LTL2FR=.FALSE.
       GOTO 120
115    CONTINUE
       IF(LTL2FR) GOTO 155
```

```
120   READ(5,5020) ILAB,(NAME(ICM,J),J=1,2),VAL(ICM),(VARY(ICM,J),J=1,2)
      IF(ILAB.EQ.IG.OR.ILAB.EQ.NCOMP) GOTO 160
      IF(ILAB.EQ.IPS) GOTO 130
      IF(ILAB.EQ.IT) GOTO 140
      WRITE(6,6010)
      STOP
130   DECIDE(ICM)=0
      GOTO 150
140   DECIDE(ICM)=1
150   START(ICM)=VAL(ICM)
      ICM=ICM+1
      IF(ICOMP) GOTO 120
      WRITE(6,6060)
      STOP
155   READ(5,5060) ILAB,ILAB1,VALUE
      IF(ILAB.EQ.IG1.OR.ILAB.EQ.NCMPLX) GOTO 159
      IP=0
      DO 156 I=1,NC
      IF(ILAB.EQ.CMNAME(I,1).AND.ILAB1.EQ.CMNAME(I,2)) IP=I
156   CONTINUE
      IF(IP.NE.0) GOTO 158
      WRITE(6,6250)
      STOP
158   K=K+1
      IOUT(K)=IP
      GOTO 155
159   INOUT=K
      MAX=5
      IF(INOUT.NE.0) MAX=0
      MAX=MAX+13
160   ICM=ICM-1
      NLNMP1=NLNM+1
      NLNMP2=NLNM+2
      NFMT1(6)=INF(10)
      IFMT1(5)=INF(9)
      IF(NLNMP1.GT.10) GOTO 162
      NFMT1(6)=INF(NLNMP1)
      IFMT1(5)=INF(NLNM)
162   NCM=ICM-NLNM
      IF(NCM.GT.0) GOTO 175
      KKK=1
      DO 165 I=1,NLNM
      IF(DECIDE(I).NE.-1) GOTO 165
      WRITE(6,6070) NAME(I,1)
      KKK=0
165   CONTINUE
      IF(KKK.EQ.0) STOP
      GOTO 245
175   NCOM=0
      DO 240 J=NLNMP1,ICM
      IP=0
      K=-1
      DO 200 I=1,NLNM
      IF(NAME(J,1).EQ.NAME(I,1)) GOTO 180
      IF(NAME(J,2).EQ.NAME(I,1)) GOTO 190
      GOTO 200
180   IF(IP.GT.0) GOTO 195
      IF(DECIDE(I).EQ.-1) IP=I
      ILAB=NAME(J,2)
      GOTO 195
190   IF(IP.GT.0) GOTO 195
      IF(DECIDE(I).EQ.-1) IP=I
      ILAB=NAME(J,1)
195   K=K+1
200   CONTINUE
      IF(IP.GT.0) GOTO 204
```

```
        WRITE(6,6020)
        IERR=.TRUE.
204     IF(K.LE.1) GOTO 206
        WRITE(6,6030)
        IERR=.TRUE.
206     IP1=0
        DO 210 I=1,NLNM
        IF(ILAB.EQ.NAME(I,1)) IP1=I
210     CONTINUE
        IF(IP1.GT.0) GOTO 214
        WRITE(6,6040)
        IERR=.TRUE.
214     IP2=0
        DO 220 I=1,NC
        IF(NAME(J,1).EQ.CMNAME(I,1).AND.NAME(J,2).EQ.CMNAME(I,2)) IP2=I
220     CONTINUE
        IF(IP2.GT.0) GOTO 230
        WRITE(6,6050)
        IERR=.TRUE.
230     NCOM=NCOM+1
        LOCCOM(NCOM,1)=IP
        LOCCOM(NCOM,2)=J
        LOCCOM(NCOM,3)=IP1
        LOCCOM(NCOM,4)=IP2
240     CONTINUE
245     IF(IERR) STOP
        LCA=0
        DO 250 I=1,NLNM
        IF(NAME(I,1).EQ.ICA) LCA=I
250     CONTINUE
260     DO 300 J=1,NC
        DO 280 I=1,NLNM
        IF(CMNAME(J,1).EQ.NAME(I,1).OR.CMNAME(J,2).EQ.NAME(I,1))
       1IALPHA(I,J)=1
280     CONTINUE
300     CONTINUE
        IP=0
        DO 302 I=1,NLNM
        IF(NAME(I,1).EQ.IEG.OR.NAME(I,1).EQ.IED) IP=I
302     CONTINUE
        IF(IP.EQ.0) GOTO 308
        K=1
        IEBIND(1)=IP
        DO 306 J=1,NC
        IF(IALPHA(IP,J).EQ.0) GOTO 306
        L=1
        IF(CMNAME(J,1).EQ.NAME(IP,1)) L=2
        DO 304 I=1,NLNM
        IF(NAME(I,1).NE.CMNAME(J,L)) GOTO 304
        K=K+1
        IEBIND(K)=I
304     CONTINUE
306     CONTINUE
        KBIND=K
308     IONS=1
        NCL=0
310     READ(5,5030) ILAB,(LABL(IONS,I),I=1,10)
        IF(ILAB.EQ.IG.OR.ILAB.EQ.NIONS) GOTO 320
        IF(LABL(IONS,1).NE.IAN) GOTO 315
        NCL=NCL+1
315     IONS=IONS+1
        GOTO 310
320     IONS=IONS-1
        IF(IONS.NE.0) GOTO 323
        NOIONS=.TRUE.
        IF(NLNM.LE.9) GOTO 322
```

```
          IFMT1(5)=IMF(10)
          GOTO 382
322       NFMT1(6)=IMF(NLNM)
          GOTO 382
323       L=0
          DO 365 I=1,IONS
          IF(LABL(I,1).EQ.IAN) GOTO 325
          L=L+1
325       DO 360 J=1,10
          ILAB=LABL(I,J)
          IF(ILAB.EQ.IBL) GOTO 365
          IF(LABL(I,1).EQ.IAN) GOTO 335
          IF(ILAB.EQ.IS) GOTO 345
          IF(ILAB.EQ.ICNST) GOTO 350
          DO 330 K=1,NLNM
          IF(ILAB.EQ.NAME(K,1)) MET(L,J)=K
330       CONTINUE
          GOTO 360
335       DO 340 K=1,NLNM
          IF(ILAB.EQ.NAME(K,1)) LCL(J-1)=K
340       CONTINUE
          GOTO 360
345       LMIS=L
          LVO=.FALSE.
          GOTO 360
350       LNCNST=L
          LV1=.FALSE.
360       CONTINUE
365       CONTINUE
          IF((.NOT.LVO.OR..NOT.LV1).OR.(L.NE.0)) GOTO 370
          WRITE(6,6380)
          STOP
370       IF(LVO.OR.LV1) GOTO 381
          WRITE(6,6350)
          STOP
381       READ(5,5040) BUFFER,PM,IONSTR
          TLGMMA=IONSTR
          IONSTR=IONSTR-BUFFER
          IF(LTL2FR) IONSTR=0.
382       IONS=IONS-NCL
          IF(LMIS.EQ.0) GOTO 385
          IONIS=MET(LMIS,1)
385       I=0
390       I=I+1
          J=1
          READ(5,5070) ILAB,LRLABL(I,J),(REGRNG(I,L),L=1,2)
          IF(ILAB.EQ.IG.OR.ILAB.EQ.NLR) GOTO 410
          LR=.TRUE.
          IF(ILAB.NE.ILR) GOTO 395
          WRITE(6,6410)
          STOP
395       IF(ILAB.EQ.IMRT) GOTO 400
          WRITE(6,6400)
          STOP
400       J=J+1
          READ(5,5090) ILAB,LRLABL(I,J)
          IF(ILAB.EQ.ILR) GOTO 400
          IF(ILAB.EQ.NOR) GOTO 390
410       IF(.NOT.LR) GOTO 427
          DO 425 I=1,5
          DO 420 J=1,10
          IF(LRLABL(I,J).EQ.IBL) GOTO 425
          DO 415 K=1,NLNM
          IF(NAME(K,1).NE.LRLABL(I,J)) GOTO 415
          LRREG(I,J)=K
          GOTO 420
```

```
415   CONTINUE
420   CONTINUE
425   CONTINUE
427   IF(KBIND.EQ.0) GOTO 447
      DO 435 I=1,10
      IF(LCL(I).EQ.0) GOTO 447
      DO 430 J=2,KBIND
      IF(IEBIND(J).EQ.LCL(I)) IEBIND(J)=-1
430   CONTINUE
435   CONTINUE
447   READ(5,5050) ISTORE
      IF(ISTORE.EQ.JST) LSTORE=.TRUE.
      BIG=0.
      DO 450 I=1,NLNM
      IF(VARY(I,2).EQ.0.) GOTO 450
      RANGE=1.+(VAL(I)-VARY(I,1))/VARY(I,2)
      IF(BIG.GE.RANGE) GOTO 450
      BIG=RANGE
      IBIG=I
450   CONTINUE
      IF(NOIONS) GOTO 469
      IF(NCL.EQ.0) GOTO 456
      DO 451 I=1,20
451   IHEAD(I)=IBL
      J=-1
      DO 452 I=1,10
      IF(LCL(I).EQ.0) GOTO 453
      J=J+2
      IHEAD(J)=NAME(LCL(I),1)
      IHEAD(J+1)=IPLUS1
452   CONTINUE
453   NCLOUT=J+1
      IHEAD(J+1)=IPAR2
      IF(IONS.EQ.0) GOTO 468
      IF(LV1) GOTO 454
      IHEAD(J+2)=NAME(MET(LNCNST,1),1)
      GOTO 456
454   IHEAD(J+2)=NAME(MET(LNIS,1),1)
456   DO 467 I=1,IONS
      IHEAD1(I,1)=NAME(MET(I,1),1)
      IHEAD1(I,2)=IPAR1
      LL=2
      DO 457 J=2,NLNM
      IF(MET(I,J).EQ.0) GOTO 460
      LL=LL+1
      IHEAD1(I,LL)=NAME(MET(I,J),1)
      LL=LL+1
      IHEAD1(I,LL)=IPLUS
457   CONTINUE
460   IHEAD1(I,LL)=IPAR2
      IF(I.NE.LNIS.AND.I.NE.LMCNST) GOTO 463
      LL=LL+1
      IHEAD1(I,LL)=IPLUS
      LL=LL+1
      IHEAD1(I,LL)=NAME(MET(I,1),1)
      LL=LL+1
      IHEAD1(I,LL)=IAN
      LL=LL+1
      IHEAD1(I,LL)=IPLUS
      LL=LL+1
      IHEAD1(I,LL)=NAME(MET(I,1),1)
      LL=LL+1
      IHEAD1(I,LL)=IOH
463   LL=LL+1
      DO 465 K=LL,19
      IHEAD1(I,K)=IBL
```

```
465   CONTINUE
467   CONTINUE
468   N=NCLOUT/2
      LFMT(6)=INF(N)
      MM=14+4*NCLOUT
      NCLOUT=MM/10
      MCL=MM-10*NCLOUT
      IF(MCL.EQ.0) MCL=1
      LFMT1(3)=INF(NCLOUT)
      LFMT1(6)=INF(MCL)
469   DO 470 I=1,22
470   IHEAD2(I)=IBL
      IF(LTL2FR) GOTO 487
      IF(DECIDE(IBIG).EQ.0) GOTO 471
      IHEAD2(1)=IT
      GOTO 472
471   IHEAD2(1)=LETBP
472   IHEAD2(2)=NAME(IBIG,1)
      K=0
      J=3
      IHEAD2(4)=IAM
      IF(.NOT.NOIONS) GOTO 474
      IHEAD2(3)=IM
      GOTO 477
474   IF(LV1) GOTO 476
      IHEAD2(3)=NAME(MET(LMCNST,1),1)
      GOTO 477
476   IHEAD2(3)=NAME(MET(LMIS,1),1)
477   DO 480 I=1,NLNM
      IF(DECIDE(I).EQ.1) GOTO 480
      J=J+2
      IHEAD2(J)=IT
      IHEAD2(J+1)=NAME(I,1)
      K=K+1
      IOUT(K)=I
480   CONTINUE
      INOUT=K
      IF(LV1) GOTO 483
      J=J+2
      IHEAD2(J)=IT
      IHEAD2(J+1)=IS1
      K=K+1
483   K=K+1
      KFMT(7)=INF(K)
      GOTO 490
487   DO 488 I=1,NLNM
488   IHEAD2(I)=NAME(I,1)
      IF(.NOT.NOIONS) IHEAD2(NLNMP1)=IS1
490   INAME=KAPP
      NOUT=0
      DO 500 I=1,NC
      INAME1=CMNAME(I,1)
      INAME2=CMNAME(I,2)
      OUTPUT=STBLTY(I)
      CALL TITLE(INAME,INAME1,INAME2,OUTPUT,NOUT)
500   CONTINUE
      IF(LTL2FR) GOTO 582
      NOUT=0
      DO 580 I=1,ICM
      IF(I.EQ.IBIG) GOTO 580
      IF(I.GT.NLNM) GOTO 555
      IF(IONS.EQ.0) GOTO 520
      IF(.NOT.LV1) GOTO 520
      DO 510 J=1,IONS
      IF(I.EQ.MET(J,1)) GOTO 580
510   CONTINUE
```

```
520   IF(DECIDE(I)) 580,530,540
530   INAME=LETP
      INAME1=NAME(I,1)
      INAME2=IBL
      GOTO 550
540   INAME=IT
      INAME1=IBL
      INAME2=NAME(I,1)
550   OUTPUT=VAL(I)
      GOTO 570
555   IF(DECIDE(I).EQ.0) GOTO 560
      INAME=IT
      GOTO 565
560   INAME=LETP
565   INAME1=NAME(I,1)
      INAME2=NAME(I,2)
      OUTPUT=VAL(I)
570   CALL TITLE(INAME,INAME1,INAME2,OUTPUT,NOUT)
580   CONTINUE
582   IF(NOIONS) GOTO 615
      IF(LTL2FR.OR.(IONS.EQ.0)) GOTO 590
      IF(LV1) GOTO 585
      WRITE(6,6360) PH,NAME(MET(LMCHST,1),1),BUFFER
      GOTO 590
585   WRITE(6,6240) TLGMMA,PH,NAME(MET(LMIS,1),1),BUFFER
590   IF(IONS.EQ.0) GOTO 595
      DO 592 I=1,IONS
      WRITE(6,6120) (IHEAD1(I,J),J=1,19)
592   CONTINUE
595   IF((NCL.NE.0) WRITE(6,LFMT) (IHEAD(I),I=1,NCLOUT)
      IF((NCL.NE.0).AND.(IONS.NE.0)) WRITE(6,LFMT1) (IHEAD(NCLOUT+1))
615   IF(LTL2FR) GOTO 620
      WRITE(6,6670) (IDASH,I=1,115)
      WRITE(6,6130) (IHEAD2(I),I=1,22)
      WRITE(6,6670) (IDASH,I=1,115)
      GOTO 627
620   IF(.NOT.NOIONS) WRITE(6,6370) BUFFER
      WRITE(6,6670) (IDASH,I=1,115)
      WRITE(6,6260) (IHEAD2(I),I=1,10)
      WRITE(6,6670) (IDASH,I=1,115)
627   DO 640 I=1,NLNM
      IF(DECIDE(I)) 633,630,635
630   FREE(I)=10.**(-VAL(I))
633   TOTAL(I)=.001
      GOTO 640
635   TOTAL(I)=VAL(I)
      FREE(I)=VAL(I)
640   CONTINUE
      IF(NCM.EQ.0) GOTO 650
      DO 645 I=NLNMP1,ICM
      IF(DECIDE(I).EQ.0) GOTO 642
      CMVAL(I)=VAL(I)
      GOTO 645
642   CMVAL(I)=10.**(-VAL(I))
645   CONTINUE
      DO 647 I=1,NCM
      A=CMVAL(LOCCOM(I,2))
      B=FREE(LOCCOM(I,3))
      C=STBLTY(LOCCOM(I,4))
      FREE(LOCCOM(I,1))=A/(B*C)
647   CONTINUE
650   METCL=0.
      CALL CALC
890   IF(.NOT.LSTORE) GOTO 920
      NSTORE=NSTORE+1
      DO 900 I=1,NLNM
```

```
          RECORD(NSTORE,I)=TOTAL(I)
900       CONTINUE
          RECORD(NSTORE,I+1)=METCL
920       NLINES=NLINES+1
          NREC=NREC+1
          IF(LTL2FR) GOTO 1045
          DO 930 I=1,INOUT
930       PRINT(I)=TOTAL(IOUT(I))
          IF(LV1) GOTO 940
          IONSTR=IONSTR+BUFFER
          WRITE(6,KFMT) NREC,VAL(IBIG),METCL,(PRINT(I),I=1,INOUT),IONSTR
          GOTO 950
940       WRITE(6,KFMT) NREC,VAL(IBIG),METCL,(PRINT(I),I=1,INOUT)
950       IF(VAL(IBIG).LE.(VARY(IBIG,1)+VARY(IBIG,2)/100.)) GOTO 1040
          VAL(IBIG)=VAL(IBIG)+VARY(IBIG,2)
          IF(NLINES.LT.45) GOTO 1100
          NLINES=0
          WRITE(6,6660)
          GOTO 1100
1040      IF(LF) CALL LFREG(NLIM,IONIS,IBIG,LCA)
          VAL(IBIG)=START(IBIG)
          GOTO 1050
1045      CALL TFOUT(NREC,NLINES,INOUT)
          IF(NLINES.LT.MAX) GOTO 1050
          NLINES=0
          WRITE(6,6660)
1050      DO 1080 I=1,ICM
          IF(VARY(I,2).EQ.0.) GOTO 1080
          IF(I.EQ.IBIG) GOTO 1080
          IF(VAL(I).LE.(VARY(I,1)+VARY(I,2)/100.)) GOTO 1060
          VAL(I)=VAL(I)-VARY(I,2)
          NLINES=0
          WRITE(6,6660)
          GOTO 1100
1060      VAL(I)=START(I)
1080      CONTINUE
          IF(.NOT.LSTORE) STOP
1090      NLINES=0
          WRITE(6,6660)
          GOTO 1190
1100      IF(NLINES.EQ.0) GOTO 490
          GOTO 627
1190      K=0
1200      READ(5,5060) ILAB,ILAB1,VALUE
          IF(ILAB.EQ.IG1.OR.ILAB.EQ.MORE) GOTO 1270
          IP=0
          DO 1210 I=1,NC
          IF(ILAB.EQ.CHNAME(I,1).AND.ILAB1.EQ.CHNAME(I,2)) IP=I
1210      CONTINUE
          IF(IP.NE.0) GOTO 1220
          WRITE(6,6250)
          STOP
1220      IF(VALUE.EQ.0.) GOTO 1230
          STBLTY(IP)=VALUE
          GOTO 1200
1230      K=K+1
          IOUT(K)=IP
          GOTO 1200
1270      LNNSTR=.TRUE.
          LSTORE=.FALSE.
          INAME=KAPP
          NREAD=0
          MAX=5
          NLINES=0
          INOUT=K
          IF(INOUT.NE.0) MAX=0
```

```
      MAX=MAX÷13
      DO 1280 I=1,NLNM
      DECIDE(I)=1
1280  IHEAD2(I)=NAME(I,1)
      IHEAD2(NLNMP1)=IDIS
      IF(LTL2FR) IHEAD2(NLNMP1)=IS1
      IF(NOIONS) IHEAD2(NLNMP1)=IBL
      DO 1290 I=NLNMP2,10
1290  IHEAD2(I)=IBL
1300  NOUT=0
      DO 1310 I=1,NC
      INAME1=CMNAME(I,1)
      INAME2=CMNAME(I,2)
      OUTPUT=STBLTY(I)
      CALL TITLE(INAME,INAME1,INAME2,OUTPUT,NOUT)
1310  CONTINUE
      WRITE(6,6670) (IDASH,I=1,115)
      WRITE(6,6260) (IHEAD2(I),I=1,10)
      WRITE(6,6670) (IDASH,I=1,115)
1330  NREAD=NREAD+1
      IF(NREAD.LE.NSTORE) GOTO 1335
      IF(ILAB.EQ.IG1) STOP
      GOTO 1090
1335  DO 1340 I=1,NLNM
      TOTAL(I)=RECORD(NREAD,I)
1340  FREE(I)=TOTAL(I)
      METCL=RECORD(NREAD,I+1)
      CALL CALC
      CALL TFOUT(NREAD,NLINES,INOUT)
      IF(NLINES.LT.MAX) GOTO 1330
      NLINES=0
      WRITE(6,6660)
      GOTO 1300
5000  FORMAT(2A4,1X,F6.3,E10.3)
5010  FORMAT(A3,A4,1X,F6.3,3(E10.3))
5020  FORMAT(A3,2A4,3(E10.3))
5030  FORMAT(A3,10(A4,1X))
5040  FORMAT(E8.2,2(E9.2))
5050  FORMAT(A4)
5060  FORMAT(2A4,E10.3)
5070  FORMAT(A3,1X,A4,F5.2,3X,F5.2)
5090  FORMAT(A3,3X,A4)
6000  FORMAT(1X,'ILLEGAL SPECIFICATION ON INPUT OF METALS',
     1' AND LIGANDS')
6010  FORMAT(1X,'ILLEGAL SPECIFICATION ON INPUT OF METAL-LIGAND',
     1' COMPLEXES')
6020  FORMAT(1X,'COMPLEX SPECIFIED BUT NO FREE METAL OR LIGAND',
     1' TO BE SET EXISTS')
6030  FORMAT(1X,'A METAL AND/OR LIGAND IS SPECIFIED MORE THAN ONCE')
6040  FORMAT(1X,'COMPLEX SPECIFIED BUT NO FREE VALUE ASSOCIATED',
     1' WITH COMPLEX IS SPECIFIED')
6050  FORMAT(1X,'EITHER NO STABILITY CONSTANT FOR SPECIFIED COMPLEX',
     1' EXISTS OR COMPLEX IS ILLEGAL')
6060  FORMAT(1X,'CANNOT SPECIFY COMPLEX WHEN NO FREE VALUE IS ',
     1'SPECIFIED TO BE SET BY COMPLEX')
6070  FORMAT(1X,'SPECIFIED FREE ',A4,' TO BE SET BY A COMPLEX',
     1' BUT NO COMPLEX EXISTS.')
6120  FORMAT(1X,'TL',1X,A3,'= 2 #',18(A4,1X))
6130  FORMAT(7X,A2,1X,A4,5X,10(A2,1X,A4,4X))
6240  FORMAT(1X,'TOTAL IONIC STRENGTH (I.S.) =',F6.3,'N. CALCULATIONS',
     1' DONE ASSUMING',1PE9.2,'M OF ',A2,'OH FOR PH-ING. I.S. OF ',
     3'BUFFER IS',1PE9.2,'M')
6250  FORMAT(1X,'ILLEGAL COMPLEX INPUT FOR TOTAL TO FREE',
     1' CALCULATIONS.')
6260  FORMAT(9X,10(8X,A4))
6350  FORMAT(1X,'CANNOT KEEP BOTH IONIC STRENGTH AND TOTAL MONOVALENT',
```

```
        1' CATION CONSTANT')
6360 FORMAT(1PE9.2,'M OF ',A2,'OH ASSUMED TO PH THE SOLUTIONS.',
     1' ALSO BUFFER ADDS',1PE9.2,'M TO THE IONIC STRENGTH.')
6370 FORMAT(1X,'BUFFER CONTRIBUTES',1PE9.2,'M TO IONIC STRENGTH')
6380 FORMAT(1X,'CATIONS WERE SPECIFIED BUT NONE WERE LABELED',
     1' WITH -IS- OR -CNST-')
6400 FORMAT(1X,'ILLEGAL INPUT IN LINEAR REGRESSION DATA')
6410 FORMAT(1X,'MUST SPECIFY WHAT LINEAR REGRESSION IS TO BE ',
     1'DONE WITH RESPECT TO FIRST')
6660 FORMAT('1')
6670 FORMAT(1X,115A1)
6680 FORMAT(' ')
     END
     SUBROUTINE TITLE(IN,IN1,IN2,OUT,NUMOUT)
     IF(NUMOUT.LT.6) GOTO 500
     NUMOUT=0
500  NUMOUT=NUMOUT+1
     GOTO (590,595,600,605,610,615),NUMOUT
590  WRITE(6,6060) IN,IN1,IN2,OUT
     GOTO 620
595  WRITE(6,6070) IN,IN1,IN2,OUT
     GOTO 620
600  WRITE(6,6080) IN,IN1,IN2,OUT
     GOTO 620
605  WRITE(6,6090) IN,IN1,IN2,OUT
     GOTO 620
615  WRITE(6,6110) IN,IN1,IN2,OUT
     GOTO 620
610  WRITE(6,6100) IN,IN1,IN2,OUT
620  RETURN
6060 FORMAT(1X,A2,2A4,'=',1PG10.3)
6070 FORMAT('+',21X,'/',A2,2A4,'=',1PG10.3)
6080 FORMAT('+',43X,'/',A2,2A4,'=',1PG10.3)
6090 FORMAT('+',65X,'/',A2,2A4,'=',1PG10.3)
6100 FORMAT('+',87X,'/',A2,2A4,'=',1PG10.3)
6110 FORMAT('+',109X,'/',A2,2A4,'=',1PG10.3)
     END
     SUBROUTINE TFOUT(NNN,LINES,NUMOUT)
     DIMENSION PFREE(20)
     COMMON /TFCALC/ STBLTY(30),IALPHA(20,30),MET(10,10),LCL(10),
    1CMCHRG(30),CHARGE(20),NLNM,ADDIN,LRNSTR,LMIS,PH,IONSTR,
    2NC,LV1,LMCNST,LTL2FR,IRBIND(10)
     COMMON /TLFR/ TOTAL(20),FREE(20),NAME(30,2),DECIDE(30),METCL,IONS
     COMMON /OUT/ NFMT1(10),IFMT1(8),IOUT(10),MAX,NOIONS,
    1BUFFER,CMNAME(30,2)
     LOGICAL NOIONS,LV1,LTL2FR,LRNSTR,LSTALL
     REAL IONSTR,METCL
     INTEGER DECIDE,CMNAME
     DATA LETP/'P '/
     DO 1360 I=1,NLNM
     IF(FREE(I).GT.0) GOTO 1350
     PFREE(I)=99.999
     GOTO 1360
1350 PFREE(I)=-1.*ALOG10(FREE(I))
1360 CONTINUE
     LSTALL=.TRUE.
     NWRITE=NLNM
     IF(NLNM.LE.9) GOTO 1362
     NWRITE=10
     LSTALL=.FALSE.
1362 LINES=LINES+1
     NOUT=0
     ADDIN=-1.*ADDIN
     IF(LTL2FR) ADDIN=ADDIN+BUFFER
     IF(NOIONS.OR..NOT.LSTALL) GOTO 1364
     WRITE(6,NFMT1) NNN,(TOTAL(I),I=1,NWRITE),ADDIN
```

```
            GOTO 1367
1364  WRITE(6,IFMT1)  NUM,(TOTAL(I),I=1,NWRITE)
1367  WRITE(6,IFMT1)  (PFREE(I),I=1,NWRITE)
      IF(NUMOUT.EQ.0) GOTO 1440
      DO 1425 J=1,NUMOUT
      CMPLX=1.
      DO 1370 I=1,NLNM
1370  CMPLX=CMPLX*FREE(I)**IALPHA(I,IOUT(J))
      CMPLX=CMPLX*STBLTY(IOUT(J))
      PCMPLX=-1.*ALOG10(CMPLX)
      IF(NOUT.LT.6) GOTO 1380
      NOUT=0
1380  NOUT=NOUT+1
      INAME=LETP
      INAME1=CNAME(IOUT(J),1)
      INAME2=CNAME(IOUT(J),2)
      GOTO (1390,1395,1400,1405,1410,1415),NOUT
1390  WRITE(6,6290) INAME,INAME1,INAME2,PCMPLX
      GOTO 1425
1395  WRITE(6,6300) INAME,INAME1,INAME2,PCMPLX
      GOTO 1425
1400  WRITE(6,6310) INAME,INAME1,INAME2,PCMPLX
      GOTO 1425
1405  WRITE(6,6320) INAME,INAME1,INAME2,PCMPLX
      GOTO 1425
1410  WRITE(6,6330) INAME,INAME1,INAME2,PCMPLX
      GOTO 1425
1415  WRITE(6,6340) INAME,INAME1,INAME2,PCMPLX
1425  CONTINUE
1440  WRITE(6,6680)
      RETURN
6290  FORMAT(6X,A2,2A4,' =',F7.3,'/')
6300  FORMAT('+',25X,A2,2A4,' =',F7.3,'/')
6310  FORMAT('+',45X,A2,2A4,' =',F7.3,'/')
6320  FORMAT('+',65X,A2,2A4,' =',F7.3,'/')
6330  FORMAT('+',85X,A2,2A4,' =',F7.3,'/')
6340  FORMAT('+',105X,A2,2A4,' =',F7.3)
6680  FORMAT(' ')
      END
      SUBROUTINE CALC
      DIMENSION KNT(20)
      COMMON /TLFR/ TOTAL(20),FREE(20),NAME(30,2),DECIDE(30),METCL,IONS
      COMMON /TFCALC/ STBLTY(30),IALPHA(20,30),MET(10,10),LCL(10),
     1CMCHRG(30),CHARGE(20),NLNM,ADDIN,LRNSTR,LNIS,PH,IONSTR,
     2NC,LV1,LNCNST,LTL2FR,IEBIND(10)
      LOGICAL LRNSTR,LV1,LTL2FR
      INTEGER DECIDE
      REAL METCL,IONSTR
      DO 600 I=1,NLNM
600   KNT(I)=0
655   NUM=0
660   NUM=NUM+1
      IF(KNT(NUM).EQ.1) GOTO 720
      SUM=0.
      DO 690 J=1,NC
      IF(IALPHA(NUM,J).EQ.0) GOTO 690
      PROD=1.
      DO 670 K=1,NLNM
670   PROD=PROD*FREE(K)**IALPHA(K,J)
      XSUBJ=PROD*STBLTY(J)
      SUM=SUM+XSUBJ
690   CONTINUE
      IF(DECIDE(NUM).EQ.1) GOTO 710
      BEFORE=TOTAL(NUM)
      TOTAL(NUM)=FREE(NUM)+SUM
      IF((ABS(BEFORE-TOTAL(NUM))/BEFORE).LE..00001) KNT(NUM)=1
```

```
          GOTO 720
  710     BEFORE=FREE(NUM)
          FREE(NUM)=FREE(NUM)*TOTAL(NUM)/(FREE(NUM)+SUM)
          IF((ABS(BEFORE-FREE(NUM))/BEFORE).LE..00001) KNT(NUM)=1
  720     IF(NUM.LT.NLNM) GOTO 660
          NUM=0
          DO 730 I=1,NLNM
          IF(KNT(I).EQ.0) GOTO 660
  730     CONTINUE
          DO 780 I=1,NLNM
  780     KNT(I)=0
          GAMMA=0.
          DO 820 J=1,NC
          CMPLX=1.
          DO 800 I=1,NLNM
          IF(IALPHA(I,J).EQ.0) GOTO 800
          CMPLX=CMPLX*FREE(I)
  800     CONTINUE
          GAMMA=GAMMA+CMPLX*STBLTY(J)*CMCHRG(J)**2
  820     CONTINUE
          DO 840 I=1,NLNM
  840     GAMMA=GAMMA+FREE(I)*CHARGE(I)**2
          DO 850 I=1,10
          IF(LCL(I).EQ.0) GOTO 852
          GAMMA=GAMMA+2.*TOTAL(LCL(I))
  850     CONTINUE
  852     CHECK=TOTAL(IEBIND(1))
          DO 854 I=2,10
          II=IEBIND(I)
          IF(II.EQ.0) GOTO 855
          IF(II.LT.0) GOTO 854
          IF(TOTAL(II).LE.CHECK) GOTO 854
          GAMMA=GAMMA+2.*(TOTAL(II)-CHECK)
  854     CONTINUE
  855     GAMMA=GAMMA+METCL
          ADDIN=IONSTR-GAMMA/2.
          IF(LRNSTR.OR.LTL2FR) RETURN
          IF(ABS(ADDIN).LT.5.0E-07) GOTO 890
          IF(IONS.EQ.0) GOTO 890
          DO 880 I=1,IONS
          SUM=0.
          DO 860 J=2,10
          IF(MET(I,J).EQ.0) GOTO 865
          SUM=SUM+TOTAL(MET(I,J))
  860     CONTINUE
  865     IF(I.NE.LMCNST) GOTO 870
          METCL=TOTAL(MET(I,1))-2.*SUM-PH
          GAMMA=GAMMA+METCL
          IONSTR=GAMMA/2.
          GOTO 880
  870     TOTAL(MET(I,1))=2.*SUM
  880     CONTINUE
          IF(.NOT.LV1) GOTO 655
          METCL=METCL+ADDIN
          TOTAL(MET(LMIS,1))=TOTAL(MET(LMIS,1))+METCL+PH
          GOTO 655
  890     RETURN
          END
          SUBROUTINE LINREG(LM,ISMET,LBIG,ICA)
          COMMON /LREG/ LNREG(5,10),REGRNG(5,2)
          COMMON /TLFR/ TOTAL(20),FREE(20),NAME(30,2),DECIDE(30),METCL,IONS
          REAL METCL,INCEPT
          DOUBLE PRECISION SUMX,SUMX2,SUMXY(20),SUMY(20)
          INTEGER DECIDE
          DATA IEG/'EGTA'/,IED/'EDTA'/
          WRITE(6,6660)
```

```
         IP=0
         DO 60 I=1,LM
         IMM=NAME(I,1)
         IF(IMM.EQ.IEG.OR.IMM.EQ.IED) IP=I
60       CONTINUE
         TOTAL(LBIG)=TOTAL(IP)
         ITEM=DECIDE(LBIG)
         DECIDE(LBIG)=1
         CALL CALC
         DECIDE(LBIG)=ITEM
         PVAL=-1.*ALOG10(FREE(LBIG))
         WRITE(6,6000) NAME(LBIG,1),NAME(IP,1),NAME(LBIG,1),PVAL
         IF(ICA.EQ.0) GOTO 70
         TOTAL(ICA)=3.0E-06
         ITEM=DECIDE(ICA)
         DECIDE(ICA)=1
         IF(ICA.EQ.LBIG) GOTO 65
         TOTAL(LBIG)=1.0E-30
65       CALL CALC
         DECIDE(ICA)=ITEM
         PVAL=-1.*ALOG10(FREE(ICA))
         WRITE(6,6005) PVAL
70       LMM1=LM-1
         DO 250 I=1,5
         WRITE(6,6050)
         IF(LNREG(I,1).EQ.0) GOTO 260
         III=LNREG(I,1)
         ITEM=DECIDE(III)
         SUMX=0.
         SUMX2=0.
         DO 80 K=1,20
         SUMY(K)=0.
         SUMXY(K)=0.
80       CONTINUE
         FREE(III)=10.**(-REGRNG(I,1))
         DECIDE(III)=0
         CALL CALC
         TL1=TOTAL(III)
         FREE(III)=10.**(-REGRNG(I,2))
         CALL CALC
         TL2=TOTAL(III)
         IF(TL2.GT.TL1) GOTO 100
         TEMP=TL2
         TL2=TL1
         TL1=TEMP
100      STEP=(TL2-TL1)/14.
         FUZZ=STEP/100.
         NN=1
         WRITE(6,6010) NAME(III,1),(REGRNG(I,J),J=1,2)
         DECIDE(III)=1
110      TOTAL(III)=TL1
         CALL CALC
         SUMX=SUMX+TOTAL(III)
         SUMX2=SUMX2+TOTAL(III)**2
         DO 140 J=2,LM
         JJJ=LNREG(I,J)
         IF(JJJ.EQ.0) GOTO 160
         IF(IONS.EQ.0) GOTO 120
         IF(JJJ.NE.ISMET) GOTO 120
         SUMY(J-1)=SUMY(J-1)+METCL
         SUMXY(J-1)=SUMXY(J-1)+METCL*TOTAL(III)
         GOTO 140
120      SUMY(J-1)=SUMY(J-1)+TOTAL(JJJ)
         SUMXY(J-1)=SUMXY(J-1)+TOTAL(JJJ)*TOTAL(III)
140      CONTINUE
160      IF(TL1.GE.(TL2-FUZZ)) GOTO 170
```

```
        TL1=TL1+STEP
        NN=NN+1
        GOTO 110
170     DO 190 J=1,LIMM
        IF(LMREG(I,J+1).EQ.0) GOTO 240
        SLOPE=(SUMXY(J)-SUMX*SUMY(J)/NN)/(SUMX2-SUMX**2/NN)
        INCEPT=(SUMY(J)-SLOPE*SUMX)/NN
        CONCA=SLOPE*TOTAL(IP)+INCEPT
        CONCB=INCEPT
        LABL=NAME(LMREG(I,J+1),1)
        LABL1=NAME(LMREG(I,1),1)
        IF(LMREG(I,J+1).NE.ISMET) GOTO 180
        WRITE(6,6030) NAME(LMREG(I,J+1),1)
        WRITE(6,6040) SLOPE,LABL1,INCEPT,CONCA,NAME(LBIG,1),
       1NAME(IP,1),CONCB,NAME(IP,1)
        GOTO 190
180     WRITE(6,6020) LABL,SLOPE,LABL1,INCEPT,CONCA,NAME(LBIG,1),
       1NAME(IP,1),CONCB,NAME(IP,1)
190     CONTINUE
240     DECIDE(III)=ITEM
250     CONTINUE
260     RETURN
6000    FORMAT(1X,'WHEN TOTAL ',A2,' EQUALS TOTAL ',A4,', P',A2,
       1' IS',1PG11.4,/)
6005    FORMAT(1X,'ASSUMING 3.000E-06M TOTAL CA CONTAMINATION, MAXIMUM',
       1' PCA IS',F7.3,/)
6010    FORMAT(1X,'LINEAR REGRESSION FOR P',A4,' =',F5.2,' TO',F5.2,/)
6020    FORMAT(6X,'TOTAL ',A4,' = ',F8.4,'(TOTAL ',A4,') +',1PE10.3,6X,
       1'GIVES:',3X,1PE10.3,'M IN ',A2,A4,' SOLUTION AND',
       21PE10.3,'M IN ',A4,' SOLUTION',/)
6030    FORMAT(6X,'TOTAL ',A4)
6040    FORMAT('+',13X,'A- = ',F8.4,'(TOTAL ',A4,') +',1PE10.3,6X,
       1'GIVES:',3X,1PE10.3,'M IN ',A2,A4,' SOLUTION AND',
       21PE10.3,'M IN ',A4,' SOLUTION',/)
6050    FORMAT(' ')
6660    FORMAT('1')
        END
```

## Instructions for Entering the Data

Before entering data into the SPECS program, preliminary calculations are necessary to obtain the apparent stability constants at the specified temperature and pH and the rms charges of the complexes and ligands. This is done by the ancillary program ABSTOAPP. The rms charges are not necessary if the ionic strength is not to be computed. Then the selected stability constants listed by the ancillary program STACONS can be used, which eliminates the risk of errors in selecting absolute stability constants.

As previously stated, the specification data can be entered with a word-processor program. This is not mandatory but merely to facilitate corrections, as explained below. For instance, if WordStar is used on an IBM-PC microcomputer with two 360-kilobyte disk drives, the WordStar program is in drive A:, but the data files should be on another diskette in drive B: because there would not be enough space for these files on diskette No. 1. After booting the computer in drive A:, WordStar is called by typing

WS

and, thereafter, the disk drive should be changed by typing

L

Then a nondocument file should be opened by typing

N

Each data file has a name of up to eight characters. Using a word-processor program allows data to be edited or copied so corrections can be made. Thus, when only a few among a large number of specifications have to be changed, it is sufficient to copy the data file into a new file and then to make the appropriate modifications in this new data file. This saves the time of reentering the unchanged specifications.

After all the data have been entered and the file saved, the program can be run. This is done in drive A: by typing SPECS or the batch file name, such as SPECSE, if the compressed mode is to be used. Then the program will request:

ENTER THE COMPLETE DATA FILENAME

"Complete" indicates that the drive should be included in the filename if an IBM-PC is used. For instance, if EXAMPLE1 is the name of the data file, the answer to the request of the computer program should be:

B:EXAMPLE1

since the data are on drive B:.

If an IBM-AT or -XT with a hard disk is used, the wordprocessor program and the specification data files are both contained in the hard disk. Then the answer to the request of the computer program should be merely:

EXAMPLE1

Note that (1) the SPECS program does not support path names, and (2) only capital letters should be used for the name of the data file as well as for entering the data.

As shown graphically in Fig. 1, the data are entered in seven categories entitled: (1) stability constants, (2) specifications of the free or total concentrations of ions and substrates, (3) specifications of the concentrations of complexes, (4) specifications of the combination of the monovalent ions, (5) ionic strength specifications, (6) linear regression specifications, and (7) calculations from total to free concentrations with change of the apparent stability constants. The following detailed instructions describe how the data are entered in these seven categories. To get familiarity with these instructions, it is suggested that the reader (1) read the instruction cursively, (2) try the first example with the explanations given in the subsequent section, and (3) return to a detailed reading of the instructions.

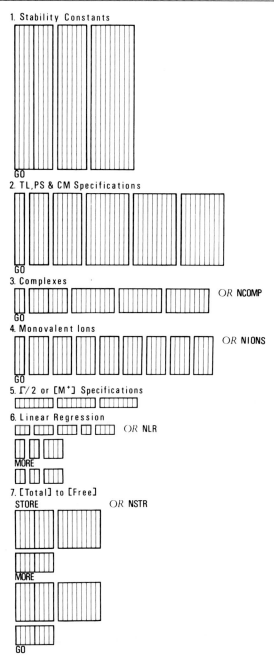

Fig. 1. Space allotted for each data entry for SPECS. Each column is one character wide.

*I. Stability Constants*

Each line must contain the name of the complex (i.e., the names of the complexing metal and ligand), the charge of the complex, and the apparent stability constants defining the complex formation. In this and all other sections of the data specifications, each ion name must be four characters long; a blank space is treated as a character by the computer. Each charge is preceded by its sign and is three decimal places long. Although the sign of the charge is respected, an error of sign would not affect the calculation of the ionic strength inasmuch as the rms charges of the ligands and complexes are used in the calculation of the ionic strength. If no computation of the ionic strength is made, charges need not be entered, i.e., the input of the rms charges could be replaced by +0.000 or the corresponding spaces may be left blank. The format for entering the data is shown graphically in Fig. 1 which indicates the number of spaces for each input. Each stability constant is three decimal places long and is in scientific notation. A space separates the complex name from the charge and the charge from the stability constant. Note that, in the following examples, both in the text and in Figs. 2, 4, and 6, the symbol ■ indicates that a blank space should be entered by pressing the space bar.

*Example.* The $K_{app}$ for the complexes between ethyleneglycol-bis($\beta$-aminoethyl ether)-*N,N'*-tetraacetic acid (EGTA) and Ca$^{2+}$ (CaEGTA) = $3.976 \times 10^6$ $M^{-1}$ with CaEGTA having a charge of $-1.999$ (as calculated by the ancillary program ABSTOAPP) is typed as:

    CA■■EGTA■−1.999■3.976E+06

When all the stability constants have been entered

    GO

is typed in the first two spaces of a line to indicate that no more stability constants are to be entered.

*II. Specifications of the Free or Total Concentrations of Ions and Substrates*

To solve the equations describing the equilibria in aqueous solutions containing multiple metals and ligands, either the total concentration or the free concentration of each metal or ligand must be known. When the concentration of a complex (e.g., MgATP) has to be kept constant, the free concentration of one of the ions participating in this complex must be specified (e.g., [free Mg$^{2+}$]; note that throughout this article brackets denote concentration of whatever is within the brackets). Then, the concentration of the complex (e.g., MgATP) is kept constant by varying the free concentration of the other ion (e.g., [free ATP]). These conditions are specified in the program through the commands TL (to specify total concentration), PS (to specify the $-\log_{10}$ of the free molar concentration, which is the p value; e.g., pCa for $Ca^{2+}$), and CM (to calculate the free concentration of the ion not specified when the complex concentration is specified).

Each line should contain the command TL, PS, or CM, the name of an ion or substrate, and the charge of the free ion or substrate. If TL or PS is specified, one should also type the initial concentration and, if desired, the final concentration and the magnitude of each step for concentration increments. If CM is specified, no concentration need be entered as the free concentration is calculated by the computer. For instance, if pMg and pMgATP are specified, the CM command should be used with ATP and the charge of free ATP should be inserted next. Each concentration, whether initial, final, or step, is three decimal places long and is in scientific notation. If PS is specified, the concentrations are in $-\log_{10}$ free molar concentration (p values). If TL is specified, the concentrations are in molar units. Whether PS or TL is specified the initial number must be greater than the final number as the computer subtracts the step from the initial and subsequent numbers (see the second example). A space separates the PS, TL, or CM specification from the substance name, the substance name from the charge, the charge from the initial concentration, the initial from the final concentration, and the final concentration from the concentration step.

*Examples.* Specifying a [total EGTA] of 10 m*M* with free EGTA having a charge of $-2.017$ is typed as:

    TL■EGTA■ $-2.107$ ■ $1.000E-02$

If pMg and pMgATP are specified, the CM commands should be used with ATP. Since free ATP has a charge of $-3.552$ at pH 7.10, this is typed as:

    CM■ATP■■ $-3.552$

Note that neither the free nor the total ATP concentration need to be specified as the CM command instructs the computer to calculate [free ATP] from the pMg, pMgATP, and the $K_{app}$ for MgATP.

Varying pCa from 9 to 4 in increments of 0.5 units with free $Ca^{2+}$ having a charge of $2+$ is typed as:

PS■CA■■■ $+2.000$ ■ $9.000E+00$ ■ $4.000E+00$ ■ $5.000E-01$

Note that when a constant ionic strength is specified (as described in subsequent sections) initial estimates of the total concentrations of monovalent cations must be entered. These estimates need not to be close to the final values, which will be found by iteration, but should not be zero. Estimates farther from the final values only increase the number of iterations, which will have a negligible effect on the computation time. The total concentration is kept constant for all monovalent cations if the

NIONS command is used and for the labeled cation if the CNST command is used. When all the PS, TL, and CM specifications have been entered
    GO
should be typed to indicate that there are no more concentration specifications.

### III. Specifications of the Concentrations of Complexes

If none of the concentrations of the complexes such as MgATP or CaEGTA is to be specified, a line typed
    NCOMP
(for "no complexes," starting at the first space of the line) should be entered. Then one can proceed to the next section without typing a GO line.

Otherwise, each line contains the command PS or TL, the name of the complex (i.e., of the complexing metal and ligand), the initial concentration of the complex, and, if it is to be varied, the final concentration and the concentration step. If PS is specified, the concentrations are in $-\log_{10}$ free molar concentration (p values). If TL is specified, the concentrations are in molar units. As for the ion and substrate specifications, if the concentration is to be varied the initial number must be greater than the final number. Each concentration, whether initial, final, or step, is three decimal places long and is in scientific notation. A space separates the PS or TL specification from the complex name and the complex name from the concentration.

*Examples.* pMgATP 2.5 is typed as:
    PS ■MG■■ATP■■2.500E+00
[CaEGTA] varying from 2 m$M$ to 1 m$M$ with increments of 250 $\mu M$ is typed as:
    TL■CA■■EGTA■2.000E−03■1.000E−03■2.500E−04
Again, this section of the input ends with a line:
    GO

### IV. Specifications of Combinations of the Monovalent Ions

This program can be used in three ways with respect to the ionic strength or the concentration of the major cation (e.g., K$^+$), which is termed M$^+$. The first way consists of keeping the ionic strength constant and is specified by the command IS (for "ionic strength"). The second consists of keeping the total concentration of the major monovalent cation M$^+$ constant and is specified by the command CNST (for "constant"). The third does not calculate ionic strength or major cation concentration and is specified by the command NIONS (for "no ions"). Thus, how (combined with which substances) the monovalent ions are added is specified as follows.

For the cations, all the ligands to which they are associated are listed beside them. For example, if $K_2EGTA$ is to be added, then total potassium should have EGTA beside it and if $Na_2ATP$ and $Na_2PC$ (where PC stands for phosphocreatine) are to be added, then total sodium should have ATP and PC beside it. If the ionic strength is to be kept constant through the addition of KCl, then IS should be listed on the same line as the list of ligands added as $K_2L$ (not necessarily last). If the [total potassium] is to be kept constant through the addition of KCl, CNST should be listed on the same line as the list of ligands added as $K_2L$. Obviously, IS and CNST may not be used together.

The anions are treated as monovalent $(A^-)$. The divalent cations added as $M^{2+}A^-_2$ are listed beside TL $A^-$. If a divalent anion is used with a divalent metal (e.g., $MgSO_4$), the actual ionic strength will be higher than the computer indicates by an amount equal to the total concentration of divalent metal added as $M^{2+}A^{2-}$ (e.g., [total Mg] added as $MgSO_4$).

If no ion specifications are to be made,

NIONS

is entered. This also cancels the next section (Section V, Ionic Strength Specifications).

Otherwise, each line for cation specifications begins with TL and contains the name of the cation and which ligand it is added to as a salt, i.e., as $M^+_2L$. One line of the cation specification contains IS or CNST. For the anions, each line begins with TL $A^-$ and contains the names of the divalent metals added as $M^{2+}A^-_2$. A space separates the TL at the beginning of each line from the cation name or the $A^-$. Each subsequent name is separated by a space.

*Example.* If EGTA is to be added as $K_2EGTA$, ATP and PC are to be added as $Na_2ATP$ and $Na_2PC$, $MgCl_2$ is to be added, and the ionic strength is to be kept constant by the addition of KCl, the instructions should be typed as follows:

TL■K■■■■EGTA■IS
TL■NA■■■ATP■■PC
TL■A−■■■MG

Again,

GO

ends the input for this section.

*V. Ionic Strength Specifications*

A single line should contain the ionic strength in molar units contributed by the pH buffer, the concentration in molar units of $M^+OH^-$ needed to adjust the pH of the solution, and the total ionic strength in molar units. The monovalent cation in the base used to adjust the pH of the solution must be the one labeled with IS or CNST in the monovalent ion specification section. If a monovalent cation has been labeled with CNST, the total ionic strength should not be entered, as this is calculated by the computer. However, entering the total ionic strength with CNST specified would cause no problems because the computer will ignore it. Each number is two decimal places long and is in scientific notation. A space separates subsequent numbers.

*Example.* Knowing from the output of the ABSTOAPP program that 30 m*M* BES contributes 7.14 m*M* to the ionic strength at pH 7.10 and 22° and that the solution to be prepared needs approximately 6 m*M* KOH for adjusting the pH to 7.10 (found experimentally from past solutions), if the total ionic strength has to be 180 m*M*, these instructions should be typed as:

7.14E−03■6.00E−03■1.80E−01

No GO command is necessary to signify the end of this section as it consists of a single line.

## VI. Linear Regression Specifications

Before even starting to read this section of the data input, the reader should know that this is the least-used section of the program and that it takes the longest to run (several seconds or minutes depending upon the difficulty of the problem). Thus, unless it is needed, this section should be eliminated by typing:

NLR

When making a series of solutions at various pCa values for an experiment, such as a determination of the force-pCa relation in skinned muscle fibers, it is simplest to make only a complete solution containing all substrates and ions, including EGTA but not calcium (EGTA solution), and a solution identical except that it contains CaEGTA instead of EGTA (CaEGTA solution). Then it is sufficient to mix these two solutions in appropriate proportions to obtain the different [total calcium] values needed for the different pCa values. Such a mixing of solutions at different pCa values was also used to produce progressive variations of [free Ca²⁺] with a microprocessor-controlled system of microinjections through microsyringes.[7] Since the [total ATP] and [total magnesium] required to obtain a given pMg and pMgATP vary when the pCa changes, the EGTA and CaEGTA solutions must have different concentrations of these two substances. To obtain the best values for [total ATP] and [total

---

[7] A. Fabiato, *J. Gen. Physiol.* **85**, 291 (1985).

magnesium] to be used in the EGTA and CaEGTA solutions, a linear regression is done over the range of pCa values to be used. This is the purpose of this section of the program, which outputs the [total ATP] and [total magnesium] to be used in the EGTA and CaEGTA solutions.

The linear regression of [total ATP] and [total magnesium] must be done with respect to [total calcium] as it is this quantity, and not [free $Ca^{2+}$], that varies linearly as the CaEGTA and EGTA solutions are mixed. Since [total calcium] is specified, [free $Ca^{2+}$] must be found to calculate for [total ATP] and [total magnesium].

The linear regression commands are WRT (for "with respect to") and LR■ON (for "linear regression on"). When a substance is being regressed (e.g., ATP or magnesium) with respect to another substance (e.g., calcium), the latter is specified by WRT and the range over which the linear regression is to be done is listed beside it in $-\log_{10}$ of the free molar concentration (e.g., pCa). The substances that are to be regressed are specified by LR■ON with one substance per line. Up to five runs of linear regressions for various ranges of the substance labeled WRT can be done by inserting MORE (for "more calculations") between runs.

If no linear regression is to be done, a line typed

    NLR

(for "no linear regression") is entered in the first three spaces.

Otherwise, the linear regression of the total concentration of various substances with respect to the total concentration of a single substance is accomplished as follows:

Each WRT line begins with WRT and contains the name of a substance and the range of the p values over which the linear regression is to be done. The range values are two decimal places long and in fixed notation. The first range value can be greater than or less than the second. A space separates the WRT command from the substance name and the substance name from the range value. The range values are separated by TO.

*Example.* If a linear regression is to be done with respect to total calcium for the pCa range of 8 to 6, this can be typed either as:

    WRT■CA■■■8.00■TO■6.00

or as:

    WRT■CA■■■6.00■TO■8.00

Each LR■ON line begins with LR■ON and contains the name of the substance to be regressed on. A space separates LR from ON and ON from the substance name. If the substance to be regressed is the monovalent cation labeled with IS, what is regressed is the total concentration of $M^+A^-$ to be added to complete the ionic strength rather than the total concentration of $M^+$.

*Example.* If a linear regression on total ATP and total magnesium and total KCl to be added to complete the ionic strength is to be done, this should be typed as:

LR ■ ON ■ ATP
LR ■ ON ■ MG
LR ■ ON ■ K

Thus, KCl, rather than K, is regressed.

If a second range of the substance labeled with WRT is to be studied, the next sequence of commands WRT and LR■ON must be preceded by MORE. This can be repeated for up to five ranges. If no new sequence is desired, GO should be typed instead of MORE.

## VII. Calculation from Total to Free Concentrations with Change of the Apparent Stability Constants

The program permits the specification of free ion concentrations and the output of the total concentrations using one set of apparent stability constants and then the use of these total concentrations to output the free ionic concentrations with a new set of apparent stability constants. The commands for the change of apparent stability constants begin with STORE. This tells the computer to store all results for conversion from total to free concentrations with the new set of apparent stability constants. The constants to be changed are entered after the STORE command. In addition, since the complex concentrations are often specified, entering the name of the complex will result in the p value of this complex being output together with the free ionic concentrations. An unlimited number of runs of computations from total to free concentrations with various apparent stability constants can be made by inserting MORE between the subsequent runs.

If no calculation from total to free concentrations is desired, type

NSTR

(for "no STORE") in the first four spaces of a line and stop there.

Otherwise, type

STORE

in the first five spaces of the line.

Each line following the STORE line contains the name of the complex (i.e., the complexing metal and ligand names), a space, and the apparent stability constant defining the complex formation. If no apparent stability constant (or an apparent stability constant of zero) is entered, the p value of that complex is calculated when the computation from total to free concentration takes place. Such a data input line is used as a command for checking, for instance, how the [MgATP] or the [CaEGTA] changes when the apparent stability constants are changed. But, if it is desired to actually change a stability constant to zero, a very small number (such as $10^{-20}$) should be entered instead of zero.

*Example.* To change the $K_{app}$ for CaEGTA to 4.410 × $10^6$ $M^{-1}$ for converting data obtained with the Allen and Blinks apparent stability constant to what would be obtained with the Schwarzenbach's apparent stability constant (see Ref. 8 for references to the original literature), and have the values of pMgATP and pCaEGTA printed along with the p values of all the other substances, one should type:

> STORE
> C A ■ ■ EGTA ■ 4 . 4 1 0 E + 0 6
> MG ■ ■ AT P
> C A ■ ■ EGTA

If another change of apparent stability constants is desired, MORE should be typed followed by the new constants. Note that all previous changes remain. Thus, to negate a change the original apparent stability constant should be reentered. If no more total to free concentration calculations are desired, GO should be typed instead of MORE.

### Examples

The following examples demonstrate the major, if not all, particular cases of the use of the SPECS program. As indicated previously, the inputs for these examples are already on diskette No. 2. Thus, for instance, the problem for the first example can be run by first typing

> SPECS

then, in answer to the question

> INPUT COMPLETE NAME OF DATA FILE

type

> B:EXAMPLE1

if two 360-kilobyte diskettes are used, or, simply,

> EXAMPLE1

if a hard disk is used.

### First Example

This problem could be that of making up solutions for experiments designed to obtain a force-pCa relation from skinned muscle fibers with the following specifications: 10 m*M* total EGTA (potassium salt), pCa varying from 9 to 4 in 0.5 pCa steps, pMg 3.0, pMgATP 2.5 (with ATP as sodium salt) in the presence of an ATP-regenerating system made of 12

---

[8] A. Fabiato, *J. Gen. Physiol.* **78**, 457 (1981).

m$M$ phosphocreatine (sodium salt, and 15 U/ml creatine phosphokinase that are ignored in the calculations), ionic strength of 180 m$M$ (with $K^+$ and $Cl^-$ as major ionic species), pH 7.1 buffered with 30 m$M$ $N,N$-bis(2-hydroxyethyl)-2-aminoethanesulfonic acid (BES) at 22°. Previous experience has indicated that about 6 m$M$ KOH is necessary to adjust the pH. It is desired to obtain the different pCa values by appropriate mixing of only three solutions at pCa 8.0, 6.0, and 4.5. Therefore, the linear regression will be done with respect to total calcium from pCa 8.0 to 6.0 and then from pCa 6.0 to 4.5. These data are to be computed with the Allen and Blinks apparent stability constant for CaEGTA (see Ref. 8 for references to the original literature). Then one wants to see what would be the effects of changing the apparent stability constant for CaEGTA to that of Schwarzenbach *et al.* (see Ref. 8 for references to the original literature), with calculation of the modification of not only the free ionic concentrations of metals and ligands but also of the [MgATP]. Finally, one wants to see the effects of changing the stability constant to that of Harafuji and Ogawa (see Ref. 8 for references to the original literature), with computation of not only the free ionic concentrations but also of the [MgATP] and the [CaEGTA].

The first step is to use the first option of the ABSTOAPP program to obtain the stability constants and the rms charges of the ligands and complexes. Then the second option of the ABSTOAPP program calculates the ionic strength contributed by the pH buffer. With this information the data can be entered as indicated in Fig. 2.

The output will be that shown in Fig. 3. Part of the output display has been deleted to save space in the second part of Fig. 3. Note that the output gives the slope and intercept for the linear regression before giving the concentrations of the substances in the two solutions. Also given is the lowest pCa value (i.e., the highest [free $Ca^{2+}$]) that can be achieved with this method, which is obtained when the total calcium concentration is equal to the total EGTA concentration.

In formatting the output, the program assumes that the PS of at least one metal ($Ca^{2+}$ here) should be specified for a range of values with a specified increment step: here pCa 9 to 4.5 with an increment step of 0.5 pCa unit. If only one pCa value were desired, then it would be necessary to type the same value twice for the initial and final values of increment and to give a step of increment different from zero. Obviously, the computer cannot obtain the second value with the specified step, and only one output is given. For instance, the following could be entered if one wanted to compute only for pCa 6.50:

PS■CA■■■+2.000■6.500E+00■6.500E+00■5.000E−01

```
CA■■EGTA■-1.999■3.976E+06
MG■■EGTA■-1.249■3.835E+01
CA■■ATP■■-1.995■4.990E+03
MG■■ATP■■-1.982■1.117E+04
K■■■ATP■■-2.954■4.377E+00
NA■■ATP■■-2.740■6.474E+00
CA■■PC■■■-0.000■1.407E+01
MG■■PC■■■-0.000■1.987E+01
GO
TL■K■■■■+1.000■1.000E-01
TL■NA■■■+1.000■3.000E-02
TL■EGTA■-2.017■1.000E-02
TL■PC■■■-1.997■1.200E-02
PS■CA■■■+2.000■9.000E+00■4.000E+00■5.000E-01
PS■MG■■■+2.000■3.000E+00
CM■ATP■■-3.552
GO
PS■MG■■ATP■■2.500E+00
GO
TL■K■■■■EGTA■IS
TL■NA■■■ATP■■PC
TL■A-■■■MG
GO
7.14E-03■6.00E-03■1.80E-01
WRT■CA■■■8.00■TO■6.00
LR■ON■ATP
LR■ON■MG
MORE
WRT■CA■■■6.00■TO■4.50
LR■ON■ATP
LR■ON■MG
GO
STORE
CA■■EGTA■4.410E+06
MG■■ATP
MORE
CA■■EGTA■3.958E+06
MG■■ATP■■1.861E+04
MG■■ATP
CA■■EGTA
GO
```

FIG. 2. Input of data for EXAMPLE1. Note that the symbol ■ indicates a blank space, which obviously should be entered by pressing the space bar (rather than by trying to create this symbol).

```
K'CA  EGTA= 3.976E+06/K'MG   EGTA= 38.3    /K'CA ATP = 4.990E+03/K'MG ATP = 1.117E+04/K'K ATP = 4.38    /K'NA ATP = 6.47
K'CA  PC = 14.1   /K'MG PC = 19.9
TL    EGTA=1.000E-02/TL   PC = 1.200E-02/P MG = 3.00    /P MG ATP = 2.50
TOTAL IONIC STRENGTH (I.S.) = .180M. CALCULATIONS DONE ASSUMING 6.00E-03M OF K OH FOR PH-ING. I.S. OF BUFFER IS 7.14E-03M
TL K  = 2 * (TL EGTA ) + TL K   A-  + TL K   OH
TL NA = 2 * (TL ATP + TL PC )
TL A- = 2 * (TL MG) + TL K A-
```

| P CA | K A- | TL CA | TL MG | TL ATP |
|------|------|-------|-------|--------|
| 1)  9.00 | 8.540E-02 | 3.815E-05 | 4.764E-03 | 3.640E-03 |
| 2)  8.50 | 8.540E-02 | 1.196E-04 | 4.761E-03 | 3.640E-03 |
| 3)  8.00 | 8.541E-02 | 3.688E-04 | 4.752E-03 | 3.641E-03 |
| 4)  7.50 | 8.543E-02 | 1.080E-03 | 4.726E-03 | 3.641E-03 |
| 5)  7.00 | 8.547E-02 | 2.769E-03 | 4.663E-03 | 3.641E-03 |
| 6)  6.50 | 8.554E-02 | 5.478E-03 | 4.563E-03 | 3.641E-03 |
| 7)  6.00 | 8.560E-02 | 7.932E-03 | 4.473E-03 | 3.642E-03 |
| 8)  5.50 | 8.562E-02 | 9.245E-03 | 4.424E-03 | 3.645E-03 |
| 9)  5.00 | 8.560E-02 | 9.771E-03 | 4.405E-03 | 3.655E-03 |
| 10) 4.50 | 8.548E-02 | 1.000E-02 | 4.399E-03 | 3.685E-03 |
| 11) 4.00 | 8.484E-02 | 1.023E-02 | 4.397E-03 | 3.782E-03 |

WHEN TOTAL CA EQUALS TOTAL EGTA, PCA IS 4.499

ASSUMING 3.000E-06M TOTAL CA CONTAMINATION, MAXIMUM PCA IS 10.106

LINEAR REGRESSION FOR PCA  = 8.00 TO 6.00

```
    TOTAL ATP =   .0002(TOTAL CA ) + 3.640E-03   GIVES:   3.642E-03M IN CAEGTA SOLUTION AND 3.640E-03M IN EGTA SOLUTION
    TOTAL MG  = -.0369(TOTAL CA ) + 4.765E-03    GIVES:   4.396E-03M IN CAEGTA SOLUTION AND 4.765E-03M IN EGTA SOLUTION
```

LINEAR REGRESSION FOR PCA  = 6.00 TO 4.50

```
    TOTAL ATP =   .0121(TOTAL CA ) + 3.540E-03   GIVES:   3.661E-03M IN CAEGTA SOLUTION AND 3.540E-03M IN EGTA SOLUTION
    TOTAL MG  = -.0361(TOTAL CA ) + 4.759E-03    GIVES:   4.397E-03M IN CAEGTA SOLUTION AND 4.759E-03M IN EGTA SOLUTION
```

K'CA EGTA= 4.410E+06/K'MG EGTA= 38.3 /K'CA ATP = 4.990E+03/K'MG ATP = 1.117E+04/K'K ATP = 4.38 /K'NA ATP = 6.47
K'CA PC = 14.1 /K'MG PC = 19.9

| | K | NA | EGTA | PC | CA | MG | ATP | D:IS |
|---|---|---|---|---|---|---|---|---|
| 1) TOTAL | 1.114E-01 | 3.128E-02 | 1.000E-02 | 1.200E-02 | 3.815E-05 | 4.764E-03 | 3.640E-03 | -4.023E-07 |
| PFREE | .954 | 1.506 | 2.018 | 1.929 | 9.045 | 3.000 | 3.548 | |
| P MG ATP = 2.500/ | | | | | | | | |
| 2) TOTAL | 1.114E-01 | 3.128E-02 | 1.000E-02 | 1.200E-02 | 1.196E-04 | 4.761E-03 | 3.640E-03 | -4.023E-07 |
| PFREE | .954 | 1.506 | 2.022 | 1.929 | 8.545 | 3.000 | 3.548 | |
| P MG ATP = 2.500/ | | | | | | | | |
| 10) TOTAL | 1.115E-01 | 3.137E-02 | 1.000E-02 | 1.200E-02 | 1.000E-02 | 4.399E-03 | 3.685E-03 | 2.533E-07 |
| PFREE | .953 | 1.504 | 4.126 | 1.930 | 4.522 | 3.000 | 3.548 | |
| P MG ATP = 2.500/ | | | | | | | | |
| 11) TOTAL | 1.108E-01 | 3.156E-02 | 1.000E-02 | 1.200E-02 | 1.023E-02 | 4.397E-03 | 3.782E-03 | 2.712E-06 |
| PFREE | .956 | 1.502 | 4.642 | 1.930 | 4.004 | 3.000 | 3.548 | |
| P MG ATP = 2.500/ | | | | | | | | |

K'CA EGTA= 3.958E+06/K'MG EGTA= 38.3 /K'CA ATP = 4.990E+03/K'MG ATP = 1.861E+04/K'K ATP = 4.38 /K'NA ATP = 6.47
K'CA PC = 14.1 /K'MG PC = 19.9

| | K | NA | EGTA | PC | CA | MG | ATP | D:IS |
|---|---|---|---|---|---|---|---|---|
| 1) TOTAL | 1.114E-01 | 3.128E-02 | 1.000E-02 | 1.200E-02 | 3.815E-05 | 4.764E-03 | 3.640E-03 | -5.840E-04 |
| PFREE | .953 | 1.505 | 2.017 | 1.929 | 9.000 | 3.042 | 3.708 | |
| P MG ATP = 2.480/P CA EGTA = 4.419/ | | | | | | | | |
| 2) TOTAL | 1.114E-01 | 3.128E-02 | 1.000E-02 | 1.200E-02 | 1.196E-04 | 4.761E-03 | 3.640E-03 | -5.843E-04 |
| PFREE | .953 | 1.505 | 2.020 | 1.929 | 8.500 | 3.042 | 3.708 | |
| P MG ATP = 2.480/P CA EGTA = 3.922/ | | | | | | | | |
| 10) TOTAL | 1.115E-01 | 3.137E-02 | 1.000E-02 | 1.200E-02 | 1.000E-02 | 4.399E-03 | 3.685E-03 | -7.257E-04 |
| PFREE | .953 | 1.504 | 4.138 | 1.931 | 4.462 | 3.055 | 3.695 | |
| P MG ATP = 2.480/P CA EGTA = 2.003/ | | | | | | | | |
| 11) TOTAL | 1.108E-01 | 3.156E-02 | 1.000E-02 | 1.200E-02 | 1.023E-02 | 4.397E-03 | 3.782E-03 | -6.070E-04 |
| PFREE | .956 | 1.501 | 4.664 | 1.929 | 3.934 | 3.059 | 3.690 | |
| P MG ATP = 2.479/P CA EGTA = 2.001/ | | | | | | | | |

FIG. 3. Output of data for EXAMPLE1.

411

## Second Example

Let us modify the specifications in the first example as follows: keep the total potassium concentration constant at 110 m$M$ (instead of maintaining the ionic strength constant), and use two values of total EGTA (10 and 5 m$M$), two values of [free Mg$^{2+}$] (pMg 3.5 and 2.5), without linear regression or calculation of free ionic concentrations from specified total concentrations.

The word-processor program permits the copy of the data file EXAMPLE1 into a new file called EXAMPLE2 (for instance, use the letter "O" command in WordStar). Then the file EXAMPLE2 is edited to obtain the display shown in Fig. 4. The output will be that shown in Fig. 5. Note that some data have been deleted to save space: only the part of the specifications that changes and the data for only one pCa value are shown. Without editing, four tables of the length shown at the top of the figure would be obtained.

```
CA■■EGTA■-1.999■3.976E+06
MG■■EGTA■-1.249■3.835E+01
CA■■ATP■■-1.995■4.990E+03
MG■■ATP■■-1.982■1.117E+04
K■■■ATP■■-2.954■4.377E+00
NA■■ATP■■-2.740■6.474E+00
CA■■PC■■■-0.000■1.407E+01
MG■■PC■■■-0.000■1.987E+01
GO
TL■K■■■■+1.000■1.100E-01
TL■EGTA■-2.017■1.000E-02■5.000E-03■5.000E-03
TL■PC■■■-1.997■1.200E-02
PS■CA■■■+2.000■9.000E+00■4.000E+00■5.000E-01
PS■MG■■■+2.000■3.500E+00■2.500E+00■1.000E+00
CM■ATP■■-3.552
GO
PS■MG■■ATP■■2.500E+00
GO
TL■K■■■■EGTA■CNST
TL■NA■■■ATP■■PC
TL■A-■■■MG
GO
7.14E-03■6.00E-03
NLR
NSTR
GO
```

FIG. 4. Input of data for EXAMPLE2.

```
K'CA EGTA= 3.976E+06/K'MG  EGTA= 38.3  /K'CA ATP = 4.990E+03/K'MG ATP = 1.117E+04/K'K  ATP = 4.38  /K'NA ATP = 6.47
K'CA PC =  14.1  /K'MG  PC = 19.9  /K'CA PC = 1.000E-02/TL  PC = 1.200E-02/P MG  = 3.50  /P MG ATP = 2.50
TL  K =  .110  /TL  EGTA= 1.000E-02/TL
6.00E-03M OF K OH ASSUMED TO PH THE SOLUTIONS. ALSO BUFFER ADDS 7.14E-03M TO THE IONIC STRENGTH.
TL = 2 * (TL EGTA ) + TL K  A-  + TL K  OH
TL K = 2 * (TL ATP + TL PC )
TL A- = 2 * (TL MG) + TL K A-
```

| | P CA | K A- | TL CA | TL MG | TL ATP | TL I.S. |
|---|---|---|---|---|---|---|
| 1) | 9.00 | 8.400E-02 | 3.914E-05 | 3.673E-03 | 1.028E-02 | 1.877E-01 |
| 2) | 8.50 | 8.400E-02 | 1.227E-04 | 3.672E-03 | 1.028E-02 | 1.877E-01 |
| 3) | 8.00 | 8.400E-02 | 3.780E-04 | 3.669E-03 | 1.028E-02 | 1.877E-01 |
| 4) | 7.50 | 8.400E-02 | 1.105E-03 | 3.660E-03 | 1.028E-02 | 1.877E-01 |
| 5) | 7.00 | 8.400E-02 | 2.821E-03 | 3.639E-03 | 1.028E-02 | 1.876E-01 |
| 6) | 6.50 | 8.400E-02 | 5.542E-03 | 3.607E-03 | 1.028E-02 | 1.875E-01 |
| 7) | 6.00 | 8.400E-02 | 7.977E-03 | 3.578E-03 | 1.029E-02 | 1.874E-01 |
| 8) | 5.50 | 8.400E-02 | 9.273E-03 | 3.562E-03 | 1.030E-02 | 1.874E-01 |
| 9) | 5.00 | 8.400E-02 | 9.808E-03 | 3.556E-03 | 1.033E-02 | 1.875E-01 |
| 10) | 4.50 | 8.400E-02 | 1.010E-02 | 3.554E-03 | 1.042E-02 | 1.878E-01 |
| 11) | 4.00 | 8.400E-02 | 1.054E-02 | 3.554E-03 | 1.073E-02 | 1.890E-01 |

```
TL  K = .110  /TL  EGTA= 5.000E-03/TL  PC = 1.200E-02/P MG  = 3.50  /P MG ATP = 2.50
```

| | P CA | K A- | TL CA | TL MG | TL ATP | TL I.S. |
|---|---|---|---|---|---|---|
| 17) | 6.50 | 9.400E-02 | 2.772E-03 | 3.580E-03 | 1.028E-02 | 1.824E-01 |

```
TL  K = .110  /TL  EGTA= 1.000E-02/TL  PC = 1.200E-02/P MG  = 2.50  /P MG ATP = 2.50
```

| | P CA | K A- | TL CA | TL MG | TL ATP | TL I.S. |
|---|---|---|---|---|---|---|
| 28) | 6.50 | 8.400E-02 | 5.286E-03 | 7.544E-03 | 3.875E-03 | 1.692E-01 |

```
TL  K = .110  /TL  EGTA= 5.000E-03/TL  PC = 1.200E-02/P MG  = 2.50  /P MG ATP = 2.50
```

| | P CA | K A- | TL CA | TL MG | TL ATP | TL I.S. |
|---|---|---|---|---|---|---|
| 39) | 6.50 | 9.400E-02 | 2.643E-03 | 7.289E-03 | 3.875E-03 | 1.641E-01 |

FIG. 5. Output of data for EXAMPLE2.

```
CA■■EGTA■-1.999■3.976E+06
MG■■EGTA■-1.249■3.835E+01
CA■■ATP■■-1.995■4.990E+03
MG■■ATP■■-1.982■1.117E+04
K■■■ATP■■-2.954■4.377E+00
NA■■ATP■■-2.740■6.474E+00
CA■■PC■■■-0.000■1.407E+01
MG■■PC■■■-0.000■1.987E+01
GO
TL■K■■■■+1.000■1.100E-01
TL■NA■■■+1.000■3.000E-02
TL■EGTA■-2.017■1.000E-02
TL■PC■■■-1.997■1.200E-02
TL■CA■■■+2.000■7.932E-03
TL■MG■■■+2.000■4.600E-03
TL■ATP■■-3.552■3.650E-03
GO
NCOMP
NIONS
NLR
STORE
MG■■ATP
GO
```

FIG. 6. Input of data for EXAMPLE3.

*Third Example*

Let us now use the program for only a calculation of free ionic concentrations from specified total concentrations using the apparent stability constants of the previous examples. Thus, let us find the free ionic concentrations and the pMgATP in a solution containing 110 m*M* total potassium, 30 m*M* total sodium, 10 m*M* total EGTA, 12 m*M* phosphocreatine, 7.932 m*M* total calcium, 4.6 m*M* total magnesium, and 3.65 m*M* total ATP.

Again the file EXAMPLE1 can be copied into a new file called EXAMPLE3 that will be edited, in order to save the time of reentering the apparent stability constants. The input data are shown in Fig. 6 and the output data in Fig. 7. The program will actually print the table shown in Fig. 7 twice, but the pMgATP will be printed only the second time. This is because the program is designed to permit printing the effect of changing the stability constants after a calculation of total from specified free ionic concentrations as shown, for instance, in the first example.

*Fourth Example*

This will compute a first approximation of the myoplasmic [total calcium] values required to obtain given myoplasmic [free Ca²⁺] values in the mammalian ventricular cardiac cell, taking into account what was known in 1982 of the intracellular steady-state buffering of calcium. This should be treated as a problem of finding the total ionic concentrations required to obtain specified pCa values in the presence of a pH of 7.1, a

```
K'CA  EGTA= 3.976E+06/K'MG  EGTA=  38.3    /K'CA  ATP = 4.990E+03/K'MG  ATP = 1.117E+04/K'K  ATP =  4.38    /K'NA  ATP =  6.47
K'CA  PC  =  14.1   /K'MG  PC  =  19.9
-----------------------------------------------------------------------------------------------------------------
           K          NA         EGTA       PC         CA         MG         ATP
-----------------------------------------------------------------------------------------------------------------
1) TOTAL   1.100E-01  3.000E-02  1.000E-02  1.200E-02  7.932E-03  4.600E-03  3.650E-03
   PFREE    .959      1.524      2.701      1.930      5.999      2.971      3.572
   P MG  ATP =  2.495/
```

Fig. 7. Output of data for EXAMPLE3.

pMg of 2.5, a pMgATP of 2.5, 140 m$M$ total potassium, and 30 m$M$ total sodium, at 22°. These were thought in 1982 to be the values existing in the myoplasm of an intact mammalian cardiac cell. Using the total concentrations listed in Table 2 of Ref. 9 and the 25 stability constants listed in Table 3 in Ref. 9, one specifies the pCa from 9.00 to 4.50 in steps of 0.01 pCa unit. The 451-line long output should contain the data shown in Table 4 of Ref. 9.

### Programs Developed by Other Investigators

Dr. Michael R. Berman (Department of Biomedical Engineering at the Johns Hopkins University, School of Medicine, 720 Rutland Avenue, Baltimore, Maryland 21205) has for a long time kindly corresponded with me with respect to the calculator programs for the Texas Instruments TI-59 programmable calculator.[2] He has adapted these programs to the Hewlett Packard Programmable Calculator HP41-CV in the form of three programs, and subsequently as a program which combines the calculator programs Nos. 2 and 3 of Ref. 2. I have verified these programs. They give exactly the same results as those obtained with either the TI-59 calculator program or the FORTRAN program described here. Subsequently, Dr. Berman has also written these programs in PASCAL for the IBM-PC. I have verified that these programs give exactly the same results as mine. Of course, Dr. Berman's programs are much more limited than the present program, since they only reproduce the calculator programs reported in Ref. 2.

Dr. Rashid Nassar (Department of Physiology, Duke University Medical Center, Durham, North Carolina 27710) also developed a PASCAL program for the IBM-PC reproducing some of the programs reported in Ref. 2. His work is only partly completed but I have verified that what he has done is accurate.[10]

Dr. Richard A. Meiss (Department of Physiology, Indiana University Medical Center, Indianapolis, Indiana 46223) has reproduced in BASIC for an IBM-PC part of the second calculator programs reported in Ref. 2. Dr. John W. Krueger (Department of Physiology, Albert Einstein College of Medicine of Yeshiva University, Bronx, New York 10461) has adapted part of the calculator programs reported in Ref. 2 to an Apple-Macintosh

[9] A. Fabiato, *Am. J. Physiol.* **245**, C1 (1983).

[10] I am most grateful to Mr. Nassib Nassar, the son of my friend Dr. Rashid Nassar, for developing a text editor (AscEdit) appropriate for entering data into SPECS. Thanks to the generosity of Mr. Nassib Nassar this text editor can be included on diskette No. 1 for the convenience of the users who would request it because they do not own a word-processor software.

microcomputer. Finally, Drs. Godt and Lindley[11] have written a program that is similar in principle to those reported in Ref. 2. They told me that it gives results identical to those reported in the examples given in Ref. 2. The purpose of this section is not to report the results of any inquiry on all related computer programs that may have been developed by others but to indicate that the rationale and computations used in the programs reported here have been extensively verified.[12]

[11] R. E. Godt and B. D. Lindley, *J. Gen. Physiol.* **80**, 279 (1982).
[12] For any problem that users of this program may have, my telephone number is (804) 786-9563.

# [32] Ionic Permeability of Isolated Muscle Sarcoplasmic Reticulum and Liver Endoplasmic Reticulum Vesicles

*By* GERHARD MEISSNER

Several permeation systems for ions and small solutes are present within the reticulum structures of cells. For example, in the endoplasmic reticulum (ER) of liver, glucose 6-phosphate hydrolysis is mediated by a two-component system consisting of a glucose-6-phosphate permease which transfers glucose-6-phosphate across the ER membrane and a phosphohydrolase–phosphotransferase (EC 3.1.3.9), localized on the luminal side of the membrane.[1] Other biologically relevant solutes and ions that cross the ER membrane include D-glucose, L-glucose, L-leucine, choline+, K+, Na+, and Cl−.[2]

Sarcoplasmic reticulum (SR) is another specialized intracellular membrane system that controls muscle contraction and relaxation by rapidly releasing and sequestering $Ca^{2+}$.[3] Rapid $Ca^{2+}$ fluxes are possible due to the presence of several high-capacity permeation systems for $Ca^{2+}$ and monovalent ions.[4] The SR membrane contains at least three and possibly four separate passive ion permeation systems: (1) a ligand-gated "$Ca^{2+}$-release channel" which is thought to play a central role in the process of

[1] L. M. Ballas and W. J. Arion, *J. Biol. Chem.* **252**, 8512 (1977).
[2] G. Meissner and R. Allen, *J. Biol. Chem.* **256**, 6413 (1981).
[3] A. N. Martonosi and T. J. Beeler, *in* "Handbook of Physiology. Sect. 10: Skeletal Muscle" (L. D. Peachey, R. H. Adrian, and S. R. Geiger, eds.), p. 417. Am. Physiol. Soc., Bethesda, Maryland, 1983.
[4] G. Meissner, *Mol. Cell. Biochem.* **55**, 65 (1983).

excitation–contraction coupling in skeletal muscle, (2) a $K^+,Na^+$ channel, (3) a $Cl^-$ channel, and (4) a $H^+(OH^-)$ permeable pathway which may or may not be synonymous with the $Cl^-$ channel. In this chapter, we describe radioisotope flux–Millipore filtration, light scattering, and membrane potential–dye measurements we have used to determine the passive $Ca^{2+}$, $K^+$, $Na^+$, $H^+$, and $Cl^-$ permeability of isolated rabbit skeletal muscle SR and rat liver ER vesicle fractions.

## Isolation of SR and ER Vesicle Fractions

During homogenization, the reticulum structures of muscle and liver are disrupted into sealed membranous vesicles. Isolation procedures have been described for skeletal muscle SR vesicle fractions that are derived from the junctional-terminal cisternae, and longitudinal sections of SR and that are capable of active $Ca^{2+}$ transport.[5–7] Rabbit skeletal muscle SR vesicle fractions used in the ion flux studies described in this chapter have been prepared using homogenization buffers of low or intermediate ionic strength. Use of sucrose solutions of low ionic strength favors the isolation of SR vesicle fractions of high purity. Three density subfractions, a light fraction derived from the longitudinal section, a heavy fraction derived from the terminal cisternae, and a mixed intermediate vesicle fraction have been prepared by differential and sucrose gradient centrifugation.[5] Vesicle fractions prepared by this method are of relatively small size ($r \approx 50$ nm). Light- and intermediate-density vesicles display little or no $Ca^{2+}$-release activity. On the other hand, a majority of these vesicles are readily permeable to $K^+$, $Na^+$, $H^+$, and $Cl^-$, making them suitable for studying the monovalent ion channels of SR. Homogenization of rabbit skeletal muscle in the presence of 0.1 $M$ NaCl yields two subpopulations of vesicles called heavy $Ca^{2+}$-release and light control vesicles.[7,7a] Heavy $Ca^{2+}$-release vesicles possess the $Ca^{2+}$-release channel and are readily permeable to $K^+$, $Na^+$, $H^+$, and $Cl^-$. A majority of these vesicles are relatively large ($r = 100$–$300$ nm; see Ref. 7b) and are recovered in the 36–45% region of sucrose gradients that contain membranes sedimenting at 2,600–35,000 $g$. Light control fractions ($r = 50$–$100$ nm) lack significant

[5] G. Meissner, *Biochim. Biophys. Acta* **389,** 51 (1975).

[6] A. Saito, S. Seiler, A. Chu, and S. Fleischer, *J. Cell Biol.* **99,** 875 (1984).

[7] G. Meissner, *J. Biol. Chem.* **259,** 2365 (1984).

[7a] It is recommended that heavy $Ca^{2+}$-release vesicles are isolated in the presence of 1 m$M$ diisopropyl fluorophosphate (DIFP), a protease inhibitor, in order to avoid partial degradation of the SR $Ca^{2+}$-release channel (F. A. Lai, H. P. Erickson, E. Rousseau, Q. Y. Liu, and G. Meissner, *Nature (London)* **331,** 315 (1988).

[7b] J. S. Smith, R. Coronado, and G. Meissner, *Biophys. J.* **50,** 921 (1986).

$Ca^{2+}$-release activity but are, like all the other SR vesicle fractions we have examined, permeable to monovalent ions. These vesicles are recovered in the 30–34% region of sucrose gradients containing membranes obtained by differential pelleting at 35,000–130,000 $g$.

Microsomal membrane fractions derived from the rough ER of rat liver were prepared by differential and sucrose gradient centrifugation[2] using the $Cs^+$ aggregation technique of Dallner.[8] The ER vesicles ($r = 50$–100 nm) are fairly impermeable to sucrose yet readily permeable to $Cl^-$. About 70% of the vesicles are permeable to choline[+] and other small solutes; the remaining 30% are relatively impermeable to these compounds.[2]

### Radioisotope Flux–Millipore Filtration Measurements

The radioisotope flux–Millipore filtration technique is useful for determining the fraction of vesicles that contain a specific permeation system as well as for measuring flux rates of relatively impermeable solutes and ions. By using a slowly permeating solute such as [$^3$H]sucrose, the internal space of all intact vesicles as well as membrane "tightness" is determined. Vesicles are incubated in media containing one or two radioactively labeled ions or solutes, diluted into isoosmolal release media, collected on Millipore filters, and rapidly rinsed. Radioactivity retained by the vesicles on the filters is measured. The radioisotope flux–Millipore filtration technique has a time resolution of 15–20 sec. In the case of the SR $Ca^{2+}$-release channel, a time resolution of at least 20 msec can be achieved using a rapid-mixing apparatus and the two channel inhibitors $Mg^{2+}$ and ruthenium red.

### Equipment

The radioisotope flux–Millipore filtration experiments require an ultracentrifuge and fixed-angle rotor for pelleting SR and ER vesicles at centrifugal forces of 100,000 $g_{av}$; a Potter–Elvehjem homogenizer (Kontes, Vineland, NJ) for the resuspension of the pelleted vesicles; an osmometer (e.g., $\mu$ Osmette, Precision Systems, Natick, MA) to adjust the osmolality of vesicle and dilution media; a filtration apparatus, attached to a house vacuum line or a vacuum pump, and consisting of a 250-ml filtering flask and a filter support (e.g., fritted glass base with stopper, Millipore Co., Bedford, MA); and a liquid scintillation counter for counting singly and doubly labeled samples.

[8] G. Dallner, *Acta Pathol. Microbiol. Scand.*, *Suppl.* No. 166 (1963).

*Experimental Procedures*

Vesicles are initially transferred to a medium of defined composition. Preferably, this medium and the subsequent incubation medium should be (1) of sufficient ionic strength (e.g., 0.1 $M$ KCl) to minimize nonspecific ion binding as well as Donnan potential effects, and (2) well buffered. Vesicles are incubated for about 1 hr at 0° in a large volume (0.5–1 mg protein/ml) of incubation medium, sedimented by centrifugation for 30–60 min at 100,000 $g_{av}$ (36,000 rpm in a Beckman 42,1 rotor), resuspended in a small volume (10–15 mg protein/ml) of incubation medium, and used immediately. If prolonged storage is unavoidable, vesicles can be quick-frozen and stored for several weeks in small aliquots at −65°. Repeated thawing and freezing increases the leakiness of the vesicles.

SR and ER vesicles are fairly permeable to most small molecules such as $^{45}$Ca$^{2+}$, [$^{3}$H]choline$^{+}$, or [$^{3}$H]sucrose. It is therefore possible to equilibrate the radioactive compounds across the vesicle membranes in a relatively short time (2–4 hr at 22° or overnight at 4°).[2,7,9] The vesicle suspension (50–200 μl) is transferred to a small container, such as a 400-μl centrifuge tube supplied with a cap. A small aliquot of incubation medium containing the radioisotope(s) is added (0.1–0.2 mCi/ml for singly labeled samples, or 0.4 mCi of a $^{3}$H-labeled compound and 0.2 mCi of the other radioisotopes for doubly labeled samples). The tube containing the sample and radioisotopes is sealed in order to minimize evaporation and mixed by gentle vortexing.

Influx and efflux rates as well as apparent isotope spaces are determined by incubating and diluting vesicles at various times. At a given time, an aliquot (5–20 μl) of the vesicles is diluted 100- to 500-fold into an isoosmolal, unlabeled release medium. Radioactivity trapped by the vesicles is determined by placing an aliquot of the diluted vesicle suspension (0.1–0.5 ml) on a filter, followed by rapid rinsing with three 1- to 2-ml aliquots of the release medium. By repeating this process at 30- to 120-sec time intervals, it is possible to determine the efflux rate of a relatively impermeable compound. The time to execute filtration and rinsing of about 10–20 sec is taken into account in the calculation of apparent isotope spaces (cf. Fig. 1).[10,11]

The time required to filter and rinse the vesicles depends on the vacuum and the amount of vesicles applied to the filters, as well as the type of filters used. Millipore filters of the MF type of mixed esters of cellulose

[9] G. Meissner and D. McKinley, *J. Biol. Chem.* **257,** 7704 (1982).
[10] D. McKinley and G. Meissner, *FEBS Lett.* **82,** 47 (1977).
[11] D. McKinley and G. Meissner, *J. Membr. Biol.* **44,** 159 (1978).

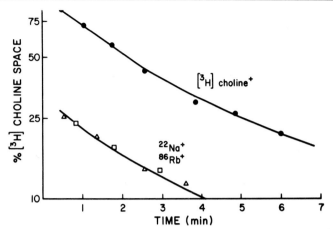

FIG. 1. Measurement of [³H]choline⁺, ²²Na⁺, and ⁸⁶Rb⁺ ion efflux rates and isotope spaces. Rabbit skeletal muscle sarcoplasmic reticulum vesicles were incubated for 24 hr at 0° in a medium containing 200 m$M$ [³H]choline-Cl, 1 m$M$ ²²NaCl or 1 m$M$ ⁸⁶RbCl, 20 μ$M$ CaCl₂, 1 m$M$ MgCl₂, and 5 m$M$ Tris-PIPES, pH 7.0. Vesicles were diluted 200 times into unlabeled media of identical composition at 4°, and the time course of [³H]choline⁺, ⁸⁶Rb⁺, and ²²Na⁺ efflux from the vesicles was determined by Millipore filtration as described in the text. (Taken with permission from Ref. 10.)

with a nominal pore size of 0.45 μm are, in our experience, well suited for these experiments. These filters have a reasonable flow rate and retain greater than 95% of native SR and ER vesicles. The filters of some suppliers retain, even at a nominal pore size of 0.20–0.22 μm, as little as one third of the vesicles.

During prolonged incubation with an organic radioactive compound (e.g., D-[³H]glucose) some of the ³H radioactivity may be incorporated into the constituents of the membrane vesicles. The amount of "bound" radioactivity can be estimated by diluting an aliquot of the vesicles into 5% trichloroacetic acid (TCA), followed by filtration.

Radioactivity retained by the vesicles on the filters by singly or doubly labeled samples may be conveniently counted in a liquid scintillation system using minivials. Total radioactivity is determined by counting a filter and a 50-μl aliquot of the diluted sample in parallel. An effective but expensive scintillation liquid that completely dissolves Millipore MF type filters and retains ions well in suspension, thereby minimizing sample variability, is the following: 60 g naphthalene, 4.2 g 2,5-diphenyloxazole (PPO), 180 mg 1,4-bis(2-(4-methyl-5-phenyloxazoyl))benzene (dimethyl-POPOP), and 70 ml water in 900 ml dioxane.

*Example: Measurement of $^{86}Rb^+$, $^{22}Na^+$, and [$^3H$]choline$^+$ Isotope Spaces and Flux Rates*

Figure 1 demonstrates the usefulness and limitations of the radioisotope flux–Millipore filtration technique in assessing the monovalent cation permeability of SR vesicles. Intermediate-density SR vesicles[5] were incubated at a protein concentration of 18 mg/ml for 24 hr at 0° in the following medium with [$^3H$]choline$^+$ (0.4 mCi/ml) and either $^{86}Rb^+$ or $^{22}Na^+$ (0.2 mCi/ml) added: 200 m$M$ choline-Cl, 20 $\mu M$ CaCl$_2$, 1 m$M$ MgCl$_2$, 1 m$M$ NaCl, 1 m$M$ RbCl, and 5 m$M$ Tris–PIPES, pH 7. Vesicles (20 $\mu$l) were diluted into 4 ml unlabeled medium of identical composition at 4°. At 30- to 45-sec intervals, 0.5-ml aliquots were placed on 0.45-$\mu$m HAWP Millipore filters followed by rapid rinsing (see above). Rb$^+$ is used rather than K$^+$ since its size and chemical properties are similar to K$^+$ and there is no readily available radioisotope of K$^+$. The [$^3H$]choline$^+$ space extrapolated back to zero time represents the internal volume of all intact SR vesicles, and typically ranges from 2–4 $\mu$l/mg protein.

Differences in [$^3H$]choline$^+$ and $^{86}Rb^+$ or $^{22}Na^+$ isotope spaces indicate the presence of two types of vesicles which differ in their permeability to Rb$^+$ and Na$^+$. About two-thirds of the vesicles (designated type I, Ref. 11) contain the K$^+$,Na$^+$ channel and will therefore release all their $^{86}Rb^+$ and $^{22}Na^+$ before the first 15–30 sec time point. Following the inflow of Tl$^+$ by a stopped-flow fluorescence quenching method, Garcia and Miller[12] have estimated that in type I vesicles both K$^+$ and Na$^+$ equilibrate in less than 3 msec across the membrane. From the remaining one-third of the vesicles (type II) which trap $^{86}Rb^+$ and $^{22}Na^+$, efflux is on a time scale of minutes which we believe is characteristic for nonmediated permeation in SR.

*Example: Measurement of Rapid $^{45}Ca^{2+}$ Efflux Rates from Heavy SR Ca$^{2+}$-Release Vesicles*

Heavy SR vesicles contain a Ca$^{2+}$-release channel which is activated by Ca$^{2+}$ and adenine nucleotides (ATP, ADP, AMP, adenosine, adenine) and inhibited by Mg$^{2+}$ and ruthenium red.[7,13–15] We have found in flux measurements that in heavy SR fractions, prepared in the presence of 0.1 $M$ NaCl, 7–9 vesicles out of 10 contain the Ca$^{2+}$ channel.[7] When placed into media at pH 7 containing 5 $\mu M$ free Ca$^{2+}$ and 5 m$M$ AMP-PCP, a

[12] A. M. Garcia and C. Miller, *J. Gen. Physiol.* **83**, 819 (1984).
[13] M. Endo, *Physiol. Rev.* **57**, 71 (1977).
[14] K. Nagasaki and M. Kasai, *J. Biochem.* (*Tokyo*) **94**, 1101 (1983).
[15] G. Meissner, E. Darling, and J. Eveleth, *Biochemistry* **25**, 236 (1986).

nonhydrolyzable ATP analog, heavy SR vesicles release $^{45}Ca^{2+}$ with a first-order rate constant ranging from 30–100 sec$^{-1}$.[15] Below are outlined two $Ca^{2+}$-release assays. In the first one, $^{45}Ca^{2+}$ release from passively loaded vesicles is measured on a time scale of 30–150 sec to determine the fraction of vesicles capable of rapid $^{45}Ca^{2+}$ release. In the second one, rapid $^{45}Ca^{2+}$ release rates are measured with an Update System 1000 Chemical Quench apparatus (Madison, WI) using quench solutions which contain $Mg^{2+}$ and ruthenium red.

A heavy SR vesicle fraction[7] is initially equilibrated for 30 min at 0° in a large volume (0.2–0.5 mg of protein/ml) of incubation medium (0.1 $M$ KCl, 1 m$M$ DIFP, 100 $\mu M$ EGTA, 100 $\mu M$ $Ca^{2+}$, and 20 m$M$ K-PIPES, pH 7.0), sedimented by centrifugation for 30 min at 100,000 $g$, and resuspended in incubation medium at a protein concentration of 2–10 mg/ml. In the rapid quench experiments, the vesicle suspensions (2 mg protein/ml) are centrifuged at 500 $g$ for 5 min prior to incubation with $^{45}Ca^{2+}$ in order to remove small amounts of aggregated material. Vesicles are passively loaded with $^{45}Ca^{2+}$ (0.2 mCi/ml) for 2 hr at 22° (5.1 m$M$ $^{45}Ca^{2+}$ in Fig. 2).

The fraction of $Ca^{2+}$-permeable vesicles is determined by transferring 5 $\mu l$ of the vesicles (5–10 mg protein/ml) to a test tube followed by addition of 1.5 ml of isoosmolal, unlabeled 20 m$M$ K-PIPES/KCl release medium. (KCl concentration is adjusted to maintain isoosmolality between vesicle and release media.) At approximately 15, 60, and 140 sec, a 0.4-ml aliquot is placed on a 0.45-$\mu$m HAWP Millipore filter, rinsed with release medium, and the radioactivity remaining with the vesicles on the filters is determined. $^{45}Ca^{2+}$ efflux is slow in a medium containing the two $Ca^{2+}$-release channel inhibitors $Mg^{2+}$ and ruthenium red at concentrations of 10 m$M$ and 10 $\mu M$, respectively (Fig. 2). This allows one to determine the amounts of $^{45}Ca^{2+}$ trapped by all intact vesicles (87 nmol/mg protein in Fig. 2). Vesicles release most of their $^{45}Ca^{2+}$ within 30 sec when diluted into a $Ca^{2+}$-release medium containing a maximally activating concentration of ~5 $\mu M$ free $Ca^{2+}$ (1 m$M$ EGTA plus 0.9 m$M$ $Ca^{2+}$). (Free $Ca^{2+}$ concentrations are calculated using a computer program based on the binding constants published by Fabiato.[16]) The small amount of $^{45}Ca^{2+}$ (15 nmol/mg protein) that remains within the vesicles 30–150 sec after dilution into the 5 $\mu M$ $Ca^{2+}$ medium reflects a subpopulation of vesicles that lack the $Ca^{2+}$-release channel.[7]

Rapid initial $^{45}Ca^{2+}$ efflux rates in $Ca^{2+}$-release media are determined using the Update System 1000 Chemical Quench Apparatus.[15] Figure 3 is a diagrammatic representation of the measurement of $Ca^{2+}$- and nucleo-

[16] A. Fabiato, *J. Gen. Physiol.* **78**, 457 (1981).

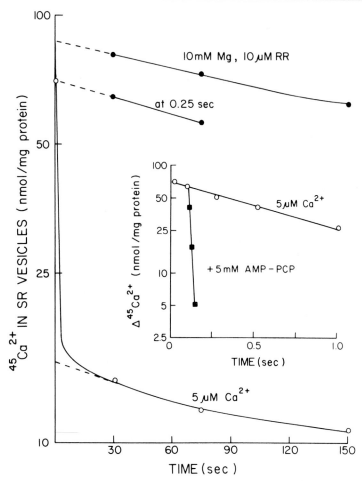

FIG. 2. Measurement of the $^{45}$Ca$^{2+}$-permeable vesicle fraction and $^{45}$Ca$^{2+}$ efflux rates. A heavy rabbit skeletal muscle SR vesicle fraction was passively loaded with $^{45}$Ca$^{2+}$ for 2 hr at 22° in a medium containing 20 mM K-PIPES, pH 7.0, 0.1 M KCl, 1 mM DIFP, 0.1 mM EGTA, and 5.1 mM $^{45}$Ca$^{2+}$. $^{45}$Ca$^{2+}$ efflux was initiated by diluting vesicles 5- to 300-fold into isoosmolal, unlabeled release media. $^{45}$Ca$^{2+}$ remaining with the vesicles is determined by Millipore filtration. An Update System 1000 Chemical Quench apparatus was used in experiments where rapid $^{45}$Ca$^{2+}$ efflux was inhibited at time intervals ranging from 25–1000 msec by the addition of the two Ca$^{2+}$-release inhibitors Mg$^{2+}$ and ruthenium red (see text). The amount of $^{45}$Ca$^{2+}$ initially trapped by all vesicles (87 nmol/protein) as well as the amount not readily released by a subpopulation of vesicles (15 nmol/mg protein) were obtained by back extrapolation to the time of vesicle dilution. In the inset, the time course of $^{45}$Ca$^{2+}$ efflux from the Ca$^{2+}$-permeable vesicle population was obtained by subtracting the amount not readily released (15 nmol/mg protein).

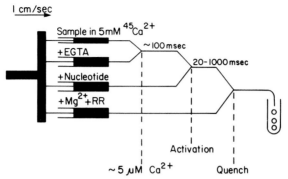

FIG. 3. Diagrammatic representation of rapid $^{45}Ca^{2+}$ efflux quench experiments using heavy SR vesicles.

tide-induced $^{45}Ca^{2+}$ efflux at time intervals ranging from 20–1000 msec. The rapid quench apparatus consists of an electronically driven ram which is programmed to push up to four syringes simultaneously 0.1–1.8 cm with a speed ranging from 0.8–8 cm/sec. Standard conditions are a ram displacement of 1.8 cm with a speed of 1 cm/sec. Lower ram speeds result in less-than-optimal mixing, while higher mixing speeds promote sample leakage. Reaction times are determined by varying the length of the "aging" hoses between the mixing chambers.

In the inset of Fig. 2, the rate of $Ca^{2+}$-induced $^{45}Ca^{2+}$ release in the presence of 5 $\mu M$ free $Ca^{2+}$ has been measured using three syringes and two Update 4-grid acrylic mixing chambers with 1.6 $\mu l$ dead volume. One 0.5-ml syringe was filled with the vesicle suspension (2 mg/ml in 20 m$M$ K-PIPES, 0.1 $M$ KCl, 1 m$M$ DIFP, 0.1 m$M$ EGTA, 5.1 m$M$ $^{45}Ca^{2+}$), one 2-ml syringe with the reaction medium, and a second 2-ml syringe with the quench solution. In the first mixing chamber, free extravesicular $Ca^{2+}$ was lowered from 5 m$M$ to 5 $\mu M$ by diluting the vesicle suspension with 4 vol of isoosmolal K-PIPES/KCl medium containing 6.25 m$M$ EGTA and 4.55 m$M$ $Ca^{2+}$. It is recommended that a reaction medium with a relatively high $Ca^{2+}$-buffering capacity be used in order to account for small fluctuations in the mixing ratio of vesicle and reaction media. Chelation of $Ca^{2+}$ by EGTA results in the release of $H^+$. It is therefore necessary to adjust the pH of the EGTA-containing solution so that after mixing the vesicles are present in a medium of defined pH. In Fig. 2 at 0.25 sec after the initial mixing step, efflux of $^{45}Ca^{2+}$ was inhibited by adding to the vesicles in the second mixing chamber a quench solution containing K-PIPES/KCl, 22.5 m$M$ $MgCl_2$, and 22.5 $\mu M$ ruthenium red. Either inhibitor alone will effectively inhibit $Ca^{2+}$-induced $^{45}Ca^{2+}$ release. However, in the presence of an additional channel activator (e.g., AMP-PCP) it is necessary to add both inhibitors to the quench solution in order to rapidly and effectively block $^{45}Ca^{2+}$ release. After the quench step, vesicles were collected, placed on a filter, and rapidly rinsed. Radioactivity remaining with the vesicles was determined by liquid scintillation counting. In the inset of Fig. 2, the time course of $^{45}Ca^{2+}$ efflux from the vesicle population capable of rapid $Ca^{2+}$ release (72 nmol $^{45}Ca$/mg protein) was obtained by adding to the 5 $\mu M$ $Ca^{2+}$-release medium at 25, 100, 250, 500, and 1000 msec the two $Ca^{2+}$-release inhibitors $Mg^{2+}$ and ruthenium red. Vesicles released $^{45}Ca^{2+}$ with a first-order rate constant of ~1 sec⁻¹.

Stimulation of $Ca^{2+}$-induced $Ca^{2+}$ release by adenine nucleotides requires the use of a four syringe–three mixing chamber arrangement (Fig. 3). After the extravesicular $Ca^{2+}$ concentration has been adjusted by transferring the vesicles into an EGTA-containing medium in a first dilution step, an activating ligand such as AMP-PCP is added to stimulate $^{45}Ca^{2+}$ release (inset of Fig. 2). A nonhydrolyzable ATP analog like AMP-PCP has to be used rather than ATP in order to avoid reuptake of the released $^{45}Ca^{2+}$ by the SR $Ca^{2+}$-pump during the relatively slow Millipore filtration step. If measurements are done at low $Ca^{2+}$ concentrations ($<10^{-7}$ $M$), it is important that the time interval between the first dilution and the activation step be at least 50–100 msec. At shorter time intervals we have observed $Ca^{2+}$ activation of nucleotide-induced $^{45}Ca^{2+}$ release, suggesting that chelation of residual amounts of excess free $Ca^{2+}$ is a rate-limiting step in the rapid quench experiments.[15,16a]

*Comments.* The radioisotope flux–Millipore filtration method is more laborious and expensive than the light-scattering and membrane potential measurements described below. However, its advantages are (1) its versatility in adjusting the ion composition of intravesicular and extravesicular spaces, and (2) measurement of unidirectional fluxes. Two major limitations of the radioisotope flux–Millipore filtration method are (1) its time resolution is rather poor (15–30 sec), unless a specific inhibitor is available and (2) internal vesicle spaces are measured so that the permeability properties of the large vesicles predominate in a heterogeneously sized vesicle population.

---

[16a] The canine cardiac $Ca^{2+}$-release channel is less sensitive to inhibition by $Mg^{2+}$ and ruthenium red than the skeletal channel. Therefore, in the presence of adenine nucleotide, an additional requirement is that the free $Ca^{2+}$ concentration is decreased to below $10^{-6}$ $M$ during the quench step [G. Meissner and J. S. Henderson, *J. Biol. Chem.* **262**, 3065 (1987)]. This can be achieved by decreasing EGTA and $Ca^{2+}$ concentrations in the reaction media and by including the $Ca^{2+}$ chelating agent EGTA in the quench solution.

## Light-Scattering Measurements

The permeability properties of SR[9,17-20] and ER (G. Meissner, unpublished studies) vesicles have been assessed by determining their osmotic behavior, as measured by light scattering. Volume changes are initiated by increasing the osmolality of the vesicle suspensions, and are measured by determining the resulting light-scattering intensity changes in a fluorometer. With a medium-priced fluorometer such as a Farrand Model 801 Fluorometer and a simple overhead stirring device, a time resolution of 2–3 sec or better is achieved. Use of a stop-flow apparatus has allowed a time resolution of ~20 msec (see [33], this volume).

### Equipment

Osmotically induced changes in vesicle size and shape are monitored at 400–500 nm at a right angle to the incoming light beam using a fluorometer equipped with a recorder and a temperature-controlled cuvette holder. Rapid mixing of the contents in the fluorometric cuvette is achieved with the use of a magnetic stirrer or an overhead stirrer. We use a small plastic stirring rod (Plumper, Calbiochem, San Diego, CA) which consists of a rod with a small paddle attached to the lower end of the rod at a right angle. The stirrer is placed into the cuvette just above the light beam. The upper end of the rod is attached to a small variable-speed motor mounted on the sample compartment cover. A small hole in the sample compartment cover allows additions to be made to the cuvette with the use of a microliter syringe. An osmometer is needed to determine the osmolality of the media.

### Experimental Procedures

Osmotically induced volume changes in the vesicles are initiated by increasing the osmolality of the vesicle medium. Vesicles present at a protein concentration of about 10 mg/ml are diluted 100- to 500-fold into a 4-ml rectangular fluorescence cell containing 2.7–3.2 ml of a filtered solution of low osmolality (e.g., 10 m$M$ K-PIPES, pH 7.0). Vesicles are equilibrated in the presence or absence of an ionophore under stirring for about 5 min or until a stable baseline is obtained. Osmolality of the vesicle medium is then rapidly increased by injecting the test solute under continued stirring.

[17] J. C. Selser, Y. Yeh, and R. J. Baskin, *Biophys. J.* **16,** 1375 (1976).
[18] T. Kometani and M. Kasai, *J. Membr. Biol.* **41,** 295 (1978).
[19] M. Kasai, T. Kanemasa, and S. Fukumoto, *J. Membr. Biol.* **51,** 311 (1979).
[20] J. R. Gilbert and G. Meissner, *Arch. Biochem. Biophys.* **223,** 9 (1983).

*Example: Effect of Changing Osmolality on the Light-Scattering
Intensity of SR and ER Vesicles*

Figure 4 shows that the addition of various test solutes results in transient changes in the light-scattering intensity of SR and ER vesicles. Vesicles were initially present at a concentration of 30 $\mu$g protein/ml in 3.15 ml of 10 m$M$ K-PIPES, pH 7.0. The extravesicular osmolality was increased by the addition of 0.35 ml of a buffered 1000 mOsm solution of

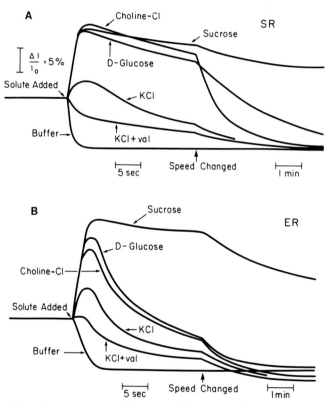

FIG. 4. Effect of increasing osmolality on the light-scattering intensity of SR (A) and ER (B) vesicles. Light control rabbit skeletal muscle SR[7] and rat liver ER[2] vesicles at a protein concentration of 10 mg/ml in 10 m$M$ K-PIPES, pH 7.0, and 0.3 $M$ sucrose were diluted in a fluorometric cuvette at 22° to a protein concentration of 30 $\mu$g/ml with 3.15 ml of 10 m$M$ K-PIPES, pH 7.0, containing or lacking 1 $\mu M$ valinomycin. After an equilibration period of 5 min, 0.35 ml of 10 m$M$ K-PIPES, pH 7, buffer or 1000 mOsm of buffered sucrose, D-glucose, choline-Cl, or KCl was added. Resulting light-scattering intensity changes were recorded at a right angle to the incoming beam at 400 nm using a Farrand Model 801 fluorometer (for details see text).

sucrose, D-glucose, choline-Cl, or KCl under rapid mixing using an over-head stirrer. In control experiments the decrease in light-scattering intensity due to vesicle dilution from 30 to 27 $\mu$g protein/ml was determined by adding 0.35 ml of 10 m$M$ K-PIPES, pH 7.

On addition of the test compounds, the vesicles shrank as water left the vesicles, thereby increasing their light-scattering properties. The signals returned to that of control vesicles as the test compounds and water entered and the vesicles expanded. The amplitude of the signal and the rate of swelling are a function of the permeability of the vesicles to the solutes. The maximal signal and the slow return rates of the light-scattering signals, seen for sucrose in Fig. 4, imply that all SR and ER vesicles are relatively impermeable to sucrose. Establishment of an equilibrium for D-glucose was much faster in ER than SR vesicles. Radioisotope space and efflux measurements have suggested that about 70% of rat liver ER vesicles are permeable to D-glucose.[2]

Net movement of salt requires that both ions move across the vesicle membranes with the slower ion determining the return rate of the light-scattering signal. Since SR[4] and ER[2] vesicles are readily permeable to Cl$^-$, the rate-limiting ions were choline$^+$ and K$^+$ in Fig. 4. A reduced light-scattering signal, as seen on addition of KCl, indicated that some of the vesicles were readily permeable to both ions, i.e., K$^+$ and Cl$^-$. The K$^+$- and Cl$^-$-permeable vesicle fraction contracted and expanded too fast to be detected on the time scale of the light-scattering experiments (2–3 sec). The further decline in the light-scattering signal, seen following KCl addition in the presence of 1 $\mu M$ valinomycin (0.7 $\mu$l 5 m$M$ valinomycin in ethanol/3.5 ml final volume), indicated the presence of vesicles that were permeable to Cl$^-$ but lacked an efficient intrinsic permeability mechanism for K$^+$ such as the K$^+$,Na$^+$ channel of SR.

*Comments.* The light-scattering method is useful in determining the overall permeability properties of vesicle fractions. Its two major advantages are that measurements are easily performed and that no prolonged vesicle incubations are required. On the other hand, interpretation of the light-scattering data is problematic. First, the physical basis of light-scattering intensity changes of small vesicle populations is not well understood.[21] Most likely, this technique primarily monitors the volume changes of the larger vesicles in a heterogeneously sized vesicle preparation. Second, the vesicles are generally present initially in media of low ionic strength which, together with the presence of fixed negative internal charges, can lead to the formation of significant Donnan potentials[22] and

[21] P. Latimer, *Annu. Rev. Biophys. Bioeng.* **11**, 129 (1982).
[22] N. Yamamoto and M. Kasai, *J. Biochem. (Tokyo)* **88**, 1425 (1980).

pH gradients.[23] Third, the differential permeation of ions results in the formation of membrane potentials which in turn influence the movement of ions across the membrane. Fourth, ions move across the membrane in pairs, so that the permeation rate of both ions must be known. Fortunately, SR[4] and ER[2] vesicles are permeable to Cl$^-$. Therefore, the permeation rates of relatively impermeable cations can be compared using salts with chloride as anion. Conversely, permeability of the vesicles to anions can be assessed by rendering all vesicles permeable to K$^+$ with valinomycin and using potassium salts. Fifth, the light-scattering signals are sensitive to changes in vesicle aggregation state, whereas the aggregation state of the vesicles is influenced by the composition of the vesicle media. Changes in vesicle aggregation state during an experiment are particularly difficult to avoid in the case of ER vesicles, and are indicated by the failure of the light-scattering signal to return to control levels (cf. Fig. 4). Finally, the refractive index of the vesicle suspension is changed by the addition of the test solute. The effect of increasing refractive index on the light-scattering signal can be assessed by diluting vesicles into media containing a permeant solute with a high refractive index such as glycerol.

### Membrane Potential Measurements with the Fluorescent Dye diO-C$_5$-(3)

The recording of membrane diffusion potentials provides another means of probing the ion permeability of isolated vesicle fractions. For example, H$^+$ permeability can be assessed by measuring the diffusion potential that is formed when vesicles are transferred from low to high pH.[23] Membrane potentials are visualized with the use of dye probes that change their absorption and/or fluorescence in response to a membrane potential change. The reviews by Cohen and Salzberg,[24] Gupta et al.,[25] and Beeler et al.[26] list the advantages and limitations of the large number of dye probes that are available for measuring potentials in membrane systems. Among these, the positive fluorescent cyanine dye 3,3′-dipentyl-2,2′-oxadicarbocyanine [diO-C$_5$-(3)] is useful for recording membrane potentials (negative inside) in SR and ER vesicles.[2,9,11,23] The dye is commercially available from Molecular Probes (4849 Pitchford Ave., Eugene, OR 97402).

---

[23] G. Meissner and R. C. Young, *J. Biol. Chem.* **255**, 6814 (1980).
[24] L. B. Cohen and B. M. Salzberg, *Rev. Physiol. Biochem. Pharmacol.* **83**, 35 (1978).
[25] R. K. Gupta, B. M. Salzberg, A. Grinvald, L. B. Cohen, K. Kamino, S. Lesher, M. B. Boyle, A. S. Waggoner, and C. H. Wang, *J. Membr. Biol.* **58**, 123 (1981).
[26] T. J. Beeler, R. H. Farmen, and A. N. Martonosi, *J. Membr. Biol.* **62**, 113 (1981).

*Equipment*

Membrane potential-induced changes in fluorescence emission of diO-$C_5$-(3) are followed in a spectrofluorometer equipped with a recorder, a temperature-controlled cell [fluorescence emission of diO-$C_5$-(3) is sensitive to small changes in temperature], and a stirring device (see Equipment, in the section Light-Scattering Measurements). Excitation is at 470 nm and emission is recorded at ~495 nm. Slits are set to give half-bandwidths of 2.5–5 nm. An ultracentrifuge with a fixed angle rotor, a Potter–Elvehjem homogenizer, and an osmometer are needed to transfer vesicles into media of well-defined composition (see Equipment, in the section Radioisotope Flux–Millipore Filtration Measurements).

*Experimental Procedures*

The fluorescent properties of diO-$C_5$-(3) have been determined with the use of unilamellar phospholipid vesicles.[11] The lipid vesicles are sufficiently impermeable to small ions to allow measurement of the dye spectra in the presence of polarized vesicles. Figure 5 shows that the addition of unpolarized lipid vesicles, prepared from SR phospholipid in 200 m$M$ potassium gluconate by sonication, shifts the emission spectrum of diO-$C_5$-(3) to longer wavelengths. In addition, an increase in fluorescence emission is noted. The fluorescence emission decreases when vesicles become negatively charged inside (theoretical Nernst potential of $-120$ mV) by dilution from 200 m$M$ $K^+$ into 1.3 m$M$ $K^+$ (plus 199 m$M$ $Na^+$) medium, in the presence of the $K^+$ ionophore valinomycin. The fluorescence emission gradually returns to that of unpolarized vesicles, as $Na^+$, the major cation in the vesicle medium, slowly moves across the membrane. This process can be accelerated by increasing the valinomycin concentration.

We have quantitated the fluorescence decreases of diO-$C_5$-(3) at 2–15° at an emission wavelength of ~495 nm, i.e., the crossover point of the solid lines in Fig. 5. At this wavelength, a rapid decrease in fluorescence emission due to membrane polarization is followed by a slow return to baseline as the potential slowly collapses. When dye responses are measured at ~495 nm, the fluorescence emission is roughly proportional to the magnitude of the developed potential (Fig. 6) as well as the fraction of vesicles which are polarized. Therefore, using an ion gradient of defined size, the proportion of vesicles permeable to a particular ion can be determined.[11]

The crossover point of the two spectra for polarized and unpolarized vesicles at 495 nm in Fig. 5 is determined by placing 3 ml of the vesicle

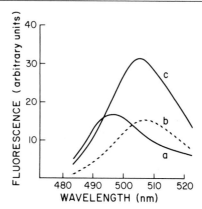

FIG. 5. Effect of membrane polarization on the fluorescence emission spectra of diO-C₅-(3). Small single-walled SR phospholipid vesicles were prepared in 200 m$M$ potassium gluconate medium by sonication.[11] Samples were kept at 2° and excitation was at 470 nm. Both slits were set to give half-bandwidths of 4 nm. Dilution media contained, after the addition of the lipid vesicles, 199 m$M$ sodium gluconate, 1.3 m$M$ potassium gluconate, 1 m$M$ magnesium gluconate, 20 $\mu M$ calcium gluconate, 10 m$M$ Tris–PIPES, pH 7.0, 8 n$M$ valinomycin, and 1.7 $\mu M$ diO-C₅-(3). Spectrum a was recorded before the addition of the lipid vesicles. Spectrum b was obtained within 3 min after lipid vesicles (final concentration 0.66 $\mu$g lipid phosphorus/ml) had been polarized by adding them to the above medium (theoretical Nernst potential of −120 mV). The concentration of valinomycin was then raised to 0.33 $\mu M$ to speed up collapse of K⁺ and Na⁺ gradients. Spectrum c was recorded when ionic gradients had decayed. Spectrum c could also be obtained by diluting lipid vesicles into potassium gluconate medium. (Taken with permission from Ref. 11).

medium (without vesicles) into a temperature-controlled fluorometric cell, followed by the addition of 2–3 $\mu$l of 2 m$M$ diO-C₅-(3) in acetone under stirring. After about 5 min, when a stable baseline is obtained (some of the dye will bind to the cuvette walls and the stirrer), the fluorescence emission maximum is determined (~500 nm at 2–15°). The emission wavelength is then decreased by 5–10 nm to give exactly 90% of the maximum response at 500 nm. Next, the amount of vesicles (15–50 $\mu$g protein/ml) is determined that does not perturb the fluorescence signal at 495 nm under nonpolarizing conditions. Initially a decrease in fluorescence emission should be observed, as small amounts of sample are added successively to the cuvette under stirring. The vesicle concentration that returns the dye signal to the initial value at 495 nm is used in subsequent membrane potential measurements. Between experiments, the cell and stirrer are rinsed with acetone to remove "bound" dye. Alternatively, a constant free dye concentration can be maintained throughout the experiments, by seeding the cell and stirrer twice with the dye before the initial

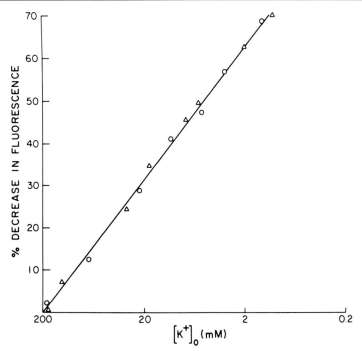

FIG. 6. Fluorescence changes of diO-C$_5$-(3) as a function of K$^+$ gradients across SR phospholipid vesicles. Lipid vesicles prepared in 200 m$M$ potassium gluconate medium were diluted into isoosmolal Na (○) or Tris (△) gluconate media containing 1.7 $\mu M$ diO-C$_5$-(3), 8 n$M$ valinomycin, and varying concentrations of potassium gluconate. The decrease in fluorescence emission observed within 5–10 sec after the addition of the lipid vesicles is plotted against the logarithm of the K$^+$ concentration in the dilution media. Decrease in fluorescence is expressed as percentage of initial fluorescence seen at 495 nm before the addition of liposomes (cf. Fig. 5). (Taken with permission from Ref. 11.)

calibration step. In this case, the cell and stirrer are briefly rinsed with distilled water between experiments.

*Example: H$^+$ Permeability of SR Vesicles*

In Fig. 7 the dye diO-C$_5$-(3) has been used to visualize H$^+$ diffusion potentials generated in SR vesicles. The potential (negative inside) was generated by transferring the vesicles from low (pH 6.2) to high pH (pH 7.8). Vesicle and dilution media contained the relatively impermeant ions Tris$^+$ and PIPES$^-$ to render the vesicles selectively permeable to H$^+$. The time course and magnitude of the fluorescence response were virtually unaffected by the addition of the H$^+$ carrier carbonyl cyanide *p*-trifluoro-

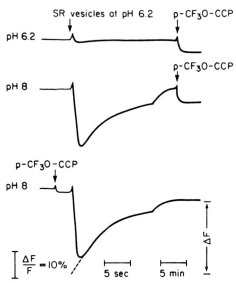

FIG. 7. Visualization of H⁺ diffusion potentials with diO-C₅-(3). SR vesicles in 480 mOsm Tris–PIPES at pH 6.2 were diluted 250-fold into an isoosmolal solution of Tris–PIPES at 15° and the indicated pH, and containing 1.5 $\mu M$ diO-C₅-(3). Additions of the H⁺-carrier carbonyl cyanide $p$-trifluoromethoxyphenylhydrazone ($p$-CF₃O-CCP) (1 $\mu$l of 1.5 m$M$ in acetone/3 ml) were at the times indicated. The addition of $p$-CF₃O-CCP partially quenched diO-C₅-(3) fluorescence. (From Ref. 23.)

methoxyphenylhydrazone, indicating that essentially all skeletal muscle SR vesicles contain a pathway that facilitates the rapid movement of protons. ER vesicles are only slightly permeable to H⁺, whereas SR phospholipid vesicles are essentially impermeable to H⁺.[23]

*Example: Cation Permeability of SR and ER Vesicles*

In Fig. 8 the dye diO-C₅-(3) has been used to determine the K⁺, Na⁺, Tris⁺, Mg²⁺, and Ca²⁺ permeability of intermediate-density SR and ER vesicles. Vesicles equilibrated in 400 mOsm K-PIPES medium, pH 7 (see Experimental Procedures in the section Radioisotope Flux–Millipore Filtration Measurements) were diluted 60-fold into isoosmolal media containing 1.5 $\mu M$ diO-C₅-(3) and the PIPES salts of K⁺, Na⁺, Tris⁺, Mg²⁺, or Ca²⁺. The vesicles elicited no significant change in fluorescence emission when diluted into K⁺ or Na⁺ media, indicating that no membrane potentials were created. Dilution of vesicles into Mg²⁺ or Ca²⁺ media created fluorescence signals of intermediate size. In Tris⁺ medium, SR but not ER vesicles elicited a significant signal. When vesicles were diluted into Na⁺, Tris⁺, Mg²⁺, or Ca²⁺ media in the presence of 0.5 $\mu M$

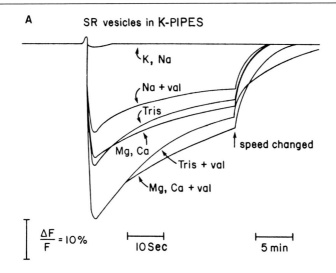

A          SR vesicles in K-PIPES

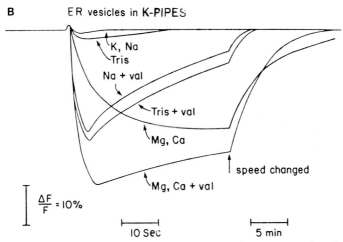

B          ER vesicles in K-PIPES

FIG. 8. Visualization of $K^+$ diffusion potentials with diO-$C_5$-(3). Intermediate-density (A) SR[5] or (B) ER[2] vesicles present in 400 mOsm K-PIPES medium, pH 7, were diluted 60-fold into isoosmolal PIPES media containing the indicated cation, 1.5 $\mu M$ diO-$C_5$-(3), and no or 0.5 $\mu M$ valinomycin. The $K^+$ of the vesicle medium served to establish an initial 60-fold $K^+$ gradient throughout the experiments. (From Ref. 2.)

valinomycin, increased fluorescence signals were observed. Fluorescence signals gradually returned to baseline as the extravesicular cations ($Na^+$, $Tris^+$, $Mg^{2+}$, or $Ca^{2+}$) slowly entered the vesicles, thereby permitting the eventual dissipation of the $K^+$ gradients.

Fluorescence data of Fig. 8 indicate a similar, although not identical, permeability behavior for SR and ER vesicles diluted into Na-, Ca-, or Mg-PIPES media. Decreases in fluorescence seen in Ca$^{2+}$ and Mg$^{2+}$ media in the absence of valinomycin point to the presence of vesicles which were more permeable to K$^+$ than to Mg$^{2+}$ and Ca$^{2+}$ and which therefore formed a negative inside potential. Increases in fluorescence signals in the presence of valinomycin showed the presence of a second population of vesicles which, in the absence of valinomycin, were similarly (or less) permeable to K$^+$ than Ca$^{2+}$ or Mg$^{2+}$. The less-than-optimal signals in Na$^+$ medium (in the presence of valinomycin) indicated that some of the vesicles were permeable to Na$^+$. These vesicles were able to exchange all of their K$^+$ for Na$^+$ within 1–2 sec, the experimental limit of detection. No membrane potential and thus fluorescence signal was formed by K$^+$,Na$^+$-permeable vesicles. Fluorescent measurements of Fig. 8 suggest that a substantial portion of SR and ER vesicles were more permeable to K$^+$ and Na$^+$ than Mg$^{2+}$ or Ca$^{2+}$.

Dye measurements also point out important differences in the permeability behavior of SR and ER vesicle fractions. In Fig. 8, SR vesicles in K-PIPES medium elicited a similar maximal fluorescence signal when diluted into Tris$^+$, Ca$^{2+}$, or Mg$^{2+}$ media containing valinomycin, suggesting that the SR vesicles were relatively impermeable to Ca$^{2+}$ and Mg$^{2+}$ as well as Tris$^+$. In contrast, as indicated by the suboptimal signal, a significant fraction of ER vesicles were more permeable to Tris$^+$ than Ca$^{2+}$ or Mg$^{2+}$. Another difference was that the initial fluorescence changes in Mg$^{2+}$ and Ca$^{2+}$ media were appreciably slower for ER vesicles (20–30 sec) than for SR vesicles (<2 sec). Despite their preferential K$^+$ permeability, a majority of liver microsomes appeared, therefore, to lack an efficient ion-conducting structure for K$^+$ such as the K$^+$,Na$^+$ channel which renders about two-thirds of sarcoplasmic reticulum vesicles highly permeable to K$^+$.

*Comments.* At low vesicle concentrations (15–50 $\mu$g protein/ml) (Figs. 7 and 8), only a fraction of the dye is bound by the membranes under nonpolarizing conditions. Formation of a membrane potential negative inside probably causes additional binding or uptake of the positively charged dye, resulting in a decrease in fluorescence emission by a mechanism not well understood. Whether the membrane potential measurements give an indication of the internal volume or the surface of the polarized vesicles is therefore unclear.

Membrane potentials, positive inside, do not appreciably affect the fluorescence emission of diO-C$_5$-(3). Conditions for the measurement of positive membrane potentials with this dye have therefore not been established. Another potential drawback is that dye probes may not only respond to transmembrane potential changes, but also to alterations in surface potential due to $Ca^{2+}$ binding or changes in ionic environment.[26] However, the optical signals of diO-$C_5$-(3), when used under the conditions described above, were found to be relatively insensitive to changes in $Ca^{2+}$ concentration.[27] On the other hand, $H^+$ diffusion potentials created in Tris–PIPES media produce a smaller dye response than those formed by $K^+$.[23]

Another point to be kept in mind is that a substantial fraction of native SR and ER vesicles are permeable to $K^+$, $H^+$, and $Cl^-$. It may be difficult therefore to avoid osmotic effects, and nearly impossible to render all vesicles selectively permeable to only one ionic species. Osmotic effects during vesicle dilution are minimized by including in vesicle and dilution media a relatively impermeable ion (PIPES$^-$, gluconate$^-$, $Mg^{2+}$; and Tris$^+$ when using SR vesicles). Further, dilution of vesicles from a high to low $K^+$ medium will not only result in the formation of a membrane potential, but also in an exchange of $K^+$ for $H^+$, until $K^+$ and $H^+$ gradients of equal size and magnitude are established. The number of $H^+$ and $K^+$ ions involved in the exchange reaction depends on the internal buffer capacity of the vesicles.[27] The rate depends on whether SR or ER vesicles are used.[23]

### Acknowledgment

This work was supported by Research Grants AR 18687 and HL 27430 from the United States Public Health Service.

[27] G. Meissner, *J. Biol. Chem.* **256**, 636 (1981).

# [33] Permeability of Sarcoplasmic Reticulum

*By* Michiki Kasai and Kazuki Nunogaki

## List of Symbols

$A$      Surface area ($cm^2$)
$A_i$      Amplitude of fluorescence change for component i (cf. Eq. 41) (arbitrary unit)
$a$      Fraction of dye leakage (dimensionless)
$b$      Constant defined as $P/P_w \bar{V}_w C_o$ [c.f. Eq. (17)] (dimensionless)
$C$, $C_o$, $C_s$, $C_m$, $C_{+o}$, $C_{+i}$, $C_{-o}$, $C_{-i}$, $\bar{C}_s$      Concentrations (mol $cm^{-3}$)
$c_i$      Fractional amplitude of component i (cf. Eq. 33) (dimensionless)
$D_{salt}$, $D_+$, $D_-$      Diffusion coefficients ($cm^2$ $sec^{-1}$)
$d$      Membrane thickness (cm)
$E$      Transmembrane potential (V)
$F$      Faraday's constant [coulomb (C) $mol^{-1}$]

$F, F_o, F_{in}, F_{ex}$   Fluorescence intensities (arbitrary unit)

$f, \Delta f_i$   Normalized fluorescence intensities (cf. Eqs. 27,40) (dimensionless)

$g$   Electric conductance [siemens (S) cm⁻²]

$J$   Solute flux (mol sec⁻¹ cm⁻²)

$K_D, K_S$   Fluorescence quenching parameters ($M^{-1}$)

$k, k_s, k_v, k_{Tris}, k_F$   Rate constants (sec⁻¹)

$m$   Protein concentration (mg ml⁻¹)

$N$   Number of channels per vesicle (dimensionless)

$P, P_{salt}, P_+, P_-, P_w$   Membrane permeabilities (cm sec⁻¹)

$p$   Single-channel permeability (cm³ sec⁻¹)

$Q, Q_o, Q_i$   Quencher concentrations (mol liter⁻¹ or $M$)

$R$   Gas constant (J K⁻¹ mol⁻¹)

$R_i, R_{i,o}, R_o, R_t$   Amounts of radioactive tracer (cpm)

$r$   Radius of vesicle (cm)

$S$   Amount of permeable solute in the cell (mol)

$T$   Absolute temperature (K)

$T$   Normalized time (cf. Eq. 17) (dimensionless)

$t$   Time (sec)

$u$   Molar mobility of solute (mol cm² sec⁻¹ J⁻¹)

$\bar{V}_w$   Molar volume of water (cm³ mol⁻¹)

$V, V_0$   Intravesicular volume (cm³)

$v_{app}, v_{iso}$   Specific intravesicular volume (cm³ mg protein⁻¹)

$v, V_{app}$   Normalized vesicular volume (cf. Eqs. 10, 17) (dimensionless)

$x$   Transmembranous distance (cm)

$z$   Valence of ion (dimensionless)

$\alpha$   Constant defined as $\sigma C_s/C_o$ (cf. Eq. 17) (dimensionless)

$\gamma$   Single-channel conductance (siemens (S))

$\rho_o$   Concentration of radioactive tracer (cpm ml⁻¹)

$\sigma$   Reflection coefficient (dimensionless)

$\tau, \tau_{salt}, \tau_+, \tau_-, \tau_i$   Permeation times (sec)

$\tau_{1/2}$   Half-permeation time (sec)

$[K]_{i,t=0}, [K]_o, [Tris]_o$   Concentrations (mol liter⁻¹ or $M$)

## Introduction

Sarcoplasmic reticulum (SR) regulates the cytoplasmic Ca²⁺ concentration of muscle cells and thereby controls muscular contraction and relaxation.[1,2] The cytoplasmic Ca²⁺ concentration is controlled by Ca²⁺ release through the Ca channel and Ca²⁺ uptake by the Ca²⁺-pump. In order to elucidate these functions on a molecular level, studies using isolated SR vesicles are essential.

It has been established that Ca²⁺ uptake is performed by Ca²⁺-ATPase molecules.[3,4] However, it is not known what kinds of ions are counter- or

[1] S. Ebashi, *Annu. Rev. Physiol.* **38**, 293 (1976).

[2] S. Ebashi and M. Endo, *Prog. Biophys. Mol. Biol.* **18**, 123, (1968).

[3] W. Hasselbach, *Biochim. Biophys. Acta* **515**, 23, (1978).

[4] M. Tada, T. Yamamoto, and Y. Tonomura, *Physiol. Rev.* **58**, 1 (1978).

cotransported together with $Ca^{2+}$ [4] and what kinds of ions are transported to neutralize the charge imbalance elicited during $Ca^{2+}$ release. This is due to the fact that the permeability of ions through SR membrane is too fast to be followed by the usual methods.[5,6]

To understand these phenomena, it is necessary to determine the ion permeability of SR vesicles. Then further information such as how much membrane potential is generated during $Ca^{2+}$ uptake and $Ca^{2+}$ release and how long it lasts can be obtained.

In this chapter, we describe the methodology to measure ion permeability of isolated SR vesicles. Since the SR vesicle is a spherical-shaped entity with a diameter of about 0.1 $\mu$m, the method described here is applicable to vesicles of different origin and different sizes including small vesicles, organelles, or cells. Further, the method to measure $Ca^{2+}$ permeability is applicable to $Ca^{2+}$ uptake or release.

## Permeation through Spherical Vesicles

### Flux, Permeability, Rate Constant, and Permeation Time

*Definition.* According to Fick's law,[7] flux of permeant molecules through a membrane is given by

$$J = - uRT \, dC/dx \qquad (1)$$

where $J$ is the flux, $u$ is the mobility, $C$ is the concentration, $R$ is the gas constant, $T$ is the absolute temperature, and $x$ is the distance measured from the outside to the inside of the membrane.

If the flow of the permeant solute is proportional to the concentration difference, putting $dC/dx = \Delta C/d$, Eq. (1) can be written as

$$J = -(uRT/d) \, \Delta C = - P \, \Delta C \qquad (2)$$

where $\Delta C$ is the concentration difference, $d$ is the thickness of the membrane, and $P$ is the permeability coefficient. Now we will consider the efflux of radioactive tracer from the spherical-shaped vesicles with radius $r$ and thickness $d$ as shown in Fig. 1. There are three assumptions: (1) the extravesicular volume is large enough to neglect the increase in the extravesicular concentration of radioactive tracer, (2) intravesicular concen-

[5] D. McKinley and G. Meissner, *J. Membr. Biol.* **44**, 159 (1978).
[6] M. Kasai, K. Nunogaki, K. Nagasaki, M. Tanifuji, and M. Sokabe, *in* "Structure and Function of Sarcoplasmic Reticulum" (S. Fleischer and Y. Tonomura, eds.), p. 537. Academic Press, New York, 1985.
[7] S. Schultz, "Basic Principles of Membrane Transport." Cambridge Univ. Press, London and New York, 1980.

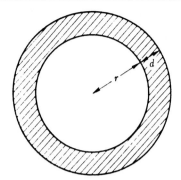

FIG. 1. Diagrammatic representation of a membrane vesicle. The vesicle is depicted as spherically shaped with the inner radius of $r$ and the membrane thickness of $d$.

tration of tracer is uniform, and (3) absorption of permeant solute to the membrane phase is negligible. Then, the amount of radioactive tracer inside the vesicles, $R_i$, is given by the following relation from Eq. (2),

$$-dR_i/dt = JA = PAR_i/V \tag{3}$$

where $A$ is the surface area of the vesicles and $V$ is the intravesicular volume. Integrating Eq. (3), we get

$$R_i = R_{i,o} \exp(-kt) \tag{4}$$
$$k = PA/V \tag{5}$$

where $k$ is the rate constant of efflux and $R_{i,o}$ is the amount of radioactive tracer inside the vesicles at $t = 0$.

If we further assume that $r \gg d$, we get $A/V = 3/r$ and then

$$k = 3P/r \tag{6}$$

Clearly, the efflux follows an exponential function from Eq. (4) and its rate constant is proportional to the permeability coefficient, $P$, and inversely proportional to the radius of vesicles, $r$.

The permeation time $\tau$ is defined by

$$\tau = 1/k \tag{7}$$

and half-permeation time $\tau_{1/2}$ is defined by the time when the radioactivity reaches half of the initial value, i.e.,

$$\tau_{1/2} = 0.693/k \tag{8}$$

In the case of influx, if the amount of intravesicular radioactive tracer at $t = 0$ is zero and the extravesicular concentration of radioactive tracer given in cpm/ml is $\rho_o$, we get the following equation by a similar procedure.

$$R_i = V\rho_o[1 - \exp(-kt)] \tag{9}$$

where $k$ is also given by $3P/r$.

*Deviation from the Exponential Function.* Sometimes the influx or efflux curves for permeant solutes do not follow the simple exponential function [see Eq. (4) or (9)]. A variety of explanations are possible: (1) effect of electric potential, i.e., membrane potential change due to the flux of permeant ions, (2) effect of absorption of permeant solutes to membrane phase or intravesicular molecules, (3) distribution of vesicular size, (4) distribution of permeability coefficient among vesicles, i.e., distribution of channel content, (5) time-dependent change of permeability coefficient, i.e., closing or opening of channels.

## Net Flux and Exchange Rate

When the radioactive tracer is placed on one side of the membrane and movement of radioactive tracer to the other side is followed, we can measure the net flux or equilibrium exchange rate depending on the constitution of the solution in the other side. If the same molecules or ions are present as the tracer used, we can measure the rate of exchange at equilibrium. On the other hand, if the molecules or ions of the same species are not present, we can measure the net flux. When the net flux of ions is measured, we have to take into consideration the fact that the anion and cation move as a pair because electroneutrality must be maintained (as described below in Light-Scattering Method).

## Intravesicular Space

Apparent intravesicular space can be defined as follows[8]:

$$V_{app} = R_{i,o}/R_o \tag{10}$$

where $R_{i,o}$ is the intravesicular radioactivity at $t = 0$ determined by Eq. (4) and $R_o$ is the total radioactivity inside and outside the vesicles. The apparent specific intravesicular space, $v_{app}$ ($\mu$l/mg protein), is given by[8]

$$v_{app} = \frac{1000R_{i,o}}{R_o m} \tag{11}$$

where $m$ is the protein concentration of the vesicle suspension (mg/ml) during the incubation. If the permeant solute does not bind to the mem-

[8] M. Kasai and H. Miyamoto, *J. Biochem. (Tokyo)* **79**, 1067 (1976).

brane phase and the intra- and extravesicular concentrations of the permeant solute are the same at $t = 0$, we can expect that $v_{app}$ corresponds to the intravesicular volume. Intravesicular volume can be determined from the exclusion volume of the nonpermeable solute such as inulin[9] or dextran.[10] Exclusion volume is the sum of the intravesicular volume and the membrane phase volume.

When the apparent space is smaller than the intravesicular volume, there are a number of possible explanations: (1) The permeant ion is not equilibrated due to slow permeation. In such a case, incubation time should be lengthened and influx measurement gives more information.[8,9] (2) Intravesicular concentration of permeant solute is not the same as the extravesicular concentration even at the equilibrium due to the Donnan effect.[9,10] In this case, apparent volume may change with changes in the ionic strength. (3) There exist fast components which cannot be followed by the tracer technique. In the case of SR vesicles this effect is dominant for K$^+$, Na$^+$, Cl$^-$, etc.[5,6]

When the apparent space is larger than the intravesicular volume, (1) binding of permeant solute to the membrane phase or intravesicular molecules[8] and (2) Donnan effect are considered.[9,10] In the case of Ca$^{2+}$ efflux measurement,[11,12] the former effect is observed.

### Radioactive Isotope Tracer Method

*Outline*

Tracer technique is useful to measure permeability as a control for other methods since it is more direct and the results are easy to interpret. However, there are some disadvantages: (1) time resolution is limited, (2) continuous measurement is difficult, and (3) it is limited for the species whose radioactive tracers are available. The tracer method can measure both net flux and equilibrium exchange rate. This method is the only one which can measure the equilibrium exchange rate.

The tracer methods are usually divided into two classes: the efflux and the influx measurements. For both, membrane vesicles must be separated from the extravesicular medium and the amount of radioactive tracer in the extravesicular medium or that of intravesicular medium determined.

[9] M. Kasai, *J. Biochem. (Tokyo)* **88**, 1081 (1980).
[10] P. F. Duggan and A. Martonosi, *J. Gen. Physiol.* **56**, 147 (1970).
[11] K. Nagasaki and M. Kasai, *J. Biochem. (Tokyo)* **94**, 1101 (1983).
[12] Y. Kirino, M. Osakabe, and H. Shimizu, *J. Biochem. (Tokyo)* **94**, 1111 (1983).

Accordingly, the practical procedure depends on the method to separate the vesicles from the extravesicular medium as follows.

1. Ultracentrifugation method. The vesicles are separated from the medium by centrifugation. This method does not have a good time resolution, but is important to estimate the extra- or intravesicular volumes of the vesicles.[9,10]

2. Dialysis method. Vesicles are loaded by the radioactive tracer and are placed in a dialyzing bag, then aliquots of the inner or outer medium of the bag are taken at appropriate time intervals, and their radioactivities are determined.[13] This method is useful to measure slow permeation such as the permeation of $Ca^{2+}$.

3. Filtration method. The vesicles are separated by filtration. SR vesicles are trapped on the Millipore filter. A time resolution of about 10 sec is readily obtained. When the method is combined with the quenching technique, faster permeation can be followed.

*Millipore Filtration Method*

*Efflux Measurement*

*Procedure*

1. Loading: SR vesicles (more than 10 mg protein/ml is preferable) are incubated with a tracer for appropriate time (usually overnight) in order to attain the equilibrium at 0°.

2. Efflux measurement: The incubated suspension is diluted more than 100 times with the solution containing no radioactive tracers. An aliquot (1 ml) is passed through a Millipore filter (pore size 0.45 $\mu$m, HAWP 025, Millipore Co., Bedford, MA) by suction. More than 99% of SR vesicles adhere to the filter. In order to remove remaining free tracers outside the vesicles, 3–5 ml of washing solution, which is similar to the dilution medium, is passed through the filter.

3. Determination of radioactivity. The filters are dried under an infrared lamp and soaked in 10 ml of a liquid scintillation counting medium [1000 ml toluene, 3 g 2,5-diphenyloxazole (PPO) and 0.3 g 2,2'-(1,4-phenylene)bis(5-phenyloxazole) (POPOP), or commercially available liquid scintillation cocktail]. The radioactivity is determined using a liquid scintillation counter.

4. Determination of total radioactivity. An aliquot (10 $\mu$l) of the diluted solution is dried on a Millipore filter and the radioactivity is determined.

[13] J. M. Vanderkooi and A. Martonosi, *Arch. Biochem. Biophys.* **147**, 632 (1971).

The total radioactivity (per milliliter), $R_o$, is 100 times of this value. This value is used for determination of the apparent intravesicular space in Eqs. (10) and (11).

APPARENT ISOTOPE SPACE. Radioactivity at any time can be expressed as the apparent isotope space, $v_{iso}$, defined by

$$v_{iso} = 1000R_t/(R_o m) \tag{12}$$

where $R_t$ is the radioactivity of the vesicles on the filter at time $t$, and the other symbols are the same as in Eq. (11). This quantity is used to get the apparent intravesicular space.

*Influx Measurement.* Usually influx measurements are more difficult than the efflux measurements. More radioactive tracer and higher specific activity are required. However, influx measurements are also necessary to determine the rectification properties of the ion permeation or reversibility of the permeation.

PROCEDURE. SR vesicles (10 mg protein/ml) are equilibrated with 5 m$M$ Tris-maleate (pH 6.8). The suspension is mixed with an equal volume of the solution containing 5 m$M$ Tris-maleate (pH 6.8), 0.1 $M$ choline chloride, and [¹⁴C]choline. Then 0.2-ml aliquots are filtered through a Millipore filter at appropriate times and are washed by passing 3 ml of the solution containing 5 m$M$ Tris-maleate (pH 6.8) and 50 m$M$ choline chloride through the filter to remove the remaining free [¹⁴C]choline. The radioactivity of the filter is assayed as in the case of efflux measurement.[14]

*Analysis and Interpretation*

From Fig. 2 it can be seen that (1) efflux rates differ, for different ions; and (2) the values extrapolated to zero time (apparent intravesicular space) are different. Efflux rate constants for various ions have been determined from the slopes of these curves according to Eq. (4). For example, the efflux rate constants obtained are $2.6 \times 10^{-3}$ sec⁻¹ for choline⁺, $1.2 \times 10^{-2}$ sec⁻¹ for $SO_4^{2-}$, and $9 \times 10^{-3}$ sec⁻¹ for K⁺. The calculated permeation times from Eq. (7) are 390 sec, 81 sec, and 111 sec, respectively. The permeation times directly obtained as the time when the isotope space becomes $1/e$ of the initial value are 96 sec for $SO_4^{2-}$ and 129 sec for K⁺. For choline there are not enough data points to obtain the permeation time directly. These values are slightly larger than those calculated from the rate constants since the efflux curves do not follow Eq. (4). Similarly, the half-permeation times directly obtained as the time when the isotope space becomes half of the initial value are 63 sec for

¹⁴ T. Taguchi and M. Kasai, *J. Biochem.* (*Tokyo*) **96**, 179 (1984).

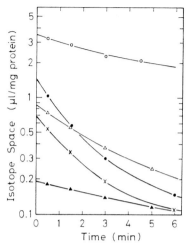

FIG. 2. Permeability measurement by tracer method.[6] SR vesicles were incubated in 0.2 Osm of a salt containing radioactive tracers, 5 m$M$ Tris-maleate (pH 6.5) and 10 mg protein/ ml at 0° for overnight. Efflux of tracer ions from SR vesicles were followed at room temperature after 100-fold dilution in the same solution used for the incubation without the labeled ions. Ordinate represents the radioactivity remaining in the vesicles, i.e., the apparent isotope space defined by Eq. (12). ○, [$^{14}$C]Choline$^+$ (choline-Cl); ●, $^{35}$SO$_4$$^{2-}$ (K$_2$SO$_4$); △, $^{22}$Na$^+$ (NaCl); ×, $^{42}$K (KCl); ▲, $^{36}$Cl$^-$ (NaCl).

SO$_4^{2-}$ and 81 sec for K$^+$. These values are also larger than those calculated from the rate constants in Eq. (8). Permeability coefficients can be calculated from Eq. (6). By assuming the radius of vesicles to be 50 nm, permeability coefficients obtained are $4.3 \times 10^{-9}$ cm/sec for choline$^+$, $2.0 \times 10^{-8}$ cm/sec for SO$_4^{2-}$, and $1.5 \times 10^{-8}$ cm/sec for K$^+$.

In the above calculation, however, the difference in the apparent intravesicular space was not taken into consideration. In the case of choline$^+$, apparent intravesicular space had almost the same value as the intravesicular volume determined by the inulin exclusion volume.[9] Thus, the permeability of all vesicles was determined by the above treatment.[6] However, in the case of other ions such as K$^+$ and Cl$^-$, apparent intravesicular spaces were smaller than the intravesicular volume. After considering many possibilities, it was concluded that a faster component exists which could not be followed by the tracer methods. The present SR preparation contained at least two types of vesicles: one permeates K$^+$ or Cl$^-$ too fast to be followed (probably contains ion channels) and the other permeates them slowly. The permeability coefficient thus determined corresponds only to that for the slowly permeating vesicles.[6] The RI tracer method was applied for various permeants, such as choline$^+$,[5,6,15] Na$^+$,[5,6] K$^+$,[6] Rb$^+$,[5] Cl$^-$,[5,6] SO$_4^{2-}$,[6,16] phosphate$^-$,[15] gluconate$^-$,[15] Ca$^{2+}$,[12,15,17–19] Mn$^{2+}$,[15] and sucrose.[15]

## Light-Scattering Method

### Background: Volume Change of Vesicles Due to Osmosis

The light-scattering method is a convenient one to measure the flux of ions or neutral molecules continuously. When an osmotic pressure gradient is formed across the membrane by the addition of solute, water first moves across the membrane because water permeability is usually higher than that of solutes and then the solute moves depending on its permeability. When an osmotic pressure gradient is applied to the membrane vesicles, movement of water and solute across the membrane causes a change in vesicular volume. If we follow the change in the vesicular volume continuously, we can calculate the permeability to water and solutes. In this section, we first describe the fundamental equation related to water and solute permeation. Next, we describe the method to follow the volume change of the vesicles and to calculate the permeability.

Jacobs[20] treated the problem of osmosis in the cell and described the change of cell volume by

$$dV/dt = P_w \bar{V}_w A[(C_0 V_0 + S)/V - C_s - C_m] \qquad (13)$$

where $V$ is the volume of solvent water in the cell at any time $t$, $A$ is the surface area of the cell membrane, $P_w$ is the water permeability given in cm/sec, $\bar{V}_w$ is the molar volume of water, $C_0$ and $V_0$ are the initial concentration of nonpermeating solute and the initial volume of solvent water within the cell, respectively, $S$ is the amount of permeating solute in the cell water at any time $t$, $C_s$ is the concentration of $S$ in the external medium and is assumed to be constant, and $C_m$ is the concentration of the nonpermeating solutes in the external medium. In the subsequent development, it is assumed that $C_m = C_0$. For the movement of solute,

$$dS/dt = PA(C_s - S/V) \qquad (14)$$

where $P$ is the permeability of the membrane for $S$ given in cm/sec.

[15] G. Meissner and D. McKinley, *J. Membr. Biol.* **30**, 79 (1976).
[16] M. Kasai, *J. Biochem. (Tokyo)* **89**, 943 (1981).
[17] J. Nagasaki and M. Kasai, *J. Biochem. (Tokyo)* **90**, 749 (1981).
[18] A. R. de Boland, R. L. Jilka, and A. N. Martonosi, *J. Biol. Chem.* **250**, 7501 (1975).
[19] H. Morii and Y. Tonomura, *J. Biochem. (Tokyo)* **93**, 1271 (1983).
[20] J. A. Johnson and T. A. Wilson, *J. Theor. Biol.* **17**, 304 (1967).

Johnson and Wilson[20] modified these equations to include the reflection coefficient $\sigma$ for the permeant solute:

$$dV/dt = P_w \bar{V}_w A[(C_0 V_0 + \sigma S)/V - \sigma C_s - C_m] \tag{15}$$
$$dS/dt = PA(C_s - S/V) + (1 - \sigma)\bar{C}_s(dV/dt) \tag{16}$$

where $\bar{C}_s$ is the mean concentration of $S$ across the membrane. These equations were solved by the perturbation method, when the amplitude of the volume change is small. For example, when the vesicle is in the swollen state and the permeable solute is added to the external medium, the change in the vesicular volume is given by

$$v = 1 - [\alpha/(1 - b)][\exp(-bT) - \exp(-T)] \tag{17}$$

where $v$ is the normalized vesicular volume, $\alpha$ is a constant related to the reflection coefficient given by $\sigma C_s/C_0$, $b$ is $P/P_w \bar{V}_w C_0$, and $T$ is the normalized time given by $P_w \bar{V}_w A C_0 t/V_0$. Eq. (17) is rewritten as

$$v = 1 - [\alpha/(1 - b)][\exp(-k_s t) - \exp(-k_v t)] \tag{18}$$

where $k_s$ is the rate constant related to the solute permeability given by

$$k_s = PA/V_0 \tag{19}$$

and $k_v$ is the rate constant of volume decrease, which is related to water permeability as follows.

$$k_v = P_w \bar{V}_w A C_0/V_0 \tag{20}$$

This relation shows that, when the permeable solute is added to the external medium of the vesicles and the permeability of the solute is smaller than that of water, the volume decreases with a time constant of $k_v$ and increases with a time constant of $k_s$. Johnson and Wilson applied this relation to the analysis of cell permeability and determined the water and solute permeabilities and the reflection coefficient.[20] We will now describe the method to determine the permeability of SR vesicles.

Volume changes in such small vesicles or cells can be followed by light-scattering[21,22] or turbidity changes.[23,24] The light-scattering intensity decreases with swelling of the vesicles and increases with shrinkage. The scattered light-intensity change is approximately proportional to the volume change.[21,22] Changes in the absorbance[24] or the reciprocal of absor-

[21] R. I. Sha'afi, G. T. Rich, V. W. Sidel, W. Bossert, and A. K. Solomon, *J. Gen. Physiol.* **50,** 1377 (1967).

[22] T. Kometani and M. Kasai, *J. Membr. Biol.* **41,** 295 (1978).

[23] A. D. Bangham, J. De Gier, and G. D. Greville, *Chem. Phys. Lipids* **1,** 225 (1967).

[24] B. De Kruijff, W. J. Gerritsen, A. Oerlemans, R. A. Demel, and L. L. M. Van Deenen, *Biochim. Biophys. Acta* **339,** 30 (1974).

bance[23] are approximately proportional to the vesicular volume change. The theoretical background for relating volume change and optical response is poor at present, although some treatments have been attempted.[25,26]

This technique has been applied to various membranes, including red blood cells,[21,27,28] liposomes,[23,24] and SR vesicles.[14,22,29,30]

## Measurement of Permeability of SR Vesicles

Since SR vesicles are highly permeable to cations and anions such as $K^+$, $Na^+$, and $Cl^-$,[5,6] fast measurement of light-scattering changes is required. Use of the stopped-flow apparatus is recommended for the measurement of SR permeability.

*Procedure.* SR vesicles are incubated in a low osmotic solution such as 5 m$M$ Tris-maleate (pH 6.5) and 0.5 mg protein/ml for more than 30 min at room temperature. After equilibration, the solution is mixed with an equal amount of solution containing an appropriate concentration of solute (usually 200 mOsm) and 5 m$M$ Tris-maleate (pH 6.5). The scattered-light intensity at 90° is followed at 400 nm with a stopped-flow spectrofluorimeter at a constant temperature. If required, some effectors or drugs such as valinomycin, $Ca^{2+}$, and caffeine, can be added to the incubated medium or mixing solution (see Fig. 3).[22]

*Remarks.* Changes in vesicular volume can also be followed using a spectrophotometer. The wavelength is usually set between 400 and 450 nm.

## Interpretation and Analysis

As shown in Fig. 3, the scattered-light intensity increases rapidly and then decreases. The fast increase in the scattered-light intensity is caused by the shrinkage of the vesicles due to the outflow of water, which is driven by the osmotic pressure difference. The later decrease in the scattered-light intensity is caused by the swelling of the vesicles due to the inflow of water accompanied by solute influx, which is driven by the chemical potential difference. If the curve can be simulated by Eq. (18),

[25] W. Yoshikawa, H. Akutu, and Y. Kyogoku, *Biochim. Biophys. Acta* **735**, 397 (1983).
[26] S. Ikegami and M. Kasai, unpublished observations, 1983.
[27] R. I. Sha'afi and C. M. Gary-bobo, *Prog. Biophys. Mol. Biol.* **26**, 103 (1973).
[28] P.A. Knauf, G. F. Fuhrmann, S. Rothstein, and A. Rothstein, *J. Gen. Physiol.* **69**, 363 (1977).
[29] M. Kasai, T. Kanemasa, and S. Fukumoto, *J. Membr. Biol.* **51**, 311 (1979).
[30] N. Yamamoto and M. Kasai, *J. Biochem.* (*Tokyo*) **92**, 465 (1982).

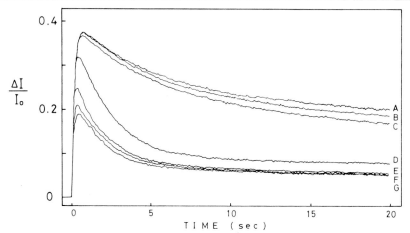

FIG. 3. Effect of valinomycin on the volume change of SR vesicles in KCl solution.[21] SR vesicles were preincubated in 2 mM KCl, 5 mM Tris-maleate (pH 6.5), and 0.4 mg SR protein/ml for more than 3 hr at room temperature. The suspension was mixed with an equal volume of 100 mM KCl and 5 mM Tris-maleate (pH 6.5) at 23° in a stopped-flow apparatus and the change in the right-angle light-scattering intensity at 400 nm was recorded. Valinomycin was added to preincubated SR suspension 30 min before the stopped-flow mixing. The concentrations of valinomycin in micrograms/milliliter were as follows: (A) 0, (B) 1.3 × 10$^{-3}$, (C) 4 × 10$^{-3}$, (D) 5 ×10$^{-2}$, (E) 1.5 × 10$^{-1}$, and (F) 5 × 10$^{-1}$, (G) 1.5. The initial light-scattering intensity ($I_0$) was determined by extrapolating the rising phase back to zero time. The change in the intensity ($\Delta I$) was obtained by subtracting the initial intensity from total intensity.

the rate constant of volume decrease, $k_v$, and the rate constant of solute permeation, $k_s$, can be determined. If we assume that the vesicles are spherical and the membrane thickness is negligible, the permeability coefficient for solute can be calculated from the rate constants according to Eq. (6). If the curve does not follow Eq. (18), the rate constant of volume change, $k_v$, can be estimated from the initial slope of the increasing phase and the apparent permeability for solute, $k_s$, from the initial gradient of decreasing phase.[30] Apparent half-permeation time, $\tau_{1/2}$, can be obtained as the time when the scattered-light intensity reaches the half-value of the maximal increment.[22]

   *Determination of Ionic Permeability.* In the case of neutral molecules, the permeability coefficient thus obtained is for the solute itself. In the case of salt ions, however, since anion and cation move as a pair while maintaining electroneutrality, the permeability coefficient thus determined is related to that of the ion pair.

According to the Nernst relation, the diffusion constant of an ion pair is given by[22]

$$1/D_{salt} = \tfrac{1}{2}(1/D_+ + 1/D_-) \tag{21}$$

where $D_{salt}$ is the diffusion constant of salt and $D_+$ and $D_-$ are those of cation and anion. Since the permeability coefficient is proportional to the diffusion coefficient, from Eq. (21),

$$1/P_{salt} = \tfrac{1}{2}(1/P_+ + 1/P_-) \tag{22}$$

where $P_{salt}$ is the permeability coefficient of salt and $P_+$ and $P_-$ are those of cation and anion. Since the permeability coefficient is inversely proportional to permeation time,

$$\tau_{salt} = \tfrac{1}{2}(\tau_+ + \tau_-) \tag{23}$$

where $\tau_{salt}$ is the permeation time of salt and $\tau_+$ and $\tau_-$ are those of cation and anion.

In the case of KCl permeation (see Fig. 3), permeation time without valinomycin was about 10 sec; it decreased with increasing valinomycin concentration and reached 0.2 sec in the presence of large amounts of valinomycin. Since valinomycin only increases the $K^+$ permeability, permeation times for each ion can be obtained from Eq. (23). The permeation times to $K^+$ and $Cl^-$ were calculated to be 20 sec and 0.4 sec, respectively, in this case.[22]

*Remarks.* The light-scattering method is easily operated and requires a small amount of sample. However, a relatively high concentration of solute is required. The detectable time range is restricted by the time constant of volume decrease related to water permeation. In the case of SR vesicles, since the time constant of volume decrease is around 0.1 sec, permeation of solute faster than 0.1 sec is difficult to determine. As shown in Fig. 4, the amplitude of light-scattering change for KCl is less than that for choline chloride, sucrose, and glycerol.[6] This indicates that faster permeating component(s) exist in KCl permeation than the rate constant of volume decrease related to water permeation (0.1 sec). That is, the SR vesicle preparation is a mixture of vesicles some of which have fast permeating channels for $K^+$ and $Cl^-$ and others not having both channels.[6] This result is consistent with that described in the radioactive isotope tracer experiment.

In this experiment, we assumed that change in the scattered-light intensity is proportional to the volume change. There are examples where the scattered-light intensity is not proportional to the volume change. In such cases, accurate permeability cannot be determined. For example, when a $Ca^{2+}$ ion binds to the membrane and causes aggregation of the

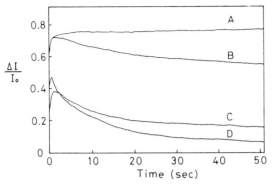

FIG. 4. Measurements of SR permeability to several solutes by the light-scattering method.[6] SR vesicles were incubated in 5 m$M$ Tris-maleate (pH 6.5) and 0.2 mg protein/ml for 1 hr at 23°. The suspension was rapidly mixed in a stopped-flow apparatus with a solution containing solute of 0.2 Osm and 5 m$M$ Tris-maleate (pH 6.5) at 23° and the change in scattered-light intensity was followed at 450 nm. (A) sucrose, (B) choline-Cl, (C) KCl, and (D) glycerol.

vesicles, an unusual increase in the scattered-light intensity was observed. Such a phenomenon was also observed when SR vesicles was treated by DIDS (4,4'-diisothiocyanostilbene-2,2'-disulfonic acid).[31] Curious behavior of the scattered-light intensity was observed in the presence of some SH reagents, such as Ag$^+$ and PCMBS ($p$-chloromercuribenzenesulfonic acid),[32] and mellitin.[32,33] When choline permeability was measured in the presence of tetraphenyl borate ion, an additional increase in the scattered-light intensity was observed due to the complex formation between tetraphenyl borate and choline.[14] In such cases, the artifact should be subtracted.

The impermeable solute ($C_m$ and $C_0$) that appeared in Eqs. (13) and (15) is the origin of the restoration force of the vesicles to swell and return to the initial volume after shrinkage. In the experiment on SR vesicles, we did not add such an impermeable solute. However, the vesicles behaved as if they had the restoration force. We consider that the vesicles have some restoration forces originating from the elasticity of the membrane and the Donnan effect.[9] These forces may have an equivalent role as the restoration force ($C_m$ and $C_0$) introduced in Eqs. (13) and (15). For larger structures such as cells, intracellular molecules might work as impermeable solutes.[20]

[31] M. Kasai, unpublished observations, 1981.
[32] K. Nunogaki, M. Suzuno, E. Kitakuni, and M. Kasai, unpublished observations, 1984.
[33] A. M. Garcia and C. Miller, *J. Gen. Physiol.* **83**, 819 (1984).

## Methods with Potential-Sensing Dyes

### Ionic Permeability and Membrane Potential

Membrane potential is related to the ionic permeability and ionic composition. According to Goldman,[7] membrane potential is given by

$$E = \frac{RT}{F} \ln \left( \frac{P_+C_{+o} + P_-C_{-i}}{P_+C_{+i} + P_-C_{-o}} \right) \qquad (24)$$

for only two kinds of ions, where $P_+$ and $P_-$ are the permeability coefficients of cation and anion, $C_{+o}$ and $C_{+i}$ are the extravesicular and intravesicular cation concentrations, and $C_{-o}$ and $C_{-i}$ are the extravesicular and intravesicular anion concentrations. This membrane potential is readily formed when the transmembrane difference in the ionic composition is applied. By changing the ionic compositions, the permeability coefficient ratio $P_+/P_-$ can be estimated.

If the ionic compositions outside and/or inside the vesicles change as a result of ion permeation, the membrane potential will change. Ionic permeability can be determined from the rate of membrane potential change. By this procedure, slow permeation of ions can be followed.

The membrane potential of SR vesicles is difficult to determine using microelectrodes since they can not be inserted into small vesicles. The potential-sensing probe is a useful tool for such small systems. One of the most popular methods uses the fluorescent potential probe.[34] There are two kinds of potential probes: penetrating and nonpenetrating dyes. Nonpenetrating dyes show a fast response (msec range), but the change of fluorescence intensity is small (less than 1%). These dyes are useful for studying the change of membrane potential during excitation.[34] The penetrating dyes show a large change in fluorescence intensity (more than 10%) but response is slow (second range). These dyes are useful for the vesicular system,[34] and have been used for the estimation of the membrane potential of red blood cells.[35]

### Measurement of Membrane Potential and Determination of Ionic Permeability

*Procedure.* SR vesicles (2.5 mg protein/ml) are incubated with 5 m$M$ K-MES (pH 6.8) and 200 m$M$ potassium gluconate. After overnight incubation, a 20-$\mu$l aliquot is diluted 100-fold into 2 ml of a medium containing 200 m$M$ Tris-gluconate, 1 $\mu M$ valinomycin, 1 $\mu$g diS-C$_3$-(5)/ml, and 5 m$M$

[34] A. S. Waggoner, *Annu. Rev. Biophys. Bioeng.* **8**, 47 (1979).
[35] J. F. Hoffman and P. C. Laris, *J. Physiol. (London)* **239**, 519 (1974).

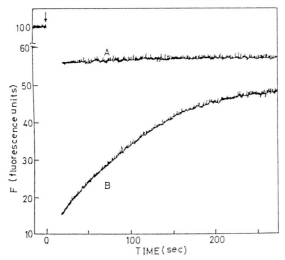

Fig. 5. Time course of fluorescence change of diS-$C_3$−(5) upon application and dissipation of the transmembrane potential.[36] At the time indicated by the arrow, a 20-$\mu$l aliquot of a suspension of SR vesicles (2.5 mg protein/ml) incubated with 5 m$M$ K-MES (pH 6.8) and 200 m$M$ potassium gluconate overnight at 0° was diluted 100-fold into 2 ml of medium containing 1 $\mu$g of diS-$C_3$−(5)/ml, 5 m$M$ Tris-MES (pH 6.8), and either 200 m$M$ potassium gluconate (A) or 200 m$M$ Tris-gluconate (B). Fluorescence emission was recorded at 670 nm with excitation at 620 nm in a fluorescence spectrophotometer at 23°. The value of the fluorescence intensity before addition of vesicles was defined at 100 fluorescence units.

Tris-MES (pH 6.8), and changes in fluorescence intensity are recorded at 670 nm with excitation at 620 nm using a fluorescence spectrophotometer.[36]

### Analysis and Interpretation

DETERMINATION OF MEMBRANE POTENTIAL. As shown in Fig. 5,[36] the fluorescence intensity is suddenly decreased after dilution. This decrease is due to the formation of an inside negative membrane potential. Since the permeability of the membrane for Tris and gluconate ions is much smaller than that for K$^+$, the membrane potential generated by this salt replacement is controlled by only the K$^+$ concentration gradient. The percent changes in the fluorescence intensity are determined by extrapolating the curves back to zero time and these values are plotted against external K$^+$ concentration in Fig. 6. Fluorescence intensity changes linearly with the logarithm of external K$^+$ concentration. This shows that

[36] N. Yamamoto and M. Kasai, J. Biochem. (Tokyo) **88,** 1425 (1980).

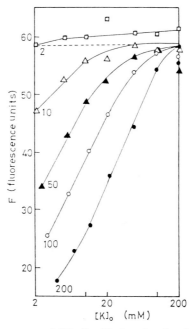

FIG. 6. Change in fluorescence of diS−C$_3$−(5) plotted against log [K]$_o$.[36] A 20-$\mu$l aliquot of a suspension of SR vesicles (2.5 mg protein/ml) incubated with 200 m$M$ mixtures of potassium gluconate and Tris-gluconate to give the K$^+$ concentration indicated in the figure was diluted 100-fold into a medium containing 1 $\mu$g of diS−C$_3$−(5)/ml, 1 $\mu$g of valinomycin/ ml, 5 m$M$ Tris-MES (pH 6.8), and 200 m$M$ mixture of potassium gluconate and Tris-gluconate to give the K$^+$ concentration indicated on the abscissa. The percent changes in the fluorescence intensity were determined by extrapolating the curves similar to one shown in Fig. 5 back to zero time in all experiments. K$^+$ concentrations of the incubation media were as follows: ●, 200 m$M$; ○, 100 m$M$; ▲, 50 m$M$; △, 10 m$M$; □, 2 m$M$.

fluorescence intensity at $t$ = 0 is proportional to the membrane potential formed. When we want to estimate the membrane potential from the fluorescence intensity, this curve is used for calibration.

DETERMINATION OF SLOW PERMEATION OF TRIS IONS. As shown in Fig. 5, fluorescence intensity gradually increases after formation of the K$^+$ diffusion potential. The later change is due to the decrease of intravesicular K$^+$ concentration. Since anion permeation is slow in this case, the efflux of K$^+$ is coupled with the influx of Tris$^+$ and the Tris$^+$ permeation is rate determining. From the rate of fluorescence change the permeation rate of Tris$^+$ can be determined. Since the fluorescence intensity is not proportional to the intravesicular K$^+$ concentration, the following relation should be used[37]:

[37] N. Yamamoto and M. Kasai, *Biochim. Biophys. Acta* **692**, 89 (1982).

$$k_{Tris} = k_F \frac{\ln([K]_{i,t=0}/[K]_o)}{[Tris]_o/[K]_{i,t=0}} \qquad (25)$$

where $k_{Tris}$ is the rate constant of $Tris^+$ influx, $k_F$ is the rate constant of the fluorescence change, $[K]_{i,t=0}$ is the intravesicular $K^+$ concentration at $t = 0$, and $[K]_o$ and $[Tris]_o$ are the extravesicular concentrations of $K^+$ and $Tris^+$, respectively.

Permeability of slowly permeating cations such as choline and TEA can be determined by the same method. Permeability of a slowly permeating anion such as gluconate can be determined by a similar method which replaces the ionic composition.[38] In this case, SR vesicles incubated in Tris–gluconate are diluted in a solution containing Tris-Cl. When cyanine dyes such as diS-C$_3$-(5) are used, conditions should be chosen in which negative membrane potential is formed, because these dyes can measure only inside negative potentials.[34] Dyes which monitor the inside positive potential are also available.[34]

### ANS Fluorescence Method

1-Anilinonaphthalene-8-sulfonic acid (ANS) is a hydrophobic probe and its fluorescence intensity increases in a hydrophobic environment.[39] When ANS binds to the membrane phase, fluorescence intensity increases markedly. Since ANS has a negative charge, its binding to the membrane phase is affected by the surface potential, which is dependent on the density of surface charge and ionic strength.[40] Surface charge can be changed with the binding of ions to the membrane phase. Accordingly, if we follow the change in ANS fluorescence, we can estimate the movement of some ions. This method has been applied to measure the permeability for $Ca^{2+}$, $K^+$,[40,41] and $H^+$ [42] and will not be described here.

### Fluorescence-Quenching Method

The highest time resolution can be obtained by using the fluorescence-quenching method for measuring the permeability of small ions. Since the quenching process occurs in the nanosecond range, the limit of measurement arises from the method of concentration jump, i.e., stopped-flow mixing.

[38] N. Yamamoto and M. Kasai, *J. Biochem. (Tokyo)* **89**, 1521 (1981).
[39] D. H. Haynes and H. Staerk, *J. Membr. Biol.* **17**, 313 (1974).
[40] V. C. K. Chiu, D. Mouring, B. D. Watson, and D. H. Haynes, *J. Membr. Biol.* **56**, 121 (1980).
[41] V. C. K. Chiu and D. H. Haynes, *J. Membr. Biol.* **56**, 203 (1980).
[42] D. H. Haynes, *Arch Biochem. Biophys.* **215**, 444 (1982).

*Features of Fluorescence Quenching*

Fluorescence of organic molecules is weakened in the presence of a certain kind of molecule, called quenchers. This phenomenon is called "fluorescence quenching" and is classified as "dynamic quenching" or "static quenching," based on its mechanism.[43,44]

In the dynamic quenching mechanism, diffusional collision of quenchers on the excited dyes shortens the lifetime of the excited state and diminishes the quantum yield of the dye fluorescence. This effect is expressed by the Stern–Volmer equation which states that the reciprocal of fluorescence intensity is proportional to quencher concentration. When the Stern–Volmer relationship does not hold, the static-quenching mechanism is also considered to operate in that diffusion of the quencher is not involved. In general, fluorescence quenching is described by the modified Stern–Volmer equation[43,44]:

$$F_o/F = (1 + K_DQ) \exp(K_SQ) \tag{26}$$

where $F$ and $F_o$ are the fluorescence intensity in the presence and in the absence of the quencher, $K_D$ and $K_S$ are the quenching constants for dynamic and static quenching, respectively, and $Q$ is the quencher concentration. $K_S$ is assumed to be zero in the case of purely dynamic quenching.

The quenching phenomena of fluorescent amino acid residues in proteins and extrinsic fluorescent labels have been used to investigate the extent of exposure to the protein surface or the microscopic environment around them. The following molecular species have been employed as quenchers: cations such as pyridinium, $Cs^+$, and $Tl^+$, anions such as $I^-$, $NO_3^-$, and $SCN^-$, and neutral molecules such as $O_2$ and acrylamide.[43,44] In the case of fluorescence quenching of charged dyes by ionic quenchers, care must be taken because the quenching profile is affected by ionic strength.[44]

*Application to Permeability Measurement*

The fluorescence-quenching phenomenon has been applied to measure ionic permeation through membrane vesicles isolated from the electric organ of *Torpedo californica* by Moore and colleagues using a water-soluble dye ANTS (8-amino-1,3,6-naphthalene trisulfonate) and $Tl^+$ as quencher.[45,46] The $Tl^+$ influx was measured with a stopped-flow instru-

[43] M. R. Eftink and C. A. Ghiron, *Anal. Biochem.* **114**, 199 (1981).
[44] S. S. Lehrer and P. C. Leavis, this series, Vol. 49, p. 222.
[45] H.-P. H. Moore and M. A. Raftery, *Proc. Natl. Acad. Sci. U.S.A.* **77**, 4509 (1980).
[46] W. C.-S. Wu, H.-P. H. Moore, and M. A. Raftery, *Proc. Natl. Acad. Sci. U.S.A.* **78**, 775 (1981).

ment and showed single-phase exponential kinetics, being accelerated by agonist for acetylcholine receptor and inhibited by its antagonist. For the measurement of SR permeability, the same procedure was used but the dye was replaced by PTS(pyrene1,3,6,8-tetrasulfonate), a dye with lower permeability than that of ANTS. The influx of $Tl^+$ and $I^-$ was measured in place of $K^+$ and $Cl^-$.[6,33,47] Here we will describe the application of the quenching method to the measurement of SR permeability.

*Quenching of PTS Fluorescence by $Tl^+$ and $I^-$.* Fluorescence of 0.1 $\mu M$ PTSNa$_4$ solutions containing 5 m$M$ K-MES (pH 6.5) is measured and concentrations of TlNO$_3$ and KNO$_3$ varied in such a manner as to keep the total concentration of TlNO$_3$ and KNO$_3$ at 0.1 $M$ to maintain the ionic strength constant. Excitation is at 373 nm or 353 nm and detection is at 403 nm. In the measurement of quenching by $I^-$, the total concentration of KNO$_3$ and KI is kept at 0.4 $M$ because of the weaker quenching effect of $I^-$.

The results are described by the modified Stern–Volmer equation. The quenching parameters obtained are, for $Tl^+$ ([TlNO$_3$] + [KNO$_3$] = 0.1 $M$), $K_D = 9.05 \times 10^{-2}$ m$M^{-1}$, $K_S = 2.51 \times 10^{-3}$ m$M^{-1}$; and, for $I^-$ ([KI] + [KNO$_3$] = 0.4 M), $K_D = 2.42 \times 10^{-2}$ m$M^{-1}$, $K_S = 5.44 \times 10^{-4}$ m$M^{-1}$. These profiles are common whether excitation is at 373 or 353 nm.

*Instrument and Zero-Time Calibration for Stopped-Flow Mixing.* Quencher influx measurement is carried out with a gas pressure-driven type stopped-flow fluorescence spectrophotometer (RA 401 and RA450, Union Giken Co., Hirakata, Osaka, Japan). The stopped-flow apparatus can be operated with either a single or a double mixing cell. In the single mixing cell, the two solutions are mixed only once; in the double mixing cell, the solution is mixed once, divided again and then remixed. Shorter dead time can be attained with the single mixing cell. With the double mixing cell, the mixed solution is more stable against the turbulent circulation of unmixed solutions after the mixing so that observation is possible for a longer period of time.

The mixing zero time is determined by the method of Tonomura *et al.*[48] for instrumental conditions used in the quencher influx measurement. In this method, the reduction reaction of DCIP (2,6-dichlorophenol-indophenol) by L-ascorbic acid is utilized under pseudo-first-order reaction conditions in which there is an excess amount of ascorbic acid with respect to DCIP. In the quencher influx measurement, the obtained zero time has been compensated for by a delay circuit before the data are stored into memory, so that the data acquisition is initiated at the mixing zero time.

[47] K. Nunogaki and M. Kasai, in preparation.
[48] B. Tonomura, H. Nakatani, M. Ohnishi, J. Yamaguchi-Ito, and K. Hiromi, *Anal. Biochem.* **84,** 370 (1978).

The values obtained for the dead time are about 0.6 msec at gas pressure of 8 kg/cm² and about 1.0 msec at 4 kg/cm² with a single mixing cell and about 4 msec at 8 kg/cm² and about 7 msec at 4 kg/cm² with a double mixing cell.[47]

### Quencher Influx Measurement

DYE INCORPORATION INTO VESICLES AND SEPARATION OF FREE DYES: SR vesicles are incubated overnight at 4°C at a protein concentration of 9 mg/ml with 1 m$M$ PTSNa$_4$ in 5 m$M$ K-MES (pH 6.5) and 100 m$M$ or 400 m$M$ KNO$_3$ (for Tl$^+$ and I$^-$ influx experiments, respectively. The incubated suspension of 350–400 μl is applied to a Sephadex G-25 column (coarse, 16 mm × 20 cm) equilibrated with the same solution as for the incubation except for PTS. The void volume containing SR vesicles (3.0–3.5 ml) is collected and diluted twofold with the filtration solution. This SR suspension is applied to the sample chamber for stopped-flow mixing.

ASSAY OF DYE LEAKAGE. One milliliter of the sample suspension is further diluted into 12 ml with the KNO$_3$ buffer solution [0.1 $M$ KNO$_3$ and 5 m$M$ K-MES (pH 6.5) for the Tl$^+$ quenching experiment and 0.4 $M$ KNO$_3$ and 5 m$M$ K-MES (pH 6.5) for the I$^-$ quenching experiment]. At appropriate times during the stopped-flow experiment, 3-ml aliquots are passed through a Millipore filter (HAWP 02500, Millipore Co., Bedford, MA) by suction. The amount of dye leaked out of SR vesicles is determined from the fluorescence of the filtrated solution.

The dye leakage increases almost linearly with time for at least 30 min after gel filtration. The rate is 0.5–0.7%/min. The leakage at the time of stopped-flow mixing is 10–15%.

STOPPED-FLOW MIXING. The sample suspension is rapidly mixed with an equal volume of quencher solution of the same ionic strength as the gel filtration medium, i.e., 100 m$M$ TlNO$_3$ or 400 m$M$ KI with 5 m$M$ K-MES (pH 6.5), in the stopped-flow spectrophotometer (RA 401 and RA 450, Union Giken Co., Hirakata, Osaka, Japan). The subsequent change in fluorescence is followed at right angles through a glass cut-off filter at 400 nm (L-40, Hoya Co., Shinjuku, Tokyo, Japan) and for excitation at 353 nm by a xenon lamp. The low-frequency noise in the photomultiplier output caused by the xenon lamp flickering is removed by a simple additional electronic system constructed with a silicone photodiode and operational amplifiers.[47]

The fluorescence signal after mixing is stored in the transient memory of the stopped-flow instrument (RA 450, Union Giken Co., Hirakata, Osaka, Japan) and later printed on paper. The data are stored at several time scales of different order, i.e., from 10 msec to 10 sec in single mixing experiments and from 0.1 sec to 100 sec in double mixing experiments.

The trace of the fluorescence change is smoothed visually by hand. A set of the signal levels at appropriate times, chosen to be almost of equal intervals on a logarithmic time scale, are measured on the curve and are put into a microcomputer (M243 mark IV, Sord Computer Corp. Masago, Chiba, Japan). The data are analyzed as described in the next section.

After mixing, the process can be monitored for longer than 100 sec with the double mixing cell. With the single mixing cell it can be monitored for about 10 sec because a ghost appears in the signal, probably due to circulation of unmixed solution in the observation cell. A pair of stopped-flow mixing measurements are done routinely with both the single mixing and the double mixing cells in order to extend the observable time range. Usually, gas pressures of 8 kg/cm² and 4 kg/cm² are used for single and double mixing measurements.

To normalize the fluorescence change due to quencher influx, the total fluorescence level is measured by mixing the sample suspension with an equal amount of $KNO_3$ buffer solution instead of quencher solution. The zero fluorescence level is measured by adding $KNO_3$ buffer solution to both chambers. Upon analysis, the total fluorescence level and the zero level are corrected for the output difference due to different refractive indices of $KNO_3$ buffer and quencher solutions and for the effect of scattering of excitation light by SR vesicles, although these effects amount only to a few percent of the total fluorescence change due to quencher influx.

An example of the trace of fluorescence change on quencher influx is shown in Fig. 7.

*Analysis*

The fluorescence intensity of PTS decreases with quencher concentration according to the Stern–Volmer law as mentioned previously:

$$F/F_o = 1/[(1 + K_DQ) \exp(K_SQ)] = f(Q) \qquad (27)$$

where symbols have already been defined except for $f(Q)$, which expresses the quenching profile as defined by Eq. (27).

In the quencher influx experiment, the fluorescence of dyes outside the vesicles ($F_{ex}$) is immediately quenched to a level $f(Q_o)$, where $Q_o$ is the extravesicular quencher concentration. The latter is considered to be constant during the inflow of quenchers since the extravesicular volume is far larger than intravesicular volume, i.e., more than 500-fold under the present experimental conditions. The fluorescence of the dyes in the vesicles ($F_{in}$) decreases as the intravesicular quencher concentration, $Q(t)$,

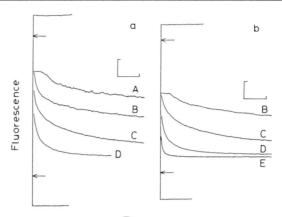

Time

FIG. 7. Measurements of Tl⁺ influx by the fluorescence-quenching method.[47] SR vesicles (9 mg protein/ml) were incubatd with 1 m$M$ PTSNa₄ in KNO₃ buffer solution [0.1 $M$ KNO₃ and 5 m$M$ K-MES (pH 6.5)] overnight at 4°. The suspension was passed through a Sephadex G-25 column to remove extravesicular dye. The eluent was diluted twofold with KNO₃ buffer solution and was rapidly mixed with an equal volume of TlNO₃ buffer solution [0.1 $M$ TlNO₃ and 5 m$M$ K-MES (pH 6.5)] in a stopped-flow apparatus at 23°. The subsequent decrease in PTS fluorescence was monitored in several time ranges: The scale bars represent (A) 2 msec, (B) 20 msec, (C) 0.2 sec, (D) 2 sec, and (E) 20 sec. The fluorescence was measured with a cutoff filter at 400 nm with excitation at 353 nm. The vertical scale indicates 0.1 V. The lowest traces are the background of the KNO₃ buffer solution and the upper traces are the total fluorescence intensity measured by mixing with KNO₃ buffer solution instead of TlNO₃ buffer solution. Upper arrows show the level of fluorescence inside the vesicles at zero time. Lower arrows show the level of fluorescence fully quenched. (a) Measurement with the single mixing cell; (b) measurement with the double mixing cell.

increases. We assume that the permeability of a vesicle is unchanged throughout the process and we get

$$Q(t) = Q_0[1 - \exp(-t/\tau)] \tag{28}$$

where $\tau$ is the permeation time of the vesicle. Eq. (28) is equivalent to Eq. (9).

For a homogeneous system, $F_{in}(t)$ can be obtained by substituting Eq. (28) into Eq. (27). Letting the fraction of the dyes outside the vesicles be denoted by $a$, total fluorescence from the sample suspension is written as

$$F(t) = F_{in}(t) + F_{ex} = (1 - a)F_0 f[Q(t)] + aF_0 f(Q_0) \tag{29}$$
$$F(\infty) = F_0 f(Q_0) \tag{30}$$

where $F_0$ is the nonquenched fluorescence level of the suspension that is experimentally measurable. In the case where $K_S = 0$, Eq. (29) can be rearranged to

$$\frac{1}{F(\infty)} - \frac{1 - a}{F(t) - aF(\infty)} = \frac{K_D Q}{F_o} \exp\left(-\frac{t}{\tau}\right) \tag{31}$$

Thus, if the inflow of quencher is a single process, one can get a straight line by plotting the logarithm of the left-hand side of Eq. (31) against time. This method was successfully employed by Moore *et al.* for analysis of the fluorescence decay of ANTS by $Tl^+$-inflow through the acetylcholine receptor.[45] It may be applicable to the quenching system with PTS and $Tl^+$ if its influx is a single process, since the quenching can be approximated by the unmodified Stern–Volmer equation. In the case of SR, however, this method is not effective because the influx of quenchers is not a single process but is a multicomponent in every case investigated and because, for $I^-$, the quenching can be expressed only by a modified Stern–Volmer equation.

Here we consider a heterogeneous system, i.e., a mixture of several types of vesicles of different permeabilities. Vesicles of type i have a permeation time of $\tau_i$ and contribute to $F_{in}$ with a fraction $c_i$ ($\Sigma_i c_i = 1$). The kinetics for the homogeneous system holds for each component:

$$Q_i(t) = Q_o[1 - \exp(-t/\tau_i)] \tag{32}$$

Summation of all components yields the total fluorescence of the suspension:

$$F(t) = (1 - a)F_o \sum_i c_i f[Q_i(t)] + aF_o f(Q_o) \tag{33}$$

$$F(\infty) = F_o f(Q_o) \tag{34}$$

We define

$$\Delta F(t) = F(t) - F(\infty) \tag{35}$$

and get

$$\Delta F(t) = (1 - a)F_o \sum_i c_i\{f[Q_i(t)] - f(Q_o)\} \tag{36}$$

$$\Delta F(0) = (1 - a)[F_o - F(\infty)] \tag{37}$$
$$= (1 - a)[F_o(1 - f(Q_o)] \tag{38}$$

Equation (37) gives the level of fluorescence at the start of mixing with quencher, which is indicated by an arrow in each trace in Fig. 7. From Eqs. (36) and (38), we obtain

$$\Delta F(t) = \sum_i A_i \, \Delta f_i(t) \tag{39}$$

by defining

$$\Delta f_i(t) = \frac{f[Q_i(t)] - f(Q_o)}{1 - f(Q_o)} \qquad (40)$$

and

$$A_i = \Delta F(0)c_i \qquad (41)$$

The problem is how to resolve $\Delta F(t)$ into each summand in Eq. (39). Experimentally, $F(t)$ and $F_o$ are obtained directly. Next, $F(\infty)$, and then $\Delta F(t)$ and $\Delta F(0)$, are obtained by calculation using Eqs. (34), (35), and (37). Since $\Delta f_i(t)$ involves $\tau_i$ only in terms of $t/\tau_i$, the plot of the logarithm of each summand against log $t$ becomes a curve with a common shape for all i, i.e., the shape of log $\Delta f_i(t)$ versus log $t$. Since $\Delta f_i(t)$ diminishes as $t$ becomes large, $\Delta F(t)$ is approximated only by the slowest component at large $t$: log $\Delta F(t)$ = log $A_s$ + log $\Delta f_s$ at large $t$. Then we can fit the data points with the curve of log $\Delta f(t)$ versus log $t$ by translating it vertically and horizontally. The extent of the translation gives $A_s$ and $\tau_s$. Then log $[\Delta F(t) - A_s \, \Delta f_s(t)]$ versus log $t$ plot is in turn subjected to the same procedure to get $A$ and $\tau$ of another component. This is repeated until the remaining amplitude is fitted with the fastest component, its amplitude $A_f$ being fixed for $\Delta F(0) - \Sigma_i A_i$.

An example of this fitting procedure is shown in Fig. 8. The parameters obtained for components are shown in Fig. 9.

## Relation between Conductance and Permeability

Recently, the method of measuring single-channel conductance of ion channels has been developed. Here we will present the interrelations between ion permeation and ionic conductance.

The relation between zero-volt conductance under symmetrical ionic conditions and permeability coefficient is given by[7]

$$g = PC[(z^2F^2)/(RT)] \qquad (42)$$

where $g$ is the electric conductance, $P$ is the permeability coefficient, $C$ is the concentration of conducting ion, $z$ is its valence, $F$ is Faraday's constant, and $R$ and $T$ have the usual meaning. Equation (42) is rewritten by using single-channel conductance $\gamma$

$$\gamma = pC[(z^2F^2)/(RT)] \qquad (43)$$

where $p$ is the permeability coefficient of the single channel.

Since $PA$ in Eq. (5) is given by $Np$, where $N$ is the number of channels in a single vesicle, the permeation time in Eq. (7) is given by

$$\tau = V/Np \qquad (44)$$

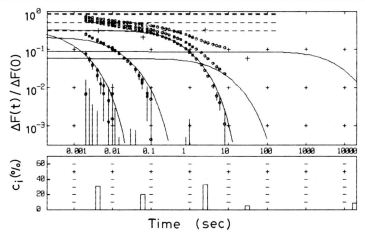

Time (sec)

FIG. 8. Analysis of the $Tl^+$ influx data obtained with a single mixing cell.[47] Upper figure: The logarithm of the normalized fluorescence $[\triangle F(t)/\triangle F(0)]$ is plotted against log $t$ (uppermost points). Successive fitting with $\triangle f(t)$ versus log $t$ curve (solid line) and its subtraction were repeated as described in the text. Lower figure: The fractional amplitude of each component is plotted against its permeation time on the same log $t$ scale as in the upper figure. Each bar in the lower figure corresponds to each solid curve of $\triangle f(t)$ versus log $t$ in the upper figure.

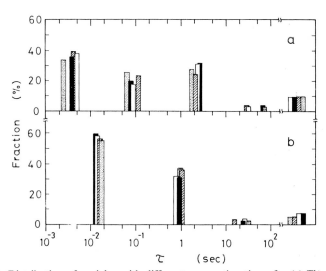

FIG. 9. Distribution of vesicles with different permeation times for (a) $Tl^+$ and (b) $I^-$ influx.[47] The permeation time and the fractional amplitude of each component were obtained from experiments similar to Fig. 7 and through analyses similar to Fig. 8. The different shadings represent different preparations.

Combining Eqs. (43) and (44), the relation between the permeation time and the single-channel conductance is given by

$$\tau = (CVz^2F^2)/(N\gamma RT) \tag{45}$$

If we assume a spherical vesicle with a radius of $r$,[6,49] we get

$$\tau = (4\pi r^3 Cz^2 F^2)/(3N\gamma RT) \tag{46}$$

Using typical values for $\gamma$ (10 pS), $C$ (0.15 $M$), and r (50 nm), we find that the permeation time is 30 msec, if the vesicle has only one channel and the channel is open all the time. Since the permeation time is inversely proportional to the single channel conductance, permeation time becomes 3 msec for $\gamma = 100$ pS.

## Cautions Concerning the Effects of Drugs

When the effect of drugs on ionic permeability is to be examined, care must be taken since problems may arise.

Drugs such as SITS (4-acetoamido-4'-isothiocyanostilbene-2,2'-disulfonate) and DIDS emit fluorescence which gradually decreases due to photobleaching under the optical condition used in the present method. The measured fluorescence change includes a superfluous decaying component.[47] If the effector itself (e.g., Cs⁺) is also a fluorescence quencher, it must be added to both sample and quencher solutions and the quenching curve must be measured in the presence of the effector. Any other molecules which may influence the quenching profile must be used with care: drugs such as decamethonium interfere with the quenching of PTS by Tl⁺, probably by binding to PTS. Some SH reagents cause the SR membrane to become leaky even to PTS, which hampers a precise estimation of the dye leakage at the instant of quencher mixing.

### Techniques for Calcium Flux Measurement

For the measurement of $Ca^{2+}$ flux, essentially the same methods used for the other ions are applicable. However, there are some specially developed methods for $Ca^{2+}$ flux measurement. Important methods include the use of tracer, use of fluorescent dyes such as ANS,[41] chlortetracycline (CTC),[11,50–52] aequorin,[53] and metallochromic dyes[54,55] such as murexide,

[49] C. Miller, *Annu. Rev. Physiol.* **46,** 549 (1984).
[50] K. Nagasaki and M. Kasai, *J. Biochem. (Tokyo)* **87,** 709 (1980).
[51] K. Nagasaki and M. Kasai, *J. Biochem. (Tokyo)* **96,** 1769 (1984).
[52] A. H. Caswell and B. C. Pressman, *Biochem. Biophys. Res. Commun.* **49,** 292 (1972).
[53] J. R. Blinks, *Ann. N.Y. Acad. Sci.* **307,** 71 (1978).
[54] A. Scarpa, this series, Vol. 56, p. 301.
[55] Y. Ogawa, H. Harafuji, and N. Kurebayashi, *J. Biochem. (Tokyo)* **87,** 1293 (1980).

arsenazo III,[56,57] and antipyrylazo III. Here we describe only the method using CTC.

## Fluorescence Characteristics of CTC

CTC exhibits a low fluorescence in aqueous solution, but its fluorescence intensity is enhanced when it interacts with divalent cations such as $Ca^{2+}$ or $Mg^{2+}$, and is further markedly enhanced when the environment is hydrophobic.[58] This phenomenon is applied to measure membrane-bound divalent cations. The fluorescence increase is proportional to the amount of divalent cation dye complex bound to the membrane phase. The CTC fluorescence technique is used for the determination of passive $Ca^{2+}$ flux[50] or $Ca^{2+}$ release[11,51,59] and active $Ca^{2+}$ uptake.[60] Further, it was used for the determination of $Mg^{2+}$ flux.[50,51] This method is useful to study $Mg^{2+}$ permeability since a radioactive tracer for $Mg^{2+}$ is not readily available.

## Measurement of Slow $Ca^{2+}$ and $Mg^{2+}$ Efflux by the CTC Method

Procedure. SR vesicles are incubated in 150 m$M$ KCl, 20 m$M$ Tris-maleate (pH 6.7), 10 m$M$ $CaCl_2$, and 5 mg protein/ml overnight at 4°. Then, a 50-$\mu$l aliquot is diluted into 40 volumes of a solution containing 150 m$M$ KCl, 20 m$M$ Tris-maleate (pH 6.7), 10 $\mu M$ CTC and a specified amount of $Ca^{2+}$ and/or effectors, and the change in fluorescence intensity is recorded at 530 nm with excitation at 390 nm using a fluorescence spectrophotometer (FS-501, Union Giken Co., Hirakata, Osaka, Japan) at 23° (see Fig. 10).[51]

Analysis. Fluorescence intensity first increases very rapidly with the influx of CTC into the vesicles. After CTC influx reaches equilibrium, fluorescence decreases due to the efflux of $Ca^{2+}$ and concomitant efflux of CTC. The $Ca^{2+}$ efflux rate can be estimated from the rate of fluorescence change. Since the fluorescence intensity is proportional to the Ca–CTC complex bound to the membrane, it is roughly proportional to the free $Ca^{2+}$ concentration inside the vesicles. However, this value is also approximately proportional to the total amount of the intravesicular $Ca^{2+}$, and the $Ca^{2+}$ efflux rate can be estimated from the rate constant of fluorescence change. Since SR vesicles have $Ca^{2+}$-binding sites inside the vesicles, $Ca^{2+}$ efflux does not follow a simple exponential curve. A three-compartment model seems to be effective for the analysis of this system.[11,12] If some effectors are present in the extravesicular medium,

[56] D. H. Kim, S. T. Ohnishi, and N. Ikemoto, J. Biol. Chem. 258, 9662 (1983).

[57] N. Ikemoto, B. Antoniu, and D. H. Kim, J. Biol. Chem. 259, 13151 (1984).

[58] A. H. Caswell and J. D. Hutchison, Biochem. Biophys. Res. Commun. 42, 43 (1971).

[59] R. L. Jilka, A. N. Martonosi, and T. W. Tillack, J. Biol. Chem. 250, 7511 (1975).

[60] A. H. Caswell and S. Warren, Biochem. Biophys. Res. Commun. 46, 1757 (1972).

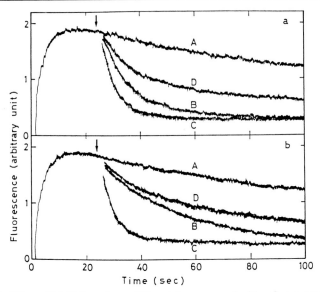

FIG. 10. Efflux of $Mg^{2+}$ from heavy SR vesicles measured with a fluorescence spectrophotometer.[51] Heavy SR vesicles were incubated in 150 m$M$ KCl, 20 m$M$ Tris-maleate (pH 6.7), 10 m$M$ MgCl$_2$, and 10 mg SR protein/ml overnight at 4°. At zero time, 20-$\mu$l aliquot of the incubated suspension was diluted into 2 ml of a solution containing 150 m$M$ KCl, 20 m$M$ Tris-maleate (pH 6.7), 0.1 m$M$ EGTA, and 20 $\mu$l CTC. Then, 23 sec after the dilution, 20 $\mu$l of the solution containing appropriate amounts of divalent cations and NaOH was added and the change in the fluorescence intensity was recorded at 530 nm with excitation at 390 nm at 23°. The added NaOH was to avoid the pH change caused by chelation of divalent cations by EGTA. The amount of NaOH was about twice that of the formed divalent cation and EGTA complex. (a) $Mg^{2+}$ efflux gated by $Ca^{2+}$ and (b) $Mg^{2+}$ efflux gated by $Sr^{2+}$. Extravesicular free divalent cation concentrations were (A) 0.25 $\mu M$, (B) 10 $\mu M$, (C) 100 $\mu M$, and (D) 1 m$M$. Arrows show the times when divalent cations were added.

we can measure $Ca^{2+}$ efflux such as $Ca^{2+}$-induced or caffeine-induced $Ca^{2+}$ release. This method is applicable to the $Mg^{2+}$ efflux measurement.[50,51]

## Measurement of Rapid Ca²⁺ and Mg²⁺ Efflux by the CTC Method

Rapid $Ca^{2+}$ and $Mg^{2+}$ efflux rates can be measured by the use of CTC in combination with the stopped-flow technique.

*Procedure.* SR vesicles are incubated in 150 m$M$ KCl, 20 m$M$ Tris-maleate (pH 6.7), 10 m$M$ CaCl$_2$, and 10 mg SR protein/ml overnight at 0°. A 0.1-ml aliquot is diluted in 2 ml of a dilution medium containing 150 m$M$ KCl, 20 mM Tris-maleate (pH 6.7), 4.5 m$M$ EGTA, and 20 $\mu M$ CTC, where $Ca^{2+}$ release is inhibited. The diluted solution is transferred to the stopped-flow apparatus and 45 sec after the dilution the solution is mixed with an equal volume of a solution containing 150 m$M$ KCl, 20 m$M$ Tris-maleate (pH 6.7), 4.5 m$M$ EGTA, 20 $\mu M$ CTC, and a specified amount of $Ca^{2+}$ and/or effectors. The fluorescence change is recorded with a cutoff filter at 470 nm by excitation at 390 nm at 23° (see Fig. 11).[11]

*Analysis.* When the $Ca^{2+}$ release rate is fast, movement of CTC does not follow the $Ca^{2+}$ movement. CTC permeation takes a few second to reach equilibrium. For very rapid $Ca^{2+}$ flux, the lag time in the fluorescence change must be corrected.[11]

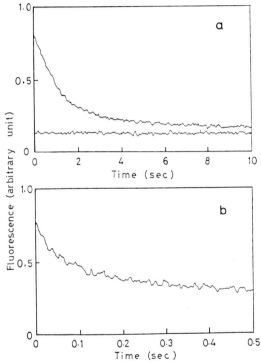

FIG. 11. The change in chloretetracycline (CTC) fluorescence caused by $Ca^{2+}$ release.[11] After preincubation in 150 m$M$ KCl, 20 m$M$ Tris-maleate (pH 6.7), 10 m$M$ CaCl$_2$, and 10 mg SR protein/ml overnight at 4°, a 0.1-ml aliquot was diluted into 2 ml of a dilution medium containing 150 m$M$ KCl, 20 m$M$ Tris-maleate (pH 6.7), 4.5 m$M$ EGTA, and 20 $\mu M$ CTC, where $Ca^{2+}$ release was inhibited. Then, 45 sec after the dilution, the solution was mixed with an equal volume of a solution containing 150 m$M$ KCl, 20 m$M$ Tris-maleate (pH 6.7), 4.5 m$M$ EGTA, 20 $\mu M$ CTC, and specified amounts of $Ca^{2+}$ and effectors, and the fluorescence change was recorded with a cutoff filter at 470 nm by excitation at 390 nm at 23°. (a) $Ca^{2+}$ release in 12.5 m$M$ caffeine and 2.5 $\mu M$ free $Ca^{2+}$. The lower trace is the record at 1.5 min after the mixing. (b) $Ca^{2+}$ release in 12.5 m$M$ caffeine, 2.5 $\mu M$ free $Ca^{2+}$, and 12.5 m$M$ ATP.

Techniques for Proton Flux Measurement

There are a number of methods for measuring the proton flux of a vesicular system. We summarize the methods to measure proton movement as follows.

1. Use of pH electrode.[61] After a pH jump, proton flux can be estimated by the change of pH outside the vesicles.

2. Use of radioactive pH-sensitive molecules. If a pH difference ($\Delta pH$) exists across the membrane, the $\Delta pH$ can be estimated from the distribution of the radioactively labeled pH-sensitive molecules, for example, [$^{14}C$]methylamine.[62] From the rate of change of $\Delta pH$, the movement of protons can be estimated.

3. Use of pH-indicating dyes. If fluorescence or absorption of dyes is pH-dependent, we can estimate the intravesicular pH. From its changing rate we can estimate the proton flux. When we use membrane-permeable dyes, such as quinacrine and 9-aminoacridine,[63] pH changes due to the flux of proton can be estimated from the dye distribution. When we use membrane-impermeable dyes, such as BCECF [bis(carboxyethyl)carboxyfluorescein][64] and pyranine(8-hydroxy-1,3,6-pyrene trisulfonate),[65,66] proton flux can be estimated from the measurement of pH inside the vesicles.

We used pyranine as intravesicular pH indicator in stopped-flow fluorophotometry.[66] The fluorescence intensity changed monophasically as expected from its pH dependency for imposed pH. The half-time for initial pH of 7.53 or 6.83 was about 6 msec. The $H^+/OH^-$ permeability was calculated to be 11 cm/sec. From the results, it was suggested that each vesicle contained large numbers of $H^+/OH^-$ channels.

4. Use of potential-sensing dyes. When the permeable ions are only protons, the permeability can be estimated from the membrane potential measurement as described above in Methods with Potential-Sensing Dyes.[67]

5. Use of ANS. As described above in ANS Fluorescence Method, ANS fluorescence is dependent on the surface potential of the membrane. Since surface potential is affected by the binding of protons to the membrane, proton flux can be estimated from the fluorescence change.[42]

[61] M. Yamaguchi and T. Kanazawa, J. Biol. Chem. 260, 4896 (1985).
[62] T. Ueno and T. Sekine, J. Biochem. (Tokyo) 89, 1239 (1981).
[63] V. M. C. Maderia, Arch. Biochem. Biophys. 185, 316 (1978).
[64] T. J. Rink, R. Y. Tsien, and T. Pozzan, J. Cell Biol. 95, 189 (1982).
[65] C. M. Biegel and J. M. Gould, Biochemistry 20, 3474 (1981).
[66] K. Nunogaki and M. Kasai, Biochem. Biophys. Res. Commun. 140, 934 (1986).
[67] G. Meissner and R. C. Young, J. Biol. Chem. 255, 6814 (1980).

[34] Measurement of Calcium Release in Isolated
Membrane Systems: Coupling between the Transverse
Tubule and Sarcoplasmic Reticulum

*By* Noriaki Ikemoto, Do Han Kim, and Bozena Antoniu

## Introduction

According to the generally accepted view concerning the excitation–contraction coupling mechanism, excitation initiated at the cell surface is propagated to the inside of the muscle cell through the transverse tubule membrane system (T tubule). The depolarization is communicated to the sarcoplasmic reticulum (SR), which in turn releases the $Ca^{2+}$ accumulated in the SR lumen. Binding of the released $Ca^{2+}$ to the $Ca^{2+}$-binding subunit of troponin located in the thin filaments releases the inhibition of the interaction between actin and myosin, which is manifested in muscle contraction. Reaccumulation of cytoplasmic $Ca^{2+}$ into the SR by the ATP-dependent $Ca^{2+}$-pump reverses the above process and muscle relaxation ensues.

The molecular mechanism by which depolarization of the T tubule membrane leads to rapid $Ca^{2+}$ release from SR is least understood. A recently developed methodology useful for studies of the T tubule-mediated $Ca^{2+}$ release from SR *in vitro* is presented here. Methods to induce $Ca^{2+}$ release by direct stimulation of SR (e.g., $Ca^{2+}$-induced or drug-induced $Ca^{2+}$ release) are presented elsewhere in this volume by Kasai and Nunogaki.[1]

## Preparation of Microsomes Enriched in the T tubule/SR Complex

Rabbit fast muscles from leg and back are homogenized in a Waring blender with 4 volumes of 2.5 m$M$ NaOH solution for 2 min (20 sec × 6 with 30 sec interval); during the homogenization the pH is adjusted to 6.8. The homogenate is then centrifuged at 10,000 $g$ for 3 min, the supernatant filtered through Whatman filter paper (No. 4), and the filtrate centrifuged at 17,000 $g$ for 30 min. The pellets are suspended in a solution containing 150 m$M$ KCl and 20 m$M$ MES, pH 6.8, and recentrifuged at 17,000 $g$ for 30 min. The pellets are suspended in the same solution at a protein concentration between 30 and 40 mg per milliliter.

---

[1] M. Kasai and K. Nunogaki, this volume [33].

METHODS IN ENZYMOLOGY, VOL. 157

According to protein and cholesterol determinations after sucrose density gradient fractionation with and without French press treatment,[2] in these preparations 1 mg SR contains about 0.02 mg free T tubule protein and 0.04 mg SR-attached T tubule protein.

## Principles and Equipment for Ca$^{2+}$-Release Assays

### Stopped-Flow Spectrophotometry of Changes in [Ca$_o^{2+}$]

In order to perform accurate measurements of the extravesicular [Ca$^{2+}$] ([Ca$_o^{2+}$]) with the use of metallochromic dyes, the dyes should meet several requirements: (1) the dye should be sensitive specifically to Ca$^{2+}$ in the [Ca$^{2+}$] range in which regulation of physiological functions occur; (2) it should not permeate across the membrane; and (3) the on rate of Ca$^{2+}$ binding to the dye should be sufficiently high to monitor the time course of rapid Ca$^{2+}$ release.[3] Arsenazo III satisfies most of these requirements,[4] though some caution is in order concerning nonspecific sensitivity of the dye to some cations other than Ca$^{2+}$ [4] and inhibitory effects of the dye on Ca$^{2+}$ release by SR when used at higher concentrations.[5]

Since relatively high concentrations of SR are required to maximize the changes in [Ca$_o^{2+}$], light-scattering changes may interfere with optical measurements unless some precautions are taken. In many dual-wavelength spectrophotometric measurements the light-scattering problem has been circumvented by collecting the difference of optical densities at two wavelengths which are sufficiently close to cancel out the light-scattering component.[4]

Thus, the most appropriate method to optically monitor the time course of rapid Ca$^{2+}$ release appears to be dual-wavelength stopped-flow spectrophotometry. An example of such a system which we have employed in our recent studies[3] is illustrated in Fig. 1. The system consists of a dual-beam stopped-flow spectrophotometer (Durrum model D-130), a log amplifier (Durrum model D-131), and a PDP-11 computer. The light that has passed through a reaction chamber containing the mixture of the contents of syringe A and syringe B is split into two by a beam splitter; the two beams are filtered through interference filters having windows at 650 nm and 680 nm, respectively. The difference of the optical densities at 650 nm and 680 nm ($\Delta OD_{650/680}$), is calculated with a log amplifier and re-

[2] A. H. Caswell, Y. H. Lau, M. Garcia, and J.-P. Brunschwig, *J. Biol. Chem.* **254**, 202 (1979).

[3] N. Ikemoto, B. Antoniu, and D. H. Kim, *J. Biol. Chem.* **259**, 13151 (1984).

[4] A. Scarpa, this series, Vol. 56, p. 301.

[5] N. Yamamoto and M. Kasai, *J. Biochem.* (*Tokyo*) **92**, 485 (1982).

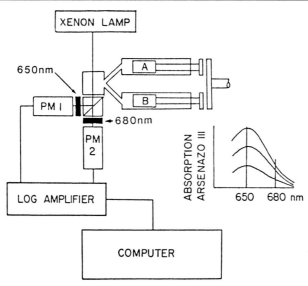

FIG. 1. Schematic representation of a dual-beam stopped-flow spectrophotometer system used for monitoring the rapid kinetics of $Ca^{2+}$ release from the isolated sarcoplasmic reticulum.

corded in the computer as a function of $t$. Sets of the data are signal-averaged, and various kinetic constants (e.g., the rate constant and the maximum amount of $Ca^{2+}$ release) are calculated by an iterative algorithm using the computer. For further details concerning the $Ca^{2+}$ indicator dyes and equipment related to the $Ca^{2+}$ measurements, we recommend the excellent article by Scarpa.[4]

*Chemical-Quench Method for the Determination of $^{45}Ca$ Movement*

Although the spectrophotometric method described above is convenient for many purposes, its capability is restricted in some types of experiments. For instance, one may wish to hold the $[Ca_0^{2+}]$ at a constant value using an EGTA–Ca buffer during the course of $Ca^{2+}$ release. Similarly, a variety of conditions which interfere with the optical measurements of $[Ca^{2+}]$ (e.g., high concentrations of ATP, pH changes) have in fact interesting effects on the $Ca^{2+}$-release functions.[6,7] One may also wish to monitor changes of the $[Ca^{2+}]$ in the intravesicular as well as in the extravesicular space. For these purposes, it is important to establish a method

---

[6] M. Endo, *Physiol. Rev.* **57**, 71 (1977).
[7] A. N. Martonosi, *Physiol. Rev.* **64**, 1240 (1984).

which permits direct determination of $Ca^{2+}$ movement across the SR membrane at a high temporal resolution.

An earlier finding that $^{45}Ca^{2+}$ transported across the SR membrane is maintained in the inside of SR vesicles upon quenching the $Ca^{2+}$-pump reaction with high concentrations of EGTA and $Mg^{2+}$[8] has led to the development of the rapid-flow chemical-quench method for studies of the transient kinetics of the $Ca^{2+}$ transport reaction.[9-12] We have used EGTA together with ruthenium red, a $Ca^{2+}$ channel blocker,[13-15] as a quenching solution, as demonstrated in several examples shown in this article.

There are at least two important criteria in choosing the chemical-quench apparatus: (1) the dead time of the system should be sufficiently short, and (2) the system should have at least three cylinders in order to mix the contents of the first and second cylinders to start the $Ca^{2+}$-release reaction and to quench the reaction with the contents of the third cylinder. Systems that have four or more cylinders are preferable since in some experiments it will be necessary to change experimental conditions before or during the $Ca^{2+}$-release reaction. The multimixing chemical-quench technique is probably the most suitable method for this type of experiment. An example of this approach is demonstrated in the next section.

## Induction and Measurements of T Tubule-Mediated Ca²⁺ Release from SR

### Loading of SR with Ca²⁺

Since the extent of $Ca^{2+}$ release depends upon the level of the intravesicular $Ca^{2+}$,[6,14] loading of the SR vesicles with a sufficient amount of $Ca^{2+}$ is a prerequisite for the $Ca^{2+}$-release experiments. $Ca^{2+}$ loading is carried out either by active loading with the aid of the active $Ca^{2+}$-pump or by passive loading by incubation of the SR vesicles with several millimolar $Ca^{2+}$ for a longer period, e.g., overnight.[14,16] Active loading is preferable for several reasons: (1) various levels of $Ca^{2+}$ loading can be achieved

[8] W. Hasselbach and M. Makinose, *Biochem. Z.* **339,** 94 (1963).

[9] J. P. Froehlich and E. W. Taylor, *J. Biol. Chem.* **250,** 2013 (1975).

[10] P. D. Boyer, L. deMeis, M. G. L. Carvalho, and D. Hachney, *Biochemistry* **16,** 136 (1977).

[11] S. Verjovski-Almeida, M. Kurzmack, and G. Inesi, *Biochemistry* **17,** 5006 (1978).

[12] N. Ikemoto, A. M. Garcia, Y. Kurobe, and T. L. Scott, *J. Biol. Chem.* **256,** 8593 (1981).

[13] S. T. Ohnishi, *J. Biochem. (Tokyo)* **86,** 1147 (1979).

[14] D. H. Kim, S. T. Ohnishi, and N. Ikemoto, *J. Biol. Chem.* **258,** 9662 (1983).

[15] B. Antoniu, D. H. Kim, M. Morii, and N. Ikemoto, *Biochim. Biophys. Acta* **816,** 9 (1985).

[16] K. Nagasaki and M. Kasai, *J. Biochem. (Tokyo)* **90,** 749 (1981).

quickly and (2) a rather lengthy exposure of the membrane preparation to high concentrations of $Ca^{2+}$ is avoided.

*For Stopped-Flow Spectrophotometry.* Since relatively high concentrations of SR protein are required, it is necessary to include an ATP-regenerating system in order to prevent spontaneous $Ca^{2+}$ release due to depletion of ATP.

*Reaction Solution A1.* (A, B, etc. refer to solutions to be mounted in syringe A, B, . . . , respectively.) SR (1.6 mg per milliliter), various concentrations of $CaCl_2$ (e.g., 150–200 nmol per milligram SR protein), 0.15 $M$ potassium gluconate, 0.5 m$M$ MgATP (disodium salt of ATP, Sigma), 2.5 m$M$ phosphoenolpyruvate, 10 units of pyruvate kinase per milliliter, 9 $\mu M$ arsenazo III, and 20 m$M$ MES (pH 6.8).

Incubation of SR in solution A1 at 22° for about 2 min leads to completion of $Ca^{2+}$ loading, and the steady-state level of $Ca^{2+}$ uptake is maintained for at least 8 min. During this period, about 10 sets of $Ca^{2+}$-release time courses are collected for signal averaging as described above.[3]

*For the Chemical-Quench Method.* By carrying out the $Ca^{2+}$-uptake reaction in the presence of higher concentrations of MgATP and lower concentrations of SR protein, it is possible to maintain the steady state of $Ca^{2+}$ uptake for a time sufficient to collect a number of data points even in the absence of an ATP-regenerating system.

*Reaction Solution A2*

0.15 $M$ potassium gluconate, SR (0.2 mg per milliliter), various concentrations of ⁴⁵$CaCl_2$ (150–200 nmol per milligram SR protein, 0.2 $\mu$Ci per milliliter), 5 m$M$ MgATP (disodium salt of ATP, Sigma), and 20 m$M$ MES (pH 6.8).

Incubation in solution A2 at 22° for 5 min leads to complete uptake of the added $Ca^{2+}$, and the steady state of $Ca^{2+}$ uptake is maintained for at least 15 min.

*Induction of $Ca^{2+}$ Release by Chemical Depolarization and Measurement of Its Time Course*

Replacement of an impermeable anion (e.g., anions of gluconate, methanesulfonate, propionate, and thiosulfonate) with a permeable anion such as $Cl^-$, or replacement of a permeable cation such as $K^+$ with an impermeable cation (e.g., cations of choline and Tris), or replacement of both anions and cations, is expected to produce an inside-negative potential across the vesicular membrane. There has been extensive discussion in the literature regarding the involvement of artifacts in $Ca^{2+}$ release produced by ionic replacement. For instance, anion replacement alone[17-19] leads to swelling of SR vesicles, and may produce artifactual Ca$^{2+}$ release due to membrane lysis.[20] One of the advantages of the stopped-flow spectrophotometric studies is that one can examine the changes in light scattering of the vesicular suspension which reflect the changes in vesicular size and shape. The light-scattering changes produced by cation replacement alone are opposite to those produced by anion replacement alone, suggesting that shrinkage of the vesicles has occurred by cation-replacement alone. Replacement of both cation and anion, when appropriate ions are selected (e.g., replacement of potassium gluconate with choline-Cl; see below), produced no light-scattering changes. Furthermore, replacement of both cation and anion should be more effective in producing membrane depolarization than replacement of either cation or anion alone.

*Stopped-Flow Spectrophotometry.* Solution A1 is loaded in syringe A of a stopped-flow apparatus (cf. Fig. 1). The content of syringe A is mixed with an equal volume of a "triggering solution" loaded in syringe B (e.g., solution B1) to induce Ca$^{2+}$ release by replacement of potassium gluconate with choline-Cl.

*Solution B1*

0.15 $M$ choline-Cl, 9 $\mu M$ arsenazo III, 20 m$M$ MES (pH 6.8)

The time course of Ca$^{2+}$ release induced by choline-Cl replacement of potassium gluconate monitored with a stopped-flow spectrophotometer system is illustrated in Fig. 2 on two different time scales. The whole time course consists of five distinguishable phases as numbered in the figure. Phase 1 is a lag phase. In phase 2, about 15 nmol Ca$^{2+}$ per milligram protein is released with a $t_{1/2} = 5$ msec. Subsequently (phase 3), there is a plateau, or in some experiments the Ca$^{2+}$ released in phase 2 is reaccumulated. In phase 4, a larger amount of Ca$^{2+}$ (e.g., 50 nmol per milligram) is released. All the Ca$^{2+}$ that has been released in both phase 2 and phase 4 is reaccumulated into SR in phase 5 (not shown). Dilution of the potassium gluconate vesicles into 0.15 $M$ potassium gluconate produces no Ca$^{2+}$ release, indicating that it is not dilution but ionic replacement that is the key factor for the induction of Ca$^{2+}$ release.

*Rapid-Flow Chemical-Quench Studies.* Solution A2 is loaded in syringe A of a multimixing apparatus (e.g., Froehlich–Berger chemical-quench flow apparatus model CQF-105, Commonwealth Technology Inc.). One part of the syringe A solution is mixed with one part of a

[17] M. Endo and Y. Nakajima, *Nature (London), New Biol.* **246**, 216 (1973).
[18] M. Kasai and H. Miyamoto, *J. Biochem. (Tokyo)* **79**, 1053 (1976).
[19] M. Kasai and H. Miyamoto, *J. Biochem. (Tokyo)* **79**, 1067 (1976).
[20] G. Meissner and D. McKinley, *J. Membr. Biol.* **305**, 79 (1976).

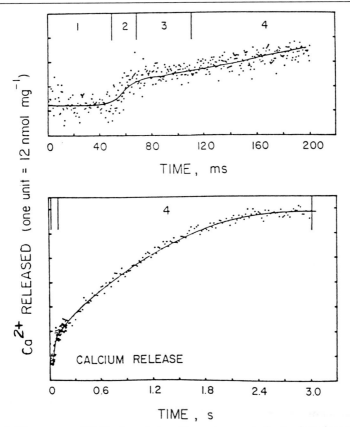

FIG. 2. Time course of Ca$^{2+}$ release induced by replacement of potassium gluconate with choline-Cl as determined by a stopped-flow spectrophotometer with the use of arsenazo III.

triggering solution loaded in syringe B. Two examples of triggering solutions are shown below; solutions B2 and B3 being for Ca$^{2+}$ release at $[Ca_o^{2+}] = 0$ and 4 $\mu M$, respectively.

*Solution B2*

0.15 $M$ choline-Cl, 1 m$M$ EGTA, 20 m$M$ MES (pH 6.8)

*Solution B3*

0.15 $M$ choline-Cl, 1 m$M$ EGTA, 0.92 m$M$ CaCl$_2$, 20 m$M$ MES (pH 6.8), pH should be adjusted to 6.8 with concentrated KOH before use

After various times the reaction is quenched by mixing with one part of a quenching solution loaded in syringe C (solution C, see below).

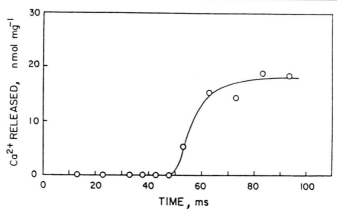

FIG. 3. Time course of $Ca^{2+}$ release induced by replacement of potassium gluconate with choline-Cl as determined by the rapid-flow chemical-quench technique.


*Solution C*

0.075 $M$ potassium gluconate, 0.075 $M$ choline-Cl, 30 m$M$ EGTA, 15 $\mu M$ ruthenium red, and 20 m$M$ MES (pH 6.8)

A portion of the reaction mixture (e.g., 0.5 ml) is rapidly (within 30 sec after quenching) filtered through Millipore filters that have been soaked in solution C and air-dried. A 12-filter Millipore manifold is convenient for completion of one set of time course such as shown in Fig. 3. A portion of the filtrate is placed on a filter paper strip (e.g., Whatman filter paper No. 2), air-dried, and the radioactivity of $^{45}Ca$ retained on the filter is counted with a scintillation counter.

In the case of continuous quench flow experiments such as those carried out with the Froehlich–Berger type apparatus, the reaction time is controlled by changing combinations of (1) speed, distance, delay time of a stepping motor, (2) internal volumes of interconnecting tubings between ball mixers, and (3) volumes of the assay solution to be collected.[21] Various reaction times are calculated using a computer program.[22] A relatively long time is required for completion of time course measurements (e.g., 12 min for 10 data points) since the changing of interconnecting tubing is rather time-consuming. Therefore, it is necessary to ensure that the state of $Ca^{2+}$ loading is maintained constant throughout the measurements. For this purpose, at both the beginning and the end of the measurements one part of solution A2 is mixed with one part of solution C, and mixed with

[21] J. P. Froehlich, J. V. Sullivan, and B. L. Berger, *Anal. Biochem.* **73**, 331 (1976).
[22] N. Ikemoto and R. W. Nelson, *J. Biol. Chem.* **259**, 11790 (1984).

TABLE I
EFFECTS OF DISSOCIATION AND REASSOCIATION OF THE
T TUBULE/SR COMPLEX[a]

| Step | Amount of $Ca^{2+}$ release (nmol per mg) | $t_{1/2}$ (msec$^{-1}$) |
|---|---|---|
| Control | 6.8 | 7.4 |
| French press treatment | 2.2 | 9.6 |
| Potassium cacodylate incubation | 7.3 | 9.1 |

[a] On the rapid phase of $Ca^{2+}$ release (phase 2, cf. Fig. 2) induced by replacement of potassium gluconate with choline-Cl.

one part of triggering solution (solution B2 or solution B3). Then the reaction mixture is filtered, and the filtrate is counted as described above to determine the background $[Ca_o^{2+}]$. If there are changes in the $[Ca_o^{2+}]$ during the collection of data points, it is necessary to carry out separate $^{45}Ca^{2+}$ loading reactions for a certain time (e.g., for 5 min) for the collection of each data point.

Figure 3 depicts an example of the time course of the rapid phase of depolarization-induced $Ca^{2+}$ release determined by the rapid-flow chemical-quench methods described above (for further details, see the legend to Fig. 3). The time course determined by the chemical-quench method at $[Ca_o^{2+}] = 4 \mu M$ shows about the same characteristics as the time course determined by the stopped-flow spectroscopic method (cf. Fig. 2A), indicating that the two methods can be used in an interchangeable fashion.

*Evidence that Chemical-Depolarization-Induced $Ca^{2+}$ from SR Is Triggered via the Attached T Tubule*

Historically, the ionic replacement methods were devised with the aim of producing membrane depolarization across the SR membrane.[17–19] However, the following evidence suggests that depolarization of the T tubule rather than depolarization of the SR membrane plays a crucial role in the ionic replacement-induced $Ca^{2+}$ release described above.[3]

As reported by Caswell et al.,[2] the attached T tubules are dissociated from SR by treatment with a French press, and they are reassociated with SR by incubation in potassium cacodylate at high concentrations (e.g. 0.4 $M$[23]). Table I depicts the results of experiments showing the effects of dissociation and reassociation of the T tubule/SR linkage on the depolarization-induced $Ca^{2+}$ release. The amount of $Ca^{2+}$ released in the rapid

[23] A. H. Caswell, N. R. Brandt, J.-P. Brunschwig, and R. M. Kawamoto, this volume [7].

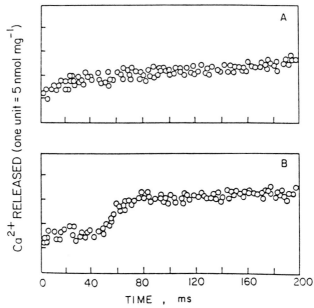

FIG. 4. Reconstitution of the T tubule/SR complex capable of depolarization-induced Ca²⁺ release from the isolated T tubule and SR fractions.

phase is decreased significantly upon dissociation of T tubules, and upon reassociation the original amount of rapid Ca²⁺ release is almost completely restored. Thus, it appears that the T tubule has to be associated with SR to induce the rapid Ca²⁺ release.

Figure 4 illustrates an example of reconstitution of an active T tubule/SR complex from the isolated T tubule and SR fractions. As shown in Fig. 4A, the purified SR alone is incapable of producing Ca²⁺ release upon replacement of potassium gluconate with choline-Cl even after incubation in potassium cacodylate. Upon incubation of a mixture of the SR with the T tubule (e.g., at 20:1 protein ratio) in 0.4 $M$ potassium cacodylate, depolarization-induced rapid Ca²⁺ release can be reconstituted, further suggesting that chemical-depolarization-induced Ca²⁺ release is triggered via T tubules attached to SR. Such reconstitution experiments will be useful for several new types of experiments, such as (1) incorporation of membrane-impermeable potential-sensitive dyes into the T tubule vesicles before association with SR, and (2) chemical modification of specific components of either the SR or the T tubule membrane with covalently reacting probes before reconstitution etc.

*Activation and Inhibition of Depolarization-Induced Ca$^{2+}$ Release*

For initial screening of reagents that activate or inhibit depolarization-induced Ca$^{2+}$ release, it is appropriate to include various concentrations of the test reagent in the triggering solution (solution B). One such experiment, in which various concentrations of ATP have been added in solution B3, is illustrated in Table II. Increasing concentrations of ATP shorten the lag phase and increase the size of Ca$^{2+}$ release. However, there is no further increase in the rate constant.

Ruthenium red and tetracaine, which inhibit Ca$^{2+}$-induced Ca$^{2+}$ release from the isolated SR,[13] inhibit depolarization-induced Ca$^{2+}$ release with $C_{1/2}$ (the concentration for half-inhibition) = 0.08–0.10 $\mu M$ and 0.07–0.11 m$M$, respectively. It is interesting that the $C_{1/2}$ is the same for several types of Ca$^{2+}$ release (depolarization-induced Ca$^{2+}$ release, Ca$^{2+}$-induced Ca$^{2+}$ release, and caffeine-induced Ca$^{2+}$ release) for each reagent. The results suggest that both ruthenium red and tetracaine interfere with a common mechanism of the different types of Ca$^{2+}$ release.

### Concluding Remarks

A number of different methods that induce Ca$^{2+}$ release from the isolated SR have been described.[6,7] In this chapter we have specifically discussed methods that permit one to induce and monitor the so-called depolarization-induced Ca$^{2+}$ release. Since physiological Ca$^{2+}$ release is expected to be very rapid ($t_{1/2}$ = several milliseconds) the method for monitoring Ca$^{2+}$ release should have a time resolution of the order of a few milliseconds or less. Stopped-flow spectrophotometry with a Ca$^{2+}$-indicator dye and the rapid-flow chemical-quench method described in this article are examples of the most promising approaches.

As demonstrated here, a preparation of microsomes enriched in the T

TABLE II
Effects of [ATP] on Depolarization-Induced
Ca$^{2+}$ Release

| [ATP] added in solution B (mM) | Lag phase (msec) | Ca$^{2+}$ released (nmol per mg) | $t_{1/2}$ (msec) |
|---|---|---|---|
| 0 | 43 | 18 | 7 |
| 2 | 15 | 35 | 20 |
| 5 | 5 | 40 | 16 |
| 15 | 0 | 40 | 10 |

tubule/SR complexes can release $Ca^{2+}$ from SR at rates comparable with those of the physiological $Ca^{2+}$ release upon chemical depolarization. To produce such $Ca^{2+}$ release, T tubules must be attached to SR, indicating that $Ca^{2+}$ release is triggered via the attached T tubule. In view of the analogy to the proposed T tubule/SR coupling *in situ,* the isolated membrane system described here would be the most suitable model for studies of physiological $Ca^{2+}$ release *in vitro.* The molecular mechanism by which depolarization of the T tubule membrane is coupled to the opening of the $Ca^{2+}$ channel of SR will be a central problem to be investigated in future studies.

Recently, there has been remarkable progress in the isolation and purification of the T tubule/SR complex.[24] It is useful to apply the methods described here to highly purified triad preparations.

### Acknowledgments

We wish to thank Drs. John Gergely and Terrence L. Scott for their comments on the manuscript. This work was supported by grants from NIH (AM 16922) and MDA.

[24] R. D. Mitchell, P. Palade, and S. Fleischer, *J. Cell Biol.* **96,** 1008 (1983) and this volume [6].

# [35] Techniques for Observing Calcium Channels from Skeletal Muscle Sarcoplasmic Reticulum in Planar Lipid Bilayers

By JEFFREY S. SMITH, ROBERTO CORONADO, and GERHARD MEISSNER

The sarcoplasmic reticulum (SR) is a specialized intracellular membrane responsible for the distribution of cytoplasmic $Ca^{2+}$. The SR membrane is freely permeable to many small monovalent ions; however, it possesses a stringently regulated system for the release and subsequent reuptake of $Ca^{2+}$.[1,2] $Ca^{2+}$ release from the SR occurs through a calcium channel which is regulated by external $Ca^{2+}$, $Mg^{2+}$, and adenine nucleotides.[3–5] The $Ca^{2+}$ release channel from SR is a large-conductance (~100

[1] G. Inesi, *Annu. Rev. Physiol.* **47,** 573 (1985).
[2] G. Meissner, *Mol. Cell. Biochem.* **55,** 65 (1983).
[3] G. Meissner, *J. Biol. Chem.* **259,** 2365 (1984).
[4] H. Morii and Y. Tonomura, *J. Biochem.* (*Tokyo*) **93,** 1271 (1983).
[5] K. Nagasaki and M. Kasai, *J. Biochem.* (*Tokyo*) **94,** 1101 (1983).

pS in 50 m$M$ Ca$^{2+}$) divalent-cation-selective channel which can be incorporated into planar lipid bilayers from heavy terminal cisternae-derived
SR fractions.[6]

The planar bilayer is a powerful tool for studying ionic currents mediated by single-ion-channel proteins. If conditions are carefully selected,
much can be learned about the biophysical and pharmacological properties of individual ion-conducting proteins without the need for rigorous
biochemical purification. There are many variations of the planar bilayer
method, most of which have been developed to increase sensitivity and to
reduce contaminating background noise (for review see Ref. 7). In this
report we describe techniques that utilize the Mueller–Rudin-type planar
lipid bilayer[8] and the vesicle fusion technique originally described by
Miller and Racker.[9]

There are several complicating factors which make observation of the
SR Ca$^{2+}$-release channel not entirely straightforward. The SR membrane
contains considerable numbers of K$^+$-and Cl$^-$-selective ion channels
which contribute substantially to the background conductance when SR
vesicles are fused into an artificial bilayer.[10] Consequently, currents originating from the Ca$^{2+}$-release channels are normally not observable in
buffers containing K$^+$, Na$^+$, or Cl$^-$. In order to visualize the currents
attributable to the SR calcium channel, buffers must be carefully designed
so that ions permeant in the contaminating K$^+$ and Cl$^-$ channels are
ommitted. A second factor crucial to the observation of these channels in
lipid bilayers is the concentration of Ca$^{2+}$ or adenine nucleotide present in
the bilayer chamber corresponding to the exterior of the vesicle or the
cytoplasm of the muscle cell. The SR calcium channel is activated by
micromolar cis Ca$^{2+}$ but is inhibited at cis Ca$^{2+}$ in excess of 1 m$M$.[3]
Therefore, free cis Ca$^{2+}$ must be maintained within this narrow range,
preferably through the use of CaEGTA buffers. Millimolar nucleotide in
the absence of activating cis Ca$^{2+}$ also activates the channel; however,
channel opening 100% of the recorded time does not occur unless both
activators are present. We have used a two-step technique for observing
the SR calcium channel which involves fusion of SR vesicles into phospholipid bilayers in a Cl$^-$ buffer followed by a perfusion step with
chloride-free buffers containing only divalent cations as the premeant
species.

[6] J. S. Smith, R. Coronado, and G. Meissner, *Nature (London)* **316,** 446 (1985).
[7] C. Miller, *Physiol. Rev.* **63,** 1209 (1983).
[8] P. Mueller and D. O. Rudin, *Curr. Top. Bioenerg.* **3,** 157 (1969).
[9] C. Miller and E. Racker, *J. Membr. Biol.* **30,** 283 (1976).
[10] C. Miller, *J. Membr. Biol.* **40,** 1 (1978).

## Preparation of SR Vesicles

Heavy SR vesicles from rabbit skeletal muscle were prepared by differential and sucrose gradient centrifugation as described previously.[3,10a] Briefly, heavy SR membranes were recovered from the 36–45% region of sucrose gradients that contained rabbit skeletal muscle membranes sedimenting at 15,000 $g$. The recovered membranes were pelleted at 105,000 $g$ for 60 min, resuspended in 0.3 $M$ sucrose, 10 m$M$ K-PIPES, pH 7.0, and stored at $-70°$.

## Selection and Preparation of Buffers

Choline chloride and Trizma base (Tris) were obtained from Sigma Chemical Co., St. Louis, MO. HEPES buffer was purchased from Research Organics Inc., Cleveland, OH. All other chemicals were reagent grade from Fisher Scientific Co., Fair Lawn, NJ. Glass-distilled deionized water was used in the preparation of all buffer solutions.

As was stated in the introduction, SR contains ion channels specific for monovalent cations, anions, and divalent cations. In light of this knowledge, it is necessary to design buffers in such a way that currents originating from one type of channel may be recorded without the interfering conductance of another channel type. We had originally tried one-step fusion and recording protocols in buffers containing only divalent cations as the permeant species. Using Tris (cis) and Ba²⁺ or Ca²⁺ (trans) with glucuronate, glutamate, methane sulfonate or HEPES as the impermeant anion, we were unable to reproducibly incorporate SR calcium channels. We now use a two-step buffer system for incorporating and recording calcium release channels from SR. SR vesicles are fused into the bilayer in an asymmetric choline chloride buffer. We have found that this buffer promotes fusion of vesicles containing calcium release channels. Fusion buffer consists of 50 m$M$ choline chloride, 5 m$M$ CaCl₂, 10 m$M$ HEPES/Tris, pH 7.4. After bilayer formation, 1 $M$ choline chloride is added to the cis chamber until cis choline chloride is equal to 250 m$M$. Choline is only very slightly permeant in the SR K⁺ channel,[11] so Cl⁻ is the only significantly permeant ion in this system.

Perfusion and recording buffers are made starting with 250 m$M$ HEPES (free acid) in water to which the appropriate base (Tris) or metal hydroxide, (Ca²⁺, Ba²⁺, Cs²⁺, etc.) is added until pH 7.4 is reached. Commercially available calcium and barium hydroxides contain insoluble impurities (mostly calcium and barium carbonates) which impart a cloudi-

[10a] G. Meissner, this volume [32].

[11] R. Coronado and C. Miller, *J. Gen. Physiol.* **79,** 529 (1982).

TABLE I
EQUIPMENT LIST FOR PLANAR LIPID BILAYER[a]

Bilayer cell
Aluminum box with hinged lid, mounted on four lead bricks
Inner tube (designed to fit an 8-in. diameter wheel)
Stir plate
Marble or epoxy resin balance table
Patch clamp
Regulated dc power supply (Analog Devices, Norwood, MA, models 920 and 950)
Dual-syringe infusion/withdrawal pump (Harvard Apparatus Company, Inc., South Natick, MA, model 915)
Oscilloscope (Hitachi Denshi Ltd., Tokyo, Japan, model VC-6015)
Eight-pole Bessel low-pass filter (Frequency Devices Inc., Haverhill, MA, model 902LPF)
*Strip chart recorder (100 Hz frequency response) (Astro Med Div Atlan-Tol Ind., West Warwick, RI, model Z-1000)
*Frequency-modulated (FM) tape recorder (Racal Recorders, Inc., Vienna, VA, model store 4DS)
*Personal computer (IBM Corp., Boca Raton, FL, model IBM-PCXT or ICM-PCAT)
*Ten-megabyte removable disk drive (INFAX Inc., Decatur, GA, model 101PC)
*Data acquisition system (Keithley das, Boston, MA, system 520-measurement and control system)

[a] Items marked with an asterisk are optional.

ness to solutions. We normally filter our $Ca^{2+}$- and $Ba^{2+}$-HEPES solutions through 0.22-$\mu$m Millipore filters to remove these visible impurities. As with the fusion buffer, we intentionally limit the number of permeant ions in our perfusion buffer so that currents mediated by the calcium release channels may be observed. $Tris^+$ permeability of the SR $K^+$ channel is very low[11] and the large organic anion HEPES does not readily permeate the $Cl^-$-selective SR channel.[2]

$Ca^{2+}$ in cis buffers is controlled with a CaEGTA buffer (usually 100 $\mu M$). Free $Ca^{2+}$ in cis buffers was determined using a computer program and absolute binding constants as given by Fabiato.[12]

Equipment and Data Acquisition

Table I is a listing of the equipment we have used for our planar lipid bilayer studies. The bilayer cell and box were constructed at a local in-

[12] A. J. Fabiato, *J. Gen. Physiol.* **78,** 457 (1981).

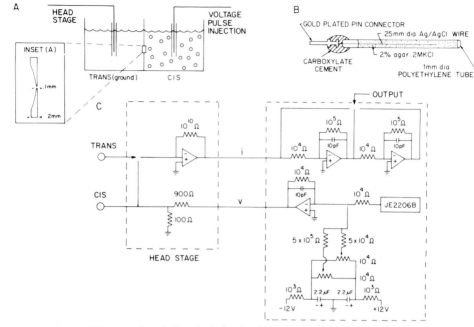

FIG. 1. Bilayer cell and electrical circuits for detection of SR calcium channels. (A) Bilayer cell for single-channel measurements. Voltages and SR vesicles are applied to the cis chamber. The trans chamber is held at virtual ground. Inset shows dimensions of cup wall and hole. (B) Ag/AgCl salt bridge electrode. Electrodes are inserted into female BNC connectors mounted on the headstage support. (C) Current amplifier and voltage pulse circuits for detection of single-channel events.

strument shop. All current amplifier and pulse injection circuits were constructed in our laboratory from available components. As an alternative, almost any commercially manufactured patch clamp unit could be used here. Prefabricated equipment was purchased from the vendors given in Table I (vendors and model numbers are given for comparative purposes only).

Figure 1A is a diagrammatic representation of the planar bilayer cell we have used for the incorporation of SR calcium channels. This cell is basically similar to the one described by Miller and Rosenberg 1979[13] and consists of two aqueous filled chambers formed by a Lexan (polycarbonate) or poly(vinylidene fluoride) cup inserted into a doubly cut-away poly(vinyl chloride) block. The bilayer cell fits into a hollow brass block situated inside an aluminum box. The box serves as a shield from atmo-

[13] C. Miller and R. Rosenberg, *Biochemistry* **18**, 1138 (1979).

spheric radio frequency noise and acoustic vibration. The aluminum box is bolted to a plastic base which in turn is fastened to four 18-kg lead bricks. The whole bilayer box/lead brick assembly rests upon a partially inflated inner tube set in the middle of an epoxy resin balance table. This setup is effective for dampening high-frequency vibrations and is a relatively inexpensive alternative to an air isolation table. Mueller–Rudin planar bilayers are formed by painting a solution of phospholipids in decane across a 0.2- to 1.2-mm diameter hole drilled into the polycarbonate cup. We have found that bilayers form most easily when the wall of the cup is thinned to 0.075–0.1 mm thickness at the point where the hole is drilled (Fig. 1A inset). The hole must be free from plastic burrs or residue which might partially occlude the hole. We normally inspect each cup under a low-power microscope to ensure that the hole is free from dirt and that cracks have not formed along the edge of the hole. Minute stress cracks radiating outward from the hole cause the bilayer to be mechanically unstable. Consequently, cups showing such signs of wear should be discarded. A detailed discussion of the physics and geometry of the planar bilayer and its support is given by White.[14]

Ag/AgCl electrodes encased in 0.2 $M$ KCl, 2% agar-filled lengths of polyethylene tubing, are inserted into each chamber for the purpose of applying and recording voltage pulses. These electrodes are fashioned from 0.25-mm diameter silver wires, soldered to miniature pin contacts. Each wire is coated with AgCl in an electrolytic cell filled with 0.1 $M$ HCl and then inserted into a length of 1-mm diameter polyethylene tubing which has been filled with 2% agar containing 0.2 $M$ KCl. Carboxylate cement is used to glue the tubing to the pin connector (Fig. 1B). This assembly is durable and may last for a month or more if between experiments the electrode tips are kept wet in a 0.2 $M$ KCl solution. Voltage pulses are applied to the cis chamber while the trans electrode is held at virtual ground.

Figure 1C is a schematic diagram of the current amplification and pulse injection circuit we have used for measuring single channel fluctuations. The circuit consists of three main parts; a head stage amplifier, an additional amplification stage, and a summing amplifier circuit for injecting voltage pulses and waveforms from a function generator. The head stage amplifier we use is a low-noise LF-157 operational amplifier (National Semiconductor, Santa Clara, CA) with a 10,000 m$\Omega$ feedback resistor. This is mounted on a small circuit board positioned next to the bilayer cell inside the aluminum box. Power for the headstage amplifier is supplied by two 9-volt nickel–cadmium batteries which are also mounted

[14] S. H. White, *Biophys. J.* **12**, 432 (1972).

inside the bilayer box. The secondary amplification stage is constructed with two low-noise LF-365N op-amps (National Semiconductor) in series and a switch for selection of the appropriate current gain (0.0001–0.01 V/pA). The summing circuit is also constructed from an LF-365N op-amp with a 10K feedback resistor and 10-fold voltage divider. Voltage clamp commands are supplied from a ±12-volt regulated power supply. Triangular or square waves are generated by an inexpensive waveform generator (JAMECO Electronics, Belmont, CA, Model JE2206B). For an in-depth discussion of circuit design for single channel recording, we refer the reader to Sigworth.[15]

Amplified currents from the bilayer are output directly to an FM tape recorder or filtered at 300 Hz (−3 dB point through an 8-pole Bessel low-pass filter) and digitized at 1 kHz for storage on hard disk. Output is also monitored on an oscilloscope and a fast strip chart recorder (100 Hz filter). Statistical analysis of single channel currents is carried out on an IBM-PCXT computer.

Fusion of SR Vesicles into Planar Lipid Bilayers
and Perfusion of the Bilayer Cell

Planar lipid bilayers containing phosphatidylethanolamine (PE) (25 mg/ml bovine brain, Avanti Biochemicals, Birmingham, AL), phosphatidylserine (PS) (15 mg/ml bovine brain, Avanti), and diphytanolyphosphatidylcholine (PC) (10 mg/ml, Avanti) in decane are formed by painting the lipid–decane solution with a plastic stick onto a 0.3-mm diameter hole in a polycarbonate cup. Bilayer formation is monitored on an oscilloscope by following the increase in capacitance during thinning. Capacitance of the membrane is measured by applying a triangular voltage pulse to the bilayer cell. Since bilayer capacitance is directly proportional to the time derivative of the applied voltage, the result is a square wave which increases in amplitude as the area of the bilayer grows. The resulting bilayers are stable in our buffer at room temperature for as long as several hours, although 45 min is more the average. We have also used PE : PS (50 : 50) in decane; however, we find that the PC-containing bilayers are more resistant to breakage and collapse, especially during the perfusion step.

Heavy SR vesicles are fused into the bilayer by adding an aliquot (1–10 μg vesicle protein/ml cis buffer) to the cis chamber containing choline chloride fusion buffer. Fusion normally ensues a few seconds to a minute or so after vesicle addition. If fusion does not occur within a few minutes,

[15] F. J. Sigworth in "Single Channel Recording," (B. Sakmann and E. Neher, eds.), pp. 3–35. Plenum, New York, 1983.

breakage and reformation of the bilayer or changing the holding potential to more negative values ($-80$ to $-100$ mV) often stimulates a fusion event. Fusion is discernible as discrete steplike conductance increases (Fig. 2A). The resulting conductance is anionic in character and reverts near $+25$ mV. Occasionally, single chloride-selective channels are fused into the bilayer (Fig. 2B). There are two common conductance states of these channels at approximately 60 and 95 pS under our buffer conditions. Immediately following fusion, both chambers are perfused to remove the permeant $Cl^-$ and unfused SR vesicles. Perfusion is accomplished using a dual-syringe infusion/withdrawal pump fitted with two 50-ml disposable syringes. The cis chamber is perfused first (about 3 ml/min, pump setting 6) with several volumes of 100 $\mu M$ CaEGTA, 125 m$M$ Tris(base)/250 m$M$ HEPES, pH 7.4, followed by perfusion of the trans chamber with 54 m$M$ Ca(OH)$_2$ or Ba(OH)$_2$/250 m$M$ HEPES, pH 7.4. During the perfusion step, buffer is pumped into the bottom of the chamber via a small Tygon hose and simultaneously withdrawn through a hose positioned at the top of the chamber. The density of the HEPES perfusate is greater than that of the choline chloride solution so that during perfusion the choline chloride is effectively displaced by the HEPES solutions. In control experiments designed to test the effectiveness of the perfusion, 6-carboxylfluorescein fluorescence is monitored as a marker for the choline chloride buffer. After perfusion, 6-CF fluorescence should be decreased by >99%, indicating a very efficient replacement by the HEPES perfusate. Figure 3 summarizes the buffer conditions used during the fusion and perfusion steps.

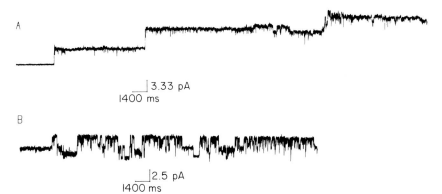

Fig. 2. Chloride-ion-specific fusion events and channels from incorporated SR vesicles. (A) Fusion steps at 0 mV in 250 m$M$ cis/50 m$M$ trans choline chloride, 5 m$M$ CaCl$_2$, 10 m$M$ Tris/HEPES, pH 7.4. (B) Single chloride channels from SR in same buffer as A and 0 mV. Major unitary conductance states are 60 and 95 pS.

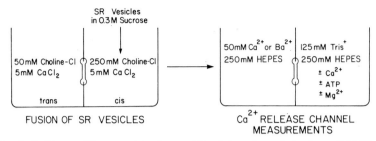

FUSION OF SR VESICLES                    Ca$^{2+}$ RELEASE CHANNEL
                                              MEASUREMENTS

FIG. 3. Buffer conditions used to observe SR calcium channels in planar lipid bilayers. SR vesicles were fused into the bilayer in the asymmetric choline chloride fusion buffer. SR calcium channels were then recorded in chloride-free HEPES buffers (perfusion/recording buffer).

The lipid bilayer becomes unstable in the presence of large current transients, which are subject to occur during the perfusion procedure. This problem is most noticeable during perfusion of the trans chamber. Any foreign object inserted into the trans chamber (i.e., perfusion tubing) serves as a conduit for these transients. In an effort to decrease this unwanted current flow through the bilayer, during perfusion the two electrodes are grounded together by means of a switch mounted on the headstage. A schematic for this switch is shown in Fig. 1C. This effectively reduces bilayer breakage during perfusion of the trans chamber. One minor drawback to this technique is that during trans-perfusion the competence of the bilayer cannot be monitored electrically.

After the perfusion step, if calcium channels are incorporated during the fusion, currents attributable to them can at this point be measured. Figure 4 shows single channel current traces recorded in 125 m$M$ Tris/250 m$M$ HEPES, pH 7.4, cis and 54 m$M$ Ca–HEPES, pH 7.4, trans following fusion of heavy SR vesicles into a PE:PS:PC 50:30:20 bilayer and perfusion with the previously described buffers. The channels are inactive in the absence of either Ca$^{2+}$ or nucleotide (Fig. 4A). When micromolar Ca$^{2+}$ is present the channels are active and appear as single events or grouped into bursts of activity (Fig. 4B). Channels may also be activated by nucleotide alone as in Fig. 4C. Complete activation occurs in the presence of both nucleotide and micromolar cis Ca$^{2+}$ (Fig. 4D).

In addition to the high-conductance calcium release channels, other channels of similar divalent cation specificity may also be incorporated from SR fractions. These small-conductance (5–10 pS in 54 m$M$ trans Ca$^{2+}$) channels may be incorporated from light- intermediate- and heavy-density SR and are not dependent on Ca$^{2+}$ or nucleotide for activity.[16] The

[16] J. S. Smith, R. Coronado, and G. Meissner, *Biophys. J.* **50**, 921 (1986).

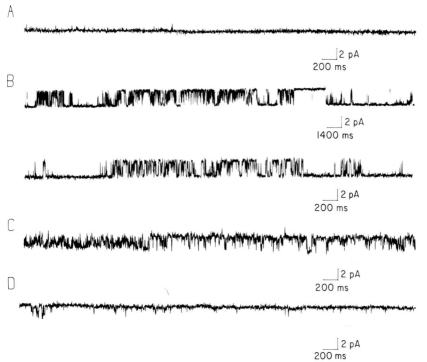

FIG. 4. Single channel recordings of SR calcium channels. Channels were recorded in 54 mM Ca(OH)$_2$/250 mM HEPES, pH 7.4, trans and 125 mM Tris (base)/250 mM HEPES, pH 7.4, cis plus(A) 5.1 mM EGTA and 0.1 mM Ca$^+$ (1 nM free Ca$^{2+}$); (B) 100 μM CaEGTA (2 μM free Ca$^{2+}$); (C) 10 mM ATP, 5.1 mM EGTA, and 0.1 mM Ca$^{2+}$ (1 nM free Ca$^{2+}$); (D) 100 μM CaEGTA (2 μM free Ca$^{2+}$), 1.75 mM ATP. Filter = 300 Hz. Sampling rate = 1 kHz.

presence of the small-conductance channels does sometimes interfere with recordings of calcium release channels, especially when many fusions have occurred in the choline chloride buffer.

### Acknowledgment

This work was supported by U.S. Public Health Service Grants AM 18687, HL 27430, and GM 32824.

# [36] Purification of Ca²⁺ Release Channel (Ryanodine Receptor) from Heart and Skeletal Muscle Sarcoplasmic Reticulum

By MAKOTO INUI and SIDNEY FLEISCHER

## Introduction[1]

In excitation–contraction coupling, the release of Ca²⁺ from the terminal cisternae of sarcoplasmic reticulum[2] triggers muscle contraction. It is by way of the triad junction that the electrical signal from transverse tubules transduces Ca²⁺ release from terminal cisternae. Ca²⁺ release occurs at the junctional face membrane of SR via Ca²⁺ release channels which structurally have been identified as the feet structures at the triad junction.

Significant progress has been made in the past several years with regard to characterizing the Ca²⁺ release machinery of SR from both skeletal muscle and heart. (1) Junctional terminal cisternae of sarcoplasmic reticulum have been isolated and characterized.[3,4] (2) Ca²⁺ permeability in terminal cisternae was found to be modulated by ryanodine in pharmacologically significant concentrations.[5] (3) Ryanodine binding in the same concentration range ($K_D \sim 50$ n$M$ for skeletal muscle and $\sim 10$ n$M$ for heart) has been localized to the terminal cisternae of SR.[4,5] (4) The ryanodine receptor has been isolated and found to be equivalent to the feet structures of the junctional terminal cisternae.[6,7] (5) SR vesicles have been fused into black lipid films and found to contain Ca²⁺ channels.[7a,7b]

[1] These studies were supported by Grants AM14632 and HL32711 from the National Institutes of Health, by a grant from the Muscular Dystrophy Association and a Biomedical Research Support Grant from NIH administered by Vanderbilt University (to SF), and by a Grant-in-Aid from the American Heart Association, Tennessee Affiliate (to MI). MI is supported by an Investigatorship of the American Heart Association, Tennessee Affiliate. We thank Ms. Laura Taylor for typing this manuscript.

[2] The abbreviations used are CHAPS, 3-[(3-cholamidopropyl)dimethylammonio]-1-propane sulfonate; JTC, junctional terminal cisternae; SDS-PAGE, sodium dodecyl sulfate-polyacrylamide gel electrophoresis; SR, sarcoplasmic reticulum.

[3] A. Saito, S. Seiler, A. Chu, and S. Fleischer, J. Cell Biol. **99**, 875 (1984).
[4] M. Inui, S. Wang, A. Saito, and S. Fleischer, submitted for publication.
[5] S. Fleischer, E. M. Ogunbunmi, M. C. Dixon, and E. A. M. Fleer, Proc. Natl. Acad. Sci. U.S.A. **82**, 7256 (1985).
[6] M. Inui, A. Saito, and S. Fleischer, J. Biol. Chem. **262**, 1740 (1987).
[7] M. Inui, A. Saito, and S. Fleischer, J. Biol. Chem. **262**, 15637 (1987).
[7a] J. S. Smith, R. Coronado, and G. Meissner, Nature (London) **316**, 446 (1985).
[7b] E. Rousseau, J. S. Smith, J. S. Henderson, and G. Meissner, Biophys. J. **50**, 1009 (1986).

These channels have characteristics of the Ca$^{2+}$ release observed in vesicles.[7c,7d] (6) Ryanodine has been found to modulate the channel in the bilayer, thereby reinforcing its role in the Ca$^{2+}$ release process.[7e,7f] (7) The ryanodine receptor incorporated into a bilayer has the Ca$^{2+}$ channel characteristics consistent with calcium release observed in isolated terminal cisternae.[7g–7j] Hence, the Ca$^{2+}$ release channel from skeletal muscle and heart has been isolated and identified in molecular and structural terms. This article describes the isolation of the Ca$^{2+}$ release channel from heart and skeletal muscle SR.

## Ryanodine-Binding Assay of Solubilized Samples

### Reagents

Tritiated ryanodine: 3 $\mu M$ [$^3$H]ryanodine[8] ($\sim$50,000 cpm/pmol) in H$_2$O

Cold ryanodine: 3 m$M$ ryanodine[9] in H$_2$O

Column solution: 1 $M$ NaCl, 0.3 $M$ sucrose, 2 m$M$ dithiothreitol, 10 mg/ml CHAPS,[10] 5 mg/ml soybean phospholipid,[11] 20 m$M$ Tris-HCl, pH 7.4. This solution is prepared by using the CHAPS-PL mixture (see p. 493).

### Procedure

[$^3$H]Ryanodine binding to solubilized samples (solubilized SR and fractions during the ryanodine receptor purification) is assayed at 24° using a Sephadex G-50 minicolumn (see below). The sample (5–600 $\mu$g of protein, depending on the purity of the receptor) is suspended in 0.2 ml

[7c] A. Chu, P. Volpe, B. Costello, and S. Fleischer, *Biochemistry* **25**, 8315 (1986).

[7d] G. Meissner, E. Darling, and J. Eveleth, *Biochemistry* **25**, 236 (1986).

[7e] E. Rousseau, J. S. Smith, and G. Meissner, *Am. J. Physiol.* **253**, C364 (1987).

[7f] K. Nagasaki and S. Fleischer, *Cell Calcium* **9**, 1 (1988).

[7g] L. Hymel, M. Inui, S. Fleischer, and H. Schindler, *Proc. Natl. Acad. Sci. U.S.A.* **85**, 441 (1988).

[7h] F. A. Lai, H. P. Erickson, E. Rousseau, Q. Liu, and G. Meissner, *Nature (London)* **331**, 315 (1988).

[7i] F. A. Lai, K. Anderson, E. Rousseau, Q. Liu, and G. Meissner, *Biochem. Biophys. Res. Commun.* **151**, 441 (1988).

[7j] L. Hymel, H. Schindler, M. Inui, and S. Fleischer, *Biochem. Biophys. Res. Commun.* **152**, 308 (1988).

[8] [$^3$H]Ryanodine (70 Ci/mmol) was prepared and purified as described in S. Fleischer *et al.*[5] Radioactive ryanodine can now be purchased from Du Pont (NEN) (Boston, MA).

[9] Ryanodine was initially obtained from the Pennick Corp. (Lindhurst, NJ). Ryanodine is currently available from Agra Systems International (Windgap, PA).

[10] CHAPS from Sigma Chemical Company (St. Louis, MO) was used in these studies.

[11] Soybean phospholipid (soybean lecithin, type IV-S) was obtained from Sigma.

(final volume) of column solution containing 25 $\mu M$ $CaCl_2$ with various concentrations of [³H]ryanodine with or without ryanodine. In the routine assay, 300 n$M$ [³H]ryanodine is included. To determine the nonspecific binding, a 1000-fold excess of cold ryanodine is included in addition to [³H]ryanodine. The reaction is started by adding an aliquot of tritiated ryanodine (20 $\mu$l) or a combination of tritiated (20 $\mu$l) and cold ryanodine (20 $\mu$l). After 40 min of incubation, bound and free [³H]ryanodine are separated on a Sephadex G-50 minicolumn. The minicolumn is prepared by pouring 1.5 ml packed volume of Sephadex G-50 (fine) (Pharmacia Inc., Piscataway, NJ) into a disposable filter column[12] (for SMA systems, Fischer Scientific, Pittsburgh, PA) and washed with 3 ml of column solution. The sample is carefully layered on top of the minicolumn, which is immediately centrifuged for 80 sec at 500 rpm (setting conditions) in a Beckman TJ-6 Centrifuge. A 100-$\mu$l aliquot of the eluate (~0.6 ml) is mixed with scintillation fluid (10 ml of ACS, Amersham Corp., Arlington Heights, IL) and radioactivity was measured by scintillation counting (standard error = 1%). For quantitative determination, the protein concentration of the eluate was estimated from a 10- to 100-$\mu$l aliquot using Amido Black 10B and a 0.45-$\mu$m Millipore filter (type HA) by the method of Kaplan and Pedersen.[13] When this method was applied to solubilized JTC of skeletal muscle SR, solubilized cardiac SR, and the purified ryanodine receptors from cardiac and skeletal muscle SR, nonspecific binding was less than 10% of the total binding under the standard assay conditions.[6]

*Notes on Methodology*

Among a variety of detergents, we find that CHAPS has a low nonspecific interaction with [³H]ryanodine and that CHAPS does not interfere with the [³H]ryanodine binding to the receptor.[6] Therefore, CHAPS was selected as a detergent. Phospholipids are essential to maintain the function of the ryanodine receptor in the solubilized state.[6] A small amount of $Ca^{2+}$ is also essential for the ryanodine binding since EGTA precludes the binding.[6] Endogenous $Ca^{2+}$ is usually sufficient. The binding is enhanced by high salt concentration. We have standardized on 1 $M$ NaCl.[6]

There can be sizeable variation from run to run in the centrifugation of the minicolumns. For example, the protein recovery in the eluate varied from 40 to 60% of the applied protein in different assays. Therefore, it is necessary to determine the protein recovery in the same eluate which is used for measuring ryanodine binding. On the other hand, the variation

---

[12] The diameter is approximately 0.9 cm.
[13] R. S. Kaplan and P. L. Pedersen, *Anal. Biochem.* **150**, 97 (1985).

within the same centrifugation run is small and usually less than 5% when columns are centrifuged at the same time.

SR vesicles also can be assayed using this method, in which case CHAPS and soybean phospholipid are omitted from the assay and column solutions. Similar results were obtained when this method was compared with the Millipore filtration assay employed previously.[5,6]

## Purification of Ryanodine Receptor from Skeletal Muscle SR

### Reagents and Preparations

CHAPS-PL mixture: 100 mg/ml CHAPS[10] and 50 mg/ml soybean phospholipid[11] in H$_2$O (sonicated until clear)

Solution A: 0.3 $M$ sucrose, 2 m$M$ dithiothreitol, 10 mg/ml CHAPS, 5 mg/ml soybean phospholipid, 0.5 μg/ml leupeptin, 20 m$M$ Tris-HCl, pH 7.4

Solution A–0.1 $M$ NaCl: Solution A containing 0.1 $M$ NaCl

Solution A–0.8 $M$ NaCl: Solution A containing 0.8 $M$ NaCl

Solution B: 0.6 $M$ KCl, 0.3 $M$ sucrose, 2 m$M$ dithiothreitol, 10 mg/ml CHAPS, 0.5 μg/ml leupeptin, 5 m$M$ potassium phosphate buffer, pH 7.4

Solution C: 0.6 $M$ KCl, 0.3 $M$ sucrose, 2 m$M$ dithiothreitol, 10 mg/ml CHAPS, 0.5 μg/ml leupeptin, 0.25 $M$ potassium phosphate buffer, pH 7.4

Solution D: 0.5 $M$ KCl, 0.3 $M$ sucrose, 2 m$M$ dithiothreitol, 10 mg/ml CHAPS, 0.5 μg/ml leupeptin, 20 m$M$ Tris-HCl, pH 7.4

Membrane fractions referrable to JTC of SR are prepared from rabbit fast twitch skeletal muscle as described previously[3,6] with some modifications. Leupeptin is included at a concentration of 0.5 μg/ml throughout the preparation procedure. The JTC are suspended in 0.3 $M$ sucrose, 0.5 μg/ml leupeptin, 5 m$M$ imidazole-HCl, pH 7.4, quick-frozen in liquid nitrogen and stored at −80°.

### Procedures

All procedures were performed in the cold at 4–6°.

*Step 1: Solubilization of Junctional Terminal Cisternae.* The JTC of SR (100 mg protein) are warmed at room temperature until thawed and suspended in 19.8 ml of 0.5 $M$ sucrose, 1.67 $M$ NaCl, 3.33 m$M$ dithiothreitol, 0.83 μg/ml leupeptin, and 33.3 m$M$ Tris-HCl, pH 7.4. Solubilization is initiated by adding 13.2 ml of CHAPS-PL mixture. The final protein concentration is about 3 mg/ml and the CHAPS/protein weight ratio is 13.3. The sample is kept for 10 min and centrifuged at 30,000 rpm for 30 min in a Beckman 70.1 Ti rotor. The supernatant (CHAPS extract) is obtained.

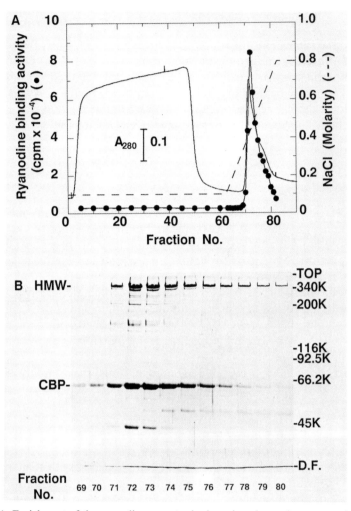

FIG. 1. Enrichment of the ryanodine receptor by heparin column chromatography. (A) The diluted CHAPS extract (~330 ml) is applied to a column (2.5 × 6 cm) of Affi-Gel heparin (Bio-Rad) preequilibrated with solution A–0.1 $M$ NaCl. The receptor is eluted at a flow rate of 3.5 ml/min with a linear NaCl gradient (NaCl concentration indicated by dashed lines) between solution A–0.1 $M$ NaCl and solution A–0.8 $M$ NaCl in 40 min. Absorbance at 280 nm (——) is monitored. Seven-milliliter fractions are collected and aliquots (125 μl) are assayed for ryanodine-binding activity (●). The peak fractions of the ryanodine-binding activity (fractions 71–77) are collected. (B) The protein pattern of the elution is checked by SDS-PAGE. SDS-PAGE electrophoresis, shown in each of the figures was carried out according to the method of U. K. Laemmli [*Nature (London)* **227,** 680 (1970)] on a 7.5% gel, employing a minislab gel electrophoresis apparatus (gel size: 75 mm length, 0.75 mm thickness). In Figs. 1 to 3, each lane contained 7.5 μl of the fraction. The gel is stained with Coomassie Blue followed by silver staining. Molecular weight standards used are $\alpha_2$-macroglobulin (nonreduced) (340K), myosin (200K), $\beta$-galactosidase (116K), phosphorylase *b* (92.5K), bovine serum albumin (66.2K), and ovalbumin (45K). HMW, high-molecular-weight ($M_r$ 360K) protein; CPP, Ca$^{2+}$-pump protein (Ca$^{2+}$-ATPase); CBP, Ca$^{2+}$-binding protein (calsequestrin); TOP, the top of the gel; D.F., the dye front.

*Step 2: Heparin Column Chromatography.* The CHAPS extract (~33 ml) is diluted with 300 ml of solution A and loaded onto an Affi-Gel heparin (Bio-Rad, Richmond, CA) column (2.5 × 6 cm) preequilibrated with solution A–0.1 $M$ NaCl. The elution profile of the chromatography is shown in Fig. 1. The chromatography is performed at a flow rate of 3.5 ml/min using an LKB Ultrachrom GTi chromatography system (Gaithersburg, MD). Seven-milliliter fractions are collected. The column is washed with solution A–0.1 $M$ NaCl until the absorbance at 280 nm comes to the base line, and elution is achieved with a linear salt gradient of solution A–0.1 $M$ NaCl to solution A–0.8 $M$ NaCl in 40 min. Aliquots (125 $\mu$l) of the fractions are assayed for ryanodine-binding activity with 300 n$M$ [$^3$H]-ryanodine. In the assay, NaCl is supplemented to make the final concentration approximately 1 $M$. The protein pattern of each fraction is checked by SDS-PAGE. The peak fractions of the ryanodine-binding activity are collected. These fractions are enriched in the high-molecular-weight protein with $M_r$ ~360K (Fig. 4).

*Step 3: Hydroxylapatite Column Chromatography.* The pooled fractions (48 ml) from an Affi-Gel heparin column are applied to a hydroxylapatite (Bio-Rad) column (1.5 × 6 cm) preequilibrated with solution B. The chromatography is performed at a flow rate of 3 ml/min using LKB Ultrochrom GTi system (Fig. 2) and 6-ml fractions are collected. The column is washed with solution B until the absorbance at 280 nm comes to the base line, and elution is achieved with a potassium phosphate linear gradient of solution B (containing 5 m$M$ potassium phosphate buffer) to solution C (containing 0.25 $M$ potassium phosphate buffer) in 40 min. Aliquots (150 $\mu$l) of the fractions are assayed for ryanodine-binding activity with 300 n$M$ [$^3$H]ryanodine in the presence of 5 mg/ml soybean phospholipid. The protein pattern of each fraction is also monitored by SDS-PAGE. The peak fractions of the ryanodine-binding activity are collected. These fractions are further enriched in the 360K protein and some other proteins with $M_r$ 330K, and traces of 175K, 63K, and 45K are also present (Fig. 4).

*Step 4: Gel-Permeation Chromatography.* The pooled fractions (30 ml) from the hydroxylapatite column are concentrated to 1 ml using a Centricon 30 (Amicon, Danvers, MA) and applied to a Superose 6 (Pharmacia) column (1.5 × 70 cm) preequilibrated and eluted in solution D (Fig. 3). The column is eluted at a flow rate of 1 ml/min and 2-ml fractions are collected. The fractions are assayed for ryanodine-binding activity with 300 n$M$ [$^3$H]ryanodine, supplemented with 5 mg/ml soybean phospholipid and 1 $M$ KCl (final concentrations). The protein pattern of the fractions is checked by SDS-PAGE. The peak fractions of the ryanodine-binding activity are pooled (20 ml) and can readily be concentrated 10-fold using the

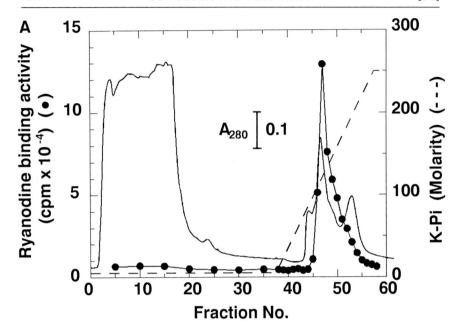

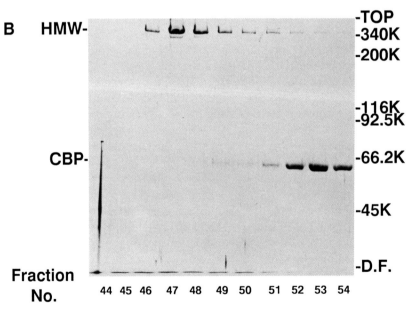

TABLE I
PURIFICATION OF RYANODINE RECEPTOR FROM JTC OF SKELETAL MUSCLE SR

| Fraction[a] | Protein[b] (mg) | Ryanodine binding[c] | | Yield (%) | Purification (-fold) |
|---|---|---|---|---|---|
| | | Specific activity (pmol/mg) | Total activity (pmol) | | |
| JTC of SR | 100 | 25.3 | 2530 | 100 | 1 |
| Solubilized JTC | 90.4 | 26.0 | 2350 | 92.9 | 1.03 |
| Heparin column chromatography | 12.8 | 135 | 1728 | 68.3 | 5.34 |
| Hydroxylapatite chromatography | 3.58 | 354 | 1267 | 50.1 | 14.0 |
| Gel-permeation chromatography | 2.26 | 392 | 886 | 35.0 | 15.5 |

[a] Protein pattern after each step is shown in the SDS-PAGE in Fig. 4.
[b] Protein concentration was determined by the method of R. S. Kaplan and P. L. Pedersen [*Anal. Biochem.* **150,** 97 (1985)] using JTC of SR as standard, which was determined by the method of Lowry *et al.* [O. H. Lowry, N. J. Rosebrough, A. L. Farr, and R. J. Randall, *J. Biol. Chem.* **193,** 265 (1951)], with bovine serum albumin as standard.
[c] Ryanodine-binding activity was determined at a [³H]ryanodine concentration of 300 n*M* using the ryanodine-binding assay for the solubilized samples.

Centricon 30. These fractions reveal a major protein band of 360K with a minor band of 330K (less than 10%) (Fig. 4). After addition of soybean phospholipid to 5 mg/ml, final concentration, the sample (0.1 to 1.0 mg protein/ml) is frozen in liquid nitrogen and stored at −80°.

*Summary of Purification*

The purification procedure of the ryanodine receptor from JTC of skeletal muscle SR is summarized in Fig. 4 and Table I. Approximately 2 mg of the purified receptor can be obtained from 100 mg of JTC. The

FIG. 2. Hydroxylapatite column chromatography of ryanodine receptor. (A) The pooled fractions (~48 ml) from an Affi-Gel heparin column are applied to a hydroxylapatite (Bio-Rad) column (1.5 × 6 cm) preequilibrated with solution B. The receptor is eluted at a flow rate of 3.0 ml/min with a linear phosphate gradient (phosphate concentration indicated by dashed lines) between solution B (5 m*M* potassium phosphate) and solution C (250 m*M* potassium phosphate) for 40 min. Absorbance at 280 nm is monitored. Six-milliliter fractions are collected. Aliquots (150 µl) of the fractions are assayed for ryanodine-binding activity (●). The peak fractions of the ryanodine-binding activity (fractions 46–50) are collected. (B) The protein pattern of the elution is monitored by SDS-PAGE. Abbreviations as defined in legend to Fig. 1.

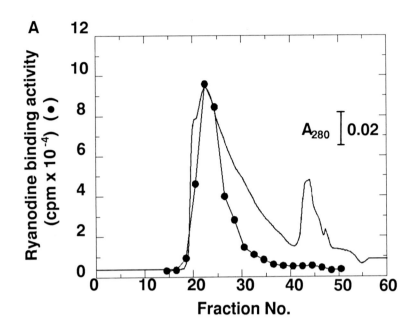

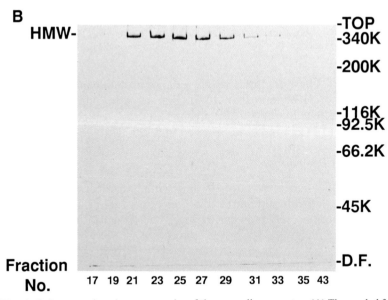

FIG. 3. Gel-permeation chromatography of the ryanodine receptor. (A) The pooled fractions (~30 ml) from an hydroxylapatite column are concentrated into 1 ml by Centricon 30 (Amicon) and applied onto a Superose 6 (Pharmacia) column (1.5 × 70 cm) preequilibrated and eluted in solution D. The column is eluted at a flow rate of 1 ml/min and 2-ml fractions are collected. Absorbance at 280 nm (——) is monitored. Fractions are assayed for ryanodine-binding activity (●). The peak fractions of the ryanodine-binding activity (fractions 21–30) are collected. (B) The protein pattern of the elution is monitored by SDS-PAGE. Abbreviations as defined in legend to Fig. 1.

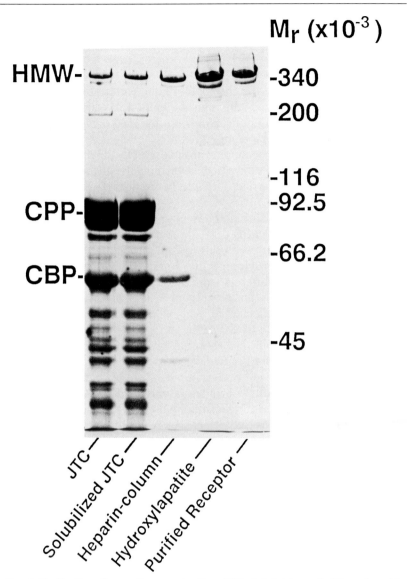

FIG. 4. Purification of ryanodine receptor from JTC of skeletal muscle SR as viewed by SDS-PAGE.

overall yield of the ryanodine-binding activity of the purified receptor is approximately 30%.

*Notes on Methodology*

Figures 1–3 illustrate the chromatographic pattern for the purification of the ryanodine receptor. Although the elution pattern is monitored by absorbance at 280 nm ($A_{280}$), protein is not the only component contributing to $A_{280}$; phospholipid and detergent are the major contributors. Especially in heparin-affinity column chromatography, high $A_{280}$ (~1.0 OD) due to phospholipid is subtracted electronically in the base line. The protein quantitation is carried out directly on the isolated fractions and is provided in Table I.

The procedure and chromatograms shown here are from a typical experiment using highly purified JTC preparation.[3] When the SR preparations with different purity are used for the purification of the ryanodine receptor, the elution profiles of the contaminating proteins from the columns are somewhat different from those presented here. Therefore, in order to obtain the highly purified receptor, it is necessary to modify the size of the fractions collected in each step of the chromatography by checking the fractions on SDS-PAGE.

Although phospholipids are essential for ryanodine binding to the receptor,[6] phospholipids are omitted in Steps 3 and 4 for several reasons. In the presence of soybean phospholipid, a proper base line cannot be achieved in hydroxylapatite chromatography in Step 3. It is difficult to concentrate the sample with phospholipids by Centricon 30 in Step 4. Normal ryanodine-binding activity can be restored when phospholipid is added to the receptor from skeletal muscle.[6]

### Purification of the Ryanodine Receptor from Cardiac SR

The ryanodine receptor can be purified from cardiac SR using the procedure described above with some modifications.[7] Cardiac microsomes prepared from dog heart ventricles by the method of Chamberlain *et al.*[14] are used for the purification. Leupeptin is included at the concentration of 0.5 μg/ml throughout the procedure for the preparation of microsomes and the receptor purification. The main difference in purifying the ryanodine receptor from cardiac and skeletal muscle SR is that phospholipid cannot be removed from the ryanodine receptor of cardiac SR without loss of function. Once phospholipid is removed, ryanodine-bind-

---

[14] B. K. Chamberlain, D. O. Levitsky, and S. Fleischer, *J. Biol. Chem.* **258**, 6602 (1983).

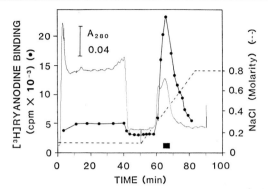

FIG. 5. Purification of the ryanodine receptor from heart using *p*-aminobenzamidine-agarose column chromatography. The pooled fraction (14 ml) from a heparin-affinity column is diluted with 100 ml of solution A and applied to a column (1.5 × 5 cm) of *p*-aminobenzamidine-agarose preequilibrated with solution A–0.1 *M* NaCl. The receptor is eluted at a flow rate of 3 ml/min with a linear salt gradient (NaCl concentration indicated by dashed lines) between solution A–0.1 *M* NaCl and solution A–0.8 *M* NaCl in 40 min. Absorbance at 280 nm (——) is monitored. Six-milliliter fractions are collected and aliquots (130 μl) are assayed for ryanodine-binding activity (●). The fractions indicated by the black bar are pooled for further purification. This chromatography shows the purification started from 100 mg cardiac microsomes.

ing activity is completely lost and is not restored by readdition of phospholipid.[7] Therefore, Steps 3 and 4 are modified as follows.

Since phospholipid interferes with the chromatography on hydroxylapatite, for Step 3, a *p*-aminobenzamidine-agarose (Sigma) column[15] is used instead of a hydroxylapatite column in the presence of 5 mg/ml soybean phospholipid. The pooled fractions from an Affi-Gel heparin column are diluted 6-fold with solution A and applied to a *p*-aminobenzamidine-agarose column (1.5 × 6 cm) preequilibrated with solution A–0.1 *M* NaCl. The column is washed with solution A–0.1 *M* NaCl and elution is achieved with a linear salt gradient from solution A–0.1 *M* NaCl to solution A–0.8 *M* NaCl in 40 min. The chromatography (Fig. 5) is performed at a flow rate of 3 ml/min using an LKB Ultrochrom GTi chromatography system. Six-milliliter fractions are collected and the peak fractions of ryanodine-binding activity are pooled.

In Step 4, gel permeation chromatography (Superose 6 column) is performed in solution D containing 5 mg/ml soybean phospholipid. The purified receptor is concentrated using a small column (1 ml of bed vol-

---

[15] The ryanodine receptor from skeletal muscle SR has similar chromatographic behavior in the presence of phospholipid on a *p*-aminobenzamidine-agarose column. The receptor is readily absorbed onto the column and eluted by increasing the salt concentration.

TABLE II
PURIFICATION OF RYANODINE RECEPTOR FROM HEART MUSCLE SR

| Fraction[a] | Protein[b] (mg) | Ryanodine binding[c] | | Yield (%) | Purification (-fold) |
| | | Specific activity (pmol/mg) | Total activity (pmol) | | |
|---|---|---|---|---|---|
| Cardiac SR | 100 | 8.64 | 864 | 100 | 1 |
| Solubilized SR | 74.2 | 9.35 | 694 | 80 | 1.1 |
| Heparin column | 1.80 | 181 | 326 | 38 | 20.9 |
| p-Aminobenzamidine column chromatography | 0.21 | 782 | 164 | 19 | 90.5 |
| Gel-permeation chromatography | 0.084 | 983[d] | 83 | 9.6 | 114 |

[a] The protein pattern after each step is shown in the SDS-PAGE gel in Fig. 1.
[b] Protein concentration was determined as described in Table I.
[c] Ryanodine binding was measured with 300 n$M$ [³H]ryanodine as described previously.[7] The binding consists of both high- and low-affinity binding.
[d] The $B_{max}$ for high-affinity binding of the purified protein was 455 pmol/mg protein (46% of the total), so that the low-affinity binding is 528 pmol/mg protein (54%).

ume) of Affi-Gel heparin. The pooled fraction (~10 ml) from a Superose 6 column is diluted 5-fold with solution A and applied to the small column of Affi-Gel heparin preequilibrated with solution A–0.1 $M$ KCl. After washing with the same solution, the receptor is eluted with solution A–0.8 $M$ KCl. In this way, the receptor can be concentrated to a small volume (approximately 1–1.5 ml).

The purification of the ryanodine receptor from cardiac microsomes is summarized in Fig. 6 and Table II. About 100 μg of the purified receptor is obtained from 100 mg of cardiac microsomes.[7] When estimated by ryanodine-binding activity with 300 n$M$ [³H]ryanodine, this method represented about 110-fold purification of the ryanodine receptor from cardiac microsomes with overall yield of about 10%.[7] The purified receptor revealed the protein band of 340K with a minor band of 300K (less than 10%) on SDS-PAGE (Fig. 6).

## Characteristics of Purified Ryanodine Receptor

Ryanodine binding is localized to the JTC of SR both in heart[4] and skeletal muscle.[5] The ryanodine-binding affinity ($K_d$) is ~50 n$M$ in skeletal muscle SR, which is in the same concentration range as its pharmacological action on Ca²⁺-release channels.[5,6] Ryanodine locks the Ca²⁺-release

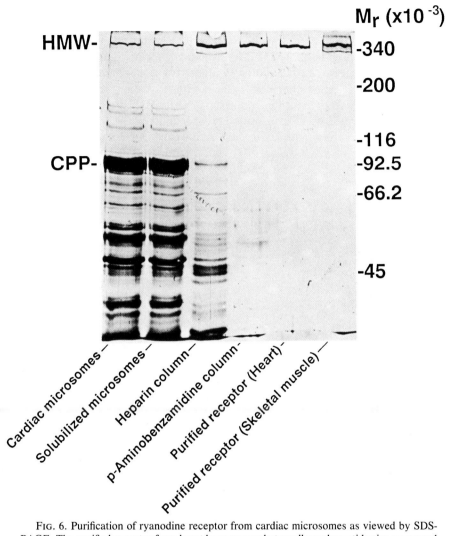

FIG. 6. Purification of ryanodine receptor from cardiac microsomes as viewed by SDS-PAGE. The purified receptor from heart has a somewhat smaller polypeptide size compared with that from skeletal muscle SR ($M_r$ 340K versus 360K).

channel in the open state. In cardiac SR, two types of ryanodine binding with $K_d$ of ~8 n$M$ and ~1 $\mu M$ are observed that are also in the same concentration ranges as its action on the Ca²⁺-release channels,[4] i.e., to lock the Ca²⁺-release channels in the open state and to close the channels, respectively. The purified receptors from cardiac and skeletal muscle SR showed about the same ryanodine-binding affinities as the original SR vesicles,[6,7] indicating that the ryanodine receptor was purified unchanged with regard to ligand binding. The maximal binding ($B_{max}$) of the purified receptor from skeletal muscle SR was about 500 pmol/mg,[6] and that from cardiac SR was ~500 pmol/mg and ~5 nmol/mg for high- and low-affinity sites, respectively.[7] The ryanodine-binding affinity is about 5-fold higher in heart than in skeletal muscle, when compared using identical binding conditions.[7] When the purified ryanodine receptor, from either heart or skeletal muscle SR, was incorporated into the lipid bilayer, Ca²⁺ activated channels were formed with Ca²⁺ release characteristics diagnostic of that in isolated JTC of SR.[7g-7j]

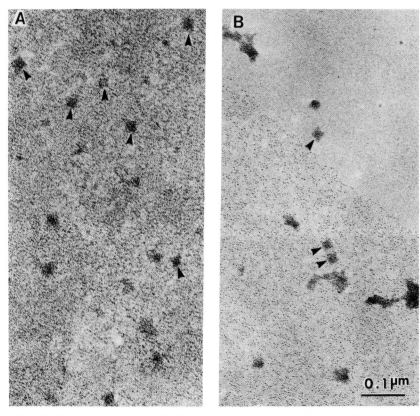

Fig. 7. Morphology of the purified ryanodine receptor from heart and skeletal muscle SR. The samples were stained with uranyl acetate using the procedure for negative staining electron microscopy.[7] Positive staining images are shown. The purified receptors from heart (A) and skeletal muscle (B) have a characteristic square-shaped appearance, with sides of 210 to 220 Å (arrowheads), which is the unique characteristic of the feet structures of the junctional face membrane of terminal cisternae of both cardiac[4] and skeletal muscle SR.[3]

The purified ryanodine receptor from skeletal muscle SR contains one major polypeptide band of $M_r$ 360K with a minor band of $M_r$ 330K (less than 10%) on SDS-PAGE. A proteinase inhibitor, leupeptin, should be included during the SR preparation and for the purification of receptor, in order to decrease proteolysis which gives rise to bands of $M_r$ 330K and 175K.[6,7] Therefore, it would appear that the two smaller bands are mainly proteolytic degradation products. The size of the receptor from cardiac SR is slightly smaller than that from skeletal muscle SR ($M_r$ 340K versus 360K), further indicating that the polypeptides are not identical.[7] On gel-permeation chromatography, the purified receptor both from cardiac SR and from skeletal muscle SR eluted in the peak which corresponds to $M_r$ of about one million.[6,7]

The morphology of the purified receptor from cardiac SR and from skeletal muscle SR, examined by electron microscopy, is observed as squares having sides of about 210–220 Å (Fig. 7). Such squares are the unique characteristic of the feet structures of the junctional face membrane of terminal cisternae,[3,4] indicating that the ryanodine receptor is identical to the foot structure. Since the molecular mass of the foot structure can be estimated to be $4.4 \times 10^6$ daltons from the dimension of the structure ($210 \times 210 \times 120$ Å), it is suggested that the foot structure of both cardiac and skeletal muscle consists of an oligomer of 12 monomers ($M_r$ 360,000 for skeletal muscle and $M_r$ 340,000 for cardiac muscle). From the ryanodine-binding data, the stoichiometry between high-affinity binding sites and the number of the feet structures is estimated to be two.[6]

# [37] Measurement of Sodium–Calcium Exchange Activity in Plasma Membrane Vesicles

By John P. Reeves

## Introduction

Na–Ca exchange is a carrier-mediated transport process in which transmembrane movements of calcium ions in one direction are directly coupled to sodium ion movements in the opposite direction. It is found primarily in plasma membranes of excitable cells such as muscle and nerve where it appears to function as a mechanism for extruding $Ca^{2+}$ from the cell. This chapter describes the techniques, and some of the pitfalls, of measuring Na–Ca exchange activity in osmotically sealed plasma membrane vesicles. The discussion is focused on sarcolemmal membranes obtained from cardiac muscle, but the techniques are applicable to other types of vesicle preparations as well. The preparation of purified plasma membrane vesicles from heart and other tissues has been discussed in other chapters in this series and will not be described here.

Na–Ca exchange activity is assayed by measuring the $Ca^{2+}$ movements which occur when transmembrane $Na^+$ concentration gradients are generated by appropriate loading and dilution procedures.[1] $Ca^{2+}$ uptake will occur when the vesicles are loaded internally with NaCl and then diluted into an isosmotic, $Na^+$-free medium containing $CaCl_2$. $Na^+$-dependent $Ca^{2+}$ efflux can be measured by equilibrating the vesicles with $CaCl_2$ in a $Na^+$-free medium and then diluting the $Ca^{2+}$-loaded vesicles into an isosmotic medium containing NaCl. The $Ca^{2+}$ movements can be detected either by filtration with $^{45}Ca^{2+}$ or by using external $Ca^{2+}$ indicators such as arsenazo III. The former approach will be described in this article; information on the use of indicators to follow Na–Ca exchange activity can be found in Caroni et al.[2] and Kadoma et al.[3]

### Sodium-Dependent Calcium Uptake

*Reagents*

Vesicle incubation medium: 160 m$M$ NaCl, 20 m$M$ 3-($N$-morpholino)propanesulfonic acid (MOPS), pH adjusted to 7.4 with Tris or NaOH

Dilution medium: 160 m$M$ KCl, 20 m$M$ MOPS, pH adjusted to 7.4 with Tris or KOH

Quenching medium: 200 m$M$ KCl, 5 m$M$ MOPS, 0.1 m$M$ EGTA, adjusted to pH 7.4 with Tris (ice-cold)

The dilution medium specified above is used in our laboratory for routine measurements of exchange activity. Other media, such as 0.3 $M$ sucrose or 160 m$M$ LiCl, can also be used. It should be noted that solutions of reagent grade LiCl should be treated with Chelex (Bio-Rad) to reduce contaminating $Ca^{2+}$ ions (20 $\mu M$ in 160 m$M$ LiCl) to tolerable levels. With distilled water of high quality, contaminating $Ca^{2+}$ in 160 m$M$ solutions of reagent grade KCl (e.g., Fisher, Baker) is 1–2 $\mu M$. The concentration of contaminating $Ca^{2+}$ can be measured with a $Ca^{2+}$ electrode (e.g., Orion) after proper calibration[4] or by using the arsenazo III technique described by Ohnishi.[5]

[1] J. P. Reeves and J. L. Sutko, *Proc. Natl. Acad. Sci. U.S.A.* **76**, 590 (1979).
[2] P. Caroni, L. Reinlib, and E. Carafoli, *Proc. Natl. Acad. Sci. U.S.A.* **77** (1980).
[3] M. Kadoma, J. Froehlich, J. Reeves, and J. Sutko, *Biochemistry* **21**, 1914 (1982).
[4] D. M. Bers, *Am. J. Physiol.* **242**, C404 (1982).
[5] S. T. Ohnishi, *Biochim. Biophys. Acta* **586**, 217 (1979).

*Assay Method.* Plasma membrane vesicles are suspended in the vesicle incubation medium at 1–5 mg protein/ml and incubated at 37° for at least 30 min, or overnight on ice, to allow the NaCl to equilibrate with the intravesicular space. $^{45}CaCl_2$ is added to the dilution medium at the desired concentration (usually 10–30 $\mu M$) and 100-$\mu$l aliquots are dispensed to 12 × 75 mm polystyrene tubes. For each assay point, a 2-$\mu$l aliquot of the vesicle suspension is placed on the side of the tube near the bottom and the reaction is initiated by vortexing the tube. It is helpful to use a vortexing apparatus such as the Thermolyne Maxi-Mix which can be run continuously and has a vibrating platform for mixing; the large area of the platform facilitates proper tube placement for vortexing and the continuous operation minimizes delay in mixing.

The reaction is terminated by the addition of 5 ml of ice-cold quenching medium with a repeating dispensing syringe. The vesicles are harvested by filtration on Whatman GF/A glass-fiber filters; other filters (e.g., Millipore-type HA filters) may be used instead of the glass-fiber filters, but the latter are relatively inexpensive and allow for easy adjustment of the flow rate through the filter. Experiments with $^{125}I$-labeled vesicles indicate that at least 80% of the vesicles are trapped by the glass-fiber filters. After the contents of the tube are poured over the filter, a second 5-ml aliquot of the quenching solution is added to the tube and its contents are poured over the filter just before the meniscus in the filtration chimney from the first rinse reaches the filter. A final 5-ml aliquot of the quenching medium is added directly to the filter, again just before the meniscus reaches the filter. It has been our experience that the reproducibility of the data is improved if the filter is not exposed to air between rinses. The radioactivity on the filter is determined by standard liquid scintillation techniques.

*Timing.* For uptake periods over 10 sec in duration, timing can be done with a stopwatch. However, cardiac sarcolemmal vesicles have a very high Na–Ca exchange activity (5–20 nmol/mg protein · sec) so that kinetic studies with these vesicles require short reaction times to approximate initial rate conditions. Reaction times as short as 1 sec can be attained by using a metronome set to beat once per second to time the mixing and quenching steps. For 1-sec uptakes, it is helpful to have the syringe containing the quench medium poised directly on the lip of the assay tube during the uptake period.

*Valinomycin.* The stoichiometry of the exchange reaction is 3 Na$^+$ per Ca$^{2+}$ [6,7] and Na–Ca exchange is, therefore, an electrogenic process. In cardiac sarcolemmal vesicles, exchange activity is restrained by the re-

[6] B. J. R. Pitts, *J. Biol. Chem.* **254**, 6232 (1979).
[7] J. P. Reeves and C. C. Hale, *J. Biol. Chem.* **259**, 7733 (1984).

sulting charge buildup[8,9] and can be stimulated by adding electrogenic ionophores, such as carbonyl cyanide $m$-chlorophenylhydrazone or valinomycin (in the presence of $K^+$), to increase the conductivity of the membranes and allow adequate charge compensation. Valinomycin can be added as a 2 m$M$ solution in dimethyl sulfoxide to either the dilution medium (0.5 $\mu M$ final concentration) or to the vesicle suspension (1–2 $\mu M$ final concentration). In the former case, the dilution medium should be used within a few minutes since the valinomycin is hydrophobic and will not remain in solution for long. When added to the membrane suspension, the valinomycin associates with the vesicles and remains effective indefinitely.

*Blanks.* To correct for $^{45}Ca^{2+}$ bound nonspecifically to the vesicles and filters, two types of blanks can be run. In the first type, Na-loaded vesicles are diluted into 160 m$M$ NaCl–20 m$M$ MOPS, pH 7.4, with the same concentration of $^{45}Ca^{2+}$ as used in the assay for Na–Ca exchange. In the second type of blank, vesicles are loaded internally with KCl (or LiCl) instead of NaCl and assayed for $Ca^{2+}$ uptake in the normal manner. Both types of blanks give comparable results. The first approach is used more commonly than the second because of the convenience of using a single batch of vesicles.

### Sodium-Dependent Calcium Efflux

*Loading Procedure.* Vesicles are suspended in 160 m$M$ KCl, 20 m$M$ MOPS, pH 7.4, and incubated with the desired concentration of $^{45}CaCl_2$ (usually 10–100 $\mu M$) for several hours at 37° or overnight on ice. If the vesicles had previously been loaded with NaCl, the $Na^+$ gradient is allowed to dissipate during a 30–45 min preincubation at 37° before adding the $^{45}CaCl_2$. In the case of cardiac sarcolemmal vesicles, the amount of $^{45}Ca^{2+}$ associated with the vesicles at equilibrium is 5- to 20-fold higher than that predicted by the intravesicular volume of the vesicles (approximately 8.5 $\mu$l/mg protein[10]). This reflects the binding of $Ca^{2+}$ to the internal surface of the membranes,[10] perhaps through an interaction with membrane phospholipids.[11] Under certain circumstances, the rate of dissociation of intravesicular $Ca^{2+}$ from these internal binding sites may be the rate-limiting step in the $Ca^{2+}$ efflux process (cf. below). Vesicles can also be loaded with $^{45}Ca^{2+}$ by means of Na–Ca exchange activity.[1]

[8] J. P. Reeves and J. L. Sutko, *Science* **208,** 1461 (1980).
[9] K. D. Philipson and A. Y. Nishimoto, *J. Biol. Chem.* **255,** 6880 (1980).
[10] R. S. Slaughter, J. L. Sutko, and J. P. Reeves, *J. Biol. Chem.* **258,** 3183 (1983).
[11] K. D. Philipson, D. M. Bers, and A. Y. Nishimoto, *J. Mol. Cell. Cardiol.* **12,** 1159 (1980).

This is less desirable than the equilibration method because the intravesicular $Na^+$ concentration is not well defined and because the amount of $Ca^{2+}$ in the vesicles tends to decline after the maximal level of accumulation has been achieved.

*Assay Method.* The procedures are similar to those for $Ca^{2+}$ uptake. A 2-$\mu$l aliquot of the $^{45}Ca$-loaded vesicles is placed on the side of the polystyrene tube containing 100 $\mu$l of a solution with the desired concentration of NaCl. The vesicles are mixed with the dilution medium, terminated, and filtered as described above for $Ca^{2+}$ uptake. The inclusion of 0.1 m$M$ EGTA in the dilution medium is recommended to preclude Ca–Ca exchange. To correct for the loss of $^{45}Ca^{2+}$ from the vesicles by passive diffusion, the vesicles are diluted into a $Na^+$-free medium (usually 160 m$M$ KCl) containing 0.1 m$M$ EGTA; the rate of passive efflux is normally only a small fraction of $Na^+$-dependent $Ca^{2+}$ efflux.

Pitfalls

*Kinetics of Sodium–Calcium Exchange.* Cardiac sarcolemmal preparations show an extreme variability in their apparent $K_m$ for $Ca^{2+}$; reported values range from 1.5 $\mu M$ to over 100 $\mu M$. Part of the variability undoubtedly reflects the responsiveness of the Na–Ca exchange system to a host of regulatory influences; these include limited proteolysis, phospholipase treatment, anionic or cationic amphiphiles, membrane phosphorylation, alkaline pH, and certain redox reactions.[12] Thus, subtle differences in preparation procedures or in the physiological state of the hearts could lead to sharp differences among the kinetic properties of different preparations. Another important influence is intravesicular $Ca^{2+}$. Recent results have shown that the $K_m$ of the Na–Ca exchange system for extravesicular $Ca^{2+}$ is markedly reduced by the presence of intravesicular $Ca^{2+}$.[13] Since cardiac sarcolemmal preparations contain considerable amounts of endogenous $Ca^{2+}$ (6.2 ± 1.0 nmol/mg protein[14]), the kinetic properties of a particular preparation are likely to be affected by such factors as its endogenous Ca content, the number or affinity of internal $Ca^{2+}$ binding sites, and the protein concentration of the vesicle suspension. The endogenous $Ca^{2+}$ content of sarcolemmal preparations can be reduced by 30–45 min of incubation at 37° in a NaCl medium containing 0.1 m$M$ EGTA, or by incubating a dilute suspension of vesicles in NaCl alone, collecting the vesicles by centrifugation, and resus-

[12] J. P. Reeves, *Curr. Top. Membr., Transp.* **25,** 77 (1985).
[13] J. P. Reeves and P. Poronnik, *Am. J. Physiol.* **252,** C17 (1987).
[14] D. K. Bartschat and G. E. Lindenmayer, *J. Biol. Chem.* **255,** 9926 (1980).

pending them in fresh NaCl solution. A more important implication of the effects of intravesicular Ca$^{2+}$ is that the kinetics of the Na–Ca exchange system will be altered as a consequence of its own activity, since this necessarily changes intravesicular Ca$^{2+}$ levels. Thus, it seems likely that the kinetics of the Na–Ca exchange system will remain complex and difficult to understand in terms of a conventional Michaelis–Menten approach.

*Ca$^{2+}$ Binding.* Much of the Ca$^{2+}$ taken up by vesicles is bound internally to sites which may originate with membrane phospholipids.[11,12] These sites act as a sink for transported Ca$^{2+}$ and amplify the extent of Ca$^{2+}$ accumulation. This is an important consideration in inhibitor studies since agents that reduce the extent of intravesicular Ca$^{2+}$ binding could produce an apparent decrease in Ca$^{2+}$ accumulation without actually affecting transport activity. For this reason, the efficacy of putative inhibitors should always be verified by measuring their effects on Na$^+$-dependent Ca$^{2+}$ efflux as well as on Ca$^{2+}$ uptake. However, it is important to bear in mind that, under certain conditions, the rate of Ca$^{2+}$ efflux may be limited by the rate of dissociation of Ca$^{2+}$ from internal binding sites. Therefore, it is best to measure efflux under conditions where this possibility is minimized, i.e., low temperature (25° or less) and high levels of Ca$^{2+}$ loading. One simple test to verify that the rate of Na$^+$-dependent Ca$^{2+}$ efflux is not limited by dissociation of Ca$^{2+}$ from internal binding sites is to dilute the Ca$^{2+}$-loaded vesicles into a KCl medium containing 0.1 m$M$ EGTA and 0.5 $\mu M$ A23187, a Ca$^{2+}$ ionophore. The rate of ionophore-mediated efflux should be considerably more rapid than Na$^+$-dependent Ca$^{2+}$ efflux.[3]

# Section II

# ATP-Driven Proton Pumps

# [38] Plasma Membrane ATPase from the Yeast
## Schizosaccharomyces pombe

By Jean-Pierre Dufour, Antoine Amory, and André Goffeau

*Schizosaccharomyces pombe* is a yeast which is well studied from both the biochemical and genetic point of view.[1] It provides a convenient material for the purification of the plasma membrane ATPase.[2] The $H^+$-translocating ATPase shares many properties with the cation-transporting ATPases such as the $Ca^{2+}$-, $Na^+/K^+$-, or $H^+/K^+$-ATPase, from mammalian tissues. The $H^+$-ATPases from the fission yeast *Schizosaccharomyces pombe*, the budding yeast *Saccharomyces cerevisiae*, and the mold *Neurospora crassa* are remarkably similar and the biochemical information gathered on any one of these enzymes is usually valid for the other two.[3] Each of these microorganisms has particular advantages. *Saccharomyces cerevisiae* offers many possibilities for genetic studies which have not yet been fully explored. In our experience,[4] however, it is still more difficult to obtain $H^+$-ATPase of high specific activity from *S. cerevisiae* than from *S. pombe* or *N. crassa*.

This chapter describes the preparation of the membrane-bound $H^+$-ATPase of *S. pombe,* the solubilization and purification of the enzyme, and the reconstitution of its $H^+$-pumping function in artificial phospholipid vesicles.

## ATPase Assay

It has to be stressed that, during the early purification steps, several other ATP hydrolytic activities can be expressed at pH 6.0, the optimum pH for the *S. pombe* plasma membrane ATPase. The major contaminant is the mitochondrial ATPase. The contributions from acid phosphatases and/or vacuolar ATPase are usually of minor importance.

The plasma membrane ATPase activity is recognized by its strict specificity for ATP, its insensitivity to 10 m$M$ sodium azide, and its total inhibition by 20 $\mu M$ vanadate or 0.5 $\mu M$ *p*-hydroxymercuribenzoate. To assess the importance of contaminant activities, one can measure ATP

[1] R. Egel, J. Kohli, P. Thuriaux, and K. Wolf, *Annu. Rev. Genet.* **14,** 77 (1980).
[2] J. P. Dufour and A. Goffeau, *J. Biol. Chem.* **253,** 7026 (1978).
[3] A. Goffeau and C. W. Slayman, *Biochim. Biophys. Acta* **639,** 197 (1981).
[4] A. Goffeau and J. P. Dufour, this volume [39].

hydrolysis in the absence of sodium azide at pH 9.0 for the mitochondrial ATPase, the hydrolysis of $\beta$-glycerophosphate at pH 5.0 for acid phosphatases, and the hydrolysis of GTP in the presence of sodium azide at pH 7.0 for the vacuolar ATPase. However, during purification of the *S. pombe* plasma membranes, it routinely suffices to perform the ATPase assays both at pH 6.0 and 9.0 in the absence and in the presence of 10 m$M$ sodium azide. The ratio of the azide-insensitive ATPase at pH 6.0 to the azide-sensitive ATPase at pH 9.0 is a convenient marker for mitochondrial contamination. This ratio increases from 0.5 to more than 20 during purification of the plasma membranes.

*Colorimetric Method*

The lowest phosphate concentration detectable with confidence is 100 $\mu M$. Since no more than 20% of the substrate should be hydrolyzed to maintain linear conditions, the lowest meaningful ATP concentration is 0.5 m$M$.

*Reagent*

ATPase stock assay medium: 7.05 m$M$ ATP sodium salt (Boehringer, Mannheim, FDR, or Grade II from Sigma Chem. Co., MO), 10.6 m$M$ MgSO$_4$, 59 m$M$ 2-($N$-morpholino)ethanesulfonic acid (MES)–KOH, pH 6.0. The solution is stored at $-20°$.

Sodium orthovanadate (22 to 25% V$_2$O$_5$) (BDH Chemical Ltd, Poole, England), 1 m$M$ stock solution in doubly distilled water kept frozen at $-20°$.

Lysolecithin (Type I, Sigma or from Avanti Polar Lipids, 2421 Highbluff Rd, Birmingham, AL 35216), 1 mg/ml stock solution freshly dissolved in doubly distilled water at room temperature.

Sodium azide, 1 $M$ stock solution kept at 4°.

Sodium dodecyl sulfate (SDS) (Merck cat. No. 822050, Darmstadt, FDR) 1% w/v.

Elon reagent: 5 g $p$-methylaminophenol sulfate (Elon) (Kodak), 15 g NaHSO$_3$ (Na$_2$S$_2$O$_5$), adjusted to 500 ml with doubly distilled water. The solution must be kept in the dark.

Ammonium molybdate reagent: Dissolve 50 g (NH$_4$)$_6$Mo$_7$O$_{24}$ · 4H$_2$O in 400 ml 10 $N$ H$_2$SO$_4$ and bring to 1 liter with doubly distilled water.

*Procedure.* The assay is performed at 30° in 1 ml final volume containing 6 m$M$ ATP, 9 m$M$ MgSO$_4$ (3 m$M$ free Mg$^{2+}$), 50 m$M$ MES–KOH (pH 6.0) (0.85 ml of ATPase assay stock medium), and 10 m$M$ sodium azide.

For the purified enzyme, the assay medium is supplemented with 50 $\mu$g/ml of lysolecithin. The reaction is started by addition of the enzyme and stopped by addition of 3 ml of SDS 1% (w/v).

Inorganic phosphate is measured as follows: 1 ml of ammonium molybdate reagent is added to each tube immediately followed by 1 ml of Elon reagent. After 15 min at room temperature, the absorbance of the final solution is measured either using a colorimeter (660 nm) or using a spectrophotometer at 720 or 800 nm (highest sensitivity). The volume of assay can be scaled down to 100 $\mu$l. In such case, the ATPase reaction is stopped by 300 $\mu$l of SDS 1% (w/v). Color development is carried as above using 250 $\mu$l of ammonium molybdate reagent and 250 $\mu$l of Elon reagent.

### Radioactive Method

For ATPase assays at concentrations of ATP lower than 0.5 m$M$, the [$^{32}$P]phosphate produced from hydrolysis of [$\gamma$-$^{32}$P]ATP is measured.[5]

#### Reagents

[$\gamma$-$^{32}$P]ATP, 10 m$M$ stock solution in 50 m$M$ MES–KOH, pH 6.0
MgSO$_4$, 100 m$M$ stock solution in 50 m$M$ MES–KOH, pH 6.0
50 m$M$ MES–KOH, pH 6.0
Charcoal (activated charcoal), 250 mg/ml 0.1 $N$ HCl. Before use, the charcoal is washed 3 times with 0.1 $N$ HCl.
Perchloric acid, 20% w/v

*Procedure.* The reaction in 1.2 ml of solution containing 3 m$M$ free Mg$^{2+}$, 50 m$M$ MES–KOH (pH 6.0), and [$\gamma$-$^{32}$P]ATP (minimum 10 $\mu M$, maximum 0.5 m$M$) is started by addition of the enzyme and stopped after 0, 5, 10, and 15 min by transferring 200 $\mu$l in 200 $\mu$l cold 20% (w/v) perchloric acid into Eppendorf tubes. Stopping time may be changed to stay in the linear response of the assay. For the purified enzyme, the assay medium is supplemented with 50 $\mu$g/ml of lysolecithin. After addition of 400 $\mu$l of charcoal, the tube is stirred for a short time and centrifuged in a table-top Eppendorf centrifuge for 5 min at room temperature. This step removes the nonhydrolyzed [$\gamma$-$^{32}$P]ATP. An aliquot of 200 $\mu$l is mixed with 10 ml of liquid scintillation mixture.

### Spectrophotometric Method

For continuous rate measurements, the ATPase activity is best measured spectrophotometrically by coupling the production of ADP to the

[5] C. Grubmeyer and H. S. Penefsky, *J. Biol. Chem.* **256**, 3718 (1981).

oxidation of NADH via the pyruvate kinase and lactate dehydrogenase reactions.

*Reagents*

ATPase stock assay medium, as for the colorimetric assay
NADH, 45 m$M$ stock solution in 10 m$M$ Tris–MES, pH 7.5
Phosphoenolpyruvate, 125 m$M$ stock solution
Lactate dehydrogenase (EC 1.1.1.27) in glycerol
Pyruvate kinase (EC 2.7.1.40) in glycerol

*Procedure.* The reaction mixture contains, in a volume of 1 ml, 6 m$M$ ATP, 9 m$M$ MgSO$_4$ (3 m$M$ free Mg$^{2+}$), 50 m$M$ MES–KOH, pH 6.0, 0.45 m$M$ NADH, 1.25 m$M$ phosphoenolpyruvate, 1.2 units of lactate dehydrogenase, and 25 units of pyruvate kinase. The requirement of the pyruvate kinase for a monovalent cation is satisfied by the KOH used to adjust the pH. The oxidation of NADH is recorded at 30° by absorbance at 340 nm. For the purified enzyme, the assay medium is supplemented with 50 μg/ml of lysolecithin.

Yeast Culture and $^{35}$S Labeling

Fifteen liters of growth medium containing 2% (w/v) yeast extract (Difco or KAT, Ohly GmbH, Hamburg, FDR) and 5.8% (w/v) glucose adjusted to pH 4.5 with HCl is inoculated with 5 × 10$^5$ cells/ml from an actively growing preculture of *S. pombe* 972h$^-$ and agitated aerobically (500 rpm) at 30° in a Virtis fermentor, model 43-100. The cells are harvested by centrifugation when the cellular density is 1.0 to 3.0 × 10$^8$ cells/ml. It is important to harvest during exponential growth in order to obtain high plasma membrane ATPase activity and to reduce mitochondrial contamination. Under these conditions, about 250 to 350 g wet weight of cells is obtained. The cells are washed three times in cold distilled water.

For $^{35}$S labeling, the cells are further incubated for 8 hr to a final density of 10$^8$ cells/ml in a 15-liter fermentor containing 40 mC$_i$ of $^{35}$SO$_4^{2-}$, 5.8% (w/v) glucose, 0.1% (w/v) KH$_2$PO$_4$, 0.05% (w/v) MgCl$_2 \cdot$6H$_2$O), 0.01% (w/v) NaCl, 0.14% (w/v) NH$_4$Cl, 0.15% (w/v) asparagine, 0.01% (w/v) CaCl$_2 \cdot$2H$_2$O, and 0.1% (v/v) of a solution of oligoelements containing 0.05% (w/v) H$_3$BO$_3$, 0.0025% (w/v) CuCl$_2$, 0.010% (w/v) KI, 0.02% (w/v) FeCl$_3 \cdot$6H$_2$O, 0.047% (w/v) MnCl$_2$, 0.040% (w/v) ZnCl$_2$, and 0.1% (w/v) (NH$_4$)$_6$Mo$_7$O$_{24} \cdot$7H$_2$O. After sterilization of the whole medium, 15 ml 0.001% (v/v) biotin in 50% (v/v) ethanol and 15 ml of a sterilized solution containing 0.01% (w/v) calcium pantothenate, 0.01% (w/v) nicotinamide, and 0.1% (w/v) *meso*-inositol are added.

## Cell Homogenization

### Principle

The yeast cells are mechanically disrupted either in a Braun MSK homogenizer (small scale) or in a continuous glass bead grinder (Dyno Mill, model KDL) (large scale).

### Reagents and Equipment

Breaking medium: 1 m$M$ MgCl$_2$ and 50 m$M$ Tris-acetate, pH 7.5, 0.25 $M$ sucrose

Phenylmethylsulfonyl fluoride (PMSF), 100 m$M$ stock solution in dimethylsulfoxide. Before use, 1 ml of PMSF stock solution is added slowly per 100 ml of the breaking medium under magnetic stirring

Glass beads, 0.45- to 0.55-mm diameter

Braun MSK homogenizer (B. Braun, Melsungen, FDR)

Dyno Mill model KDL (W. A. Bachofen, 4005 Basel, Utengasse 15, Switzerland)

*Procedure Using Dyno Mill Glass Beads Grinder (Large Scale).* The cell suspension (40 g wet weight in 60 g of cold breaking medium) is passed twice through the 600-ml disrupting chamber containing 520 ml of glass beads at a flow rate of 235 ml/min with an agitator shaft speed of 3000 rpm. The disrupting chamber is refrigerated at −20°. The eluant is collected in a flask kept in crushed ice through a glass cooling coil (approx. 1 m length, approx. 8 mm internal diameter) maintained in ice and CaCl$_2$ to have the temperature of the eluting suspension close to 4°. After the second passage, the residual homogenate is recovered with 400 ml of cold breaking medium pumped through the disrupting chamber under stirring.

*Procedure Using Braun MSK Homogenizer (Small Scale).* Ten grams wet weight of cells is suspended in 25 ml of cold breaking medium and poured into the 75-ml-capacity homogenizer glass flask containing 15 g glass beads. This suspension is homogenized three times for 1 min at full speed (2800 oscillations per minute) under cooling by expansion of liquid carbon dioxide. The homogenate is decanted and the beads rinsed with 10 ml breaking medium.

It has been verified that for both methods more than 80% of the cells are broken.

## Isolation of Plasma Membranes

*Principle*

For successful solubilization of the plasma membrane $H^+$-ATPase activity, it is important to start from plasma membrane fractions of high specific ATPase activity: from 10 to 20 $\mu$mol phosphate $min^{-1}$ $mg^{-1}$ for azide-insensitive activity and less than 0.1 $\mu$mol phosphate $min^{-1}$ $mg^{-1}$ for azide-sensitive activity, both at pH 6.0. Such plasma membranes are obtained by selective pH precipitation of mitochondria in a crude membrane fraction followed by a stripping treatment with Triton X-100.

*Reagents and Equipment*

Suspension medium: 1 m$M$ ATP sodium salt (Grade II, Sigma), 1 m$M$ EDTA sodium salt, and 10 m$M$ Tris-acetate, pH 7.5
Precipitation medium: the same as the suspension medium, pH 5.2
Triton X-100, 1% 2/v (Koch-Light Ltd, Haverhill, Suffolk, England), stock solution
1 $N$ NaOH
1 $N$ $CH_3COOH$
Glass homogenizers ($\sim$10 ml, $\sim$20 ml, and $\sim$50 ml chamber volume) with tight-fitting Teflon pestle

*Procedure*

*Large-Scale Purification.* The purification procedure is carried out at 4°. All $g$ values are calculated at the middle height of the volumes centrifuged. The cell walls and the remaining unbroken cells are eliminated by two centrifugations at 1000 $g$ and one centrifugation at 3000 $g$, each for 5 min in a Sorvall GS-3 rotor. The supernatants, collected with minimal disturbance by using a Pasteur pipet connected to a vacuum vial kept in crushed ice, are subsequently centrifuged at 15,000 $g$ for 40 min in a Sorvall GSA rotor. After suction of the supernatant, the crude membranes pellet is suspended in the suspension medium (20–40 mg of protein/ml) using a glass homogenizer with a Teflon pestle at 1300 rpm. After homogenization, the crude membrane suspension can eventually be kept frozen at $-20°$.

The reproducibility of the acid precipitation depends on tight control of all parameters (time, rotor, centrifuge, speed, volumes). The crude membranes are brought to 5 mg of protein/ml in the precipitation medium with magnetic stirring at 4°. After a final adjustment to pH 5.2 with acetic acid, the suspension is quickly centrifuged in a Sorvall GSA rotor using centrifuge bottles containing between 140 and 150 ml each. For this centrifugation, the tachometer is preset at 6500 rpm and, 2.5 min after the start, the centrifugation is stopped with the brake on. The centrifugation includes 1.5 min of acceleration to reach 6500 rpm, 1 min at 6500 rpm, and 4 min of deceleration (Sorvall RC-5B centrifuge). The combined supernatants are rapidly neutralized to pH 7.5 with NaOH with magnetic stirring. Altogether, the acid precipitation step should not exceed 20 min. The supernatant is divided into 220-ml aliquots in Beckman R19 centrifuge bottles (if required the volume can be adjusted with the suspension medium) and centrifuged at 16,000 rpm (26,000 $g$) in the Beckman R19 rotor for 36 min.

After suspension of each pellet with 4 ml of suspension medium, each bottle is rinsed with another 4 ml of suspension medium. The suspensions are pooled and homogenized at 1300 rpm in a glass homogenizer with the Teflon pestle. This yields a suspension of 5 to 10 mg of protein/ml. At this stage, the plasma membrane suspension can be kept frozen at −20°.

The plasma membrane suspension is further purified by stripping the membrane with Triton X-100 as follows. The plasma membranes are mixed at a final concentration of 0.84 mg of protein/ml with 0.39 mg of Triton X-100/ml in the suspension medium at 4° and quickly transferred to Beckman R30 centrifuge tubes (20 to 25 ml/tube).

After centrifugation at 28,000 rpm (74,000 $g$) for 1.5 hr, each pellet is suspended in 2 ml of suspension medium and homogenized manually using the glass homogenizer with the Teflon pestle. The purified plasma membranes are stored at −20°. The Triton X-100 treatment yields a 30 to 50% increase in the total ATPase units recovered (pellet + supernatant).

*Small-Scale Purification.* The preparation of the crude membranes follows the same scheme as that for large-scale procedure but uses the Sorvall SS34 rotor. The crude membranes are suspended to 3 mg protein/ml of the precipitation medium. After acid treatment at pH 5.2, the protein aggregates are removed by centrifugation at 7500 rpm (7500 $g$) for 45 sec in a Sorvall HB4 rotor (15 ml/tube). The centrifugation, from the start to the complete stop of the rotor, takes 5 min. The 7500 $g$ supernatant is brought to pH 7.5 with NaOH and centrifuged at 22,000 rpm (45,000 $g$) for 12.5 min in the Beckman R42 rotor.

Stripping conditions with Triton X-100 are identical in both methods. After stripping, the membranes are recovered by centrifugation for 1 hr at 35,000 rpm (92,000 $g$) in a Sorvall AG41 rotor. The Triton X-100 treatment yields a 30 to 50% increase in the total ATPase units recovered (pellet + supernatant).

Solubilization

*Principle*

The main problem encountered is to extract the plasma membrane ATPase with minimal loss of the activity. This is achieved by addition of a natural detergent, egg lysolecithin. The conditions for solubilization have to be carefully controlled, especially the lysolecithin to protein ratio.[2]

Other factors to be considered are lysolecithin concentration, pH, ionic strength, time of contact between the detergent and the membranes, presence of stabilizing agents, and temperature. Lysolecithin concentration should ideally be kept below 3 mg/ml. Above this limit, the recovery of ATPase units (sum of the units in the pellet and supernatant) decreases. However, such conditions are not convenient because volumes that are too large must be handled and concentration is required for the subsequent sucrose gradient centrifugation. This leads to loss of up to 50% of the ATPase units. As a compromise, a lysolecithin concentration of 10 mg/ml is routinely used at a ratio of 4 mg of lysolecithin per milligram of protein. Under such conditions, about 50% of the proteins and total ATPase units is solubilized routinely.

Below or above pH 7.5, the solubilized ATPase is irreversibly denatured. Ionic strength (up to 100 mM Tris) does not have any effect. The presence of 1 mM EDTA and/or 1 mM ATP increases significantly the solubilization of the ATPase activity and proteins and doubles the specific ATPase activity. Phenylmethylsulfonyl fluoride (1 mM), a protease inhibitor, does not enhance the solubilization although there is no inhibition of the enzyme. Temperature also markedly affects the solubilization. The optimal solubilization occurs between 10 and 15°, both for total ATPase units and specific ATPase activity.

Lysolecithin also appears to be an efficient detergent for solubilization of *N. crassa*,[6] *S. cerevisiae*[4] and *Dictyostelium discoideum*[7] plasma membranes. For each system, optimizing the yield of solubilized proteins and/ or enzyme activities requires a systematic study of all the aforementioned parameters.

*Reagents*

Suspension medium: 1 mM ATP sodium salt (Grade II, Sigma), 1 mM EDTA sodium slat, and 10 mM Tris-acetate, pH 7.5.
Lysolecithin: egg lysolecithin is purchased either from Sigma (Type I) or from Avanti Polar Lipids (2421 Highbluff Rd, Birmingham,

[6] R. Addison and G. A. Scarborough, *J. Biol. Chem.* **256**, 13165 (1981).
[7] R. Pogge-von Strandmann, R. R. Kay, and J. P. Dufour, *FEBS Lett.* **175**, 1811 (1984).

AL 35216). It should be stored under dry conditions at $-20°$. Lysolecithin stock solution (20 mg/ml) is freshly prepared in the suspension medium at room temperature with magnetic stirring. The solution should be clear.

*Procedure.* Sixty milligrams of purified plasma membranes proteins is suspended in 24 ml of the suspension medium containing 240 mg of lysolecithin. After 10 min incubation at $15°$ with manual stirring every 30 sec for 5 sec, the suspension is centrifuged for 1 hr at 36,000 rpm (100,000 g) in a Beckman R42 rotor at $15°$. The supernatant, containing around 50% of the proteins and ATPase units, is stored on ice and layered on the sucrose gradient as quickly as possible. The extract may appear translucent after storage on ice.

## Purification

### Principle

Centrifugation on a continuous sucrose density gradient has so far been the only successful purification step for the *S. pombe* plasma membrane ATPase. After purification through the sucrose gradient, the ATPase activity (measured by the colorimetric assay) is stimulated up to 20 times by merely mixing the enzyme with phospholipid micelles or vesicles of sufficient fluidity and hydrophobicity.[8,9] It appears that the enzyme is purified in an active oligomeric form and that irreversible inactivation occurs during the assay when tested in the absence of added lipids.

### Reagents and Equipment

Polyproplyene cryotubes (38 × 12.5 mm) (A/S NUNC, Kamstrup, 4000 Roskilde, Denmark)

Sucrose, 6% (w/w) solution containing 6 g sucrose, 9.8 ml of 10 m$M$ ATP sodium salt, 10 m$M$ EDTA sodium salt, and 100 m$M$ Trisacetate, pH 7.5, adjusted to 100 g with doubly distilled water

Sucrose, 30% (w/w) solution containing 30 g sucrose, 8.87 ml of 10 m$M$ ATP sodium salt, 10 m$M$ EDTA sodium salt, and 100 m$M$ Tris-acetate, pH 7.5, adjusted to 100 g with doubly distilled water

These solutions are stored at $-20°$. The gradients are made up from 17.5 ml and 15.8 ml of the 6 and the 30% solutions, respectively, using a gradient former.

---

[8] J. P. Dufour and A. Goffeau, *J. Biol. Chem.* **255,** 10591 (1980).
[9] J. P. Dufour and T. Y. Tsong, *J. Biol. Chem.* **256,** 1801 (1981).

*Procedure.* The lysolecithin extract is divided into aliquots of a maximum of 3.6 ml, each layered on a 33.3-ml linear sucrose gradient containing 6 to 30% sucrose (w/w). After centrifugation for 16.5 hr at 24,000 rpm (76,000 $g$) at 4° in a Beckman SW 27 rotor, the sucrose gradient is divided into 28 fractions (approximately 1.3 ml each). The white opaque zone usually observed in the bottom of the gradient corresponds to aggregates of lysolecithin (fractions 23 to 25 in Fig. 1).

The three to four fractions of the highest ATPase activity (fractions 13 to 16 in Fig. 1) are pooled, divided into 0.5-ml aliquots in the polypropylene cryotubes (NUNC), and stored at −80° after freezing in liquid nitrogen. The specific activities and the yields obtained during a typical purification are presented in Table I.

## Comments on Yields and Purity of the Purified Plasma Membrane ATPase

A 15-liter yeast culture harvested at $1.2 \times 10^8$ cells/ml yields 250 g wet weight of yeast cells. Homogenization followed by differential centrifugations provides approximately 4.5 g of crude plasma membranes. After acid precipitation and Triton X-100 treatment, 60 to 120 mg of purified plasma membranes is obtained which, after solubilization with lysolecithin, requires 6 to 12 sucrose gradients for further purification. Pooling of the three to four fractions of the highest ATPase activity from six sucrose gradients yields about 7.5 mg of purified enzyme. As judged by silver nitrate and Coomassie Blue staining after SDS-polyacrylamide gel elec-

TABLE I

ACTIVITIES AND YIELDS DURING PURIFICATION OF THE PLASMA MEMBRANE ATPase FROM *Schizosaccharomyces pombe*

| | Specific activity[a] ($\mu$mol phosphate/min · mg) | | | Yield (%) | |
| --- | --- | --- | --- | --- | --- |
| Component | pH 6.0 azide-insensitive | pH 9.0 azide-sensitive | pH 6.0/pH 9.0 | Protein | Activity (pH 6.0) |
| Crude membranes | 1.1 | 1.3 | 0.8 | 100 | 100 |
| Plasma membranes | 7.0 | 0.5 | 14.0 | 5.0 | 29.2 |
| Purified plasma membranes | 10.3 | 0.2 | 51.5 | 3.5 | 30.3 |
| Purified solubilized enzyme | 45 | nd[b] | — | 0.24 | 9.1 |

[a] Specific activities are determined using the colorimetric ATPase assay at 30° and the Folin protein assay (bovine serum albumin as standard).
[b] nd, Not detectable.

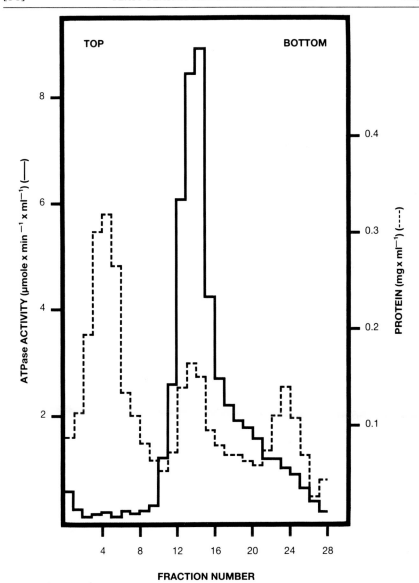

FRACTION NUMBER

FIG. 1. Distribution of ATPase and protein after centrifugation of *Schizosaccharomyces pombe* plasma membrane lysolecithin extract on a sucrose density gradient. An aliquot of 3.6 ml of *S. pombe* plasma membrane lysolecithin extract was layered on a linear sucrose density gradient containing 6 to 30% sucrose (w/w) in 33.3 ml of solubilization buffer and centrifuged for 16.5 hr at 24,000 rpm (76,000 $g$) at 4° in a Beckman SW 27 rotor. The ATPase assays were carried out at 30° in the presence of lysolecithin using the colorimetric method. Protein concentrations were determined by the Folin test using bovine serum albumin as a standard.

trophoresis, the purity of the final ATPase is higher than 80%. Starting from the yeast culture, the whole purification procedure can be carried out within 1 week.

## Concentration

### Principle

Highly concentrated purified ATPase solution, up to 5 mg of protein/ml is needed in many instances (e.g., for incorporation into lipid vesicles using the cholate dialysis method), for ligand-binding studies or for crystallization assays.

*Procedure.* When collected from the sucrose gradient, the concentration of proteins in the purified ATPase is 150 to 200 μg/ml. This dilute solution is best concentrated by ultrafiltration using an Amicon 8 MC concentration cell equipped with an Amicon XM50 membrane under nitrogen pressure at 4°. The use of Amicon CF cones is not recommended because of slow flow rate and formation of a cake in the bottom.

The purified enzyme can also be quick-frozen in liquid nitrogen and lyophilized. The 0.5-ml aliquots of the ATPase peak from the sucrose gradient can be directly lyophilized in the polypropylene cryotubes. No more than 10% of the ATPase units is lost. However, use of larger volumes tends to decrease the recovery of ATPase units. In lyophilized form, the enzyme can be stored at −80° for considerable time (up to 1 year) without loss of activity. As a powder, the enzyme can withstand room temperaturs for a limited period (2–3 days, sufficient for transport by air mail). To recover ATPase activity, a minimum of 100 μl of water is added per tube. This corresponds approximately to a five times concentration factor.

### Change of Buffer

Change of buffer can be carried out either before or after concentration by continuous ultrafiltration using the Amicon 8 MC concentration cell (XM50 membrane) at 4° or by fast gel filtration.[10] In this case, a disposable 1-ml plastic tuberculin syringe plugged with glass wool or a porous polyethylene disk (1.6-mm thick, 35-μm pore size, 4-mm diameter, Bel-Art Products, 6 Industrial Rd, Pequannock, NJ 07440) is filled up to 1 ml with Sephadex G-50 (medium or coarse) swelled in the desired buffer. The column is allowed to stand until no further liquid drained from

---

[10] H. S. Penefsky, *J. Biol. Chem.* **252,** 2891 (1977).

it (about 5 min) and put on the edge of a glass test tube which is centrifuged at 1000 $g$ and 4° for 1.5 min in a clinical centrifuge equipped with a swinging bucket rotor. In a second step, 60 $\mu$l of the enzyme solution is layered on the gel and centrifuged in the same conditions, using a clean test tube to collect the effluent.

From 80 up to 95% of the ATPase and protein units is recovered in the bottom of the glass tube in an equivalent or slightly higher volume of the new buffer.

Active ATPase can be obtained by centrifugation through sucrose gradient using the procedure described above, with the exception that no ATP and EDTA are added to the sucrose solutions. However, when frozen under these conditions the ATPase activity is markedly reduced.

## Incorporation into Lipid Vesicles

### Principle

Reconstitution of the H⁺-pumping properties of the purified ATPase is obtained by incorporation into phospholipid vesicles. In such proteoliposomes, the plasma membrane ATPase translocates protons from the outside to the inside, generating a membrane potential $\Delta\psi$ (positive inside) as well as a $\Delta$pH (acid inside).[11-13] The relative values of $\Delta\psi$ and $\Delta$pH depend on the experimental conditions, especially on the leakiness of the proteoliposomes to protons as well as to other cations or anions present.

Reconstitution can be carried out by at least two techniques: the freeze–thaw or the cholate dialysis methods.[14,15] Both procedures have specific advantages and drawbacks. The main advantages of the freeze–thaw method are the use of amounts of protein varying from micrograms to several milligrams, its simplicity and rapidity, and the possibility of storing the reconstituted proteoliposomes at −80° for as long as 6 months. However, the average coupling efficiency measured by stimulation of the incorporated ATPase activity by protonophore is lower for the freeze–thaw method (50%) than for the cholate dialysis method (120%). It must also be stressed that the success of the reconstitution is strongly dependent on the lipids and buffers. For example, the presence of chloride and/or nitrate has to be avoided.

---

[11] A. Villalobo, M. Boutry, and A. Goffeau, J. Biol. Chem. 256, 12081 (1981).

[12] J. P. Dufour and T. Y. Tsong, Biophys. J. 37, 96 (1982).

[13] J. P. Dufour, A. Goffeau, and T. Y. Tsong, J. Biol. Chem. 257, 9365 (1982).

[14] M. Kasahara and P. C. Hinkle, J. Biol. Chem. 252, 7384 (1977).

[15] Y. Kagawa and E. Racker, J. Biol. Chem. 246, 5477 (1971).

*Reagents and Equipment for the Freeze–Thaw Method*

Asolectin (95% purified soy phosphatides) (Associated Concentrates, Woodside, Long Island, NY 11377) or soybean lecithins (Type II S, Sigma)

Reconstitution medium: 25 m$M$ K$_2$SO$_4$ and 10 m$M$ MES–KOH, pH 6.0

Dilution medium: 50 m$M$ K$_2$SO$_4$ and 20 m$M$ MES–KOH, pH 6.0

Polypropylene cryotubes (38 × 12.5 mm) (NUNC)

Bath sonicator, special cylindrical Ultrasonic Tank, and Generator GS122 SP 1 (Laboratory Supplies Co., 20 Jefry Lane, Hicksville, NY 11801)

*Reagents and Equipment for the Cholate Dialysis Method*

Soybean lecithins (Type II S, Sigma)

Reconstitution medium stock solution: 100 m$M$ K$_2$SO$_4$, 20 m$M$ MgSO$_4$, and 10 m$M$ HEPES-KOH, pH 6.1

Cholate stock solution, 50 mg cholic acid/ml adjusted to pH 7.5 with KOH

Dialysis tubing: Visking, size 1-8/32″ (Medicell International Ltd, 239 Liverpool Road, London N1 1 LX, England)

Dialysis medium: 50 m$M$ K$_2$SO$_4$, 10 m$M$ MgSO$_4$, and 5 m$M$ HEPES-KOH, pH 6.1

*Procedure for the Freeze–Thaw Method.* One hundred milligrams of phospholipid is weighed in a 30-ml Corex tube. Two milliliters of the reconstitution medium is added and the lipids are allowed to swell for 10 min at room temperature. The tube is placed in the bath sonicator as described by Hokin and Dixon.[16] After sonication for 5 to 20 min at room temperature, the translucent suspension is put on ice and used within a few hours. Three milligrams of such asolectin vesicles are mixed with the purified ATPase at an asolectin/protein ratio higher than or equal to 0.7 mg/μg in a final volume of 150 μl. A minimum of 1 m$M$ sucrose (usually brought with the enzyme sample) is required to prevent fusion and/or aggregation of the vesicles.

For example, to reconstitute 500 μl of purified plasma membrane ATPase (160 μg/ml) one will mix 1.2 ml of asolectin vesicles with 800 μl of the reconstitution medium and 500 μl of the dilution medium. The mixture is divided into 125-μl aliquots in polypropylene cryotubes kept on ice. An aliquot of 25 μl of the purified enzyme is then added to each tube, mixed, and immediately frozen in liquid nitrogen. In this state the

---

[16] L. E. Hokin and J. F. Dixon, *in* "Na,K-ATPase, Structure and Kinetics" (J. C. Skou and J. G. Nørby, eds.), p. 48. Academic Press, New York, 1979.

proteoliposomes can be kept at $-80°$ for several months. Before use, the proteoliposomes are thawed for 1.5 min at 30° and used in 130-$\mu$l aliquots for subsequent assays, e.g., proton movements using fluorescent $\Delta$pH probes.[13]

*Procedure for the Cholate Dialysis Method.* One hundred milligrams of soybean lipids is weighed in a glass conical bottom tube followed by 2.5 ml of the reconstitution medium, 1 ml of cholate stock solution, and 0.5 mg of purified ATPase protein (in a maximum of 1 ml of the sucrose gradient medium). After adjustment to 5 ml with doubly distilled water, the suspension maintained in ice-water is sonicated by using a Virsonic Cell Disrupter model 16-850 (The Virtis Company, Gardiner, NY) sonicator with a standard titanium microtip (5/32-in. diameter) at 35% power level for 10 min with frequent on/off periods until clarification is obtained. After sonication, the mixture is dialyzed against 500 ml of dialysis medium with four buffer changes in a 500-ml graduated cylinder with magnetic stirring at 4° using a Crowe–Englander microdialyzer (A.H. Thomas Co., Vine St. at Third, P.O. Box 779, Philadelphia, PA 19105).

## General Properties of the Plasma Membrane ATPase from *Schizosaccharomyces pombe*

The purified ATPase is isolated in an oligomeric form.[17] Addition of phospholipids activates the enzyme and disaggregates spontaneously the ATPase oligomer into smaller molecular forms which associate with lipid vesicles.[8,9] The kinetic properties of the enzyme are very similar in the membrane-bound and solubilized forms.[17] The enzyme requires $Mg^{2+}$ and shows high substrate specificity for ATP. Other ester phosphates such as ITP, GTP, CTP, *p*-nitrophenyl phosphate or adenylylimidophosphate are not hydrolyzed. The optimal pH for activity is 5.7 to 6.0 and the enzyme exhibits a strict Michaelis–Menten kinetic with a $K_m$ for MgATP of 0.3–3.8 m$M$, depending on the experimental conditions. The ATPase activity can be inhibited by vanadate, dicyclohexylcarbodiimide, Dio-9, miconazole, diethylstilbestrol, *p*-hydroxymercuribenzoate, and octylamino esters.[18] Upon SDS-polyacrylamide gel electrophoresis, the preparations display a single polypeptide band of $M_r$ 100,000–105,000.[2] This subunit forms a phosphorylated catalytic intermediate during ATP cleavage, the phosphorylated amino acid being an aspartate residue.[19,20] [18]O-labeling

---

[17] J. P. Dufour and A. Goffeau, *Eur. J. Biochem.* **105**, 145 (1980).
[18] J. P. Dufour, M. Boutry, and A. Goffeau, *J. Biol. Chem.* **255**, 5735 (1980).
[19] A. Amory, F. Foury, and A. Goffeau, *J. Biol. Chem.* **255**, 9353 (1980).
[20] A. Amory and A. Goffeau, *J. Biol. Chem.* **257**, 4723 (1982).

methods and phosphorylated intermediate measurement suggest that a similar mechanism seems to apply for the $H^+$-translocating ATPase and other well-characterized cations translocating ATPases ($Na^+,K^+$- or $Ca^{2+}$-ATPases).[21]

[21] A. Amory, A. Goffeau, D. B. McIntosh, and P. D. Boyer, *J. Biol. Chem.* **257**, 12509 (1982).

# [39] Plasma Membrane ATPase from the Yeast *Saccharomyces cerevisiae*

*By* ANDRÉ GOFFEAU and JEAN-PIERRE DUFOUR

When applied exactly to *Saccharomyces cerevisiae*, the procedure reported for the purification of the plasma membrane ATPase from *Schizosaccharomyces pombe*[1] yields enzyme of low specific activity: less than 1 $\mu$mol min$^{-1}$ mg$^{-1}$. Several modifications have to be brought to the *S. pombe* procedure to obtain purified ATPase of specific activity above 15 $\mu$mol min$^{-1}$ mg$^{-1}$ from *S. cerevisiae*. These modifications are detailed in this article. All other conditions are similar to those used for *S. pombe*.[1]

## ATPase Assay

### Principle

The ATPase assay described for *S. pombe*[1] is applicable to the *S. cerevisiae* activity with the following modifications. For membrane-bound preparations, the addition of 40 $\mu M$ Triton X-100 stimulates ATPase activity by 30 to 70%. The solubilized purified ATPase is inhibited during the assay by lysolecithin concentrations higher than 10 $\mu$g/ml. However, 125 to 500 $\mu$g/ml asolectin, combined with 25 to 50 $\mu$g/ml lysolecithin in the reaction mixture, stimulates ATPase activity up to 110%. In addition, the presence of 50 $\mu$g/ml bovine serum albumin stimulates the ATPase activity by 50%.

Due to the low *S. cerevisiae* ATPase activity, we found it useful to carry out the assay at 35° where ATPase activity is 36% higher than at 30° for a 32-min assay. After purification through a sucrose gradient, we use

[1] J. P. Dufour, A. Amory, and A. Goffeau, this volume [38].

METHODS IN ENZYMOLOGY, VOL. 157

the colorimetric microassays in order to reduce the amount of enzyme required per assay.

*Reagents*

ATPase assay stock medium as for *S. pombe*[1]

NaN$_3$ 1 *M* as for *S. pombe*[1]

Bovine albumin, 5 mg/ml (Sigma, fatty acid free)

Lysolecithin, 1 mg/ml (Sigma, Type I)

Sonicated asolectin (95% purified soy phosphatides) (Associated Concentrates, Woodside, Long Island, NY), 5 mg/ml

Sodium dodecyl sulfate, ammonium molybdate, and Elon reagents as for *S. pombe*[1]

Triton X-100 10 m*M* (Koch–Light Laboratories Ltd., Colnbrook Bucks, England, puriss.)

*Procedure for Assay of Membrane-Bound ATPase*

To 850 μl of ATPase assay stock medium, 10 μl of 1 *M* NaN$_3$, 10 μl of 10 m*M* Triton X-100, and 125 μl of water are added. The reaction is started with 5 μl of membrane fractions and incubated at 35° for 8 to 32 min according to activity. All membrane fractions are centrifuged for 10 min at 132,000 *g* in a Beckman airfuge just prior to the assay; the pellets are suspended in the original volume of 10 m*M* imidazole-HCl, pH 7.5. The assay is stopped as described for *S. pombe*.[1]

*Procedure for Microassay of the Purified ATPase*

To 85 μl of ATPase assay stock medium, 1 μl of 1 *M* NaN$_3$, 1 μl of 5 mg/ml bovine albumin, 2.5 μl of 1 mg/ml lysolecithin, 5 μl of 5 mg/ml asolectin, and 3 μl of water are added. The reaction is started with 5 μl of purified ATPase from the sucrose gradient and incubated at 35° for 8 to 32 min according to activity. After addition of 0.3 ml of 1% (w/v) sodium dodecyl sulfate, 0.25 ml of ammonium molybdate reagent, and 0.25 ml of Elon reagent, and then 15 min incubation at room temperature, the absorbance is measured at 800 nm.

Culture

Strain: *Saccharomyces cerevisiae* Σ 1278b

Culture medium: glucose 5.8% (w/v) and yeast extract 2% (w/v) from KAT (Ohly, Hamburg). The pH is not adjusted (pH 5.2)

Growth: A 15-liter fermentor is inoculated with 2.5 × 10$^5$ cells/ml from an active preculture and agitated for 16 hr at 30° with aeration. The final density is 3 to 5 × $10^8$ cells/ml. About 350 to 450 g wet weight of cells is harvested by centrifugation and washed twice with cold 100 m$M$ glucose and 50 m$M$ imidazole-HCl, pH 7.5. The final temperature of the yeast suspension during washing is approximately 10°.

### Isolation of Plasma Membranes

*Principle*

After preparation of crude membranes, the mitochondrial and ribosomal contaminants are reduced by acid precipitation at pH 5.2 in the presence of 1 m$M$ MgCl$_2$ and by treatment with 0.039% Triton X-100.

*Reagents*

Breaking medium: 1 m$M$ MgCl$_2$, 0.25 $M$ sucrose, 0.1 $M$ glucose, 50 m$M$ imidazole (UCB, 1395)-HCl pH 7.5, and 100 $\mu M$ sodium orthovanadate (BDH, 30194)

Suspension medium: 1 m$M$ MgCl$_2$, 10 m$M$ imidazole-HCl pH 7.5, and 100 $\mu M$ sodium orthovanadate

Precipitation medium: the same as the suspension medium, pH 5.2

Phenylmethylsulfonyl fluoride (Sigma, P-7626), 100 m$M$ in dimethyl sulfoxide

DL-Dithiothreitol (Sigma, D-0632) solid

*Procedure.* Just before use, 10 ml of 100 m$M$ phenylmethylsulfonyl fluoride and 770 mg solid dithiothreitol are slowly added to 1 liter of breaking medium with magnetic stirring at 4°. Breakage of the cells, subsequent centrifugations, acid precipitation at pH 5.2, and the Triton treatment (Table I) are carried out exactly as described for *S. pombe*[1] except that after cell homogenization and the first three low-speed centrifugations, the crude membrane pellet is centrifuged for 1 hr at 36,000 rpm in a Sorvall AG 41 rotor (100,000 $g$). Freezing and thawing of membrane preparations should be avoided.

### Solubilization and Purification

*Principle*

After solubilization with lysolecithin, the *S. cerevisiae* ATPase is obtained in a form which sediments during centrifugation in the sucrose gradient at a lower rate than the *S. pombe* enzyme. The centrifugation has thus to be adapted for optimal separation of the *S. cerevisiae* ATPase from the bulk protein peak.

TABLE I

ACTIVITIES AND YIELDS DURING PURIFICATION OF THE PLASMA MEMBRANE ATPase
FROM *Saccharomyces cerevisiae*

| | Specific activity[a] ($\mu$mol phosphate/min mg) | | | Yield (%) | |
| | pH 6.0 azide-insensitive | pH 9.0 azide-sensitive | pH 6.0/pH 9.0 | Protein | Activity (pH 6.0) |
| Component | | | | | |
|---|---|---|---|---|---|
| Crude membranes | 0.3 | 0.9 | 0.2 | 100 | 100 |
| Plasma membranes | 6.7 | <0.1 | >67 | 5.1 | 127 |
| Plasma membranes treated with Triton | 8.1 | <0.1 | >81 | 3 | 94 |
| Purified soluble enzyme | 16 | nd[b] | — | 0.18 | 9.6 |

[a] Specific activities are determined with the colorimetric ATPase assay at 35° and the Folin protein assay.
[b] nd, Not detectable.

### Reagents

Suspension medium as for *S. pombe*[1]
Sucrose gradient solutions as for *S. pombe*[1]
Lysolecithin as for *S. pombe*[1]

*Procedure.* After the second Triton treatment, the purified plasma membranes are used in aliquots of 60 mg protein at the concentration of 10 mg/ml in the solubilization medium.

The solubilization by lysolecithin and purification through sucrose gradient centrifugation are carried out exactly as for *S. pombe*[1] except that the centrifugation time must be extended to 20.5 hr in order to obtain the separation shown in Fig. 1. The four fractions of highest specific activity correspond to the ATPase peak itself (fractions 18 and 19) and the heavy (left) side of the peak (fractions 20 and 21) in Fig. 1. These fractions are mixed and 0.5-ml aliquots are transferred in polypropylene cryotubes (NUNC), frozen in liquid nitrogen, and stored at −80°.

### Yields

A 15-liter fermentor yields 350 to 450 g wet weight of cells. This provides approximately 3 to 4 g crude membrane proteins. After the acid precipitation step and the two Triton treatments, a total of about 60 mg of purified plasma membranes is recovered which, after solubilization, provides sufficient material to load six sucrose gradients. From each sucrose

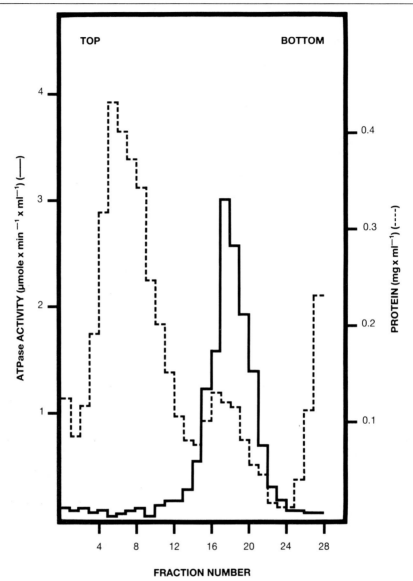

FIG. 1. Distribution of ATPase and protein after centrifugation of *Saccharomyces cere-visiae* plasma membrane lysolecithin extract on a sucrose density gradient. An aliquot of 3.6 ml of *S. cerevisiae* plasma membrane lysolecithin extract was layered on a linear sucrose density gradient containing 6 to 30% sucrose (w/w) in 33.3 ml of solubilization buffer and centrifuged for 20.5 hr at 24,000 rpm (76,000 $g$) at 4° in a Beckman SW 27 rotor. The ATPase assays were carried out at 35° in the presence of the mixture asolectin–lysolecithin–bovine serum albumin using the colorimetric method. Protein concentrations were determined by the Folin test using bovine serum albumin as a standard.

gradient, the four fractions of highest specific ATPase activity yield 600 to 800 μg protein. A total of about 4 mg of purified enzyme can thus be obtained from one fermentor. The whole procedure from culture to sucrose gradient can be carried out in 1 week.

## Comments on Purity

Even though the procedure described above yields useful ATPase preparation of appreciable specific activity, the *S. cerevisiae* purified ATPase is of lower quality than the *S. pombe* enzyme. Its specific activity is two to three times lower. It has been assessed that more than 80% of the *S. pombe* purified ATPase is composed of the ATPase subunit. Similar observations on sodium dodecyl sulfate-polyacrylamide gel electrophoresis lead to the conclusion that the *S. cerevisiae* purified ATPase contains no more than 60% of intact ATPase subunit. The latter is of slightly lower mobility in *S. cerevisiae* ($M_r$ 107,000) than in *S. pombe* ($M_r$ 104,000). In sodium dodecyl sulfate-polyacrylamide gel electrophoresis from *S. cerevisiae* ATPase, a contaminant glycoprotein is located just above the intact ATPase subunit. Just below intact ATPase, a band of variable concentration is observed which is most probably an ATPase proteolytic product. These three bands are easily confused, especially by automated scanning procedures. Several other components of high mobility seem to be hydrolytic components of the ATPase subunit, as indicated by immunoreactions with antibody obtained against the eluted $M_r$ 107,000 ATPase subunit. Prolonged incubation in the cold in the presence of 1% (w/v) sodium dodecyl sulfate leads to total degradation of the ATPase intact subunit into several smaller components of definite size.

## [40] H$^+$-ATPase from Plasma Membranes of Saccharomyces cerevisiae and Avena sativa Roots: Purification and Reconstitution

By RAMÓN SERRANO

The plasma membranes of fungi and plants contain a similar ATPase which represents a novel type of proton pump.[1] Two convenient sources for these enzymes are bakers' yeast (*Saccharomyces cerevisiae*) and oat

[1] R. Serrano, *Curr. Top. Cell. Reg.* **23**, 87 (1984).

(*Avena sativa*) roots. The purified plasma membrane ATPases from these organisms can be reconstituted into proteoliposomes that catalyze ATP-driven proton transport.

## Analytical Methods

### *Measurement of Protein Concentration*

Protein is determined by the dye-binding method of Bradford[2] as modified by Read and Northcote.[3] Bovine serum album serves as standard.

### *Assay of Plasma Membrane ATPase Activity*[4]

#### *Reagents*

Incubation buffer: 50 m$M$ 2-($N$-morpholino)ethanesulfonic acid adjusted to either pH 5.7 (yeast) or 6.5 (oat roots) with Tris, 5 m$M$ MgSO$_4$, 50 m$M$ KNO$_3$, 5 m$M$ sodium azide, and 0.2 m$M$ ammonium molybdate. Store at 4°C.

ATP: sodium ATP (vanadate-free) is dissolved to 0.1 $M$, adjusted to pH 7, and stored frozen.

Phosphate reagent: 0.5% sodium dodecyl sulfate, 0.5% ammonium molybdate, and 2% (v/v) sulfuric acid. Store at room temperature.

Ascorbic acid: dissolve to 10% and store frozen in small aliquots.

Diethylstilbestrol: dissolve in methanol to 0.1 $M$ and store at −20°.

Soybean phospholipids: crude soybean phospholipids (Sigma phosphatidylcholine Type II-S) are dissolved with 3 ml diethyl ether per gram of lipid and the insoluble residue removed by low-speed centrifugation. The supernatant is precipitated with 18 ml acetone per gram of lipid and the pellet dried under vacuum. The purified phospholipids are resuspended to 2% in water and sonicated[5] to clarity. They are stored at 4° for up to 1 week.

*Definition of Unit and Specific Activity.* A unit of activity is defined as the amount of enzyme which catalyzes the hydrolysis of 1 $\mu$mol of ATP per minute under the assay conditions described below. Specific activity is expressed as units per milligram protein.

---

[2] M. M. Bradford, *Anal. Biochem.* **72,** 248 (1976).

[3] S. M. Read and D. H. Northcote, *Anal. Biochem.* **116,** 53 (1981).

[4] R. Serrano, *Mol. Cell. Biochem.* **22,** 51 (1978).

[5] Sonication can be performed in a cylindrical bath sonicator of the type provided by Laboratory Supplies Co., Hicksville, NY (tank model T80-80-1 RS). Alternatively, the cup horn accessory offered by Heat Systems Ultrasonics, Farmingdale, NY, is also convenient.

*Procedure.* The enzyme (from 5 to 50 milliunits in less than 50 $\mu$l) is added to 1 ml of incubation buffer and preincubated for 2–3 min at 30°. The reaction is started with 20 $\mu$l of 0.1 $M$ ATP and stopped after 10 min at 30° with 2 ml of phosphate reagent. The presence of sodium dodecyl sulfate in this acidic reagent prevents the precipitation of protein and lipids. Color is developed with 20 $\mu$l of 10% ascorbic acid and the absorbance at 750 nm read after 5–10 min. A phosphate standard with 0.2 $\mu$mol P$_i$ gives a reading of 0.2–0.4.

The assay can be performed with crude preparations by determining the activity sensitive to 0.2 m$M$ diethylstilbestrol. Azide inhibits the mitochondrial ATPase,[4] molybdate inhibits acid phosphatase,[6] and nitrate inhibits the vacuolar ATPase.[7] Some ATP hydrolysis may be caused by alkaline phosphatases[7] and apyrases[8] which are resistant to these inhibitors but also to the plasma membrane ATPase inhibitor diethylstilbestrol. Vanadate is not as convenient as diethylstilbestrol because it inhibits acid and alkaline nonspecific phosphatases[9] in addition to the plasma membrane ATPase.

The assay of the yeast membrane ATPase should include 0.4 mg of sonicated soybean phospholipids after solubilization by Zwittergent-14 (see below). Exogenous phospholipids are inhibitory to the oat root ATPase solubilized by lysolecithin.

*Assay of $^{32}$P$_i$-ATP Exchange*[10]

*Reagents*

5× buffer: 50 m$M$ MgSO$_4$ and 0.25 $M$ 2-($N$-morpholino)ethanesulfonic acid adjusted to pH 6.0 with Tris

$^{32}$P$_i$: $^{32}$P$_i$ (carrier-free) is boiled in 1 $N$ HCl for 4–6 hr to hydrolyze pyrophosphate and polyphosphates. It is diluted with 0.25 $M$ phosphoric acid to a specific activity of about 1000 counts min$^{-1}$ nmol$^{-1}$ and adjusted to pH 6.0 with Tris

0.5 $N$ perchloric acid

2.5% ammonium molybdate

Isobutanol

Gramicidin D: (Sigma) dissolve to 2 mg/ml with methanol and store at $-20°$

1799 (2,2′bis(hexafluoracetonyl)acetone): it can be obtained from Dr.

---

[6] J. D'Auzac, *Phytochemistry* **14,** 671 (1975).
[7] E. J. Bowman and B. J. Bowman, *J. Bacteriol.* **151,** 1326 (1982).
[8] F. Vara and R. Serrano, *Biochem. J.* **197,** 637 (1981).
[9] R. L. Van Etten, P. P. Waymack, and D. M. Rehkop, *J. Am. Chem. Soc.* **96,** 6782 (1974).
[10] F. Malpartida and R. Serrano, *J. Biol. Chem.* **256,** 4175 (1981).

P. G. Heytler, Du Pont de Nemours, Wilmington, DE. It is dissolved in methanol to 10 m$M$ and stored at $-20°$

*Procedure.* Fifty microliters of reconstituted proteoliposomes (see below) are mixed with 100 $\mu$l of water and 50 $\mu$l of 5× buffer. Two microliters of either 2 mg/ml gramicidin or 10 m$M$ 1799 may be included in control experiments where the proton gradient is dissipated by the ionophores. After 2–3 min of preincubation at 30° the reaction is started with 20 $\mu$l of 0.25 $M$ $^{32}$P[P$_i$] and 25 $\mu$l of 0.1 $M$ ATP. After 10 min at 30° the reaction is stopped with 2 ml of 0.5 $N$ perchloric acid and after addition of 1 ml of 2.5% ammonium molybdate the nonesterified phosphate is extracted three times with 3 ml of isobutanol. Aliquots of 1.5 ml of the aqueous phase are counted by Cerenkov radiation.

*Measurement of Intravesicular Acidification by the Quenching of Acridine Dye Fluorescence*[11–13]

*Reagents*

Incubation buffer: 10 m$M$ 2-($N$-morpholino)ethanesulfonic acid, 5 m$M$ MgSO$_4$, and 50 m$M$ nitric acid, adjusted to pH 6.0 with Tris.
Acridine dye: the dye 9-amino-6-chloro-2-methoxyacridine can be obtained from Dr. R. Kraayenhof, Free University of Amsterdam, The Netherlands. It is dissolved in water to 1 m$M$ and stored frozen.

*Procedure.* Reconstituted proteoliposomes (20–100 $\mu$l, see below) are added to 2 ml of incubation buffer containing 2 $\mu$l of 1 m$M$ acridine dye. Two microliters of either 2 mg/ml gramicidin D or 10 m$M$ 1799 may be included in control experiments where the proton gradient is dissipated by the ionophores. 1799 is more convenient than other commercial uncouplers because it is colorless. Measurements are performed at 30° in a fluorescence spectrophotometer with temperature-regulated circulating water. Exciter and analyzer wavelengths are set at 415 and 485 nm, respectively. The reaction is started with 20 $\mu$l of 0.1 $M$ ATP, which produced an instantaneous artifactual quenching of about 10%. The time-dependent quenching reflecting intravesicular acidification is completed in 2–4 min. The extent of quenching is expressed as a percentage of the initial fluorescence, after subtracting the instantaneous quenching produced by ATP.

[11] F. Malpartida and R. Serrano, *FEBS Lett.* **131,** 351 (1981).
[12] F. Vara and R. Serrano, *J. Biol. Chem.* **257,** 12826 (1982).
[13] R. Serrano, *Biochem. Biophys. Res. Commun.* **121,** 735 (1984).

*Purification of the Yeast Plasma Membrane ATPase*[4,14]

*Materials*

Baker's yeast: either commercial baker's yeast or laboratory strains of *Saccharomyces cerevisiae* grown to either exponential or stationary phases can be employed with similar results. The cells are extensively washed with water before homogenization

0.5 *M* Tris adjusted to pH 8.5 with HCl

0.5 *M* EDTA adjusted to pH 8

0.2 *M* phenylmethylsulfonyl fluoride, dissolve in methanol and store at −20°

TED buffer: 10 m*M* Tris adjusted to pH 7.5 with HCl, 0.2 m*M* EDTA, and 0.2 m*M* dithiothreitol. Store frozen

Zwittergent-14 (Calbiochem): 3-(tetradecyldimethylammonium)-1-propane sulfonate. Dissolve at 5% and store frozen

Phospatidlyserine (Sigma, brain extract type III). Resuspend at 2% in water and sonicate[5] to clarity.

*Procedure.* Unless otherwise indicated, all the operations are carried out at 0–4°. Relative centrifugal forces refer to the average radius of the tubes. The TED buffer is present in all the sucrose and glycerol solutions. Pellets are resuspended manually with a glass homogenizer.

*Preparation of Homogenates.* Portions of 10–30 g of cells (fresh weight) are diluted with water to 84 ml and transferred to 200-ml container of a Vibrogen Cell Mill (E. Bühler, Tübingen, FDR). After addition of 5 ml 0.5 *M* Tris, 1 ml of 0.5 *M* EDTA, 0.25 ml of 0.2 *M* phenylmethylsulfonyl fluoride, and 170 ml of glass beads (0.5 mm), the shaker is operated for 2.5 min with circulating cold water as coolant. The beads are removed by filtration under vacuum through a glass-sintered filter and washed with 50 ml 20% glycerol in TED buffer.

*Differential Centrifugation.* The homogenate is first centrifuged for 10 min at 700 *g* to remove debris and the supernatant further centrifuged for 20 min at 20,000 *g*. This second pellet, enriched in plasma membranes, is resuspended with about 50 ml of 20% glycerol and 0.1 ml of 0.2 *M* phenylmethysulfonyl fluoride per every 100 g of starting yeast.

*Sucrose Gradient Centrifugation.* Fourteen milliliters of the 20,000 *g* pellet is applied to a discontinuous gradient made of 8 ml 53% (w/w) sucrose and 16 ml 43% (w/w) sucrose. After centrifugation for 6 hr at 25,000 rpm in a Beckman SW 27 rotor, the purified plasma membranes are

---

[14] F. Malpartida and R. Serrano, *Eur. J. Biochem.* **116**, 413 (1981).

recovered at the 43/53 interface. The band is collected with a Pasteur pipet, diluted with 4 volumes of water, and pelleted by centrifugation for 20 min at 80,000 g. The purified membranes are resuspended with about 20 ml of 20% glycerol per every 100 g of starting yeast. In the case of commercial baker's yeast and of stationary-phase grown cells the large amount of mitochondria on top of the 43% layer results in some contamination of the plasma membrane band. A second sucrose gradient is then recommended. Four milliliters of membranes from the first gradient is applied to a discontinuous gradient made of 3 ml 53% (w/w) sucrose and 5 ml 43% (w/w) sucrose. After centrifugation for 4 h at 38,000 rpm in a Beckman SW40 Ti rotor, the plasma membrane band is collected, diluted, pelleted, and resuspended as above.

*Solubilization.* The purified plasma membranes are diluted to 4 mg protein/ml in 20% glycerol containing 3.2 mg/ml Zwittergent-14 and 0.65 mg/ml phosphatidylserine and the mixture sonicated[5] for 1 min. The temperature of the sample should not rise above 30° at this step. After centrifugation for 30 min at 80,000 g, the pellet is discarded.

*Glycerol Gradient and Guanidine Treatment.* Portions of 4 ml of solubilized enzyme are applied to a discontinuous gradient made of solubilized enzyme are applied to a discontinuous gradient made of 10 ml 50% (v/v) glycerol and 10 ml 20% (v/v) glycerol. After centrifugation for 5 hr at 130,000 g in a Beckman 60Ti angular rotor, the pellet is resuspended with 5 ml of 30% glycerol per every 100 g of starting yeast. Guanidine is added to 0.5 M and after 10 min, portions of 4 ml are applied over 20 ml of 40% (v/v) glycerol. After centrifugation for 5 hr at 130,000 g in a Beckman 60Ti rotor, the pellet is resuspended with about 1 ml of 20% glycerol per every 100 g of starting yeast and stored at −70°.

### Purification of the Oat Root Plasma Membrane ATPase[12,13]

*Materials*

Oat seeds: no significant differences have been found between several varieties. Seeds are stored in a cold room and washed with water for 20–30 min before sowing

Vermiculite (type 3)

Homogenization buffer: 30% (w/w) sucrose, 0.25 M Tris, 25 mM EDTA, and 5 mM dithiotreitol, adjusted to pH 8.5 with HCl. Store frozen

Lysolecithin (lysophosphatidylcholine, Sigma type I, from egg yolk): dissolve at 5% (about 0.1 M) and store frozen. After thawing it must be heated briefly at 70–90° to dissolve

*Plant Growth.* Plastic trays of 50 × 30 cm are covered with 1 liter of vermiculite and about 50 g of oat seeds are extended and covered with another liter of vermiculite. The trays are irrigated with 1.3 liters of water and incubated at 26–28° in the dark. After 4 days they are further irrigated with 0.6 liters of water and the roots harvested on the seventh day.

## Procedure

Unless otherwise indicated, all the operations are carried out at 0–4°. Relative centrifugal forces refer to the average radius of the tubes. The TED buffer (see above) is present in all the sucrose and glycerol solutions. Pellets are resuspended manually with a glass homogenizer.

*Preparation of Homogenates.* The roots are cut with scissors close to the seeds and freed of vermiculite by washing with water. After blotting the excess water with a filter paper, from 40 to 60 g of roots can be obtained from every tray. They are cut with scissors into small pieces and homogenized with 5 ml homogenization buffer and 50 $\mu$l 0.2 $M$ phenylmethylsulfonyl fluoride. Homogenization can be effected manually with a mortar and pestle or, more conveniently, with a Moulinex juice extractor. In this later case, the extracted juice is collected over the homogenization buffer. The homogenate is filtered through cheesecloth to remove fibrous residues.

*Differential Centrifugation.* The homogenate is first centrifuged for 10 min at 2,000 $g$ to remove debris and the supernatant further centrifuged for 20 min at 20,000 $g$. This second pellet, enriched in plasma membranes, is resuspended with about 30 ml of 20% glycerol and 60 $\mu$l 0.2 $M$ phenylmethysulfonyl fluoride per every kilogram of starting roots.

*Sucrose Gradient Centrifugation.* Six milliliters of the 20,000 g pellet is applied to a discontinuous gradient made of 2 ml 46% (w/w) sucrose and 4 ml 33% (w/w) sucrose. After centrifugation for 5 hr at 38,000 rpm in a Beckman SW40 Ti rotor the purified plasma membranes are recovered at the 33/46 interface. The band is collected with a Pasteur pipet, diluted with 4 volumes of water, and pelleted by centrifugation for 20 min at 80,000 $g$. The pellet is resuspended with about 10 ml of 20% glycerol per every kilogram of starting roots.

*Extraction with Triton X-100 and KCl.* The purified plasma membranes are diluted to 1 mg protein/ml in 20% glycerol containing 0.5 $M$ KCl and 4 mg/ml Triton X-100 (about 6 m$M$). After 10 min of incubation, the sample is centrifuged for 45 min at 80,000 $g$ and the pellet, enriched in plasma membrane ATPase, is resuspended with about 5 ml of 20% glycerol per kilogram of starting roots.

*Solubilization and Glycerol Gradient.* The extracted membranes are diluted to 1 mg protein/ml with 20% glycerol and placed at room temperature. Concentrated lysolecithin is added to 6 mg/ml (about 12 m$M$) and, after 10 min at room temperature, the mixture is centrifuged in the cold for 30 min at 80,000 $g$ and the pellet is discarded. Four milliliters of solubilized enzyme is applied to a linear glycerol gradient [34 ml, from 25 to 50% (v/v)] prepared in quick-seal tubes for the Beckman vertical VTi 50 rotor. One convenient way of preparing the gradients is to layer 17 ml of 25% glycerol over 17 ml of 50% glycerol. After slowly tilting the tubes to a horizontal position, the gradient is allowed to diffuse for 2 hr and the tubes slowly put back in a vertical position. The tubes are centrifuged for 16 hr at 45,000 rpm with slow acceleration and without brake. Fractions of 2 ml are collected from the top with a Büchler Auto Densi-Flow apparatus and aliquots of 50 $\mu$l assayed for plasma membrane ATPase activity. The two or three peak fractions in the middle of the gradient are pooled, diluted with 1 volume of water, and the enzyme is concentrated by overnight centrifugation at 130,000 $g$. The pellet is resuspended with about 1 ml of 20% glycerol per every kilogram of starting roots and stored at $-70°$.

## Reconstitution of Proteoliposomes[10–13]

The freeze–thaw–sonication procedure of Kasahara and Hinkle[15] is modified as follows. The purified ATPases are diluted to 0.3 protein/ml with 10% glycerol. Soybean phospholipids (see above) are resuspended to 9 mg/ml (oat roots) or 30 mg/ml (yeast) in TED buffer and sonicated[5] to clarity. Equal volumes (0.2–1 ml) of enzyme and lipid solutions are mixed in nitrocellulose tubes (Beckman 5/8 × 4 inches) and 1 $M$ MgSO$_4$ is added to a final concentration of 5 m$M$ (yeast) or 25 m$M$ (oat roots). The tubes are placed for 1 hr in a freezer at $-70°$ and thawed in a water bath at room temperature. The turbid suspension is clarified by sonication[5] for 2 min at 15° and the whole cycle of freeze–thaw–sonication is repeated once. Proteoliposomes can be stored frozen for several days.

## Evaluation of the Purification Procedures

A summary of the purification of the yeast and oat root plasma membrane ATPases is shown in Table I. The evaluation of the purifications is complicated by the activating and inactivating effects of detergents. Detergent activation results from unmasking latent enzyme molecules present in closed vesicles while detergent inactivation may be caused by delipidation of the enzymes. In oat root homogenates the plasma membrane ATPase is partially latent and the presence of 0.1 mg/ml lysole-

---

[15] M. Kasahara and P. C. Hinkle, *J. Biol. Chem.* **252**, 7384 (1977).

TABLE I
PURIFICATIONS OF THE PLASMA MEMBRANE ATPases FROM BAKERS' YEAST AND
OAT ROOTS

| Fraction | Protein (mg) | Plasma membrane ATPase | |
|---|---|---|---|
| | | units | units mg$^{-1}$ |
| Oat roots (1 kg fresh weight) | | | |
| Homogenate | 480 | 44 | 0.09 |
| 20,000 g pellet | 60 | 36 | 0.6 |
| Sucrose gradient | 21 | 25 | 1.2 |
| Triton–KCl treatment | 9 | 20 | 2.2 |
| Solubilization and glycerol gradient | 1 | 6 | 6.0 |
| Bakers' yeast (100 g fresh weight) | | | |
| Homogenate | 5200 | 104 | 0.02 |
| 20,000 g pellet | 600 | 96 | 0.16 |
| Sucrose gradient | 50 | 75 | 1.5$^a$ |
| Solubilization and glycerol gradient | 7 | 42 | 6.0 |
| Guanidine treatment | 4 | 32 | 8.0 |

$^a$ Assayed after activation with lysolecithin, as described in text.

cithin in the assay medium results in optimal activation (from 1.5- to 2-fold). Taking this factor into account, the Triton X-100 and KCl treatment causes some inactivation (20–30%). There is no inactivation during solubilization by lysolecithin, probably because this natural detergent satisfies the lipid requirements of the plant enzyme. However, a large inactivation (40–50%) occurs during the glycerol gradient and only about one-third of the missing activity can be recovered by adding lysolecithin during the assay. With the uncertainties introduced by all these phenomena, the corrected yield and purification factor for the oat root plasma membrane ATPase are 7–10% and 30- to 40-fold, respectively.

In yeast homogenates, detergents produce very little activation (less than 1.3-fold) of the plasma membrane ATPase. However, latency is induced during the sucrose gradient employed for plasma membrane purification, and the ATPase activity of the purified membranes can be increased from 3- to 6-fold by detergents under optimal conditions. Optimal detergent activation of purified yeast plasma membranes is achieved by preincubating at 0.2 mg protein/ml in 20% glycerol containing 0.2 mg/ml lysolecithin. Fifty microliters of this mixture is diluted into the ATPase assay buffer containing 0.4 mg of sonicated soybean phospholipids. Direct treatment with detergents after dilution in the assay buffer results in much lower activation and omission of the phospholipids results in inactivation. Taking this effect into account, it can be estimated that there is a 30–40% inactivation of the yeast plasma membrane ATPase during solubilization by Zwittergent-14 and the glycerol gradient step. This is probably caused by delipidation, which cannot be compensated by the soybean lipid added in the assay. The estimated yield and purification factor for the yeast plasma membrane ATPase are about 45% to 600-fold, respectively.

The purified enzymes are stable up to several weeks at −70° in a medium with glycerol, EDTA, and dithiothreitol (see above).

### Enzymatic Properties[1]

The purified plasma membrane ATPase from yeast and oat roots contain a major polypeptide of about 105 kDa which represents more than 70–80% of the protein in the preparations. Some minor polypeptides of lower molecular weight are usually present in variable and less than stoichiometric amounts. They may represent contaminants or result from proteolytic degradation of the enzyme. Visualization of the 105-kDa polypeptide during electrophoresis in the presence of sodium dodecyl sulfate is favoured by two modifications of the usual procedure for sample preparation: the sample is first precipitated with tricholoroacetic acid before dissolving in sodium dodecyl sulfate and heating is limited to 30–40° instead of boiling. The trichloroacetic acid inactivates endogenous proteases which are activated if the samples are directly dissolved in sodium dodecylsulfate, and boiling induces aggregation of the enzyme, mostly in crude fractions.

The 105-kDa polypepeide forms and acyl phosphate intermediate in the course of ATP hydrolysis. With purified ATPase, the maximum level of intermediate formed at saturating ATP concentrations is 0.6–0.7 nmol per milligram protein, representing a phosphorylation of about 10% of the polypeptides. It is not clear if this low level of phosphorylation results from inactivation of the enzyme during the purification or whether it is a consequence of the relative values of the kinetic constants for the phosphorylation and dephosphorylation steps. This second possibility is more likely because the estimated inactivation during purification is much lower than 90% (see above).

The plasma membrane ATPases require phospholipids for activity. The purified yeast ATPase is activated 3- to 6-fold by soybean phospholipids. The oat root ATPase is also activated by exogenous phospholipids after delipidation with cholate and ammonium sulfate[12] but the enzyme purified with lysolecithin is not activated by exogenous lipids, probably because the lysolecithin itself activates.

The yeast and oat root plasma membrane ATPases requires magnesium for activity. This physiological cofactor can be substituted by $Mn^{2+}$

or Co$^{2+}$ but not by Ca$^{2+}$, which is inhibitory in the presence of Mg$^{2+}$. Both enzymes are very specific for ATP as substrate, the activity with any other nucleotide or phosphate ester being less than 5% that with ATP. The pH optima are 6.5 (oat roots) and 5.7 (yeast) and the $K_m$ for ATP at their respective pH optima are about 0.3 m$M$ (oat roots) and about 1 m$M$ (yeast).

The plasma membrane ATPases are sensitive to the following inhibitors (concentrations giving 50% inhibition in parentheses): vanadate (5–15 $\mu M$), diethylstilbestrol (40–60 $\mu M$), dicyclohexylcarbodiimide (10–30 $\mu M$), p-chloromercuriphenyl sulfonic acid (1–3 $\mu M$), and Cu$^{2+}$ (1–3 $\mu M$). The inhibition by dicyclohexylcarbodiimide of both the yeast and oat root enzymes involves the covalent modification of the 105-kDa polypeptide.[16] Molybdate (up to 0.5 m$M$), azide (up to 10 m$M$), oligomycin (up to 20 $\mu$g/ml), and ouabain (up to 1 m$M$) produced no significant inhibition.

The monovalent cations K$^+$, Rb$^+$, and NH$_4^+$ produce a small activation (about 1.4-fold) of the oat root ATPase, with half-maximal effects at 2–4 m$M$. Na$^+$ is much less effective and Li$^+$ is without effect. The activation by K$^+$ disappears if the assays are performed at pH greater than 7. The yeast enzyme is not significantly affected by monovalent cations under the usual assay conditions. However, some activation by high concentrations of K$^+$ has been observed in assays performed at very low pH.[17]

### Transport Properties[1]

The proton pumping activity of the purified yeast and oat root ATPases has been demonstrated by the quenching of acridine dye fluorescence in reconstituted proteoliposomes. The addition of ATP results in a 50–80% quenching which can be prevented by weak bases like imidazole (20 m$M$), proton ionophores like 1799 (20 $\mu M$), and gramicidin D 1 $\mu$g/ml), and by ATPase inhibitors like vanadate (0.2 m$M$) and diethylstilbestrol (60 $\mu M$). Omission of nitrate in the assay medium decreases the quenching 3-fold in the case of the yeast enzyme and more than 10-fold with the oat root ATPase. This suggest that proton transport is electrogenic and therefore requires a permeable anion such as nitrate to proceed at maximal rates. The proteoliposomes reconstituted with the plant enzyme seems to have a lower ion permeability than those formed with the yeast ATPase. Potassium is not required for proton transport and it does not affect the quenching observed with the yeast enzyme. In the case of the plant ATPase, potassium produces a small increase in quenching

[16] A. Cid, Tesina de Licenciatura, Fac. Med., Univ. Complutense de Madrid, 1984.
[17] P. H. J. Peters and G. W. F. H. Borst-Pauwels, *Physiol. Plant.* **46**, 330 (1979).

(about 1.4-fold), in accordance with the activation of the ATPase activity discussed above. The potassium effect on the fluorescence quenching proceeds without detectable lag, suggesting that it is caused by an action on the active site of the enzyme facing the external medium. A lag in stimulation should be expected if it were caused by a proton–potassium exchange activity of the enzyme because then the potassium would first have to diffuse into the proteoliposomes. Nevertheless, the existence of a minor proton–potassium exchange activity in addition to the major electrogenic proton transport cannot be dismissed by these experiments, and some evidence for this exchange has been presented for both the yeast[18] and plant[19] enzymes.

In the case of the yeast plasma membrane ATPase, proton transport has also been indirectly demonstrated by measuring a $^{32}P_i$–ATP exchange activity sensitive to proton ionophores. The rate of this reversed reaction (10–20 nmol $min^{-1}$ mg $protein^{-1}$) is three orders of magnitude lower than the rate of ATP hydrolysis but it can be measured with high sensitivity. The complete inhibition caused by proton ionophores suggests that the proton gradient generated during ATP hydrolysis by the proteoliposomes drives back the enzyme to catalyze a small rate of ATP synthesis.

### Acknowledgments

This work was supported by grants from the Spanish Comisión Asesora de Investigación Científica y Técnica and Fondo de Investigaciones Sanitarias. I want to thank Eulalia Moreno and Consuelo Montesinos for the preparation of membranes.

[18] A. Villalobo, *Can. J. Biochem.* **62**, 865 (1984).
[19] D. P. Briskin and R. T. Leonard, *Proc. Natl. Acad. Sci. U.S.A.* **79**, 6922 (1982).

## [41] Purification of Yeast Vacuolar Membrane $H^+$-ATPase and Enzymological Discrimination of Three ATP-Driven Proton Pumps in *Saccharomyces cerevisiae*

By Etsuko Uchida, Yoshinori Ohsumi, and Yasuhiro Anraku

### Introduction

In the yeast *Saccharomyces cerevisiae,* vacuoles are the largest organelles, occupying about 25% of the cell volume, and they are postulated to function as lysosomes and as a storage compartment in which the bulk of basic amino acids, *S*-adenosylmethionine, and polyphosphates are localized.[1] These observations have suggested the existence of some specific transport systems in the vacuolar membranes. Recently, we established a procedure for preparing right-side-out vacuolar membrane vesicles of high purity from cells of the yeast *Saccharomyces cerevisiae*[2] and showed that the vesicles catalyze active transport of 10 amino acids[3,4] and Ca$^{2+}$[5] which are driven by an electrochemical potential difference of protons formed by ATP hydrolysis.[6] Subsequent studies on ATP hydrolysis by vacuoles and vacuolar membrane vesicles indicated the presence of a new type of ATPase in the membranes.[6]

Recently, this vacuolar membrane-bound ATPase was characterized as a H$^+$-translocating ATPase and is the third type of ATP-driven H$^+$-pump found in the yeast *S. cerevisiae*.[7]

This chapter describes the purification and properties of vacuolar membrane H$^+$-ATPase and its enzymological discrimination from other ATP-driven proton pumps, F$_1$F$_0$-ATPase from mitochondria, and H$^+$-translocating ATPase from the plasma membrane of *S. cerevisiae*.

### Preparation of Vacuolar Membrane Vesicles

Vacuoles are the largest organelles in yeast cells, but they contain only 0.2% of total cell protein. To analyze the functions of vacuoles and vacuolar membrane ATPase, it is crucial to obtain pure materials in hand. The following procedures are established for this purpose.

#### Strain and Culture Condition

Haploid strain X2180-1A from the Yeast Genetic Stock Center, Berkeley, is used. Cells are grown in medium containing 1% yeast extract (Difco), 2% polypeptone, 2% glucose at 30° with reciprocal shaking at 120 strokes/min. On a large scale, cells are grown in a 10-liter Magnaferm fermentor (New Brunswick Scientific Co., Inc.) that is aerated at an air flow of 6 liters/min and an agitation speed of 200 rpm. These culture conditions are critical to obtain pure vacuolar membrane vesicles. Cells grown under the conditions indicated above can be converted easily to spheroplasts without harsh pretreatment.

[1] A. Wiemken, M. Schellenberg, and K. Urech, *Arch. Microbiol.* **123**, 23 (1979).
[2] Y. Ohsumi and Y. Anraku, *J. Biol. Chem.* **256**, 2079 (1981).
[3] T. Sato, Y. Ohsumi, and Y. Anraku, *J. Biol. Chem.* **259**, 11505 (1984).
[4] T. Sato, Y. Ohsumi, and Y. Anraku, *J. Biol. Chem.* **259**, 11509 (1984).
[5] Y. Ohsumi and Y. Anraku, *J. Biol. Chem.* **258**, 5614 (1983).
[6] Y. Kakinuma, Y. Ohsumi, and Y. Anraku, *J. Biol. Chem.* **256**, 10859 (1981).
[7] E. Uchida, Y. Ohsumi, and Y. Anraku, *J. Biol. Chem.* **260**, 1090 (1985).

*Vacuolar Membrane Vesicles*

The following buffer solutions are required for routine operation.

1 $M$ sorbitol

Buffer A: 10 m$M$ tris(hydroxymethyl)aminomethane (Tris)/2-($N$-morpholino)ethanesulfonic acid (MES) (pH 6.9), 0.1 m$M$ MgCl$_2$, 12% Ficoll-400

Buffer B: 10 m$M$ Tris/MES (pH 6.9), 0.5 m$M$ MgCl$_2$, 8% Ficoll-400

Buffer C: 10 m$M$ Tris/MES (pH 6.9), 5 m$M$ MgCl$_2$, 25 m$M$ KCl

2× Buffer C

Exponentially growing cells (30 liters, $4 \times 10^7$ cells/ml) are harvested by centrifugation at 4500 $g$ for 3 min, washed twice with distilled water at room temperature, and suspended in 1 $M$ sorbitol at a density of $2 \times 10^8$ cells/ml. To this suspension, zymolyase (20T) is added to a final concentration of 1 unit/ml, and the mixture is incubated at 30° for 90 min with gentle shaking. This treatment converts most (>95%) of the cells to spheroplasts, which are then recovered by centrifugation and washed twice with 1 $M$ sorbitol with centrifugation (2200 $g$, 5 min). The pellet is suspended in about 10 volumes of buffer A, homogenized in a loosely fitting Dounce homogenizer, and centrifuged in a swinging bucket rotor at 4500 $g$ for 10 min.

For isolation of vacuoles, 20-ml volumes of supernatant are transferred to centrifuge tubes and 10 ml of buffer A is layered on top. The tubes are centrifuged in a swinging bucket rotor RPS27-2 of a Hitachi model 70P centrifuge at 51,900 $g$ for 30 min at 4°. The white layer on top of the tubes, which contains most of the vacuoles, is collected and resuspended in buffer A with an homogenizer. Then, 15-ml volumes of crude vacuoles are transferred to centrifuge tubes and 15 ml of buffer B is placed on top. After recentrifugation under the conditions described above, the vacuoles are recovered from the top of the tubes almost free from contaminating lipid granules, other membranous organelles, or unbroken spheroplasts. During these steps, the purity of preparations is checked by phase-contrast microscopy. The vacuoles thus obtained are converted to vacuolar membrane vesicles by diluting them first with an equal volume of 2× buffer C and then 2 volumes of buffer C. The vesicles are recovered by centrifugation (37,000 $g$, 20 min) as an opalescent pellet. The final pellet can be kept at −80° for several months without significant loss of ATPase activity.

### Properties of Vacuolar Membrane Vesicles

The distributions of marker enzymes ($\alpha$-mannosidase for vacuoles, glucose-6-phosphate dehydrogenase for cytosol, succinate dehydro-

TABLE I
MARKER ENZYME ACTIVITIES IN VACUOLAR MEMBRANE VESICLES[a]

| Enzyme | Spheroplast lysate | | Vacuolar membrane vesicles | | Ratio | Recovery |
| | A | a | B | b | B/A | (b/a × 100) |
|---|---|---|---|---|---|---|
| $\alpha$-Mannosidase | 0.036 | 97 | 1.62 | 24.9 | 45.0 | 27 |
| Glucose-6-phosphate dehydrogenase | 44.0 | 118800 | 0.1 | 1.5 | 0.002 | 0.002 |
| Succinate dehydrogenase | 5.9 | 15930 | 0.1 | 0.15 | 0.002 | 0.002 |
| NADH–cytochrome-$c$ reductase | 3.1 | 8370 | 0.01 | 0.15 | 0.005 | 0.002 |
| Chitin synthase | 1.2 | 3320 | 0.1 | 1.5 | 0.08 | 0.04 |
| Protein (mg) | 2700 | | 15.4 | | | |

[a] A and B, specific activity (nmol/min/mg protein); a and b, total activity (nmol/min).

genase for mitochondria, NADH–cytochrome $c$ reductase for microsomes, and chitin synthase for plasma membrane) are examined by measuring enzyme activities in the vacuolar membrane vesicles and spheroplast lysate fractions. Table I shows that the recovery of $\alpha$-mannosidase activity in the vacuolar membrane vesicles is about 27% and that its specific activity in this fraction increases 45-fold. The recoveries of the four other marker enzyme activities are found to be less than 0.1% of their total activities. These results indicate that the vacuolar membrane vesicles obtained are virtually free from mitochondria and other organelles. At the last step of the procedure, the preparation is almost free of marker constituents of vacuoles, such as arginine (78%), polyphosphates (90%), and alkaline phosphatase (83%). The alkaline phosphatase activity and $\alpha$-mannosidase activity are activated by detergents such as Triton X-100, but $Mg^{2+}$-ATPase activity is rather inhibited. This suggests that $Mg^{2+}$-ATPase has a catalytic site exposed to the cytoplasm.

The orientation and size of vacuolar membrane vesicles are studied by electron microscope using a freeze–fracturing technique. Many small particles are seen on the concave face of the vesicles, while the convex face is fairly smooth. The concave face corresponds to the exoplasmic fracturing face of intact vacuoles. This means that almost all the vesicles are right-side-out. The vesicles, which are much smaller than intact vacuoles, are perfectly spherical and 0.2–1.6 $\mu$m in diameter.

## Characterization of Vacuolar Membrane ATPase

*Assay of ATPase Activity*

Vacuolar membrane ATPase is assayed at 30° in a final volume of 0.1 ml containing 25 m$M$ MES/Tris (pH 6.9), 5 m$M$ ATP-2Na, 5 m$M$ MgCl$_2$, and enzyme. For assay of the enzyme activity after solubilization, sonicated soybean phospholipids prepared by the method of Kagawa and Racker[8] are added to the reaction mixture at a final concentration of 0.1 mg/ml. When indicated, the inhibitor in ethanol solution or in dimethyl sulfoxide solution is added to the assay mixture at a final concentration of solvent less than 0.2% (v/v). The reaction is started by addition of Mg$^{2+}$-ATP and, after 5–20 min of incubation, is stopped by adding 0.1 ml of 5% (w/v) sodium dodecyl sulfate (SDS). Inorganic phosphate liberated is measured by the method of Ohnishi,[9] and 1 unit of enzyme is defined as the amount liberating 1 $\mu$mol of inorganic phosphate per minute under the standard conditions described above. For determination of the $K_m$ value, the activity of ATPase is assayed by measuring the release of $^{32}P_i$ from [$\gamma$-$^{32}$P]ATP (1–5 Ci/mol) in the standard mixture as described above. After 1 min at 30°, the reaction is stopped by adding 0.061 ml of 13.2% trichloroacetic acid (TCA), 1 m$M$ phosphoric acid, and 10 m$M$ ATP, and then 0.3 ml of 4% perchloric acid. The radioactivity recovered as inorganic phosphate is counted in a liquid scintillation counter as described by Martin and Doty.[10]

Plasma membranes and submitochondrial particles from the same strain, X2180-1A, are prepared by the method of Serrano[11] and Takeshige *et al.*,[12] respectively. Mitochondrial F$_1$-ATPase is extracted and purified by sucrose density gradient centrifugation as described by Takeshige *et al.*[12] The reaction mixtures for assays of plasma membrane ATPase and mitochondrial F$_1$-ATPase are similar to that for the assay of vacuolar membrane ATPase, except for the pH of the medium. The pH of the reaction mixture for assay of plasma membrane ATPase is pH 5.5, obtained with 25 m$M$ MES/Tris buffer, and that for assay of mitochondrial F$_1$-ATPase is pH 9.0, obtained with 25 m$M$ Tris-HCl buffer.

[8] Y. Kagawa and E. Racker, *J. Biol. Chem.* **246,** 5477 (1971).
[9] T. Ohnishi, *Anal. Biochem.* **69,** 261 (1975).
[10] J. B. Martin and D. M. Doty, *Anal. Chem.* **21,** 965 (1949).
[11] R. Serrano, *Mol. Cell. Biochem.* **22,** 51 (1978).
[12] K. Takeshige, B. Hess, M. Bohm, and H. Zimmermann-Telschow, *Hoppe-Seyler's Z. Physiol. Chem.* **357,** 1605 (1976).

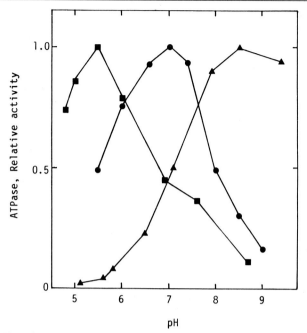

FIG. 1. pH profiles of three ATPases of yeast, *Saccharomyces cerevisiae*. Assays were carried out using 25 m$M$ MES/Tris buffer, 5 m$M$ MgCl$_2$, and 5 m$M$ ATP-2Na. (●) Vacuolar membrane vesicles, (■) plasma membranes, (▲) mitochondria. (From Serrano.[11])

## Properties of Vacuolar Membrane ATPase

The enzyme requires $Mg^{2+}$ ion for ATP hydrolysis. The optimal ratio of ATP to $Mg^{2+}$ of 1 indicates that an $ATP–Mg^{2+}$ complex is substrate for the enzyme.[6] $Ca^{2+}$ ion has no stimulatory effect on the activity. The optimal pH of the enzyme is determined to be pH 7.0, which is apparently different from that of mitochondrial $F_1$-ATPase (pH 8.5) and plasma membrane ATPase (pH 5.5) (Fig. 1). The enzyme hydrolyzes ATP and three other ribonucleoside triphosphates, GTP, UTP, and CTP, with this order of preference, but does not hydrolyze ADP and *p*-nitrophenyl phosphate (pNPP) (see Table III). The $K_m$ value for ATP is determined as 0.2 m$M$, which is 8-fold smaller than that of the ATPase of plasma membrane of *S. cerevisiae*.[13] Effects of various inhibitors and cations are summarized in a later section.

---

[13] G. R. Willsky, *J. Biol. Chem.* **254,** 3326 (1979).

*Evidence that Mg$^{2+}$-ATPase of Vacuoles Is*
*an H$^+$-Translocating ATPase*

The Mg$^{2+}$-ATPase activities of vacuoles and vacuolar membrane vesicles are stimulated 3- and 1.5-fold, respectively, by the protonophore uncoupler 3,5-di-*tert*-butyl-4-hydroxylbenzylidine malononitrile (SF6847) and the K$^+$/H$^+$ antiporter ionophore nigericin, indicating that these reagents decrease the proton gradient formed by ATP hydrolysis.[6] Valinomycin has no effect on the activities.

ATP hydrolysis-dependent formation of a proton gradient is demonstrated by recording the change in quenching of quinacrine or 9-amino acridine fluorescence. The fluorescence signal of quinacrine is quenched by incubating the vacuolar membrane vesicles with ATP, reflecting up-

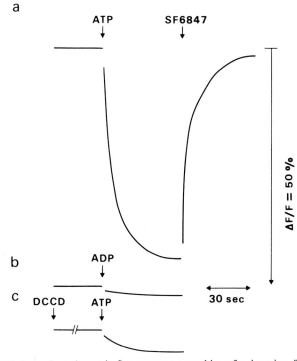

Fig. 2. ATP-dependent change in fluorescence quenching of quinacrine. The method of Tsuchiya and Rosen[14] was used with a modification. The reaction mixture (2.5 ml), which contained 50 m$M$ Tricine/NaOH (pH 7.5), 5 m$M$ MgCl$_2$, 2 $\mu M$ quinacrine, and 180 $\mu$g of protein of membrane vesicles, was incubated in a quartz cuvette at 25° for 2–5 min. To this mixture, ATP (0.5 m$M$) was added to start quenching. Additions were made where indicated by the arrows at the following final concentrations: (a) 0.5 m$M$ ATP, (b) 0.5 m$M$ ADP; (c) vacuolar membrane vesicles were treated with 0.1 m$M$ DCCD for 20 min at 25° before addition of 0.5 m$M$ ATP.

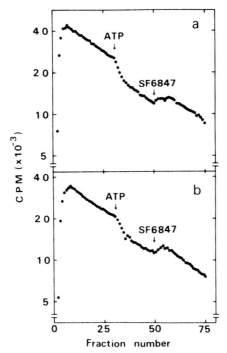

FIG. 3. ATP-dependent uptake of (a) [¹⁴C]methylamine and (b) KS¹⁴CN by vacuolar membrane vesicles. Flow dialysis was done by the method of Ramos *et al.*[15] The upper chamber (1 ml) contained 1.5 mg of protein of membrane vesicles in buffer C. Circulation of buffer C in the lower chamber was conducted with a peristaltic pump at a rate of 4.2 ml/min at 25°. Immediately after addition of 30 $\mu M$ [¹⁴C]methylamine (40 mCi/mmol) or 150 $\mu M$ KS¹⁴CN (15 mCi/mmol) to the upper chamber, 4 m$M$ ATP was added to start the reaction. Fractions of 2.1 ml were collected and their radioactivity was determined.

take of protons and formation of ΔpH (Fig. 2),[14] which indicates that the uptake of protons is coupled with ATP hydrolysis by Mg²⁺-ATPase. ATP-dependent alkalinization of reaction mixture containing vesicles and ATP in 2 m$M$ glycylglycine buffer (pH 6.3) is also observed with a pH electrode.

The electrochemical potential difference of protons across the vacuolar membrane generated upon ATP hydrolysis is determined quantitatively by the flow-dialysis method with [¹⁴C]methylamine for measuring the formation of ΔpH and with KS¹⁴CN for measuring the membrane potential (Fig. 3).[15] The $\Delta\bar{\mu}H^+$ thus calculated is 180 mV, with contribu-

[14] T. Tsuchiya and B. P. Rosen, *J. Biol. Chem.* **251**, 962 (1976).

[15] S. Ramos, S. Schuldiner, and H. R. Kaback, *Proc. Natl. Acad. Sci. U.S.A.* **73**, 1892 (1976).

tion of 1.7 pH units, interior acid, and of a membrane potential of 75 mV, interior positive.

### Purification of Vacuolar Membrane H$^+$-ATPase

The following buffer solutions are required for routine operation.

Washing buffer: 10 m$M$ Tris-HCl (pH 7.5), 1 m$M$ ethylenediamine-tetraacetic acid (EDTA)

Solubilizing buffer: 10 m$M$ Tris-HCl (pH 7.5), 1 m$M$ EDTA, 2 m$M$ dithiothreitol (DTT), 0.5 m$M$ phenylmethanesulfonyl fluoride (PMSF), and 10% glycerol

Detergent: 10% (w/v) zwitterionic detergent (ZW3-14); $N$-tetradecyl-$N,N$-dimethyl-3-ammonio-1-propane sulfonate

Glycerol density gradient: 8 ml 20–50% (v/v) glycerol in a solution of 10 m$M$ Tris-HCl (pH 7.5), 1 m$M$ EDTA, 2 m$M$ DTT, 0.5 m$M$ PMSF, 0.005% (w/v) ZW3-14

*Step 1: EDTA Wash.* Vacuolar membrane vesicles (approximately 30 mg of protein), prepared as described above in the section Preparation of Vacuolar Membrane Vesicles, are suspended in washing buffer at a protein concentration of 1 mg/ml and homogenized by 5 strokes of a Dounce homogenizer. The suspension is centrifuged at 37,000 $g$ for 30 min and the supernatant is discarded. Unless otherwise noted, all preparations are carried out at 0–4°. This EDTA wash is repeated three times with recovery of 70% of the total protein and no loss of ATPase activity.

*Step 2: Solubilization.* The EDTA-washed membranes are suspended at a protein concentration of 5 mg/ml in a solubilizing buffer. To this suspension, 10% (w/v) ZW3-14 is added dropwise under vigorous stirring to a final weight ratio of the detergent to protein of 1.0. The mixture is kept at 4° for 15 min with gentle shaking and then centrifuged at 100,500 $g$ for 60 min in a Hitachi RP70 rotor. The supernatant is collected as the solubilized fraction.

*Step 3: Glycerol Density Gradient.* Aliquots (0.25 ml) of the supernatant obtained at Step 2 are layered on top of 20–50% glycerol density gradient, and centrifuged at 180,000 $g$ for 8 hr in a Hitachi RP65 rotor at 4°. Then fractions are collected from the bottom of the tubes using a Perista minipump.

Figure 4 shows the distribution profile of protein and activities of H$^+$-ATPase, $\alpha$-mannosidase, acid phosphatase, and alkaline phosphatase. H$^+$-ATPase sediments faster than the bulk of protein and separates well from acid phosphatase and alkaline phosphatase. Under the conditions used, mitochondrial F$_1$-ATPase sediments around fraction 13, three fractions upper. The peak fraction has a specific activity of 18 units/mg of

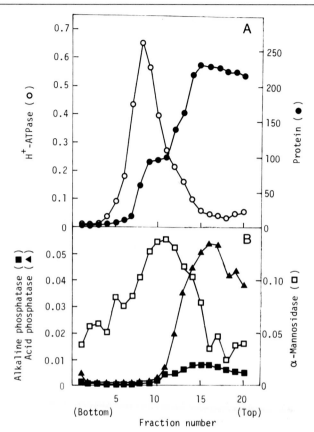

FIG. 4. Distributions of protein and enzyme activities in a glycerol gradient. The solubi-lized fraction in Step 2 (0.25 ml) was applied to a glycerol density gradient. Centrifugation was at 180,000 g for 8 hr and the distributions of protein and activities of marker enzymes were determined. (●) Protein (μg/ml), (○) H+-ATPase (unit/ml), (□) α-mannosidase (mil-liunit/ml), (■) alkaline phosphatase (unit/ml), (▲) acid phosphatase (unit/ml).

protein. Fractions with activity of more than 16 units/mg of protein are pooled and used as purified vacuolar membrane H+-ATPase.

Table II summarizes the purification of vacuolar H+-ATPase. More than 38-fold purification is attained with recovery of 5% of the activity of the vacuolar membranes. It is important to use vacuolar membrane vesi-cles that are free from mitochondria and other membranous organelles. Therefore, to obtain vacuolar membrane vesicles of high purity, we rec-ommend starting the purification with cells harvested in the middle of the exponential phase of growth, or at a density of 4 × 10^7 cells/ml.

TABLE II
PURIFICATION OF VACUOLAR MEMBRANE H⁺-ATPase

| | Specific activity (unit/mg) | | | |
|---|---|---|---|---|
| Purification step | $H^+$-ATPase | $\alpha$-Mannosidase | Acid phosphatase | Alkaline phosphatase |
| Vacuolar membrane vesicles | 0.48 (1.0)[a] | 0.009 | 0.27 | 0.45 |
| EDTA-washed membranes | 3.3  (6.9)[a] | 0.002 | 0.09 | 0.10 |
| Solubilized fraction | 4.6[b]  (9.6)[a] | —[c] | —[c] | —[c] |
| Glycerol density gradient peak | 18[b]  (38)[a] | 0.008 | 0.009 | 0.02 |

[a] Purification factor.
[b] Activity determined in the presence of sonicated soybean phospholipids (0.1 mg/ml).
[c] Not assayed.

Washing with EDTA is an effective treatment before solubilization, since it washes off most of the acid and alkaline phosphatases, which are contaminating soluble enzymes, and also strips off proteins loosely bound to the membranes, resulting in appreciable increase in specific activity. The specific activity of H⁺-ATPase is increased about 7-fold at this step with elimination of 30% of the total protein (Table II).

The zwitterionic detergent ZW3-14 is the best for solubilization; detergents such as Triton X-100, cholic acid, Sarkosyl, and Tween 80 are not effective. The weight ratio of the detergent to protein is critical for solubilization. In Step 2, in which the protein concentration of EDTA-washed membranes is adjusted to 5 mg/ml, the detergent at a concentration of 7 m$M$ or less is not effective for solubilization, while at 16 m$M$ it solubilizes more than 70% of the membrane protein, but the solubilized ATPase activity is rapidly inactivated. At the optimal concentration of ZW3-14 of 14 m$M$, or a weight ratio of the detergent to protein of 1.0, approximately 70% of the membrane protein and a maximum of 55% of the H⁺-ATPase activity are recovered in the supernatant fraction (Table II).

Further purification of the solubilized enzyme by several conventional techniques is not successful. No appreciable results are obtained by column chromatography through DE-52 (Whatman) or DEAE-Sepharose (Pharmacia), gel filtration with high-performance liquid chromatography (SW-3000, Toyo Soda), affinity chromatography with Blue Sepharose CL-6B (Pharmacia), or isoelectric focusing electrophoresis. The concentration of the detergent ZW3-14 in the glycerol density gradient is critical for

TABLE III

SUBSTRATE SPECIFICITY OF PURIFIED VACUOLAR MEMBRANE H⁺-ATPase AND
H⁺-ATPases OF VACUOLAR MEMBRANES, PLASMA MEMBRANES, AND MITOCHONDRIA

| | Relative activity (%)[a] | | | |
| Substrate | Purified vacuolar membrane $H^+$-ATPase | Vacuolar membrane $H^+$-ATPase | Plasma membrane $H^+$-ATPase | Mitochondrial[b] ATPase |
| --- | --- | --- | --- | --- |
| ATP | 100 | 100 | 100 | 100 |
| GTP | 68 | 56 | 29 | 147 |
| UTP | 22 | 26 | 15 | 33 |
| CTP | 19 | 22 | 12 | 1.6 |
| ADP | 0.5 | 11 | 5.9 | 8.0 |
| AMP | 0.2 | 3.7 | 4.1 | 3.2 |
| pNPP | 0.9 | 8.9 | 2.8 | — |

[a] Activity was measured with 1 m$M$ substrate, 1 m$M$ $MgCl_2$, 25 m$M$ MES/Tris, pH 6.9 (purified vacuolar membrane $H^+$-ATPase and vacuolar membrane $H^+$-ATPase) or pH 6.0 (plasma membrane $H^+$-ATPase).
[b] Data from Peters and Borst-Pauwels.[16]

recovery and separation of the enzyme activity. Addition of KCl (25–300 m$M$) to the gradient reduces the recovery.

## Enzymatic Properties of the Purified Vacuolar Membrane $H^+$-ATPase

### Activation by Phospholipids

The purified vacuolar membrane $H^+$-ATPase requires phospholipids for maximal activity. The activity is stimulated 3-fold by phospholipids at more than 0.1 mg/ml. Under the standard assay conditions in the presence of phospholipids, the $K_m$ value for ATP is determined to be 0.21 m$M$. ADP inhibits the enzyme activity competitively and its $K_i$ value is determined to be 0.31 m$M$. The purified enzyme hydrolyzes ATP, GTP, UTP, and CTP, with this order of preference (Table III).[16] ADP, AMP, and pNPP are not hydrolyzed. The optimal pH is determined to be 6.9.

### Inhibitors

The activity of purified enzyme is not inhibited at all by antiserum against mitochondrial $F_1$-ATPase, which inhibits the $F_1$-ATPase activity

[16] P. H. J. Peters and G. W. F. H. Borst-Pauwels, *Physiol. Plant.* **46**, 330 (1979).

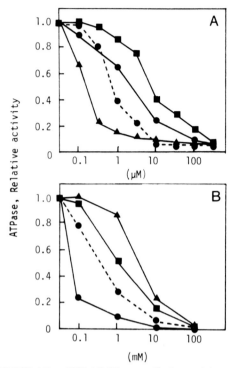

FIG. 5. Effects of DCCD (A) and EDAC (B) on purified vacuolar membrane H⁺-ATPase, and H⁺-ATPases in vacuolar membranes, plasma membranes, and submitochondrial particles. Enzyme samples were incubated with DCCD or EDAC at the concentrations indicated for 25 min at 30° before starting the reaction. Assays were carried out as described in the section Characterization of Vacuolar Membrane ATPase. (●--●) Purified vacuolar membrane H⁺-ATPase, (●—●) vacuolar membrane vesicles, (■) plasma membranes, and (▲) submitochondrial particles.

in submitochondrial particles completely. The activity of the purified vacuolar membrane H⁺-ATPase is not affected by the mitochondrial $F_1$-ATPase inhibitor protein.[17] These results clearly indicate that the vacuolar membrane H⁺-ATPase is a unique ATPase and differs from mitochondrial $F_1$-ATPase. The purified vacuolar membrane H⁺-ATPase is not cold-labile.

$N,N'$-Dicyclohexylcarbodiimide (DCCD) is a potent inhibitor of various H⁺-translocating ATPases. The $K_i$ values for DCCD of the purified vacuolar enzyme and ATPases in vacuolar membrane vesicles, submitochondrial particles, and plasma membranes are determined to be 0.8, 2, 0.2, and 8 $\mu M$, respectively (Fig. 5A). Thus the vacuolar membrane

[17] H. Matsubara, K. Inoue, T. Hashimoto, K. Yoshida, and K. Tagawa, *J. Biochem.* (*Tokyo*) **94**, 315 (1983).

ATPase is approximately 10-fold less sensitive to DCCD than mitochondrial F$_1$F$_0$-ATPase and 4-fold more sensitive than plasma membrane ATPase. The vacuolar membrane H$^+$-ATPase is more sensitive to the hydrophilic derivative of DCCD, 1-ethyl-3-(3-dimethylaminopropyl)carbodiimide (EDAC), than the plasma membrane H$^+$-ATPase or mitochondrial F$_1$F$_0$-ATPase (Fig. 5B). EDAC at 0.1 m$M$ inhibits the vacuolar membrane H$^+$-ATPase, but reduces the activities of the plasma membrane and mitochondrial enzymes less than 10%. Therefore, EDAC is a specific inhibitor of vacuolar membrane H$^+$-ATPase at this concentration.

The effects of several other inhibitors of H$^+$-translocating ATPases are examined and the results are summarized in Table IV. The activity of purified vacuolar membrane H$^+$-ATPase is strongly inhibited by 7-chloro-4-nitrobenzoxazole (NBD-Cl) and tributyltin, which are known to interact specifically with the $\beta$-subunit of F$_1$F$_0$-ATPase. The activity is also inhibited by diethylstilbestrol (DES) and quercetin, but not by sodium azide or oligomycin.

Of the specific inhibitors of plasma membrane H$^+$-ATPase, sodium vanadate and miconazole nitrate, which are known to inhibit the dephos-

TABLE IV

Effects of Inhibitors on Purified Vacuolar Membrane H$^+$-ATPase and H$^+$-ATPases from Vacuolar Membranes, Plasma Membranes, and Mitochondria from *S. cerevisiae*

| | Relative activity (%)[a] | | | |
|---|---|---|---|---|
| Inhibitor (m$M$) | Purified vacuolar membrane H$^+$-ATPase | Vacuolar membrane H$^+$-ATPase | Plasma membrane H$^+$-ATPase | Mitochondrial[b] ATPase |
| None | 100 | 100 | 100 | 100 |
| DCCD (0.001) | 38 | 63 | 86 | 12 |
| EDAC (0.1) | 77 | 23 | 95 | 100 |
| NBD-Cl (0.1) | 23 | 27 | 79 | 6 |
| Tributyltin (0.1) | 14 | 45 | 33 | 15 |
| Sodium azide (2.0) | 95 | 110 | 105 | 4 |
| Oligomycin (0.047) | 96 | 74 | 74 | 10 |
| DES (0.1) | 30 | 48 | 16 | 95 |
| Quercetin (0.1) | 37 | 67 | 30 | 100 |
| Sodium vanadate (0.1) | 95 | 96 | 50 | 100 |
| Miconazole nitrate (0.3) | 66 | 52 | 5 | 86 |

[a] Assays were carried out under standard conditions with the indicated inhibitors. In assays with purified vacuolar membrane H$^+$-ATPase, sonicated soybean phospholipids (0.1 mg/ml) were added to the reaction mixture.

[b] Submitochondrial particles were used as an enzyme source.

phorylation and phosphorylation of plasma membrane $H^+$-ATPase from *Schizosaccharomyces pombe*,[18] do not inhibit purified vacuolar membrane $H^+$-ATPase noticeably (Fig. 6). This suggests that vacuolar membrane $H^+$-ATPase is different from the plasma membrane $H^+$-ATPase. Interestingly, however, both ATPases are inhibited similarly by DES and quercetin.

The activity of purified vacuolar membrane $H^+$-ATPase is stimulated by KCl and inhibited by $KNO_3$ (Fig. 6) and KSCN. $Cu^{2+}$ and $Zn^{2+}$ are potent inhibitors of the vacuolar enzyme. The activity of purified vacuolar membrane $H^+$-ATPase is insensitive to ammonium molybdate (0.1 m$M$), which is an inhibitor of acid phosphatase.[19]

Effects of various ions on three $H^+$-ATPases are summarized in Table V. As shown in Fig. 6, three simple anions, azide, vanadate, and nitrate, are very effective compounds for enzymological discrimination of three types of $H^+$-ATPases of yeast.

## Polypeptide Composition of Purified Vacuolar Membrane $H^+$-ATPase

The polypeptide compositions of samples during purification are analyzed by gel electrophoresis in the presence of SDS (Fig. 7).[20] Lanes 2, 3, and 4 show the polypeptide compositions of vacuolar membrane vesicles, EDTA-washed membranes, and the peak fraction from the glycerol density gradient, respectively. Lane 5 shows the peak fraction of purified mitochondrial $F_1$-ATPase, where the $\alpha$ ($M_r$ 68,000), $\beta$ (52,000) and $\gamma$ (27,000) subunits are seen.

Lane 4 shows no polypeptide with $M_r$ 105,000, which is a component of plasma membrane $H^+$-ATPase and also no polypeptides corresponding to the $\alpha$, $\beta$, and $\gamma$ subunits of the $F_1$-ATPase. Two major polypeptides, a and b, respectively, of $M_r$ 89,000 and 64,000 and one small polypeptide c with a diffuse band are copurified during the preparation. The ratio of polypeptide a to polypeptide b seemed to be constant in different preparations at Step 3 but the ratio of polypeptide a to those of other small polypeptides of about $M_r$ 55,000, such as those seen in lane 4, changes during purification and differs in different preparations at Step 3. Upon column chromatography with Blue Sepharose CL-6B, these contaminant polypeptides can be eliminated without loss of enzyme activity.

DCCD is a potent inhibitor of vacuolar membrane $H^+$-ATPase, as shown in Fig. 5A. To identify a DCCD-binding polypeptide of vacuolar

[18] A. Amory and A. Goffeau, *J. Biol. Chem.* **257**, 4723 (1982).
[19] R. A. Leigh and R. R. Walker, *Planta* **150**, 222 (1980).
[20] U. K. Laemmli, *Nature (London)* **227**, 680 (1970).

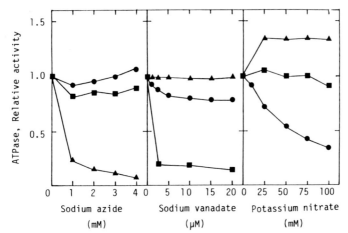

FIG. 6. Discrimination of three H⁺-ATPases from *S. cerevisiae* by sensitivity to azide, vanadate, and nitrate. Enzyme samples were incubated with sodium azide, sodium vanadate, or potassium nitrate at the concentrations indicated for 20 min at 30° before starting the reaction. Assays were carried out at optimal pH of each ATPase. (●) Vacuolar membrane vesicles, (■) plasma membranes, and (▲) submitochondrial particles.

TABLE V

ION SENSITIVITIES OF THREE H⁺ATPases OF *S. cerevisiae*

| | Relative activity (%)[a] | | |
| Salt (m$M$) | Vacuolar membrane H⁺-ATPase | Plasma membrane H⁺-ATPase | Mitochondrial[b] ATPase |
|---|---|---|---|
| None | 100 | 100 | 100 |
| KCl (10) | 117 | 120 | 118 |
| (100) | 128 | 91 | 154 |
| NH₄Cl (10) | 141 | 127 | 105 |
| ZnCl₂ (0.2) | 49 | 76 | 85 |
| CuCl₂ (0.1) | 30 | 6 | 114 |
| HgCl₂ (0.01) | 32 | 2 | 79 |
| LaCl₃ (1.0) | 69 | 38 | 72 |

[a] Assays were carried out with 25 m$M$ MES/Tris, pH 6.9 (vacuolar membrane H⁺-ATPase), pH 6.0 (plasma membrane H⁺-ATPase), and pH 8.9 (mitochondrial ATPase) in the presence of ions indicated.

[b] Submitochondrial particles were used as an enzyme source.

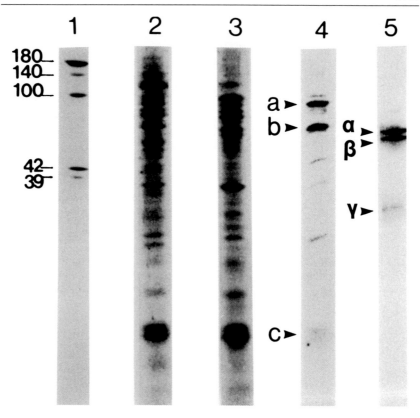

FIG. 7. SDS-polyacrylamide gel electrophoresis of purified vacuolar membrane H+-ATPase. Analytical polyacrylamide gel electrophoresis in the presence of SDS was carried out using the system of Laemmli with 15% polyacrylamide gel.[20] The gel was stained with Coomassie Brilliant Blue R-250. Lane 1 shows the positions of molecular weight markers of RNA polymerase from a thermophilic bacterium; lane 2, vacuolar membranes; lane 3, EDTA-washed membranes; lane 4, purified vacuolar membrane H+-ATPase. Arrowheads indicate the positions of polypeptide a ($M_r$ 89,000), b ($M_r$ 64,000), and c ($M_r$ 19,500) as constituent subunits of the enzyme. Lane 5, mitochondrial $F_1$-ATPase.

membrane H+-ATPase, several samples are subjected to fluorography of [14C]DCCD-binding components (Fig. 8).

When purified plasma membranes are used (lane 3), the polypeptide of $M_r$ 105,000 is labeled predominantly, as reported by Sussman and Slayman.[21] Lane 2 shows the DCCD-binding polypeptides of submitochondrial particles; two polypeptides corresponding to the $\beta$-subunit and the

[21] M. R. Sussman and C. W. Slayman, *J. Biol. Chem.* **258**, 1839 (1983).

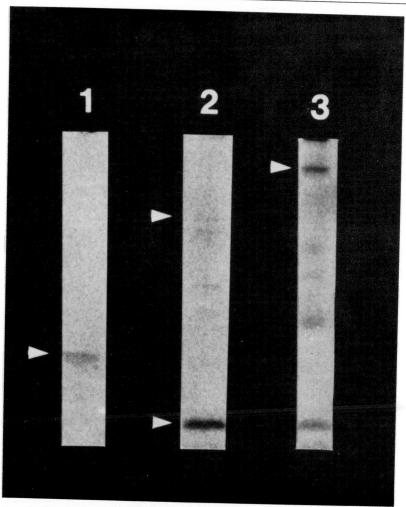

Fɪɢ. 8. Fluorogram of SDS-polyacrylamide gel showing [¹⁴C]DCCD-binding polypeptides (indicated by arrowheads) of purified vacuolar membrane H⁺-ATPase (polypeptide c of $M_r$ 19,500; lane 1), mitochondrial $F_1$-ATPase ($\beta$-subunit of $M_r$ 52,000 and DCCD-binding protein of $M_r$ 8,000; lane 2), and plasma membrane ATPase ($M_r$ 105,000; lane 3). Samples were subjected to gel electrophoresis and bands were located by fluorography. (See details in Ref. 7.)

8,000-Da DCCD-binding polypeptide are labeled specifically. On the other hand, a DCCD-binding protein of purified vacuolar membrane $H^+$-ATPase is found to have an apparent molecular weight of 19,500 (lane 1). Accordingly, polypeptide c in Fig. 7 is concluded to be a DCCD-binding protein (Fig. 8).

These results indicate that the vacuolar membrane $H^+$-ATPase has a complex structure like mitochondrial $F_1$-ATPase. However, it has no polypeptides corresponding to the $\alpha$, $\beta$, and $\gamma$ subunits, and a unique structure that differs from those of mitochondrial $F_1$-ATPase and plasma membrane $H^+$-ATPase of *S. cerevisiae*.

# [42] Purification of Vacuolar Membranes, Mitochondria, and Plasma Membranes from *Neurospora crassa* and Modes of Discriminating among the Different $H^+$-ATPases

*By* EMMA JEAN BOWMAN and BARRY J. BOWMAN

Three $H^+$-translocating ATPases have been described in membranes of *Neurospora crassa*. The mitochondrial ATPase catalyzes the synthesis of ATP and has a complex structure consisting of 11 different polypetides, typical of the family of $F_0F_1$-ATPases.[1] The plasma membrane ATPase serves as an electrogenic pump, hydrolyzing ATP to generate an electrochemical proton gradient which can be used to drive a series of $H^+$-cotransport systems. This enzyme, an integral membrane protein, appears to be a dimer[2] of a large polypeptide of $M_r$ 104,000.[3,4] The vacuolar ATPase hydrolyzes ATP and generates a pH gradient (and possibly a membrane potential) that is coupled to the uptake of basic amino acids into the vacuolar interior. This enzyme has not yet been completely purified, but partial purification suggests that polypeptides of approximate $M_r$ 70,000, 60,000, and 15,000 are associated with activity.[5] We describe here the preparation of membrane fractions highly enriched for each of these

[1] W. Sebald, *Biochim. Biophys. Acta* **463**, 1 (1977).
[2] B. J. Bowman, *in* "The Na Pump" (I. M. Glynn and J. C. Ellory, eds.), p. 739. Company of Biologists, Cambridge, England, 1985.
[3] B. J. Bowman, F. Blasco, and C. W. Slayman, *J. Biol. Chem.* **256**, 12343 (1981).
[4] R. Addison and G. A. Scarborough, *J. Biol. Chem.* **256**, 13165 (1981).
[5] E. J. Bowman and B. J. Bowman, *in* "Biochemistry and Function of Vacuolar Adenosine-Triphosphatase in Fungi and Plants" (B. P. Marin, ed.), p. 131. Springer-Verlag, Berlin and New York, 1985.

ATPases and the assay procedures used to quantitate the amount of activity specifically due to each enzyme.

## Preparation of Plasma Membranes, Mitochondria, and Vacuolar Membranes

Our primary purpose in isolating membranes from *Neurospora* has been to characterize the $H^+$-translocating ATPases. Therefore, it has been important to obtain homogeneous preparations with a single type of ATPase. After solubilization the ATPases tend to copurify, making mixed-membrane preparations unsuitable for enzyme purification. The first critical step is the choice of a method for cell breakage. We have found two methods to be useful: (1) treatment of mycelia with crude $\beta$-glucuronidase to weaken cell walls, followed by gentle disruption in a glass–Teflon tissue grinder[6,7]; and (2) mechanical disruption of mycelia by mixing them vigorously with glass beads in a glass-bead blender, the Bead-Beater.[8,9] Either procedure can be used to prepare all three types of membranes. However, the $\beta$-glucuronidase procedure is preferred for plasma membranes because use of the Bead-Beater results in contamination by mitochondrial ATPase, presumably due to increased mitochondrial breakage. By contrast, Bead-Beater preparations of vacuolar membranes are superior because the greater rapidity of the procedure gives higher yields of this labile organelle. Mitochondria are typically taken from the Bead-Beater procedure.

### $\beta$-Glucuronidase Procedure

Wall digestion medium (WDM): 590 m$M$ sucrose, 5 m$M$ Na$_2$EDTA, 50 m$M$ NaH$_2$PO$_4 \cdot$ H$_2$O, adjusted to pH 6.8 with KOH
Sucrose, 680 m$M$ (in distilled water)
Mitochondrial preparation medium (MPM): 330 m$M$ sucrose, 1 m$M$ Na$_2$EDTA, bovine serum albumin, 0.3% (w/w), adjusted to pH 7.1 with KOH
EGTA–Tris: ethylene glycol bis($\beta$-aminoethyl ether)-$N$-$N'$-tetraacetic acid (EGTA), 1 m$M$ adjusted to pH 7.5 with Tris base
The essential features of the $\beta$-glucuronidase procedure are outlined in Fig. 1.

[6] A. M. Lambowitz, C. W. Slayman, C. L. Slayman, and W. D. Bonner, Jr., *J. Biol. Chem.* **247**, 1536 (1972).

[7] E. J. Bowman, B. J. Bowman, and C. W. Slayman, *J. Biol. Chem.* **256**, 12336 (1981).

[8] C. L. Cramer, J. L. Ristow, T. J. Paulus, and R. H. Davis, *Anal. Biochem.* **128**, 384 (1983).

[9] E. J. Bowman, *J. Biol. Chem.* **258**, 15238 (1983).

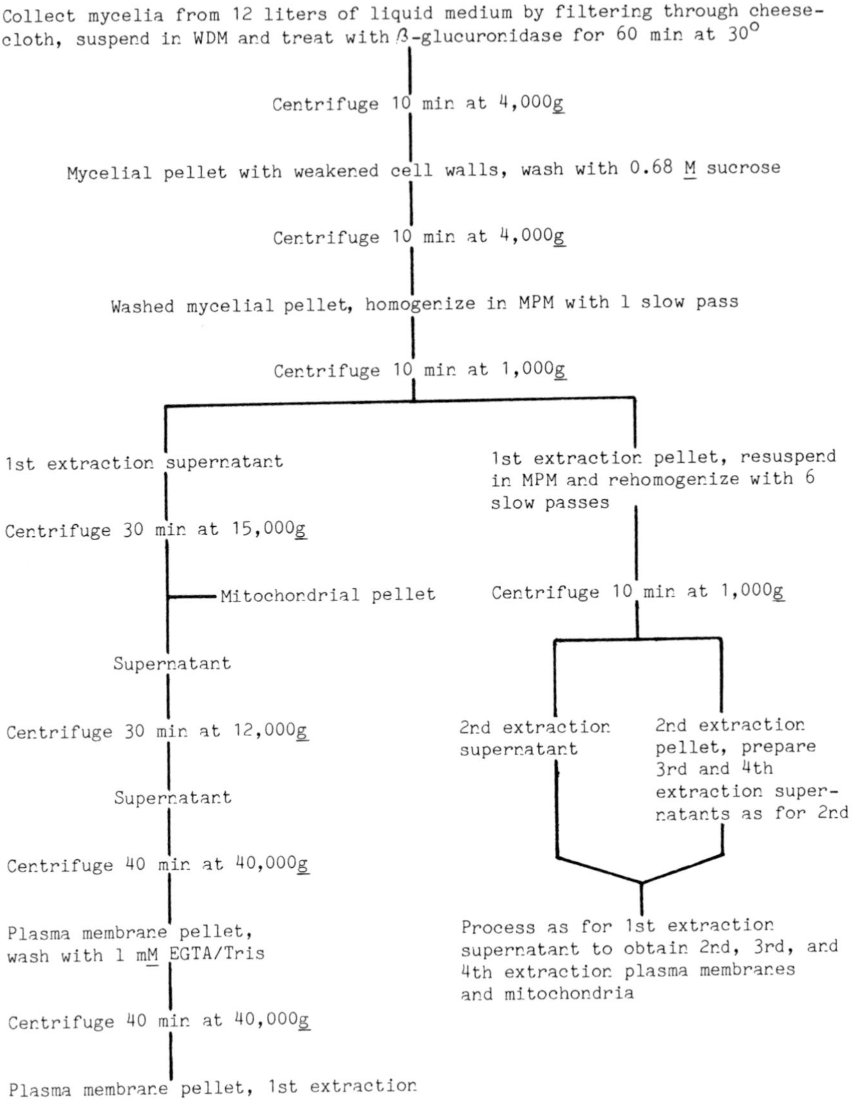

Collect mycelia from 12 liters of liquid medium by filtering through cheesecloth, suspend in WDM and treat with β-glucuronidase for 60 min at 30°

Centrifuge 10 min at 4,000g

Mycelial pellet with weakened cell walls, wash with 0.68 M sucrose

Centrifuge 10 min at 4,000g

Washed mycelial pellet, homogenize in MPM with 1 slow pass

Centrifuge 10 min at 1,000g

1st extraction supernatant

Centrifuge 30 min at 15,000g

Mitochondrial pellet

Supernatant

Centrifuge 30 min at 12,000g

Supernatant

Centrifuge 40 min at 40,000g

Plasma membrane pellet,
wash with 1 mM EGTA/Tris

Centrifuge 40 min at 40,000g

Plasma membrane pellet, 1st extraction

1st extraction pellet, resuspend
in MPM and rehomogenize with 6
slow passes

Centrifuge 10 min at 1,000g

2nd extraction
supernatant

2nd extraction
pellet, prepare
3rd and 4th
extraction super-
natants as for 2nd

Process as for 1st extraction
supernatant to obtain 2nd, 3rd, and
4th extraction plasma membranes
and mitochondria

FIG. 1. Purification of plasma membranes by the β-glucuronidase procedure.

*Cell Harvest.* Twelve liters of Vogel's minimal medium[10] supplemented with 2% sucrose is inoculated with 6- to 14-day-old conidia ($10^9$/
liter) and grown in a 25° water bath with vigorous aeration from a labora-

[10] H. J. Vogel, *Am. Nat.* **98**, 435 (1964).

tory compressed air line. (The procedure can be readily scaled up or down to use 4 to 24 liters of culture.) Mycelia are harvested after approximately 13 hr at a dry weight of 0.5–0.6 mg/ml (acetone pad[11]) by pouring the culture through four layers of cheesecloth in a large Büchner funnel. The mycelial pad is rinsed with distilled water and blotted with filter paper to remove excess water. The mycelial pad from 12 liters is divided into three portions of approximately 16 g (wet weight) and loosely shredded by hand. Each portion is placed in a 125-ml Erlenmeyer flask containing 33 ml of WDM and 70 μl of 2-mercaptoethanol. The three flasks are capped tightly with Parafilm and vigorously shaken by hand until the mycelia are evenly dispersed.

*β-Glucuronidase Digestion.* β-Glucuronidase (Type H1, Sigma Chemical Co.) should be prepared about 30 min before needed by rehydrating 1.2 g in 12 ml of WDM. This crude digestive enzyme is then added to the dispersed mycelia (4 ml/flask) and mixed gently by hand until the brown color is evenly distributed. The flasks are incubated for 1 hr at 30° either with gentle mixing every 15 min or with constant slow shaking. (For optimal results the exact amount of β-glucuronidase must be empirically determined: too little gives low yields and low specific activity ATPase, while too much results in contamination of the plasma membrane fractions by broken mitochondria.)

The treated mycelia are transferred to three 250-ml centrifuge bottles and centrifuged at 4000 $g$ for 10 min (GSA rotor, 5000 rpm). For this and all subsequent steps, samples are kept at 0–4° or on ice. The supernatants are poured off and discarded. To wash the pellets, 100 ml of 0.68 $M$ sucrose is added to each bottle without disturbing the pellet, and the bottles are returned to the centrifuge with the pellets near the center of the rotor. After centrifugation at 4000 $g$ for 10 min the supernatants are discarded. The pellets contain mycelia with weakened cell walls.

*Homogenization.* Individual mycelial pellets are coarsely suspended in 160 ml of MPM, using a glass stirring rod. Aliquots of approximately 40 ml are then poured into a 55-ml glass tissue grinder and homogenized with one slow pass (~1 min) of a motor-driven (~1100 rpm) Teflon pestle. (Potter–Elvehjem-type glass tissue grinders with smooth-tipped Teflon pestles came from Arthur H. Thomas Co., Philadelphia, PA.) The combined homogenate is divided among three 250-ml bottles and centrifuged at 1000 $g$ for 10 min (GSA rotor, 2500 rpm). Each 1000 $g$ pellet is coarsely suspended in 33 ml of MPM and saved for reextraction (see below). The supernatants are decanted through cheesecloth (to catch any loose pellet

[11] R. H. Davis and F. J. de Serres, this series, Vol. 17A, p. 79.

material) into clean bottles and centrifuged at 15,000 $g$ for 30 min (GSA rotor, 9500 rpm).

*Mitochondrial Fraction.* The 15,000 $g$ supernatants are collected into a beaker covered with a layer of cheesecloth. The pellets, which contain predominantly mitochondria, are gently suspended in 3–4 ml MPM, frozen in liquid nitrogen, and stored at $-70°$.

*Plasma Membrane Fraction.* The 15,000 $g$ supernatants are dispensed into 12 50-ml tubes and centrifuged at 12,000 $g$ for 30 min (SA600 rotor, 9000 rpm or SS34 rotor, 10,000 rpm). The supernatants are decanted off the small pellets (which contain a mixture of mitochondria and plasma membranes and are discarded) into clean tubes and centrifuged at 40,000 $g$ for 40 min (SA600 rotor, 16,500 rpm or SS34 rotor, 18,500 rmp) to pellet plasma membranes. The supernatants are completely removed by aspiration and discarded. Because the pellets are small and can slide to the bottom, the aspiration step should be done immediately after the centrifuge stops. The plasma membrane pellets are suspended with a Pasteur pipet in EGTA–Tris (1 ml/tube) and transferred to a 55-ml glass tissue grinder. EGTA–Tris is added to a volume of 40 ml, and the membrane suspension is washed by homogenizing (10 passes) at high speed with the motor-driven pestle. The plasma membranes are repelleted by centrifugation at 40,000 $g$ for 40 min. After complete removal of the supernatant by aspiration, the plasma membrane pellet is suspended in ~2ml (for 12 liters of mycelia) of EGTA–Tris, homogenized by hand in a 4-ml tissue grinder, dispensed in ~0.5 ml aliquots to microfuge tubes, frozen in liquid nitrogen, and stored at $-70°$.

*Reextraction of Mycelial Pellets.* The initial mycelial pellets from the first extraction, which were set aside in MPM (99 ml total), should be homogenized again. The procedure for purification of mitochondria and plasma membranes is identical to that for the first extraction except (1) six passes (instead of one) are made with the pestle during homogenization and (2) the first spin in 50-ml tubes (12,000 $g$ in the SA600 rotor, see Plasma Membrane Fraction above) is omitted. Third and fourth extractions in 75 and 33 ml MPM, respectively, are also worth doing. Since the volume of homogenate decreases with each extraction, the 1000 $g$ supernatants for second, third, and fourth extractions can be processed at the same time starting with the 30-min centrifugation step to pellet mitochondria.

*Membrane Yields.* From 12 liters of cells we obtain about 4 mg of plasma membrane protein with an ATPase specific activity of 2–4 U/mg in the first extraction. For reasons not understood, second, third, and fourth extractions plasma membranes consistently have higher ATPase specific activities, 3–7 U/mg. The relative yield in protein from the four extractions is typically ~$1.0:2.0:1.3:0.7$. The plasma membrane ATPase preparations are contaminated by endoplasmic reticulum. The endoplasmic reticulum can be separated from the plasma membranes by a sucrose gradient step before pelleting the plasma membranes,[12] but this results in lower yields of the plasma membranes. For assay or purification of the plasma membrane ATPase,[3] removal of the endoplasmic reticulum, which contains no detectable ATPase activity, is not necessary and is not routinely done.

Mitochondria obtained from all four extractions have similar ATPase specific activities of $1.0-2.0$ U/mg. Protein yields are high, typically 125 mg both in the first extraction and in the combined second, third, and fourth extractions. The mitochondria are contaminated with 2–5% vacuoles on a protein basis and can be further purified on sucrose gradients as described in the Bead-Beater procedure below.

*Bead-Beater Procedure*

Vacuolar preparation medium (VPM): 1 $M$ sorbitol, 1 m$M$ Na$_2$EDTA, 10 m$M$ N-2-hydroxyethylpiperazine-N'-2-ethanesulfonic acid (HEPES), adjusted to pH 7.5 with NaOH

Na$_2$ATP, 100 m$M$, adjusted to pH 7.5 with Tris base

Chymostatin stock solution, 20 mg/ml dimethyl sulfoxide

Sucrose gradient solutions: sucrose, 10% and 30%, 1 m$M$ Na$_2$EDTA, 2 m$M$ Na$_2$ATP, 2 $\mu$g/ml chymostatin, 10 m$M$ HEPES, adjusted to pH 7.5 with NaOH

EGTA–Tris/chymostatin/ATP (ECA): EGTA–Tris (1 m$M$, pH 7.5, from $\beta$-glucuronidase procedure), 2 $\mu$g/ml chymostatin (diluted from stock solution), 2 m$M$ Na$_2$ATP (diluted from pH 7.5 stock solution)

The essential features of the Bead-Beater procedure are outlined in Fig. 2.

*Equipment.* The Bead-Beater was bought from Biospec Products (Bartlesville, OK) with standard-sized chambers (340 ml). The ice jacket was found not to be necessary for our short beating times. Glass beads, lead-free, 0.3–0.5 mm diameter, were also obtained form Biospec Products. The initial beads were usable as delivered. Replacement beads, which have a gray tint of unknown identity, have unfortunately required extensive washing before use. We now cycle new beads through an overnight soak in 4 $N$ HCl, a detergent wash, and 2–3 sham homogenizations with *Neurospora* mycelia before using the beads to isolate vacuolar membranes. Without this washing, the vacuolar membrane ATPase activity is

[12] C. E. Borgeson and B. J. Bowman, *J. Bacteriol.* **156,** 362 (1983).

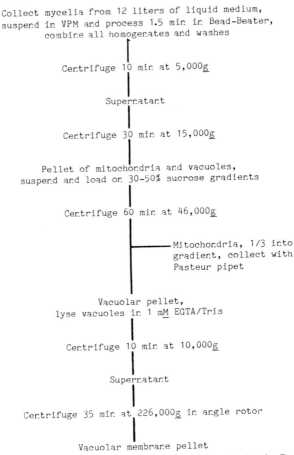

FIG. 2. Purification of vacuolar membranes and mitochondria by the Bead-Beater procedure.

quite low. Other fractions are less affected. Once cleaned, the beads can be reused.

*Mycelial Harvest.* *Neurospora* cultures are prepared as in the β-glucuronidase procedure except that the mycelia are harvested at a dry weight of 0.8–1.0 mg/ml (14.5–15 hr) and rinsed with VPM while still in the Büchner funnel.

*Homogenization.* All procedures are carried out on ice or at 0–4°. Glass beads (170 ml) are put in a Bead-Beater chamber with cold VPM. VPM is added to about two-thirds full. ATP (2 ml, 100 m$M$) and chymostatin (20 μl of stock solution) are added. The mycelial pad from 12 liters is divided into three parts, equal to ~3 g dry weight each. One third is placed in the chamber, which is then completely filled with VPM so that no air bubbles remain with the impeller in place. The mycelia are homogenized for 1.5 min, and the homogenate is decanted into a beaker in ice. The glass beads are rinsed twice (35 ml VPM containing 350 $\mu$l 100 m$M$ ATP and 3.5 $\mu$l chymostatin stock) by stirring with a glass rod and decanting off the liquid. The remaining two mycelial pads are treated in the same way. The three homogenates and six washes are combined in the beaker. The pooled homogenate is then divided among six 250-ml bottles and centrifuged at 1000 $g$ for 10 min (GSA rotor, 2500 rpm) to pellet unbroken mycelia and cell debris.

*Mitochondria and Vacuoles.* The low-speed supernatants are decanted into clean bottles and centrifuged at 15,000 $g$ for 30 min (GSA rotor, 9500 rpm). The 15,000 $g$ supernatants are poured off and either discarded or used to prepare plasma membranes (see below). Supernatant that remains in the bottles and loose material from the pellets should be aspirated and discarded. The pellets contain mitochondria (95–98% of the protein) and vacuoles (2–5% of the protein), which can be separated on sucrose gradients prepared the day before or during the 30-min centrifugation. (Gradients consist of a 5-ml cushion of 30% sucrose solution pipetted into a 16-ml centrifuge tube with a 6-ml 10–30% linear sucrose gradient poured over the cushion.) Using a small Teflon pestle, the organellar pellets are suspended in approximately 3.0 ml of VPM supplemented with ATP (2 m$M$) and chymostatin ( 2 $\mu$g/ml) to give a total volume of 6–8 ml, transferred to a 10-ml glass tissue grinder, and gently homogenized by hand. The suspension is layered onto sucrose gradients (four for a 12-liter culture) and centrifuged at 46,000 $g$ for 1 hr (SS34 rotor, 19,500 rpm). Mitochondria form a prominent band located about one-third down the gradients. They are collected with a Pasteur pipet, placed in microfuge tubes, frozen in liquid nitrogen, and stored at $-70°$. The remaining liquid in the gradients is completely aspirated off, leaving a small pellet of vacuoles. To save whole vacuoles, the pellets are suspended in a ~0.5 ml VPM (+2 m$M$ ATP and 2 $\mu$g/ml chymostatin), frozen, and stored at $-70°$.

*Vacuolar Membranes.* To prepare vacuolar membranes the vacuolar pellets are suspended in ECA with a small pestle, transferred to a 55-ml glass tissue grinder, brought to 30 ml with more ECA, and homogenized for 10 passes with a motor-driven pestle at top speed. The homogenate is centrifuged at 10,000 $g$ for 10 min (SS34 rotor, 9000 rpm). The pellets, containing residual cell wall and mitochondrial fragments, are discarded. The supernatants are divided into four 10-ml Oak Ridge-type centrifuge tubes and centrifuged at 226,000 $g$ for 35 min (Beckman type 50Ti rotor, 50,000 rpm). (Using a swinging bucket rotor at this step results in lower activity of vacuolar membrane ATPase). The supernatants are removed by aspiration and discarded. The pellets of vacuolar membranes are suspended in 0.3 ml ECA, transferred to a 1-ml tissue grinder, homogenized by hand, divided among three to four microfuge tubes, frozen in liquid nitrogen, and stored at −70°.

*Plasma Membranes.* The 15,000 $g$ supernatant obtained after pelleting mitochondria and vacuoles is centrifuged at 46,000 $g$ for 1 hr. (This can be done at the same time the sucrose gradients are centrifuged.) After aspirating off the supernatants, the unwashed plasma membrane pellets are suspended in 40 ml of ECA and homogenized in a 55-ml tissue grinder with 10 passes at top speed of the motor. The membranes are repelleted in 50-ml tubes by centrifuging at 46,000 $g$ for 40 min. Supernatants are completely removed by aspiration. The plasma membrane pellet is suspended in 0.5–1.0 ml ECA, homogenized by hand in a 4-ml tissue grinder, and stored at −70° as above.

*Membrane Yields.* From 12 liters of mycelia we obtain 1.0–2.0 mg of protein in the vacuolar membranes, which have a vacuolar ATPase specific activity of 2–4 U/mg. We do not aim for quantitative recovery of gradient purified mitochondria, but can easily collect 150 mg of protein with mitochondrial ATPase specific activities of 1.2–2.2 U/mg. If all the 15,000 $g$ supernatant is used for plasma membrane isolation, ∼25 mg of protein is obtained in the plasma membranes, which have a plasma membrane ATPase specific activity of 3–6 U/mg. This plasma membrane preparation is useful for routine assays and comparisons with the two other membrane fractions; however, because of some contamination from mitochondrial membranes we recommend the $\beta$-glucuronidase procedure for preparing purified plasma membrane ATPase.

## Discriminating among the Three H⁺-ATPases

### Activity Assays

*Rationale.* The most straightforward procedure for distinguishing among the three proton pumps is to measure their ATPase activities. The three membrane ATPases are characterized by different pH optima and specific inhibitor sensitivities. The pH optima for the plasma membrane, vacuolar membrane, and mitochondrial ATPases are, respectively, 6.7,[3] 7.4,[13] and 8.3.[14] The most useful inhibitors include vanadate, specific for the plasma membrane ATPase; azide, specific for the mitochondrial ATPase in its soluble or membrane-bound form; and oligomycin, specific

[13] E. J. Bowman and B. J. Bowman, *J. Bacteriol.* **151**, 1326 (1982).
[14] S. E. Mainzer and C. W. Slayman, *J. Bacteriol.* **133**, 584 (1978).

for the membrane-bound mitochondrial ATPase.[15] Although an inhibitor with high affinity for the vacuolar membrane ATPase has not been identified, the vacuolar activity is inhibited selectively by high concentrations of KSCN or KNO$_3$ and is completely resistant to both vanadate and azide.[9]

*Assay Procedure.* ATPase activities in membrane fractions are measured at 30° by the liberation of inorganic phosphate as follows: (1) For the assay of plasma membrane ATPase, the reaction mixture contains 5 mM Na$_2$ATP, 5 mM MgCl$_2$, 15 mM NH$_4$Cl, 3 mM phosphoenolpyruvate (monopotassium salt), 25 μg pyruvate kinase, 5 mM KN$_3$ (to inhibit residual mitochondrial ATPase), and 10 mM piperazine-*N,N'*-bis(2-ethanesulfonic acid) (PIPES) buffer, adjusted to pH 6.7 with Tris base. (2) For the assay of mitochondrial ATPase, the reaction mixture contains 5 mM Na$_2$ATP, 5 mM MgCl$_2$, 15 mM NH$_4$Cl, 3 mM phosphoenolpyruvate, 25 μg pyruvate kinase, and 10 mM PIPES buffer, adjusted to pH 8.3 with Tris base. (3) The vacuolar membrane ATPase reaction mixture consists of 5 mM Na$_2$ATP, 5 mM MgSO$_4$, 10 mM NH$_4$Cl, 5 mM KN$_3$, 0.1 mM Na$_3$VO$_4$ (to inhibit residual plasma membrane ATPase and vacuolar alkaline phosphatase), and 10 mM PIPES buffer, pH adjusted to 7.4 with Tris base. (4) For each determination 0.5 ml of reaction mixture is dispensed to disposable plastic tubes (to avoid contamination with phosphate) and preincubated for 10 min at 30°. The ATPase reaction is started by addition of membranes (3–10 μg protein/assay tube) at timed intervals and stopped 10–30 min later with 1.25 ml of Fiske–Subbarow reagent[16] containing 0.5% sodium dodecyl sulfate. Ten minutes after stopping the last tube, $A_{660\,nm}$ is read on a spectrophotometer.

*Comments.* Pyruvate kinase is employed to regenerate ATP in the plasma membrane and mitochondrial assays to avoid inhibition of the ATPases by ADP. The regenerating system is not used in the vacuolar membrane mixture because of the possibility of inorganic phosphate release from phosphoenolpyruvate by vacuolar phosphatases; fortunately, no inhibition of vacuolar membrane ATPase is seen when as much as 14% of the ATP is hydrolyzed during an assay. ATPase activities in our membrane fractions show 90–98% inhibition when diagnostic inhibitors are added to the reaction mixtures: 20 μM Na$_3$VO$_4$ for the plasma membrane ATPase, 5 mM KN$_3$ or oligomycin (10 μg/μl from an ethanolic stock solution) for the mitochondrial ATPase, and 100 mM KSCN or KNO$_3$ for the vacuolar membrane ATPase. Similarly, preincubation of any of

[15] B. J. Bowman, S. E. Mainzer, K. E. Allen, and C. W. Slayman, *Biochim. Biophys. Acta* **512,** 13 (1978).
[16] C. H. Fiske and Y. Subbarow, *J. Biol. Chem.* **66,** 375 (1925).

the membrane fractions with 100 $\mu M$ $N,N'$-dicyclohexylcarbodiimide (DCCD) results in 90–95% inhibition of ATPase activity.

### Criteria for Purity of Membrane Fractions

*Enzyme Markers.* With a primary aim of studying and purifying the $H^+$-translocating ATPases of *Neurospora*, we consider the three ATPases, which can be assayed specifically as described above, the best enzyme markers for plasma membranes, mitochondria, and vacuolar membranes. On the basis of ATPase specific activity, all three membrane fractions are at most 5% contaminated by the other two ATPase-containing membranes. Chitin synthetase serves as an additional marker for plasma membranes,[7] and cytochrome oxidase and succinate dehydrogenase are standard mitochondrial markers. Although other enzyme markers for vacuolar membranes are uncertain, several enzymes are localized in whole vacuoles and remain partially associated with the membranes: these include 5′-AMPase, $\alpha$-mannosidase, and protease.[13] Furthermore, acid-soluble, sedimentable arginine constitutes an excellent marker for intact vacuoles.[17] Using these marker enzyme activities, we find no more than 5% contamination of any of the three membrane fractions by the other two. However, as mentioned above, our plasma membrane fractions are contaminated with endoplasmic reticulum.[12] We identify the endoplasmic reticulum with the enzyme activity phosphatidylcholine glyceride transferase and have thus far failed to detect ATPase activity. To our knowledge other membrane fractions, which may well contain $H^+$-ATPases, have not yet been identified in extracts of *Neurospora crassa*.

*[14C]DCCD Labeling.* All three ATPases are inhibited by DCCD, which blocks proton translocation. The subunits of mitochondrial and plasma membrane ATPases to which DCCD binds have been identified. Consequently, incubating membrane fractions with radioactive DCCD and locating labeled protein bands by fluorography proves to be a sensitive method for detecting contamination by ATPase-containing membranes. Mitochondrial membranes can be identified by the presence of a DCCD-binding protein of $M_r$ 8000 in which the labeling is prevented by preincubation with venturicidin.[9,18] Weaker labeling of the $\beta$-subunit of the mitochondrial ATPase ($M_r$ 56,000) can also be detected. The presence of plasma membranes is indicated by the labeling (though weaker than for mitochondria membranes) of its $M_4$ 104,000 polypeptide.[9,19] In vacuolar

[17] L. E. Vaughn and R. H. Davis, *Mol. Cell. Biol.* **1,** 797 (1981).
[18] W. Sebald and J. Hoppe, *Curr. Top. Bioenerg.* **12,** 1 (1981).
[19] M. R. Sussman and C. W. Slayman, *J. Biol. Chem.* **258,** 1839 (1983).

membranes, specific peptides have not yet been conclusively associated with the membrane ATPase. However, no labeling by DCCD of bands of $M_r$ either 8,000 or 104,000 can be detected in our current preparations. (In previous preparations, this procedure was instrumental in detecting contamination by mitochondria when mitochondrial ATPase activity was not detectable.[9]) However, a single band of $M_4$ ~15,000 is labeled in both vacuolar membranes[9] and partially purified vacuolar ATPase preparations[20]; this band is not visible in mitochondria or plasma membranes.

*Immunological Cross-reactivity.* Using rabbit antisera prepared in other laboratories[21] and donated to us, we have found that the three H$^+$-ATPases show no immunological cross-reactivity by Western blot analysis,[22] but that antisera to individual ATPases provide an extremely sensitive method for detecting contamination of membrane fractions.[20] Antiserum to the plasma membrane ATPase of *Neurospora*, which cross-reacts with a protein of $M_r$ 104,000 but not with lower $M_r$ bands, reveals small amounts of plasma membrane contamination in mitochondrial and vacuolar membrane preparations. Antiserum raised against the F$_1$ complex (or the $\beta$-subunit) of the *E. coli* ATPase cross-reacts strongly and specifically with a protein of $M_r$ 56,000 (the $\beta$-subunit of the ATPase) in mitochondria; this antiserum reveals essentially no mitochondrial ATPase contamination in either vacuolar membranes or snail enzyme plasma membranes, but does show contamination of Bead-Beater plasma membranes. Finally, antiserum to the $M_r$ ~70,000 protein of corn vacuolar membranes shows strong cross-reactivity with a band of identical $M_r$ in *Neurospora* vacuolar membranes and reveals barely detectable amounts of this band in plasma membrane and mitochondrial fractions.

### Acknowledgment

This work was supported by NIH Grants GM-28703 and RR-08132.

[20] E. J. Bowman, unpublished experiments, 1984.
[21] Antibody against the *Neurospora* plasma membrane ATPase was donated by Dr. Carolyn Slayman of Yale University. Antibody to the *E. coli* F$_1$-ATPase or the $\beta$-subunit came from Dr. R. D. Simoni of Stanford University. Antibody to the $M_r$ ~70,000 protein of corn vacuolar membranes was contributed by Dr. L. Taiz of the University of California, Santa Cruz.
[22] R. Rott and N. Nelson, *J. Biol. Chem.* **256**, 9224 (1981).

## [43] Large-Scale Purification of Plasma Membrane H+-ATPase from a Cell Wall-Less Mutant of *Neurospora crassa*

*By* GENE A. SCARBOROUGH

### Introduction

The plasma membrane of the filamentous fungus, *Neurospora crassa,* contains an electrogenic, proton-translocating ATPase which functions to generate a transmembrane electrochemical protonic potential difference that is utilized by substrate-specific porters to energize the cellular accumulation or extrusion of a variety of specific ions, nutrients, and metabolites. Two original published methods for the purification of this ATPase[1,2] involved isolation of plasma membranes, solubilization of the ATPase with detergents, and subsequent purification of the enzyme by Sepharose CL-6B chromatography and/or glycerol density gradient sedimentation. Both of these methods yielded reasonably pure ATPase, but they were both fairly time consuming and the yields were in the range of only a few hundred micrograms. These shortcomings prompted the development of a large-scale isolation procedure[3] which, along with some minor improvements, is described in this article. With this procedure, 50- to 100-mg quantities of highly purified, high specific activity H+-ATPase can be readily and reproducibly prepared in a relatively small amount of time.

### Principles

The cell breakage problem associated with the use of wild-type *Neurospora* is obviated by the use of a cell wall-less mutant. Cells of the cell wall-less mutant are treated with concanavalin A (Con A), which coats and stabilizes the plasma membrane. The Con A-treated cells are then homogenized in the presence of a high concentration of deoxycholate which solubilizes most of the cellular constituents, including certain plasma membrane proteins, but does not solubilize the complex between Con A and at least some of its receptors or the 100,000-Da H+-ATPase. Thus, simple centrifugation of the deoxycholate lysate results in a pellet that is greatly enriched in the plasma membrane H+-ATPase. The pelleted

[1] B. J. Bowman, F. Blasco, and C. W. Slayman, *J. Biol. Chem.* **256,** 12343 (1981).
[2] R. Addison and G. A. Scarborough, *J. Biol. Chem.* **256,** 13165 (1981).
[3] R. Smith and G. A. Scarborough, *Anal. Biochem.* **138,** 156 (1984).

material is subsequently washed and treated with α-methylmannoside to remove the bulk of the Con A, and is then treated with lysolecithin, which solubilizes the $H^+$-ATPase but does not so efficiently dissolve the major remaining Coomassie Blue-stainable contaminating proteins. The solubilized ATPase is then further purified by glycerol density gradient centrifugation, a step which is extremely effective because of the propensity of the ATPase to form high-molecular-weight aggregates and the tendency of most of the contaminants to remain as lower-molecular-weight species.

### Procedures

#### Growth of Cells

The cells used are a variant strain of the cell wall-less *fz;sg;os*-1 strain of *Neurospora,* designated *fz;sg;os*-1 V.[4] The procedure does not work nearly as well with *fz;sg;os*-1. The growth medium is Vogel's medium N[5] (without chloroform as a preservative) supplemented with 2% (w/v) mannitol, 0.75% (w/v) yeast extract (Difco), and 0.75% (w/v) nutrient broth (Difco), sterilized by autoclaving. Cells are routinely maintained by transfer of 1 ml of a 1- to 3-day 50-ml culture contained in a 250-ml Erlenmeyer flask into 50 ml of fresh growth medium and rotary shaking (150 rpm) at 30° for 1–3 days. Two days before the ATPase isolation is to be carried out, three 500-ml portions of growth medium contained in 1-liter flat-bottomed boiling flasks are each inoculated with a 50-ml overnight culture followed by growth overnight as above. The resulting cultures are then transferred to three 5-liter portions of growth medium contained in 12-liter flat-bottomed boiling flasks, and grown for about 22 hr with rotary shaking (100 rpm, 30°) with a stream of sterile $O_2$ (approximately 1.5 liters/min) passing sequentially over the surface of each of the cultures. The optical density (650 nm, 1 cm) of such cultures is about 12, as calculated from the optical density of an appropriate dilution.

#### Isolation of the H⁺-ATPase

The cells are harvested by centrifugation (650 g, 15 min, 4°) in 1-liter plastic bottles and resuspended by swirling to a total volume of 2 liters with ice-cold buffer A (0.05 $M$ Tris containing 0.01 $M$ $MgSO_4$ and 0.25 $M$ mannitol, pH 7.5 with HCl, filtered through an ethanol-washed AAWP Millipore filter). All resuspension steps in the procedure are best accom-

---

[4] G. A. Scarborough, *Exp. Mycol.* **9,** 275 (1985). The variant strain is obtainable from the author or the Fungal Genetics Stock Center, University of Kansas Medical Center, Kansas City, KS 66103.

[5] H. J. Vogel, *Am. Nat.* **98,** 435 (1964).

plished by first resuspending pellets in an approximately equal volume of the resuspension solution and then adding the remaining volume. The resulting cell suspension is then apportioned equally into four 1-liter plastic bottles and the cells are then pelleted again (290 $g$, 15 min, 4°), resuspended in 900 ml of ice-cold buffer A (225 ml per bottle) pelleted once more (290 $g$, 15 min, 4°), and resuspended in 1200 ml of room temperature (~22°) buffer A (300 ml per bottle). The cells are then treated with Con A by mixing the contents of each bottle with 370 ml of room temperature buffer A containing 0.7 mg/ml of Con A (Calbiochem, not corrected for salt content), incubating the resulting mixture at room temperature for 10 min, and chilling it on ice for 10 min. A copious precipitate of agglutinated cells forms almost immediately after the Con A solution and cell suspension are mixed. The agglutinated cells are then pelleted by brief centrifugation (200 $g$, 1 min, 4°), gently resuspended in 900 ml of ice-cold buffer A (225 ml per bottle), and pelleted again (200 $g$, 6 min, 4°), after which they are resuspended and quickly combined into a total of 2.7 liters of cold lysis buffer in the vessel of a 3.8-liter Plexiglas homogenizer[6] embedded in ice in a cold room. The lysis buffer is prepared just before use by mixing 324 ml of room temperature 10% (w/v) sodium deoxycholate (Sigma) solution (Millipore filtered as above) with 2376 ml of ice-cold buffer B (0.01 $M$ Tris containing 1.5 $\mu$g/ml of chymostatin (Sigma), pH 7.5 with HCl).

After 50 passes of the homogenizer during a 10-min period, the somewhat viscous cell lysate is then centrifuged (13,500 $g_{max}$, 30 min, 4°) to sediment the $H^+$-ATPase-enriched plasma membrane sheets. The membranes are then washed by alternate resuspension and centrifugation (14,000 $g_{max}$, 15 min, 4°), first with 1.5 liters of a 0.6% (w/v) deoxycholate solution prepared just before use by mixing 90 ml of the room temperature 10% sodium deoxycholate solution described above with 1410 ml of ice-cold buffer B, and then with 1.5 liters of ice-cold buffer B alone. All membrane resuspensions are performed with the aid of an artist's 0.5-in. nylon paint brush. The washed membranes are then resuspended in a total of 450 ml of room temperature 0.5 $M$ α-methylmannoside (Sigma, recrystallized from 85° water) in buffer B, and incubated at 30° for 5 min to dissociate the bulk of the Con A from the membranes. The membrane

---

[6] The homogenizer is readily fashioned from commercially available Plexiglas rod and tubing. The vessel consists of a Plexiglas tube (10.1 cm inside diameter, 11.3 cm outside diameter, 62 cm length) plugged at one end with a Plexiglas disk 1.5 cm in thickness. The pestle consists of a Plexiglas rod (2.6 cm diameter, 61 cm length) attached concentrically to a second piece of Plexiglas rod (9.8 cm diameter, 7 cm length). A cross bar (1.2 cm diameter, 10 cm length) at the end opposite the larger piece of rod serves as a handle. With the vessel filled with distilled water, the drop time of the pestle is about 50 sec at room temperature.

suspension is then diluted to 1500 ml with ice-cold buffer B and centrifuged again (14,000 $g$, 30 min, 4°). At this point the white, plasma membrane-derived pellets are physically separated from a small underlying dark pellet utilizing small amounts of buffer B and the paint brush, resuspended in a total of 250 ml of ice-cold buffer B, and the resulting suspension is centrifuged again (14,000 $g$, 30 min, 4°) to sediment the H$^+$-ATPase-enriched membranes. The membranes are then resuspended in room temperature ATPase solubilization buffer [1 m$M$ disodium ATP containing 1 m$M$ MgSO$_4$, 0.1 m$M$ Na$_3$ VO$_4$, 2 $\mu$g/ml chymostatin, pH 7.5 with Tris, and 15 mg/ml egg lysolecithin (Sigma)] to a final volume of 62 ml, and the resulting opaque white suspension is incubated at room temperature for 15 min with occasional agitation, after which it is centrifuged (15,000 $g$, 10 min, room temperature).

The still somewhat opaque supernatant fluid, which contains the bulk of the H$^+$-ATPase, is then adjusted to 0.2% (w/v) sodium deoxycholate using the abovementioned 10% sodium deoxycholate solution, and approximately 5.1-ml portions are then layered over 12 ice-cold 30-ml linear glycerol gradients [20–40% (w/v) glycerol in 2 m$M$ disodium ATP containing 2 m$M$ EDTA, 2 $\mu$g/ml chymostatin, 0.2% (w/v) sodium deoxycholate, and 1 m$M$ dithiothreitol, pH 6.8 with Tris] using a peristaltic pump in a cold room. The gradients are then centrifuged (31,000 rpm, 18 hr, 4°) in a Beckman 50.2 Ti fixed angle rotor, which sediments the ATPase to a region near the bottom of the gradients. One of the gradients is then fractionated and assayed for ATPase activity. It is unnecessary and potentially misleading, because of detergent effects, to assay for ATPase activity before this point. Fractions containing significant amounts of ATPase activity are then pooled into a centrifuge tube the same as that used for the gradients, and the resulting fluid volume is used as an indicator of the ATPase-containing portion of the other 11 gradients. The upper portions of each of the 11 remaining gradients are then aspirated and discarded accordingly, and the remaining portions of each tube are decanted and pooled, and the pooled fractions (usually about 160 ml) are assayed for ATPase activity and protein content, and then stored at −20°. The yield of protein is between 50 and 100 mg.

The ATPase preparation is approximately 85% pure as judged by quantitative densitometry of Coomassie Blue R-250-stained sodium dodecyl sulfate-containing polyacrylamide gel electrophoresis gels, and the specific activity is 20–30 $\mu$mol/mg protein/min. This is a minimum range as the enzyme activity can be increased by about 50% by the inclusion of an additional 0.03% (w/v) sodium deoxycholate in the assay medium. These values are also not altered to take into account the reported[1] overestimation of the ATPase protein content by the Lowry protein assay. The isolation procedure has been carried out more than 50 times with essentially identical results every time. The ATPase appears to be stable indefinitely when stored under the above conditions. Virtually no loss of activity could be measured in two different preparations assayed occasionally over a period of 1 year. Turbidity due to the formation of small spherical particles (as judged by light microscopy) usually develops in the preparation, sometimes as early as during the gradient harvesting process, and sometimes much later during storage. The particles do not contain much, if any, ATPase and can be removed by centrifugation at 12,000 $g$ for 10 min.

### Analytical Procedures

ATPase activity is measured as follows. Five microliters of the individual or pooled gradient fractions are added to a mixture containing 5 $\mu$l of 0.2 $M$ disodium ATP/MgSO$_4$ (pH 6.8 with Tris), 5 $\mu$l of 0.1 $M$ NaN$_3$ (pH 6.8 with Tris), 40 $\mu$l of 0.1 $M$ 4-morpholinoethanesulfonic acid (pH 6.8 with Tris), 4 $\mu$l of 0.5 $M$ (NH$_4$)$_2$SO$_4$, 2 $\mu$l of 0.1 $M$ EDTA (pH 6.8 with Tris), 10 $\mu$l of a sonicated suspension of Folch Fraction I from bovine brain (Sigma, 5 mg/ml in H$_2$O), and 29 $\mu$l of H$_2$O, and the resulting assay mixture is incubated for 5 min at 30°, after which the reaction is terminated by the addition of 100 $\mu$l of 5% (w/v) sodium dodecyl sulfate solution. The liberated inorganic phosphate in the samples is then determined essentially according to the method of Stanton.[7] Protein content is determined by the procedure of Lowry et al.[8] after precipitation by the deoxycholate–trichloroacetic acid method of Bensadoun and Weinstein[9] with bovine serum albumin as a standard. Quantitative amino acid analyses indicate that this method of protein estimation provides a reasonably accurate indication of the amount of ATPase present. A procedure for highly efficient reconstitution of the H$^+$-ATPase and assays for protonophore-stimulated ATP hydrolysis and ATP-dependent proton translocation by the reconstituted ATPase-bearing proteoliposomes have also been described.[10]

### Isolation of Essentially Homogeneous ATPase

For chemical analyses and other investigations that do not require catalytically active enzyme, essentially pure, inactive ATPase is prepared as follows. The gradient-purified ATPase solution ($\sim$10 mg of protein) is centrifuged as described above to remove any turbidity that has accumu-

[7] M. G. Stanton, Anal. Biochem. 22, 27 (1968).
[8] O. H. Lowry, N. J. Rosebrough, A. L. Farr, and R. J. Randall, J. Biol. Chem. 193, 265 (1951).
[9] A. Bensadoun and D. Weinstein, Anal. Biochem. 70, 241 (1976).
[10] G. A. Scarborough and R. Addison, J. Biol. Chem. 259, 9109 (1984).

lated, and the supernatant fluid is then dialyzed against 100–200 volumes of cold (6°) water containing 1.5 $\mu$g/ml chymostatin and 0.1 m$M$ phenylmethylsulfonyl fluoride (PMSF, Sigma) over a period of 3 days with several changes, whereupon virtually all of the ATPase precipitates. The ATPase is then pelleted by centrifugation (12,000 $g$, 15 min, 4°), resuspended in 5 ml of room temperature disaggregation buffer [0.05 $M$ Tris, pH 6.8 with H$_3$PO$_4$, containing 4% (w/v) sodium dodecyl sulfate, 2% (v/v) 2-mercaptoethanol, 2 m$M$ EDTA, 20% (v/v) glycerol, 4 $\mu$g/ml chymostatin, and 0.2 m$M$ PMSF, heated to the boiling point and then cooled], and the resulting mixture is gently stirred at room temperature for 2 hr. The solution is then centrifuged (116,000 $g$ at $r_{max}$, 1 hr, room temperature) and the resulting supernatant fluid is applied to a 2.6 cm × 92 cm column of Sephacryl S-300 (Sigma) preequilibrated at room temperature with elution buffer [0.05 $M$ Tris, pH 6.8 with H$_3$PO$_4$, containing 0.1% (w/v) sodium dodecyl sulfate, boiled 5 min, cooled, and then adjusted to 1 m$M$ 2-mercaptoethanol, 0.1 m$M$ PMSF, and 1.5 $\mu$g/ml chymostatin]. The column is then eluted at a flow rate of 1 ml/min and the eluate is monitored for absorbance at 280 nm. The first peak emerging near the void volume is discarded, and the second, much larger peak, which contains the ATPase, is collected, dialyzed against water (6°) for 24 hr with three changes, and lyophilized.

### Acknowledgment

This work was supported by U.S. Public Health Service National Institutes of Health Grant GM 24784.

## [44] H$^+$-ATPase from Vacuolar Membranes of Higher Plants

By Alan B. Bennett, Roger A. Leigh, and Roger M. Spanswick

### Introduction

The biochemical study of ion transport in plant cell membranes has been advanced by the development of methods to isolate sealed membrane vesicles from a variety of plant tissues.[1–3] The results obtained with

[1] H. Sze, *Proc. Natl. Acad. Sci. U.S.A.* **77**, 5904 (1980).
[2] F. M. DuPont, D. L. Giorgi, and R. M. Spanswick, *Plant Physiol.* **70**, 1694 (1982).
[3] A. B. Bennett, S. D. O'Neill, and R. M. Spanswick, *Plant Physiol.* **74**, 538 (1984).

preparations of plant membrane vesicles have demonstrated the presence of two distinct $H^+$-translocating ATPases of nonmitochondrial origin.[3,4] Because both the plasma membrane and tonoplast (vacuolar membrane) maintain substantial gradients of $\Delta\bar{\mu}_{H^+}$ it was anticipated that $H^+$-ATPases from both membranes contributed to the activity measured in relatively crude membrane preparations. The presence of multiple ATPases, however, confounded attempts to characterize the activities present and to assign a single $H^+$-ATPase activity to a particular subcellular membrane. In order to clarify our view of the biochemical basis of $H^+$ transport across both the plasma membrane and tonoplast of plant cells, it was necessary to develop procedures for clearly separating these two membranes into distinct fractions and in sufficient quantity to allow detailed characterization of the respective $H^+$-ATPases. The storage tissue (swollen hypocotyl) of red beet has proved to be a good source of membranes because (1) the tissue is commercially available year-round, (2) the densities of tonoplast and plasma membranes are sufficiently different to allow good separation by density gradient centrifugation, and (3) relatively easy methods are available to isolate and purify intact vacuoles from this tissue.[5]

In this report we will describe procedures to isolate and assay the $H^+$-ATPase from red beet tonoplast. Intact vacuoles isolated from red beet storage tissue have been important in establishing the major characteristics of the tonoplast $H^+$-ATPase,[6] whereas tonoplast membrane vesicles isolated from tissue homogenates can be prepared in large quantities and have been useful in characterizing $H^+$ transport catalyzed by this $H^+$-ATPase.[3,7,8] Here we will review both methods of membrane preparation and illustrate the salient properties of the $H^+$-ATPase from red beet tonoplast. The characteristics of the red beet tonoplast $H^+$-ATPase are similar to those described for other higher plant tonoplast $H^+$-ATPases.[2,4,9]

### Materials and Methods

*Plant Material.* Red beets can be obtained commercially or grown in the field or greenhouse. If purchased, care must be taken to ensure that they were freshly harvested since membrane densities shift upon prolonged storage. We purchase red beets with fresh leafy tops to ensure

[4] K. A. Churchill, B. Holaway, and H. Sze, *Plant Physiol.* **73**, 921 (1983).
[5] R. A. Leigh and D. Branton, *Plant Physiol.* **58**, 656 (1976).
[6] R. R. Walker and R. A. Leigh, *Planta* **153**, 140 (1981).
[7] A. B. Bennett and R. M. Spanswick, *Plant Physiol.* **74**, 545 (1984).
[8] R. J. Poole, D. P. Briskin, Z. Kratky, and R. M. Johnstone, *Plant Physiol.* **74**, 549 (1984).
[9] B. Marin, *Plant Physiol.* **73**, 973 (1983).

freshness. The leafy tops can be removed and the storage tissue stored at 4° in moist vermiculite for up to 1 month without affecting membrane properties.

### Solutions—Procedure 1

Buffer 1A: 250 m$M$ sucrose containing 70 m$M$ Tris-Cl, 4 m$M$ DTT, 3 m$M$ Na$_2$EDTA, 0.5% polyvinylpyrrolidine (PVP-40), 0.1% bovine serum albumin (BSA), pH 8.0

Buffer 1B: 250 m$M$ sucrose containing 10 m$M$ Tris-MES, 2 m$M$ DTT, pH 7.0

Sucrose solutions: 16%, 26%, 34%, 40% (w/w) sucrose containing 10 m$M$ Tris-MES, 2 m$M$ DTT, pH 7.0

### Solutions—Procedure 2

Buffer 2A: 1 $M$ sorbitol containing 50 m$M$ Tris-MES, 5 m$M$ EDTA, 4 m$M$ DTT, 0.5% polyvinylpyrrolidine (PVP-40), pH 8.0

Buffer 2B: 1.2 $M$ sorbitol containing 25 m$M$ Tris-MES, 1 m$M$ EDTA, 2 m$M$ DTT, pH 7.0

Sodium diatrizoate solutions: 10%, 15% (w/v) sodium diatrizoate containing 1.2 $M$ sorbitol, 25 m$M$ Tris-MES, 1 m$M$ EDTA, 2 m$M$ DTT, pH 7.0. Sodium diatrizoate may be obtained in crystalline form (Sigma) or obtained as a sterile 76% (w/v) solution (Renografin-76, Squibb) from a medical or veterinary hospital pharmacy.

### Membrane Preparation—Procedure 1

The isolation of tonoplast vesicles from red beet tissue homogenates is outlined in Fig. 1. All buffers, glassware, and rotors are cooled to 4° and the homogenate is maintained on ice throughout the procedure. Red beet storage tissue is peeled and quartered before homogenization. Harsh methods of tissue disruption are required because of the rigidity of the tissue; we have found that homogenization for 1 min in a blender at a medium setting gives good results. The use of a juice extractor, as previously described for the isolation of red beet plasma membranes,[10] did not preserve H$^+$ transport activity. The presence of BSA and polyvinylpyrrolidone in the homogenization buffer (buffer 1A) causes foaming during homogenization. The foam is allowed to settle for approximately 20 min in the refrigerator before filtering the homogenate through cheesecloth.

We routinely use a Beckman SW 28 rotor for all centrifugations, although other rotors should be suitable. Membrane pellets can be resuspended with a pipet tip although we have found that more complete dis-

---

[10] D. P. Briskin and R. J. Poole, *Plant Physiol.* **71**, 350 (1983).

**PROCEDURE I**

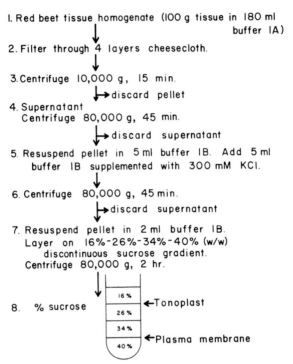

1. Red beet tissue homogenate (100 g tissue in 180 ml
                                           buffer IA)

2. Filter through 4 layers cheesecloth.

3. Centrifuge 10,000 g, 15 min.
                      →discard pellet

4. Supernatant
    Centrifuge 80,000 g, 45 min.
                              →discard supernatant

5. Resuspend pellet in 5 ml buffer IB. Add 5 ml
    buffer IB supplemented with 300 mM KCl.

6. Centrifuge 80,000 g, 45 min.
                            →discard supernatant

7. Resuspend pellet in 2 ml buffer IB.
    Layer on 16%-26%-34%-40% (w/w)
    discontinuous sucrose gradient.
    Centrifuge 80,000 g, 2 hr.

8. % sucrose

16 %   ←Tonoplast
26 %
34 %
40 %   ←Plasma membrane

FIG. 1. Flow chart describing the preparation of tonoplast vesicles from homogenates of red beet storage tissue.

persal of the pellet is obtained with an artist's paint brush. A salt (150 mM KCl) wash of the crude membrane fraction is accomplished by slowly adding an equal volume of buffer 1B supplemented with 300 mM KCl to the resuspended membrane pellet (step 5, Fig. 1). The discontinuous sucrose gradient is conveniently prepared by successively underlaying 6 ml each of 16%, 26%, 34%, and 40% (w/w) sucrose into a 34-ml polycarbonate centrifuge tube. After centrifugation, the tonoplast fraction can be removed from the 16%/26% (w/w) interface with a Pasteur pipet. This fraction is routinely diluted with an equal volume of 10 mM Tris-MES, 2 mM DTT, pH 7.0, and repelleted at 80,000 $g$ for 30 min. The pellet is then resuspended in 1 ml of buffer 1B. The protein concentration in this fraction is typically 1–1.5 mg/ml and can be used directly in assays of ATP hydrolysis or $H^+$ transport.

*Membrane Preparation—Procedure 2*

The method originally devised for the isolation of intact vacuoles from red beet relied on the use of a slicing machine that rapidly processed large amounts of beet tissue.[5] Because such specialized slicing machines are not readily available, the slicing procedure has been modified by simply finely slicing the tissue by hand with a single-edge razor blade. Red beet storage tissue (500 g) is first sliced into approximately 5-mm slices with a large knife and placed in a shallow dish resting on a bed of crushed ice. The slices are covered with cold buffer 2A (1 liter) and the cells allowed to plasmolyze for 30 min, after which the tissue is finely sliced with a single-edge razor blade. The finely sliced tissue in buffer 2A is filtered through cheesecloth and the filtrate carried through the isolation procedure outlined in Fig. 2. Pellets should be gently resuspended by pipetting with a plastic pipet tip having a large-diameter tip opening. Intact vacuoles float

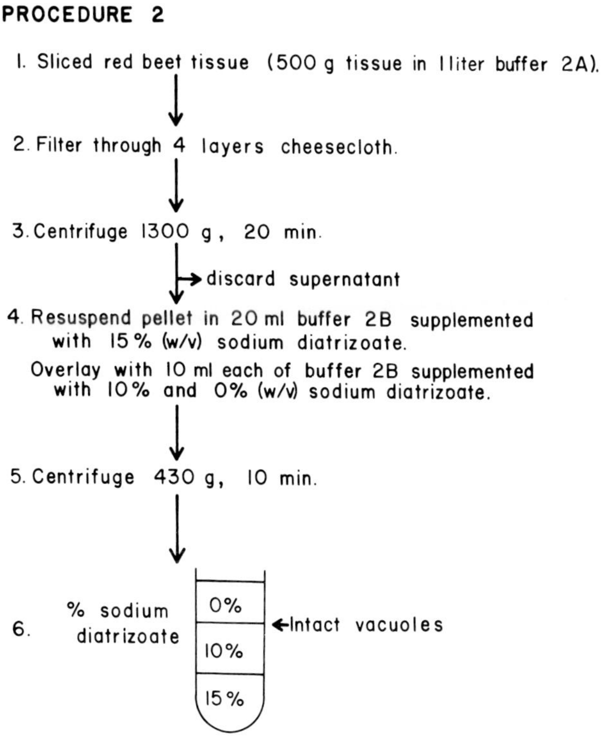

**PROCEDURE 2**

1. Sliced red beet tissue (500 g tissue in 1 liter buffer 2A).

   ↓

2. Filter through 4 layers cheesecloth.

   ↓

3. Centrifuge 1300 g, 20 min.

   → discard supernatant

4. Resuspend pellet in 20 ml buffer 2B supplemented with 15% (w/v) sodium diatrizoate.
   Overlay with 10 ml each of buffer 2B supplemented with 10% and 0% (w/v) sodium diatrizoate.

   ↓

5. Centrifuge 430 g, 10 min.

   ↓

6.   % sodium diatrizoate   | 0% | ←Intact vacuoles
                            | 10% |
                            | 15% |

FIG. 2. Flow chart describing the preparation of intact vacuoles from red beet storage tissue.

to the 0%/10% (w/v) sodium diatrizoate interface and can be collected with a Pasteur pipet. These vacuoles may be assayed intact. However, for many purposes, including measurements of $H^+$ transport, we found it necessary to lyse the vacuoles by diluting them 10-fold into 2 m$M$ DTT, 25 m$M$ Tris-MES, pH 7.0. The diluted vacuolar membranes are then pelleted at 80,000 $g$ for 30 min and resuspended in buffer 1B. Yield of membrane protein by this procedure is about 10% of that obtained by procedure 1.

### Assay of ATP Hydrolysis

ATPase activity is routinely assayed in a volume of 0.5 ml containing 3 m$M$ Tris-ATP, 3 m$M$ MgSO$_4$, 50 m$M$ KCl, and 30 m$M$ Tris-MES, pH 7.0. If sensitivity of the ATPase to NO$_3^-$ is to be determined, the basic reaction medium is supplemented with either 25 m$M$ K$_2$SO$_4$ or 50 m$M$ KNO$_3$. The assay is initiated by the addition of 1–10 $\mu$g membrane protein and allowed to incubate for 30 min at 28°. Phosphate is determined by the method of Ames.[11] Because sorbitol interferes with the Ames method of phosphate determination, if vacuoles in buffer 2B are to be assayed directly for ATPase activity the method of phosphate determination must be modified as previously described.[6]

### Assay of $H^+$ Transport

$H^+$ transport activity is assayed as quenching of the fluorescent amine dyes, quinacrine or acridine orange.[12] Membrane vesicles (75–100 $\mu$g membrane protein), MgSO$_4$ (3 m$M$ final concentration), appropriate monovalent salts, and quinacrine or acridine orange (10 $\mu M$ or 5 $\mu M$ final concentration, respectively) are added to an assay buffer of 250 m$M$ sucrose and 10 m$M$ Tris-MES, pH 7.0, to give a final volume of 1.5 ml. Fluorescence is measured in a temperature-controlled cell (28°) at excitation/emission wavelengths of 425/500 nm for quinacrine and 472/525 nm for acridine orange. Measurements of ATP-dependent quenching of fluorescence are initiated by the addition of 9 $\mu$l of 0.5 $M$ Tris-ATP, pH 7.0.

### Characteristics of the Tonoplast $H^+$-ATPase

Characterization of the density of tonoplast membranes was important in designing preparative procedures for isolating tonoplast membrane ves-

[11] B. N. Ames, this series, Vol. 8, p. 115.
[12] A. B. Bennett and R. M. Spanswick, *J. Membr. Biol.* **71**, 95 (1983).

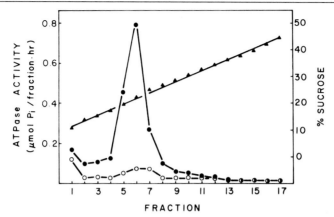

FIG. 3. Distribution of ATPase activity associated with isolated vacuolar membranes on a continuous sucrose gradient. ATPase activity was assayed in the presence of 3 m$M$ MgSO$_4$, 3 m$M$ Tris-ATP, 50 m$M$ KCl, 2 $\mu M$ gramicidin, and the absence (●) or presence (○) of 50 m$M$ KNO$_3$. Sucrose concentrations (%, w/w) are also shown (▲). [Reproduced with permission from *Plant Physiol.* **74**, 538 (1984).]

icles from crude membrane fractions. Figure 3 shows the distribution on a continuous sucrose gradient of ATPase activity associated with membranes derived from purified red beet vacuoles. The vacuoles were isolated by procedure 2 (described above) and osmotically lysed prior to density gradient centrifugation. ATPase associated with purified vacuolar membranes formed a single peak at a density of 1.10 g/cm$^3$. This single peak of ATPase activity was completely inhibited by NO$_3^-$ (Fig. 3). The characteristic of NO$_3^-$ inhibition has become a distinguishing feature of this ATPase.[3,6,13]

The distribution of ATPase activities associated with a crude membrane fraction (10,000–80,000 $g$ membrane pellet) on a continuous sucrose gradient also showed a peak of NO$_3^-$-inhibited ATPase activity at 1.10 g/cm$^3$ (Fig. 4). This low density peak of ATPase activity is well separated from a peak of vanadate-inhibited ATPase which is thought to be the plasma membrane H$^+$-ATPase[3,14] and is well separated from the mitochondrial marker, cytochrome-$c$ oxidase. These results indicate that tonoplast has a density near 1.10 g/cm$^3$ and that a relatively pure tonoplast fraction can be prepared on discontinuous sucrose gradients such as that shown in Fig. 1.

The sensitivity of the tonoplast H$^+$-ATPase to NO$_3^-$ has been useful in distinguishing this enzyme from the plasma membrane H$^+$-ATPase that is

[13] S. D. O'Neill, A. B. Bennett, and R. M. Spanswick, *Plant Physiol.* **72**, 837 (1983).

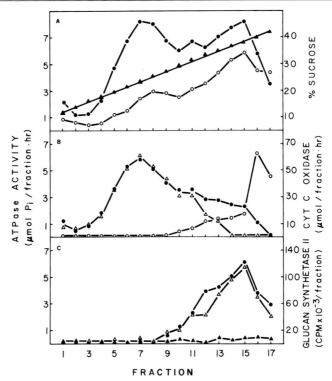

FIG. 4. Distribution of ATPase activities, cytochrome-c oxidase, and glucan synthase II associated with red beet microsomal membranes on a continuous (12–45%, w/w) sucrose gradient. (A) ATPase activity assayed in the presence of 3 m$M$ MgSO$_4$, 3 m$M$ Tris-ATP, 50 m$M$ KCl, and in the absence (○) or presence (●) of 2 $\mu M$ gramicidin. Sucrose concentration (%, w/w) is also shown (▲). (B) Gramicidin-stimulated ATPase ($\Delta$G ATPase, ●) is calculated as the difference in ATPase activity assayed as in (A) in the absence or presence of gramicidin. NO$_3^-$-sensitive ATPase ($\Delta$NO$_3^-$ ATPase, △) is calculated as the difference in ATPase activity assayed in the presence of 3 m$M$ MgSO$_4$, 3 m$M$ Tris-ATP, 50 m$M$ KCl, 2 $\mu M$ gramicidin, and in the absence or presence of 50 m$M$ KNO$_3$. Cytochrome-c oxidase activity is also shown (○). (C) Vanadate-sensitive ($\Delta$vanadate ATPase, △) or azide-sensitive ATPase ($\Delta$azide ATPase, ▲) activity assayed in the presence of 3 m$M$ MgSO$_4$, 3 m$M$ Tris-ATP, 50 m$M$ KCl, 2 $\mu M$ gramicidin, and in the absence or presence of either 20 $\mu M$ vanadate or 1 m$M$ azide. Glucan synthase II activity (●) is also shown. [Reproduced with permission from *Plant Physiol.* **74**, 538 (1984).]

insensitive to NO$_3^-$ but inhibited by vanadate.[14–16] Table I illustrates the sensitivity of tonoplast membranes, isolated by either procedure 1 or

---

[14] S. D. O'Neill and R. M. Spanswick, *J. Membr. Biol.* **79**, 245 (1984).

[15] S. D. O'Neill and R. M. Spanswick, *Plant Physiol.* **75**, 586 (1984).

[16] S. R. Gallagher and R. T. Leonard, *Plant Physiol.* **70**, 1335 (1982).

TABLE I
EFFECT OF INHIBITORS ON ATPase ACTIVITY ASSOCIATED WITH MEMBRANES[a]

| | ATPase activity | | | |
| | Tonoplast (procedure 1) | | Tonoplast (procedure 2) | |
| Treatment | Specific activity ($\mu$mol P$_i$/mg · min) | % | Specific activity ($\mu$mol P$_i$/mg · min) | % |
|---|---|---|---|---|
| Control | 0.201 | — | 0.464 | — |
| +Gramicidin (2 $\mu M$) | 0.596 | 100 | 0.898 | 100 |
| +Vanadate (50 $\mu M$) | 0.551 | 92 | 0.863 | 96 |
| +Molybdate (100 $\mu M$) | 0.623 | 105 | 0.810 | 90 |
| +KNO$_3$ (50 $\mu M$) | 0.158 | 26 | 0.208 | 23 |

[a] Membranes were collected from 16/26% (w/w) sucrose interface (procedure 1) or from purified vacuoles (procedure 2). Relative activities are expressed as a percentage of the activity in the presence of gramicidin. Data from Ref. 3.

procedure 2, to NO$_3^-$ and its insensitivity to vanadate. The slight inhibition of ATPase activity by vanadate most likely represents contamination of the tonoplast fractions with plasma membrane. Plasma membranes collected from the 34/40% (w/w) sucrose interface (procedure 1, Fig. 1) shows the opposite inhibitor sensitivity, being 80% inhibited by vanadate and only 10% inhibited by NO$_3^-$.[3] Tonoplast ATPase activity was stimulated 200–300% by gramicidin (Table I), indicating that these membranes are present as sealed vesicles capable of maintaining ion gradients.[1,2] Molybdate inhibition of ATPase activity has been used as a criterion to identify acid phosphatase activity associated with plant membranes.[6,16] Although the presence of acid phosphatase has been reported to be associated with red beet tonoplast membranes,[6] the low inhibition of ATPase activity by molybdate (Table I) indicated that these membrane preparations were largely free of acid phosphatase activity. Since the presence of acid phosphatase may vary with the source of plant material, assays can be routinely carried out in the presence of molybdate (0.1 m$M$) since the activity of acid phosphatase may obscure characteristics of the H$^+$-ATPase.

H$^+$ transport activity catalyzed by the tonoplast H$^+$-ATPase showed inhibitor sensitivity that was similar to the sensitivity when ATP hydrolysis was assayed (Fig. 5). Whether tonoplast was prepared by procedure 1 (Fig. 5A) or procedure 2 (Fig. 5B), ATP-dependent H$^+$ transport activity

A          B

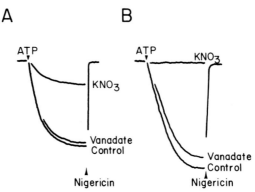

FIG. 5. ATP-dependent quenching of quinacrine fluorescence with tonoplast membrane vesicles prepared by either procedure 1 (A) or procedure 2 (B). Fluorescence assays were as described in Materials and Methods in the presence of 50 m$M$ KCl. Inhibitors were added as indicated at the following concentrations: vanadate, 50 $\mu M$; KNO$_3$, 50 m$M$. [Reproduced with permission from *Plant Physiol.* **74**, 538 (1984).]

was largely insensitive to vanadate but 90–100% inhibited by NO$_3^-$. Other inhibitors tested indicated that H$^+$ transport and ATP hydrolysis in these membrane fractions was insensitive to oligomycin but inhibited by $N,N'$-dicyclohexylcarbodiimide (DCCD) and diethylstilbestrol (DES).[3]

Because ion stimulation of transport ATPases has often been used as a diagnostic feature of various cellular ATPases, stimulation of the plant tonoplast H$^+$-ATPase by monovalent and divalent salts has been well characterized.[2,3,6,8,12,13] Table II illustrates the sensitivity of tonoplast membranes, isolated by either procedure 1 or procedure 2, to monovalent salts. All of the Cl$^-$ salts stimulate ATPase activity to a similar extent, regardless of the accompanying cation. This indicates that the tonoplast H$^+$-ATPase, in contrast to the plant plasma membrane H$^+$-ATPase, is stimulated by anions and relatively insensitive to monovalent cations. The preferred anion is Cl$^-$ followed by Br$^-$, acetate, HCO$_3^-$, and SO$_4^{2-}$.[3]

Stimulation of ATP hydrolysis catalyzed by the tonoplast H$^+$-ATPase by Cl$^-$ is observed even in the presence of gramicidin,[3,12] indicating that this anion effect does not result from Cl$^-$ acting as a permeant anion, with consequent effects on the membrane potential generated by the H$^+$ transport activity of the H$^+$-ATPase.[12] This direct effect of anions in stimulating the H$^+$-ATPase is illustrated in Fig. 6 where H$^+$ transport was monitored in the presence of KCl, KBr, and K$_2$SO$_4$, with and without valinomycin. In the presence of K$^+$, the addition of valinomycin accelerates the initial rate of H$^+$ transport. Even in the presence of valinomycin, however, Cl$^-$ and Br$^-$ support a greater extent of H$^+$ transport than does

TABLE II
EFFECT OF MONOVALENT SALTS ON ATPase ACTIVITY ASSOCIATED WITH MEMBRANES[a]

| | ATPase activity | | | |
|---|---|---|---|---|
| | Tonoplast (procedure 1) | | Tonoplast (procedure 2) | |
| Salt addition | Specific activity ($\mu$mol P$_i$/mg · min) | Salt stimulation (% of KCl stimulation) | Specific activity ($\mu$mol P$_i$/mg · min) | Salt stimulation (% of KCl stimulation) |
| MgSO$_4$ | 0.322 | | 0.756 | |
| +KCl | 0.604 | 100 | 1.43 | 100 |
| +NaCl | 0.639 | 113 | 1.48 | 107 |
| +RbCl | 0.605 | 100 | 1.44 | 101 |
| +LiCl | 0.611 | 102 | 1.32 | 84 |
| +CsCl | 0.573 | 89 | 1.38 | 92 |
| +KBr | 0.574 | 89 | 1.25 | 73 |
| +K-acetate | 0.516 | 67 | 1.13 | 55 |
| +K$_2$SO$_4$ | 0.386 | 20 | 0.82 | 9 |
| +KNO$_3$ | 0.170 | −59 | 0.40 | −52 |

[a] Membranes were collected from 16/26% (w/w) sucrose interface (procedure 1) or from purified vacuoles (procedure 2). Data from Refs. 3 and 6.

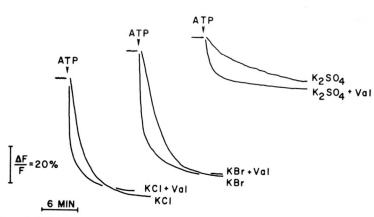

FIG. 6. Effect of anions and valinomycin on ATP-dependent quenching of quinacrine fluorescence with tonoplast vesicles prepared by procedure 1. Fluorescence assays were as described in Materials and Methods in the presence of either 50 mM KCl, 50 mM KBr, or 25 mM K$_2$SO$_4$. Valinomycin (Val), when present, was 0.5 $\mu$M. [Reprinted with permission from *Plant Physiol.* **74**, 538 (1984).]

$SO_4^{2-}$. Since valinomycin relieves electrical constraints on $H^+$ transport, this result indicates that $Cl^-$ directly activates the ATPase, apart from any role in stimulating $H^+$ transport as a permeant anion.

## Discussion

The $H^+$-ATPase associated with higher plant vacuolar membranes has been well characterized in membrane vesicles derived from tissue homogenates and in membranes derived from purified intact vacuoles. The main characteristics that distinguish the tonoplast $H^+$-ATPase from other cellular ATPases are (1) sensitivity to $NO_3^-$, but not to vanadate, azide, or oligomycin and (2) stimulation by $Cl^-$ and lack of sensitivity to monovalent cations. Biophysical characterization of this $H^+$-ATPase has demonstrated that $H^+$-ATPase is electrogenic[12] and that the stoichiometry of $H^+/ATP$ coupling is 2.[7] Other studies have also suggested that the $H^+$-ATPase may be intimately associated with an anion channel,[12,17,18] although this suggestion has been disputed by other investigators.[19]

The tonoplast $H^+$-ATPase has been solubilized and functionally reconstituted into artificial membrane vesicles,[20] and has been purified. Purification suggests that the $H^+$-ATPase is an oligomeric protein comprised of at least three subunits with apparent molecular masses of 60, 70, and 16 kDa.[21,22,23] In general, the characteristics of the plant vacuolar $H^+$-ATPase are similar to those described for $H^+$ ATPase identified in vacuolar membranes of *Neurospora crassa*[24] and *Saccharomyces cerevisiae*.[25] We anticipate that the membrane preparations described in this report will be useful in further characterizing the plant vacuolar $H^+$-ATPase and in establishing its relationship with anion-sensitive $H^+$-ATPases from other organisms.

[17] A. Hager and M. Helmle, *Z. Naturforsch.* **36**, 997 (1981).
[18] R. R. Lew and R. M. Spanswick, *Plant Physiol.* **77**, 352 (1985).
[19] K. A. Churchill and H. Sze, *Plant Physiol.* **76**, 490 (1984).
[20] A. B. Bennett and R. M. Spanswick, *J. Membr. Biol.* **75**, 21 (1984).
[21] S. Mandala and L. Taiz, *Plant Physiol.* **78**, 327 (1985).
[22] S. Randall and H. Sze, *J. Biol. Chem.* **261**, 1364 (1986).
[23] M. Manolson, P. Rea, and R. Poole, *J. Biol. Chem.* **260**, 12273 (1985).
[24] E. J. Bowman, *J. Biol. Chem.* **258**, 15238 (1983).
[25] E. Uchida, Y. Ohsumi, and Y. Anraku, *J. Biol. Chem.* **260**, 1090 (1985).

## [45] Preparation and Acidification Activity of Lysosomes and Lysosomal Membranes

By DONALD L. SCHNEIDER and JEAN CHIN

### Introduction

Lysosomes are organelles of a membrane system which is known collectively as the vacuolar apparatus. This system includes lysosomes, coated vesicles, the Golgi apparatus, and endocytic vesicles, all of which participate in the essential cellular functions of endocytosis and exocytosis. The molecular events responsible for these functions are subjects of considerable current interest, especially now that proton pump activity, first found in lysosomes, is apparently present throughout the entire vacuolar apparatus. One would like to know whether the same molecular species of proton pump exists in all parts of the apparatus or whether each part has a distinct proton pump. The regulation of the proton pump(s) is another important area of research. In this chapter, we present methods for the isolation of lysosomes and lysosomal membranes from rat liver and from phagocytic cells and for the assay of acidification activity. In addition, the properties of the lysosomal ATP-driven proton pump will be considered.

### Preparation of Lysosomes from Rat Liver

This technique takes advantage of the buoyancy of lipoproteins to effect a very clean (and otherwise difficult) separation of Triton–lipoprotein-filled lysosomes (also known as tritosomes) from mitochondria. Rats are injected with the nonhemolytic agent Triton WR-1339 and, over the course of the next $3\frac{1}{2}$ days, the 10-fold increase in plasma lipoproteins leads to accumulation of the Triton and lipoproteins in the liver lysosomes. A large granule fraction containing primarily mitochondria and lysosomes is prepared by differential centrifugation, and then, the lysosomes are floated up from the mitochondria by discontinuous sucrose gradient centrifugation.[1-3]

[1] A. Trouet, Arch. Int. Physiol. Biochem. **72,** 698 (1964).
[2] F. Leighton, B. Poole, H. Beaufay, P. Baudhuin, J. W. Coffey, S. Fowler, and C. de Duve, J. Cell Biol. **37,** 482 (1968).
[3] J. Burnside and D. L. Schneider, Biochem. J. **204,** 525 (1982).

*Reagents*

Triton WR-1339 (from Ruger Chemical, Irvington, NJ, or from Sigma, St. Louis, MO, as Tyloxapol): A 17% solution in saline is prepared by stirring at room temperature; complete dissolution may require stirring for about an hour. Although this stock solution of Triton WR-1339 is stable at ambient temperature, it should be stored frozen to avoid microbial growth that can be toxic to rats.

Sucrose, 0.25 *M*, unbuffered. Prepare about 100 ml per rat and store in the refrigerator.

Sucrose, 14.3% (w/w), density 1.06 g/ml, or 18% (w/v).

Sucrose, 34.5% (w/w), density 1.155 g/ml, or 44% (w/v).

Sucrose, 45% (w/w), density 1.21 g/ml, or 57% (w/v).

Sucrose, 60% (w/w), density 1.295 g/ml, or 76.5% (w/v).

*Procedure*

The preparation can easily be scaled up to a dozen rats or down to one; the desired number of rats are weighed individually and, under light ether anesthesia, each is injected intraperitoneally with 1 ml of Triton WR-1339 stock solution per 100 g body weight (dose: 850 mg of Triton WR-1339 per kilogram rat body weight). After $3\frac{1}{2}$ days, decapitate the rats and rapidly excise and chill the livers by placing them in a tared beaker of ice-cold 0.25 *M* sucrose. Weigh the livers and, after mincing to 1-cm cubes, homogenize the livers one at a time in a size C Potter tissue grinder with motor-driven, serrated Teflon pestle in a volume of 25 ml of 0.25 *M* sucrose. A single stroke is sufficient. The homogenates are transferred to 50-ml centrifuge tubes and spun at 1500 *g* for 3 min. The supernatants are set aside on ice and the pellets are rehomogenized in the same volume and recentrifuged. The nuclear pellets are resuspended and a representative aliquot is saved. The combined supernatants are mixed well to give the extract (E) fraction; the volume is recorded. A 1-ml aliquot is saved, and the remainder of the extract is centrifuged at 40,000 *g* (Sorvall SS34 rotor, 19,000 rpm) for 12 min. Decant the supernatants and set aside on ice. The pellets are resuspended in half the original volume of 0.25 *M* sucrose, using a hand-operated Potter tissue grinder with a smooth Teflon pestle. After recentrifugation, the supernatants are decanted, and the mitochondrial–lysosomal pellets are resuspended in 45% sucrose using a hand-operated Potter tissue grinder. To assure that the density of the resuspended pellets is adequately high, a volume of 60% sucrose approximately equal to that of the pellets is added and, after very thorough mixing, a drop of the suspension is checked in a test tube containing a few milliliters of 34.5% sucrose in which it should sink. For large preparations, the total volume should be at least 7 ml per rat. A 0.5-ml aliquot is set aside and the remainder of the resuspended mitochondrial–lysosomal fraction is applied to the gradient. Meanwhile, the supernatants are combined as the microsomal + soluble fraction (PS), and a 1-ml aliquot is saved.

*Sucrose Gradients.* These are layered from the top down by adding, through a long, broken-tip Pasteur pipet, successive solutions to the bottom of the centrifuge tube. The Beckman SW 27 rotor with its capacity of 35 ml per tube provides a convenient compromise between capacity and centrifugation time, but other rotors may be used with appropriate adjustments. To avoid damage to organelles, excessive speeds are to be avoided.[4] Layer into each tube in the following order:

(1) 5 ml of 14.3% sucrose,
(2) 15 ml of 34.5% sucrose,
(3) 13 ml of sample in 45% sucrose, and
(4) 2 ml of 60% sucrose.

When switching from one solution to the next, begin adding the next solution as the first enters the constricted part of the pipet so as to avoid (1) excessive mixing and (2) air bubbles that can disrupt the gradient. The gradients are centrifuged at 100,000 $g$ (24,000 rpm) for 2 hr. Afterward the lysosomes at the 14.3–34.5% interface are apparent from the rich golden yellow-brown color (presumably due to aging pigment, lipofuscin). Since the lysosomes are above the mitochondria and other contaminants, it is important to unload the gradient from the top. This can be done in two ways. By using an inverted funnel (a handmade hollowed-out rubber stopper is perfectly adequate), one can pump heavy 60% sucrose into the bottom and float up the gradient. The lysosomes are collected in a volume of about 5 ml per gradient; all other parts of the gradient are combined as residue fraction. Alternatively, one can pierce the side of the tube with a syringe needle attached to a syringe and draw off the interface.

*Analysis of the Preparation*

The lysosomal fraction will be enriched 40- to 60-fold relative to homogenate. If the rats are fasted overnight before sacrificing, the enrichment will be about 40; whereas without fasting, the enrichment will be about 60-fold. These differences presumably occur due to a greater protein content in lysosomes of fasted rats that are more active in protein degradation.[5] Any number of marker enzymes are suitable, and a particu-

[4] R. Wattiaux, S. Wattiaux-De Coninck, and M.-F. Ronveaux-Dupal, *Eur. J. Biochem.* **22**, 31 (1971).

[5] D. Ray, E. Cornell, and D. Schneider, *Biochem. Biophys. Res. Commun.* **71**, 1246 (1976).

larly convenient one is $N$-acetyl-$\beta$-D-glucosaminidase (NA$\beta$Gase)[6] assayed with $p$-nitrophenyl-$N$-acetyl-$\beta$-D-glucosaminide, available from Sigma. In view of the sucrose content of gradient fractions and its interference in protein determinations, precipitation with trichloroacetic acid is advisable. The activity and protein values of the homogenate can be calculated from the values of the extract and nuclear fractions as recommended by de Duve et al.[7] By assaying all of the fractions one can also make useful calculations concerning recoveries and yields, a detailed description of which is available.[8] The overall yield of lysosomes is typically 30% or 10 mg of protein per rat.

### Preparation of Lysosomes from Phagocytic Cells

Many cells take up inert, buoyant, latex beads and through endocytosis transfer the beads to their lysosomes. With latex, lysosomes from such diverse cell types as polymorphonuclear leukocytes,[9] amoebas,[10] and even fibroblastic L cells[11] can be isolated. Using again the principle of separation by flotation gradient centrifugation, a very clean separation of latex-filled lysosomes (more accurately referred to as phagolysosomes to denote the uptake of particulate material) from mitochondria is achieved. The preparation involves ingestion of the beads, homogenization, and flotation gradient centrifugation to isolate the lysosomes.

#### Reagents

Latex beads: These are available from Sigma or Difco as aqueous suspensions. It is preferable to use beads about 1 $\mu$m in diameter, about the same size as bacteria (whence "fake bacteria"), because small beads (0.2 $\mu$m) are not ingested efficiently and large ones (>1.5 $\mu$m) promote the secretion of lysosomal enzymes and render the lysosomes fragile to manipulations. It is also to be noted that the beads are shipped in detergent solution to promote dispersion. Before using the latex, detergent is removed by diluting with 10 volumes of saline and centrifuging at 20,000 $g$ (Sorvall SS34 rotor, 15,000 rpm) for 20 min. The pelleted beads are gently resuspended in a small volume of saline using a hand-operated Potter tissue

[6] J. Findlay, G. A. Levvy, and C. A. Marsch, *Biochem. J.* **69,** 467 (1958).

[7] C. de Duve, B. C. Pressman, R. Gianetto, R. Wattiaux, and F. Appelmans, *Biochem. J.* **60,** 604 (1955).

[8] C. de Duve, *J. Cell Biol.* **50,** 20D (1971).

[9] D. R. Crawford and D. L. Schneider, *J. Biol. Chem.* **258,** 5363 (1983).

[10] E. D. Korn, this series, Vol. 31, p. 686.

[11] A. L. Hubbard and Z. A. Cohn, *J. Cell Biol.* **64,** 461 (1975).

grinder. To effect monodispersion, the washed beads are briefly subjected to gentle bath sonication just prior to use.

Ingestion media: This can be either the growth media or phosphate-buffered saline if the cells will tolerate the lack of serum proteins and nutrients. Since uptake depends on the adsorptive properties of the beads, the presence of serum proteins may decrease ingestion and the ultimate yield of lysosomes.

Phosphate-buffered saline (PBS): 150 m$M$ sodium chloride, 10 m$M$ sodium phosphate, pH 7.4.

Albumin-PBS: Dissolve 1 g of bovine serum albumin in 100 ml of PBS and readjust the pH to 7.4.

Sucrose solutions (w/v): 10%, 20%, and 30%, 100 ml each, chilled to 0–4°.

Sucrose solution (w/w), 60%, chilled.

Sucrose solutions, 0.25 and 2 $M$.

*Procedure*

Ingestion of latex is effectively carried out with quite concentrated solutions of beads and cells. One gram wet weight of packed cells is dispersed in 100 ml of ingestion media containing 0.1 g wet weight of packed latex. After incubation, at either 30 or 37° for 20 min with gentle shaking, the mixture is thoroughly chilled. The cells are collected by low-speed centrifugation, washed twice with 30 ml of cold albumin-PBS, washed twice with chilled PBS, and resuspended in a small volume of 0.25 $M$ sucrose to give 2 ml.

*Homogenization.* The resuspended cells are transferred to a Dounce tissue grinder (14 ml size, type A, tight) and diluted to 50 m$M$ sucrose by addition of 8 ml of ice-cold water. Cell disruption is achieved by 10 strokes of a tight pestle and isoosmolarity is restored by adding 1.2 ml of 2 $M$ sucrose.

*Sucrose Gradient.* Layer in order from top to bottom by adding through a long, broken-tip Pasteur pipet in an SW 27 tube the following:

(1) 3 ml of 10% sucrose,
(2) 10 ml of 20% sucrose,
(3) 10 ml of 30% sucrose,
(4) 10 ml of a 50:50 mixture of homogenate and 60% sucrose, and
(5) 2 ml of 60% sucrose.

Add successive solutions when the previous one enters the constricted part of the pipet. The gradients are centrifuged at 60,000 $g$ (18,000 rpm) for 45 min. The lysosomes are collected from the 10–20% interface by pumping 60% sucrose into the bottom and floating the gradient up through an inverted funnel.

*Analysis of the Preparation*

The NA$\beta$Gase activity, discussed above, is a convenient marker. Lysosomelike granules or phagolysosomes are relatively abundant in amoebae and neutrophils, and the phagolysosomal fraction will be at most 10-fold enriched in marker activity relative to homogenate. Fibroblasts are less rich in lysosomes, and an enrichment of more than 20 is expected. With the latex procedure, the yields are commonly low, usually about 10% of total.

Other comments are in order regarding analysis of phagolysosomal preparations. First, the neutrophils, known to contain different kinds of granules, carry out a sequence of fusion events between the forming phagolysosome and tertiary granules, specific granules, and azurophilic granules. When incubating latex beads with cells for short times, 6 min or less, it is important to be mindful that the azurophilic granules, marked by NA$\beta$Gase, may not yet have fused with the forming phagolysosomes. Under such conditions, gelatinase for the tertiary granules or lactoferrin for the specific granules is a better marker. However, especially for these granules, fusion may precede closure of the forming phagolysosome from the outside, and considerable secretion occurs that selectively depletes phagolysosomes and may lead to an underestimation of enrichment. Second, protein determination of latex-laden phagolysosomes is facilitated by centrifugation of the assay solutions, after addition of all reagents (Lowry C and Folin) but prior to reading the absorbance, at 15,000 $g$ for 10 min to eliminate the turbidity of the latex beads.

### Lysosomes from Other Cells

At the present time no single technique is applicable in every instance. Ones that have a wide spectrum of uses and merit consideration are centrifugation in metrizamide,[12] centrifugation in Percoll,[13] and free flow electrophoresis.[14]

### Assay of Acidification Activity

In our laboratory, we routinely use the radioactive methylamine assay with rapid gel filtration[15] because it is very sensitive, easy to do batchwise

---

[12] R. Wattiaux, S. Wattiaux-De Coninck, M.-F. Ronveaux-Dupal, and F. Dubois, *J. Cell Biol.* **78**, 349 (1978).

[13] L. H. Rome, A. J. Garvin, M. M. Allientta, and E. F. Neufeld, *Cell (Cambridge, Mass.)* **17**, 143 (1979).

[14] R. Stahn, K. P. Maier, and K. Hannig, *J. Cell Biol.* **46**, 576 (1970).

[15] D. L. Schneider, *J. Biol. Chem.* **256**, 3858 (1981).

with large numbers of samples (we routinely do 48 assays at once, about 2 hr from start to finish), and the activity value obtained is ΔpH that is more obviously related to acidity than a rate value such as nmol H$^+$ per min · mg. The theory of using permeant weak bases to probe intravesicle pH is well founded as discussed in the literature.[16,17]

*Reagents*

Methylamine, [14]C-labeled, is commercially available. The hydrochloride is not volatile; a stock aqueous solution, 0.1 mCi per milliliter, is convenient and stable when stored frozen at −20°. The chemical concentration of methylamine in the assay is low, 1–10 $\mu M$, and one should be mindful that millimolar levels of permeant weak bases, including methylamine, ammonium sulfate, etc., are inhibitory and must be avoided when measuring acidification with any assay.

ATP, 45 m$M$, pH 7.0, stock solution stored at −20°.

MOPS [3-($N$-morpholino)propanesulfonic acid], 0.2 $M$, pH 7.

DTT (dithiothreitol), 1 $M$, aqueous solution, stored at −20°.

EDTA (ethylenediaminetetraacetic acid) and EGTA (ethylene glycol bis($\beta$-aminoethyl ether)-$N,N,N',N'$-tetraacetic acid) solutions, 200 and 50 m$M$, respectively, pH 7.

Magnesium chloride, 0.2 $M$.

Potassium chloride and sodium chloride solutions, 1 $M$ each.

BSA (bovine serum albumin), 20 mg/ml aqueous solution, pH 7, sterilized by membrane filtration, stored at 4°.

Phospholipid solutions, crude soybean (a mixture consisting primarily of phosphatidylcholine and phosphatidylethanolamine) or brain phosphatidylserine, 20 mg/ml aqueous dispersions obtained by homogenizing in a Potter tissue grinder and bath sonicating, dialyzing against 40 volumes of 1 m$M$ Tris, 4 changes, 4 hr each. These are stored at −20°.

Sephadex G-50: 6 g swollen overnight at 4° in 100 ml of 20 m$M$ potassium MOPS, pH 7, 60 m$M$ sucrose, 60 m$M$ potassium chloride, 1 m$M$ EDTA, 2 m$M$ magnesium chloride, 0.5 m$M$ EGTA, is enough for 60 assays.

*Procedure*

Tuberculin syringes, 1-ml size, are outfitted with fritted disks, and filled with the swollen Sephadex G-50 previously equilibrated at room

[16] C. de Duve, T. De Barsy, A. Trouet, P. Tulkens, and F. Van Hoff, *Biochem. Pharmacol.* **23**, 2495 (1974).

[17] D. J. Reijngoud and J. M. Tager, *Biochem. Biophys. Acta* **472**, 419 (1977).

temperature. Immediately prior to use, these are centrifuged at 200 $g$ for 3 min.

The sample is preincubated with buffer and [$^{14}$C]methylamine (500,000 cpm, about 10 nmol or 5 $\mu M$) at room temperature for 10 min (volume 180 $\mu$l), and the assay is initiated by addition of 20 $\mu$l of either ATP (+ATP) or EDTA (−ATP control). The final concentrations are 20 m$M$ MOPS, pH 7, 60 m$M$ sucrose, 50 m$M$ potassium chloride, 12.5 m$M$ sodium chloride, 0.5 m$M$ EGTA, 0.1 mg/ml BSA, 75 $\mu$g/ml phospholipid, 9 m$M$ magnesium chloride, and 4.5 m$M$ of either ATP or EDTA. After incubation (usually 10 min), 150 $\mu$l of each mixture is transferred to a tuberculin syringe filled with Sephadex (prespun) and centrifuged at 200 $g$ for 3 min. The eluate, collected in a clean 13 × 100 mm tube during centrifugation, is diluted to 250 $\mu$l with water (add about 125 $\mu$l), and protein is precipitated by addition of 40 $\mu$l of 40% trichloroacetic acid. After centrifugation, 250 $\mu$l of the supernatant is removed and counted for radioactivity, and the residue is analyzed for protein by the Lowry method.[18]

*Calculations.* The counts are first corrected for methylamine flow-through, then the ATP-dependent proton gradient is calculated by dividing the cpm/mg value of the +ATP sample by that of the −ATP one as shown in the tabulation below. The cpm corrected value is obtained by subtracting an appropriate amount based on protein, e.g., the 0.0243 value for lysos-ATP is in part due to bovine serum albumin (7.25/0.0091 = 797 cpm/mg flow-through, 76.6 $\mu$g sample + 20 $\mu$g BSA or 20.7% BSA); therefore, the correction is (0.0243)(797)(0.207) = 4.01 cpm. The ATP-dependent $\Delta$pH is obtained by taking the log, log 12.99 = 1.11 = $\Delta$pH. Since the outside pH is 7.0, the pH inside is 5.89. It should be noted that this method assumes insignificant volume changes, an assumption that has been verified for lysosomes, 4 $\mu$l/mg ± ATP.[15]

| Sample | Net cpm | mg protein | cpm corrected | cpm/mg | Proton gradient |
|---|---|---|---|---|---|
| Flow-through | 7.25 | 0.0091 | — | | |
| Lysos−ATP | 141.6 | 0.0243 | 137.6 | 5,662.5 | |
| Lysos+ATP | 1739.6 | 0.0236 | 1735.7 | 73,547 | 12.99× |

A number of other methods for measuring acidification activity exist.[19,20]

[18] O. H. Lowry, N. J. Rosebrough, A. L. Farr, and R. J. Randall, *J. Biol. Chem.* **193**, 265 (1951).
[19] X.-S. Xie, D. K. Stone, and E. Racker, this volume [49].
[20] N. Nelson, S. Cidon, and Y. Moriyama, this volume [48].

Preparation of Membranes with Acidification Activity

Freshly prepared lysosomes are mixed with proteinase inhibitors[21] and disrupted by freezing. Proteinase action is minimized by adding 1 m$M$ Tris free base, 50 m$M$ potassium phosphate, pH 7, 50 $\mu M$ chymostatin, 25 $\mu M$ pepstatin, 3 $\mu M$ leupeptin, 5 mg/ml BSA, 2 m$M$ $p$-aminobenzamidine, and 0.5 $\mu M$ aprotinin. It is important to use a complete mixture of inhibitors because a limited mixture will stabilize the residual proteinase activities[22] and thereby labilize the acidification activity. Freezing at $-20°$ is satisfactory for disruption and one overnight storage. When stored at $-80°$ with the proteinase inhibitors, the acidification activity is stable for at least 6 months. To isolate the membranes, the thawed lysosomes are diluted and either collected by discontinuous sucrose gradient centrifugation[21] or pelleted by centrifuging at 100,000 $g$ for 35 min. The pelleted membranes are resuspended by hand in a small volume of 0.25 $M$ sucrose using a Potter tissue grinder. Significantly, the pelleted membranes are active in acidification only if the proteinase inhibitor mixture is used. Also, it should be pointed out that the dilution is required: in 20% sucrose as isolated, the membranes are quite slow sedimenting as their equilibrium position is at a sucrose density of 1.12 (25% sucrose). Furthermore, excessive dilution, to less than density 1.03 (8% sucrose), will lead not only to pelleting of lysosomal membranes but also some of the lysosomal contents. For example, a significant part of the total lysosomal cholesterol (assayed enzymatically[23]), more than half in the case of tritosomes, appears on linear sucrose gradients at density 1.05 with about one third of the phospholipid (assayed as described[24]). The cholesterol is presumably from endocytosed lipoprotein cholesterol inasmuch as these gradient fractions contain no ATPase, acidification, nor acid 5′-nucleotidase activities. One might also point out that the Triton WR-1339 does not sediment but fractionates with the soluble proteins.

Additional modifications are useful for the removal of latex beads from phagolysosomal membranes. The disrupted preparation is diluted to less than 15% sucrose (w/w) and layered over a cushion of 15% sucrose; by centrifugation at 100,000 $g$ for 35 min, the membranes are pelleted while the latex beads accumulate above the 15% sucrose cushion. These are completely removed by inverting the tubes and wiping the walls with a swab before righting the tubes.

[21] D. L. Schneider, *J. Biol. Chem.* **258,** 1833 (1983).
[22] D. L. Schneider, *Intracell. Protein Catabolism, Proc. Int. Symp., 5th,* p. 291 (1985).
[23] C. A. Allain, L. S. Poon, C. S. G. Chan, W. Richmond, and P. C. Fu, *Clin. Chem.* **20,** 470 (1974).
[24] J. Chin and K. Bloch, *J. Biol. Chem.* **259,** 11735 (1984).

Concerning the isolation of lysosomal membranes, some miscellaneous points should be made. The addition of millimolar amounts of magnesium is often beneficial for the pelleting of membranes active in acidification. The effect is not an aggregative one, but may be related to a suppression of proteinase activity[22] or perhaps due to the stabilization of a nonsedimentable factor. In addition, unless the ionic strength is at least 0.2 $M$, significant amounts of the soluble hydrolases like NA$\beta$Gase will cofractionate with the membranes.[25]

### Properties of the Lysosomal Proton Pump

Acidification activity is inhibited by treatment of preparations with dicyclohexylcarbodiimide, diisothiocyanostilbenedisulfonic acid, and N-ethylmaleimide but not by oligomycin, ouabain, or vanadate. Thus, it is clearly different from mitochondrial[26] and fungal plasma membrane[27] proton pumps. In addition, acidification activity is inhibited by the ionophores carbonylcyanide $p$-trifluoromethoxyphenylhydrazone (FCCP)[21] and monensin.[28] The relationship of the proton pumps in the various regions of the vacuolar apparatus is less clear, inasmuch as they all have similar sensitivities to dicyclohexylcarbodiimide and N-ethylmaleimide and insensitivities to oligomycin, ouabain, and vanadate.[15,29–33] Suggestion that the lysosomal pump may be unique is based on (1) a less strict specificity for ATP (GTP supports acidification activity nearly as well),[15,34] and (2) an insensitivity to duramycin.[35] However, there are alternative explanations for the observed data, and protein purification studies are necessary. Concerning the studies on reconstitution of acidification carried out in Racker's laboratory,[36] which indicate that the

[25] D. L. Schneider, J. Burnside, F. R. Gorga, and C. J. Nettleton, *Biochem. J.* **176,** 75 (1978).

[26] E. Racker, "A New Look at Mechanisms in Bioenergetics." Academic Press, New York, 1976.

[27] B. J. Bowman, F. Blasco, and C. W. Slayman, *J. Biol. Chem.* **256,** 17343 (1981).

[28] J. A. Fink, M. J. Cahilly, and D. L. Schneider, *Biochem. Arch.* **1,** 37 (1985).

[29] S. Ohkuma, Y. Moriyama, and T. Takano, *Proc. Natl. Acad. Sci. U.S.A.* **79,** 2758 (1982).

[30] M. Forgac, L. Cantley, B. Wiedenmann, L. Altstiel, and D. Branton, *Proc. Natl. Acad. Sci. U.S.A.* **80,** 1300 (1983).

[31] D. K. Stone, X.-S. Xie, and E. Racker, *J. Biol. Chem.* **258,** 4059 (1983).

[32] F. Zhang and D. L. Schneider, *Biochem. Biophys. Res. Commun.* **114,** 620 (1983).

[33] J. Glickman, K. Croen, S. Kelley, and Q. Al-Awqati, *J. Cell Biol.* **97,** 1303 (1983).

[34] C. J. Galloway, G. E. Dean, M. Marsh, G. Rudnick, and I. Mellman, *Proc. Natl. Acad. Sci. U.S.A.* **80,** 3334 (1983).

[35] D. K. Stone, X.-S. Xie, and E. Racker, *J. Biol. Chem.* **259,** 2701 (1984).

[36] X.-S. Xie, D. K. Stone, and E. Racker, *J. Biol. Chem.* **259,** 11676 (1984).

coated-vesicle pump consists of a unique set of polypeptides that fit the $F_1F_0$ paradigm in which the ATPase ($F_1$ portion) is more readily solubilized than the proton channel ($F_0$ portion), it is worth noting that both the coated vesicle and the lysosomal ATPases are stimulated by dithiothreitol and phosphatidylserine.

To date, there are no known inhibitors specific for the lysosomal proton pump. Although micromolar concentrations of trifluoperazine inhibit acidification and ATPase activities ($IC_{50} = 50 \mu M$), the involvement of calmodulin is unlikely inasmuch as 50 $\mu M$ levamisole, bromotetramisole, and calmidazolium are without effect, and, therefore, the inhibitory effect of trifluoperazine is presumably nonspecific.

The $N$-ethylmaleimide sensitivity of the lysosomal proton pump requires comment. Although the majority of the ATPase activity in rat liver lysosomes is membranous, only 5.5% of the membranous ATPase is inhibited by $N$-ethylmaleimide. Moreover, the majority of the $N$-ethylmaleimide-sensitive ATPase activity in lysosomes is soluble (greater than 80%) and not membrane bound. Thus, we are faced with the awesome probability that the proton pump is a very minor protein component of lysosomes.

# [46] Analysis of Endosome and Lysosome Acidification in Vitro

By Cynthia J. Galloway, Gary E. Dean, Renata Fuchs, and Ira Mellman

The low internal pH of endocytic organelles plays a critical role in maintaining the orderly traffic of receptors and receptor-bound ligands during endocytosis.[1,2] In endosomes, acidic pH promotes the dissociation of many ligands from their receptors, allowing the receptor to recycle to the cell surface and the ligand to be transported to lysosomes. In lysosomes, acidic pH facilitates degradation of internalized macromolecules by lysosomal hydrolases, many of which have acidic pH optima. In addition to these beneficial functions, low intravesicular pH is used by a variety of pathogenic agents to enter the cell. For example, many enveloped viruses and bacterial toxins penetrate endosomal or lysosomal mem-

[1] A. Helenius, I. Mellman, D. Wall, and A. Hubbard, *Trends Biochem. Sci.* **8,** 245 (1983).
[2] M. S. Brown, R. G. W. Anderson, and J. L. Goldstein, *Cell (Cambridge, Mass.)* **32,** 663 (1983).

METHODS IN ENZYMOLOGY, VOL. 157

branes as a result of acid pH-catalyzed membrane fusion or insertion events.

Acidic pH in endocytic organelles is due in part to the activity of ATP-dependent proton pumps present in the membranes of both endosomes and lysosomes. These proton ATPases are capable of maintaining the internal pH of these organelles at about 5.0 compared to a cytoplasmic pH of about 7.0.[3–5] While they have yet to be isolated and studied in detail, the endosomal and lysosomal proton pumps appear to be related to each other as well as to proton pumps found in a variety of other organelles of the vacuolar apparatus: acidic endocrine secretion granules, synaptic vesicles, clathrin-coated vesicles, and Golgi membranes.[6–8] All of these proton ATPases are electrogenic and require influx of permeant anions into the vesicle to dissipate the positive interior membrane potential generated by proton transport and allow the continued influx of protons. However, proton transport does not appear to require direct molecular coupling to any other ions. All vacuolar $H^+$-ATPases are inhibited by low concentrations of sulfhydryl alkylating reagents, such as N-ethylmaleimide and 4-chloro-7-nitrobenzo-2-oxa-1,3-diazole, but not by inhibitors of other ion-transport ATPases, such as orthovanadate, oligomycin, ouabain, azide, or efrapeptin.[6,9]

These characteristics differentiate the proton ATPases of endocytic and secretory organelles from the two well-characterized classes of ion-transport ATPases: (1) those enzymes which contain phosphorylated intermediates in their reaction cycle, like the plasma membrane $Na^+$,-$K^+$-ATPase, $Ca^{2+}$-ATPase of the sarcoplasmic reticulum, and $K^+$,-$H^+$-ATPase of the gastric parietal cell, which are inhibited by the transition state analog orthovanadate and (2) the proton pumps of mitochondria, chloroplasts, and bacteria, which normally function at ATP synthetases when coupled to electron transport, and which are inhibited by efrapeptin, oligomycin, and azide.[10] Thus, the vacuolar proton pumps seem to comprise a unique class of ATPase present throughout the functionally continuous membrane systems of the endocytic and exocytic pathways.

[3] S. Ohkuma and B. Poole, *Proc. Natl. Acad. Sci. U.S.A.* **75**, 3327 (1978).
[4] B. Tycko and F. Maxfield, *Cell (Cambridge, Mass.)* **28**, 643 (1982).
[5] A. Roos and W. F. Boron, *Physiol. Rev.* **61**, 296 (1981).
[6] G. Rudnick, in "Physiology of Membrane Disorders" (T. E. Andreoli, D. D. Fanestil, J. F. Hoffman, and S. G. Schultz, eds.), 2nd Ed. Plenum, New York, 1985.
[7] M. Forgac, L. Cantley, B. Weidenman, L. Altsteil, and D. Branton, *Proc. Natl. Acad. Sci. U.S.A.* **80**, 1300 (1983).
[8] J. Glickman, K. Croen, S. Kelly, and Q. Al-Awqati, *J. Cell Biol.* **97**, 1303 (1983).
[9] C. J. Galloway, G. E. Dean, M. Marsh, G. Rudnick, and I. Mellman, *Proc. Natl. Acad. Sci. U.S.A.* **80**, 3334 (1983).
[10] H. E. Ives and F. C. Rector, Jr., *J. Clin. Invest.* **73**, 285 (1984).

## Approaches to Studying Organelle Acidification *in Vitro*

Two general methods have been used to study the ATP-dependent acidification of isolated endocytic vesicles. The first relies on the uptake of lipophilic weak bases in response to a pH gradient. For example, a radiolabeled probe such as [14C]methylamine, which is unprotonated and permeant at neutral pH, is allowed to equilibrate across the vesicle membrane. If acidification occurs after ATP addition, the new equilibrium favors accumulation of the probe which becomes impermeant when protonated. Accordingly, acidification is directly related to the measured increase in intravesicular [14C]methylamine.[7,11,12] Certain fluorescent weak bases, such as acridine orange and 9-aminoacridine, have also been used to study endocytic vesicle acidification *in vitro*. Since these dyes exhibit predictable concentration-dependent shifts in their absorption and fluorescence spectra, the kinetics of their accumulation into acidic membrane vesicles can be monitored using an absorbance or fluroescence spectrophotometer.[8,13,14] While weak base partitioning is technically simple to use, and has yielded much important data, it has a major disadvantage. Weak bases partition across *any* membrane in response to a pH gradient, making interpretation of results dependent on the homogeneity of the membrane fraction being studied. Thus, even small amounts of contaminating organelles with a large intravesicular volume or potent acidification activity may account for a disproportionate fraction of the observed signal.

The second general approach avoids this problem by selectively labeling endocytic organelles with pH-sensitive probes prior to vesicle isolation. This method, originally introduced by Ohkuma and Poole,[3] involves allowing cells to internalize extracellular fluid which contains a macromolecule that is coupled to a pH-sensitive fluorochrome. Fluorescein has proved to be a particularly effective fluorescent reagent since its emission spectrum and quantum yield are both titrable functions of pH in the physiological range (pH 4 to 8). While fluorescein can be conveniently coupled to markers of both fluid-phase and receptor-mediated endocytosis, we have generally relied on derivatized fluid-phase markers (such as high-molecular-weight dextran) to ensure that all endocytic compartments are labeled equally. Fluorescein isothiocyanate-dextran (FD) is also resistant to degradation by lysosomal hydrolases. The probe can be localized in lysosomes or endosomes by controlling the time and tempera-

[11] J. P. Reeves and T. Reames, *J. Biol. Chem.* **256**, 6047 (1981).
[12] D. L. Schneider, *J. Biol. Chem.* **256**, 3858 (1981).
[13] D. K. Stone, X.-S. Xie, and E. Racker, *J. Biol. Chem.* **258**, 4059 (1983).
[14] R. W. Van Dyke, C. J. Steer, and B. F. Scharschmidt, *Proc. Natl. Acad. Sci. U.S.A.* **81**, 3108 (1984).

ture of incubation with cells. After labeling, the cells are homogenized, fractions enriched in the organelles are prepared, and the ATP-dependent acidification of the FD-containing vesicles is examined. This technique eliminates the need for complete purification of the organelles required by weak base partitioning since it limits the analysis only to a functionally defined compartment, rather than a vesicle population which may be both functionally and structurally heterogeneous. This becomes crucial when considering an organelle like an endosome whose definition rests solely on function, and for which fractions of sufficient purity and yield are difficult to obtain.

### Selective Labeling of Endosomes and Lysosomes by Endocytosis

We have successfully studied the acidification of endosomes and lysosomes isolated from a variety of monolayer and suspension cell culture lines as well as from rat liver. In this section, we will summarize the protocol for labeling endosomes and lysosomes using both spinner cultures of the mouse macrophage line J774, and monolayer cultures of Chinese hamster ovary (CHO) cells.[9,15]

Fluorescein isothiocyanate-coupled dextran (Sigma MW 70,000) is prepared as a 25 mg/ml stock solution in phosphate-buffered saline. Ten milliliters of this stock must be dialyzed for 24–48 hr in the cold against 2 liters of phosphate-buffered saline to remove toxic, uncoupled fluorescein and low-molecular-weight glucose polymers. J774 cells are harvested from spinner culture by centrifugation (300 $g$, 5 min at 4°) and resuspended at up to $5 \times 10^8$ cells per 20 ml of warm culture medium. To label endosomes, enough warm FD stock solution is added to the cells to make 2 mg/ml final FD concentration. After 5 min at 37° with gentle agitation, the cells are diluted to 200 ml in cold HEPES–saline (10 m$M$ HEPES, 100 m$M$ NaCl, 20 m$M$ KCl, pH 7.4) to stop endocytosis. The cells are then washed three times with 25 ml of cold HEPES–saline by centrifugation. Other cell lines can also be used to study endosomal acidification, but because their rate and volume of fluid-phase uptake is usually less than that exhibited by macrophages, longer labeling periods and higher concentrations of FD may be required. For example, to label adherent cultures of CHO cells, culture medium from 10 confluent 150-mm dishes is replaced with 37° medium containing 3–6 mg/ml FD. After 30 min at 37°, the plates are washed with cold HEPES–saline. Because transfer of content from endosomes to lysosomes in most cells is prevented at 17°,

[15] A. R. Robbins, C. Oliver, J. L. Bateman, S. S. Krag, C. J. Galloway, and I. Mellman, *J. Cell. Biol.* **99**, 1296 (1984).

endosomes may also be labeled by incubation with FD for up to an hour at this temperature.[9,16] Cells are removed from the plates by scraping with a Teflon scraper in 10 ml of cold HEPES–saline and washed three times by centrifugation at 4° with cold HEPES–saline.

Lysosomes are labeled by the same protocol, except that uptake is not stopped with cold buffer. Rather, the FD-containing medium is removed and replaced with warm FD-free medium. Macrophages are pelleted (300 $g$, 5 min), resuspended in 100 ml of warm media, and returned to culture for an additional 30 min prior to harvest. CHO cells are washed once and incubated in FD-free medium for 60–90 min, conditions which ensure that all of the marker is transferred from endosomes to lysosomes.

### Separation of Endosomes and Lysosomes Using Percoll Gradients

Following the desired labeling protocol, cells are harvested and homogenized in such a way as to minimize disruption of FD-containing organelles yet maximize cellular lysis. After washing with cold HEPES–saline, J774 cells are resuspended in 7 ml of homogenizing buffer (10 m$M$ HEPES, 250 m$M$ sucrose, 1 m$M$ EDTA, pH 7.4, with NaOH) and homogenized at 4° using 10 strokes of a stainless steel dounce homogenizer (Konte, Vineland, NJ). For CHO cells, the cells scraped from plates are resuspended in 8 ml of 10 m$M$ triethanolamine, 10 m$M$ acetate, 250 m$M$ sucrose, 1 m$M$ EDTA, pH 7.4 (TEA buffer), incubated for 10 min at 4°, and then homogenized with 10 strokes of a tight glass dounce (Konte, type B, Vineland, NJ). Both procedures usually disrupt >90% of the cells as monitored by phase-contrast microscopy.

Separation of endosomes from lysosomes is achieved by centrifuging the homogenate in self-forming Percoll density gradients. Percoll, a polyvinlypyrrolidone-coated colloidal silica, is generally obtained from Sigma. The homogenate is first centrifuged at 750 $g$ for 10 min to remove nuclei and unbroken cells. The resulting postnuclear supernatant is then diluted to 24.5 ml with cold homogenizing buffer (TEA buffer for CHO cells) and 10.5 ml of stock Percoll (40 ml Percoll + 10 ml 2.5 $M$ sucrose) is added to bring the final volume to 35.0 ml. The sample is loaded into a Quick-seal centrifuge tube (Beckman) over a 4.0 ml cushion of 2.5 $M$ sucrose and centrifuged in a Beckman L8-70 ultracentrifuge using a type 70Ti rotor at 16,000 rpm for 2 hr with the brake off. An alternative centrifugation procedure, which provides a better separation on some cell lines such as CHO cells and requires a less lengthy run time, uses a vertical rotor (Beckman type VTi 50) at 18,000 rpm for 1 hr with slow acceleration

[16] A. W. Dunn, A. L. Hubbard, and N. N. Aaronson, Jr., *J. Biol. Chem.* **255**, 5971 (1979).

and deceleration on the L8-70 ultracentrifuge. After centrifugation, 1-ml fractions are collected from the bottom by tube puncture using a Beckman fraction recovery system. Notice that in this procedure, no provisions have been made to completely remove the organelles from soluble cytoplasmic molecules. However, if desired, membranes may be pelleted at 100,000 $g$ for 1 hr and washed before mixing with Percoll. Also, the sucrose cushion may be omitted.

Because there are no known marker enzymes for endosomes, their position in the gradient can be defined only by measuring the amount of fluorescein fluorescence in each fraction. To maximize the sensitivity of FD detection, 0.25-ml aliquotes are diluted into 2.5 ml of HEPES–saline, pH 7.4, containing 0.1% Triton X-100. The fluorescence is then measured in each fraction using a fluorescence spectrophotometer with excitation wavelength at 485 nm and the emission wavelength at 515 nm. Excitation and emission slit widths are set to 10 nm.

As shown in Fig. 1, endosomes should be well separated from the major peak of lysosomal enzyme activity. In most cells, a minor fraction of the lysosomal enzyme activity will also be present near the endosomal peak, and probably represents activity in Golgi membranes. We have

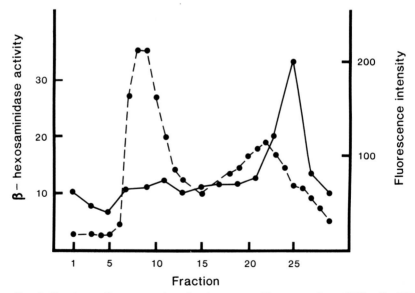

FIG. 1. Density gradient separation of endosomes and lysosomes from CHO cells. FD-filled endosomes (●—●) migrate to the low-density region, while lysosomes (●--●) containing $\beta$-hexosaminidase activity migrate to the high-density region. After a chase period, the fluorescent peak will migrate with the lysosomal enzyme peak.

generally found it convenient to measure either $\beta$-hexosaminidase or $\beta$-glucuronidase activities using $p$-nitrophenol-coupled substrates.[9]

### Fluorescence Assay of Vesicular Acidification

The fluorescence spectrum of FD varies as a predictable function of pH. At neutral pH the excitation spectrum has two well-defined peaks at 490 and 450 nm, the intensity of which decline with decreasing pH. The fluorescence intensity ratio of these two peaks has often been used to provide a measure of intracellular pH that is independent of the quantity of the marker internalized. This method is generally used to express intraorganellar pH in systems where contributions to the shorter wavelength region of the spectrum from light scattering and autofluorescence are minimal.[3,4,17,18] However, when measuring fluorescence from turbid solutions of membrane vesicles containing Percoll, contributions from light scattering distort the excitation spectrum making ratio measurements inaccurate. Therefore, changes in intravesicular pH are more reliably measured by comparing the intensity of fluorescence after acidification to the intensity of fluorescence at a standard pH value.[19] A standard curve of fluorescence intensity at different pH values, making pH 7.4 equal to 100% of relative fluorescence intensity, can be constructed (Fig. 2) and used to estimate pH changes from the relative decrease in fluorescence (see below).

To study ATP-dependent acidification, 0.1- to 0.25-ml samples of FD-loaded endosomes or lysosomes (up to 1 mg protein) are diluted to a final volume of 3 ml with 120 m$M$ KCl, 1 m$M$ EDTA, 5 m$M$ MgSO$_4$, and 10 m$M$ HEPES, pH 7.4. After gentle mixing, the sample is transferred to a fluorometer cuvette and examined using a Perkin–Elmer model LS-5 fluorometer or its equivalent. Excitation and emission wavelengths are set at 485 and 515 nm. Slit widths are usually 10 nm. An arbitrary baseline signal is established for each sample by adjusting the instrument's gain control such that the chart recorder pen rests at 50–80% of full scale deflection.

When the baseline has stabilized, Na$^+$-ATP is rapidly added to a final concentration of 5 m$M$ (from a 500 m$M$ stock, pH 7.0). Intravesicular acidification is indicated by a gradual but rapid decrease in the fluorescence signal, which reaches a minimum after 5–10 min (see Fig. 3). When the chart recorder pen has again stabilized, the K$^+$/H$^+$ ionophore nigericin is added at 0.5 $\mu M$ (nigericin is obtained from Calbiochem and pre-

[17] F. R. Maxfield, *J. Cell Biol.* **95**, 676 (1982).
[18] M. J. Geisow, *Exp. Cell Res.* **150**, 29 (1984).
[19] S. Ohkuma, Y. Moriyama, and T. Takano, *Proc. Natl. Acad. Sci. U.S.A.* **79**, 2758 (1982).

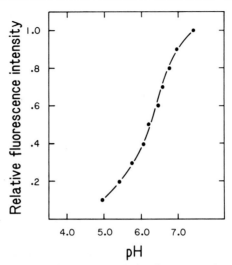

FIG. 2. A standard curve of pH versus relative fluorescence intensity for FD. To construct the curve, the fluorescence of FD is measured at various pH values, and then maximum relative intensity is set at pH 7.4.

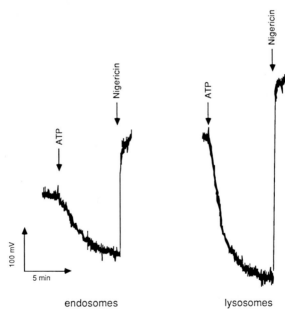

endosomes                                    lysosomes

FIG. 3. ATP-dependent acidification of endosomes and lysosomes from CHO cells. Addition of ATP quenches the fluorescence of FD-containing organelles. The quenching is reversed when the proton gradient across the membrane is released using the ionophore nigericin in a $K^+$-containing buffer.

pared as a 1 m$M$ stock solution in ethanol) to dissipate the transmembrane pH gradient generated due to ATP-driving proton pumping. Because nigericin mediates electroneutral exchange of $K^+$ for protons, the inophore equilibrates intravesicular pH with that of the suspending medium when the medium contains $K^+$. Nigericin should cause an immediate increase in fluorescence to a level equal to or greater than the initial baseline (Fig. 3). This result constitutes proof that acidification has occurred because nigericin can only dissipate a pH gradient across a membrane. If the fluorescence after nigericin addition exceeds the baseline, then the vesicles were probably slightly acidic as isolated; this can easily be tested by adding nigericin before ATP. A weak base such as ammonium chloride (10 m$M$) may also be used to dissipate the pH gradient. If neither agent causes an increase in fluorescence after ATP addition, then the concentration of protons inside the vesicles is not higher than the concentration outside the vesicles.

The fluorescence signal obtained after nigericin addition provides an internal standard to relate to the magnitude of fluorescence quenching to to ATP. This is necessary when working with vesicles which have a low internal pH before ATP addition. In this case, it is often valuable to express the extent of acidification due to ATP as a fraction of the total pH gradient (i.e., fluorescence quenching due to any preexisting pH gradient + that generated by ATP). Using the standard curve (Fig. 2), it is possible to relate fluorescence quenching to pH. However, these data do not necessarily yield absolute pH measurements. Not all FD-containing vesicles in the fraction may acidify equally, nor can the contribution of extravesicular FD or light scattering be accurately quantitated.[9] Therefore, although it is valid to express the degree of acidification in pH units, the absolute pH values before and after ATP addition must not be taken literally.

This protocol can be used in many experimental designs. For instance, to test agents which may inhibit vacuolar acidification, the potential inhibitor is added to the cuvette containing the FD-labeled membranes and any possible nonspecific effect on fluorescence is noted. The effect of the inhibitor on the rate and extent of fluorescence quenching after ATP addition is observed. To examine any possible anion requirement of acidification, chloride in the assay medium can be replaced by gluconate, an organic impermeant anion. Studies of nucleotide specificity and competitive inhibition require low-noise signals in order to compare the initial rates of acidification. ADP and AMP-PNP (a nonhydrolyzable ATP analog) will not support acidification but will act as competitive inhibitors. A photoaffinity analog of ATP, 3'-$O$-(4-benzoyl)benzoic-ATP, is a potent competitive inhibitor and may be used as a photoaffinity reagent to label the proton pump. These methods have also been used to characterize cell lines exhibiting nonlethal or temperature-sensitive defects in endosomal acidification.[14]

### Acidification of Highly Purified Endosomes

The same basic approach has recently allowed us to study the acidification and membrane transport properties of highly purified endosomes isolated from rat liver by free-flow electrophoresis. Selective labeling of rat liver endosomes with FD was accomplished by *in situ* perfusion with 5 mg/ml FD in Hanks' buffer for 5–10 min at 37°. Livers were then flushed for 5 min with cold FD-free Hanks' buffer and homogenized using a Dounce homogenizer in cold TEA buffer. Subsequent isolation steps are described in detail elsewhere.[20–22] To summarize, centrifugation in discontinuous sucrose gradients is used to prepare crude "Golgi" fractions which contain most of the internalized FD and also demonstrate ATP-dependent acidification activity. These Golgi fractions are then subjected to free-flow electrophoresis at 4° in TEA buffer using a Bender and Hobein (Munich, FDR) ElphorVap 21, according to a modification of the method of Harms *et al.*[23] An anodally migrating peak containing most of the FD activity is obtained which is well separated from the major protein-containing peak as well as from the major peak of the Golgi marker enzyme galactosyltransferase. The presence of marker enzymes for mitochondria and lysosomes is negligible. The most purified endosome fractions are 200–300 times enriched in ATP-dependent acidification activity, and contain up to 5 mg of endosomal protein. In contrast, the Percoll gradient procedure described for tissue culture cells may result in only a 2- to 5-fold enrichment in endosomal acidification activity. Thus, using FD as a marker, we have been able to obtain relatively large amounts of highly purified, functionally competent endosomes.

These highly purified preparation offer several advantages. Among the most important, endosomes purified in this way are exceedingly stable, and can be kept 48–72 hr at 4° without loss of acidification activity. Therefore, isolated vesicles can be incubated in isosmotic buffers for the time necessary to equilibrate internal and external ions (usually 2–16 hr at 4°). The ability to use endosomes with a controlled internal ionic environ-

[20] M. Marsh, S. Schmid, H. Kern, E. Harms, I. Mellman, and A. Helenius, *J. Cell Biol.* **104,** 875 (1987).

[21] S. L. Schmid, R. Fuchs, P. Mâle, and I. Mellman, *Cell* **52,** 73 (1988).

[22] S. L. Schmid and I. Mellman, *in* "Cell Free Analysis of Membrane Transport," (D. J. Morré, K. E. Howell, and G. M. W. Cook, eds.). A.R. Liss, New York, in press.

[23] E. Harms, H. Kern, and J. A. Schneider, *Proc. Natl. Acad. Sci. U.S.A.* **77,** 6139 (1980).

ment permits detailed study of the ion requirements and electrogenicity of the proton pump as well as of the ion permeability characteristics of the endosomal membrane. For instance, replacing internal $K^+$ with $Na^+$ (and vice versa) has shown that proton transport into endosomes is not directly coupled to efflux of a particular cation. Replacing external $Cl^-$ with gluconate inhibits acidification unless valinomycin is present to potentiate the efflux of internal $K^+$; therefore, endosomal acidification appears to be electrogenic. Be observing the rate at which the pH gradient dissipates after ATP removal, we have shown that the endosomal membrane has a significant proton conductance which is, however, less than its conductance for other cations ($K^+ > Na^+ > H^+$). All of these data were obtained using the FD assay, eliminating the possibility that the observed signals might have been due to minor or undetected contaminants of the final endosomal fraction. Nevertheless, we have been able to confirm our findings using acridine orange and the voltage-sensitive dye diS-C$_3$(5).

In summary, the FD approach, which relies on selective labeling and functional definition of endocytic organelles, is a simple, sensitive, and versatile method which can be applied to the study of crude or highly purified fractions of endocytic vesicles isolated from a wide variety of sources.

# [47] Proton ATPases in Golgi Vesicles and Endoplasmic Reticulum: Characterization and Reconstitution of Proton Pumping

By Anson Lowe and Qais Al-Awqati

The discovery that endocytic vesicles are acidified by a proton-translocating ATPase in urinary epithelia[1-4] and elsewhere[5-9] and that lysosomes also contain them[8,9] led us to search for these proton ATPases in other intracellular organelles such as Golgi vesicles and endoplasmic re-

[1] T. E. Dixon and Q. Al-Awqati, *Proc. Natl. Acad. Sci. U.S.A.* **76**, 3135 (1979).
[2] S. Gluck, C. Cannon, and Q. Al-Awqati, *Proc. Natl. Acad. Sci. U.S.A.* **79**, 4327 (1982).
[3] S. Gluck, S. Kelly, and Q. Al-Awqati, *J. Biol. Chem.* **257**, 9230 (1982).
[4] S. Gluck and Q. Al-Awqati, *J. Clin. Invest.* **73**, 1704 (1984).
[5] M. Forgac, L. Cantley, B. Wiedemann, L. Altstiel, and D. Branton, *Proc. Natl. Acad. Sci. U.S.A.* **80**, 1300 (1983).
[6] D. K. Stone, X.-S. Xie, and E. Racker, *J. Biol. Chem.* **258**, 4059 (1983).
[7] C. J. Galloway, G. E. Dean, M. Marsh, G. Rudnick, and I. Mellman, *Proc. Natl. Acad. Sci. U.S.A.* **80**, 3334 (1983).
[8] S. Ohkuma, Y. Moriyama, and T. Tatsuy, *Proc. Natl. Acad. Sci. U.S.A.* **79**, 2758 (1982).
[9] D. L. Schneider, *J. Biol. Chem.* **256**, 3858 (1981).

ticulum.[10,11] The methods of preparation of these organelles are essentially those used by others with some minor modification as described below.

### Preparation of Golgi Membranes

Golgi membranes are prepared according to the procedure of Wibo *et al.*[12] Livers of starved rats are excised, weighed, finely minced with a razor blade, and then homogenized in 0.5 *M* sucrose using 20 strokes in a Dounce-type homogenizer fitted with a loose pestle in 2–3 ml homogenizing buffer/g tissue. Homogenizing buffer is 0.5 *M* sucrose which contains (as do all other sucrose solutions) 37.5 m*M* Tris-maleate, pH 6.5, 1% dextran T-500, 5 m*M* $MgCl_2$, and 1 m*M* dithiothreitol, unless otherwise specified. The homogenate is centrifuged for 10 min at 2100 rpm in a Sorvall centrifuge using the SA-600 rotor (680 *g*). The supernatant is decanted and the pellet is resuspended in homogenizing buffer, homogenized with six strokes in a Dounce-type homogenizer (loose pestle), and centrifuged for 10 min at 1800 rpm in a SA-600 rotor (500 *g*). The pooled supernatants are centrifuged for 7.9 min at 20,000 rpm in a Beckman ultracentrifuge using the SW-27 rotor (73,000 $g_{max}$) and the pellet is resuspended in homogenizing buffer and subjected to another centrifugation under the same conditions. The resulting pellet (microsomal fraction) is suspended in 37.4% sucrose to a final volume of 1 ml/g liver tissue, and a discontinuous sucrose density gradient is made in a 36-ml cellulose nitrate centrifuge tube as follows: 6.4 ml 50.4% sucrose, 16.1 ml sample in 37.4% sucrose, 7.7 ml 30.4% sucrose, 5.8 ml 16% sucrose. The gradient is centrifuged for 2.5 hr at 24,000 rpm, in a Beckman ultracentrifuge using the SW-27 rotor (104,000 $g_{max}$). Golgi membranes which collected at the 16%–30.1% interface were aspirated with a syringe, diluted in 0.1 *M* sucrose, and centrifuged for 30 min at 15,000 rpm in a Sorvall centrifuge. The pellets are resuspended in 0.1 *M* sucrose and used either immediately for assays or are frozen at −70° in the appropriate buffer containing 1 m*M* dithiothreitol and 3 m*M* ATP. When thawed they are diluted in 20 ml of the appropriate buffer and pelleted again before use.

When compared to the homogenate the Golgi vesicles are found to contain less than 0.2% of the activity of $Na^+,K^+$-ATPase, oligomycin-sensitive ATPase, glucose-6-phosphatase and *N*-acetylglucosaminidase. However, they contain 30% of the galactosyltransferase, a marker now

[10] J. Glickman, K. Croen, S. Kelly, and Q. Al-Awqati, *J. Cell. Biol.* **97**, 1303 (1983).
[11] R. Rees-Jones and Q. Al-Awqati, *Biochemistry* **23**, 2236 (1984).
[12] M. Wibo, D. Thines-Sempoux, A. Amar-Costesec, H. Beaufay, and D. Godelaine, *J. Cell Biol.* **89**, 456 (1981).

known to be present largely in the trans Golgi. There is 140-fold enrichment in the specific activity of this enzyme in the Golgi fraction.

## Preparation of Smooth and Rough Endoplasmic Reticulum

Rat liver microsomes are prepared according to the procedure of Adelman et al.[13] All low-speed centrifugations are performed in a Sorvall Model RC-5B; and the high-speed centrifugations in a Beckman Model L5-50 ultracentrifuge. All steps are performed at ~4°. Briefly, fresh livers are excised, placed in ice-cold 0.25 $M$ sucrose, cut into five pieces each, blotted on filter paper, and forced through a stainless steel screen with a mesh size of 1 × 1 mm into 1 $M$ sucrose (2 ml/gram of tissue). The mixture is homogenized with eight strokes of a motor-driven Teflon pestle and glass homogenizer at 1500 rpm (Caframo type RZR50, Warton, Ontario), filtered through gauze, mixed with an equal volume of 2.5 $M$ sucrose, and centrifuged for 45 min at 100,000 $g$ to pellet nuclei. The postnuclear supernatant is mixed with one-half its volume of water and centrifuged for 15 min at 22,000 $g$ to bring down the mitochondria. The postmitochondrial supernatant is saved, and the mitochondrial pellet is washed twice by resuspending in 12 ml of a 9 : 1 mixture of 0.5 $M$ sucrose : inhibitory supernatant and centrifuging for 15 min at 17,000 $g_{max}$. To prepare the ribonuclease inhibitory supernatant, we centrifuge the 0.25 $M$ sucrose homogenate for 20 min at 40,000 $g$ in the Beckman ultracentrifuge using the SW-40 rotor. The supernatant from that spin is recentrifuged for 2 hr at 100,000 $g$. The supernatant is then divided into 10-ml aliquots and stored at −20° until used.

The combined postmitochondrial supernatant is centrifuged for 15 min at 17,000 $g$ again for the final postmitochondrial supernate. The mitochondrial pellets are combined and saved. Smooth and rough microsomes are then obtained from a sucrose density gradient where 11 ml of postmitochondrial supernate is underlayed with 2 ml of 2 $M$ sucrose : inhibitory supernatant at 3 : 1 (final molarity ~1.5 $M$) and 0.5 ml of a solution containing 2 $M$ sucrose, 50 m$M$ Tris-HCl, 25 m$M$ KCl, 5 m$M$ MgCl$_2$, pH 7.5, and centrifuged for 18–20 hr at 200,000 $g$. Rough membranes band at the interface between steps of 1.265 and 1.202 densities while the smooth membranes band between densities of 1.202 and 1.085. The membranes are aspirated separately using a syringe and steel cannula, diluted with twice their volume in transport medium [0.25 $M$ sucrose, 150 m$M$ KCl, 6 m$M$ MgCl$_2$, 2 m$M$ Tris base, 2 m$M$ 2-($N$-morpholino)ethanesulfonic acid (MES), titrated to pH 7.0 with HCl], and pelleted by centrifuging for 1 hr

---

[13] M. R. Adelman, G. Blobel, and D. D. Sabatini, *J. Cell Biol.* **56**, 191 (1973).

at 46,800 $g_{max}$ in a Sorvall centrifuge using the SM-24 rotor. These pellets are resuspended in 10 ml of transport medium per fraction (smooth and rough) and portions are either used immediately in transport experiments or frozen and stored at $-20°$ for subsequent enzyme assays. Dithiothreitol is added to all solutions to a final concentration of 1 m$M$. The rough and smooth membranes are depleted in oligomycin-sensitive ATPase containing only 2% of the homogenate activity. They are enriched in glucose-6-phosphatase containing 16% of the homogenate activity. The proton-pumping activity in the rough membranes is readily demonstrable initially but rapidly decays over the following 4–6 hr.

*Enzyme Assays*

*ATPase Assays.* The final composition of the ATPase assays was 150 m$M$ KCl, 30 m$M$ Tris, 30 m$M$ MES, 6 m$M$ MgCl$_2$, pH 7.0, in 1 ml. Samples of membrane are incubated with inhibitors for at least 1 hr at 0° before the assays are performed. The reaction is started by the addition of 100 $\mu$l of 30 m$M$ Tris-ATP, pH 7.0, to 900 $\mu$l of assay mixture. They are then incubated for 20 min at 25° and the reaction is stopped by the addition of 100 $\mu$l of 100% trichloroacetic acid (TCA). The samples are allowed to sit on ice for 20 min., and phosphate assays are either done immediately or the samples were frozen and the assays done the following day. The ATPase samples are centrifuged at 2500 $g$ for 10 min. Two hundred microliters of the supernatant is transferred to another set of tubes on ice. The following reagents are then added serially using a repetitive pipetter: 1 ml 2% ascorbic acid in 10% TCA (made freshly), 0.5 ml 1% ammonium molybdate, 1 ml 2% sodium citrate, 2% sodium arsenite, 2% acetic acid. The samples are then vortexed, allowed to stand 20 min, and then the $A_{700}$ read. Na$^+$,K$^+$-ATPase is measured essentially as above except that 20 m$M$ Na is present in the assay mixture. The phosphate release before and after the addition of ouabain is considered to be the Na$^+$,K$^+$-ATPase activity.

Glucose-6-phosphatase activity is measured by the method of Baginski *et al.*[14]; galactosyltransferase is measured using ovalbumin as acceptor[15]; $N$-acetylglucosaminidase activity is measured by the method of Ray *et al.*[16]; and protein is measured using the Bradford method with bovine serum albumin as standard.[17]

[14] E. S. Baginski, P. P. Foa, and B. Zak, *in* "Methods of Enzymatic Analysis" (H. U. Bergmeyer, ed.), p. 876. Academic Press, New York, 1974.

[15] G. N. Andersson and L. C. Eriksson, *J. Biol. Chem.* **256**, 9633 (1981).

[16] D. Ray, E. Cornell, and D. Schneider, *Biochem. Biophys. Res. Commun.* **71**, 1246 (1976).

[17] M. M. Bradford, *Anal. Biochem.* **72**, 248 (1976).

*Transport Assays*

The final composition of the transport buffer is 150 m$M$ KCl, 6 m$M$ MgCl$_2$, 2 m$M$ Tris, 2 m$M$ MES, pH 7.0. Membrane fractions taken from the sucrose gradients are suspended in transport buffer, pelletted at 22,600 $g$ for 1 hr, and resuspended in a small amount of transport buffer for use in the assays. Oligomycin (final concentration 5 $\mu$g/ml) is added to the plasma membrane samples and these are allowed to stand at 0° for at least 1 hr before use. The samples are pipetted into a cuvette containing 3–7 $\mu M$ acridine orange in 1.5 ml of transport buffer. The optimum protein concentration varies among different preparations from 20–1000 $\mu$g/ml. The cuvette is then placed in a dual-wavelength spectrophotometer[18] (Univ. of Pennsylvania Instrument Shop) and the $A_{492}$–$A_{520}$ (an index of the amount of acridine orange in free solution) monitored continuously on a chart recorder. After the baseline reaches a steady level, transport is initiated by the addition of ATP to a final concentration of 0.4 m$M$. Initial rates of acridine orange uptake into vesicles are calculated by taking the steepest tangent of the curve after the initial drop caused by quenching from the ATP.

The assay is based on the fact that acridine orange is a weak base. Since only its unprotonated form is permeable, the steady-state distribution ratio of the acridine orange concentration ([AO]) across the vesicle membrane will be equal to the H$^+$ concentration ([H$^+$]) ratio,

$$[H^+]_i/[H^+]_o = [AO]_i/[AO]_o \tag{1}$$

where subscripts i and o refer to inside and outside the vesicle. The protonated form is less permeable and hence will accumulate in the vesicles when their contents are acidified. Acridine orange tends to aggregate at high concentration with a consequent change in its absorption characteristics. Using differential absorption tuned to the absorption characteristics of the monomeric form (i.e., that in the bulk medium), one can follow the disappearance of the dye from the medium and use it as a reasonable index of the development of a pH difference across the membrane. The magnitude of the steady-state pH difference across the membrane depends on the ATP-driven proton flux and the leakage of protons out of the vesicles. The latter is given by the proton conductance of the membrane times the proton electrochemical gradient. The initial rate of uptake of acridine orange, however, should reflect H$^+$ transport before a significant pH difference develops; hence, it will largely be due to the ATP-driven H$^+$ influx provided the permeability of the unprotonated acridine orange is not rate-limiting. The "instantaneous" discharge of acri-

[18] B. Chance, V. Legallias, J. Sorge, and N. Graham, *Anal. Biochem.* **66**, 498 (1975).

dine orange from the vesicles on addition of uncouplers (Fig. 1) indicates that the permeability of the unprotonated dye is not limiting. The initial rate is also affected by the buffering power of the vesicle contents. It is also assumed that the various treatments used have little effect on the

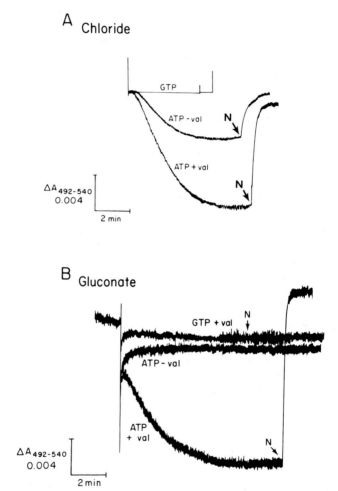

FIG. 1. $H^+$ transport in Golgi vesicles measured as the uptake of acridine orange in a dual-wavelength spectrophotometer as described in the text. Transport was initiated by the addition of ATP or GTP as labeled to a final concentration of 0.4 m$M$. Oligomycin was present in all assays in a final concentration of 50 $\mu$g/ml. Valinomycin (val), when present, was in a final concentration of 1 $\mu M$. Nigericin was added at the end (labeled N) to a final concentration of 1.5 $\mu M$. Each assay contained 150 $\mu$g of protein. In A, the media were composed of chloride salts whereas in B they were composed of gluconate salts. The noise in the tracing is largely a function of the stirring rate. (From Ref. 10.)

buffering capacity of the vesicles. We found that this initial rate varied linearly with the amount of vesicle protein added. As can be seen below, the steady-state uptake is also expected to be a linear function of the protein added. Since the absorbance is proportional to the acridine orange concentration, it can be shown readily that, in the steady state,

$$[H^+]_i/[H]_o = (V_o/V_i)(D/(1 - D)) \qquad (2)$$

where $V_o$ and $V_i$ are the extra- and intravesicular volume, respectively, and $D$ is the fractional decrease in the absorption signal from its initial level. The protein concentration is proportional to $V_i/V_o$. Since $D$ is $\ll 1$, $D/(1 - D)$ is $\approx D$. When more vesicles are placed in the apparatus, the pH across each membrane will not change; hence, there should be a direct relation between the protein concentration and the steady-state signal. However, the decline in the signal is also due to binding of acridine orange to the vesicle interior. Although we found that the steady-state decrease was a linear function in a number of preparations, this was not a universal finding, hence the calculation should be done for each new preparation.

### Reconstitution of H⁺-ATPase

Reconstitution of the H⁺-ATPase into soybean phospholipid vesicles was performed as described by Bennet and Spanswick.[19] The bovine kidney vesicles are thawed and washed with 15 volumes 5 m$M$ Tris-MES, pH 7.0, and 1 m$M$ ethylene glycol bis($\beta$-aminoethyl)-$N,N'$-tetraacetic acid (EGTA) by centrifugation at 40,000 $g$ for 1 hr in a RC-5B centrifuge using the SM-24 rotor. The supernatant is discarded and the pellet is suspended in the Tris-EGTA buffer. Glycerol is added to the vesicle suspension to a final concentration of 40% (v/v). The vesicles are then solubilized with 0.7% deoxycholate (w/v); final concentration of protein is 2 mg/ml. Before use, the deoxycholate is filtered through charcoal and recrystallized from ethanol. The mixture is allowed to incubate at 0° for 15 min before centrifugation at 150,000 $g$ for 1 hr in a Beckman ultracentrifuge using a Ti50 rotor. The supernatant is then used for reconstitution using soybean phospholipids. These phospholipids are prepared from asolectin as described by Kagawa and Racker and stored in ether at −70°. An aliquot of phospholipid is dried with argon gas, suspended in Tris-EGTA buffer to a final concentration of 5 mg/ml, and sonicated with a probe sonicator (Branson Co., Shelton, CT) in an ice bath until clear. The phospholipids, 0.5 ml, are loaded on top of a 25 cm × 1 cm Sephadex G-200 column. After the phospholipids have entered, 1–2 ml of the solubilized

---

[19] A. B. Bennet and R. M. Spanswick, *J. Membr. Biol.* **75**, 21 (1983).

protein is layered on top. The column is eluted at room temperature with the Tris-EGTA buffer at 20 ml/hr until a cloudy fraction is collected. The vesicles are then placed on ice and used the same day. The reconstituted proton pumping activity has all the characteristics of that in the native vesicles. The transport is electrogenic, inhibited by N-ethylmaleimide but not by oligomycin or vanadate.

## Characteristics of the Proton ATPase

The proton pump of these vesicles generates a pH difference and a membrane potential. The rate of transport of an electrogenic $H^+$ pump is affected by the net transmembrane electrochemical gradient ($\Delta\bar{\mu}_{H^+}$).

$$\Delta\bar{\mu}_{H^+} = RT \ln H_i/H_o + zF \, \Delta\psi \tag{3}$$

In an open circuited system, the proton pump generates an electrochemical gradient sufficient to bring net $H^+$ transport to zero. In the absence of $H^+$ leaks, the proton pump can shut itself off. The maximum gradient that can be generated by the pump is given by:

$$(\Delta\bar{\mu}_{H^+})_{J_{H=0}} = n \, \Delta G_{ATP} \tag{4}$$

where $n$ is the ATP : $H^+$ stoichiometry, $\Delta G_{ATP}$ is the free energy of ATP hydrolysis under the actual conditions, and $J_H$ is the $H^+$ flux by the pump. The electrochemical gradient will be composed of an electrical term and a pH term. If one collapses the electrical term, e.g., by adding a potassium conductor like valinomycin, the $\Delta$pH will increase so that at the steady state all the electrochemical gradient is due to a pH difference (Fig. 1A). If the native electrical conductance of the membrane is reduced, e.g., by removal of permeant anions, a large membrane potential can develop with a few turnover cycles of the pump and can generate a sufficient adverse membrane potential to shut off the $H^+$ pump before any pH difference can develop (Fig. 1B). (Using a simple electrostatic model one can calculate that the electrogenic transport of only 623 ions can lead to the generation of 100 mV.)

The ATPases from these vesicles are inhibited by low concentrations of N-ethylmaleimide ($K_i \simeq 20 \ \mu M$) and by dicyclohexyl carbodiimide ($K_i \simeq 100 \ \mu M$). However, it is resistant to oligomycin, aurovertin, efrapeptin, and vanadate. The purified vesicles have a high ATPase activity but most of it is due to a phosphatase activity. N-Ethylmaleimide inhibits only a few percent of its activity. Solubilization and reconstitution in asolectin improves this and as much as 40% of the activity can be inhibited by N-ethylmaleimide. Only ATP can be used for $H^+$ pumping, other nucleotides are inactive.

This ATPase is similar to that found in clathrin-coated vesicles, lysosomes, chromaffin granules, and other endocytic vesicles. However, this similarity is based on physiological and pharmacological criteria rather than by analysis of the polypeptides that mediate ATP-driven $H^+$ transport. This similarity imposes a serious problem for the investigator since contamination of any of these fractions by lysosomes or clathrin-coated vesicles could produce results similar to those that have been described above. The ultimate test of the location of ATP-driven proton pumping is microscopic identification of these pH or potential gradients. Anderson *et al.*[20] have recently developed an ingenious new method which promises to resolve this problem. They synthesized a primary amine linked to dinitrophenol which will accumulate in acid compartments. When the cells are fixed with glutaraldehyde, the amine is covalently attached to any protein in the vicinity. Using antidinitrophenyl antibodies, the site of accumulation of this reagent can be localized by immunoelectron microscopy. Preliminary results (R. G. W. Anderson, personal communication) suggest that the trans Golgi is acid. It would be important to collapse the potential difference, generated by the proton pump to enhance the development of a pH gradient, to identify the other sites where a functional proton pump is present. The principle of this method might also serve to develop probes to measure a membrane potential by the use of negatively charged permeant probes that can be fixed by glutaraldehyde or other fixatives.

[20] R. G. W. Anderson, J. R. Falck, J. L. Goldstein, and M. S. Brown, *Proc. Natl. Acad. Sci. U.S.A.* **81,** 4838 (1984).

# [48] Chromaffin Granule Proton Pump

*By* NATHAN NELSON, SHULAMIT CIDON, and YOSHINORI MORIYAMA

A continuous process of catecholamine uptake takes place in chromaffin granules of adrenal medullae cells.[1] The driving force for this uptake is a proton-motive force generated by ATP hydrolysis which is catalyzed by an ATPase located in the granule membrane.[1] Though ATPase activity of chromaffin granule membranes was discovered over two decades ago,[2]

[1] D. Njus, J. Knoth, and M. Zallakian, *Curr. Top. Bioenerg.* **11,** 107 (1981).
[2] N. Kirshner, *J. Biol. Chem.* **237,** 2311 (1962).

METHODS IN ENZYMOLOGY, VOL. 157

until recently the identity of the enzyme responsible for the formation of proton-motive force was not clear.[3,4] The revelation of the true identity of the chromaffin granule proton-ATPase was hampered by contamination of mitochondrial membranes in the granule membrane preparation and the instability of the genuine enzyme. By overcoming these two obstacles, the proton-ATPase of chromaffin granule membrane was identified, solubilized, and reconstituted into phospholipid vesicles and isolated with a reasonable purity.[5-7] It is difficult to define the final structure of a membrane protein complex.[8] The proton-ATPase of chromaffin granules may be defined as the minimal structure, isolated from chromaffin granule membranes, that, upon reconstitution into phospholipid vesicles, will catalyze the reaction of ATP-dependent proton uptake. It is quite likely that only part of this structure will be sufficient to catalyze the ATPase activity of this protein complex.

### Preparation of Chromaffin Granule Membranes

Chromaffin granules contain catecholamines and ATP at high concentrations; therefore it is convenient to isolate them on the basis of their heavy density. Depending on the required purity of the membranes the preferred isolation method can be more or less rigorous.

#### Reagents

SME: 0.3 $M$ sucrose, 10 m$M$ (MOPS) 4-morpholinepropanesulfonic acid (pH 7.5), and 5 m$M$ EDTA (pH 7.5)
Phenylmethylsulfonyl fluoride (PMSF), 1 $M$ in dimethyl sulfoxide
0.1 $M$ of each of PMSF, L-1-tosylamino-2-phenylethyl chloromethyl ketone (TPCK), and $p$-aminobenzamidine in dimethyl sulfoxide
Pepstatin A, 1 mg/ml in ethanol

#### Procedure

Bovine adrenal glands are obtained from the local slaughterhouse. Twenty to 40 adrenal medullae are separated from the cortex by pinching the connecting tissue with pointed forceps.[9] The isolated adrenal medul-

---

[3] D. K. Apps, J. G. Pryde, and R. Sutton, *Ann. N.Y. Acad. Sci.* **402**, 134 (1982).
[4] S. Cidon and N. Nelson, *J. Bioenerg. Biomembr.* **141**, 499 (1982).
[5] S. Cidon and N. Nelson, *J. Biol. Chem.* **258**, 2892 (1983).
[6] S. Cidon, H. Ben-David, and N. Nelson, *J. Biol. Chem.* **258**, 11684 (1983).
[7] S. Cidon and N. Nelson, *J. Biol. Chem.* **261**, 9222 (1986).
[8] N. Nelson and S. Cidon, *J. Bioenerg. Biomembr.* **16**, 11 (1984).
[9] D. I. Meyer and M. M. Burger, *J. Biol. Chem.* **254**, 9854 (1979).

lae are kept on ice in 150 ml of a freshly prepared ice-cold solution of SME supplemented with 1.1 m$M$ PMSF, 0.1 m$M$ TPCK, 0.1 m$M$ $p$-aminobenzamidine, and 2 $\mu$g/ml of pepstatin A. These protease inhibitors are added throughout the purification of the membranes. The medullae are briefly (2–4 sec) homogenized in a Waring blender at low speed. The mixture is filtered through four layers of gauze, the filtrate kept on ice, and the solid material is further homogenized in about 100 ml of the same solution by a glass–Teflon homogenizer. The homogenate is filtered through four layers of gauze and the filtrate is combined with the previous one. The suspension is centrifuged at 1000 $g$ for 15 min.[9] The pellet is discarded and the supernatant is centrifuged at 10,000 $g$ for 20 min. The pellet is gently suspended in about 30 ml of SME containing the protease inhibitors. The suspension is then applied on top of sucrose layers in SW-27 cellulose nitrate tubes. The buffer is 20 m$M$ MOPS (pH 7.5) containing all of the protease inhibitors at a tenth of their concentration in homogenization solution. The bottom layer fo 15 ml contains 1.8 $M$ sucrose, and the top layer of 10 ml contains 1.2 $M$ sucrose. Five milliliters of the suspension is applied on each tube and centrifuged in an SW-27 rotor at 20,000 rpm at 2° overnight. The solution is removed by aspiration and the pellet containing the chromaffin granules is collected in a minimal volume of a solution containing 10 m$M$ MOPS (pH 7.5), 1 m$M$ ATP, and the protease inhibitors at a tenth of their concentration in the homogenization solution. After homogenization in a glass–Teflon homogenizer the suspension is diluted with about 1 liter of the same solution. After centrifugation at 3000 $g$ for 10 min, the pellet is discarded and the supernatant centrifuged at 200,000 $g$ for 60 min. The pellet is homogenized in about 10 ml of a solution containing SME, 1 m$M$ ATP, 25% glycerol, and protease inhibitors at a tenth of their concentration in the homogenization solution. This preparation of chromaffin granule membranes can be kept at −80° and used for measurements of ATPase, ATP-dependent proton uptake, and the isolation of the $N$-ethylmaleimide (NEM)-sensitive ATPase.

If purer membranes are required they can be further purified on sucrose layers or Percoll gradient. For that purpose the 25% glycerol is omitted from the above solution and the membrane suspension is applied over layers of 1.8 and 0.7 $M$ sucrose in a buffer containing 10 m$M$ MOPS (pH 7.5), 1 m$M$ ATP, and protease inhibitors. Then it is centrifuged in an SW-27 rotor at 20,000 rpm overnight. The chromaffin granule membranes are collected from the top of the 0.7 $M$ sucrose layer. The mitochondrial membranes are present on top of the 1.8 $M$ sucrose layer. The suspension is diluted 3-fold, centrifuged at 200,000 $g$ for 60 min, and the pellet is collected and stored as was described above. These preparations have remained active after storage at −80° for over a year.

Assay of ATP-Dependent Proton Uptake

Several methods can be used for measuring ATP-dependent proton uptake into isolated chromaffin granules and resealed membranes after lysis by osmotic shock. We used measurements of absorbance changes of acridine orange or fluorescence quenching of 9-amino-6-chloro-2-methoxyacridine as indicators for proton uptake into the membrane vesicles.

*Reagents*

Reaction medium: 0.3 $M$ sucrose, 5 m$M$ Tricine–NaOH (pH 8), and 40 m$M$ KCl
Acridine orange, 1 m$M$
9-Amino-6-chloro-2-methoxyacridine (ACMA), 0.1 m$M$ in ethanol
Valinomycin, 1 mg/ml in ethanol
MgATP, 0.1 $M$ (pH 7.5)
Carbonyl cyanide $p$-trifluoromethoxyphenylhydrazone (FCCP), 1 m$M$ in ethanol

*Procedure*

Into 3 ml of the reaction medium, chromaffin granule membranes containing 100 to 200 $\mu$g of protein are added and mixed. When required, 1 $\mu$l of the valinomycin solution is added and the mixture is equilibrated at room temperature for 10 min. For assay by absorbance changes, 10 to 20 $\mu$l of the acridine orange solution is added (check the optimal amount for every source of acridine orange). Place the cuvette in an Aminco DW dual-wavelength spectrophotometer. The sample is equilibrated for 3 min and 15 $\mu$l of the MgATP solution is added and mixed. The absorbance changes at 492 nm with a reference at 540 nm are recorded. When required, 1 to 5 $\mu$l of the FCCP solution is added and the increase in absorbance is recorded the same way. Figure 1 depicts a typical experiment as was described above and shows that the ATP-dependent proton uptake by chromaffin granules is sensitive to $N$-ethylmaleimide.

When dual-wavelength spectrophotometry is not available, one can use the same procedure except that the acridine orange is replaced by 10 $\mu$l of ACMA and the fluorescence quenching in the presence of MgATP is followed in a spectrofluorometer with an excitation wavelength of 412 nm and emission at 480 nm.

Resolution and Reconstitution of the $N$-Ethylmaleimide-Sensitive
ATP-Dependent Proton Uptake of Chromaffin Granule Membranes

Reconstitution of the purified enzyme serves as the most decisive test for obtaining the minimal structure of this enzyme. For that goal and

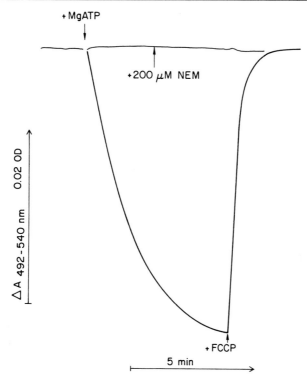

FIG. 1. Effect of *N*-ethylmaleimide (NEM) on ATP-dependent proton uptake by chromaffin granule membranes. The membranes were prepared and proton uptake was assayed as was described in the text. The ATP-induced changes in the absorbance of acridine orange were followed using an Aminco DW-2a spectrophotometer. Chromaffin granule membranes containing about 100 $\mu$g protein were used in the presence of 1 $\mu$g valinomycin. When specified, 1 nmol of FCCP was added. *N*-Ethylmaleimide was added prior to the assay as was described in the text.

especially for the identification of the enyzme, we initiated studies on resolution and reconstitution of the NEM-sensitive proton-ATPase of the chromaffin granule membrane.[5,6]

### Reagents

20% solution of *n*-octyl-$\beta$-D-glucopyranoside

20% potassium cholate (pH 7.8): the cholic acid is treated with charcoal and crystallized twice from hot 70% ethanol

Brain lipids: prepared by homogenizing bovine brain in 10 volumes of ethanol. The homogenate is filtered through four layers of gauze and the filtrate is centrifuged at 20,000 *g* for 30 min. The supernatant is evaporated to dryness in a rotary evaporator. The residue is homogenized in 4 volumes of water and centrifuged at 20,000 $g$ for 30 min. The supernatant is discarded and the pellet is dried under vacuum.

*Escherchia coli* phospholipids: prepared as described by Newman and Wilson.[10]

*Procedure*

Most of the chromaffin granule membrane preparations tested so far are suitable for resolution and reconstitution studies. The only factor that interferes is the presence of oxidized catecholamines, manifested as membrane preparation with a brown color. Sodium bromide-treated membranes in which the mitochondrial ATPase is inactivated are suitable for these experiments.[5,6] In the latter case, the presence of 5 m$M$ dithiothreitol or 5 m$M$ thioglycerol during the treatment and solubilization of the membrane are necessary. The membranes can be solubilized by either 1% cholate or 0.5% cholate and 1% octylglucoside, and the enzyme can be partially purified in glycerol gradients containing phospholipids and cholate. The active fractions are reconstituted by cholate dilution.[11] A protocol for such experiments is as follows: To 0.5 ml of chromaffin granule membranes containing about 5 mg of protein, potassium cholate and octylglucoside are added to give final concentrations of 0.5% and 1%, respectively. After incubation on ice for 10 min, the suspension is centrifuged at 50,000 $g$ for 15 min. About 200 $\mu$l of the supernatant is applied on a glycerol gradient of 7 to 40% in a solution containing 0.3 $M$ sucrose, 10 m$M$ 3-($N$-morpholino)propanesulfonic acid (pH 7.5), 5 m$M$ EDTA, 5 m$M$ thioglycerol, 1 $\mu$g/ml pepstatin A, 0.5% potassium cholate, and 0.5 mg/ml *E. coli* phospholipids. The gradient is centrifuged in a SW-60 rotor at 57,000 rpm for 5 hr. Fractions are collected and samples of 25 $\mu$l are assayed for ATP-dependent proton uptake as was described in the previous section.

Figure 2[12,13] depicts the results of such an experiment. Most of the ATP-dependent proton uptake activity appeared in fraction 8. In subsequent experiments it was observed that, at longer times of centrifugation, the activity appeared lower in the gradient. Brain lipids successfully replaced the *E. coli* phospholipids, but soybean phospholipids were not effective.

As shown in Fig. 2, the fractions were also assayed for the presence of the mitochondrial proton-ATPase by electrophoresis followed by immune

[10] M. J. Newman and T. H. Wilson, *J. Biol. Chem.* **255,** 10583 (1980).
[11] E. Racker, P. Chien, and A. Kandrach, *FEBS Lett.* **57,** 14 (1975).
[12] N. Nelson, this series, Vol. 97 [46].
[13] N. Nelson, this series, Vol. 118 [24].

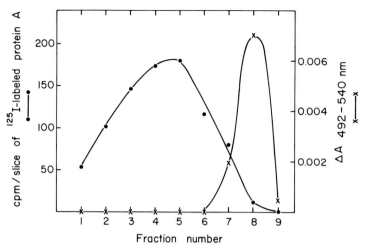

Fraction number

FIG. 2. Resolution, partial purification, and reconstitution of the proton-ATPase of chromaffin granules. Chromaffin granule membranes containing 10 mg protein per milliliter were treated with 0.5% potassium cholate and 1% octylglucoside, and separated on a glycerol gradient as described in the text. Samples of 25 $\mu$l of each fraction were assayed for ATP-dependent proton uptake and for the presence of the mitochondrial proton-ATPase. For the latter, the fractions were electrophoresed in the presence of SDS, electrotransferred to a nitrocellulose filter, decorated by an antibody against $\beta$-subunit of the proton-ATPase of *E. coli,* and the amount of radioactivity of [125]I-labeled protein A in the corresponding spots was assayed as previously described.[12,13]

decoration with antibody against the $\beta$-subunit.[12] It is apparent that the proton uptake activity was detected in different fractions from those in which the mitochondrial proton-ATPase was present.

## Purification of *N*-Ethylmaleimide-Sensitive ATPase of Chromaffin Granule Membranes

Our previous studies and the experiment described in the previous section clearly indicated that the proton-ATPase of the chromaffin granule membrane is not homologous to the mitochondrial enzyme.[4–7] On the other hand, several of its properties indicated its resemblance to similar enzymes that may function in lysosomes,[14] plant vacuoles,[15] synaptosomes,[6,16] platelet dense granules,[17] and clathrin-coated vesicles.[18,19] The

[14] D. L. Schneider, *J. Biol. Chem.* **256**, 3858 (1981).

[15] A. B. Bennett and R. M. Spanswick, *J. Membr. Biol.* **71**, 95 (1983).

[16] L. Toll and B. D. Howard, *J. Biol. Chem.* **255**, 1787 (1980).

[17] H. Fishkes and G. Rudnick, *J. Biol. Chem.* **257**, 5671 (1982).

[18] M. Forgac, L. Cantley, B. Wiedenmann, L. Altstiel, and D. Branton, *Proc. Natl. Acad. Sci. U.S.A.* **80**, 1300 (1983).

[19] D. K. Stone, X.-S. Xie, and E. Racker, *J. Biol. Chem.* **258**, 4059 (1983).

instability of these enzymes hampered their purification. This property was probably due to the susceptibility of these enzymes to proteases and their special requirement for phospholipids for maintaining activity. Recently, it was demonstrated that the enzyme for clathrin-coated vesicles is stabilized by phosphatidylserine.[20]

By the combination of appropriate protease inhibitors and phosphatidylserine we stabilized the proton-ATPase of chromaffin granules, and thereby purified it into a well-defined protein complex.

### Reagents

L-$\alpha$-Phosphatidylserine suspension, 1%, prepared by sonication in 10 m$M$ MES-Tris (pH 7.0)

10% polyoxyethylene 9-lauryl ether ($C_{12}E_9$) (Sigma)

20% potassium cholate (pH 7.8)

Hydroxylapatite (HTP) purchased from Bio-Rad Laboratories

DEAE-cellulose (DE-23) was purchased from Whatman and was washed and equilibrated as previously described[21]

Thioglycerol, 1 $M$

ATP, 0.1 $M$, adjusted to pH 7 with NaOH

L-$\alpha$-Phosphatidylserine suspension, 1%, prepared by sonication in 10 m$M$ MES-Tris (pH 7.0)

ATPase activity is assayed as previously described[22] except that the reaction mixture contains 80 m$M$ Tricine (pH 8), 4 m$M$ ATP, 4 m$M$ MgCl$_2$, and about 2 $\times$ 10$^5$ cpm of [$\gamma$-$^{32}$P]ATP.

### Procedure

The purification procedure is carried out at temperatures of 0 to 4°. About 10 ml of chromaffin granule membranes, containing about 10 mg protein per milliliter, is treated with 1% $C_{12}E_9$. After 10 min incubation on ice, the suspension is centrifuged at 200,000 $g$ for 30 min. The supernatant is applied on an hydroxylapatite column (1 $\times$ 13 cm) that is equilibrated with a buffer containing 10 m$M$ MES-Tris (pH 7.0), 5 m$M$ thioglycerol, and 0.1% $C_{12}E_9$. The column is washed with 5 ml of the same buffer and the enzyme eluted with a linear sodium phosphate (pH 7.0) gradient from 0 to 0.2 $M$ in the same buffer. Each chamber contains 20 ml solution. About 50 fractions of 1 ml are collected, and as soon as possible 10 $\mu$l of the 1% phosphatidylserine solution is added to each fraction and mixed by a Vortex.

[20] X.-S. Xie, D. Stone, and E. Racker, *J. Biol. Chem.* **259,** 11676 (1984).
[21] N. Nelson, D. W. Deters, H. Nelson, and E. Racker, *J. Biol. Chem.* **248,** 2049 (1973).
[22] N. Nelson, this series, Vol. 69 [27].

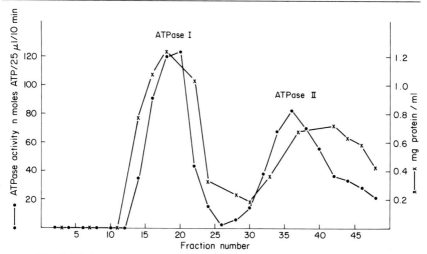

FIG. 3. Elution pattern of ATPase activity from an hydroxylapatite column.

Figure 3 shows the elution profile of the hydroxylapatite column. Two peaks with ATPase activity are eluted from the column. The activity of both of them is sensitive to $N$-ethylmaleimide. The purification of ATPase II is not continued. ATPase I is purified as follows.

Fractions 14 to 22 are combined and potassium cholate added to give a final concentration of 0.5%. About 1.3 ml of this solution is applied on each of six SW-41 tubes containing linear glycerol gradients of 10 to 30%. The gradient is formed in solution containing 10 m$M$ MES-Tris (pH 7.0), 1 m$M$ ATP, 5 m$M$ thioglycerol, 0.3% $C_{12}E_9$, and 0.1 mg/ml phosphatidylserine. After overnight centrifugation at 38,000 rpm, 14 fractions of 0.8 ml are collected from each tube. The most active fractions (6 to 9) are pooled and applied to a DEAE-cellulose column (1 × 13 cm) that is equilibrated with a solution containing 10 m$M$ MES-Tris (pH 7), 5 m$M$ thioglycerol, and 0.3% $C_{12}E_9$. The column is washed with 5 ml of the same buffer, and the enzyme is eluted with 20 ml of the same buffer supplemented with 0.2 $M$ sodium phosphate (pH 7.0). About 30 fractions of 1 ml are collected and 10 $\mu$l of the 1% phosphatidylserine suspension is added to each as soon as each fraction is eluted from the column. The active fractions (24 to 28) are pooled and applied to a second glycerol gradient that is formed and centrifuged similar to the first one. About 18 fractions of 0.6 ml are collected from each tube and fractions 12 to 14 found to be the most active ones. This preparation can be kept at $-80°$ with very little or no loss of activity.

Table I summarizes the purification procedure. Most of the activity which is not sensitive to $N$-ethylmaleimide was present in the pellet after

TABLE I
PURIFICATION OF N-ETHYLMALEIMIDE-SENSITIVE ATPase OF CHROMAFFIN GRANULES

| Purification steps | Total protein (mg) | Specific activity ($\mu$mol ATP mg protein/min) | Total activity | Recovery (%) |
|---|---|---|---|---|
| Chromaffin granule membranes | 147 | 0.171 | 25.1 | 100 |
| Supernatant of $C_{12}E_9$ | 115 | 0.203 | 23.3 | 93 |
| HTP column fractions 14 to 22 | 34.2 | 0.273 | 9.34 | 37 |
| Glycerol gradient fractions 6 to 9 | 12.8 | 0.686 | 8.78 | 35 |
| DEAE-cellulose fractions 24 to 28 | 3.1 | 1.09 | 3.38 | 14 |
| Glycerol gradient fractions 12 to 14 | 0.1 | 17.6 | 1.76 | 7 |

$C_{12}E_9$ treatment. This activity was enhanced by the detergent treatment and therefore the combined ATPase activity of the pellet and the supernatant was greater than the total activity of the chromaffin granule membranes.

Figure 4 shows a sodium dodecyl sulfate-polyacrylamide gel of the fractions from the second glycerol gradient. In the active fractions 12 to 14 five main protein bands are copurified. Using protein markers, apparent molecular weights of 115K, 72K, 57K, 39K, and 17K were measured on 10% polyacrylamide gel in the presence of SDS. These apparent molecular weights nearly correspond to the ones detected earlier.[4,5] The 17K polypeptide is the chromaffin granule dicyclohexylcarbodiimide-binding protein.[7,23]

Figure 5 shows that both the 115K and 39K polypeptides bind N-[$^{14}$C]ethylmaleimide. However, when [$^{14}$C]NEM with higher specific activity was used and the NEM concentration was decreased to 2 $\mu M$ the 72K subunit was strongly labeled.[24] The 115K protein also binds [$^{14}$C]dicyclohexylcarbodiimide while, except for the proteolipid, the three other polypeptides do not interact with this chemical.[7]

## Purification of a Reconstitutively Active H$^+$-ATPase

Previous attempts to reconstitute the purified ATPase were hampered by two main factors. Recently, we observed that the mixture of protease

[23] R. Sutton and D. K. Apps, *FEBS Lett.* **130**, 103 (1981).
[24] Y. Moriyama and N. Nelson, *J. Biol. Chem.* **262**, 9175 (1987).

8  9  10  11  12  13  14  15  16  17

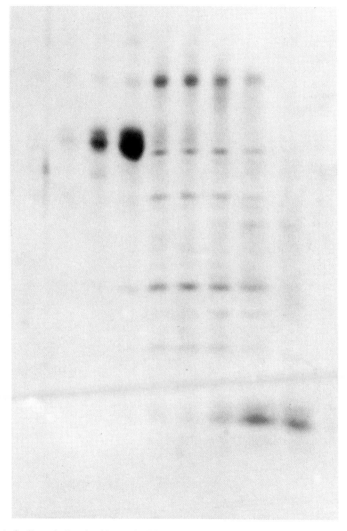

FIG. 4. Sodium dodecyl sulfate gel of fractions from the second glycerol gradient. Samples of 40 $\mu$l of each fraction were dissociated, electrophoresed, stained by Coomassie Brilliant Blue, and destained as previously described.[12]

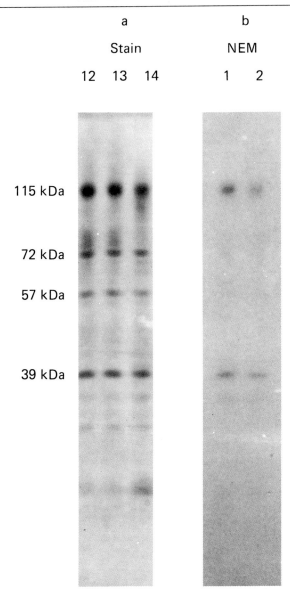

FIG. 5. Labeling of the purified ATPase of chromaffin granules by $N$-[$^{14}$C]ethylmaleimide. The enzyme was purified while thioglycerol was omitted from all the purification steps. Samples of 50 $\mu$l of the purified enzyme were incubated in a total volume of 0.1 ml with 2 $\mu$Ci $N$-[$^{14}$C]ethylmaleimide (Amersham, 2–10 mCi/mmol), in a solution containing 0.3 $M$ sucrose, about 20% glycerol, 5 m$M$ EDTA, and 10 m$M$ MOPS (pH 7.5); the NEM concentration was about 1 m$M$. After incubation at room temperature, the reaction was terminated by the addition of 25 $\mu$l of 5-fold concentrated dissociation buffer containing 10%

inhibitors used in the previous study inhibits the proton pumping activity of the enzyme.[24] Therefore, in this procedure only pepstatin A and leupeptin are used. A second observation was that the enzyme lost its activity with time, and thus the new purification procedure is completed in a single day. After the purification is accomplished the enzyme becomes more stable and it can be stored at −85° and maintains its original activity.

### Reagents

0.2 $M$ 2($N$-morpholino)ethanesulfonic acid adjusted to pH 7.0 by Tris (MES-Tris)
10% polyoxyethylene 9-lauryl ether ($C_{12}E_9$) (Sigma)
20% cholic acid adjusted to pH 7.0 by NaOH
Hydroxylapatite (HTP) purchased from Bio-Rad Laboratories
Thioglycerol, 1 $M$
ATP, 0.1 $M$, adjusted to pH 7 with NaOH
Pepstatin A, 2 mg/ml in ethanol
Leupeptin, 5 mg/ml in ethanol

### Procedure

Chromaffin granule membranes are prepared from bovine adrenal glands as described above, except that the only protease inhibitors used during the preparation are pepstatin A at 2 $\mu$g/ml and leupeptin at 5 $\mu$g/ml. The membranes are suspended at protein concentration of 5 mg/ml in a solution containing 0.3 $M$ sucrose, 10 m$M$ MOPS (pH 7), 1 m$M$ ATP, 25% glycerol, 2 $\mu$g/ml pepstatin A, and 5 $\mu$g/ml leupeptin. The membranes are kept frozen at −85°. To about 5 ml of chromaffin granule membranes, containing 5 mg of protein/ml, $C_{12}E_9$ is added as a 10% solution to give a final concentration of 1%. The suspension is vortexed and centrifuged at 250,000 $g$ for 1 hr. The supernatant is applied to an hydroxylapatite column (2.5 × 1.5 cm) that is equilibrated with a solution containing 10 m$M$ MES-Tricine (pH 7.0), 5 m$M$ monothioglycerol, and 0.1% $C_{12}E_9$. The column is washed with the same buffer, and the enzyme is collected in about 5 ml of the flow-through fraction. This fraction is

---

2-mercaptoethanol. Aliquots of 50 $\mu$l were electrophoresed on 10% SDS gels stained with Coomassie Blue or exposed to X-ray film after equilibration with 1 $M$ sodium salicylate. (a) Stained fractions 12 to 14 of the glycerol gradient. (b) Radioactive bands obtained following $N$-[$^{14}$C]ethylmaleimide treatment. 1, Without addition of cold NEM; 2, in the presence of 0.5 m$M$ cold NEM.

diluted by an equal volume of the elution buffer containing 2% sodium cholate, and solid ammonium sulfate is added to give a final concentration of 25% saturation. The suspension is centrifuged at 250,000 g for 20 min and the floating precipitate is separated by a careful suction of the solution below it. The pellet is suspended in 1 ml of the same buffer and layered on two linear glycerol gradients of 10 to 30% containing 10 mM MES-Tricine

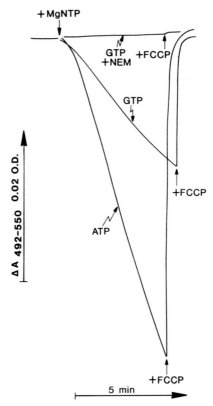

FIG. 6. Effect of GTP and NEM on the proton uptake activity of the purified chromaffin granule ATPase after reconstitution into vesicles. Reconstitution was carried out by dilution of 5 μl of the purified enzyme containing 1.4 μg protein into 1 ml buffer containing 10 mM MOPS-Tris (pH 7.0), 0.1 M KCl, and 0.1 μg valinomycin. After 10 min incubation at room temperature, 15 μl of 1 mM acridine orange was added followed by addition of 10 μl of 0.1 M MgATP or MgGTP. When specified, 100 μM NEM was added 2 min prior to the addition of MgGTP, as well as 1 μl of 1 mM carbonyl cyanide p-trifluoromethoxyphenylhydrazone (FCCP). Proton-uptake activity was followed by absorption change at 492–540 nm using an Aminco DW-2 spectrophotometer.

(pH 7.0), 5 m$M$ monothioglycerol, 1 m$M$ ATP, and 0.25% $C_{12}E_9$. The gradients are centrifuged at 60,000 rpm (450,000 $g$) in a SW-60 rotor for 4 hr at a temperature of 4°. A total of about 20 fractions are collected from each gradient and assayed for ATP-dependent proton uptake activity by dilution into the assay buffer and following acridine orange absorbance changes. For reconstitution, the sample is diluted at least 50-fold and incubated for 10 min at room temperature prior to the addition of acridine orange. This preparation is quite similar in its polypeptide composition to the purified ATPase that was previously described (Fig. 4). Its ATPase specific activity is about 5 $\mu$mol/mg protein/min which is somewhat lower than the previous one; it also contains higher amounts of contaminating cytochrome $b_{561}$. An overall purification of 50- to 100-fold was obtained for the ATPase and ATP-dependent proton uptake activities. A complete correlation was observed between the distribution of the above-mentioned polypeptides and the reconstitutable activity of ATP-dependent proton uptake.[24] Accumulation of protons in the reconstituted vesicles is dependent on the presence of potassium and valinomycin during the reconstitution of the vesicles (Fig. 6). This is due to the complete elimination of the chloride channel during the purification of the enzyme. Apparently the proton uptake activity is very sensitive to positive membrane potential inside the vesicles. Inclusion of NEM abolishes the proton-uptake activity of the enzyme. As shown in Fig. 6, GTP can serve as a substrate for the ATPase. The proton uptake activity is also dependent on the presence of $Cl^-$ or $Br^-$ outside the vesicles. Sulfate inhibits the ATP-dependent proton uptake by competing with $Cl^-$. Nitrate can replace $Cl^-$ in this site but it is a potent inhibitor when present inside the vesicles.[24,25]

Nucleotide binding sites were detected on subunits I, II, and IV.[26] Subunit II (72K) plays a major role in the catalytic activity of the enzyme. As with every other protein complex, the integrity of each one of the polypeptides in the preparation can be proven only when the specific function of each polypeptide is revealed, or when inactivation of the gene coding for this polypeptide prevents the activity of the enzyme.[8]

[25] Y. Moriyama and N. Nelson, *Biochem. Biophys. Res. Commun.* **149**, 140 (1987).
[26] Y. Moriyama and N. Nelson, *J. Biol. Chem.* **262**, 14723 (1987).

## [49] Proton Pump of Clathrin-Coated Vesicles

By Xiao-Song Xie, Dennis K. Stone, and Efraim Racker

An ATP-driven proton pump activity is present in clathrin-coated vesicles which is distinguished from the well-characterized proton pump of mitochondria by its insensitivity to oligomycin, azide, and efrapeptin.[1,2] We describe here solubilization, isolation, and reconstitution of the clathrin-coated vesicle proton-translocating complex. By sucrose density gradient centrifugation determination, the molecular weight of the proton pump is 530K. In addition, the purification of a 116K ATPase from clathrin-coated vesicles is described. This enzyme, unlike the proton pump, is sensitive to vanadate (10 $\mu M$) and hence is unrelated to the 530K ATPase. Some of the procedures have been published elsewhere.[1,3–5]

### Assay Methods

*Proton Pump Activity[6]*

*Reagents*

Stock solution I: 3 m$M$ MgCl$_2$, 85 m$M$ KCl, 30 m$M$ histidine, final pH 7.0

ATP: 200 m$M$, pH 7.0, adjusted with NaOH

Acridine orange (Kodak): 700 $\mu M$ in H$_2$O

Oligomycin (Sigma): 0.5 mg/ml in ethanol

1799 (DuPont): bis(hexafluoroacetonyl) acetone, 1 m$M$ in ethanol

Valinomycin (Sigma): 0.1 m$M$ in ethanol

Stock solution II: 3 m$M$ MgCl$_2$, 100 m$M$ KCl, 20 m$M$ Na-Tricine, pH 7.0, 3 m$M$ sodium azide, 1 n$M$ valinomycin

*Procedure for Measuring Proton Pump Activity in Clathrin-Coated Vesicles.* Add to a 3-ml cuvette 1.5 ml of stock solution (kept at room

[1] D. K. Stone, X.-S. Xie, and E. Racker, *J. Biol. Chem.* **258**, 4059 (1983).

[2] M. Forgac, L. Cantley, B. Wiedenmann, L. Altstiel, and D. Branton, *Proc. Natl. Acad. Sci. U.S.A.* **80**, 1300 (1983).

[3] X.-S. Xie, D. K. Stone, and E. Racker, *J. Biol. Chem.* **258**, 14834 (1983).

[4] X.-S. Xie, D. K. Stone, and E. Racker, *J. Biol. Chem.* **259**, 11676 (1984).

[5] X.-S. Xie and D. K. Stone, *J. Biol. Chem.* **261**, 2492 (1986).

[6] P. Dell'Antone, R. Colonna, and G. F. Azzone, *Biochim. Biophys. Acta* **234**, 441 (1971).

temperature), 2 $\mu$l of oligomycin, and 7.5 $\mu$l of acridine orange. After mixing, vesicles in the range from 10 to 500 $\mu$g are added to the curvette, which is placed in an Aminco dual-wavelength spectrophotometer with gentle magnetic stirring until the 492–540 nm absorbance is stabilized. Ten microliters of ATP is added to initiate the reaction and 5 $\mu$l of 1799 to collapse the pH gradient after completion of the assay. To measure proton pump activity in the absence of a chloride transporter or in a reconstituted system, 3 $\mu$l of valinomycin is added to collapse the membrane potential by allowing $K^+$ to move out of the vesicles.

The activity of the pump is expressed in arbitrary optical density units, either in terms of the initial slope or the extent of ATP-induced decrease in absorbance, under the assay conditions described above. The slope and extent of ATP-generated $\Delta$ absorbance are linear with clathrin-coated vesicles over a range of 10 to 500 $\mu$g protein.

*ATPase Assay*

*Reagents*

Stock solution I: 10 m$M$ ATP, 10 m$M$ MgCl$_2$, 80 m$M$ histidine buffer, pH 7.0, 1 m$M$ ouabain, and 3 m$M$ NaN$_3$

Phosphatidylserine (Avanti): 0.5 $\mu$g/ml in 0.1% polyoxyethylene 9-lauryl ether ($C_{12}E_9$) (Sigma)

$N$-ethylmaleimide (NEM): 150 m$M$ in distilled water

Sodium dodecyl sulfate (SDS): 20% in distilled water

Molybdate reagent I: dilute 200 ml of 2.5% ammonium molybdate in 5 $N$ H$_2$SO$_4$ with 700 ml of distilled water

ANS: 52.6 g Na$_2$S$_2$O$_5$, 2 g Na$_2$SO$_3$, and 1 g 1-amino-2-naphthol-4-sulfonic acid are dissolved in 400 ml distilled water by heating to 90°

Stock solution II: stock solution I containing [$\gamma$-$^{32}$P]ATP (2500 cpm/nmol)

Perchloric acid, 1.25 $N$ in distilled water

Molybdate reagent II: 5% ammonium molybdate (w/v) in distilled water

Isobutanol–benzene 1 : 1 (v/v)

*Colorimetric Assay.* The sample to be assayed is incubated with 4 $\mu$l of phosphatidylserine at room temperature for 10 min (not necessary for crude preparation), and then diluted to 50 $\mu$l with distilled water with or without 2 $\mu$l of $N$-ethylmaleimide. The reaction is started by adding 50 $\mu$l of stock solution I. After incubation at 37° for 30 min, the reaction is stopped by adding 40 $\mu$l SDS. Then 0.9 ml of molybdate reagent I and 50 $\mu$l of ANS are added and mixed well by vortexing. Absorbance at 740 nm is measured for each sample after incubation at room temperature for 20 min.

*Radioactive Assay.* The reaction conditions are the same as described above, except that stock solution II is used instead of stock solution I. To terminate the reaction, 0.75 ml of perchloric acid is added after incubation. The $^{32}P_i$ released is extracted by adding 0.25 ml of molybdate reagent II and 2 ml of isobutanol–benzene, followed by vortexing for 15 sec. Aliquots of 200 $\mu$l of the organic phase are taken from each tube and counted in a liquid scintillation counter.

DEFINITION OF UNITS AND SPECIFIC ACTIVITY. A unit of ATPase activity is defined as the amount of enzyme hydrolyzing 1 $\mu$mol of ATP per minute under the assay condition specified above. Specific activity is expressed as units per milligram of protein.

## Purification of [γ-$^{32}$P]ATP

Commercial [γ-$^{32}$P]ATP frequently contains more than 2% of $^{32}P_i$ and is not suitable for the measurement of low ATPase activities. The following simple procedure has been developed to reduce the $^{32}P_i$ contamination in commercial [γ-$^{32}$P]ATP preparation.

### Reagents

Activated charcoal (C-4386 Sigma)
EDTA, 10 m$M$ in distilled water
Ethanol, 50% aqueous solution
Acetone–EDTA: 40% acetone, 0.5 m$M$ EDTA in distilled water

*Treatment of Charcoal.* Wash 4 g of activated charcoal (Sigma-HCl washed) twice with 50 m$M$ of EDTA, followed by 50 ml of ethanol. The charcoal pellet is finally washed with 10 ml of absolute ethanol, and dried at 100°.

*Procedure.* Add 5 mg of washed charcoal to 5 ml of a solution of [γ-$^{32}$P]ATP (1 × 10$^7$ cpm/ml or more) containing 2 m$M$ NaP$_i$ and 200 m$M$ NaCl. Centrifuge for 2 min at 2000 $g$ and test the supernatant. At least 80% of radioactivity should be adsorbed. The pellet is washed with 5 ml of EDTA, and then eluted twice with 1 ml of acetone–EDTA. The eluates, containing more than 50% of adsorbed counts and less than 0.5% of $^{32}P_i$ are combined, evaporated to 40% of the volume, and stored at −20°.

## $^{32}P_i$-ATP Exchange

### Reagents

Stock solution: 20 m$M$ ATP, 10 m$M$ MgCl$_2$, 20 m$M$ K$^{32}$P$_i$ (5 × 10$^6$ cpm/$\mu$mol), pH 7.0

Perchloric acid: 1.25 $N$ in distilled $H_2O$
Ammonium molybdate: 5% in distilled $H_2O$
Isobutanol–benzene 1 : 1 (v/v)
$H_2O$-saturated isobutanol
Ethyl ether

$^{32}P_i$-ATP Exchange Assay. Coated vesicles (50 $\mu$g) are diluted with distilled water to 0.1 ml, and mixed with 0.1 ml of the stock solution. After incubation at 37° for 30 min, the reaction is terminated by adding 4.0 ml of 1.25 $N$ perchloric acid, followed by the addition of 5 ml isobutanol–benzene and 1.0 ml of ammonium molybdate. The mixture is vortexed vigorously for 30 sec. After removing the upper organic phase by aspiration, the aqueous phase is reextracted twice with 5 ml isobutanol, followed by a final extraction with 3 ml of ethyl ether. One milliliter of the clear aqueous phase is removed to measure radioactivity in a liquid scintillation counter.

## Measurement of Chloride Uptake[3,7,8]

Principle. Coated vesicles contain a transporter through which chloride or bromide pass, thereby balancing the electrogenic proton translocation. Chloride transport can be functionally dissociated from the proton pump.[3] It is facilitated either by ATP-driven proton movement, or by an inside positive membrane potential generated by $K^+$ moving with valinomycin from the outside to the inside of the vesicles.

## Reagents

Stock solution: 5 m$M$ MgSO$_4$, 30 m$M$ histidine, pH 7.0, 40 m$M$ sucrose, 50 m$M$ potassium gluconate, 3.7 m$M$ K$^{36}$Cl (8.5 × 10$^5$ cpm/$\mu$mol), and either 3 m$M$ ATP or 3 $\mu M$ valinomycin
4,4′-Diisothiocyanostilbene-2,2-disulfonic acid disodium salt trihydrate (DIDS) (Pierce), 1 m$M$ in $H_2O$
Dowex AG1-X8 anion-exchange resin (Bio-Rad)
Sucrose–BSA: 180 m$M$ sucrose containing 3.3 mg bovine serum albumin/ml (Sigma)
Sucrose: 180 m$M$ in distilled $H_2O$

Preparation of Anion-Exchange Column. Dowex AG1-X8, 2 ml for each column, is converted from the chloride form to formate form,[8] and washed with 3 ml of sucrose-BSA.

[7] A. Soumarmon and E. Racker, Front. Biol. Energ. 1, 555 (1978).
[8] O. D. Gasko, A. F. Knowles, H. G. Shertzer, E. M. Suolinna, and E. Racker, Anal. Biochem. 72, 57 (1976).

*Assay Procedure.* The reaction is initiated by adding 30 $\mu$l of coated vesicles (20 mg/ml) to 70 $\mu$l of stock solution, with or without 1 $\mu$l DIDS as an inhibitor. After incubation at room temperature for 5 min, 90 $\mu$l of the reaction mixture is passed through the Dowex AG1-X8 anion-exchange column, followed by washing with 3 ml of sucrose. The eluant is collected, mixed with 10 ml of Liquiscint, and counted in a liquid scintillation counter. The activity is expressed as nmol $Cl^-$/min/mg protein.

## Measurement of $S^{14}CN$ Uptake

*Principle.* The ATP-driven proton pump in coated vesicles is electrogenic and the uptake of $S^{14}CN$, a permanent anion, is used to register the generation of an inside positive membrane potential.[3,9]

### Reagents

Stock solution: 140 m$M$ sucrose, 30 m$M$ histidine, pH 7.0, 5 m$M$ $MgSO_4$, 3 $\mu$g oligomycin/ml, 1 m$M$ ouabain, 0.083 m$M$ $S^{14}CN$ (specific activity 56 mCi/mmol) with or without 3 m$M$ ATP
Dowex AG1-X8 anion-exchange resin
Sucrose–BSA: 180 m$M$ sucrose containing 3.3 mg BSA/ml
Sucrose: 180 m$M$

*Experimental Procedure.* The procedure is basically the same as that for $^{36}Cl$ uptake measurement, except that only a 1-min incubation is used before passing the mixture through the anion-exchange resin. $S^{14}CN$ uptake is expressed as pmol $S^{14}CN$/min/mg of protein.

### Preparations

#### Preparation of Clathrin-Coated Vesicles from Bovin Brain

Clathrin-coated vesicles are prepared from bovine brain basically according to the procedures of Pearse[10] and Nandi *et al.*,[11] with some modifications that facilitate large-scale preparations. Defatted bovine brains (20 to 30) are briefly rinsed with an ice-cold solution of 0.1 $M$ Na-MES, pH 6.5, 1 m$M$ EGTA, 0.5 m$M$ $MgCl_2$, and 3 m$M$ $NaN_3$ (buffer A). Five kilograms of the cleaned tissue is homogenized in 4500 ml of buffer A in a Waring blender with three 10-sec bursts at maximum speed. The homogenate is centrifuged in a GSA (Sorvall) rotor for 50 min at 20,000 $g$ and the

[9] H. Rottenberg, this series, Vol. 55, pp. 547–569.
[10] B. M. F. Pearse, *J. Mol. Biol.* **97**, 93 (1975).
[11] P. K. Nandi, G. Irace, P. P. VanJaarsveld, R. E. Lippoldt, and H. Edelhoch, *Proc. Natl. Acad. Sci. U.S.A.* **79**, 5881 (1982).

supernatant is centrifuged at 140,000 $g$ for 1 hr to sediment the clathrin-coated vesicles. The pellets are resuspended with a Dounce homogenizer in 800 ml of buffer A. The suspension is centrifuged at 10,000 $g$ for 10 min to eliminate aggregated material which can be used as a crude preparation of coated vesicles, and the supernatant is spun again at 140,000 $g$ for 1 hr. The pellets are resuspended in 300 ml of buffer A and the low speed–high speed spins are repeated once more. The last pellets are resuspended in buffer A to give a final volume of 114 ml. Samples of 19 ml of the suspension are loaded onto centrifuge tubes which are half-filled with 8% sucrose in buffer A prepared with $D_2O$, and centrifuged at 80,000 $g$ (25,000 rpm, SW28 rotor) for 2 hr at 15–20°. The pellets containing enriched coated vesicles are resuspended in buffer A at a protein concentration of 20–25 mg/ml, divided into small aliquots, frozen in liquid nitrogen, and stored at −80°. All procedures are carried out at 0–4° unless otherwise stated. The procedure, at the scale of 20 to 30 brains, yields 5 to 6.5 mg of enriched coated vesicles per 100 g of wet brain tissue. The proton pump activity of the frozen preparation of coated vesicles is stable for at least 6 months.

## Preparation of Crude Brain Lipids[4]

Crude brain lipids are obtained by extracting 100 ml of bovine brain homogenate (the same as that used for the coated vesicle preparation) with 20 volumes of ethanol at room temperature overnight with magnetic stirring. The mixture is filtered to remove ethanol-insoluble materials, dried under reduced pressure, and stored at −20°. The dried lipids (150 mg) are suspended in 1 ml of 1% cholate, 100 m$M$ KCl, and 2 m$M$ Tris-MES, pH 7.0, and sonicated under nitrogen in a bath-type sonicator for 15 min.

## Purification of $H^+$-Translocating Complex from Coated Vesicles[5]

### Reagents

Tris-HCl: 3 $M$, pH 8.5
Cholate: 20% in distilled water, pH 8.0 (sodium salt)
Polyoxyethylene 9-lauryl ether ($C_{12}E_9$, Sigma): 20% in distilled water
Column solution: 0.1% $C_{12}E_9$, 10 m$M$ Na-Tricine, pH 7.0
Sodium phosphate: 0.3 $M$ in column solution
Dithiothreitol: 1 $M$ in distilled water

### Procedure

*Step 1: Solubilization.* To 74 ml of crude coated vesicles (16.9 mg/ml) 25 ml of ice-cold 3$M$ Tris-HCl, pH 8.5, is added and incubated on ice for

30 min to strip the vesicles of clathrin.[12] The mixture is centrifuged at 150,000 $g$ for 1 hr and the pellet is resuspended in 110 ml of 0.5% cholate, pH 8.0, and incubated on ice for 30 min. After centrifugation of the mixture at 150,000 $g$ for 1 hr, the NEM-sensitive ATPase is in the pellet. The pellet is extracted at 0° for 1 hr with 140 ml of 10 m$M$ Na–Tricine, pH 7.0, and 0.75% $C_{12}E_9$, and centrifuged at 150,000 $g$ for 90 min. Glycerol is added to the clear supernatant to a final concentration of 10% (v/v).

*Step 2: Ammonium Sulfate Fractionation.* To 140 ml of the supernatant, 16 g of $(NH_4)_2SO_4$ is added, and the mixture is stirred on ice for 20 min, followed by centrifugation at 85,000 $g$ for 30 min at 0°. To the clear supernatant 18 g of $(NH_4)_2SO_4$ is added, and after stirring for 20 min on ice, the mixture is centrifuged at 85,000 $g$ for 30 min at 0°. The pellet is dissolved in 100 ml of column buffer containing 10% glycerol (v/v).

*Step 3: Hydroxylapatite Chromatography.* The solutilized ATPase is loaded onto two identical hydroxylapatite columns (20 ml) equilibrated with column solution containing 10% glycerol (v/v). After washing the column with the same buffer, the enzyme is eluted with a gradient of $NaP_i$ from 0 to 0.3 $M$ (total volume of 150 ml for each column) in column buffer with 10% glycerol.

*Step 4: Glycerol Gradient I.* The reconstitutively active fractions from HTP column (50 ml) are pooled and precipitated by adding 15.65 g of $(NH_4)_2SO_4$, mixing, and centrifuging at 85,000 $g$ for 30 min. The precipitate is dissolved in 2 ml of column buffer. Aliquots of 0.5 ml are loaded on top of four 12-ml glycerol gradients (10–30%) prepared in 0.1% $C_{12}E_9$, and 10 m$M$ Na-Tricine, pH 7.0. The tubes are centrifuged in a SW41 rotor at 170,000 $g$ for 18 hr at 0–4°, and the fractions collected from the bottom after piercing the tubes.

*Step 5: Glycerol Gradient II.* The reconstitutively active fractions are pooled (usually about 8 ml) and precipitated again with $(NH_4)_2SO_4$ as described in Step 4. The precipitate is dissolved in 0.5 ml of 100 m$M$ dithiothreitol, 0.05% $C_{12}E_9$, 10 m$M$ Na-Tricine, pH 7.0, and incubated on ice for 30 min. Each aliquot of 0.25 ml is loaded on top of a glycerol gradient prepared as described in Step 4, except that 10 m$M$ dithiothreitol is present in the gradient. The tubes are centrifuged and fractionated as described in Step 4. SDS-PAGE, performed under reducing conditions, reveals the presence of eight peptides of apparent MW 116K, 70K, 58K, 40K, 38K, 34K, 33K, and 15K. Shown in Table I is a summary of the purification of the complex, and in Fig. 1 a gel illustrating components of the pump.

---

[12] H. A. Sober, ed., "Handbook of Biochemistry," pp. C10-C25. The Chemical Rubber Co., Cleveland, 1968.

TABLE I
PURIFICATION OF THE PROTON-TRANSLOCATING COMPLEX
FROM CLATHRIN-COATED VESICLES

| Purification step | Total protein (mg) | Total activity (units)[a] | Specific activity (units/mg) | Purification (-fold) |
|---|---|---|---|---|
| Clathrin-coated vesicles | 1,250 | 93.7 | 0.07 | |
| $C_{12}E_9$ extract | 175 | 78.2 | 0.45 | 6.4 |
| HTP eluate | 15 | 60 | 4.0 | 57 |
| Glycerol gradient I | 2.2 | 16.2 | 7.2 | 103 |
| Glycerol gradient II | 0.24 | 3.3 | 13.5 | 193 |

[a] Units are defined as $\mu$mol $P_i$ ml$^{-1}$ min$^{-1}$.

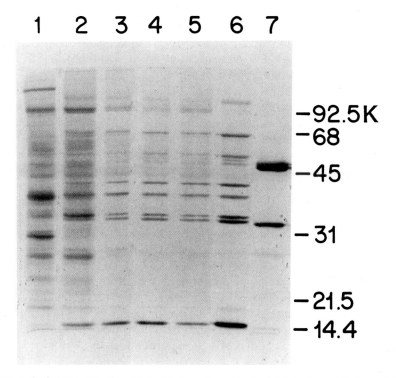

FIG. 1. SDS-polyacrylamide gel electrophoresis and silver staining of the clathrin-coated vesicle proton-translocating complex. Lane 1, 2 $\mu$g of $C_{12}E_9$ extract; lane 2, 2 $\mu$g of hydroxylapatite #1 eluate; lane 3, 1 $\mu$g of glycerol gradient 1; lanes 4 and 5, 1 $\mu$g of hydroxylapatite #2 eluate; lane 6, 1 $\mu$g of glycerol gradient purified ATPase; lane 7, 1 $\mu$g of purified bovine heart mitochondrial $F_1$.

*Reconstitution of the Proton Pump*

Sonicated brain lipids, containing 900 $\mu$g of lipid in 6 $\mu$l, are mixed with 20 $\mu$l of ATPase (containing 25% glycerol and from 0.2 to 2 $\mu$g of protein) and 1.0 $\mu$l of 20% sodium cholate. The mixture is vortexed, and 1 $\mu$l 100 m$M$ MgCl$_2$ is added. After vortexing, the mixture is incubated at 0° for 30 min and frozen at $-70°$ for at least 15 min. After thawing, proton pumping is assessed by adding the proteoliposomes to 1.5 ml of proton pump stock solution II containing 2.5 $\mu$M acridine orange, and measuring ATP-generated changes in the absorbance of acridine orange at $\Delta A_{492-540}$ in an Aminco DW2C dual-wavelength spectrophotometer.

*Molecular Weight Determination*

*Reagents*

H$_2$O gradient solutions: 0.02% C$_{12}$E$_9$, 10 m$M$ Tris-MES, pH 7.0, with 5% sucrose or 20% sucrose (w/v) in H$_2$O

D$_2$O gradient solutions: 0.02% C$_{12}$E$_9$, 10 m$M$ Tris-MES, pH 7.0, with 5% sucrose or 20% sucrose (w/v) in D$_2$O (Aldrich)

Bovine liver catalase (Pharmacia)

*Procedure.* To 2.2 ml of isolated ATPase (280 $\mu$g/ml) is added 0.69 g of (NH$_4$)$_2$SO$_4$. After incubation for 20 min at 0°, the mixture is centrifuged at 85,000 $g$ for 45 min. The pellet is resuspended in 0.1% C$_{12}$E$_9$, 10 m$M$ Tris-MES, pH 7.0, to 0.1 ml and is mixed with 0.1 ml (8.0 mg/ml) of bovine liver catalase in the identical buffer. One-half of the solution is loaded on each of two 12.5-ml sucrose gradients (5–20%) prepared in D$_2$O and H$_2$O. After centrifugation at 170,000 $g$ in a SW40 rotor for 14 hr at 4°, the distance traveled by the green catalase is measured, and the tubes are pierced at the bottom and collected in 48 fractions. The distance traveled by the proton ATPase is determined by measuring ATPase activity of the fractions and the distance traveled by the catalase is confirmed by protein determination. The proton ATPase migrates 68.7 mm in D$_2$O and 83.0 mm in H$_2$O; catalase migrates 42 mm in H$_2$O and 26.5 mm in D$_2$O. By using existing data for catalase (MW of 232K and $s_{20,w}$ 11.3)[13] and assuming a partial specific volume of 0.73 for the proton pump,[13,14] it is determined by prescribed formulas[14,15] that the protein–detergent–lipid complex has an $s_{20,w}$ of 18.4 and a calculated MW of the protein moiety of the complex of 530K.

[13] B. R. Oakley, D. R. Kirsch, and N. R. Morris, *Anal. Biochem.* **105,** 361 (1980).

[14] W. Schaffner and C. Weissmann, *Anal. Biochem.* **56,** 502 (1973).

[15] R. E. Gibson, R. D. O'Brien, S. J. Edelstein, and W. R. Thompson, *Biochemistry* **15,** 2377 (1976).

*Purification of a 116K ATPase from Coated Vesicles*[4]

A second NEM-sensitive ATPase is present in clathrin-coated vesicles which has a MW of 116K. This ATPase, unlike the 530K proton pump, is inhibited by vanadate. Its function is unknown.

### Reagents

Tris-HCl: 1 $M$, pH 8.5
Cholate: 20% in distilled water, pH 8.0 (potassium salt)
Polyoxyethylene 9-lauryl ether ($C_{12}E_9$, Sigma): 20% in distilled water
Tris-MES: 1 $M$, pH 7.0
Column solution: 0.1% $C_{12}E_9$, 10 m$M$ Tris-MES, pH 7.0
NaCl: 0.25 $M$ in column solution
Ammonium sulfate: saturated aqueous solution
Sodium phosphate: 0.2 $M$ in column solution
Phosphatidylserine (Avanti): 25 mg/ml, in column solution, sonicated
   for 10 min under nitrogen
EDTA: 0.1 $M$, pH 7.0 (sodium salt)
SDS: 1% in distilled water
Dithiothreitol: 1$M$
Thioglycerol: (Evans Chemetics) 1 $M$ in distilled water

### Procedure

*Step 1: Solubilization.* To 30 ml of crude coated vesicles (25 mg/ml), 90 ml of ice-cold 1 $M$ Tris-HCl, pH 8.5, is added, and incubated in ice for 30 min to have the vesicles stripped of clathrin. The mixture is centrifuged at 150,000 $g$ for 1 hr and the pellet is resuspended in 60 ml of 1% cholate, pH 8.0, and incubated in ice for 30 min. The NEM-sensitive ATPase remains in the pellet after centrifugation of the mixture at 150,000 $g$ for 1 hr. The pellet is extracted at 0° for 1 hr with 120 ml of 10 m$M$ Tris-MES, pH 7.0, and 0.1% or 0.75% $C_{12}E_9$. The extraction with 0.75% $C_{12}E_9$ gives a higher yield, but lower specific activity. A preparation with one major band in SDS-PAGE was obtained with the 0.1% $C_{12}E_9$ extract (~40% yield) starting with 750 mg of crude coated vesicles.

*Step 2.* The $C_{12}E_9$ extract of the ATPase is loaded onto a DEAE-Sepharose column (2.5 × 7 cm) that has been equilibrated with a solution of 0.1% $C_{12}E_9$, 10 m$M$ Tris-MES, pH 7.0. After washing of the column with the same buffer until the absorbance at 280 nm is back to the baseline, the ATPase is eluted with 100 ml of 0.25 $M$ NaCl.

*Step 3. Ammonium Sulfate Fractionation.* To 50 ml of pooled active fractions from Step 2, 23.5 ml of room-temperature saturated $(NH_4)_2SO_4$

is added dropwise and the mixture is kept on ice for 20 min, followed by centrifugation at 85,000 $g$ for 30 min at 0–3°. To the clear supernatant 3.4 ml of saturated $(NH_4)_2SO_4$ is added dropwise at 0°, and the mixture is centrifuged at 100,000 $g$, allowing the temperature of the centrifuge head to increase from 0 to 15° (approximately 1 hr). The centrifugation is then continued for 20 min at 15°. The supernatant is put on ice, and 14 ml of saturated $(NH_4)_2SO_4$ is added dropwise. After 30 min at 0° the mixture is centrifuged at 100,000 $g$ for 30 min at 4°and the precipitate is dissolved in 20 ml of 0.1% $C_{12}E_9$, 10 m$M$ Tris-MES, pH 7.0. The $(NH_4)_2SO_4$ concentration in this fraction is less than 20 m$M$, as determined by conductivity measurement.

*Step 4: Hydroxylapatite Chromatography.* The ATPase preparation from Step 2 is loaded onto a hydroxylapatite column (1.5 × 4 cm) equilibrated with 0.1% $C_{12}E_9$, 10 m$M$ Tris-MES, pH 7.0. After washing the column with 10 ml of the same buffer, the enzyme is eluted with a gradient of NaP$_i$ from 0 to 0.2 $M$ (total volume of 60 ml). Phosphatidylserine (0.1 mg/ml) is immediately added to each fraction.

*Step 5: Glycerol Gradient I.* The most active fractions from the HTP column (15 ml) are pooled and precipitated by adding 15 ml of saturated $(NH_4)_2SO_4$. The precipitate is dissolved in 1.0 ml of 0.1% $C_{12}E_9$, 10 m$M$ Tris-MES, pH 7.0. Aliquots of 0.33 ml are loaded on top of a 12-ml glycerol gradient (10–30%) prepared in 0.05% $C_{12}E_9$, 1 m$M$ EDTA, 10 m$M$ Tris-MES, pH 7.0. The fourth tube is loaded with the same buffer without enzyme in it, serving as a control for protein determinations. The four tubes are centrifuged at 170,000 $g$ for 20 hr at 0–4°, and the fractions collected from the bottom after piercing the tubes.

TABLE II
PURIFICATION OF THE 116K ATPase FROM COATED VESICLES

| Purification step | Total protein (mg) | Total activity (units)[a] | Specific activity (units/mg)[a] | Purification (-fold) |
|---|---|---|---|---|
| Clathrin-coated vesicles | 750 | 45 | 0.06 | — |
| 0.1% $C_{12}E_9$ extract | 71 | 18 | 0.25 | 4.2 |
| DEAE-eluate | 31 | 12.5 | 0.4 | 6.6 |
| $(NH_4)_2SO_4$ precipate | 8.9 | 8.0 | 0.9 | 15 |
| Hydroxylapatite eluate | 1.7 | 3.5 | 2.1 | 35 |
| Glycerol gradient I fractions | 0.23 | 1.4 | 6.2 | 103 |
| Glycerol gradient II fractions | 0.008 | 0.8 | 10 | 167 |
| Glycerol gradient III fractions | 0.005 | 0.21 | 42 | 700 |

[a] Units are defined as $\mu$mol P$_i$ ml$^{-1}$ min$^{-1}$.

*Step 6: Glycerol Gradient II.* The most active fractions, usually about 1.2 ml, are pooled and precipitated again with $(NH_4)_2SO_4$ as described above. The precipitate is dissolved in 0.6 ml of 0.01% SDS, 50 m$M$ dithiothreitol, 0.05% $C_{12}E_9$, 1 m$M$ EDTA, 10 m$M$ Tris-MES, pH 7.0, and incubated in ice for 30 min. Each aliquot of 0.3 ml is loaded on top of a glycerol gradient prepared as described in Step 5 except that 10 m$M$ of thioglycerol is present in the gradient. The two tubes of gradient are centrifuged and fractionated again as described in Step 5.

*Step 7: Glycerol Gradient II.* This step is the same as Step 6, except that (1) 0.02% SDS is used instead of 0.01% SDS; (2) the sample volume is 0.2 ml, and only one tube of gradient is used. To the purified ATPase 0.01 mg phosphatidylserine/ml is added and the preparation is stored at $-80°$ (stable for at least 1 month).

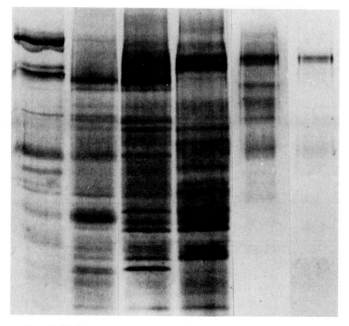

FIG. 2. SDS-polyacrylamide gel electrophoresis and silver staining of various fractions during the purification of the 116K ATPase of coated vesicles. Lane 1, clathrin-coated vesicles (1 $\mu$g); lane 2, $C_{12}E_9$ (0.1%) extract, (1 $\mu$g); lane 3, ammonium sulfate fraction (1 $\mu$g); lane 4, hydroxylapatite chromatography fraction (1 $\mu$g); lane 5, fraction from glycerol gradient I (300 ng); lane 6, fraction from glycerol gradient III (50 ng).

Shown in Table II is a summary of the purification of the 116K ATPase and in Fig. 2, an SDS-PAGE illustrating the ATPase in various states of purification.

### Acknowledgment

This investigation was supported by PHS grant CA08964, awarded by the National Cancer Institute, DHHS, NIH grant DK33627, and by the Perot Family Foundation.

# Section III

# ATP-Driven K$^+$ Pumps

# [50] Preparation of Gastric H$^+$,K$^+$-ATPase

## By EDD C. RABON, WHA BIN IM, and GEORGE SACHS

## The Gastric H$^+$,K$^+$-ATPase

### Preparation of Hog Gastric H$^+$,K$^+$-ATPase

In this procedure, large yields of the membrane-bound H$^+$,K$^+$-ATPase are obtained from the fundic mucosa of hogs. Initially, stomachs are procured from a convenient meat-packing plant. At the plant, the fundic mucosa from 10 stomachs, selected for the presence of pink coloration and distinct fundic mucosal folds, are isolated and packed in layers of wet ice for transport back to the preparation area. This quantity of tissue will provide about 100 mg of the gradient-resolved microsomal H$^+$,K$^+$-ATPase.

In the laboratory, the external muscularis, stripped from the epithelial cell layers, is discarded. The remaining intact epithelial mucosa, flattened on a glass plate set either on a bed of ice or in a cold room, is wiped vigorously to remove surface mucus and debris. The flattened tissue is then flooded with a solution of saturated NaCl, incubated for 1 or 2 min, and again thoroughly wiped dry. This step is intended to reduce mucus contamination by lysis of mucous cells.[1] The epithelial cell layers are then scraped free of the internal muscularis with a blunt metal spatula and placed in ice-cold homogenization buffer. This buffer is a solution of 0.25 $M$ sucrose and 5 m$M$ PIPES/Trizma, pH 6.8. About 120 g of these scrapings are added to 400 ml of the buffer, minced well with scissors, and homogenized by a motor-driven Teflon pestle rotating at 2300 rpm in a Potter–Elvehjem homogenizer. The homogenate is diluted to 750 ml by further addition of homogenization buffer, mixed, and placed in 250-ml bottles in ice awaiting centrifugation. As an optional procedure, floating mucus-containing material is siphoned from the homogenate and the remaining suspension filtered through cheesecloth.

In an abbreviated differential centrifugation pattern, a pellet is first collected from the crude homogenate by centrifugation at 11,000 rpm in a Sorvall GSA rotor for 45 min. Because of the rich microsomal yield, the unrestricted availability of hog stomachs, and the significant mucus contamination of this preparation, this initial pellet is usually discarded without further washing. Microsomal pellets are then collected from the su-

---

[1] J. G. Forte, T. K. Ray, and J. L. Poulter, *J. Appl. Physiol.* **32**, 714 (1972).

pernatant by centrifugation at 30,000 rpm in a Beckman type 30 rotor for 1 hr. The microsomal pellets can be refrigerated overnight or conveniently frozen at $-20°$ and later thawed for final purification by density-gradient and free-flow electrophoresis techniques.

For density-gradient purification, microsomal pellets are resuspended to 30 ml in the homogenization buffer. The resuspended microsomes are layered onto a step gradient which, in the Beckman Z-60 zonal rotor, consists of an inner cushion of the homogenization buffer, 125 ml of 7% Ficoll (w/w) in 0.25 $M$ sucrose and 175 ml of 1.1 $M$ sucrose, both with 5 m$M$ PIPES/Trizma, pH 6.8. The microsomes are resolved into light (GI on the 0.25 $M$ sucrose/7% Ficoll interface) and heavy (GII on the 7% Ficoll/1.1 $M$ sucrose interface) microsomal fractions by centrifugation at 59,000 rpm for 2.5 hr. A pellet enriched in mitochondrial membranes is discarded.

The lighter material (GI) can comprise up to 30% of the total purified microsomal population and exhibits a 95% latent K$^+$-stimulated ATPase activity of approximately 160 $\mu$mol mg$^{-1}$hr$^{-1}$. Basal Mg$^{2+}$-dependent ATPase activity is less than 5% of the K$^+$-stimulated ATPase activity. The characteristic ionophoretic enhancement of K$^+$-stimulated ATPase activity of the GI membrane fraction is shown in Table I and results from the almost unanimity of vesicles oriented with the lumen accessible K$^+$ activation site. Because of its high relative peptide purity (greater than 70% of total protein) and the large percentage of vesicles containing the tightly coupled H$^+$,K$^+$-ATPase, this fraction is most often used for transport studies.

The heavier membrane fraction (GII) comprises the majority of the microsomal membranes and contains both broken and intact membrane

TABLE I

IONOPHORETIC ENHANCEMENT OF K$^+$-STIMULATED ATPase ACTIVITY IN THE MICROSOMAL FRACTIONS OF HOGS$^a$

| Hog microsomal fractions | K$^+$-stimulated ATPase activity ($\mu$mol/hr · mg protein) | | | |
| --- | --- | --- | --- | --- |
| | Mg$^{2+}$ | K$^+$ | K$^+$/valinomycin | K$^+$/nigericin |
| GI | 6.3 | 4.2 | 33.5 | 160 |
| GII | 13.8 | — | — | 80.6 |

$^a$ K$^+$-dependent ATP hydrolysis was measured in an aliquot of gastric microsomes containing 10 $\mu$g membrane-bound protein in media composed of 0.25 $M$ sucrose, 10 m$M$ KCl, 40 m$M$ Trizma-Cl (pH 7.4), 2.0 m$M$ MgCl, and 2.0 m$M$ ATP at 37°.

vesicles. The K⁺-stimulated ATPase activity of this fraction is more typically 80–100 $\mu$mol mg⁻¹hr⁻¹ and exhibits a basal Mg²⁺-dependent component of less than 20% of total Mg²⁺- and K⁺-dependent ATPase activities. The specific activity of the GII material can be increased significantly by further purification utilizing free-flow electrophoresis.

For this procedure, mucus contamination of the GII fraction is first reduced by twice washing the membranes in the separation buffer consisting of 8 m$M$ Trizma base, 8 m$M$ acetic acid, 0.25 $M$ sucrose, and 0.1 m$M$ MgATP adjusted to pH 7.4 with NaOH. The washed sample, resuspended to a final protein concentration of 8–10 mg ml⁻¹, is then injected into a free-flow electrophoresis apparatus at 1 ml/hr. A second buffer required for this procedure, the electrophoresis buffer, is composed of 100 m$M$ Trizma base and 100 m$M$ acetic acid adjusted to pH 7.4 with NaOH. During electrophoresis of this sample, the potential across the chamber is maintained at 120 ± 10% V/cm, the current at 147 mA, the temperature at 7°, and the separation buffer flow at 4 ml per fraction per hour. The GII microsomal material is resolved by free-flow electrophoresis into three fractions, designated FI, FII, and FIII. The membrane fraction most displaced toward the anode is designated FI and displays the highest specific activity of the three fractions.[2] The K⁺-stimulated activity of the FI material consistently equals or slightly exceeds that of the GI membrane fractions.

Enzyme activity in these fractions can be preserved by storage at −80° in various buffers. The addition of 20% glycerol or 30% sucrose prior to freezing will preserve ionic integrity in vesicles intended for future transport studies.

### Preparations of Rabbit Gastric H⁺,K⁺-ATPase

The rabbit membrane-bound enzyme preparations have been developed to provide transport competent vesicles enriched in K⁺-stimulated ATPase activity. Two preparations have been developed for this purpose.[3,4] In either case, polyacrylamide gels of these preparations exhibit a pattern of peptide heterogeneity greater than that derived from hog and provide enzyme of lower specific activity. In both, the basal oligomycin-insensitive Mg²⁺ component of total ATPase activity equals or exceeds the K⁺-stimulated component. The K⁺-stimulated component ranges

[2] G. Sacconari, H. B. Stewart, D. Shaw, M. Lewin, and G. Sachs, *Biochim. Biophys. Acta* **465**, 311 (1977).
[3] J. M. Wolosin and J. G. Forte, *J. Biol. Chem.* **256**, 3149 (1981).
[4] J. Cuppoletti and G. Sachs, *J. Biol. Chem.* **259**, 14952 (1984).

from 10 to 40 $\mu$mol mg$^{-1}$hr$^{-1}$. As with the rat preparation detailed below, either method can be used to isolate gastric membranes from animals maintained in the resting or secreting state of acid secretion.

## Procedure

This preparation provides membranes enriched in K$^+$-stimulated ATPase from both a 10,000 $g$ pellet and the microsomal fraction obtained for the 10,000 $g$ supernatant.[3] The stomach from a New Zealand white rabbit is opened along the greater curvature and washed with a homogenization buffer containing 120 m$M$ mannitol, 40 m$M$ sucrose, 5 m$M$ PIPES/Trizma, pH 6.7, and 1 m$M$ EDTA. The fundus is isolated from the stomach and flattened on an ice-cold glass plate. The tissue is wiped free of mucus and the epithelia removed by scraping with a glass microscope slide. The scrapings from a single stomach are minced well with scissors in 60 ml of the homogenization buffer and then gently homogenized by 15 strokes of a loose-fitting, motor-driven, Teflon pestle rotating at 300 rpm in a Potter–Elvehjem homogenizer. A pellet, collected by centrifugation at 800 $g$ for 3 min, is washed in an additional 30 ml of the homogenization buffer and the two 800 $g$ supernatants combined. The combined supernatants are titrated to pH 7.4 by addition of Trizma base and a pellet collected by centrifugation at 10,000 $g$ for 10 min. The microsomal pellet is then collected from the 10,000 $g$ supernatant by centrifugation at 100,000 $g$ for 1 hr. Each of these pellets is then resuspended in 0.3 $M$ sucrose and 5 m$M$ PIPES/Trizma, pH 7.4, and layered onto a two-step Ficoll/sucrose gradient consisting of 9 and 17% Ficoll (w/w) in 0.3 $M$ sucrose with 5 m$M$ PIPES/Trizma, pH 7.4. Membrane fractions are recovered at the 9 and 17% Ficoll interfaces following 3 hr of centrifugation at 24,000 rpm in a Beckman SW-28 swinging bucket rotor. The higher K$^+$-stimulated ATPase activity derived from the 10,000 $g$ pellet is isolated at the 17% Ficoll/sucrose interface. Typical preparations isolated from the secreting rabbit exhibit a K$^+$-stimulated ATPase component of 10 to 35 $\mu$mol mg$^{-1}$hr$^{-1}$. A fraction containing a somewhat higher K$^+$-stimulated ATPase activity is derived from the microsomal fraction collected at the 9% Ficoll/sucrose interface. The specific activity of this fraction ranges from 25 to 40 $\mu$mol mg$^{-1}$hr$^{-1}$.

## Procedure

This preparation provides K$^+$-stimulated, ATPase-enriched membranes derived from the microsomal fraction of the rabbit preparation.[4] The rabbit fundic mucosa is isolated as in procedure 1, and homogenized with 20 strokes of a motor-driven Teflon pestle rotation at 3000 rpm in 100 ml of homogenization buffer consisting of 0.3 $M$ sucrose in 4 m$M$ PIPES/ Trizma, pH 7.4. A pellet is removed by centrifugation at 13,000 $g$ for 15 min and a microsomal fraction obtained by centrifugation of the supernatant at 100,000 $g$ for 60 min. These microsomes are suspended in the homogenization buffer and applied to a step gradient consisting of 4% Ficoll (w/v) and 12% Ficoll (w/v) in 0.3 $M$ sucrose and 4 m$M$ PIPES/ Trizma, pH 7.4. The H$^+$,K$^+$-ATPase-enriched fraction is collected at the 12% Ficoll interface following 3 hr of centrifugation at 24,000 rpm. The K$^+$-stimulated component of ATPase activity is 25–40 $\mu$mol mg$^{-1}$hr$^{-1}$. The enyzme activity and transport competency of both rabbit preparations can be preserved by storage in 30% sucrose (w/v) at $-80°$.

## Preparation of Rat Gastric H$^+$,K$^+$-ATPase

Male Sprague-Dawley rats weighing about 230 g are fasted overnight. For preparation of the gastric membranes in the stimulated state of acid secretion, the rats are injected subcutaneously with carbachol (350 $\mu$g/kg) or histamine (30 mg/kg). For the membranes in the resting state, the animals are injected intraperitoneally with cimetidine (100 mg/kg). All the injections are prepared in 0.9% saline and the dose of the secretagogues or cimetidine has been chosen to give the maximal response on the dose–response curve. The animals are sacrificed 1 hr later by cervical dislocation. The stomachs are opened along the greater curvature and washed with buffer 1 containing 250 m$M$ sucrose, 2 m$M$ MgCl$_2$, 1 m$M$ EGTA, and 2 m$M$ HEPES-Tris, pH 7.4. The top one-third of the fundic mucosal layer is removed and the rest of the mucosa enriched with the parietal cells is collected from 12 rat stomachs by scraping with a glass slide. All of the remaining procedures are carried out at 0–4°. The mucosal tissues are suspended in 40 ml of buffer 1 and homogenized with 20 strokes of a motor-driven (1500 rpm) Teflon pestle in a Potter–Elvehjem homogenizer. The homogenate is centrifuged at 20,000 $g$ for 15 min. The supernatant is layered, 15 ml each, over a gradient of two $^2$H$_2$O media: the top layer (density, 1.07) consists of 10 ml of 40% $^2$H$_2$O buffer containing 270 m$M$ sucrose and the rest of the ion components of buffer 1, and the bottom layer (density 1.13) consists of 5 ml of 99.9% $^2$H$_2$O buffer containing 300 m$M$ sucrose and the usual ion ingredients. The gradients are centrifuged in a SW 24.1 rotor at 24,000 rpm for 30 min. No distinctive interfacial bands are observed, most likely due to formation of a continuous gradient as a result of $^2$H$_2$O diffusion. The heavy microsomes refer to the membranous materials sedimented to form a pellet. The light micro-

TABLE II
SPECIFIC ACTIVITY OF K⁺-STIMULATED ATPase IN THE LIGHT
AND THE HEAVY GASTRIC MICROSOMES FROM THE RATS
TREATED WITH CIMETIDINE, HISTAMINE, AND CARBACHOL

| | K⁺-stimulated ATPase activity ($\mu$mol/hr · mg protein) | |
| Treatments | Light microsomes | Heavy microsomes |
| --- | --- | --- |
| Cimetidine | 52.5 ± 1.4 | 23.1 ± 1.9 |
| Histamine | 32.8 ± 3.6 | 18.1 ± 1.9 |
| Carbachol | 19.5 ± 1.6 | 24.6 ± 2.3 |

[a] K⁺-dependent ATP hydrolysis was measured in the membranes rendered permeable to K⁺ by freeze–thawing and in the media containing 40 m$M$ Tris-acetate (pH 7.4), 2 m$M$ MgCl$_2$, 2 m$M$ ATP, with or without 7 m$M$ NH$_4$Cl and 7 m$M$ KCl at 37°.

somes are recovered from the gradient media by a second centrifugation at 170,000 $g$ for 35 min. The membranes were suspended in 2 ml of buffer 1 to give a final concentration of protein of 7 to 10 mg/ml. The specific activity of K⁺-stimulated ATPase in these membranes is shown in Table II. These membrane vesicles are inside-out and able to accumulate H⁺ in exchange for K⁺ in the presence of ATP, valinomycin-independently in the secretagogue-stimulated heavy microsomes, but ionophore-dependently in the light or the resting heavy microsomes.[5]

In general, isolated rat gastric membranes are fairly unstable as compared to the hog membranes. It has frequently been observed that the rat membranes, when prepared in the absence of 1 m$M$ EGTA, lose K⁺-dependent ATPase activity with a $t_{1/2}$ of about 30 min at 37°[6] and are unable to develop a pH gradient because of membrane leakiness to hydrogen ion.[7]

[5] W. B. Im, D. P. Blakeman, J. M. Fieldhouse, and E. C. Rabon, *Biochim. Biophys. Acta* **772,** 167 (1984).
[6] W. B. Im and D. P. Blakeman, *Biochim. Biophys. Acta* **692,** 355 (1982).
[7] W. B. Im and D. P. Blakeman, *Biochem. Biophys. Res. Commun.* **108,** 635 (1982).

# [51] Genetics of Kdp, the K⁺-Transport ATPase of *Escherichia coli*

*By* JAMES W. POLAREK, MARK O. WALDERHAUG, and
WOLFGANG EPSTEIN

## Introduction

Kdp is a complex of three membrane proteins that form a $K^+$-transport ATPase in *Escherichia coli* (for review see Ref. 1). Kdp belongs to the $E_1$-$E_2$ class of transport ATPases, as shown by the formation of an acyl phosphate intermediate and by homology to the $Ca^{2+}$-ATPase[2,3] and the $Na^+,K^+$-ATPase[4,5] of animal cells, and to $H^+$-ATPases of yeast and fungi.[6-8] Genetic analysis has been important from the outset in the identification and characterization of the Kdp transport ATPase, in contrast to eukaryotic transport ATPases where genetics has only recently added to the extensive information obtained by biochemical techniques. The difficulty of analyzing the structure and function of membrane systems by biochemical methods makes a genetic approach especially promising.

In this article we describe the application of a variety of widely used techniques of bacterial genetics to the study of Kdp. Analogous genetic methods are already available for lower eukaryotes such as yeast, but at present only a few are readily applied to higher eukaryotes. The rapid progress of genetics gives promise that the ease and extent of genetic dissection in bacteria and yeast will soon be possible in higher eukaryotes. The biochemical characterization of Kdp is the subject of another article in this volume.[9]

[1] W. Epstein, *Curr. Top. Membr. Transp.* **23**, 153 (1985).
[2] J. E. Hesse, L. Wiedzorek, K. Altendorf, A. S. Reicin, E. Dorus, and W. Epstein, *Proc. Natl. Acad. Sci. U.S.A.* **81**, 4746 (1984).
[3] D. H. MacLennan, C. J. Brandl, B. Korczak, and N. Michael Green, *Nature (London)* **316**, 696 (1985).
[4] G. E. Shull, A. Schwartz, and J. B. Lingrel, *Nature (London)* **316**, 691 (1985).
[5] K. Kawakami, S. Noguchi, M. Noda, H. Takahashi, T. Ohta, M. Kawamura, H. Nojima, K. Nagano, T. Hirose, S. Inayama, H. Hayashida, T. Miyata, and S. Numa, *Nature (London)* **316**, 733 (1985).
[6] Serrano, R., M. C. Kielland-Brandt, and G. R. Fink, *Nature (London)* **319**, 689 (1986).
[7] R. Addison, *J. Biol. Chem.* **261**, 14896 (1986).
[8] K. M. Hager, S. M. Mandala, J. W. Davenport, D. W. Speicher, E. J. Benz, and C. W. Slayman, *Proc. Natl. Acad. Sci. U.S.A.* **83**, 7693 (1986).
[9] A. Siebers, L. Wieczorek, and K. Altendorf, this volume [52].

Description of Kdp

Kdp has a high affinity for K$^+$ ($K_m = 2 \mu M$) and serves primarily as an efficient scavenger for K$^+$ when this ion is present at low concentration in the medium. Kdp is expressed only during growth at low K$^+$ concentrations where other uptake paths (e.g., the Trk transport system, see below) cannot satisfy the cells' needs for K$^+$. Five *kdp* genes, clustered near min 16 on the current map of *E. coli*,[10] are required for Kdp activity (Fig. 1). Three of these genes, *kdpA*, *kdpB*, and *kdpC* form the *kdpABC* operon that codes for the three subunits of the Kdp transport complex. An adjacent operon containing the *kdpD* and *kdpE* genes codes for two proteins necessary for the expression of the *kdpABC* operon.[11] Insufficient intracellular K$^+$ appears to lead to a reduction in turgor pressure, the osmotic pressure difference across the cell envelope. Reduced turgor pressure turns on expression of the *kdpABC* operon by a mechanism which requires the products of the *kdpD* and *kdpE* genes. Because Kdp is but one of at least two K$^+$ transport systems, it is possible to delete part or all of the *kdp* gene cluster with the sole result that the strain requires elevated medium K$^+$ concentrations for growth. In addition to deletions, a variety of point mutations and insertion mutations have been obtained in the course of a genetic analysis of the organization and function of the *kdp* gene products.

Isolation of Mutants

The isolation and screening of *kdp* mutants is based on the elevated K$^+$ concentration of the medium required for growth of strains with impaired transport. The growth rate of wild-type strains is essentially independent of medium K$^+$ concentration, while a *kdp* mutant such as strain FRAG-5 requires 70 $\mu M$ K$^+$ to achieve half-maximal growth rate (Fig. 2). The K$^+$ transport needs of *kdp* mutants are met by a separate K$^+$ transport activity called Trk which is expressed constitutively.[12,13] Strains with impaired Trk activity grow slowly at low K$^+$ concentrations in the absence of Kdp activity, and therefore are readily identified in *kdp* mutants. Of particular utility for studies of Kdp are strains with two mutations, *trkA* and *trkD*. The severe transport defect of strains such as TK2205 with *trkA*, *trkD*, and *kdp* mutations results in a requirement of about 20 m$M$ K$^+$

[10] B. J. Bachmann, *Microbiol. Rev.* **47**, 180 (1983)
[11] J. Polarek, Thesis, Univ. of Chicago, 1985.
[12] D. B. Rhoads, F. B. Waters, and W. Epstein, *J. Gen. Physiol.* **67**, 325 (1976).
[13] G. L. Helmer, L. A. Laimins, and W. Epstein, *in* "Membranes and Transport" (A. N. Martonosi, ed), Vol. 2, p. 123. Plenum, New York, 1982.

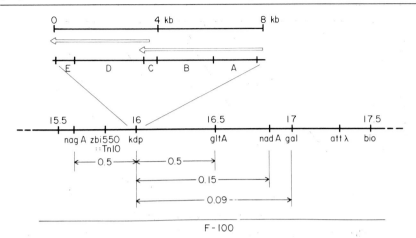

FIG. 1. A diagrammatic representation of the arrangement of the *kdp* genes and of their location relative to nearby markers on the chromosome of *E. coli*. At the top the *kdp* genes are shown, drawn to scale based on the DNA sequence of the *kdpABC* operon and an analysis of clones of the *kdpD* and *kdpE* genes. The two arrows indicate the two transcripts from the cluster; not shown is the fact that there is significant readthrough of the *kdpABC* transcript into the *kdpDE* operon.[11] The broad line in the middle represents the chromosome, with loci and minute markers as shown on the most recent map of *E. coli*.[10] The Tn10 insertion has been placed midway between *kdp* and *nagA* because it is about equally transduced with these two markers. The numbers between arrowpoints are P1*kc* cotransduction frequencies. The line at the bottom indicates the region carried by the F-100 episome.

to achieve half-maximal growth rate. When this strain is made $kdp^+$ (e.g., by transduction) to create strain TK2240, growth becomes like that of the wild-type and is independent of external $K^+$ (Fig. 2). In the *trkA trkD* mutant background, the dramatic effect of *kdp* mutations on the potassium dependence of the growth rate facilitates selection and scoring of mutants.

A convenient strain to use in obtaining *kdp* mutations is TK2240: *trkA405 trkD1* (also $F^-$ *thi rha lacZ nagA*). Standard high-$K^+$ medium containing 115 m$M$ $K^+$ and referred to as K115 contains: 46 m$M$ $K_2HPO_4$, 23 m$M$ $KH_2PO_4$, 8 m$M$ $(NH_4)_2SO_4$, 0.2 m$M$ $MgSO_4$, 4 $\mu M$ $FeSO_4$. KO medium is similar, but with $Na^+$ salts replacing the $K^+$ salts of the medium. KO contains about 20 $\mu M$ $K^+$ due primarily to contamination of the sodium phosphates. Media containing up to 10 m$M$ $K^+$ are prepared by adding KCl to KO; higher $K^+$ concentrations are achieved by mixing suitable proportions of KO and K115. Vitamin $B_1$ at 1 mg/liter, and glucose at 2 g/liter for liquid medium or 10 g/liter for agar plates complete the medium. Growth and selection are conveniently performed at 37° or at 30°.

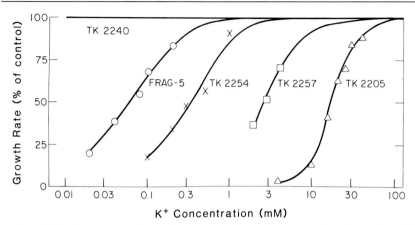

FIG. 2. The dependence of growth rate on medium K⁺ concentration for strains with mutations in transport genes. The horizontal line at the top, drawn for strain TK2240 (*kdp⁺ trkA trkD*), is a typical growth curve for all strains that are wild-type for *kdp*, regardless of other mutations. Loss of Kdp function alone has only a slight effect, shown for strain FRAG-5. The other strains all have low Trk activity due to *trkA trkD* mutations. When Kdp function is also lost (TK2205, *kdpABC5*), about 20 m$M$ K⁺ is required to achieve half-maximal growth. The other two *trkA trkD* strains carry mutations in which the affinity of Kdp for K⁺ is reduced. TK2254 (*kdpA54*) has a relatively normal $V_{max}$ but a much higher $K_m$ for K⁺, while TK2257 (*kdpA57*) is an example of a relatively infrequently isolated mutation in which an increase in $K_m$ and a marked decrease in $V_{max}$ occur.

We have isolated *kdp* mutants by using any one of a number of different mutagens. A wide variety of mutational lesions, including point mutations and deletions, is obtained by mutagenesis of cells grown in minimal medium (K115) by irradiation with ultraviolet light to a survival of about 5%. When cells are grown from a light inoculum ($10^3$ cells/ml) in K115 containing 50 $\mu$g/ml of the base analog mutagen 2-aminopurine, a relatively high proportion of amber mutations is obtained. Any mutagen can be used; however, very powerful ones such as *N*-methyl-*N*-nitro-*N*-nitrosoguanidine are not necessary and should not be used if strains with multiple mutations are to be avoided. After mutagenesis with ultraviolet light we usually allow the cells to recover for 2 hr in complex medium (KML: 10 g/liter tryptone, 5 g/liter yeast extract, 10 g/liter KCl) to maximize recovery, and then allow them to grow in minimal medium for a number of generations to allow for segregation of the mutations and full expression of the mutant phenotype.

A very effective technique for enriching *kdp* mutants is penicillin selection[14] in which nongrowing mutants survive treatment with the antibi-

[14] B. D. Davis, *J. Am. Chem. Soc.* **70**, 4267 (1948).

otic. For the permissive medium, a K$^+$ concentration is chosen where the desired class of mutants grows rapidly. The nonpermissive condition used for penicillin treatment is one where the parental strain grows well and therefore will be efficiently killed, while the desired mutants grow at a negligible rate and thus survive. To obtain *kdp* null mutants (mutants with no Kdp transport activity) of strain TK2240, K115 is used as permissive medium, while medium containing up to 5 m$M$ K$^+$ is a suitable nonpermissive medium.

To begin selection, mutagenized cells are grown into the exponential phase of growth in K115 medium. Then the cells are transferred, by brief centrifugation and washing at room temperature, to the nonpermissive medium. There is often a lag in growth after the shift, because the Kdp system must be derepressed before rapid growth ensues. When rapid growth resumes, as determined from periodic measurements of turbidity, antibiotic is added. Penicillin at 1000 units/ml, ampicillin at 100 $\mu$g/ml, carbenicillin at 150 $\mu$g/ml, cefoxitin at 100 $\mu$g/ml, or other antibiotics of the penicillin–cephalosporin class are all effective. We find it best to wait at least 60 min at 37°, or 90 min at 30°, after the shift to the nonpermissive medium even if cells are growing rapidly earlier. There is always residual growth of mutants after the shift, so that mutant survival is highest if antibiotic is not added until they have practically ceased growing. The addition should be done rapidly without interrupting growth because any slowing of growth will reduce killing of the parental strain. Antibiotics should not be added as K$^+$ salts if the resulting K$^+$ concentration is higher than desired for the nonpermissive condition. Growth is allowed to continue for at least 4 hr and until lysis has occurred. Longer selection is often best, and overnight selection at 30° is very effective. The selection is terminated by diluting the culture 1:1 with ice-cold distilled water to enhance killing of wall-weakened but still viable cells. After centrifugation and washing twice to remove antibiotic, cells are suspended in permissive medium to which 0.2% (w/v) of casamino acids (acid hydrolysate of casein) is added. Recovery of the survivors of penicillin treatment is more rapid and reproducible in medium containing casamino acids. After the survivors have grown to a sufficient density, the antibiotic selection step is repeated once, this time with casamino acids added to all media. Casamino acids are included to avoid the long lag of cells adapting to unsupplemented medium. The nonpermissive medium for the second selection must be adjusted so that total K$^+$ (from casamino acids plus that of the salts) does not exceed that desired. Salt-free, vitamin-free casein hydrolysate (ICN Biochemicals, Cleveland, OH) has had sufficiently low K$^+$ for use. A mixture of amino acids can be used if casamino acids of low K$^+$ content are not available. After recovery from the second selection, survivors are spread on plates made with K115 medium and replica plated to nonpermissive medium to identify mutants.

Penicillin selection is especially successful for transport mutants, because the parental cells that are killed cannot cross feed the desired mutants. In selections for auxotrophs or mutants with catabolic blocks, the parental strain can release metabolic intermediates before or after lysis, and these intermediates often support some growth of the desired mutants. This cannot occur for mutants where the defect is in uptake of a nutrient already present in the medium. The selection is thus very powerful.

The selection conditions can be altered to obtain *alloiophysic kdp* mutants, mutants to altered, partial function of Kdp. The types we have isolated to date are ones with a reduced affinity for $K^+$ so that Kdp activity in a strain with *trkA* and *trkD* mutations does not support growth at 0.1 m$M$ but does at 5 or 10 m$M$ $K^+$. For this type of selection the permissive medium is adjusted to contain 5 or 10 m$M$ $K^+$ and the nonpermissive medium contains 0.05 or 0.1 m$M$ $K^+$; otherwise the selection is like that for *kdp* null mutations. Figure 2 shows growth curves for two mutants of this type, referred to as "$K_m$" mutants. Using temperature rather than $K^+$ concentration to distinguish permissive from nonpermissive medium yields temperature-sensitive mutants. In this selection with strain TK2240, K1 medium is used at 25 to 30° as the permissive condition, while the same medium at 42° is nonpermissive. In this selection, it is wise to wait at least 90 min after shift to the high temperature to allow the transport activity of a temperature-sensitive Kdp system to fall to negligible levels.

A wide range of *kdp* mutations can be isolated through penicillin selection. For example, insertions of bacteriophage λ into four of the *kdp* genes have been obtained in a strain deleted for the chromosomal λ attachment site (*att*λ) and otherwise identical to TK2240.[15] Since the strain lacks the *att*λ site, infection with λ and selection for lysogens yield strains where the phage has integrated into sequences resembling that of *att*λ.[16] Penicillin selection allows for the isolation of strains where λ has inserted into the *kdp* genes. Similarly, insertions into *kdp* genes of bacteriophage Mu and of its derivatives which form fusions to the gene for bacterial β-galactosidase (*lacZ*) are also readily enriched by penicillin selection.[17] Kdp mutants that arise spontaneously may also be obtained by penicillin selection.

[15] D. B. Rhoads, L. Laimins, and W. Epstein, *J. Bacteriol.* **135**, 445 (1978).
[16] K. Shimada, R. A. Weisberg, and M. E. Gottesman, *J. Mol. Biol.* **63**, 483 (1972).
[17] L. A. Laimins, D. B. Rhoads, and W. Epstein, *Proc. Natl. Acad. Sci. U.S.A.* **78**, 464 (1981).

Moving *kdp* Mutations by Transduction

To analyze the *kdp* genotype of a strain and to construct strains with particular *kdp* genotypes for mapping, complementation analysis, or cloning, it is useful to be able to move *kdp* mutations from one strain to another. A convenient way to move them is cotransduction by bacteriophage P1 with a nearby marker. Three useful markers close to *kdp* are *nagA*, *gltA*, and *nadA*. The first two are about 50% cotransduced while the last is about 15% linked. We tend to avoid *gltA* because point mutations in this gene have a rather high apparent reversion rate due to the higher growth rate of the revertants. We have recently isolated two Tn10 insertions useful in moving *kdp* mutations. One of these, *zbg-550*::Tn*10*, is about midway between *nagA* and *kdp* and can be used to bring in *nagA* in a first step; in the second step *nagA*⁺ is selected and the desired *kdp* mutation is introduced while the Tn10 insertion is eliminated. The other, *kdpA123*::Tn*10*, allows the introduction of a very stable *kdp* mutation into any strain. A selection for tetracycline sensitivity[18] can eliminate the antibiotic resistance if desired, and at the same time often deletes *kdp* sequences.

Isolation of Deletions as Chromosomal Mutations

Fine structure mapping and a variety of genetic manipulations are facilitated by deletion mutations. The four sites of bacteriophage λ insertion provide an easy way to obtain deletions. The strains in question carry phage with the thermoinducible *cI857* mutation. When the strains are grown at 42° induction of the prophage kills the cell. However, cells in which a spontaneously arising deletion has deleted part or all of the prophage will survive at the high temperature. Deletions obtained in this way often extend into chromosomal sequences adjoining the prophage. This method has allowed us to isolate *kdp* deletions beginning at the sites of phage insertion and extending partway, or completely out of the *kdp* genes in one or both directions. A few representative deletions obtained in this way are shown in Fig. 3. Large deletions extending clockwise from *kdp* into or beyond *gltA* represent several percent of the survivors of such selections with any of the λ insertion mutants. Deletions extending as far as *nagA* in the counterclockwise direction have not been obtained, suggesting an essential gene lies between *kdp* and *nagA*. Deletions removing just an internal region of *kdp* are not frequent, apparently because there are preferred deletion endpoints outside the *kdp* genes.

[18] S. R. Maloy and W. D. Nunn, *J. Bacteriol.* **145,** 1110 (1981).

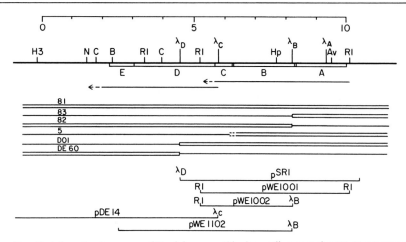

FIG. 3. A fine structure map of the *kdp* genes. The heavy line near the top represents the *kdp* region, with the extent of each gene shown by a box. The scale at the top is in kilobases. The sites of a few restriction enzymes are shown using the following abbreviations: Av, *Ava*I; B, *Bam*HI; C, *Cla*I; H3, *Hind*III; N, *Nru*I; R1, *Eco*RI. The sites of insertion of bacteriophage λ are shown, subscripts identifying the *kdp* gene into which the phage inserted. The two lines with dashed arrowheads designate the two transcription units. Seven useful deletions are shown below, the deleted region indicated by the space enclosed by double lines. All of these, except *kdpABC5*, were isolated from strains with λ insertions. At the bottom are a few selected *kdp* clones, whose end points are either a restriction site or a λ insertion site. The ends of pSR1 and pDE14 are not known precisely; they are sites of recombination between λ and chromosomal sequences that generated transducing phages λ*pkdp-1* and λ*DE-14*, respectively.

## Moving Mutations onto F-100

Mapping and complementation analysis of the *kdp* region are conveniently done by conjugal crosses with the F-100 episome. The high efficiency of transfer of the episome enhances the sensitivity of mapping crosses, and its stability facilitates complementation studies. Nearby markers on the episome (*nagA, gltA, nadA, gal, bio*) are useful in selecting for transfer of the episome; most of the strains we use are *nagA* for this reason. For storage and transfer of the episome, two *recA* strains are in common use in our laboratory: TJC111 (F⁻ *metB nadA rpsL recA12*), and TK2611 (F⁻ *gltA pyrF trp thi rha recA1 rpsL trkDl*). TJC111 is preferred for routine use because it is more robust and faster growing than strain TK2611. Transfer of F-100 into these strains can be selected by complementation of *nadA* for TJC111 or of *gltA* for TK2611 in crosses where the donors can be counterselected by auxotrophic markers, or by streptomycin if the donor is *rpsL⁺*. Both of these *recA* strains, TJC111 and TK2611, are auxotrophic for markers not complemented by F-100, and so can be readily counterselected in crosses where they serve as F-100 donors. The *recA* mutations assure that mutant episomes are not altered by recombination with chromosomal *kdp* sequences.

Mutations can be introduced into the episome by recombination of the episome with the chromosome of the mutant. Many recombinational events appear to represent gene conversion because strains with the same mutation on episome and chromosome are common. Thus one way of placing mutations on the episome is to introduce the episome, perform one cycle of penicillin selection for the mutant phenotype, and then identify the desired homozygous diploids by their *kdp*-defective phenotype and their ability to donate the episome to a suitable tester strain. An easier method takes advantage of an F-100 episome carrying the D01 *kdp-gltA* deletion (Fig. 3). This episome is transfered into a *rpsL*⁺ strain with the desired *kdp* mutation, and the resulting diploid is purified. This diploid then serves as donor in a cross with strain TK2611 in which *gltA*⁺ *rpsL* exconjugants are selected. This selection will yield episomes that have acquired the *gltA*⁺ allele from the chromosome of the *kdp* mutant. To generate *gltA*⁺ episomes, one cross-over to the right of the *gltA* marker (Fig. 1) and one to the left of the end of deletion D-01 in *kdpD* must have occurred. The result is that the episome acquires the chromosomal sequences from beyond *gltA* on the right to past the middle of *kdpD* on the left. In practice, the cross-over point on the left usually occurs past the *kdpE* extremity of the *kdp* gene cluster, and about half of the resulting episomes pick up the *nagA* mutation where this is a chromosomal marker in the original diploid. We generally screen for episomes which retain the *nagA*⁺ marker to be able to use it in selecting for transfer of the episome.

### Cloning *kdp* Genes

Clones of *kdp* genes in multicopy vectors have been useful in the analysis of Kdp. The *kdp* genes were initially cloned *in vivo* by the isolation of a λ transducing phage originating from the insertion in *kdpD*. This phage, λp*kdp*-1, has the *kdpABC* operon, its promoter, and a fragment of *kdpD*. Phages carrying part or all of the *kdpDE* operon have also been isolated in this way from strains with λ inserted in *kdpC* or *kdpB*. From these phages, smaller regions ending at restriction enzyme sites have been cloned into pBR322. The *kdpABC* operon has also been isolated by shotgun cloning of *E. coli* DNA and selecting for complementation of TK2205 (isogenic with TK2240, but *kdpABC5*) to growth on KO medium. From these clones, derivatives carrying selected parts of the region were prepared; a few of these are shown at the bottom of Fig. 3. All plasmids we have constructed in this way are stably maintained, suggesting none express the cloned genes at a level that interferes with growth.

Parts of the *kdp* region, generally small fragments ranging from 20 to several hundred base pairs, have been cloned into one of the derivatives of phage M13 widely used for DNA sequencing by the dideoxy method.[19] These phages are very convenient for fine-structure mapping because the double-stranded replicative form of M13 will recombine at modest frequency with regions homologous to the inserted DNA.

### Mapping and Complementation Analysis

Several of the deletions shown in Fig. 3 can be used to map mutations to one of the *kdp* genes by testing for complementation or recombination. A convenient scheme is outlined in Table I. The mutation to be studied will be in a recipient (F⁻) strain which includes a marker (*nagA, nadA,* or another marker) which allows for an estimate of the number of recipients which have received the F-100 episome in a cross. The crosses are performed semiquantitatively, so that one obtains an estimate of the number of recipients that have received the episome (as the number of Nag⁺ colonies if this is the marker used), as well as those that have become wild-type for Kdp function (Kdp⁺) by growth at very low K⁺ concentrations. Complementation, in which episomal genes compensate for the defect in *trans,* is indicated by equality between the numbers of Nag⁺ and Kdp⁺ colonies. Complementation occurs when the mutational defect is in a gene present and expressed on the episome. At the other extreme are crosses where no Kdp⁺ colonies arise (i.e., none above the level of detection determined by the reversion frequency of the mutation being mapped). In this case the episome is deleted for the site of the mutation. The intermediate result is seen when Kdp⁺ colonies are obtained. but at a much lower frequency than Nag⁺ colonies. This result is due to recombination, and indicates the deletion does not remove the site of the mutation, but does not allow expression of the gene in which the mutation lies. For example, the *kdpABC5* deletion is useful to identify *kdpC* mutations. The deletion ends near the *kdpB–kdpC* border and removes proximal regions including the promoter of the *kdpABC* operon. While this deletion does not complement any *kdpC* mutations, it recombines with all of them. As noted in the table, some ambiguities will arise because the end points of most of the deletions are not known precisely. Ambiguities in assigning mutations to a given gene can be resolved by complementation analysis. The precision of mapping with deletions is limited by the uncertainty of the end points of the deletions.

Clones of parts of the *kdp* gene cluster obtained by recombinant DNA methods *in vitro* are also useful in mapping and complementation analy-

[19] J. Messing, R. Crea, and P. H. Seeburg, *Nucleic Acids Res.* **9,** 309 (1981).

TABLE I
MAPPING MUTATIONS TO A SPECIFIC *kdp* GENE BY EPISOMAL
CROSSES WITH DELETION MUTATIONS

|                              | Result obtained for mutation in *kdp* gene[a] | | | | |
| Deletion episome[b]          | A | B    | C    | D    | E |
| --- | --- | --- | --- | --- | --- |
| *kdpABC5*                    | — | —[c] | R[c] | C    | C |
| *kdpA83*                     | — | R[d] | R    | C    | C |
| *kdpBCDE82*                  | C | —[d] | —    | —    | — |
| *kdpDE60*                    | C | C    | C    | R[e] | — |

  [a] Symbols used: C, episome complements mutation; R, episome recombines with mutation; —, episome neither complements nor recombines with mutation.
  [b] The extent of these deletions is shown in Fig. 3.
  [c] Because the end point of *kdpABC5* is not known precisely, either some *kdpB* may, or some *kdpC* mutations may not recombine with it.
  [d] Deletions 82 and 83 end in *kdpB*, about 50 bases from the start codon for its product. Therefore, mutations very early in *kdpB* will not recombine with deletion 83, but will recombine with deletion 82.
  [e] *kdpDE60* removes *kdpE* and the distal 40% of *kdpD*. It will recombine with mutations in the proximal part of *kdpD*.

sis. To facilitate their use in screening large numbers of mutations it is convenient to force integration of *kdp* clones in pBR322 onto the F-100 episome so that transfer is efficient and can be done by replica plating, rather than having to transform each strain individually. To integrate pBR322-derived plasmids into an F-episome we take advantage of the inability of plasmids, like pBR322, that have a *colE1* replication origin to replicate autonomously in *polA* mutants.[20] Transformants to drug resistance coded by the plasmid represent plasmids which have integrated into the chromosome or into an episome, such integration occuring most frequently by recombination between homologous sequences. For mapping *kdp* mutants we use a *polA* mutant which carries *trkA* and *trkD* mutations, and a total deletion of the *kdp* region, such as deletion 81 (Fig. 3). Into this strain we introduce a F-100 episome deleted for the *kdp* gene or region where we wish to map mutations. For example, a set of plasmids with partial deletions of *kdpA* obtained from pWE1001 *in vitro* would be introduced into an F-100 episome with the *kdpABC5* deletion. The only homol-

[20] D. T. Kingsbury and D. R. Helinski, *J. Bacteriol.* **114,** 1116 (1973).

ogy between plasmid and episome is the approximately 1 kb between the end of deletion 5 near the *kdpB-kdpC* junction and the *Eco*R1 site early in *kdpD* (Fig. 3). Transformants formed through integration by homology are unstable because they have an insertion of pBR322 sequences between directly repeated regions of homology. Recombination will cause loss of the integrated plasmid with its particular deletion. We stabilize these transformants by storing in a *recA* host such as TJC111.

### Sequencing *kdp* Mutations

The determination of the change in primary structure of the *kdp* proteins in a given mutant is readily determined by sequencing the DNA in the region of the mutation. The first step in sequence analysis is the mapping of the mutation to a small region. Mapping proceeds as outlined above. The first step is to identify the gene, and then a region of a gene, by testing for complementation and recombination as outlined in Table I. A somewhat more precise location is then determined by the use of clones with *in vitro* generated deletions that divide each gene into several regions. The final step is to test recombination with a collection of M13 clones of the wild-type *kdp* sequence, clones generated in the process of sequencing the region.[2] The end result of this process is to identify the location of the mutation to within about 100 bp.

We currently use a combination of *in vivo* and *in vitro* methods to clone mutations for sequencing. We begin by using a mini-Mu cloning vector[21] *in vivo*. A derivative of TK2247 that is lysogenic for Muc_ts and carries a suitable mini-Mu plasmid such as pEG5005 is transduced to *nad*⁺ with a P1 lysate of the mutant to be sequenced, and recombinants carrying the mutation are identified by replica plating. This strain is purified, and then grown and induced to produce phage at 42°. The lysate, which has both phage particles and particles containing mini-Mu plasmids with adjoining chromosomal sequences, is used to infect a Mu lysogen in which the desired *kdp* clones can be selected. For example, to clone mutations in *kdpA* we would use a *kdpB* mutant and select for cells able to grow on low-K⁺ medium by complementation of the host *kdpB* mutation by *kdpB*⁺ on the plasmid. The colonies are screened to make sure they have the correct phenotype for antibiotic resistance expected for clones. In the case of pEG5005, this means the colonies will be resistant to kanamycin but sensitive to ampicillin. Plasmid preparations are made from these strains, and those carrying the entire *kdpABC* operon are identified by the 4.9-kb fragment generated by digestion with restriction endonuclease *Eco*RI. In the second step, the mini-Mu plasmid and another plas-

---

[21] E. G. Groisman and M. J. Casadaban, *J. Bacteriol.* **168,** 357 (1986).

mid (we use pJD100, an $EcoRV-PvuII$ deletion derivative of pBR322) are digested with $EcoRI$ and ligated. The ligation mixture is transformed into a strain like that used to select the mini-Mu plasmid (i.e., a $kdpB$ mutant if a $kdpA$ mutation is being cloned), and colonies that are resistant to ampicillin and grow in low-$K^+$ medium are selected. The selection should yield only plasmids that have the $amp$ gene and replication origin from pJD100 and the $kdpABC$ region from the mini-Mu plasmid. The final step is sequencing the double-stranded plasmids,[22] using as sequencing primers oligonucleotides that hybridize to a region within about 250 bp of the mutation. For a gene the size of $kdpA$, with a total of 1671 bp, one would need only 6 or 7 primers to be able to sequence mutations in this gene. To sequence the mutations on the other strand, a set of primers that hybridize to that other strand can be used.

The same approach can be used for mutations in $kdpD$ and $kdpE$, except that the subcloning from the mini-Mu into the plasmid to be used for sequencing requires a different choice of restriction enzymes and a different plasmid. We have found the method described here, in which one takes advantage of the ability to select for plasmids carrying specific regions, a very quick way to move mutations from the chromosome to a plasmid where they can be sequenced.

## Conclusions

This overview has emphasized applications of classical genetic methods (mutagenesis, mapping, complementation) in the analysis of Kdp. These techniques are rapid and labor-saving, relying on enzyme systems *in vivo* to generate and analyze mutants. *In vitro* recombinant DNA methods greatly expand the possibilities for genetic manipulation and analysis. Devising new selections for other types of alloiophysic mutants, such as ones with a primary defect in the $V_{max}$ of transport, as well as for further genetic analysis of the mechanism whereby the $kdpD$ and $kdpE$ gene products translate a change in turgor into a signal to turn on $kdpABC$ operon expression, are tasks for which classical genetic methods will be very useful.

The genetic analysis outlined here has complemented biochemical approaches[9] in the study of Kdp. We look to a time when prokaryotic and eukaryotic transport systems will have undergone thorough analysis by genetic and biochemical methods, so that their similarities as well as their differences can be exploited to gain a better understanding of the structure of these systems and of how they effect transport.

[22] E. Y. Chen and P. H. Seeburg, DNA 4, 165 (1965).

# [52] K⁺-ATPase from *Escherichia coli*: Isolation and Characterization

By ANNETTE SIEBERS, LESZEK WIECZOREK, and KARLHEINZ ALTENDORF

## Introduction

The Kdp-ATPase is one of at least two systems involved in potassium ($K^+$) transport in *Escherichia coli*.[1,2] In contrast to the constitutive Trk system, it is a high-affinity transport system ($K_m = 2$ $\mu M$, $V_{max} = 0.15$ $\mu$mol mg$^{-1}$ min$^{-1}$) which is expressed when drastic changes in turgor occur or when $K^+$ is limiting in the medium.[3,4]

The Kdp complex consists of three polypeptides located in the cytoplasmic membrane[5] and is responsible for the $K^+$-stimulated ATPase activity associated with everted membrane vesicles.[6,7] The formation of a phosphorylated intermediate during ATP hydrolysis[8] suggests a functional homology to other eukaryotic ion transport ATPases. In addition, regions of strong structural homology with ATPases of the $E_1E_2$-type have been identified.[9,10,11,12] Although the genetic organization of *kdp* is well established,[13,14] only little is known about its expression and regulation.[15]

This report covers the description of the procedures for the purification and the phosphorylation of the Kdp-ATPase.

[1] W. Epstein and B. S. Kim, *J. Bacteriol.* **108**, 639 (1971).

[2] W. Epstein, *Curr. Top. Membr. Transp.* **23**, 153 (1985).

[3] D. B. Rhoads, F. B. Waters, and W. Epstein, *J. Gen. Physiol.* **67**, 325 (1976).

[4] L. A. Laimins, D. B. Rhoads, and W. Epstein, *Proc. Natl. Acad. Sci. U.S.A.* **78**, 464 (1981).

[5] L. A. Laimins, D. B. Rhoads, K. Altendorf, and W. Epstein, *Proc. Natl. Acad. Sci. U.S.A.* **75**, 3216 (1978).

[6] W. Epstein, V. Whitelaw, and J. Hesse, *J. Biol. Chem.* **253**, 6666 (1978).

[7] L. Wieczorek and K. Altendorf, *FEBS Lett.* **98**, 233 (1979).

[8] W. Epstein, L. A. Laimins, and J. E. Hesse, *Int. Congr. Biochem., 11th, Toronto* p. 449 (1979).

[9] J. E. Hesse, L. Wieczorek, K. Altendorf, A. S. Reicin, E. Dorus, and W. Epstein, *Proc. Natl. Acad. Sci. U.S.A.* **81**, 4746 (1984).

[10] R. Serrano, M. C. Kielland-Brandt, and G. R. Fink, *Nature (London)* **319**, 689 (1986).

[11] R. Addison, *J. Biol. Chem.* **261**, 14896 (1986).

[12] G. E. Shull and J. B. Lingrel, *J. Biol. Chem.* **261**, 16788 (1986).

[13] D. B. Rhoads, L. A. Laimins, and W. Epstein, *J. Bacteriol.* **135**, 445 (1978).

[14] J. W. Polarek, M. O. Walderhaug, and W. Epstein, this volume [51].

[15] W. Epstein, *FEMS Microbiol. Rev.* **39**, 73 (1986).

General Methods

K$^+$-stimulated ATPase activity of the membrane-bound and purified Kdp-ATPase is measured according to the method of Fiske and Subbarow[16] in a continuous-flow apparatus.[17] The protein is preincubated for 1 min at 37° in 50 m$M$ Tris-HCl, pH 7.8, 1 m$M$ MgCl$_2$, 1 m$M$ KCl, and the reaction started with 0.75 m$M$ ATP. In studies with inhibitors the following conditions are employed (modified assay buffers and incubation times are indicated): (1) orthovanadate, 2 min; (2) $N$, $N'$-dicyclohexylcarbodiimide, 30 min; (3) $N$-ethylmaleimide, 50 m$M$ Tris-HCl, pH 8.0, 30 min; (4) fluoresceinisothiocyanate, 50 m$M$ Tris-HCl, pH 8.8, 30 min.

Protein is determined by the method of Lowry et al.[18] with the modifications by Hartree[19] or, if detergent is present, by Dulley and Grieve.[20] For protein quantification in column fractions the method of Bradford[21] is applied.

Sodium dodecyl sulfate (SDS)-polyacrylamide gel electrophoresis is performed on slab gels (11% acrylamide) according to Lugtenberg et al.[22] The proteins are visualized with Coomassie Brilliant Blue G250; gels are stained and destained following the procedure of Weber and Osborn.[23]

The detergent Aminoxid is extracted with chloroform according to the procedure of Horikawa and Ogawara[24] with some minor modifications: SDS is added to 0.25 ml of the sample containing protein and detergent, giving a final concentration of 2%. The sample is vortexed, 1 ml chloroform is added, and the aqueous and chloroform phases mixed by vortexing for 2 min. The homogeneous mixture is centrifuged for 3 min in an Eppendorf centrifuge for phase separation. An aliquot of the upper aqueous phase is directly prepared for gel electrophoresis.

Bacterial Strains and Growth Conditions

For the isolation of the wild-type Kdp-ATPase Escherichia coli TK 2240-40 (thi rha lacZ nagA trkA405 trkD1 F' kdp$^+$), kindly provided by Dr. W. Epstein, is grown at 37° in a nominally K$^+$-free minimal medium,

[16] C. H. Fiske and Y. Subbarow, J. Biol. Chem. 66, 375 (1925).
[17] A. Arnold, H. U. Wolf, B. P. Ackermann, and H. Bader, Anal. Biochem. 71, 209 (1976).
[18] O. H. Lowry, N. J. Rosebrough, A. L. Farr, and R. J. Randall, J. Biol. Chem. 193, 265 (1951).
[19] E. F. Hartree, Anal. Biochem. 48, 422 (1972).
[20] J. R. Dulley and P. A. Grieve, Anal. Biochem. 64, 136 (1975).
[21] M. M. Bradford, Anal. Biochem. 72, 248 (1976).
[22] B. Lugtenberg, J. Meijers, R. Peters, P. van der Hoek, and L. van Alphen, FEBS Lett. 58, 254 (1975).
[23] K. Weber and M. Osborn, J. Biol. Chem. 244, 4406 (1969).
[24] S. Horikawa and H. Ogawara, Anal. Biochem. 97, 116 (1979).

designated KO, containing 46 m$M$ Na$_2$HPO$_4$, 23 m$M$ NaH$_2$PO$_4$, 8 m$M$ (NH$_4$)$_2$SO$_4$, 0.4 m$M$ MgSO$_4$, 1 m$M$ sodium citrate, 6 $\mu$$M$ Fe$^{II}$SO$_4$, 1 $\mu$g/ ml thiamine, 1% glucose.

The K$^+$ content of this medium is usually about 20 $\mu$$M$ or less, depending on the quality of the deionized water and the reagents used. For large-scale cultivation, 100 liters of KO is inoculated with 4 liters of a stationary overnight culture grown in KO.5—KO supplemented with 0.5 m$M$ KCl—resulting in an OD$_{610}$ of about 0.1. The cells are grown at 37° with vigorous aeration. The pH is adjusted to 7.0 with 20% NaOH during growth and cell density is measured every 30 min. Depending on the amount of residual K$^+$ contamination, cells will cease growth after 90 to 120 min, usually with OD$_{610}$ between 0.25 and 0.35. Upon reaching the very early stationary growth phase, 50 $\mu$$M$ KCl is added and the fermentation continued. Cells will now resume growth until K$^+$ is limiting again. This time 200 $\mu$$M$ KCl is added and the fermentation is continued until the cells again reach early stationary growth phase, which normally occurs at a final OD$_{610}$ of about 2.0. The culture is then cooled down to 10–15° and the cells are collected by centrifugation. The resulting cell paste is packed in sealable bags, quickly frozen in liquid nitrogen, and stored at −80°. The yield is approximately 300 g cell paste per 100 liters.

This procedure of large-scale *kdp* induction can be applied to all *E. coli* strains. Good results have been achieved with *E. coli* K-12 (λ), *E. coli* ML 308-225, and also with *kdp* mutant strains. However, best yields of Kdp-ATPase have always been achieved with strains diploid for the *kdp* genes.

### Preparation of Everted Vesicles

*Principle*

Frozen cells are thawed, washed once, and disrupted by a single passage through a Ribi cell fractionator (Sorvall). In the first step of the purification procedure, the everted vesicles are washed twice with low-ionic-strength buffer to remove peripherally membrane-associated proteins.

*Materials*

Buffer A: 50 m$M$ N-2-hydroxyethylpiperazine-N'-2-ethanesulfonic acid (HEPES), adjusted with Tris base to pH 7.5 (HEPES-Tris); 10 m$M$ MgCl$_2$

Buffer B: 50 m$M$ HEPES-Tris (pH 7.5), 10 m$M$ MgCl$_2$, 1 m$M$ 1,4-dithiothreitol (DTT), 0.5 m$M$ phenylmethylsulfonyl fluoride (PMSF) in methanol, 10% glycerol, 0.1 mg DNase I/ml

Buffer C: 1 m$M$ HEPES-Tris (pH 7.5); 3 m$M$ ethylenediamino-tetraacetic acid (EDTA), adjusted with NaOH to pH 7.5; 0.5 m$M$ PMSF (in methanol)

*Procedure*

All of the following steps are performed at 4° unless stated otherwise. Fifty grams of frozen *E. coli* cells is thawed and washed with 400 ml buffer A. Cells are collected by centrifugation (9000 $g$, 10 min), resuspended in 200 ml buffer B, and then disrupted by one passage in a Ribi cell fractionator (Sorvall) at 20,000 psi (138 MPa), 5 to 10°, under a constant stream of nitrogen. The resulting suspension is centrifuged at 9000 $g$ for 10 min to remove intact cells. The supernatant fraction containing the everted vesicles is then centrifuged at 190,000 $g$ for 60 min to sediment the membranes. The resulting pellet is carefully resuspended in buffer C (400 ml) using a homogenizer and incubated for 10 min in an ice-bath with gentle stirring. The vesicle suspension is centrifuged again (190,000 $g$, 60 min) and the washing procedure repeated.

At this stage, the vesicle pellet can either be stored at 4° on ice over-night without significant loss of activity or can directly be used for the solubilization of the Kdp-ATPase.

### Solubilization and Purification

*Principle*

The Kdp-ATPase is solubilized from the membrane by treating the everted vesicles with the nonionic detergent Aminoxid WS 35. Following an ultracentrifugation, the extract containing the Kdp-ATPase is purified by two successive ion-exchange chromatography steps, including a pH shift from 7.5 to 6.4. The enzyme is then concentrated and further purified by gel filtration.

*Materials*

Aminoxid WS 35: 1-alkoyl($C_7$-$C_{17}$)amino-3-dimethylaminopropane 3-$N$-oxide (Th. Goldschmidt AG, D-4300 Essen, FRG)
DEAE-Sepharose CL-6B (Pharmacia, Freiburg, FRG)
Fractogel TSK HW-65 (F) (Merck, Darmstadt, FRG)
Buffer D: 20 m$M$ HEPES-Tris (pH 7.5), 0.5 m$M$ PMSF (in methanol), 0.2% Aminoxid
Buffer E: 20 m$M$ 3-($N$-morpholino)propanesulfonic acid (MOPS)-Tris (pH 6.4), 0.5 m$M$ PMSF (in methanol), 0.2% Aminoxid

*Procedure*

*Step 1: Solubilization.* The vesicle pellet is carefully resuspended in 50 ml 100 m$M$ HEPES-Tris (pH 7.5). The suspension is then adjusted to 10 mg/ml protein with 100 m$M$ HEPES-Tris (pH 7.5), 200 m$M$ Na$_2$SO$_4$, 0.5 m$M$ PMSF, 1 m$M$ DTT, 1% Aminoxid.

The mixture is incubated in an ice-bath for 25 min with gentle stirring and centrifuged (190,000 $g$, 90 min). The supernatant containing the solubilized Kdp-ATPase is diluted 1 : 5 with a prechilled solution of 0.5 m$M$ PMSF in water and applied on an ion-exchange chromatography column.

*Step 2: First Column.* A DEAE-Sepharose CL-6B column (5 × 7.5 cm) is equilibrated with buffer D and the extract applied at a flow rate of 180 ml/hr. The column is then washed with at least two column volumes of buffer D. The bound protein is eluted with a 0 to 800 m$M$ linear KCl gradient (800 ml) in buffer D at 140 ml/hr; fractions of 10 ml are collected (Fig. 1).

*Step 3: Dialysis.* Aliquots of the fractions of the first main peak are analyzed for vanadate-sensitive ATPase activity and subjected to SDS-gel electrophoresis. Fractions eluted between 220 and 270 m$M$ KCl contain all three Kdp subunits and are pooled and dialyzed for 8 hr into 100 volumes of buffer E, with one buffer change. Precipitated material is removed by centrifugation (33,000 $g$, 20 min) and discarded.

*Step 4: Second Column.* A DEAE-Sepharose CL-6B column (2.5 × 7.5 cm) is equilibrated with buffer E. The dialyzed enzyme fraction of Step 3 is loaded onto the column at a flow rate of 140 ml/hr, and the column is washed with at least two column volumes of buffer E. Bound protein is eluted with a 0 to 500 m$M$ linear NH$_4$Cl gradient (400 ml) in buffer E at a flow rate of 100 ml/hr and collected in fractions of 3 ml. The single protein peak contains the Kdp-ATPase; it is normally eluted between 120 m$M$ and 180 m$M$ NH$_4$Cl (Fig. 2). Aliquots of the fractions are tested by ATPase activity measurements and SDS-gel electrophoresis. Fractions are pooled (Fig. 2) and either stored in liquid nitrogen or subjected to further purification.

*Step 5: Gel Filtration.* The combined fractions of Step 4 are concentrated by ultrafiltration (YM100 membranes, Amicon) to a final volume of approximately 4 ml. Since Aminoxid is also concentrated by this procedure and high concentrations interfere with SDS-gel electrophoresis, it has to be removed by chloroform extraction prior to electrophoretic separation (see General Methods). A column packed with Fractogel TSK HW-65 (F) (2.5 × 90 cm) is equilibrated with buffer D, and the sample (4 ml) is applied onto the column. Fractions of 4 ml are collected at a flow rate of 190 ml/hr and pooled after analysis of the single protein peak by

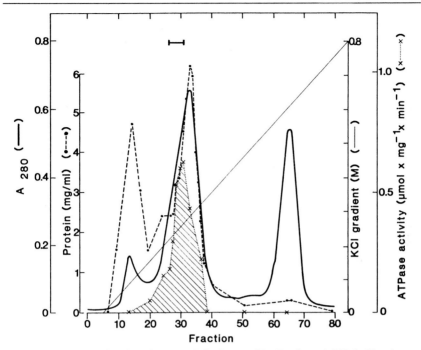

FIG. 1. Elution profile of the first DEAE-Sepharose CL-6B column (pH 7.5). The absorption profile at 280 nm (——) and the linear KCl gradient from 0–0.8 $M$ (——) are depicted. Protein in the fractions was determined by the method of Bradford[21] (●– –●); results indicate that the second main peak contains no protein but only detergent micelles. The ATPase activity over the whole fractionation range was measured in the absence and presence of 0.5 m$M$ vanadate using the automated assay (see "General Methods"); the vanadate-sensitive part of total ATPase activity was monitored (x·····x) and the corresponding peak area is hatched. The bar designates fractions containing the Kdp-ATPase that were pooled for further purification.

ATPase activity measurements and SDS-gel electrophoresis. The protein solution is concentrated by ultrafiltration (YM100 membranes, Amicon) and stored in liquid nitrogen.

*Summary of the Purification Procedure*

The protein yield and specific activities of the Kdp-ATPase at the individual purification steps are summarized in Table I. The washing of membranes with EDTA at low ionic strength proved to be an efficient initial step, removing about 40% of the peripheral or adsorbed membrane proteins. Under our solubilization conditions, Aminoxid WS 35 extracted 50% of the residual membrane-bound proteins.

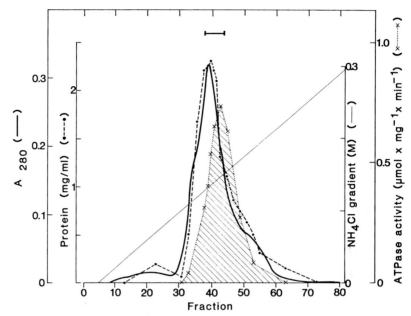

FIG. 2. Elution profile of the second DEAE-Sepharose CL-6B column (pH 6.4). Proteins were eluted with a linear NH₄Cl gradient from 0–0.5 $M$. For explanation of the symbols, see the legend to Fig. 1.

TABLE I

PURIFICATION OF Kdp-ATPase FROM *E. coli* TK 2240-40

| Step | Total protein (mg) | Specific activity[a] ($\mu$mol mg⁻¹ min⁻¹) |
|---|---|---|
| Vesicles | 2716 | nd |
| Washed vesicles | 1657 | nd |
| Solubilized protein | 845 | 0.16 |
| 1. Column (pH 7.5) | 186 | 0.43 |
| 2. Column (pH 6.4) | 45 | 0.82 |
| Gel filtration | 12 | 0.74 |

[a] The specific activity of the Kdp-ATPase was determined in the absence and presence of 0.5 m$M$ vanadate as described in "General Methods." The values represent the vanadate-sensitive part of total ATPase activity. nd, Not detectable (for explanation consult the text).

In everted vesicles of Kdp wild-type strains the determination of the Kdp-ATPase activity—defined as $K^+$-stimulated activity—is impossible (see Table I) for two reasons: (1) The activity of the $F_1F_0$-ATPase is 20-fold higher than that of the Kdp-ATPase, thereby masking any stimulatory effect of $K^+$ on the Kdp-ATPase. (2) Due to the high affinity of the Kdp-ATPase for $K^+$, the contaminating $K^+$ in the assay medium (about 3 $\mu M$) already causes a stimulation of the enzymatic activity. However, a reliable determination of the Kdp-ATPase activity can be obtained for everted membrane vesicles derived from a strain (1) without functional $F_1F_0$-ATPase ($\Delta unc$), and (2) at the same time bearing a mutation in $kdpA$ lowering the affinity of the system for $K^+$. In such a strain, everted membrane vesicles exhibit a $K^+$-stimulated ATPase activity of about 0.05 $\mu$mol mg$^{-1}$ min$^{-1}$. From studies with these $kdpA$ $\Delta unc$ mutants it can be deduced that in everted vesicles of the examined wild-type strain TK2240-40 the values obtained for the vanadate-sensitive portion of ATPase reflect an overestimation of the genuine amount of Kdp-associated activity. After solubilization of membranes, however, the Kdp activity is correctly represented by the vanadate-sensitive part of total ATPase activity (Table I).

The total ATPase activity of the preparation is extremely dependent on the extent of $kdp$ induction. Significant fluctuations in specific enzymatic activity can be observed, for unknown reasons, even within the same batch of cells. The highest activity is found after Step 4 of the purification procedure; it decreases slightly in the subsequent gel filtration step.

The specific enrichment of the Kdp-ATPase after the individual purification steps is demonstrated by the gel presented in Fig. 3. SDS-gel electrophoresis shows three major polypeptides with recalibrated apparent molecular masses of 47 kDa (KdpA), 79 kDa (KdpB), and 22 kDa (KdpC). Based on DNA sequencing data, the exact molecular weights have been determined to be 59,189 for KdpA, 72,112 for KdpB, and 20,267 for KdpC.[9] The physical association of the three proteins is maintained throughout the purification procedure. Column fractions are pooled according to ATPase activity measurements and identification of Kdp proteins by SDS-gel electrophoresis. The result is a compromise between acceptable protein yield and satisfactory purity. The grade of purity is at least 90% as judged by densitometric scans of silver-stained SDS gels. This estimation takes into account that some of the protein bands not corresponding to the three Kdp subunits represent proteolytic degradation products of KdpB, as can be verified by immunoblot analysis.

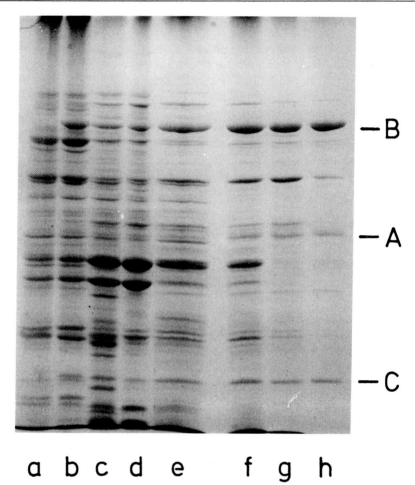

FIG. 3. Purification of the Kdp-ATPase: SDS-gel electrophoresis.[22] (a, b) References: cytoplasmic membranes prepared according to M. J. Osborn and R. Munson (this series, Vol 31, p. 642) from repressed (a) and derepressed (b) cells, 25 µg each; (c) everted (Ribi press) vesicles, 30 µg; (d) washed everted vesicles, 30 µg; (e) solubilized protein, 25 µg; (f) protein pool after the first ion-exchange column (pH 7.5), 12 µg; (g) protein pool after the second ion-exchange column (pH 6.4), 10 µg; (h) protein pool after gel filtration, 8 µg. The position of the three Kdp subunits is indicated by A, B, and C.

## Properties of the Purified Kdp-ATPase

The purified Kdp-ATPase has an apparent $K_m$ of 10 µM for K$^+$ stimulation and a $K_m$ of about 80 µM for ATP. The enzyme is characterized by a high specificity for K$^+$; other monovalent cations such as Rb$^+$, Cs$^+$, Li$^+$,

$NH_4^+$, $Na^+$ at 1 m$M$ concentration do not act stimulatory. Cooperative effects of ions together with $K^+$ are not observed. Whereas these findings are in accordance with measurements on the vesicle level,[6,7] the inhibition of the $K^+$-ATPase activity by $Tl^+$ is contradictory to $K^+$ transport experiments *in vivo*. In *E. coli* cells, $Tl^+$ is accepted as a substrate by the Kdp transport system.[25] The Kdp-ATPase has a requirement for $Mg^{2+}$ which can be replaced to some extent by $Mn^{2+}$ or $Co^{2+}$. $Ca^{2+}$, on the other hand, is completely ineffective in stimulation and even inhibits the enzyme to 80% when present in equimolar concentration with $Mg^{2+}$. In line with the narrow substrate specificity of the $K^+$-ATPase is the fact that *p*-nitrophenylphosphate, an artificial substrate that is hydrolyzed by other ATPases of the $E_1E_2$-type,[26,27,28] is not dephosphorylated by the Kdp-ATPase.

The cardiac glycoside ouabain, a specific inhibitor of the $Na^+,K^+$-ATPase, has no effect on the Kdp-ATPase in concentrations up to 2 m$M$. The $K_i$ values for different inhibitors are: 2 $\mu M$ for *o*-vanadate, a potent inhibitor of all ATPases forming a phosphorylated intermediate; 60 $\mu M$ for $N, N'$-dicyclohexylcarbodiimide (DCCD); 0.1 m$M$ for the thiol reagent *N*-ethylmaleimide (NEM); and 3.5 $\mu M$ for fluoresceinisothiocyanate (FITC). In eukaryotic ion-motive ATPases, FITC labels a specific lysine residue which seems to form part of the ATP-binding site.[10] In contrast to eukaryotic systems, with the bacterial Kdp-ATPase neither the inhibition by FITC nor by NEM can be abolished in the presence of high concentrations of protecting ADP.

## Phosphorylation

The energy required for the active transport of $K^+$ against concentration gradients up to $4 \times 10^6$ stems from the hydrolysis of ATP.[29] During the energy transduction process the terminal phosphate group of the ATP molecule is transferred to the Kdp-ATPase. A cycle of phosphorylation and dephosphorylation steps occurs, probably accompanied by conformational changes in the Kdp complex. In this section, a method for quantifying the amount of phosphoprotein formed by purified Kdp-ATPase is described.

[25] P. D. Damper, W. Epstein, B. P. Rosen, and E. N. Sorensen, *Biochemistry* **18**, 4165 (1979).
[26] B. Rossi, F. de Assis Leone, C. Gache, and M. Lazdunski, *J. Biol. Chem.* **254**, 2302 (1979).
[27] S. Bandopadhyay, P. K. Das, M. V. Wright, J. Nandi, D. Bhattacharyyay, and T. K. Ray, *J. Biol. Chem.* **262**, 5664 (1987).
[28] D. J. Horgan and R. Kuypers, *Anal. Biochem.* **166**, 183 (1987).
[29] D. B. Rhoads and W. Epstein, *J. Biol. Chem.* **252**, 1394 (1977).

*Strain and Growth Conditions*

Phosphorylation experiments are performed with *E. coli* TK 2242-42 (*thi rha lacZ nagA trkA405 trkD1 kdpA42* F' *kdpA42*), a diploid *kdpA* mutant strain.[6]

The $K_m$ of the purified mutant Kdp-ATPase for K⁺ is 6 m$M$ compared to 10 $\mu M$ for the wild-type enzyme. This reduced affinity for K⁺ makes it possible to examine the effects of K⁺ on the formation and turnover of the phospho intermediate, since the influence of K⁺ contamination in the incubation solutions can be neglected. In contrast to the induction protocol for TK 2240-40, strain TK 2242-42 is grown in 100 liters of KO.5. As soon as the culture has reached an $OD_{610}$ of about 2.0, cells are harvested and the resulting cell paste is shock-frozen and stored at −80°.

*Enzyme*

The mutant Kdp-ATPase is prepared according to the procedure described for the wild-type enzyme. The purified enzyme fraction of Step 4 is used in all phosphorylation experiments.

*Materials and Solutions*

Glass microfiber filters, type GF/F (Whatman, Maidstone, England)
Microfiber glass prefilters, type AP 25 (Millipore, Molsheim, France)
Solution 1 (phosphorylation buffer): 50 m$M$ HEPES-Tris (pH 7.8), 1 m$M$ MgCl₂
Solution 2 (stop solution): 35% (w/v) trichloroacetic acid (TCA)
Solution 3 (carrier solution): 100 m$M$ ATP, 100 m$M$ phosphate (P$_i$)
Solution 4 (filter preincubation solution): 10 m$M$ ATP, 10 m$M$ P$_i$
Solution 5 (washing solution): 2 m$M$ ATP, 10 m$M$ P$_i$, 5% (w/v) TCA

*Principle*

The partially purified Kdp-ATPase is incubated with [γ-³²P] ATP, the reaction stopped by acid precipitation, and both ATPase activity and formation of the phospho intermediate are measured. The latter is monitored by collecting the denatured protein on glass microfiber filters (Whatman) and counting the filter-bound radioactivity. ATPase activity is determined as the release of ³²P$_i$ after adsorption of nonhydrolyzed [γ-³²P]ATP to charcoal.[7]

*Procedure*

This protocol follows essentially the procedure described by Ray and Forte.[30]

[30] T. K. Ray and J. G. Forte, *Biochim. Biophys. Acta* **443**, 451 (1976).

The phosphorylation reaction is carried out at 37° in solution 1 in the presence or absence of 100 m$M$ KCl in a total volume of 0.2 ml. Fifty micrograms of the purified Kdp-ATPase is preincubated for 5 min. The reaction is started with 10 $\mu$l of a solution containing 2 m$M$ ATP and 3 $\mu$Ci [$\gamma$-$^{32}$P]ATP. The labeling is stopped after 10 sec by the addition of 0.3 ml ice-cold solution 2. The mixture containing the acid-precipitated protein is chilled on ice for at least 15 min and an aliquot of 0.1 ml is removed for the ATPase assay. To the residual 0.4 ml of the stopped reaction mix 0.044 ml solution 3 is added to give a final concentration of 10 m$M$ ATP and P$_i$. The precipitate is collected by filtration on glass microfiber filters (Whatman) using a suction device. The filters are pretreated by incubation in solution 4 at 60° for at least 2 hr in order to minimize unspecific binding of radioactivity. The TCA-precipitate is washed extensively, three times with 2 ml ice-cold solution 5, followed by three times with 2 ml ice-cold ethanol. Filters are dried and bound radioactivity is monitored after the addition of 4 ml Quickszint 212 scintillation cocktail (Zinsser, Frankfurt, FRG). Controls without enzyme and with previously acid-denatured Kdp-ATPase are run in parallel with each assay to determine the unspecific binding of radioactivity to the filters and to denatured protein, respectively. The calculated values reflect the steady-state level of $^{32}$P-labeled intermediate and are expressed in picomoles phospho intermediate per milligram protein.

The ATPase assay is continued by adding 1 ml of 5% TCA (w/v) to the 0.1-ml test volume containing the denatured protein. The precipitate is centrifuged (15,500 $g$, 2 min) and the supernatant is treated with charcoal. The suspension is kept at room temperature for 30 min and revortexed occasionally. The total volume is filtered through a Pasteur pipet stuffed with a microfiber glass prefilter (Millipore), discarding the first few drops. Gentle nitrogen pressure can be applied to accelerate the filtration. The radioactivity of 0.1 ml of the clear filtrate is counted and the ATPase activity calculated (micromoles P$_i$ liberated per milligram protein per minute).

*Properties*

The steady-state level of the phosphoprotein depends largely on the enzyme preparation. Usually values of about 500 pmol/mg protein are found, but up to 1 nmol phospho intermediate/mg protein can be observed.

K$^+$, the substrate of the Kdp-ATPase, stimulates the dephosphorylation step within the phosphorylation cycle. However, the extent of K$^+$-dependent dephosphorylation of the phospho intermediate varies between 30 and 80% for different enzyme preparations. It seems that ATP hydrolysis and K$^+$ binding are not completely coupled in the isolated Kdp-ATPase; this may be an artifact of the purification procedure.

The pH-lability profile of the phosphoprotein and its sensitivity to hydroxylamine suggest that an alkali-labile acyl phosphate is involved.[31] Therefore, electrophoretic separation is possible only on gel systems running at acidic pH values. Good results are obtained with the gel system described by Lichtner and Wolf [32] that resolves proteins according to their molecular weights at pH 2.4 and 4°. Two modifications have been introduced: We have used 9% acrylamide gels and varied the concentrations of the polymerization catalysts to achieve the recommended polymerization time of about 1 hr. With this gel system, the KdpB protein was clearly identified as the phosphorylated subunit of the purified Kdp-ATPase, proving that KdpB is involved in ATP hydrolysis.[31]

### Acknowledgments

The authors would like to thank Dr. Tilly Bakker-Grunwald for critically reading and Mrs. Gudrun Wallis and Johanna Petzold for typing the manuscript. This work was supported by the Deutsche Forschungsgemeinschaft (SFB 171) and the Fonds der Chemischen Industrie.

[31] A. Siebers, unpublished results.
[32] R. Lichtner and H. U. Wolf, *Biochem. J.* **181,** 759 (1979).

## [53] Purification of the ATPase of *Streptococcus faecalis*

By MARC SOLIOZ and PETER FÜRST

### Introduction

The vanadate-sensitive ATPase of *Streptococcus faecalis* is a novel bacterial ion pump, probably involved in K$^+$-transport.[1] The gene for this ATPase has recently been cloned.[2] The enzyme belongs to a new and growing class of prokaryotic ion-motive ATPases. Members of this class of enzymes exhibit a certain degree of sequence homology to eukaryotic ion-motive ATPases and undoubtedly bear an evolutionary relationship to them (see Ref. 3 for review). This relationship also manifests itself in a

[1] P. Fürst and M. Solioz, *J. Biol. Chem.* **261,** 4302 (1986).
[2] M. Solioz, S. Mathews, and P. Fürst, *J. Biol. Chem.* **262,** 7358 (1987).
[3] M. Solioz, *J. Membr. Biol.* in press (1988).

number of properties these ATPases, whether eukaryotic or prokaryotic, have in common: (1) they form an acyl phosphate intermediate as part of the reaction cycle, (2) they are inhibited by micromolar concentrations of vanadate, and (3) they contain a major, and often single, polypeptide component of 60 to 120 kDa. The *S. faecalis* ATPase may serve as a useful model system for the study of such ion-motive ATPases since it can readily be prepared in homogeneous form in milligram quantities without the use of specialized equipment. The purified enzyme can be transferred to a variety of detergents for reconstitution into phospholipid vesicles which are useful for transport studies.

### General Comments

1. The isolation procedure is outlined below, starting with 24 liters of culture. Scaling up the procedure has, in our hands, never yielded an ATPase preparation as pure as can be obtained following the protocol below. If smaller preparations are to be conducted, the volumes of the columns and the gradients for elution should not be reduced for maximal purity.

2. If large fermenting facilities are available, the cells can be grown in large batches and frozen as a cell paste. Aliquots can then be thawed, washed with $MgCl_2$, and further processed as described.

3. It is imperative to keep all biological materials at 0 to 4° beyond the lysozyme treatment, and to employ the protease inhibitors detailed in the procedure. Failure to do so results in degradation of the ATPase, e.g., multiple peaks of ATPase activity of varying sensitivity to vanadate elute from the columns.

4. Since the high Triton concentration used in the isolation interferes with the commonly used protein assays, we do not normally monitor the protein concentrations during purification. If desired, protein can be estimated by densitometric scanning of polyacrylamide gels, run in the presence of sodium dodecyl sulfate and stained with Coomassie Blue. Using defined amounts of bovine serum albumin applied to the same gel for standardization of the method, the ATPase content is underestimated 25-fold, as determined by quantitative amino acid analysis. The specific activity of the purified ATPase is 2 U/mg, as determined by the latter method.[4]

5. If column fractions are not assayed and processed within a few hours after collection, they should be stored frozen at −70°. Under these conditions, they are stable for several weeks.

[4] P. Fürst and M. Solioz, *J. Biol. Chem.* **260**, 50 (1985).

## Materials

The following list of materials includes, in parentheses, the required amounts for a standard preparation starting with 24 liters of culture, and the recommended storage temperatures for ready use and at which the reagents are stable for at least a few weeks.

Bacteria: *Streptococcus faecalis* ATCC9790 can be obtained from the American Type Culture Collection, Rockville, MD. For permanent storage, a 10-ml culture in the growth medium below is grown to stationary phase, supplemented with 0.5 volume sterile 50% (v/v) glycerol, and 1-ml aliquots are stored at −80°.

Growth medium (24 liters): the following are dissolved in 24 liters of distilled or deionized water: 120 g yeast extract (Difco Laboratories, Detroit, MI, or BBL, Cockeysville, MD), 240 g Trypticase peptone (BBL) or Bacto-Tryptone (Difco), and 256 g $Na_2HPO_4 \cdot 2H_2O$. 240 g of glucose is suspended in 400 ml of water and sterilized in separate flasks together with the growth medium for 1 hr at 121°. The glucose dissolves during sterilization and is added to the medium afterward.

$MgSO_4$ (5 liters, room temperature): 2 m$M$ $MgSO_4$.

GMK buffer (500 ml, room temperature): 400 m$M$ KCl, 50 m$M$ glycylglycine, 2 m$M$ $MgSO_4$, pH 7.2 with KOH.

PMSF (10 ml, −20°): 100 m$M$ phenylmethylsulfonyl fluoride in anhydrous dimethyl sulfoxide. In aqueous media, PMSF looses its activity as a protease inhibitor within hours.

DNase (10 ml, −20°): 1 mg/ml DNase I (DN-100, Sigma Chemical Company, St. Louis, MO) in 0.9% NaCl.

PIM (2 ml, −20°): protease inhibitor mixture containing 100 m$M$ each of the following in anhydrous dimethyl sulfoxide: phenylmethylsulfonyl fluoride, L-1-($p$-toluenesulfonyl)-amido-2-phenylethyl chloromethyl ketone, L-5-amino-1-($p$-toluenesulfonyl)amidopentyl chloromethyl ketone, 1,10-phenanthroline (all from Merck, Darmstadt, FRG), $p$-aminobenzamidine (Sigma Chemical Company). Some of these inhibitors loose their activity within hours in aqueous media.

D buffer (500 ml, 4°): 10 m$M$ 4-(2-hydroxyethyl)-1-piperazineethanesulfonic acid, 5 m$M$ $MgCl_2$, 5% (v/v) glycerol, 1 m$M$ $\beta$-mercaptoethanol, 0.1 m$M$ ethylenediaminetetraacetic acid, adjusted to pH 7.5 with KOH.

DT buffer (2 liters, 4°): D buffer containing 2% (w/v) Triton X-100.

DHT buffer (20 ml, 4°): D buffer containing 20% (w/v) Triton X-100.

Succinic acid (20 ml, room temperature): 0.5 $M$ in 0.02% $NaN_3$.

Asolectin suspension (3 ml, 4°): 150 mg of crude soybean phospholipids (Asolectin, Associated Concentrates, Woodside, NY) is suspended in 3 ml of water with a sonifier probe, under nitrogen and in an ice bath, until the suspension is transparent.

CM buffer (500 ml, 4°): DT buffer containing 50 mM succinic acid and 0.1 mg/ml asolectin, added as an aqueous suspension. The pH is adjusted at 0° with KOH to 4.5.

Mix (100 ml, −20°): the coupled enzyme assay to measure ATPase activity is conducted with this "mix," containing all the necessary components. To 99 ml of DT buffer are added: 1 mM $Na_2ATP$, 0.5 mM phosphoenolpyruvate, 0.2 mM NADH, 1 ml asolectin suspension. Following readjustment of the pH to 7.5 with KOH, 100 U each of pyruvate kinase and lactate dehydrogenase (no. 128 155 and 127 230, respectively, from Boehringer Mannheim GMBH, Mannheim, FRG) are added. This mix can be frozen and thawed repeatedly without loss of activity.

Miscellaneous materials: lysozyme (0.5 g, −20°, Sigma Chemical Company), $Na_3VO_4$ (10 mg, Fischer Scientific Company, Fair Lawn, NJ), DEAE-cellulose, microgranular, preswollen (35 g, DE-52, Whatman, Maidstone, England), hydroxylapatite (7 g, Bio-Rad, Richmond, CA), CM-Sephadex C-25 (1 g, Pharmacia Fine Chemicals, Uppsala, Sweden).

ATPase Assay

ATPase activities are determined with a coupled enzyme assay that links the generation of ADP to the oxidation of NADH, which can be monitored spectrophotometrically at or near 340 nm. Measurements are best conducted with a dual-wavelength spectrophotometer, using the wavelength couple 550/366 nm. Signals are recorded with a strip chart recorder at a chart speed of 1 cm/min and a sensitivity of 0.1 absorbance units for a full-scale deflection. A stable single-beam spectrophotometer can also be employed. To assay ATPase activity, 990 $\mu$l of mix are equilibrated at 37° in the spectrophotometer. The reaction is then initiated by the addition of 10 $\mu$l of sample. When a stable rate has been attained, vanadate sensitivity can be checked by adding 100 $\mu M$ $Na_3VO_4$ as a 10 mM stock solution. Full inhibition is only displayed after 1 to 2 min and should be greater than 90% for the ATPase eluting from any of the three columns. The assay is calibrated by adding a defined amount of ADP to the equilibrated assay mix. A change in optical density of 0.1 absorbance units should correspond to approximately 30 nmol of ADP.[5]

[5] G. Hugentobler, I. Heid, and M. Solioz, *J. Biol. Chem.* 258, 7611 (1983).

Purification Procedure

*Growth of Cells*

Although fermentors can be employed, *S. faecalis* may be conveniently grown in Erlenmeyer flasks, filled to capacity, and thermostated at 37° in a water bath. Slow stirring with magnetic stirring bars assures an even temperature distribution. Three liters of the growth medium is used for a preculture that is inoculated with either 1 ml from the frozen stock cultures, or 1 ml from a stationary 10-ml culture (such a culture can be stored for repeated use at 4° for several weeks without loss of viability). The preculture is allowed to grow to stationary phase (10 to 16 hr). The remaining 21 liters of medium is then inoculated with the preculture and growth is monitored by diluting samples 10-fold with water and measuring the optical density at 550 nm. When the diluted samples have reached 0.19 to 0.21 OD (late logarithmic phase, 3 to 4 hr after inoculation), the cells are harvested by repeated batchwise centrifugation at room temperature for 10 min at 5000 $g$. The cells are washed once with 3.5 liters and once with 1.5 liters of MgSO₄ by thorough resuspension and renewed centrifugation at room temperature. A yield of 50 to 60 g of wet cells will be obtained.

*Isolation of Membranes*

The wet cells are suspended in 6 ml GMK buffer per gram of wet cell weight. Protoplasts are formed by incubating this slurry with 4 mg of lysozyme per gram of wet cells, in the presence of 1 m$M$ PMSF, for 1 hr at 37° with slow magnetic stirring. Although an isotonic buffer is employed, a fraction of the cells lyses and the cell suspension becomes highly viscous. If this indication of good protoplast formation is not observed after 1 hr, 50% more lysozyme should be added and the incubation continued for another 30 min. All subsequent steps must strictly be conducted at 0 to 4° and with the use of precooled buffers. The protoplasts are collected by centrifugation for 12 min at 23,000 $g$ and the resulting fluffy pellet transferred to D buffer to obtain a total volume corresponding to 2.2 ml per gram of original wet cell weight. After the addition of 50 $\mu$l/ml of DNase and 5 $\mu$l/ml of PIM, the solution is homogenized with a motor-driven Potter–Elvehjem-type homogenizer at 1000 rpm. Complete suspension of the cells requires quite forceful homogenization. The cells are then broken with a French press at 10,000 psi, followed by centrifugation for 12 min at 23,000 $g$ to remove unbroken cells. An increase in the yield of membranes by 50% is achieved if the pellets are again suspended by homogenization as above, in half the previous volume of D buffer, and sedimented by centrifugation as before. The membranes contained in the combined supernatants are finally collected by centrifugation for 1 hr at 90,000 $g$. The gray to yellow translucent pellets are suspended in 20 ml of D buffer by homogenization. The resulting protein concentration as determined with the Bradford protein assay,[6] using bovine serum albumin as a standard, should be 10 to 20 mg/ml. Stored at $-70°$, the membranes are stable for several months. The ATPase activity can not reliably be estimated in this preparation due to the many-fold higher activity of the $F_1F_0$-ATPase and, possibly, the presence of other vanadate-sensitive activities.

*Extraction of Membranes*

Membranes are suspended in D buffer to a protein concentration of 8.6 mg/ml. The suspension is stirred on ice and 0.005 volume of PIM is added, followed by the addition of 0.3 volumes of DHT buffer. After stirring for 30 min, residual membranous material is collected by centrifugation for 45 min at 90,000 $g$. The resulting supernatant contains from 0.5 to 1.2 U/ml of vanadate-sensitive ATPase activity, and should be processed further immediately.

*DEAE-Cellulose Chromatography*

Fines are removed from 40 g of DEAE-cellulose by two cycles of suspension in 250 ml DT-buffer/sedimentation for 30 min. The slurry is degassed for 5 min with an aspirator, and the pH adjusted to 7.5 with HCl. A column of 2.5 × 10 cm is poured and equilibrated with 100 ml of DT buffer. The Triton extract is applied to the column at a flow rate of 70 ml/hr. Elution is achieved with a 540-ml linear gradient from 0 to 250 m$M$ KCl in DT buffer, at the same flow rate. Fractions of 5 ml are collected and assayed for ATPase activity as described above (these and all following fractions exhibit vanadate-sensitive ATPase activity only in the presence of phospholipids, which are contained in the assay mix). The ATPase elutes around fraction 40 (approximately 100 m$M$ KCl), but trails over 15 to 20 fractions. At around fraction 80, vanadate-insensitive ATPase (probably $F_1F_0$-ATPase) starts to elute. Fractions containing more than 0.2 U/ml of vanadate-sensitive ATPase activity are pooled for further purification.

*Hydroxlapatite Chromatography*

Fines are removed from 8 g of hydroxylapatite as described above. The column material is also degassed, but the somewhat lower pH should

[6] M. M. Bradford, *Anal. Biochem.* **72,** 248 (1976).

not be corrected. A column of 1.6 × 12 cm is prepared and equilibrated with 50 ml of DT buffer at a flow rate of 35 ml/hr, as is used throughout the run. The combined fractions from the DEAE-cellulose column are applied and elution performed with 300 ml of a linear gradient of 0 to 100 m$M$ NaP$_i$ (100 m$M$ NaH$_2$PO$_4$, pH readjusted with KOH).[7] The tubes to be used for collection of 5-ml fractions are supplemented with 10 $\mu$l of aso-lectin suspension to stabilize the eluting enzyme. The K⁺-ATPase is the only ATPase eluting from this column, and usually appears in fractions 20 to 35. For maximal purification of the ATPase, it is essential that the fractions are analyzed for their purity by polyacrylamide gel electrophore-sis in sodium dodecyl sulfate. To this end, 50 to 100 $\mu$l of the active fractions is applied to a 10% gel, prepared as described,[8] and stained with Coomassie Blue. Only the fractions of sufficient purity (see Fig. 1, lane 4) are pooled for further purification, and 30 to 50% of the activity may have to be discarded at this point.

### CM-Sephadex Chromatography

One gram of CM-Sephadex C-25 is swollen in CM-buffer, degassed, and the pH adjusted to 4.5 with succinic acid. A column of 1 × 6 cm is prepared and equilibrated with 10 ml of CM buffer, pH 4.5. The pooled fractions from the hydroxylapatite column are acidified to pH 4.5 with succinic acid, and then applied to the column at a flow rate of 35 ml/hr. A linear 200-ml gradient from pH 4.5 to 7 in CM buffer is used for elution. Fractions of 3 ml are collected and assayed for ATPase activity, which appears in approximately fractions 20 to 40. The purity of the ATPase is analyzed by gel electrophoresis as described above, but em-ploying silver staining[9] for the visualization of the protein bands. Most fractions should just exhibit one band of $M_r$ 78,000, corresponding to the K⁺-ATPase (Fig. 1, lane 5). These fractions can be stored without neutral-ization at −70° for several months.

### Concentrating the ATPase and Exchanging the Detergent

The following steps can be used to concentrate the ATPase and, if desired, to exchange the Triton X-100 for any other detergent. Pooled fractions from the CM-Sephadex column are adjusted to pH 4.5 with 0.5 $M$ succinic acid. The ATPase is then adsorbed to a small CM-Sephadex column, equilibrated with CM buffer, pH 4.5 (a bed volume of 1 ml suf-fices to concentrate over 10 mg of purified ATPase). Concentrated en-

---

[7] Do not precool this gradient since the sodium phosphate may precipitate.

[8] U. K. Laemmli and M. Favre, *J. Mol. Biol.* **80**, 575 (1973).

[9] C. R. Merril, M. L. Bunan, and D. Goldmann, *Anal. Biochem.* **110**, 201 (1981).

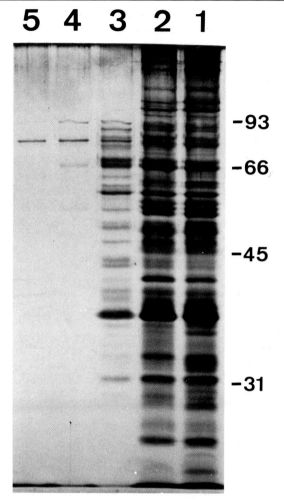

FIG. 1. Protein patterns at different stages of purification of the ATPase. Samples were resolved by polyacrylamide gel electrophoresis in sodium dodecyl sulfate,[4] followed by silver staining.[5] The probes applied to the gel were: lane 1, 10 μg of membrane protein; lanes 2 to 5, membrane extract, pooled fractions from the hydroxylapatite column, and the peak fraction from the CM-Sephadex column, respectively, each sample containing a total of 0.004 U of vanadate-sensitive ATPase activity. All the samples are from the preparation outlined in Table I. The two-digit numbers indicate the positions of protein standards of the corresponding molecular masses in kilodaltons.

zyme is eluted with CM buffer, supplemented with 100 m*M* 4-(2-hydroxy-ethyl)-1-piperazineethanesulfonic acid, adjusted to pH 7.5 with NaOH or KOH. The active fractions eluting from the column are stable if stored at −70°.

To exchange the detergent, the elution of the ATPase is preceded by extensive washing of the column with 20 column volumes of 50 m*M* 2-(*N*-morpholino)ethanesulfonic acid, adjusted to pH 4.5 with NaOH and containing the detergent of choice in a concentration that is above its critical micellar concentration. Well suited is decylmaltoside (Calbiochem) at a concentration of 0.1%. In this detergent, the ATPase is exceedingly stable and can be reconstituted into proteoliposomes by detergent dialysis.[10] Removal of the Triton from the column can be monitored conveniently via the strong UV absorption of this detergent. Elution of the concentrated ATPase is finally effected with 100 m*M* 4-(2-hydroxyethyl)1-piperazineethanesulfonic acid, adjusted to pH 7.5 with NaOH or KOH and containing, for example, 0.1% decylmaltoside.

### Summary of the Purification and Properties of the K⁺-ATPase

Table I and Fig. 1 summarize the results of a typical preparation. Although we do not routinely monitor the protein concentration in the course of the purification, due to the difficulties it represents, we can rule out extensive inactivation of the enzyme by two lines of reasoning: first, the total recovery of ATPase activity in every purification step beyond the Triton-extraction amounts to at least 70% and often approaches 100% and, second, the purified ATPase incorporates phosphate in the form of the acyl phosphate intermediate, to the extent of 0.6 mol of phosphate per mole of 78-kDa polypeptide.[11] Thus, at least 60% of the enzyme molecules in the final ATPase preparation have retained full functional integrity.

The purified K⁺-ATPase consists of a single polypeptide component of $M_r$ 78,000 that represents at least 90% of the total protein in the final preparation (Fig. 1, lane 5). The enzyme exhibits maximal rate of ATP hydrolysis at pH 7.3, but is active over a remarkably broad pH range, retaining over 50% of the maximal activity between pH 5.6 and 8.6. The best substrate is $Mg^{2+}$-ATP, with a $K_m$ of 60 $\mu M$ and a $V_{max}$ of 2 $\mu$mol/min · mg. $Mg^{2+}$-ITP and $Mg^{2+}$-dATP are hydrolyzed at 57 and 24%, respectively, of the rate observed for $Mg^{2+}$-ATP. Vanadate, an apparently

[10] H. Alpes, K. Allmann, H. Plattner, J. Reichert, R. Riek, and S. Schulz, *Biochim. Biophys. Acta* **862,** 294 (1986).
[11] P. Fürst, unpublished observations.

TABLE I
PURIFICATION OF ATPase OF *S. faecalis*[a]

| Purification step | Total protein (mg) | Vanadate-sensitive ATPase activities | | Yield (%) |
|---|---|---|---|---|
| | | Total (U) | Specific (U/ml) | |
| Membranes | 496 | 192[b] | 6.0[b] | 100 |
| Triton extract | nd[c] | 77 | 1.2 | 40 |
| DEAE-column pool | nd[c] | 41 | 0.84 | 21 |
| Hydroxylapatite pool | nd[c] | 31 | 1.0 | 16 |
| CM-Sephadex pool | nd[c] | 24 | 0.57 | 12 |

[a] The preparation was started with 64.5 g wet weight of cells.
[b] This value is most likely an overestimate (see text for details).
[c] nd, Not determined.

universal inhibitor of $E_1E_2$-ATPases, reduces the activity of the ATPase of *S. faecalis* by 50% at a concentration of 3 $\mu M$. All of these properties have been described previously.[5]

### Acknowledgments

We thank H. Alpes for introducing us to decylmaltoside. This work was supported by Grant 3.591-0.84 of the Swiss National Science Foundation.

Addendum

## Addendum to Article [30]

By JOËL LUNARDI, PAUL DeFOOR, and SIDNEY FLEISCHER

*Calculation of Phospholipid Transfer (Exchange) Based on Tracer Methodology*

*Liposomes*

Dioleyl-PC, 3400 $\mu$g phospholipid phosphorus (PLP)
1-Palmitoyl-2-oleyl [9,10,$^3$H]PC (tracer amount)
[*carboxyl*-$^{14}$C] Triolein (99 mCi/mmol; tracer amount)
The radioactivity of the mixture is 23.3 [dpm $^3$H]/nmol PC and the ratio of [dpm $^3$H]/[dpm $^{14}$C] = 3.1.

*Sarcoplasmic Reticulum*

15.9 mg of SR protein; 21.4 $\mu$g PL/mg of SR protein; 340 $\mu$g PLP

*Results*

Time 0: 21.4 $\mu$g P/mg SR = 690 nmol PL/mg SR
Time 2H: 23.6 $\mu$g P/mg SR; 8208 [dpm $^3$H]/mg SR; 460 [dpm $^{14}$C]/mg SR

*Calculations*

$$([^3H]PC)_{total} = 8208\ [dpm\ ^3H]/mg\ SR/23.3[dpm\ ^3H]/nmol\ PC$$
$$= 352.3\ nmol\ [^3H]PC/mg\ SR$$
$$([^3H]PC)_{sticking} = [dpm\ ^{14}C] \times [^3H/^{14}C]\ liposomes/23.3$$
$$= 460 \times 3.1/23.3 = 61.2\ nmol\ [^3H]PC$$
$$([^3H]PC)_{transferred} = ([^3H]PC)_{total} - ([^3H]PC)_{sticking}$$
$$= 352.2 - 61.2\ = 291\ nmol\ [^3H]PC$$

Catalyzed exchange (% PL)
$$= ([^3H]PC)_{transferred}/(SR\ phospholipids)_{total} \times 100\%$$
$$= (291/690) \times 100\% = 42.2\%\ of\ the\ PL\ exchanged$$

This value is to be compared with 40.8% transfer based on fatty-acid analysis determined in the same experiment. With oleyl-PC liposomes, both calculation methods gave essentially the same results.

It may also be noted that the amount of sticking [$^3$H]PC calculated from [$^3$H/$^{14}$C] in liposomes (61.2 nmol [$^3$H]PC) is in good agreement with the "extra" phospholipid phosphorus determined on the SR after the exchange (23.6 − 21.4 = 2.2 $\mu$g phosphorus or 71.0 nmol PL/mg SR).

# Author Index

Numbers in parentheses are footnote reference numbers and indicate that an author's work is referred to although the name is not cited in the text.

# M

Van Winkle, W. B., 126
Vanaman, T. C., 328, 330, 336, 339(13)
Vanderkoi, J. M., 207
Vanderkooi, J. M., 443
VanJaarsveld, P. P., 638
Vara, F., 535, 536, 538(12), 540(12)
Varecka, L., 4
Varga, S., 272
Vaughn, L. E., 572
Vergara, C., 69, 74(2), 370
Verjovski-Almeida, S., 172, 175, 186(37), 188(37), 198, 203, 251, 256(8), 262, 472
Verma, A. K., 341, 348, 350
Vianna, A. L., 107, 118(3), 119(3), 122(3), 191
Vibert, P., 287
Vieyra, A., 200
Villa-Komaroff, L., 290
Villalobo, A., 525, 544
Vincenzi, F. F., 340, 341(4)
Vinkler, C., 248
Vogel, H. J., 564, 575
Volpe, P., 29, 43, 59, 64(17), 65(17), 68(17), 92, 313

**W**

Wada, A., 240
Waggoner, A. S., 430, 452
Wakabayashi, S., 236
Waku, K., 15
Walderhaug, M. O., 668
Walker, R. R., 558, 580, 584(6), 585(6), 587(6), 588(6)
Wall, D., 601
Wallace, R. B., 297
Wang, C. H., 430
Wang, C., 271
Wang, C.-T., 310, 314, 319
Wang, J. A., 340, 345(5)
Wang, J. H., 151
Wang, S., 92, 100, 102(11), 103(11), 106(11), 502, 503(17)
Wang, T., 97, 111, 119(33)
Warren, G. B., 370
Warren, S., 465
Watanabe, A. M., 85, 86(1), 87(1), 89(1), 90(1), 91(1), 92, 109, 126, 152, 360
Watanabe, T., 149, 216, 233, 236(2), 239

Waters, F. B., 656, 668
Watterson, D. M., 328, 330, 331(2, 3, 8, 10), 332, 336, 337, 338, 339(13)
Wattiaux, R., 594, 596
Wattiaux-De Coninck, S., 593, 596
Waugh, R. A., 282
Wautier, J.-L., 351
Waymack, P. P., 535
Webb, J. G., 89, 90(11)
Weber, A., 162
Weber, K., 124, 357, 669
Wegener, A. D., 126, 137, 145(67), 148, 149(67, 83), 150, 361, 363(7, 8), 365(7, 8), 366(7, 8), 368(7, 8)
Wehrle, J., 248
Weidenman, B., 602, 603(7)
Weidenmann, B., 625, 634
Weinstein, D., 345, 346(14), 354, 578
Weir, W. G., 16
Weisberg, R. A., 660
Weissmann, C., 140, 144(77), 364, 365(11), 642
Westcott, K. R., 339
Westerhoff, H., 248
White, S. H., 484
Whitelaw, V., 668, 678(7)
Wibo, M., 612
Wieczorek, L., 655, 667(9), 668, 675(9), 678(6)
Wiedenmann, B., 600, 611
Wiedzorek, L., 655, 666(2)
Wiemken, A., 545
Wier, W., 164
Wiggins, M. E., 330, 331(10)
Wilderspin, A. F., 254
Wildhaber, I., 283
Will, H., 126
Willams, R. J. P., 4
Willison, J. H. M., 282
Willsky, G. R., 549
Wilson, T. A., 446, 447(20), 451(20)
Wilson, T. H., 624
Wilson, W. W., 126, 148, 149(83), 361, 363(8), 365(8), 366(8), 367(8), 368(8)
Winegrad, S., 15
Winkler, F., 179, 215
Winkler, H., 283
Witt, H. T., 247
Wolf, H. U., 341, 669, 680
Wolf, K., 513

# Subject Index

# N

Na$^+$,K$^+$-adenosine triphosphatase
  (ATPase), 29, 31
  active cation pumping of, 240–251
  active Rb$^+$ uptake induced by electric
    field, 245–246
  activity, in sealed membrane vesicles,
    measurement of, 32
  Adopp(NH)P binding to, measurement,
    239
  ATP, ADP, or P$_i$ binding to, measure-
    ment, double-membrane filtration
    method, 238–239
  E$_1$ $\rightleftharpoons$ E$_2$ scheme for, 206
  electrogenic transport of rubidium ion
    by, 244–248
  human erythrocyte, electric field induced
    active cation pumping, 241
  ion transport, 31
  monovalent cation binding to, during
    ATP hydrolysis, 233–236
  ouabain-sensitive, 31
    in skeletal muscle plasma membrane,
      30
  of plasma membrane, 190
  stimulated Rb$^+$ uptake mediated by,
    245–248
Na$^+$/Ca$^{2+}$ exchange, 31, 505–510
  affinity for Ca$^{2+}$, 7, 9
  Ca$^{2+}$ transport via, in skeletal muscle
    sarcolemmal vesicles, 35–36
  electrogenicity of, 7
  energy-independent, 9–11
  in excitable plasma membranes,
    6–7
  maximal transport capacity, 9
  mitochondrial, 9
    kinetic properties of, 10
  of plasma membranes, kinetic properties
    of, 10
  properties of, 7
Na$^+$ channel, 5
Na$^+$/H$^+$ exchange, 31
Nassar, Rashid, computer program for
  ionic concentration of aqueous solutions, 416
*Neurospora crassa*
  cell wall-less mutant, 574
  growth, 575

H$^+$-ATPase, 513, 562–573
  activity assays, 570–572
  discrimination among, 570–573
  immunological cross-reactivity, 573
  membrane fractions
    [$^{14}$C]DCCD labeling, 572–573
    enzyme markers, 572
  membranes, isolation
    bead-beater procedure, 567–570
    $\beta$-glucuronidase procedure, 563–567
  mitochondria, preparation, 563–570
  mitochondrial H$^+$-ATPase, 562
  plasma membrane
    preparation, 563–570
    solubilization, 520
  plasma membrane H$^+$-ATPase, 562
    large-scale purification, 574–579
  plasma membrane Na$^+$-ATPase
    analytical procedures for, 578
    essentially homogeneous, isolation of,
      578–579
    isolation of, 575–578
  vacuolar membrane H$^+$-ATPase, 562,
    590
  vacuolar membranes, preparation, 563–
    570

# O

Oat. *See Avena sativa*
Occluded Ca$^{2+}$, 228–233
  measurement, centrifuge column proce-
    dure, 229, 233–234
Octaethylene glycol dodecyl monoether.
  *See* C$_{12}$E$_8$
Oligonucleotide, synthetic
  labeled, purification of, 298–299
  labeling of, 298
Organelle acidification, studying, *in vitro*,
  603–604
Osmosis, 446–448
Oxalate-facilitated calcium uptake, 112
  assay for, 114–115

# P

P$_i$
  apparent $K_m$ for, at different pH values,
    193
  phosphorylation by, 192–194